GEOMATHEMATICS

Geomathematics provides a comprehensive summary of the mathematical principles behind key topics in geophysics and geodesy, covering the foundations of gravimetry, geomagnetics, and seismology. Theorems and their proofs explain why physical realities in geoscience are the logical mathematical consequences of basic laws. The book also derives and analyzes the theory and numerical aspects of established systems of basis functions and presents an algorithm for combining different types of trial functions. Topics cover inverse problems and their regularization, the Laplace/Poisson equation, boundary-value problems, foundations of potential theory, the Poisson integral formula, spherical harmonics, Legendre polynomials and functions, radial basis functions, the Biot–Savart law, decomposition theorems (orthogonal, Helmholtz, and Mie), basics of continuum mechanics, conservation laws, modelling of seismic waves, the Cauchy–Navier equation, seismic rays, and travel-time tomography. Each chapter ends with review questions, with solutions for instructors available online, providing a valuable reference for graduate students and researchers.

VOLKER MICHEL is a mathematics professor at the University of Siegen, where he founded the geomathematics group. He is also an editor-in-chief of the *International Journal on Geomathematics* and a member of the editorial boards of the *Journal of Geodesy* and *Mathematics of Computation and Data Science*. Previous works include *Lectures on Constructive Approximation* (2013) and *Multiscale Potential Theory* (2004) with Willi Freeden. Michel has organized several conferences on inverse problems and geomathematics.

GEOMATHEMATICS

Modelling and Solving Mathematical Problems in Geodesy and Geophysics

VOLKER MICHEL
University of Siegen

CAMBRIDGE
UNIVERSITY PRESS

University Printing House, Cambridge CB2 8BS, United Kingdom

One Liberty Plaza, 20th Floor, New York, NY 10006, USA

477 Williamstown Road, Port Melbourne, VIC 3207, Australia

314–321, 3rd Floor, Plot 3, Splendor Forum, Jasola District Centre,
New Delhi – 110025, India

103 Penang Road, #05–06/07, Visioncrest Commercial, Singapore 238467

Cambridge University Press is part of the University of Cambridge.

It furthers the University's mission by disseminating knowledge in the pursuit of education, learning, and research at the highest international levels of excellence.

www.cambridge.org
Information on this title: www.cambridge.org/9781108419444
DOI: 10.1017/9781108297882

First published 2022

Printed in the United Kingdom by TJ Books Limited, Padstow Cornwall

A catalogue record for this publication is available from the British Library.

ISBN 978-1-108-41944-4 Hardback

Additional resources for this publication at www.cambridge.org/9781108419444.

To my parents

Rita Michel (1938–2015)

and Walter Michel (1937–2019)

Contents

1
Introduction

Ich behaupte aber, dass in jeder besonderen Naturlehre nur so viel eigentliche Wissenschaft angetroffen werden könne, als darin Mathematik anzutreffen ist.
[But I maintain that in every special natural doctrine only so much science proper is to be met with as mathematics.]

Immanuel Kant, 1786

Eratosthenes of Cyrene, who lived basically in the third century BC, was one of those first mathematicians whose knowledge and abilities at these early stages of human civilization were remarkable. Besides his method for seeking prime numbers, he particularly also contributed to the measurement of the Earth by, for example, determining its circumference. In this respect, he might have been the first geomathematician, or at least one of the first. Many more followed him, and definitely Carl Friedrich Gauß must be mentioned here, as he can be seen as the greatest genius in mathematical history. His works and their influence are widespread in mathematics, and they are also of essential importance in various applications, in particular and (in the author's possibly biased point of view) first of all in Earth sciences, especially geomagnetics and potential theory. The awareness, which reaches back to classical antiquity, that mathematics is the foremostly required skill and toolbox for understanding the objects and processes that surround us has been preserved up to the present. It has nicely and more generally been put in a nutshell by the preceding quotation, which is from Kant (1786); for the English translation, see Kant (1883). Over the centuries, Earth sciences and mathematics have both advanced. While the achievements at the time of Eratosthenes and his fellows are nowadays parts of the curricula at schools, many modern challenges in geosciences are equally challenges to twenty-first century mathematics.

The importance of mathematics for the understanding of the entire phenomenologies which are associated to the Earth was recognized and highlighted by the initiative Mathematics of Planet Earth (MPE), which was launched by UNESCO in 2013. Since then, the interest in geomathematics has grown extensively and many publications have occurred in the wake of MPE. Certainly, the Earth is as complex as it is versatile. Therefore, one single book cannot cover the whole mathematics which models processes occurring in the Earth, at its surface, and in the atmosphere. This new book that you are currently reading concentrates on specific topics: gravitation, magnetics, and seismology, though also these fields could easily fill books of the same size on their own. Moreover, this book is not only devoted to the modelling of these physical areas, but also to the mathematical foundations which are necessary for understanding and solving the occurring challenges. This comprises, in particular, a selection of basis function systems, including an algorithm for best basis choice, and the theory of inverse problems and their regularization.

Given the limitations that naturally occur when an author tries to squeeze these topics into approximately 500 pages, one needs to concentrate on certain essentials. A particular focus was put on the interconnections. There are mathematical tools which are essential in more than one of the three considered Earth sciences. Nevertheless, each chapter can be read on its own, but cross references are set where theorems and concepts are needed which have been derived in other contexts. Looking back at approximately a quarter of a century of being a scientist, the author has noticed that often the solution of one's currently urgent problems is located beyond one's own nose. Moreover, for opening the right drawer, one needs to know what is inside each drawer – otherwise the search can become very time consuming. Therefore, gravity experts, geomagnetics scientists, and seismologists are particularly encouraged to look not only into those chapters which have headings that are familiar in their own disciplines. Another focus was set on clear and rigorous deductions. For example, in the chapter on gravitation, we start with nothing more than Sir Isaac Newton's law of gravitation, which leads us to an integral. Then we observe the properties of this integral and their consequences. This brings us automatically, for instance, to the modelling with spherical harmonics and radial basis functions. Many tools and common facts, which are often accepted without scrutinizing them, occur here as logical consequences of elementary starting points in the modelling.

This book addresses several groups of readers: mathematicians who are interested in the theoretical foundations of certain areas of Earth sciences but also those who need tools for solving applied problems with mathematical means; geoscientists who look for a mathematical reference which explains why common physical realities are actually mathematical facts; and also those who look for new inspirations and alternative approaches for solving various topical problems in their fields. The monograph has been written for students as well as researchers who will both find their own particular benefit from it. It was also very important for the author to produce a book where proofs and other derivations are comprehensible to a wide audience. Numerous graphical illustrations support the understanding of complicated mathematical concepts. Moreover, for finding important keywords more easily, many of them have been set in boldface.

Acknowledgements: Last but not least, the author, who switches here to the first person, feels the strong need to thank many people who made this book possible and who helped to bring it to what it has become. The largest gratitude is owed Kornelia Mielke, who coped with my handwriting and translated it into LaTeX for major parts of this book. Without her, the book would be far from finished yet. Certainly, I am also grateful to the people at Cambridge University Press for giving me the opportunity to publish my book under their famous brand and for being extremely patient with me, because I should have actually finished this work years ago.

Moreover, I am beholden to those members of my group and students who have been proofreading various intermediate versions of this book. In particular, I want to mention

Amna Ishtiaq, Max Kontak, Bianca Kretz, Sarah Leweke, Naomi Schneider, and Katrin Seibert. The latter two also earn my gratitude regarding some of the numerical calculations and the corresponding plots. Furthermore, there are also some other people without whom this book might have never been written: those many students in Kaiserslautern and Siegen who attended some of my master courses and were missing accompanying textbooks, which was the reason why they asked me to fill this gap.

And now definitely last but surely not least: with special emphasis, I want to thank Willi Freeden, my academic teacher, mentor, and friend, for awakening my lifelong enthusiasm for geomathematics.

2
Required Mathematical Basics

2.1 Some Important Definitions

We introduce here some notations and mathematical nomenclatures which we will need frequently within the book. First of all, $\mathbb{R}$ denotes, as usual, the set of all real numbers, whereas $\mathbb{C}$ stands for the set of all complex numbers, $\mathbb{Z}$ for the set of all integers, and $\mathbb{N}$ for the set of all positive integers with $\mathbb{N}_0 := \mathbb{N} \cup \{0\}$. Furthermore, $\mathbb{R}^+$ represents the set of all positive real numbers, where $\mathbb{R}_0^+ := \mathbb{R}^+ \cup \{0\}$, etc.

Definition 2.1.1 Let $x^0 \in \mathbb{R}^n$ and $R \in \mathbb{R}^+$ be given.

(a) The (open) **ball** with centre x^0 and radius R is defined by $B_R(x^0) := \{x \in \mathbb{R}^n \mid |x - x^0| < R\}$ such that $\overline{B_R(x^0)} = \{x \in \mathbb{R}^n \mid |x - x^0| \leq R\}$ is the closed ball.
(b) The corresponding **sphere** is $S_R(x^0) := \{x \in \mathbb{R}^n \mid |x - x^0| = R\}$.
(c) The **unit sphere** in $\mathbb{R}^3$ is denoted by $\Omega := S_1(0)$.

Here, $|\cdot|$ represents the **Euclidean norm** with $|x| := \sqrt{\sum_{j=1}^n x_j^2}$, $x \in \mathbb{R}^n$.

The following definition is based on Walter (1990).

Definition 2.1.2 Let $R \subset \mathbb{R}^n$ and $x_1, \ldots, x_k \in \mathbb{R}^n$, $k \in \mathbb{N}$.

(a) We call $\overline{x_1 x_2} := \{(1-t)x_1 + t x_2 \mid 0 \leq t \leq 1\}$ a **line segment**.
(b) The **polygonal chain** $\overline{x_1 x_2 \cdots x_k}$ is the composition of the line segments through $x_1, x_2, \ldots, x_k$, that is, $\overline{x_1 x_2 \cdots x_k} := \overline{x_1 x_2} \cup \overline{x_2 x_3} \cup \cdots \cup \overline{x_{k-1} x_k}$.
(c) The set R is called **connected**, if every arbitrary pair of points $x, y \in R$ can be connected by a polygonal chain which remains in R, that is, there exist $z_1, \ldots, z_j \in R$ (for some $j \in \mathbb{N}$) such that $\overline{x z_1 \ldots z_j y} \subset R$; see also Figure 2.1.
(d) The set R is called **open**, if the following holds true: for every $x \in R$, there exists $\varepsilon > 0$ such that $B_\varepsilon(x) \subset R$.
(e) The set R is called a **region**, if R is non-empty, open, and connected.
(f) The set R is called **bounded**, if there exists $r > 0$ such that $R \subset B_r(0)$.
(g) A point $x \in \mathbb{R}^n$ is called an **accumulation point** of R, if, for every $\varepsilon > 0$, the intersection $(B_\varepsilon(x) \setminus \{x\}) \cap R$ is non-void.
(h) The set R is called **closed**, if R contains all of its accumulation points.
(i) The set R is called **compact**, if it is closed and bounded.

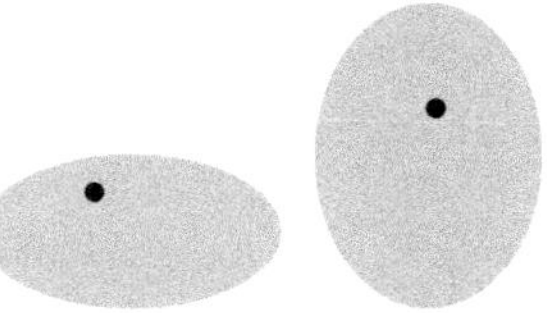
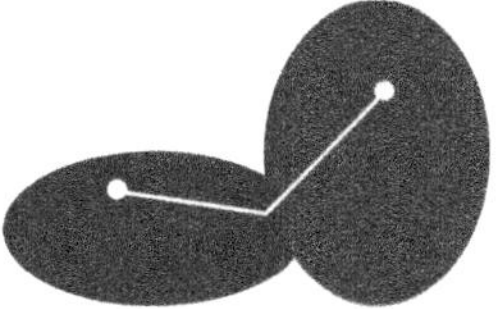

Figure 2.1 The light grey set (left) is not connected. For example, the two black points cannot be connected by a polygonal chain which remains in the set. The dark grey set (right) is connected – if it is also open (i.e., it does not contain its boundary), then it is a region. Any choice of two points in the set can be connected by a polygonal chain within the set.

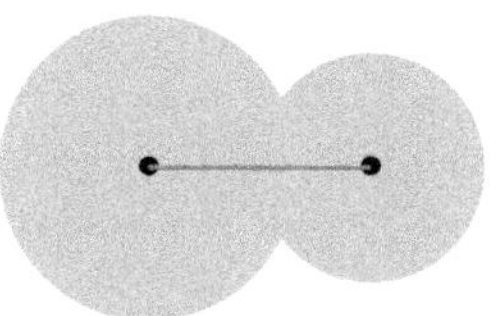
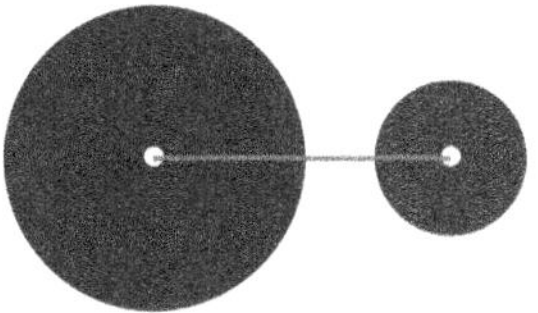

Figure 2.2 The light grey set (left) is a union of two balls. The distance of their two centres is less than the sum of the two radii. Hence, this set is connected. In the case of the dark grey set (right), the distance is larger than the sum of the two radii and the set is not connected.

For example, every ball $B_R(x_0)$ is a region (for $x, y \in B_R(x_0)$, we always have $\overline{xx_0y} \subset B_R(x_0)$). However, the union of two balls $B_{R_0}(x_0) \cup B_{R_1}(x_1)$ is not a region, if $|x_0 - x_1| > R_0 + R_1$, since there is, for example, no polygonal chain connecting x_0 and x_1 (see Figure 2.2 for an illustration).

Besides the Euclidean norm, there also exist other commonly used operations for vectors in $\mathbb{R}^n$.

Definition 2.1.3 Let $x, y \in \mathbb{R}^n$. Then $x \cdot y := \langle x, y\rangle_{\mathbb{R}^n} := \sum_{j=1}^n x_j y_j$ represents the **Euclidean inner product**, the **Euclidean scalar product**, or the **dot product** of x and y. Vectors $x_1, \ldots, x_k \in \mathbb{R}^n$ are called **orthogonal**, if $x_i \cdot x_j = 0$ for all i, j with $i \neq j$. They are called **orthonormal** if they are orthogonal and additionally satisfy $|x_i| = 1$ for all i. Note that $|x| = \sqrt{x \cdot x}$ for all $x \in \mathbb{R}^n$. Moreover, in the case $n = 3$,

$$x \times y := \begin{pmatrix} x_2 y_3 - x_3 y_2 \\ x_3 y_1 - x_1 y_3 \\ x_1 y_2 - x_2 y_1 \end{pmatrix} \in \mathbb{R}^3$$

is called the **cross product** or the **vector product** of x and y.

Furthermore, the **tensor product** (**dyadic product**) of two vectors $x \in \mathbb{R}^n$ and $y \in \mathbb{R}^m$ is the $n \times m$-matrix $x \otimes y := (x_i y_j)_{\substack{i=1,\ldots,n \\ j=1,\ldots,m}}$.

Eventually, for vectors $w, z \in \mathbb{C}^n$, the common inner product is defined by $w \cdot z := \langle w, z\rangle_{\mathbb{C}^n} := \sum_{j=1}^n w_j \overline{z_j}$, where $\overline{z_j}$ is the complex conjugation of z_j.

Theorem 2.1.4 *The vector product can be represented by using the* ***Levi-Cività (alternating) symbol***

$$\varepsilon_{ijk} := \begin{cases} 1, & \textit{if } (i,j,k) \textit{ is an even permutation of } (1,2,3), \\ -1, & \textit{if } (i,j,k) \textit{ is an odd permutation of } (1,2,3), \\ 0, & \textit{else} \end{cases}$$

as $x \times y = \left(\sum_{j,k=1}^{3} \varepsilon_{ijk}\, x_j\, y_k\right)_{i=1,2,3}$. *Moreover, the following identities hold true:*

$$\sum_{i=1}^{3} \varepsilon_{ijk}\varepsilon_{imn} = \delta_{jm}\delta_{kn} - \delta_{jn}\delta_{km}, \qquad \sum_{i,j=1}^{3} \varepsilon_{ijk}\varepsilon_{ijn} = 2\delta_{kn},$$

$$\sum_{i,j,k=1}^{3} (\varepsilon_{ijk})^2 = 6,$$

where δ_{ij} *with* $\delta_{ij} := 1$ *for* $i = j$ *and* $\delta_{ij} := 0$ *for* $i \neq j$ *is the* ***Kronecker delta****.*

This theorem is easy to verify such that we omit the proof here.

We summarize here a few basic properties of the vector product, which can be easily proved by using the Levi-Cività symbol.

Theorem 2.1.5 *Let* $a,b,c,d \in \mathbb{R}^3$*. Then* $a \times b = -b \times a$ *and*

$$(a \times b) \cdot c = (b \times c) \cdot a = (c \times a) \cdot b.$$

Moreover, we have the ***expansion theorem***

$$a \times (b \times c) = (a \cdot c)b - (a \cdot b)c$$

and the ***Lagrange identity***

$$(a \times b) \cdot (c \times d) = (a \cdot c)(b \cdot d) - (b \cdot c)(a \cdot d).$$

Definition 2.1.6 The standard orthonormal basis in $\mathbb{R}^3$ is denoted by $\varepsilon^1 := (1,0,0)^{\mathrm{T}}$, $\varepsilon^2 := (0,1,0)^{\mathrm{T}}$, and $\varepsilon^3 := (0,0,1)^{\mathrm{T}}$. Note that the notations e^1, e^2, e^3 are also common in the literature.

We will need the well-known Landau symbols.

Definition 2.1.7 (Landau Symbol) Let f and g be two functions (where g is scalar valued) for which a limit $x \to a$ with a particular $a \in \mathbb{R} \cup \{-\infty, +\infty\}$ is declared. We say that $f(x) = \mathcal{O}(g(x))$ as $x \to a$, if $\frac{f(x)}{g(x)}$ is bounded in a neighbourhood of a (for $a \notin \mathbb{R}$, the neighbourhood is $]-\infty, b[$ or $]b, +\infty[$, respectively, for some $b \in \mathbb{R}$). We say that $f(x) = o(g(x))$ as $x \to a$, if $\lim_{x\to a}(f(x)/g(x)) = 0$. Analogously, the Landau symbols can be defined for the left-hand and the right-hand limit, that is, $x \to a+$ or $x \to a-$ for $a \in \mathbb{R}$. Moreover, limits to vectors $a \in \mathbb{R}^n$ are also possible.

Example 2.1.8 Simple examples are as follows: $\sin x = \mathcal{O}(x)$ as $x \to 0$, since $\lim_{x\to 0}(x^{-1}\sin x) = 1$. Moreover, $\sin x = o(x)$ as $x \to +\infty$ and, consequently, $\sin x = \mathcal{O}(x)$ as $x \to +\infty$, since $\lim_{x\to +\infty}(x^{-1}\sin x) = 0$.

There are some other concepts that we need.

Definition 2.1.9 Let $D \subset \mathbb{R}^n$ be an arbitrary set. Then the **characteristic function**, which is also called the **indicator function**, $\chi_D \colon \mathbb{R}^n \to \mathbb{R}$ corresponding to D is defined by $\chi_D(x) := 1$ for $x \in D$ and $\chi_D(x) := 0$ for $x \notin D$.

Definition 2.1.10 The **Gauß brackets** are defined by

$$\lceil x \rceil := \min\{n \in \mathbb{Z} \,|\, x \leq n\}, \quad \lfloor x \rfloor := \max\{n \in \mathbb{Z} \,|\, n \leq x\}$$

for $x \in \mathbb{R}$. They represent rounding up and down.

Some of the properties which have been defined in this section can also be defined in more general spaces than $\mathbb{R}^n$, as we will see in Section 2.5. Moreover, we clarify that, throughout this book, log always stands for the **natural logarithm**, which is also known as ln.

2.2 A Short Course on Tensors

We will only need some simple aspects of tensors as tools, mainly in the seismological modelling in Chapter 7. The following introduction is based on Dahlen and Tromp (1998, appendix A), where further details also can be found.

Definition 2.2.1 A **tensor** of **order** $q \in \mathbb{N} \setminus \{0\}$ is a functional $T \colon (\mathbb{R}^3)^q \to \mathbb{R}$, which is linear in all q arguments (which are also called the **slots** of the tensor). This means that

$$T\left(\lambda_1 x^{(1)}, \ldots, \lambda_q x^{(q)}\right) = \left(\prod_{j=1}^{q} \lambda_j\right) T\left(x^{(1)}, \ldots, x^{(q)}\right)$$

and

$$\begin{aligned} &T\left(x^{(1)}, \ldots, x^{(k-1)}, x^{(k)} + y, x^{(k+1)}, \ldots, x^{(q)}\right) \\ &\quad = T\left(x^{(1)}, \ldots, x^{(k-1)}, x^{(k)}, x^{(k+1)}, \ldots, x^{(q)}\right) \\ &\qquad + T\left(x^{(1)}, \ldots, x^{(k-1)}, y, x^{(k+1)}, \ldots, x^{(q)}\right) \end{aligned}$$

for all $\lambda_1, \ldots, \lambda_q \in \mathbb{R}$, all $x^{(1)}, \ldots, x^{(q)}, y \in \mathbb{R}^3$, and all $k = 1, \ldots, q$. Furthermore, **scalar multiplication** and **addition** of tensors are defined in the usual way for mappings, that is, $(\lambda_1 T_1 + \lambda_2 T_2)(x) := \lambda_1 T_1(x) + \lambda_2 T_2(x)$ for all $\lambda_1, \lambda_2 \in \mathbb{R}$ and all $x \in (\mathbb{R}^3)^q$, if T_1 and T_2 are both tensors of the same order $q \in \mathbb{N} \setminus \{0\}$. Eventually, a **tensor of order 0** is a real number (that is, the slots are considered as non-existent).

The tensor product which we encountered in Definition 2.1.3 is only a particular case of the following definition. Note that we will soon have a closer look on the link between tensors, vectors, and matrices.

Definition 2.2.2 Let S and T be tensors of orders $p \in \mathbb{N}$ and $q \in \mathbb{N}$, respectively. Then the **tensor product** or **dyadic product** $S \otimes T$ is a tensor of order $p + q$ which is given by

$$(S \otimes T)\left(x^{(1)}, \ldots, x^{(p+q)}\right) := S\left(x^{(1)}, \ldots, x^{(p)}\right) T\left(x^{(p+1)}, \ldots, x^{(p+q)}\right)$$

for all $x^{(1)}, \ldots, x^{(p+q)} \in \mathbb{R}^3$.

Note that the tensor product is not commutative, that is, we have, in general, $S \otimes T \neq T \otimes S$.

Definition 2.2.3 Within Section 2.2, $\left(b^{(1)}, b^{(2)}, b^{(3)}\right)$ denotes an arbitrary choice of an orthonormal basis of $\mathbb{R}^3$ (with respect to the Euclidean inner product), where the matrix whose columns are built by these vectors has a positive determinant (that is the basis is **right-handed**). Such a system is called a **Cartesian axis system**.

An example of such a system is given by the standard orthonormal basis $\varepsilon^1, \varepsilon^2, \varepsilon^3$; see Definition 2.1.6.

Definition 2.2.4 Let T be a tensor of order $q \in \mathbb{N} \setminus \{0\}$. Then the 3^q **components** of T with respect to $(b^{(1)}, b^{(2)}, b^{(3)})$ are defined by

$$T_{i_1 \dots i_q} := T\left(\left(b^{(i_1)}\right), \dots, \left(b^{(i_q)}\right)\right)$$

for all tuples $(i_1, \dots, i_q) \in \{1, 2, 3\}^q$.

Usually, we choose our Cartesian axis system as one of our first steps when we do any modelling. In this respect, there is a fixed and clearly defined basis which underlies all the considerations. Therefore, tensors are usually represented by their components (since they are one-to-one associated to the tensors, once we have chosen our Cartesian axis system). With this in mind, we can also consider **vectors and matrices as particular cases of tensors**, as the following example demonstrates.

Example 2.2.5 Let us have a closer look at the cases of orders 1 or 2.

(a) Let c be a tensor of order 1. Then c has three components, namely $c_1 = c\left(b^{(1)}\right)$, $c_2 = c\left(b^{(2)}\right)$, $c_3 = c\left(b^{(3)}\right)$. Remember that every $x \in \mathbb{R}^3$ can be represented by $x = \sum_{j=1}^{3} \left(x \cdot b^{(j)}\right) b^{(j)}$, where '$\cdot$' is the Euclidean dot product (see Definition 2.1.3). Hence, the linearity of the tensor c yields

$$c(x) = c\left(\sum_{j=1}^{3} \left(x \cdot b^{(j)}\right) b^{(j)}\right) = \sum_{j=1}^{3} \left(x \cdot b^{(j)}\right) c\left(b^{(j)}\right) = \sum_{j=1}^{3} \left(x \cdot b^{(j)}\right) c_j.$$

Moreover, after having chosen our Cartesian axis system, we would usually write $x_j := x \cdot b^{(j)}$. Thus, if we associate the tensor c to a vector $c \in \mathbb{R}^3$ with the components $c = (c_1, c_2, c_3)^{\mathrm{T}}$, then 'the tensor c and the vector c' satisfy $c(x) = x \cdot c = x^{\mathrm{T}} c$. In particular, $c\left(b^{(j)}\right) = b^{(j)} \cdot c = c_j$. Furthermore, every vector can be uniquely associated to a tensor of order 1 (again, provided that the Cartesian axis system has already been chosen).

(b) Let A be a tensor of order 2, which, consequently, has $3^2 = 9$ components $A_{jk} = A\left(b^{(j)}, b^{(k)}\right)$, $j, k = 1, 2, 3$. This suggests that A can be associated to a matrix, where we would usually write $a_{jk} := A_{jk}$ for the components. If we represent $x, y \in \mathbb{R}^3$ with respect to the Cartesian axis system $\left(b^{(1)}, b^{(2)}, b^{(3)}\right)$ as $x = (x_1, x_2, x_3)^{\mathrm{T}}$ and $y = (y_1, y_2, y_3)^{\mathrm{T}}$, then we get, due to the bilinearity,

$$A(x,y) = \sum_{j,k=1}^{3} x_j y_k A\big(b^{(j)}, b^{(k)}\big) = \sum_{j,k=1}^{3} x_j y_k a_{jk}.$$

Hence, the tensor A is the quadratic form associated to the matrix A, that is, $A(x,y) = x^{\mathrm{T}} A y$. In analogy to the preceding case, we get here a one-to-one relation between tensors of order 2 and matrices (also here, provided that the Cartesian axis system is fixed).

(c) The tensor product which we know from Definition 2.1.3 is, indeed, a particular case of the tensor product in Definition 2.2.2. If $x = (x_1, x_2, x_3)^{\mathrm{T}}$ and $y = (y_1, y_2, y_3)^{\mathrm{T}}$ are arbitrary vectors in $\mathbb{R}^3$ (or tensors of order 1), then Definition 2.2.2 defines the tensor product as the tensor of order 2 whose components satisfy

$$(x \otimes y)_{jk} = (x \otimes y)\big(b^{(j)}, b^{(k)}\big) = x(b^{(j)}) y\big(b^{(k)}\big) = x_j y_k$$

for all $j, k = 1, 2, 3$.

Note also that the Levi-Cività alternating symbol, see Theorem 2.1.4, represents components of a tensor of order 3.

Definition 2.2.6 We define the trace of a tensor and its generalization.

(a) The **trace** of a tensor T of order 2 (or, in the sense explained previously, a matrix T) is defined by $\operatorname{tr} T := \sum_{j=1}^{3} T\big(b^{(j)}, b^{(j)}\big) = \sum_{j=1}^{3} T_{jj}$. Hence, $\operatorname{tr} T$ is a tensor of order 0, that is, a real number.

(b) The **contraction** of a tensor T of order $q \in \mathbb{N} \setminus \{0, 1\}$ upon the rth and the sth slot of T is defined by

$$\operatorname{tr}_{rs} T := \sum_{j=1}^{3} T\Big(\ldots, \underbrace{b^{(j)}}_{\text{slot } r}, \ldots, \underbrace{b^{(j)}}_{\text{slot } s}, \ldots\Big),$$

provided that $r, s \in \{1, \ldots, q\}$ with $r \neq s$ and '...' means that the corresponding slots remain untouched.

Obviously, $\operatorname{tr}_{rs} T$ is a tensor of order $q - 2$ and, if T is a tensor of order 2, then $\operatorname{tr}_{12} T = \operatorname{tr} T$.

Definition 2.2.7 We define now the transpose of a tensor, first for the case of order 2, where we recognize the usual transpose of a matrix, and then for general tensors.

(a) The **transpose** of a second-order tensor T is defined as the second-order tensor T^{T}, which satisfies

$$T^{\mathrm{T}}(x,y) := T(y,x) \quad \text{for all } x, y \in \mathbb{R}^3. \tag{2.1}$$

T is called **symmetric**, if $T = T^{\mathrm{T}}$.

(b) The transpose of a tensor T of order $q \in \mathbb{N} \setminus \{0,1\}$ with respect to the rth and the sth slot is defined via

$$\Pi_{rs}\, T\Big(\ldots, \underbrace{x}_{\text{slot } r}, \ldots, \underbrace{y}_{\text{slot } s}, \ldots\Big) := T\Big(\ldots, \underbrace{y}_{\text{slot } r}, \ldots, \underbrace{x}_{\text{slot } s}, \ldots\Big)$$

for all $x, y \in \mathbb{R}^3$, where again '…' stands for untouched slots.

In terms of components, (2.1) means that $T^{\mathrm{T}}_{jk} = T_{kj}$ for all $j,k = 1,2,3$.

Definition 2.2.8 Let S and T be tensors of orders $p \in \mathbb{N}$ and $q \in \mathbb{N}$, respectively, where $q \le p$. Then the **double-dot product** $S : T$ is the tensor of order $p-q$, which is defined by its components as

$$(S:T)_{j_1\ldots j_{p-q}} := \sum_{k_1,\ldots,k_q=1}^{3} S_{j_1\ldots j_{p-q}k_1\ldots k_q}\, T_{k_1\ldots k_q}.$$

Note that an analogous definition for complex-valued matrices (or complex tensors of order 2) is given in Example 2.5.2, part c.

Example 2.2.9 The double-dot product contains particular cases which are known from vector and matrix calculations.

(a) Let S be a tensor of order 2 and t be a tensor of order 1; then the double-dot product yields the tensor of order 1 with $(S:t)_j = \sum_{k=1}^{3} S_{jk}t_k$. In other words, if S is a matrix and t is a vector, then the vector $S:t$ is the usual matrix-vector multiplication St.
(b) Let s and t be tensors of order 1. Then $s:t$ is the tensor of order 0 which is given by $s:t = \sum_{k=1}^{3} s_k t_k$. The analogous terms are here: if s and t are both vectors, then the real number $s:t$ is given by the Euclidean inner product $s \cdot t$.
(c) Let x and y be tensors of order 1 and let Λ be the tensor of order 3 whose components are given by the Levi-Cività alternating symbol ε_{ijk} (see Theorem 2.1.4). Then $\Lambda : (x \otimes y)$ is a tensor of order $3-(1+1) = 1$. Its components satisfy

$$\big(\Lambda : (x \otimes y)\big)_i = \sum_{j,k=1}^{3} \varepsilon_{ijk}(x \otimes y)_{jk} = \sum_{j,k=1}^{3} \varepsilon_{ijk} x_j y_k = (x \times y)_i.$$

Hence, $\Lambda : (x \otimes y)$ stands for the vector product of x and y.

In general, the following proposition is easy to verify.

Theorem 2.2.10 *Within the set of all tensors of equal order $q \in \mathbb{N}$, the double-dot product satisfies the properties of an inner product, as it will be defined in Definition 2.5.1.*

Note that the usual multiplication of matrices S and T fulfils $ST = \big(\sum_{j=1}^{3} S_{ij}T_{jk}\big)_{i,k=1,2,3} = \mathrm{tr}_{23}\,(S \otimes T)$.
We need another notation in the context of tensors.

Definition 2.2.11 Let x be a tensor of order 1 and let A be a tensor of order 2. We define the **vector product** or **cross product** of x and A as the tensor of order 2 which is given by $(x \times A)_{jk} := \sum_{l,m=1}^{3} \varepsilon_{jlm} x_l A_{mk}$, $A \times x := -x \times A$.

This means, if A_1, A_2, A_3 are the rows of the matrix associated to A, then we formally apply the familiar formula for the vector product to get

$$x \times A = \begin{pmatrix} x_2 A_3 - x_3 A_2 \\ x_3 A_1 - x_1 A_3 \\ x_1 A_2 - x_2 A_1 \end{pmatrix},$$

where each entry such as $x_2 A_3 - x_3 A_2$ is a row vector of the obtained matrix ($x_2 A_3$ has the form 'scalar times row vector').

2.3 Derivatives: Notations and More

We start here with the introduction of some common differential operators.

Definition 2.3.1 The **nabla operator** ∇ maps differentiable functions $F \colon D \to \mathbb{R}$, $D \subset \mathbb{R}^n$, to their gradient

$$\text{grad } F(x) := \nabla F(x) := \left(\frac{\partial F}{\partial x_1}(x), \ldots, \frac{\partial F}{\partial x_n}(x) \right)^{\mathrm{T}}, \quad x \in D.$$

For differentiable vectorial functions $f \colon D \to \mathbb{R}^m$ with $f = (f_1, \ldots, f_m)^{\mathrm{T}}$, we call (as usual) $\nabla f(x) := \left(\frac{\partial f_j}{\partial x_k}(x)\right)_{\substack{j=1,\ldots,m \\ k=1,\ldots,n}}$, $x \in D$, the **Jacobian matrix** of f. Note that there is also the notation $\nabla \otimes f := (\nabla f)^{\mathrm{T}}$ for the transposed Jacobian matrix. If $n = m$, then we call the trace of ∇f, that is, $\text{div } f(x) := (\nabla \cdot f)(x) := \sum_{j=1}^{n} \frac{\partial f_j}{\partial x_j}(x)$, $x \in D$, the **divergence** of f. Furthermore, if $n = m = 3$, then

$$\text{curl } f(x) := \nabla \times f(x) := \begin{pmatrix} \frac{\partial}{\partial x_2} f_3 - \frac{\partial}{\partial x_3} f_2 \\ \frac{\partial}{\partial x_3} f_1 - \frac{\partial}{\partial x_1} f_3 \\ \frac{\partial}{\partial x_1} f_2 - \frac{\partial}{\partial x_2} f_1 \end{pmatrix}, \quad x \in D,$$

is called the **curl** of f. Finally, for twice differentiable functions $F \colon D \to \mathbb{R}$, $D \subset \mathbb{R}^n$, the **Hessian** of F is the second-order derivative

$$(\nabla \otimes \nabla) F(x) := \left(\frac{\partial^2 F}{\partial x_j \partial x_k}(x) \right)_{j,k=1,\ldots,n}, \quad x \in D.$$

The curl can also be represented in terms of the Levi-Cività symbol (see Theorem 2.1.4) as

$$\text{curl } f(x) = \left(\sum_{j,k=1}^{3} \varepsilon_{ijk} \frac{\partial}{\partial x_j} f_k(x) \right)_{i=1,2,3}.$$

Note that no requirement on D was formulated in Definition 2.3.1. Usually, one would suggest that D is open. On the other hand, also derivatives on boundary points can (under appropriate conditions) be defined, for example, in the sense of a left and a right derivative in the 1D case. Moreover, also derivatives on manifolds, such as the sphere, can be meant. For this reason, we state such conditions on D only if they are unavoidable and assume otherwise that D is 'sufficiently harmless'.

Sometimes it is necessary to make clear to which variable a differentiation operator applies, for example, if $F(x \cdot y)$ with $F\colon \mathbb{R} \to \mathbb{R}$, $x, y \in \mathbb{R}^n$, is to be differentiated with respect to x. In this case (but not necessarily only in this case), we add the variable as an index to the operator, that is, we write $\nabla_x F(x \cdot y)$, $\operatorname{div}_x f(x \cdot y)$, and so on.

Next, we introduce classes of functions with different levels of smoothness.

Definition 2.3.2 Let $D \subset \mathbb{R}^n$ and $E \subset \mathbb{R}^m$. By $\mathrm{C}^{(k)}(D,E)$, we denote the space of all $F\colon D \to E$ for which all derivatives of order up to k exist and are continuous. In particular, $\mathrm{C}(D,E)$ represents the space of all continuous functions $F\colon D \to E$. Furthermore, $\mathrm{C}^{(\infty)}(D,E)$ contains all infinitely differentiable functions from D to E. Moreover, following Freeden et al. (1998), we use the following abbreviations:

$$\mathrm{C}^{(k)}(D) := \mathrm{C}^{(k)}(D,\mathbb{R}), \qquad \mathrm{C}(D) := \mathrm{C}(D,\mathbb{R}),$$
$$\mathrm{c}^{(k)}(D) := \mathrm{C}^{(k)}(D,\mathbb{R}^3), \qquad \mathrm{c}(D) := \mathrm{C}(D,\mathbb{R}^3),$$
$$\mathbf{c}^{(k)}(D) := \mathrm{C}^{(k)}(D,\mathbb{R}^{3\times 3}), \qquad \mathbf{c}(D) := \mathrm{C}(D,\mathbb{R}^{3\times 3}).$$

For intervals $[a,b]$, we write $\mathrm{C}[a,b] := \mathrm{C}([a,b])$ and $\mathrm{C}^{(k)}[a,b] := \mathrm{C}^{(k)}([a,b])$.

Definition 2.3.3 The **Laplace operator** Δ on $\mathrm{C}^{(2)}(D)$ is defined as

$$\Delta_x F(x) := \sum_{j=1}^{n} \frac{\partial^2}{\partial x_j^2} F(x), \quad x \in D, F \in \mathrm{C}^{(2)}(D),$$

where $D \subset \mathbb{R}^n$. Functions $F \in \mathrm{C}^{(2)}(D)$ which satisfy $\Delta_x F(x) = 0$ for all $x \in D$ are called **harmonic**. For vectorial functions, the Laplace operator is applied componentwise and the definition of harmonicity is analogous.

There are some commonly known and easy to verify rules concerning the differential operators that we have discussed.

Theorem 2.3.4 *Let $F \in \mathrm{C}^{(2)}(D)$, $f \in \mathrm{c}^{(2)}(D)$, $G \in \mathrm{C}^{(1)}(D)$, and $g \in \mathrm{c}^{(1)}(D)$, where $D \subset \mathbb{R}^3$ is open. Then the following holds true on D:*

$$\operatorname{curl}\operatorname{grad} F = 0, \tag{2.2}$$
$$\operatorname{div}\operatorname{curl} f = 0, \tag{2.3}$$
$$\operatorname{div}\operatorname{grad} F = \Delta F, \tag{2.4}$$
$$\Delta f = \operatorname{grad}\operatorname{div} f - \operatorname{curl}\operatorname{curl} f, \tag{2.5}$$
$$\operatorname{div}(Gg) = (\operatorname{grad} G) \cdot g + G \operatorname{div} g. \tag{2.6}$$

According to (2.2), gradient fields are curl-free. However, not every curl-free field is a gradient field. This converse implication requires conditions on the domain. For example, on sets which are open and simply connected, all curl-free and continuously differentiable vector fields are gradient fields. A **simply connected** set is one where every two points of the set can be connected by a curve which remains within the set and, additionally, every closed curve (i.e., the initial and final point are identical) can be continuously deformed into a single point, again without leaving the set during the deformation. For instance, every ball in $\mathbb{R}^n$ is simply connected, but $\mathbb{R}^2 \setminus \{0\}$ is not simply connected (circle lines $S_R(0) \subset \mathbb{R}^2$ cannot be continuously deformed into a point because of the missing point 0 – this is like knotting a lasso around a stick and trying to contract the lasso completely; the stick makes this impossible).

Definition 2.3.5 Let $f \in \mathbf{c}^{(1)}(D)$ with $D \subset \mathbb{R}^3$. Then the **divergence** of f is defined as the vector field $g \in \mathrm{c}(D)$ which is given by

$$\operatorname{div} f(x) := g(x) := \begin{pmatrix} \operatorname{div} f_1(x) \\ \operatorname{div} f_2(x) \\ \operatorname{div} f_3(x) \end{pmatrix}, \qquad x \in D,$$

where f_1, f_2, f_3 denote the rows of f.

Theorem 2.3.6 (Leibniz Rule) *Let the functions F and G be m-times differentiable in $x_0 \in \mathbb{R}$, where $m \in \mathbb{N}_0$. Then the product FG is also m-times differentiable in x_0 and the mth derivative $(FG)^{(m)}$ satisfies*

$$(FG)^{(m)}(x_0) = \sum_{j=0}^{m} \binom{m}{j} F^{(j)}(x_0)\, G^{(m-j)}(x_0).$$

Remember that the binomial coefficient is defined as $\binom{m}{j} := \frac{m!}{j!(m-j)!}$ for integers j, $m \in \mathbb{N}_0$ with $j \leq m$. This definition can be generalized by using the **gamma function** $\Gamma \colon \mathbb{R}^+ \to \mathbb{R}$, which is given by $\Gamma(x) := \int_0^\infty \mathrm{e}^{-t}\, t^{x-1}\, \mathrm{d}t$, $x > 0$. It is known that $\Gamma(n+1) = n!$ for all $n \in \mathbb{N}_0$ and $\Gamma(x+1) = x\Gamma(x)$ for all $x \in \mathbb{R}^+$.

The gamma function can be continued to a metamorphic function on $\mathbb{C}$ which satisfies the functional equation $\Gamma(x+1) = x\Gamma(x)$ in every $x \in \mathbb{C}$ except the non-positive integers; see, for example, Beals and Wong (2016, Theorem 2.1.1).

Definition 2.3.7 For $x, y \in]-1, +\infty[$ with $x - y > -1$, the **binomial coefficient** is defined by

$$\binom{x}{y} := \frac{\Gamma(x+1)}{\Gamma(y+1)\Gamma(x-y+1)}.$$

2.4 Some Theorems on Integration

This section is mainly devoted to convergence theorems for integrals. In other words, we have a function depending on $(x, y) \in D \times E$. We integrate the function with respect to one of its dependencies (e.g., x) and apply a limit (where this limit can also be the limit included in a differentiation) to the other dependence (e.g., y). Moreover, the well-known theorem by Gauß and Green's formulae are recapitulated.

Remark 2.4.1 Note that all integrals in this book are **Lebesgue integrals**. Furthermore, we will use the word 'measurable' occasionally in this book in the context of sets or functions. For further details, we refer to, for example, McShane (1974) or any other textbook on measure theory or Lebesgue integrals. Nevertheless, it is not necessary to be familiar with these mathematical subjects to understand this book. Measurable sets represent a very large class of sets to which all geoscientifically relevant subsets of $\mathbb{R}^3$ belong. Moreover, wherever we write 'measurable function', the reader may replace this by 'integrable function'. Furthermore, the concept of 'almost everywhere' or 'for almost every …' means 'with the exception of a set of measure zero'. Examples of sets of measure zero are all countable (and, in particular, all finite) sets. For instance, a function on $\mathbb{R}$ which is discontinuous at every integer $n \in \mathbb{Z}$ (and only there) is almost everywhere continuous.

Some of the most important theorems in mathematical physics are Gauß's law and Green's identities, which will be summarized here. For proofs, see Mikhlin (1970, pp. 210–212) and Walter (1990, pp. 291–294) (see also Freeden and Gerhards, 2013, for the conditions on the occurring functions).

Theorem 2.4.2 *Let $G \subset \mathbb{R}^2$ be open and bounded and let $g,h\colon \overline{G} \to \mathbb{R}$ be given continuous functions (with $g(x_1,x_2) < h(x_1,x_2)$ for all $(x_1,x_2) \in G$) that define the set*

$$V := \left\{x \in \mathbb{R}^3 \mid (x_1,x_2) \in G, g(x_1,x_2) < x_3 < h(x_1,x_2)\right\}. \tag{2.7}$$

Moreover, let ∂V be the surface of V and let n be the outer unit normal field on ∂V. Then the following holds true:

(a) ***(Gauß's Law)*** *If $f : \overline{V} \to \mathbb{R}^3$ is continuously differentiable on V and continuous on $\overline{V}$, then*

$$\int_V \operatorname{div} f(x)\,\mathrm{d}x = \int_{\partial V} f(x) \cdot n(x)\,\mathrm{d}\omega(x),$$

provided that the occurring integrals exist.

(b) ***(First and Second Green's Identity)*** *If $u, v, w\colon \overline{V} \to \mathbb{R}$ are continuously differentiable functions on $\overline{V}$, where v and w are additionally twice continuously differentiable in V with Lebesgue integrable derivatives Δv and Δw on V, then*

$$\int_V u(x)\Delta v(x) + (\nabla u(x)) \cdot (\nabla v(x))\,\mathrm{d}x = \int_{\partial V} u(x)\frac{\partial v}{\partial n}(x)\,\mathrm{d}\omega(x)$$

and

$$\int_V w(x)\Delta v(x) - v(x)\Delta w(x)\,\mathrm{d}x = \int_{\partial V} \left(w\frac{\partial v}{\partial n} - v\frac{\partial w}{\partial n}\right)(x)\,\mathrm{d}\omega(x),$$

where $\frac{\partial v}{\partial n}$ stands for the directional derivative of v in the direction n, that is, $\frac{\partial v}{\partial n} = n \cdot \nabla v$.

It is worth mentioning that Gauß's theorem (i.e., Gauß's law) is sometimes also called **Green's theorem**.

Remark 2.4.3 (Domain for Gauß's Law and Green's Identities) The domain V for which the identities in Theorem 2.4.2 are valid can be defined in many different ways such that Gauß's law and Green's identities are not restricted to (2.7). In particular, V may also be the interior of a **regular surface**; see Canuto and Tabacco (2010, section 9.5.2) and Mikhlin (1970, chapter 10 §5–6). For the definition of a regular surface, see Definition 2.6.4.

Moreover, in analysis, it is often critical to check if the interchanging of some limits, such as a (common) limit and an integration, is allowed. Fortunately, there are a couple of useful theorems for this purpose. We will cite some of them here (from Bauer, 1990, p. 95, and McShane, 1974, p. 168).

Theorem 2.4.4 (Dominated Convergence Theorem I) *Let $D \subset \mathbb{R}^n$ be a measurable set and $(f_n)_{n\in\mathbb{N}}$ be an almost everywhere convergent sequence of measurable functions on D with $\int_D |f_n(x)|^p \,\mathrm{d}x < +\infty$ for all $n \in \mathbb{N}$ and for one $p \in [1, +\infty[$. If there exists a function $g\colon D \to \mathbb{R}_0^+$ such that $\int_D |g(x)|^p \,\mathrm{d}x < +\infty$ and $|f_n(x)| \leq g(x)$ for all $x \in D$, then $(f_n)_{n\in\mathbb{N}}$ converges almost everywhere to a measurable function $f\colon D \to \mathbb{R}$. Every such function f is p-integrable, that is, $\int_D |f(x)|^p \,\mathrm{d}x < +\infty$, and*

$$\int_D |f_n(x) - f(x)|^p \,\mathrm{d}x \xrightarrow[n\to\infty]{} 0.$$

A useful variant is the following theorem (from Atiyah et al., 2001b, p. 260).

Theorem 2.4.5 (Dominated Convergence Theorem II) *Let $D \subset \mathbb{R}^n$ be a measurable set and $(f_n)_{n\in\mathbb{N}}$ be a sequence of integrable functions on D which converges pointwise almost everywhere on D to the function $f\colon D \to \mathbb{R}$. If there exists a function $g\colon D \to \mathbb{R}_0^+ \cup \{+\infty\}$ such that $\int_D g(x)\,\mathrm{d}x < +\infty$ and $|f_n(x)| \leq g(x)$ for all $n \in \mathbb{N}$ and $x \in D$, then f is integrable and*

$$\int_D f(x)\,\mathrm{d}x = \lim_{n\to\infty} \int_D f_n(x)\,\mathrm{d}x.$$

Another theorem of this kind is dedicated to the case of a monotone sequence of functions; see Walter (1990, pp. 331–332).

Theorem 2.4.6 (Beppo Levi's Monotone Convergence Theorem) *Let $D \subset \mathbb{R}^n$ be a measurable set and let $(f_n)_{n\in\mathbb{N}}$ be a monotonically increasing sequence of measurable non-negative functions and let $f(x) := \lim_{n\to\infty} f_n(x)$ for all $x \in D$, where $f\colon D \to \mathbb{R} \cup \{+\infty\}$. Then f is measurable and*

$$\int_D f(x)\,\mathrm{d}x = \lim_{n\to\infty} \int_D f_n(x)\,\mathrm{d}x. \tag{2.8}$$

Corollary 2.4.7 *Let $D \subset \mathbb{R}^n$ be a measurable set and let $(f_n)_{n\in\mathbb{N}}$ be a sequence of measurable non-negative functions. Then the series $\sum_{n=1}^{\infty} f_n$ is measurable and*

$$\int_D \sum_{n=1}^{\infty} f_n(x)\,\mathrm{d}x = \sum_{n=1}^{\infty} \int_D f_n(x)\,\mathrm{d}x. \tag{2.9}$$

In particular, we can say for each of (2.8) and (2.9) that if any of the two sides is finite, then the other one must be finite too.

Theorem 2.4.8 (Interchanging Differentiation and Integration I) *Let $D \subset \mathbb{R}^n$ be a measurable subset and $[a,b]$ be a given interval with $a,b \in \mathbb{R}$ and $a < b$. Moreover, let $\tau \in [a,b]$ be given. If $F\colon D \times [a,b] \to \mathbb{R}$ satisfies the following conditions*

(a) for every $t \in [a,b]$, the function $D \ni x \mapsto F(x,t)$ is integrable,
(b) there exists an integrable function $G\colon D \to \mathbb{R}$ such that

$$\left| \frac{F(x,t) - F(x,\tau)}{t-\tau} \right| \leq G(x)$$

for all $t \in [a,b] \setminus \{\tau\}$ and all $x \in D$,
(c) for almost every $x \in D$, the partial derivative $\frac{\partial F}{\partial t}(x,t)$ exists for $t = \tau$,

then the function $I(t) := \int_D F(x,t)\,\mathrm{d}x$ is differentiable in $t = \tau$ and

$$I'(\tau) = \int_D \left. \frac{\partial F}{\partial t}(x,t) \right|_{t=\tau} \mathrm{d}x. \tag{2.10}$$

For a proof, see, for example, McShane (1974, pp. 216–217). Note that sets of measure zero may be removed from a domain of integration without any consequences for the value of the integral. Therefore, it does not matter that $\frac{\partial F}{\partial t}$ only exists almost everywhere on D in the integral (2.10). Furthermore, if $\tau = a$ or $\tau = b$, then the right-hand side and left-hand side limit, respectively, is meant.

A similar proposition is the following theorem from Bauer (1990, pp. 103–104) and Craven (1982, Proposition 4.4.8).

Theorem 2.4.9 (Interchanging Differentiation and Integration II) *Let $U \subset \mathbb{R}^d$ be open and $D \subset \mathbb{R}^n$ be measurable. If $F\colon U \times D \to \mathbb{R}$ satisfies the following properties*

(a) the mapping $D \ni y \mapsto F(x,y)$ is integrable on D for every $x \in U$,
(b) the mapping $U \ni x \mapsto F(x,y)$ is partially differentiable with respect to x_i in every $x \in U$ and for all $y \in D$,
(c) there exists a non-negative integrable function $h\colon D \to \mathbb{R} \cup \{+\infty\}$ such that $\left|\frac{\partial F}{\partial x_i}(x,y)\right| \leq h(y)$ for all $(x,y) \in U \times D$,

then the function $U \ni x \mapsto \int_D F(x,y)\,\mathrm{d}y$ is partially differentiable with respect to x_i in every $x \in U$, the function $D \ni y \mapsto \frac{\partial F}{\partial x_i}(x,y)$ is integrable on D for every $x \in U$, and

$$\frac{\partial}{\partial x_i} \int_D F(x,y)\,\mathrm{d}y = \int_D \frac{\partial}{\partial x_i} F(x,y)\,\mathrm{d}y$$

for every $x \in U$.

Note that 'integrable' includes here that the corresponding function has a *finite* integral, which is equivalent to a finite integral of the absolute value of the function.

Furthermore, absolutely continuous functions have some nice properties. The first theorem is obvious and the second one (see also Jones, 1993, p. 550) requires more effort regarding its proof.

Definition 2.4.10 A function $f\colon [a,b] \to \mathbb{R}$, $a,b \in \mathbb{R}$, $a < b$, is called **absolutely continuous**, if, for every $\varepsilon > 0$, there exists $\delta > 0$ such that $\sum_{j=1}^{m} |f(y_j)-f(x_j)| < \varepsilon$ for each finite set of points $a \leq x_1 < y_1 \leq x_2 < y_2 \leq \cdots \leq x_m < y_m \leq b$ with $\sum_{j=1}^{m}(y_j - x_j) < \delta$.

Theorem 2.4.11 *Every absolutely continuous function is continuous. Every continuously differentiable function $f\colon [a,b] \to \mathbb{R}$ is also absolutely continuous.*

Theorem 2.4.12 (Generalized Fundamental Theorem of Calculus) *A given function $f\colon [a,b] \to \mathbb{R}$ is absolutely continuous if and only if there exists a function $g\colon [a,b] \to \mathbb{R}$ with $\int_a^b |g(t)|\,\mathrm{d}t < +\infty$ and*

$$f(x) = f(a) + \int_a^x g(t)\,\mathrm{d}t \quad \textit{for all } x \in [a,b].$$

In this case, f is differentiable almost everywhere in $[a,b]$ and $f'(x) = g(x)$ for almost all $x \in [a,b]$.

2.5 Selected Topics of Functional Analysis

We summarize here some fundamental concepts, definitions, and theorems of functional analysis. This part of mathematics helps us, for example, to investigate the convergence of sequences of functions, to decide whether a function is 'close' to another one, and to use the concept of an orthonormal basis not only for vectors in $\mathbb{R}^n$ but also for functions. For further details on functional analysis and for the proofs which will be omitted here, see, for instance, Heuser (1992) and Yosida (1995).

From Definition 2.1.3, we already know the Euclidean inner product $x \cdot y = \langle x, y\rangle = \sum_{j=1}^{n} x_j y_j$, $x, y \in \mathbb{R}^n$. It provides us with the possibility to measure angles between vectors in $\mathbb{R}^n$ by $\sphericalangle(x,y) = \arccos[\langle x,y\rangle(|x|\,|y|)^{-1}] \in [0,\pi]$, $x,y \in \mathbb{R}^n \setminus \{0\}$, where $|z| = \sqrt{\langle z,z\rangle}$ for all $z \in \mathbb{R}^n$. In particular, $x,y \in \mathbb{R}^n$ are **orthogonal**, if $\sphericalangle(x,y) = \frac{\pi}{2}$, that is, $\langle x,y\rangle = 0$. Furthermore, the inner product can be used to **project** a vector x onto the straight line $\{\lambda y \mid \lambda \in \mathbb{R}\}$ corresponding to the vector $y \in \mathbb{R}^n \setminus \{0\}$. This is done by the following (see Figure 2.3):

$$\frac{\langle x,y\rangle}{|y|^2}\, y = \left\langle x, \frac{y}{|y|}\right\rangle \frac{y}{|y|}. \tag{2.11}$$

This concept of an inner product was generalized essentially, since many of its features are based on only a few properties of the inner product.

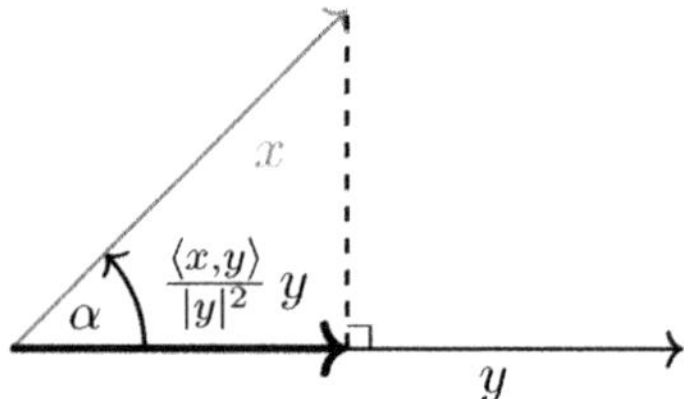

Figure 2.3 The projection of the vector x (grey) onto the straight line $\{\lambda y \mid \lambda \in \mathbb{R}\}$ associated to the vector y is given by $\langle x, y\rangle |y|^{-2} y$ (thick vector). The angle α between x and y satisfies $\cos\alpha = \langle x, y\rangle(|x|\,|y|)^{-1}$.

Definition 2.5.1 Let $\mathcal{X}$ be a vector space over the field $\mathbb{K} \in \{\mathbb{R}, \mathbb{C}\}$. If $\mathcal{X}$ is equipped with a mapping $\langle\cdot,\cdot\rangle\colon \mathcal{X} \times \mathcal{X} \to \mathbb{K}$ such that the following listed properties are satisfied, then $\mathcal{X}$ is called an **inner product space** or a **pre-Hilbert space** and $\langle\cdot,\cdot\rangle$ is called an **inner product** or a **scalar product**. The properties are as follows: $\langle\cdot,\cdot\rangle$ has to be

(IP1) positive definite: $\langle x,x\rangle \geq 0$ for all $x \in \mathcal{X}$, where $\langle x,x\rangle = 0$, if and only if $x = 0$.

(IP2) conjugately symmetric: $\langle x, y\rangle = \overline{\langle y,x\rangle}$ for all $x, y \in \mathcal{X}$, where $\overline{\langle y,x\rangle}$ stands here for the complex conjugate of $\langle y,x\rangle \in \mathbb{C}$.

(IP3) bilinear: $\langle \lambda x+\mu y, z\rangle = \lambda\langle x, z\rangle+\mu\langle y, z\rangle$ for all $x, y, z \in \mathcal{X}$ and all $\lambda, \mu \in \mathbb{K}$.

Note that the complex conjugation causes that $\langle x, \lambda y + \mu z\rangle = \overline{\langle \lambda y + \mu z, x\rangle} = \overline{\lambda\langle y,x\rangle + \mu\langle z,x\rangle} = \overline{\lambda}\langle x, y\rangle + \overline{\mu}\langle x,z\rangle$ for all $x, y, z \in \mathcal{X}$ and all $\lambda, \mu \in \mathbb{K}$. Furthermore, if $\mathbb{K} = \mathbb{R}$, that is, if $\mathcal{X}$ is a real vector space, then $\langle x, y\rangle = \langle y,x\rangle$ for all $x, y \in \mathcal{X}$.

Example 2.5.2 Here are some examples of pre-Hilbert spaces.

(a) On $\mathcal{X} = \mathbb{R}^n$ (where $\mathbb{K} = \mathbb{R}$), the Euclidean inner product is, as we know, given by $x \cdot y = \langle x, y\rangle = \sum_{j=1}^n x_j y_j$, $x, y \in \mathbb{R}^n$.

(b) In the complex case $\mathcal{X} = \mathbb{C}^n$ (where $\mathbb{K} = \mathbb{C}$), we have $x \cdot y := \langle x, y\rangle := \sum_{j=1}^n x_j \overline{y_j}$, $x, y \in \mathbb{C}^n$. Note that the complex conjugation is alternatively placed at the first argument in some literature (in particular in physics).

(c) For tensors/matrices $A, B \in \mathbb{C}^{n\times m}$, we can define an inner product, which is also called the **double-dot product**, by $A : B := \langle A, B\rangle := \sum_{j=1}^n \sum_{k=1}^m a_{jk}\overline{b_{jk}}$, where a_{jk} and b_{jk} are the components of A and B, respectively.

(d) If $\mathcal{X} = \mathrm{C}(D)$ is the set of continuous real functions on a compact domain $D \subset \mathbb{R}^n$, then $\langle F, G\rangle_2 := \int_D F(x)G(x)\,\mathrm{d}x$, $F, G \in \mathrm{C}(D)$, is an inner product. Analogously, for vectorial real functions $f, g \in \mathrm{c}(D)$, we can define an inner product by $\langle f, g\rangle_2 := \int_D f(x) \cdot g(x)\,\mathrm{d}x$, where '$\cdot$' is here the dot product from earlier. Eventually, for tensorial real functions $\mathbf{f}, \mathbf{g} \in \mathbf{c}(D)$, an inner product is given by $\langle \mathbf{f}, \mathbf{g}\rangle_2 := \int_D \mathbf{f}(x) : \mathbf{g}(x)\,\mathrm{d}x$.

(e) For complex-valued functions, the spaces $\mathrm{C}^{(k)}(D, E)$, $\mathrm{C}(D, E)$, $\mathrm{C}^{(k)}(D)$, $\mathrm{C}(D)$, $\mathrm{c}^{(k)}(D)$, $\mathrm{c}(D)$, $\mathbf{c}^{(k)}(D)$, and $\mathbf{c}(D)$ can be defined in analogy to Definition 2.3.2. Corresponding inner products would be, if D is compact, given by

$$\langle F,G\rangle_2 := \int_D F(x)\overline{G(x)}\,\mathrm{d}x, \quad F,G \in \mathrm{C}(D),$$

$$\langle f,g\rangle_2 := \int_D f(x)\cdot g(x)\,\mathrm{d}x = \sum_{j=1}^{3}\int_D f_j(x)\overline{g_j(x)}\,\mathrm{d}x,\ f,g \in \mathrm{c}(D),$$

$$\langle \mathbf{f},\mathbf{g}\rangle_2 := \int_D \mathbf{f}(x) : \mathbf{g}(x)\,\mathrm{d}x = \sum_{j,k=1}^{3}\int_D f_{jk}(x)\overline{g_{jk}(x)}\,\mathrm{d}x,\ \mathbf{f},\mathbf{g} \in \mathbf{c}(D).$$

Definition 2.5.3 Let $(\mathcal{X},\langle\cdot,\cdot\rangle)$ be a pre-Hilbert space. Two vectors $x,y \in \mathcal{X}$ are called **orthogonal** if $\langle x,y\rangle = 0$. A subset $\mathcal{S} \subset \mathcal{X}$ is called an orthogonal system if every arbitrary pair of distinct elements of $\mathcal{S}$ is orthogonal (i.e., $x,y\in\mathcal{S},\ x\neq y \Rightarrow \langle x,y\rangle = 0$). Moreover, an **orthonormal system** is an orthogonal system $\mathcal{S}\subset\mathcal{X}$, where $\langle x,x\rangle = 1$ for all $x \in \mathcal{S}$. If the elements of $\mathcal{S}$ can be associated to an index such that $\mathcal{S} = \{x_\alpha\}_{\alpha\in A}$, then we can represent the property of an orthonormal system by the Kronecker delta

$$\langle x_\alpha, x_\beta\rangle = \delta_{\alpha\beta} = \begin{cases} 0, & \alpha\neq\beta,\\ 1, & \alpha=\beta.\end{cases}$$

Example 2.5.4 For geomathematical applications, functions on spheres are particularly relevant. For this reason, we have a closer look here at the corresponding geometry.

The **unit sphere** on $\mathbb{R}^{n+1}$ is defined as $\mathbb{S}^n := \{\xi \in \mathbb{R}^{n+1} \mid |\xi| = 1\}$. Note that we defined earlier that Ω is used for $\mathbb{S}^2$. We start with the easier case of a one-dimensional sphere, that is, a circle.

(a) Points $\xi \in \mathbb{S}^1$ can be uniquely represented by the angle $\varphi \in [0,2\pi[$, where any other interval of length 2π which includes exactly one boundary point can be used alternatively. This gives rise to the **polar coordinates** in $\mathbb{R}^2$: every $x \in \mathbb{R}^2$ is representable as

$$x(r,\varphi) = \begin{pmatrix} r\cos\varphi \\ r\sin\varphi\end{pmatrix}, \quad r\in\mathbb{R}_0^+, \quad \varphi\in[0,2\pi[,$$

where $r = 1$ corresponds to $\mathbb{S}^1$ and the polar coordinates are unique for $r\neq 0$. Let us now write

$$\xi(\varphi) = \begin{pmatrix}\cos\varphi\\ \sin\varphi\end{pmatrix} =: \varepsilon^r(\varphi)$$

for $\xi\in\mathbb{S}^1$. By differentiating this with respect to φ, we get

$$\varepsilon^\varphi(\varphi) := \begin{pmatrix}-\sin\varphi\\ \cos\varphi\end{pmatrix}.$$

Note that every vector $\xi\in\mathbb{S}^1$ is always the outer unit normal to $\mathbb{S}^1$ in the corresponding point ξ. In combination with the orthogonality relation $\xi(\varphi)\cdot\varepsilon^\varphi(\varphi) = 0$ for all φ, it becomes clear that ε^φ represents a unit tangential vector in the very same point ξ, as shown in Figure 2.4.

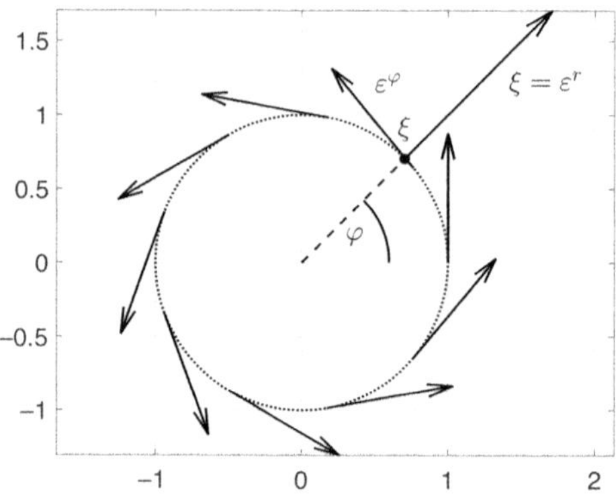

Figure 2.4 The unit circle $\mathbb{S}^1$ can be parameterized by one angle φ. The associated vector field ε^φ yields a tangential field on the sphere $\mathbb{S}^1$. Note that the position vector $\xi \in \mathbb{S}^1$ is automatically the outer unit normal ε^r at the position ξ.

(b) In $\mathbb{R}^3$, the sphere $\mathbb{S}^2$ can be constructed as follows: we cut the sphere into slices S_z associated to z-coordinates such that

$$S_z := \left\{(x,y,z) \in \mathbb{R}^3 \,|\, x^2+y^2+z^2 = 1\right\}, \quad z \text{ fixed.}$$

We write $t := z$ for the z-coordinate and call it the **polar distance**. Then S_z is a circle with centre $(0,0,t)^{\mathrm{T}}$ and radius $\sqrt{1-t^2}$. Hence,

$$\mathbb{S}^2 = \left\{ \left. \begin{pmatrix} \sqrt{1-t^2}\cos\varphi \\ \sqrt{1-t^2}\sin\varphi \\ t \end{pmatrix} \right| \varphi \in [0,2\pi[,\ t \in [-1,1] \right\}.$$

Commonly, a second angle is used for representing the polar distance. This angle is called the **latitude** and can, for example, be chosen such that $t = \sin\theta$, $\theta \in [-\pi/2, \pi/2]$, or $t = \cos\vartheta$, $\vartheta \in [0,\pi]$. Since $x = r\xi$, $r \in \mathbb{R}_0^+$, $\xi \in \mathbb{S}^2$ may be written for all $x \in \mathbb{R}^3$, we obtain the **polar coordinates** in $\mathbb{R}^3$ by

$$x(r,\varphi,t) = \begin{pmatrix} r\sqrt{1-t^2}\cos\varphi \\ r\sqrt{1-t^2}\sin\varphi \\ rt \end{pmatrix}, \quad r \in \mathbb{R}_0^+, \varphi \in [0,2\pi[, t \in [-1,1],$$

where this representation is unique for $x \neq 0$. Thus, φ is called the **longitude**. In analogy to the case of $\mathbb{S}^1$, we can differentiate this representation again with respect to the parameters:

$$\frac{\partial}{\partial r}x(r,\varphi,t) = \begin{pmatrix} \sqrt{1-t^2}\cos\varphi \\ \sqrt{1-t^2}\sin\varphi \\ t \end{pmatrix}, \qquad \frac{\partial}{\partial \varphi}x(r,\varphi,t) = \begin{pmatrix} -r\sqrt{1-t^2}\sin\varphi \\ r\sqrt{1-t^2}\cos\varphi \\ 0 \end{pmatrix},$$

$$\frac{\partial}{\partial t}x(r,\varphi,t) = \begin{pmatrix} -\frac{rt}{\sqrt{1-t^2}}\cos\varphi \\ -\frac{rt}{\sqrt{1-t^2}}\sin\varphi \\ r \end{pmatrix}.$$

The latter two vectors are not normalized such that we define the following unit vectors in local coordinates:

$$\varepsilon^r(\varphi,t) := \frac{\partial}{\partial r}x(r,\varphi,t) = \begin{pmatrix} \sqrt{1-t^2}\cos\varphi \\ \sqrt{1-t^2}\sin\varphi \\ t \end{pmatrix},$$

$$\varepsilon^\varphi(\varphi) := \frac{1}{r\sqrt{1-t^2}}\frac{\partial}{\partial \varphi}x(r,\varphi,t) = \begin{pmatrix} -\sin\varphi \\ \cos\varphi \\ 0 \end{pmatrix},$$

$$\varepsilon^t(\varphi,t) := \frac{\sqrt{1-t^2}}{r}\frac{\partial}{\partial t}x(r,\varphi,t) = \begin{pmatrix} -t\cos\varphi \\ -t\sin\varphi \\ \sqrt{1-t^2} \end{pmatrix}.$$

It is easy to verify that these three vectors are orthonormal, where $\varepsilon^r \times \varepsilon^\varphi = \varepsilon^t$. They serve as geoscientifically relevant directions on the sphere $\mathbb{S}^2 = \Omega$, since ε^r represents the upward direction, ε^φ stands for the east direction, and ε^t is the north direction, as shown in Figure 2.5.

Remember that the Euclidean norm and the Euclidean inner product are linked via $|x| = \sqrt{\langle x,x\rangle}$, $x \in \mathbb{R}^n$. A norm in Euclidean space can be used to measure the length of a vector. Note that

$$\frac{\langle x,x\rangle}{|x|^2}x = \left\langle x, \frac{x}{|x|}\right\rangle \frac{x}{|x|}$$

is the projection of x onto itself. Its length is, indeed,

$$\left|\frac{\langle x,x\rangle}{|x|^2}x\right| - \langle x,x\rangle\frac{|x|}{|x|^2} = |x|.$$

This concept of a norm can be generalized to vector spaces.

Definition 2.5.5 Let $\mathcal{X}$ be a vector space with respect to the field $\mathbb{K} \in \{\mathbb{R},\mathbb{C}\}$. If $\|\cdot\|\colon \mathcal{X} \to \mathbb{R}$ is a mapping with the following listed properties, then $\|\cdot\|$ is called a **norm** and $(\mathcal{X}, \|\cdot\|)$ is called a **normed space**. The properties are as follows: a norm $\|\cdot\|$ has to

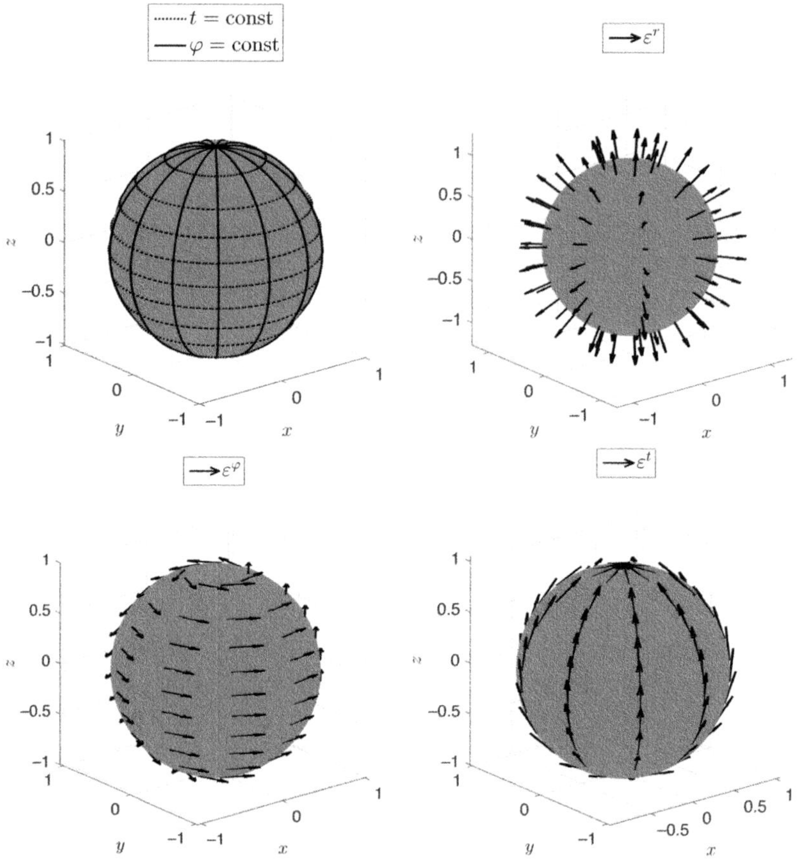

Figure 2.5 The unit sphere $\mathbb{S}^2$ needs two parameters (here φ and t). The associated vector fields ε^φ and ε^t are tangential and point eastwards and northwards, respectively. Note that the position vector $\xi \in \mathbb{S}^2$ is automatically the outer unit normal ε^r at the position ξ.

(N1) be **positive definite**: $\|x\| \geq 0$ for all $x \in \mathcal{X}$, where $\|x\| = 0$, if and only if $x = 0$,

(N2) be **homogeneous**: $\|\lambda x\| = |\lambda| \cdot \|x\|$ for all $\lambda \in \mathbb{K}$ and all $x \in \mathcal{X}$,

(N3) satisfy the **triangle inequality**: $\|x + y\| \leq \|x\| + \|y\|$ for all $x, y \in \mathcal{X}$.

Theorem 2.5.6 *Let $(\mathcal{X}, \langle \cdot, \cdot \rangle)$ be an inner product space. Then the mapping $\|\cdot\| : \mathcal{X} \to \mathbb{R}$ defined by $\|x\| := \sqrt{\langle x, x \rangle}$, $x \in \mathcal{X}$, satisfies the properties of a norm. It is called the **induced norm**.*

Theorem 2.5.7 *Let $(\mathcal{X}, \| \cdot \|)$ be a normed space. The norm $\| \cdot \|$ is an induced norm if and only if the **parallelogram identity** holds true:*

$$\|x + y\|^2 + \|x - y\|^2 = 2\|x\|^2 + 2\|y\|^2 \textit{ for all } x, y \in \mathcal{X}.$$

Theorem 2.5.8 (Cauchy–Schwarz Inequality) *If $(\mathcal{H}, \langle\cdot,\cdot\rangle)$ is a given pre-Hilbert space with the induced norm $\|x\| = \sqrt{\langle x,x\rangle}$, $x \in \mathcal{H}$, then $|\langle x,y\rangle| \le \|x\| \cdot \|y\|$ for all $x, y \in \mathcal{H}$.*

Example 2.5.9 Besides the induced norms associated to the inner products in Example 2.5.2, we mention here some further relevant examples:

(a) If $D \subset \mathbb{R}^n$ is compact, then we can define

$$\|F\|_\infty := \|F\|_{\mathrm{C}(D)} := \max_{x\in D} |F(x)|, \quad F \in \mathrm{C}(D),$$

$$\|f\|_\infty := \|f\|_{\mathrm{c}(D)} := \max_{x\in D} |f(x)|, \quad f \in \mathrm{c}(D),$$

$$\|\mathbf{f}\|_\infty := \|\mathbf{f}\|_{\mathbf{c}(D)} := \max_{x\in D} |\mathbf{f}(x)|, \quad \mathbf{f} \in \mathbf{c}(D),$$

as norms on $\mathrm{C}(D)$, $\mathrm{c}(D)$, and $\mathbf{c}(D)$. Note that $|\mathbf{f}(x)| = \sqrt{(\mathbf{f}(x)) : (\mathbf{f}(x))}$ uses the induced norm of the double-dot product. The term $\|\cdot\|_\infty$ is also called the **maximum norm**.

(b) If $D \subset \mathbb{R}^n$ is compact and $1 \le p < +\infty$ is a fixed parameter, then we can define the following alternative norms:

$$\|F\|_p := \left(\int_D |F(x)|^p \,\mathrm{d}x\right)^{1/p}, \quad F \in \mathrm{C}(D),$$

$$\|f\|_p := \left(\int_D |f(x)|^p \,\mathrm{d}x\right)^{1/p}, \quad f \in \mathrm{c}(D),$$

$$\|\mathbf{f}\|_p := \left(\int_D |\mathbf{f}(x)|^p \,\mathrm{d}x\right)^{1/p}, \quad \mathbf{f} \in \mathbf{c}(D).$$

Note that $\|\cdot\|_2$ is the induced norm associated to $\langle\cdot,\cdot\rangle_2$ in Example 2.5.2, part e.

(c) Let $k \in \mathbb{N}$. Then we can equip $\mathrm{C}^{(k)}(D)$ with the following norm $\|F\|_{\mathrm{C}^{(k)}(D)} := \sum_{j=0}^{k} \|F^{(j)}\|_\infty$, $F \in \mathrm{C}^{(k)}(D)$, where $F^{(j)}$ is the jth derivative of F, provided that $D \subset \mathbb{R}^n$ is compact.

(d) The sequence space ℓ^p with $1 \le p < +\infty$ consists of all (complex or real) sequences $(a_n)_{n\in\mathbb{N}_0} \subset \mathbb{K}$ such that $\sum_{n=0}^{\infty} |a_n|^p < +\infty$. Its elements are called p-summable (and square-summable, if $p = 2$). The space can be equipped with the norm $\|(a_n)\|_p := \|(a_n)\|_{\ell^p} := \left(\sum_{n=0}^{\infty} |a_n|^p\right)^{1/p}$. This norm is induced if and only if $p = 2$ (note the similarity to the case in Example 2.5.2, part b). More precisely, $\|\cdot\|_2$ is induced by the inner product $\langle (a_n),(b_n)\rangle_2 := \sum_{n=0}^{\infty} a_n\overline{b_n}$, $(a_n),(b_n) \in \ell^2$, where $\overline{b_n}$ stands again for the complex conjugate of b_n.

Norms can also be used to measure the distance of two elements $x, y \in \mathcal{X}$ by $\|x - y\|$. This leads us to the definition of metric spaces.

Definition 2.5.10 Let $\mathcal{X} \ne \emptyset$ be an arbitrary set. If there exists a mapping $d\colon \mathcal{X} \times \mathcal{X} \to \mathbb{R}$ with the following listed properties, then d is called a **metric** and $(\mathcal{X}, d)$ is called a **metric space**. The required properties of a metric are as follows: d has to

(M1) be **positive definite**: $d(x, y) \geq 0$ for all $x, y \in \mathcal{X}$, where $d(x, y) = 0$, if and only if $x = y$,

(M2) be **symmetric**: $d(x, y) = d(y, x)$ for all $x, y \in \mathcal{X}$,

(M3) satisfy the **triangle inequality**: $d(x, y) \leq d(x, z) + d(z, y)$ for all $x, y, z \in \mathcal{X}$.

Theorem 2.5.11 *Let $(\mathcal{X}, \|\cdot\|)$ be a normed space. Then the mapping $d\colon \mathcal{X} \times \mathcal{X} \to \mathbb{R}$ with $d(x, y) := \|x - y\|$ is a metric. It is called the* ***induced metric***.

Example 2.5.12 We mention here two examples of metric spaces:

(a) The inner product $\langle F, G\rangle_2 = \int_D F(x)\overline{G(x)}\,\mathrm{d}x$ for continuous functions $F, G\colon D \to \mathbb{C}$ on a compact domain $D \subset \mathbb{R}^n$ was introduced before. Its induced norm $\|F\|_2 = \left(\int_D |F(x)|^2\,\mathrm{d}x\right)^{1/2}$, $F \in \mathrm{C}(D, \mathbb{C})$, induces the metric $d(F, G) = \left(\int_D |F(x) - G(x)|^2\,\mathrm{d}x\right)^{1/2}$, $F, G \in \mathrm{C}(D, \mathbb{C})$.

(b) Every non-void set $\mathcal{X}$ can be equipped with the **discrete metric**, which is defined as $d(x, y) := 1$ for $x \neq y$ and $d(x, y) := 0$ for $x = y$. It is easy to verify that this is a metric. To check if it is also an induced metric, we first observe that metric spaces do not require a vector space, in contrast to normed spaces. Moreover, if $\mathcal{X}$ is a vector space with at least one non-trivial element $x \neq 0$, then, if there existed a norm which induced d, we would get $2|\lambda| \cdot \|x\| = \|2\lambda x\| = \|\lambda x - (-\lambda x)\| = d(\lambda x, -\lambda x) = 1$ for all $\lambda \in \mathbb{K} \setminus \{0\}$, which is impossible. The discrete metric is, therefore, never induced, except for the trivial case where $\mathcal{X} = \{0\}$.

Many basic concepts of analysis such as convergence and continuity as well as topological terms such as open and closed sets actually only need the possibility to measure distances. We can, therefore, generalize them to metric spaces. Note that we automatically also define these concepts for normed spaces and pre-Hilbert spaces via the induced metric.

Definition 2.5.13 Let $(\mathcal{X}, d)$ be a metric space. Moreover, let (a_n) be an arbitrary sequence in $\mathcal{X}$, let $\alpha, x \in \mathcal{X}$ be arbitrary elements of $\mathcal{X}$, and let $\mathcal{S} \subset \mathcal{X}$ be an arbitrary subset of $\mathcal{X}$. We define the following:

(a) The sequence (a_n) **converges** to α if the real sequence $(d(a_n, \alpha))_n$ converges to $0 \in \mathbb{R}$. In other words: for every $\varepsilon > 0$, there exists $n_0 > 0$ such that, for all $n \geq n_0$, we have $d(a_n, \alpha) < \varepsilon$. We write $\lim_{n\to\infty} a_n = \alpha$ or, more precisely, ${}^{d}\lim_{n\to\infty} a_n = \alpha$.

(b) The sequence (a_n) is a **Cauchy sequence** if the following holds true: for every $\varepsilon > 0$, there exists $n_0 > 0$ such that, for all $n, m \geq n_0$, we have $d(a_n, a_m) < \varepsilon$.

(c) The metric space $\mathcal{X}$ is called **complete** if every Cauchy sequence in $\mathcal{X}$ converges to an element of $\mathcal{X}$. In the particular cases of complete normed spaces, one refers to **Banach spaces**, and complete pre-Hilbert spaces are called **Hilbert spaces**.

(d) The element $x \in \mathcal{S}$ is called an **interior point** of the set $\mathcal{S}$ if there exists an $\varepsilon > 0$ such that the **ball** $B_\varepsilon(x) := \{y \in \mathcal{X} \mid d(x, y) < \varepsilon\}$ is a subset of $\mathcal{S}$. The set of all interior points of $\mathcal{S}$ is denoted by $\operatorname{int}\mathcal{S}$ or $\mathring{\mathcal{S}}$.

(e) The set $\mathcal{S}$ is **open** if every element of $\mathcal{S}$ is an interior point of $\mathcal{S}$, that is, $\mathcal{S} = \operatorname{int} \mathcal{S}$.
(f) The set $\mathcal{S}$ is **closed** if, for every sequence $(a_n) \subset \mathcal{S}$ which is convergent in $\mathcal{X}$, we have ${}^d\lim_{n\to\infty} a_n \in \mathcal{S}$.
(g) The **closure** of the set $\mathcal{S}$ is defined by

$$\overline{\mathcal{S}}^d := \overline{\mathcal{S}} := \bigcap \{\mathcal{C} \subset \mathcal{X} \mid \mathcal{S} \subset \mathcal{C} \text{ and } \mathcal{C} \text{ is closed}\}.$$

(h) The **boundary** of the set $\mathcal{S}$ is defined by $\partial \mathcal{S} := \overline{\mathcal{S}} \setminus \mathring{\mathcal{S}}$.
(i) The point $x \in \mathcal{X}$ is called an **accumulation point** of $\mathcal{S}$ if, for every $\varepsilon > 0$, we have $(B_\varepsilon(x) \setminus \{x\}) \cap \mathcal{S} \neq \emptyset$.
(j) The set $\mathcal{S}$ is **dense** in $\mathcal{X}$, if $\overline{\mathcal{S}} = \mathcal{X}$.
(k) The set $\mathcal{S}$ is called **compact** if every sequence in $\mathcal{S}$ has a subsequence which converges to an element of $\mathcal{S}$.
(l) The set $\mathcal{S}$ is called **relatively compact** if $\overline{\mathcal{S}}$ is compact.
(m) The set $\mathcal{S}$ is called **bounded** if there exists an element $x_0 \in \mathcal{X}$ and a radius $r > 0$ such that $\mathcal{S} \subset B_r(x_0)$.
(n) The **distance** of the point x to the set $\mathcal{S}$ is defined by $\operatorname{dist}(x, \mathcal{S}) := \inf_{y\in\mathcal{S}} d(x, y)$.
(o) The **diameter** of the set $\mathcal{S}$ is defined as $\operatorname{diam} \mathcal{S} := \sup\{d(x, y) \mid x, y \in \mathcal{S}\}$.

If we use a metric d which is induced, for example, by a norm $\|\cdot\|$, then we may also write ${}^{\|\cdot\|}\lim_{n\to\infty} a_n$ instead of ${}^d\lim_{n\to\infty} a_n$ and $\overline{\mathcal{S}}^{\|\cdot\|}$ instead of $\overline{\mathcal{S}}^d$.

Theorem 2.5.14 *Let $(\mathcal{X}, d)$ be a metric space. Then the following holds true:*

(a) A set $\mathcal{S} \subset \mathcal{X}$ is open, if and only if $\mathcal{X} \setminus \mathcal{S}$ is closed.
(b) Finite *intersections of open sets are open, that is, if $\mathcal{S}_1, \mathcal{S}_2, \ldots, \mathcal{S}_n \subset \mathcal{X}$ are open, then $\bigcap_{j=1}^n \mathcal{S}_j$ is open.*
(c) Arbitrary *unions of open sets are open, that is, if $\{\mathcal{S}_\alpha\}_{\alpha\in A} \subset \mathcal{X}$ is a (not necessarily countable) system of open sets, then $\bigcup_{\alpha\in A} \mathcal{S}_\alpha$ is open.*
(d) Arbitrary *intersections of closed sets are closed.*
(e) Finite *unions of closed sets are closed.*
(f) $\emptyset$ and $\mathcal{X}$ are always open and closed.
(g) A set is closed if and only if it contains all of its accumulation points.
(h) A compact set is always closed and bounded.

This also implies that closures are always closed.

Theorem 2.5.15 *Let $(\mathcal{X}, \|\cdot\|)$ be a normed space. Then the statement 'every closed and bounded subset of $\mathcal{X}$ is also compact' is true if and only if $\mathcal{X}$ is finite dimensional.*

Theorem 2.5.16 *Let $(\mathcal{X}, d)$ be a complete metric space and $\mathcal{S}$ be a subset. If $\mathcal{S}$ is closed, then $(\mathcal{S}, d)$ is also a complete metric space.*

Note that, in every metric space $\mathcal{X}$, the ball $B_R(x) = \{y \in \mathcal{X} \mid d(x, y) < R\}$ is open. Moreover, in every *normed* space $\mathcal{X}$, the closure of this ball is given by $\overline{B_R(x)} = \{y \in \mathcal{X} \mid d(x, y) \leq R\} = \{y \in \mathcal{X} \mid \|x - y\| \leq R\}$. This is not true in some metric spaces (with the discrete metric, we obtain $B_R(x) = \{x\} = \overline{B_R(x)}$ for $0 \leq R \leq 1$, but $\{y \in \mathcal{X} \mid d(x,y) \leq 1\} = \mathcal{X} \neq \overline{B_1(x)}$, at least, if $\mathcal{X}$ has more than one element).

We also have to consider **complete spaces**, since they play an important role in some parts of this book.

Definition 2.5.17 Let $(\mathcal{X}, d)$ be a metric space.

(a) A subset $\mathcal{S} \subset \mathcal{X}$ is called **nowhere dense** if the closure $\overline{\mathcal{S}}$ does not contain any interior point, that is, if $\operatorname{int} \overline{\mathcal{S}} = \emptyset$.
(b) $\mathcal{X}$ is said to be of the **first category** if there exists a countable system $\{\mathcal{S}_k\}_{k \in \mathbb{N}}$ of nowhere dense sets $\mathcal{S}_k$ such that $\mathcal{X} = \bigcup_{k \in \mathbb{N}} \mathcal{S}_k$. Otherwise, $\mathcal{X}$ is said to be of the **second category**.

Theorem 2.5.18 (Baire–Hausdorff Theorem) *Every non-empty and complete metric space is of the second category.*

Sets of the first category are relatively small. Sometimes, such sets are also alternatively called 'meagre'. So, there is not very much to find in there. For instance, $\mathbb{Q} = \bigcup_{r \in \mathbb{Q}} \{r\}$ is of the first category and every $\{r\}$, $r \in \mathbb{Q}$, is nowhere dense, while $\mathbb{R}$ is of the second category.

Example 2.5.19 We consider some of the examples of normed spaces introduced in the preceding discussion. Certainly, $\mathbb{K}^n$ is complete for $\mathbb{K} \in \{\mathbb{R}, \mathbb{C}\}$ as well as $\mathbb{K}^{n \times m}$, which can be handled like $\mathbb{K}^{n \cdot m}$.

(a) If $D \subset \mathbb{R}^n$ is compact, then the spaces $(\mathrm{C}(D), \|\cdot\|_\infty)$, $(\mathrm{c}(D), \|\cdot\|_\infty)$, and $(\mathbf{c}(D), \|\cdot\|_\infty)$ are Banach spaces (see Example 2.5.9, part a).
(b) If $D \subset \mathbb{R}^n$ is compact, then $(\mathrm{C}^{(k)}(D), \|\cdot\|_{\mathrm{C}^{(k)}(D)})$ is a Banach space (see Example 2.5.9, part c).
(c) In general, the $\|\cdot\|_p$-norm (again, for $1 \leq p < +\infty$) for continuous functions (see Example 2.5.9b) does not generate a complete space. For this reason, one uses the space $\mathrm{L}^p(D, \mathbb{C}^m)$, where $D \subset \mathbb{R}^n$ is measurable. For its construction, it has to be taken into account that $\int_D |F(x)| \, \mathrm{d}x = 0$ only implies that $F(x) = 0$ for almost every $x \in D$. Therefore, $\mathrm{L}^p(D, \mathbb{C}^m)$ consists of all $F\colon D \to \mathbb{C}^m$ which are integrable with $\int_D |F(x)|^p \, \mathrm{d}x < +\infty$. However, functions which are equal almost everywhere, that is, pairs of functions $F, G\colon D \to \mathbb{C}^m$ with $\int_D |F(x) - G(x)| \, \mathrm{d}x = 0$, are identified with each other. Note that this uses the algebraic concept of equivalence classes, but the details are omitted here. Furthermore, we write $\mathrm{L}^p[a,b] := \mathrm{L}^p([a,b], \mathbb{R})$ for intervals $D = [a,b]$. In analogy to the spaces of continuous functions (see Definition 2.3.2 and Example 2.5.9), we write $\mathrm{L}^p(D) := \mathrm{L}^p(D, \mathbb{R})$, $\mathrm{l}^p(D) := \mathrm{L}^p(D, \mathbb{R}^3)$, and $\mathbf{l}^p(D) := \mathrm{L}^p(D, \mathbb{R}^{3 \times 3})$. Accordingly, if we wish to distinguish if the norm $\|\cdot\|_p$ belongs to scalar, vectorial, or tensorial functions, we may also write $\|\cdot\|_{\mathrm{L}^p(D)}$, $\|\cdot\|_{\mathrm{l}^p(D)}$, and $\|\cdot\|_{\mathbf{l}^p(D)}$, respectively. Analogously, we can write $\langle\cdot,\cdot\rangle_{\mathrm{L}^2(D)}$, $\langle\cdot,\cdot\rangle_{\mathrm{l}^2(D)}$, and $\langle\cdot,\cdot\rangle_{\mathbf{l}^2(D)}$. The resulting spaces $\mathrm{L}^p(D, \mathbb{C}^m)$ with $1 \leq p < +\infty$ are Banach

spaces such that $(\mathrm{L}^2(D,\mathbb{C}^m), \langle\cdot,\cdot\rangle_2)$ is a Hilbert space (see Example 2.5.2, parts d and e, for the formula of the inner product).

(d) For $1 \le p < +\infty$, the spaces $(\ell^p, \|\cdot\|_p)$ are Banach spaces and $(\ell^2, \langle\cdot,\cdot\rangle_2)$ is a Hilbert space; see Example 2.5.9, part d.

(e) Let $D \subset \mathbb{R}^n$ be measurable. Then we define the **essential supremum** as follows: a measurable function $F\colon D \to \mathbb{C}^m$ is called **essentially bounded** if there exists a constant M such that $|F(x)| \le M$ for almost every $x \in D$. Then

$$\begin{aligned}\|F\|_\infty &:= \sup_{x\in D} \operatorname{ess} |F(x)| \\ &:= \inf\{M \ge 0 : |F(x)| \le M \text{ for almost every } x \in D\}.\end{aligned}$$

As in the case of $\mathrm{L}^p(D,\mathbb{C}^m)$ with $1 \le p < +\infty$, we have to identify again functions which are almost everywhere identical. Collecting all essentially bounded and measurable functions from D into $\mathbb{C}^m$ in this sense, we get $\mathrm{L}^\infty(D,\mathbb{C}^m)$, which is together with the essential supremum a Banach space.

In geomathematics, we are often in the situation that we want to approximate an unknown function by a numerically calculated function (or a sequence of such functions). For this purpose, norms and their induced metrics are an important and useful tool. However, the decision about the accuracy of the approximation or the convergence of the sequence can essentially depend on the chosen norm. Let us discuss this with the help of some examples.

Since $\|F - G\|$ tells us how close the function F is with respect to the function G and $\|(F-G)-0\| = \|F-G\|$, we restrict our attention to the case where the zero function is approximated. Consider the sequence $(F_n)_{n\in\mathbb{N}}$ of functions $F_n\colon [-1,1] \to \mathbb{R}$ with $F_n(t) := \max(1-n|t|,0)$, $t \in [-1,1]$. In the $\mathrm{L}^p[-1,1]$-norm, we obtain

$$\begin{aligned}\|F_n - 0\|_p = \|F_n\|_p &= \left(\int_{-1}^1 |F_n(t)|^p \,\mathrm{d}t\right)^{1/p} = \left(2\int_0^{1/n} (1-nt)^p \,\mathrm{d}t\right)^{1/p} \\ &= \left[-\frac{2}{n(p+1)}(1-nt)^{p+1}\Big|_{t=0}^{t=1/n}\right]^{1/p} = \sqrt[p]{\frac{2}{n(p+1)}}.\end{aligned}$$

Consequently, $(F_n)_{n\in\mathbb{N}}$ converges to the zero function as $n\to\infty$ in every $\mathrm{L}^p[-1,1]$-norm, that is, $\lim_{n\to\infty}\|F_n\|_p = 0$ for all $p \in [1,+\infty[$. However, in the maximum norm, we obtain $\|F_n\|_\infty = 1$ for all $n \in \mathbb{N}$, because the maximum of F_n is always $F_n(0) = 1$. This means that the sequence remains on the unit sphere of the Banach space $(\mathrm{C}[-1,1], \|\cdot\|_\infty)$ and never approaches the zero element. Note that (F_n) also does not converge pointwise to the zero function, because $F_n(0) = 1$ for all $n \in \mathbb{N}$. It converges pointwise to the function $F\colon [-1,1] \to \mathbb{R}$ with $F(0) = 1$ and $F(t) = 0$ for $t \neq 0$, which is not continuous; see also Figure 2.6.

Furthermore, let us consider the sequence $(G_n)_{n\in\mathbb{N}}$ in $\mathrm{C}^{(\infty)}[-1,1]$ with $G_n(t) := n^{-1}\cos(n^2\pi t)$, $t \in [-1,1]$. In this case, we have $\|G_n\|_\infty = n^{-1} \to 0$ as $n \to \infty$ and, due to

$$\|G_n\|_p = \left(\int_{-1}^1 |G_n(t)|^p \,\mathrm{d}t\right)^{1/p} \le \left(\int_{-1}^1 \|G_n\|_\infty^p \,\mathrm{d}t\right)^{1/p} = \|G_n\|_\infty \sqrt[p]{2},$$

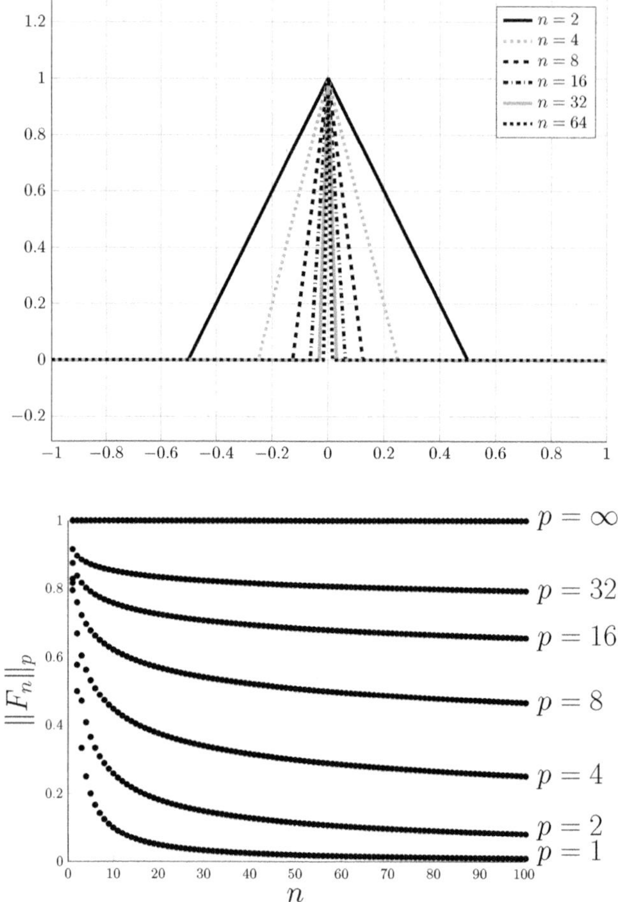

Figure 2.6 The approximation quality and the convergence of a sequence of functions (F_n) such as $F_n(t) := \max(1 - n|t|, 0)$, $n \in \mathbb{N}$, can be measured by the metric induced by a norm in terms of $\|F_n - F\|$. However, the result can essentially depend on the chosen norm. This is illustrated by two examples. In this first example, the sequence (F_n) of functions (top) converges to the zero function in all $\|\cdot\|_p$-norms *but neither* pointwise *nor* uniformly (i.e., in the $\|\cdot\|_\infty$-norm). Moreover, the smaller p, the 'weaker' is the requirement for convergence and the faster the sequence converges (bottom). For the second example, see Figure 2.7.

also $\lim_{n\to\infty} \|G_n\|_p = 0$ for all $p \in [1, +\infty[$. However, if we consider the derivative $G_n'(t) = -n\pi \sin(n^2\pi t)$, $t \in [-1, 1]$, then $\|G_n\|_{C^{(1)}[-1,1]} = \|G_n\|_\infty + \|G_n'\|_\infty = n^{-1} + n\pi$ such that $\|G_n\|_{C^{(1)}[-1,1]} \to \infty$ as $n \to \infty$; see also Figure 2.7.

Hence, there is, indeed, a major difference between the various kinds of convergences which we get for different choices of norms. In general, we can say that integration smooths a function, whereas differentiation roughens a function. Hence, integration-based norms more easily 'excuse' small outliers in contrast to differentiation-based norms.

Concerning the norms which are important for us, we can formulate the following result.

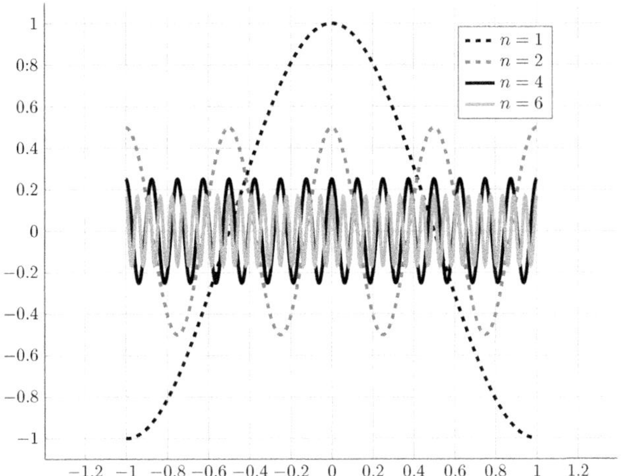

Figure 2.7 In this second example, the sequence of functions $G_n(t) := n^{-1}\cos(n^2\pi t)$, $t \in [-1,1]$, $n \in \mathbb{N}$, converges to the zero function in all $\|\cdot\|_p$-norms *as well as* pointwise and uniformly. However, it diverges in the $\|\cdot\|_{C^{(1)}[-1,1]}$-norm.

Theorem 2.5.20 *Let $D \subset \mathbb{R}^n$ be a given domain.*

(a) *If D is measurable with a finite Lebesgue measure (i.e., $\lambda(D) < +\infty$) and $1 \le p \le q < \infty$, then $\|F\|_p^p \le \lambda(D) + \|F\|_q^q$ for all $F \in \mathrm{L}^q(D)$. In particular, $\mathrm{L}^q(D) \subset \mathrm{L}^p(D)$.*

(b) *If D is compact, then $\|F\|_p \le \sqrt[p]{\lambda(D)}\,\|F\|_\infty$ for all $p \in [1,+\infty[$ and all $F \in \mathrm{C}(D)$. In particular, $\mathrm{C}(D) \subset \mathrm{L}^p(D)$ for all $p \in [1,+\infty[$.*

(c) *If D is compact (note that D should here be reasonably harmless such that derivatives $F^{(j)}$ make sense), then $\|F\|_\infty \le \|F\|_{\mathrm{C}^{(j)}(D)} \le \|F\|_{\mathrm{C}^{(k)}(D)}$ for all $F \in \mathrm{C}^{(k)}(D)$ and all $j,k \in \mathbb{N}_0$ with $j \le k$.*

Proof Each part has a relatively easy proof.

(a) If $F \in \mathrm{L}^q(D)$ and $p \le q$, then we consider $S_1 := \{x \in D \mid |F(x)| \le 1\}$ and $S_2 := D \setminus S_1$. By subdividing D into the disjoint union $D = S_1 \dot\cup S_2$, we obtain

$$\|F\|_p^p = \int_D |F(x)|^p\,\mathrm{d}x \le \int_{S_1} 1\,\mathrm{d}x + \int_{S_2} |F(x)|^q\,\mathrm{d}x \le \lambda(D) + \int_D |F(x)|^q\,\mathrm{d}x.$$

(b) If $F \in \mathrm{C}(D)$, then the monotonicity of the integral yields that

$$\|F\|_p = \left(\int_D |F(x)|^p\,\mathrm{d}x\right)^{1/p} \le \left(\int_D \|F\|_\infty^p\,\mathrm{d}x\right)^{1/p}$$
$$= \|F\|_\infty \left(\int_D 1\,\mathrm{d}x\right)^{1/p} = \|F\|_\infty \sqrt[p]{\lambda(D)}.$$

(c) Finally, the inequality for the norm in $\mathrm{C}^{(k)}(D)$ is trivial. □

Hence, if D is compact, then we have, for $1 \leq p \leq q < \infty$ and $j,k \in \mathbb{N}_0$ with $j \leq k$, the inclusion

$$\mathrm{C}^{(k)}(D) \subset \mathrm{C}^{(j)}(D) \subset \cdots \subset \mathrm{C}(D) \subset \mathrm{L}^q(D) \subset \mathrm{L}^p(D)$$

and the implications (for $n \to \infty$)

$$\begin{aligned} \|F_n - F\|_{\mathrm{C}^{(k)}(D)} \to 0 &\Rightarrow \|F_n - F\|_{\mathrm{C}^{(j)}(D)} \to 0 \\ &\Rightarrow \|F_n - F\|_\infty \to 0 \Rightarrow \|F_n - F\|_p \to 0. \end{aligned}$$

Moreover, again for $n \to \infty$,

$$\|F_n - F\|_\infty \to 0 \Rightarrow |F_n(x) - F(x)| \to 0 \text{ for all } x \in D.$$

In general, these implications are not valid in the opposite direction, as we demonstrated for some examples.

In this context, it is also useful to know that $\mathrm{L}^p(D)$-functions can be approximated arbitrarily well by continuous functions, as the following theorem shows (for a proof, see McShane, 1974, p. 229).

Theorem 2.5.21 *Let $D \subset \mathbb{R}^n$ be measurable and $p \in [1, +\infty[$. If $F \in \mathrm{L}^p(D)$ and $\varepsilon > 0$, then there exists a continuous function $G\colon \mathbb{R}^n \to \mathbb{R}$ with $\|F - G\|_p = \left(\int_D |F(x) - G(x)|^p \,\mathrm{d}x\right)^{1/p} < \varepsilon$. In particular, if D is compact, then $\mathrm{C}(D)$ is a dense subset of $\mathrm{L}^p(D)$, that is, $\overline{\mathrm{C}(D)}^{\|\cdot\|_{\mathrm{L}^p(D)}} = \mathrm{L}^p(D)$.*

Even more can be shown. The following theorem from Voigt and Wloka (1975, Theorems 3.2.6 and 3.6.2) uses the concept of manifolds. We omit a discussion of the precise definition and only mention that the sphere is a valid example here.

Theorem 2.5.22 *Let $D \subset \mathbb{R}^n$ be an open set and M be an n-dimensional and compact $\mathrm{C}^{(\infty)}$-manifold. Then $\mathrm{C}^{(\infty)}(D) \cap \mathrm{L}^2(D)$ is dense in $\mathrm{L}^2(D)$ and $\mathrm{C}^{(\infty)}(M)$ is dense in $\mathrm{L}^2(M)$.*

We will also make use of the well-known Stone–Weierstraß approximation theorem (see Dieudonné, 1960, (7.4.1)).

Theorem 2.5.23 (Stone–Weierstraß Theorem) *Let $D \subset \mathbb{R}^n$ be a compact subset. Then the set $\mathrm{Pol}(D)$ of all polynomials in n variables restricted to D is dense in $\left(\mathrm{C}(D), \|\cdot\|_{\mathrm{C}(D)}\right)$.*

Let us come back to the most specialized spaces, the Hilbert spaces. We have already come across some examples of orthonormal systems. In $\mathbb{R}^3$, the standard basis $\{\varepsilon^1, \varepsilon^2, \varepsilon^3\}$ is obviously also an orthonormal system. Moreover, the representation of an arbitrary vector $x \in \mathbb{R}^3$ in this basis is very easy: $x = \left(\langle x, \varepsilon^1\rangle, \langle x, \varepsilon^2\rangle, \langle x, \varepsilon^3\rangle\right)^{\mathrm{T}} = \sum_{j=1}^3 \langle x, \varepsilon^j\rangle \varepsilon^j$, where $\langle\cdot,\cdot\rangle$ is the Euclidean inner product (see Definition 2.1.3). In Example 2.5.4, we discussed a common local orthonormal system, which can be used to separate a vectorial field into a normal (i.e., vertical), a North, and an East component: $f\colon \Omega \to \mathbb{R}^3$ can be decomposed into $f = \langle f, \varepsilon^r\rangle \varepsilon^r + \langle f, \varepsilon^\varphi\rangle \varepsilon^\varphi + \langle f, \varepsilon^t\rangle \varepsilon^t$, where $\langle\cdot,\cdot\rangle$ is still the Euclidean inner product.

As we clearly see in the first example, we have that $x = 0$ holds if and only if $\langle x, \varepsilon^j \rangle = 0$ for all j. Indeed, this holds in a more general context: if $(\mathcal{X}, \langle \cdot, \cdot \rangle)$ is a Hilbert space with finite dimension $n \in \mathbb{N}$ and $\{y_1, \ldots, y_n\} \subset \mathcal{X}$ is an orthonormal system (and, consequently, also a basis), then every $x \in \mathcal{X}$ can be represented as

$$x = \sum_{j=1}^{n} \langle x, y_j \rangle y_j. \tag{2.12}$$

Due to the bilinearity of the inner product, the zero vector $x = 0$ obviously satisfies $\langle x, y_j \rangle = 0$ for all j. Vice versa, (2.12) immediately yields that $x = 0$, if $\langle x, y_j \rangle = 0$ for all j.

These nice and in many contexts extremely useful properties are not restricted to the finite-dimensional case.

Definition 2.5.24 Let $(\mathcal{X}, \langle \cdot, \cdot \rangle)$ be a pre-Hilbert space. A system of vectors $\{x_\alpha\}_{\alpha \in A} \subset \mathcal{X}$ is called **complete** if the following holds true: the only vector $f \in \mathcal{X}$ which satisfies $\langle f, x_\alpha \rangle = 0$ for all $\alpha \in A$ is $f = 0$.

Note that there is an essential difference between complete sets, where all Cauchy sequences are convergent (see Definition 2.5.13), and complete systems of vectors, which are defined in Definition 2.5.24.

In the finite-dimensional case, it is clear that a system of n orthonormal vectors is a basis if n is the dimension of the space. This system is then also complete. However, in the infinite-dimensional case, this argumentation is not possible, because 'infinite is not infinite'. For example, the functions $F_n(t) := \sqrt{\pi}^{-1} \sin(nt)$, $t \in [0, 2\pi]$, $n \in \mathbb{N}$, constitute an (infinite but countable) orthonormal system in the infinite-dimensional space $\mathrm{L}^2[0, 2\pi]$. However, this system is neither a basis nor complete: every constant function is orthogonal to each F_n, because $\sqrt{\pi}^{-1} \int_0^{2\pi} c \sin(nt) \, dt = 0$ for all $c \in \mathbb{R}$ and all $n \in \mathbb{N}$. A well-known complete orthonormal system is instead given by

$$F_{0,1}(t) := \frac{1}{\sqrt{2\pi}}, \qquad F_{n,1}(t) := \frac{1}{\sqrt{\pi}} \cos(nt), \quad n \in \mathbb{N},$$

$$F_{n,2}(t) := \frac{1}{\sqrt{\pi}} \sin(nt), \quad n \in \mathbb{N},$$

$t \in [0, 2\pi]$; see, for example, Chui (1992, pp. 36–43).

The concept of complete orthonormal systems, indeed, generalizes the concept of a basis to the case of arbitrary dimensions, as the following fundamental theorem shows.

Theorem 2.5.25 *Let $(\mathcal{X}, \langle \cdot, \cdot \rangle)$ be a Hilbert space and $\{x_\alpha\}_{\alpha \in A}$ be an arbitrary orthonormal system of vectors in $\mathcal{X}$. Then the following holds true:*

(1) *For every $f \in \mathcal{X}$, we have $\langle f, x_\alpha \rangle \neq 0$ for at most a countable number of indices $\alpha \in A$, and the **Bessel inequality** $\sum_{\alpha \in A} |\langle f, x_\alpha \rangle|^2 \leq \|f\|^2$ holds true, where $\|\cdot\|$ is the induced norm of $\mathcal{X}$.*

(2) *The system* $\{x_\alpha\}_{\alpha\in A}$ *is complete if and only if one of the following* equivalent *criteria is satisfied:*

(a) *Every* f *can be expanded into a* ***Fourier series*** $f = \sum_{\alpha\in A}\langle f, x_\alpha\rangle x_\alpha$.

(b) *The* ***Parseval identity I*** *holds true:* $\langle f, g\rangle = \sum_{\alpha\in A}\langle f, x_\alpha\rangle\overline{\langle g, x_\alpha\rangle}$ *for all* $f, g \in \mathcal{X}$.

(c) *The* ***Parseval identity II*** *holds true:* $\|f\|^2 = \sum_{\alpha\in A}|\langle f, x_\alpha\rangle|^2$ *for all* $f \in \mathcal{X}$.

(d) *We have* $\overline{\operatorname{span}\{x_\alpha \mid \alpha \in A\}}^{\|\cdot\|} = \mathcal{X}$.

The latter property can be regarded as a basis property. In approximation theory, systems with this property are called closed (which should not be mixed up with the concept of closed sets; see Definition 2.5.13).

Definition 2.5.26 Let $(\mathcal{X}, \|\cdot\|)$ be a normed space. A system of vectors $\{x_\alpha\}_{\alpha\in A} \subset \mathcal{X}$ is called **closed (in the sense of the approximation theory)** if

$$\overline{\operatorname{span}\{x_\alpha \mid \alpha \in A\}}^{\|\cdot\|} = \mathcal{X}. \tag{2.13}$$

If $\mathcal{X}$ is additionally a Hilbert space, then an orthonormal system $\{x_\alpha\}_{\alpha\in A} \subset \mathcal{X}$ is called an **orthonormal basis** if and only if it satisfies (2.13). More generally, an orthogonal system $\{x_\alpha\}_{\alpha\in A} \subset \mathcal{X}$ which does *not* include the zero vector is called an **orthogonal basis** if and only if it satisfies (2.13).

A proof of the following theorem can be found, for example, in Davis (1975, p. 263).

Theorem 2.5.27 *Let* $(\mathcal{X}, \langle\cdot,\cdot\rangle_{\mathcal{X}})$ *be an inner product space and* $\mathcal{S} \subset \mathcal{X}$. *Then* $\mathcal{S}$ *is complete in* $\mathcal{X}$ *(in the sense of Definition 2.5.24) if and only if it is closed in* $\mathcal{X}$ *in the sense of the approximation theory.*

We conclude the general discussions on complete orthonormal systems with the following theorem, which is a consequence of Zorn's lemma.

Theorem 2.5.28 *Let* $(\mathcal{X}, \langle\cdot,\cdot\rangle)$ *be an arbitrary Hilbert space which consists of more than only the zero vector. If* $\{x_\alpha\}_{\alpha\in A} \neq \emptyset$ *is an orthonormal system in* $\mathcal{X}$, *then it can be supplemented by further vectors to obtain an orthonormal basis. In particular, every such Hilbert space has at least one orthonormal basis.*

The second part of the theorem is an easy consequence of the first one: if $x \in \mathcal{X}$ with $x \neq 0$ (such a vector must exist in $\mathcal{X}$), then $\{x/\|x\|\}$ is a (simple) orthonormal system. It has only one element, but this is enough to apply the first part of the theorem and supplement this system to a complete system.

Another important concept in functional analysis in general but also in geomathematics is the operator theory, of which we will summarize the required essentials here.

Definition 2.5.29 A mapping $\mathcal{T}: \mathcal{X} \to \mathcal{Y}$ between two metric spaces $\mathcal{X}$ and $\mathcal{Y}$ is called an **operator**. If $\mathcal{X}$ is a normed space with the norm $\|\cdot\|: \mathcal{X} \to \mathbb{K}$, then an operator $\mathcal{F}\colon \mathcal{X} \to \mathbb{K}$ is (also) called a **functional**. We usually write $\mathcal{T}x$ instead of $\mathcal{T}(x)$ for the images of an operator $\mathcal{T}$.

Definition 2.5.30 Let $(\mathcal{X}, d_{\mathcal{X}})$ and $(\mathcal{Y}, d_{\mathcal{Y}})$ be two metric spaces and $\mathcal{T}\colon \mathcal{X} \to \mathcal{Y}$ be an operator. Then $\mathcal{T}$ is **continuous** in $x \in \mathcal{X}$ if, for every sequence $(x_n) \subset \mathcal{X}$ which converges to x, the sequence $(\mathcal{T}x_n)$ converges in $\mathcal{Y}$ to $\mathcal{T}x$. $\mathcal{T}$ is called continuous if it is continuous in every $x \in \mathcal{X}$.

As we know from real analysis, the previous sequential continuity is equivalent to the ε-δ-criterion: $\mathcal{T}\colon \mathcal{X} \to \mathcal{Y}$ is continuous in $x \in \mathcal{X}$ if and only if, for every $\varepsilon > 0$, there exists $\delta > 0$ such that all $\xi \in \mathcal{X}$ with $d_{\mathcal{X}}(x, \xi) < \delta$ satisfy $d_{\mathcal{Y}}(\mathcal{T}x, \mathcal{T}\xi) < \varepsilon$.

Moreover, $\mathcal{T}\colon \mathcal{X} \to \mathcal{Y}$ is continuous if and only if the **preimage** (also called the **inverse image**) $\mathcal{T}^{-1}(\mathcal{V})$ of every open set $\mathcal{V} \subset \mathcal{Y}$ is open in $\mathcal{X}$. This equivalent criterion for continuity is also called the **topological continuity**.

For linear operators, the case of continuous operators plays a particular role.

Definition 2.5.31 Let $(\mathcal{X}, \|\cdot\|_{\mathcal{X}})$ and $(\mathcal{Y}, \|\cdot\|_{\mathcal{Y}})$ be two normed spaces and $\mathcal{T}\colon \mathcal{X} \to \mathcal{Y}$ be a given operator.

(a) $\mathcal{T}$ is called **linear** if $\mathcal{T}(\lambda x + \mu y) = \lambda \mathcal{T}x + \mu \mathcal{T}y$ for all $x, y \in \mathcal{X}$ and all $\lambda, \mu \in \mathbb{K}$ (where $\mathcal{X}$ and $\mathcal{Y}$ are both $\mathbb{K}$-vector spaces).

(b) $\mathcal{T}$ is called **bounded** if

$$\|\mathcal{T}\|_{\mathcal{L}} := \|\mathcal{T}\|_{\mathcal{L}(\mathcal{X},\mathcal{Y})} := \sup_{\substack{x \in \mathcal{X} \\ x \neq 0}} \frac{\|\mathcal{T}x\|_{\mathcal{Y}}}{\|x\|_{\mathcal{X}}}$$

is *finite*. The set of all bounded linear operators $\mathcal{T}\colon \mathcal{X} \to \mathcal{Y}$ is denoted by $\mathcal{L}(\mathcal{X}, \mathcal{Y})$. Moreover, $\mathcal{L}(\mathcal{X}) := \mathcal{L}(\mathcal{X}, \mathcal{X})$.

Theorem 2.5.32 *Let $(\mathcal{X}, \|\cdot\|_{\mathcal{X}})$, $(\mathcal{Y}, \|\cdot\|_{\mathcal{Y}})$, and $(\mathcal{Z}, \|\cdot\|_{\mathcal{Z}})$ be normed spaces. Then the following holds true:*

(a) A linear operator $\mathcal{T}\colon \mathcal{X} \to \mathcal{Y}$ is bounded if and only if it is continuous.

(b) The vector space $(\mathcal{L}(\mathcal{X}, \mathcal{Y}), \|\cdot\|_{\mathcal{L}})$ is a normed space. It is complete, if $\mathcal{Y}$ is complete.

*(c) The **operator norm** $\|\cdot\|_{\mathcal{L}}$ has the following equivalent formulae:*

$$\begin{aligned}\|\mathcal{T}\|_{\mathcal{L}(\mathcal{X},\mathcal{Y})} &= \sup\left\{\|\mathcal{T}x\|_{\mathcal{Y}} \mid x \in \mathcal{X},\ \|x\|_{\mathcal{X}} \le 1\right\} \\ &= \sup\left\{\|\mathcal{T}x\|_{\mathcal{Y}} \mid x \in \mathcal{X},\ \|x\|_{\mathcal{X}} = 1\right\}.\end{aligned}$$

(d) If $\mathcal{S} \in \mathcal{L}(\mathcal{Y}, \mathcal{Z})$ and $\mathcal{T} \in \mathcal{L}(\mathcal{X}, \mathcal{Y})$, then $\mathcal{S}\mathcal{T} \in \mathcal{L}(\mathcal{X}, \mathcal{Z})$ and

$$\|\mathcal{S}\mathcal{T}\|_{\mathcal{L}(\mathcal{X},\mathcal{Z})} \le \|\mathcal{S}\|_{\mathcal{L}(\mathcal{Y},\mathcal{Z})} \cdot \|\mathcal{T}\|_{\mathcal{L}(\mathcal{X},\mathcal{Y})}.$$

Theorem 2.5.33 (Banach–Steinhaus Theorem) *Let $(\mathcal{X}, \|\cdot\|_{\mathcal{X}})$ as well as $\left(\mathcal{Y}, \|\cdot\|_{\mathcal{Y}}\right)$ be Banach spaces and $(\mathcal{T}_n) \subset \mathcal{L}(\mathcal{X}, \mathcal{Y})$ be a given sequence. This sequence converges pointwise (i.e., $\|\mathcal{T}_n x - \mathcal{T}x\|_{\mathcal{Y}} \to 0$ as $n \to \infty$ for each $x \in \mathcal{X}$) to a* *continuous and linear operator* $\mathcal{T}: \mathcal{X} \to \mathcal{Y}$ *if and only if both of the following criteria are fulfilled:*

(a) $\sup_n \|\mathcal{T}_n\|_{\mathcal{L}} < +\infty$.
(b) *There exists a dense subset* $\mathcal{S} \subset \mathcal{X}$ *such that, for each fixed* $x \in \mathcal{S}$, *the sequence* $(\mathcal{T}_n x)_n$ *converges in* $\mathcal{Y}$.

Example 2.5.34 We consider two operators:

$$\mathcal{S}: \mathrm{C}[0,1] \to \mathrm{C}[0,1] \qquad\qquad \mathcal{T}: \mathrm{C}^{(1)}[0,1] \to \mathrm{C}[0,1]$$

$$F \mapsto \mathcal{S}F := \int_0^{\cdot} F(t)\,\mathrm{d}t, \qquad\qquad F \mapsto \mathcal{T}F := F',$$

where the maximum norm $\|\cdot\|_\infty$ is chosen for both involved spaces. Obviously, both operators are linear. Moreover,

$$\|\mathcal{S}F\|_\infty = \max_{x\in[0,1]} \left| \int_0^x F(t)\,\mathrm{d}t \right| \le \max_{x\in[0,1]} \int_0^x \underbrace{|F(t)|}_{\le \|F\|_\infty}\,\mathrm{d}t \le \|F\|_\infty \cdot 1$$

such that $\mathcal{S}$ is bounded (and, consequently, continuous) with $\|\mathcal{S}\|_{\mathcal{L}} \le 1$. If we set $F \equiv 1$, then $\|\mathcal{S}F\|_\infty = 1 = \|F\|_\infty$ such that $\|\mathcal{S}\|_{\mathcal{L}} = 1$.

However, if we choose $F_n(t) := \sin(n\pi t)$, $t \in [0,1]$, $n \in \mathbb{N}$, then $(\mathcal{T}F_n)(t) = n\pi\cos(n\pi t)$. Consequently, $\|F_n\|_\infty = 1$ and $\|\mathcal{T}F_n\|_\infty = n\pi$ for all $n \in \mathbb{N}$, that is, $\|\mathcal{T}\|_{\mathcal{L}} = \infty$.

Hence, note that linear operators where at least one of the involved spaces is infinite dimensional need not necessarily be continuous. It suffices, indeed, that $\mathcal{X}$ is infinite dimensional such that *not all* linear operators $\mathcal{T}: \mathcal{X} \to \mathcal{Y}$ are continuous: let $\mathcal{F}: \mathrm{C}^{(1)}[0,1] \to \mathbb{R}$ be the functional given by $\mathcal{F}F := F'(0)$. Then our previous counterexample works again: $|\mathcal{F}F_n| = n\pi$ for all $n \in \mathbb{N}_0$.

Definition 2.5.35 Let $(\mathcal{X}, d_{\mathcal{X}})$ and $(\mathcal{Y}, d_{\mathcal{Y}})$ be two metric spaces and let $\mathcal{T}: \mathcal{X} \to \mathcal{Y}$ be a mapping. This mapping $\mathcal{T}$ is called an **open mapping**, if the image $\mathcal{T}(\mathcal{U})$ of every open set $\mathcal{U} \subset \mathcal{X}$ is an open set in the subspace $\mathcal{T}(\mathcal{X}) \subset \mathcal{Y}$.

Theorem 2.5.36 (Open Mapping Theorem) *Let* $(\mathcal{X}, \|\cdot\|_{\mathcal{X}})$ *and* $(\mathcal{Y}, \|\cdot\|_{\mathcal{Y}})$ *be two Banach spaces. Every* $\mathcal{T} \in \mathcal{L}(\mathcal{X},\mathcal{Y})$ *with* $\mathcal{T}(\mathcal{X}) = \mathcal{Y}$ *is an open mapping.*

Lemma 2.5.37 *If* $(\mathcal{X}, \|\cdot\|_{\mathcal{X}})$ *and* $(\mathcal{Y}, \|\cdot\|_{\mathcal{Y}})$ *are two Banach spaces, then the Cartesian product* $\mathcal{X} \times \mathcal{Y}$ *equipped with* $\|(x,y)\|_{\mathcal{X}\times\mathcal{Y}} := \sqrt{\|x\|_{\mathcal{X}}^2 + \|y\|_{\mathcal{Y}}^2}$, $x \in \mathcal{X}$, $y \in \mathcal{Y}$, *is a Banach space.*

Theorem 2.5.38 (Closed Graph Theorem) *Let* $(\mathcal{X}, \|\cdot\|_{\mathcal{X}})$ *and* $(\mathcal{Y}, \|\cdot\|_{\mathcal{Y}})$ *be two Banach spaces and let* $\mathcal{T}: \mathcal{X} \to \mathcal{Y}$ *be a linear mapping. Then* $\mathcal{T}$ *is continuous if and only if its* ***graph****, that is,* $\operatorname{graph}\mathcal{T} := \{(x, \mathcal{T}x) \mid x \in \mathcal{X}\}$, *is a closed set in* $\mathcal{X} \times \mathcal{Y}$.

In several applications in the geosciences but also in other disciplines such as medical imaging, **inverse problems** play an important role. In functional analytic language, we have an operator $\mathcal{T}: \mathcal{X} \to \mathcal{Y}$, and the inverse problem is as follows: given $G \in \mathcal{Y}$, find $F \in \mathcal{X}$ such that $\mathcal{T}F = G$. We will further investigate such problems in detail, when we are at the right point in this book. One aspect could or should be addressed already here: if $\mathcal{T} \in \mathcal{L}(\mathcal{X}, \mathcal{Y})$, is there a continuous inverse, that is, does $\mathcal{T}^{-1} \in \mathcal{L}(\mathcal{Y}, \mathcal{X})$ or, at least, $\mathcal{T}^{-1} \in \mathcal{L}(\mathcal{T}(\mathcal{X}), \mathcal{X})$ exist?

Functional analysis provides us with some general answers. Certainly, the inverse of a linear operator is also linear (if it exists).

Theorem 2.5.39 *Let $(\mathcal{X}, \|\cdot\|_{\mathcal{X}})$ and $(\mathcal{Y}, \|\cdot\|_{\mathcal{Y}})$ be two normed spaces and $\mathcal{T}: \mathcal{X} \to \mathcal{Y}$ be a linear operator. Then $\mathcal{T}$ has a continuous inverse $\mathcal{T}^{-1} \in \mathcal{L}(\mathcal{T}(\mathcal{X}), \mathcal{X})$, if and only if there exists a constant $C > 0$ such that $C\|x\|_{\mathcal{X}} \leq \|\mathcal{T}x\|_{\mathcal{Y}}$ for all $x \in \mathcal{X}$.*

A direct conclusion is the following theorem.

Theorem 2.5.40 *Let $(\mathcal{X}, \|\cdot\|_{\mathcal{X}})$ and $(\mathcal{Y}, \|\cdot\|_{\mathcal{Y}})$ be two normed spaces and $\mathcal{T}: \mathcal{X} \to \mathcal{Y}$ be a linear operator. Then $\mathcal{T}$ does* not *have a continuous inverse (from $\mathcal{T}(\mathcal{X})$ to $\mathcal{X}$) if and only if there exists a sequence $(x_n) \subset \mathcal{X}$ such that $\|x_n\|_{\mathcal{X}} = 1$ for all n and $(\mathcal{T}x_n)$ converges to 0 in $\mathcal{Y}$.*

The next theorem is an immediate consequence of the open mapping theorem (Theorem 2.5.36) if one discovers that a (topologically) continuous inverse $\mathcal{T}^{-1}$ is connected to an open mapping $\mathcal{T}$.

Theorem 2.5.41 (Inverse Mapping Theorem) *Let $(\mathcal{X}, \|\cdot\|_{\mathcal{X}})$ as well as $(\mathcal{Y}, \|\cdot\|_{\mathcal{Y}})$ be given Banach spaces and $\mathcal{T} \in \mathcal{L}(\mathcal{X}, \mathcal{Y})$ be a given operator, which is additionally bijective. Then its inverse $\mathcal{T}^{-1}: \mathcal{Y} \to \mathcal{X}$ is continuous.*

Example 2.5.42 Let us have a look again at Example 2.5.34. The operator $\mathcal{S}$ is invertible on its image, that is, $(\mathcal{T} \circ \mathcal{S})F = F$ for all $F \in \mathrm{C}[0,1]$. $\mathcal{S}$ is linear and continuous but its inverse $\mathcal{S}^{-1} = \mathcal{T}$ is not continuous. This is not a contradiction to Theorem 2.5.41, because $\mathcal{S}$ is not surjective: all functions in its image $\mathcal{S}(\mathrm{C}[0,1])$ are continuously differentiable and vanish at $x = 0$, that is, the image is a proper subspace of $\mathrm{C}[0,1]$. Moreover, we can refer to Theorem 2.5.40 and use (F_n) from Example 2.5.34: we have $\|F_n\|_\infty = 1$ for all n and $\|\mathcal{S}F_n\|_\infty = (n\pi)^{-1} \to 0$ as $n \to \infty$.

On the other hand, $\mathcal{T}$ is not injective. However, on the subspace $\mathcal{X} := \{F \in \mathrm{C}^{(1)}[0,1] \mid F(0) = 0\}$, it is invertible: $(\mathcal{S} \circ \mathcal{T})(F) = F$ for all $F \in \mathcal{X}$. Moreover, $F(s) = \int_0^s F'(t)\,\mathrm{d}t = \int_0^s (\mathcal{T}F)(t)\,\mathrm{d}t$, $s \in [0,1]$, such that $|F(s)| \leq \int_0^s \|\mathcal{T}F\|_\infty\,\mathrm{d}t$, $s \in [0,1]$, and $\|F\|_\infty \leq \|\mathcal{T}F\|_\infty$. This corresponds to Theorem 2.5.39 and confirms that $\mathcal{S} = (\mathcal{T}|_{\mathcal{X}})^{-1}$ is continuous. Nevertheless, one could get the idea to refer still to the inverse mapping theorem and say: let us look at $\tilde{\mathcal{S}}: \mathrm{C}[0,1] \to \mathcal{X}$ with $\tilde{\mathcal{S}}F := \mathcal{S}F$ for all $F \in \mathrm{C}[0,1]$.

Then $\tilde{\mathcal{S}}$ is bijective. So, doesn't it have a continuous inverse then? Actually, no, because $\mathcal{T}$ is still the inverse. This is again no contradiction to Theorem 2.5.41, because $(\mathcal{X}, \|\cdot\|_\infty)$ is not complete: let

$$G_n(t) := \sqrt{\left(t - \frac{1}{2}\right)^2 + \frac{1}{n^2}} - \frac{\sqrt{n^2+4}}{2n}, \quad t \in [0,1], \quad n \in \mathbb{N}.$$

Then $G_n \in \mathcal{X}$ for all $n \in \mathbb{N}$, because the radicand is away from zero and $G_n(0) = 0$. Moreover, since

$$\left|\sqrt{\left(t - \frac{1}{2}\right)^2 + \frac{1}{n^2}} + \left|t - \frac{1}{2}\right|\right| \geq \frac{1}{n}$$

and

$$\left|\sqrt{\left(t - \frac{1}{2}\right)^2 + \frac{1}{n^2}} - \left|t - \frac{1}{2}\right|\right| \cdot \left|\sqrt{\left(t - \frac{1}{2}\right)^2 + \frac{1}{n^2}} + \left|t - \frac{1}{2}\right|\right|$$
$$= \left(t - \frac{1}{2}\right)^2 + \frac{1}{n^2} - \left(t - \frac{1}{2}\right)^2 = \frac{1}{n^2},$$

we have

$$\left|G_n(t) - \underbrace{\left(\left|t - \frac{1}{2}\right| - \frac{1}{2}\right)}_{=:G(t)}\right| \leq \frac{1}{n} + \sqrt{\frac{1}{4} + \frac{1}{n^2}} - \frac{1}{2}$$

for all $t \in [0, 1]$ and every $n \in \mathbb{N}$. Hence, $\|G_n - G\|_\infty \to 0$, that is, G is the uniform limit of (G_n). Thus, (G_n) is a Cauchy sequence in $(\mathcal{X}, \|\cdot\|_\infty)$, but it does not converge in this space, because G is not differentiable in $t = 1/2$ and, consequently, $G \notin \mathcal{X}$.

The existence, or better the non-existence, of continuous inverses is associated to compact operators, particularly for practical problems.

Definition 2.5.43 A linear operator $\mathcal{T}: \mathcal{X} \to \mathcal{Y}$ between normed spaces $(\mathcal{X}, \|\cdot\|_\mathcal{X})$ and $(\mathcal{Y}, \|\cdot\|_\mathcal{Y})$ is called **compact** if the image $\mathcal{T}\mathcal{U}$ of the unit sphere $\mathcal{U} := \{x \in \mathcal{X} \mid \|x\|_\mathcal{X} = 1\}$ is relatively compact in $\mathcal{Y}$. The space of all compact linear operators $\mathcal{T}: \mathcal{X} \to \mathcal{Y}$ is denoted by $\mathcal{K}(\mathcal{X}, \mathcal{Y})$. Again, $\mathcal{K}(\mathcal{X}) := \mathcal{K}(\mathcal{X}, \mathcal{X})$.

Theorem 2.5.44 *It is $\mathcal{K}(\mathcal{X}, \mathcal{Y}) \subset \mathcal{L}(\mathcal{X}, \mathcal{Y})$ for all normed spaces $\mathcal{X}$ and $\mathcal{Y}$, where $\mathcal{K}(\mathcal{X}, \mathcal{Y})$ is a vector subspace of $\mathcal{L}(\mathcal{X}, \mathcal{Y})$.*

Theorem 2.5.45 *Let $\mathcal{X}, \mathcal{Y}$, and $\mathcal{Z}$ be normed spaces.*

(a) If $\mathcal{S} \in \mathcal{L}(\mathcal{Y}, \mathcal{Z})$ and $\mathcal{T} \in \mathcal{K}(\mathcal{X}, \mathcal{Y})$, then $\mathcal{S}\mathcal{T} \in \mathcal{K}(\mathcal{X}, \mathcal{Z})$.
(b) If $\mathcal{S} \in \mathcal{K}(\mathcal{Y}, \mathcal{Z})$ and $\mathcal{T} \in \mathcal{L}(\mathcal{X}, \mathcal{Y})$, then $\mathcal{S}\mathcal{T} \in \mathcal{K}(\mathcal{X}, \mathcal{Z})$.

Theorem 2.5.46 *Let $\mathcal{X}$ be a normed space and $\mathcal{Y}$ be a* Banach space. *If $(\mathcal{T}_n) \subset \mathcal{K}(\mathcal{X}, \mathcal{Y})$ is a sequence which converges to $\mathcal{T} \in \mathcal{L}(\mathcal{X}, \mathcal{Y})$ in the sense that $\lim_{n\to\infty} \|\mathcal{T} - \mathcal{T}_n\|_\mathcal{L} = 0$, then $\mathcal{T} \in \mathcal{K}(\mathcal{X}, \mathcal{Y})$. In other words, $\mathcal{K}(\mathcal{X}, \mathcal{Y})$ is a closed subspace of $\mathcal{L}(\mathcal{X}, \mathcal{Y})$.*

Theorem 2.5.47 *Let $\mathcal{X}$ and $\mathcal{Y}$ be normed spaces and $\mathcal{T}: \mathcal{X} \to \mathcal{Y}$ be a linear operator. If $\mathcal{X}$ is of finite dimension or if the image $\mathcal{T}(\mathcal{X})$ is of finite dimension and $\mathcal{T}$ is continuous, then $\mathcal{T}$ is compact.*

By combining Theorems 2.5.46 and 2.5.47, we immediately get the following proposition.

Theorem 2.5.48 *Let $\mathcal{X}$ be a normed space and $\mathcal{Y}$ be a* Banach space *and let $(\mathcal{T}_n) \subset \mathcal{L}(\mathcal{X},\mathcal{Y})$ be a sequence of operators with finite-dimensional images which converges to $\mathcal{T} \in \mathcal{L}(\mathcal{X},\mathcal{Y})$ uniformly, that is, $\lim_{n\to\infty} \|\mathcal{T}_n - \mathcal{T}\|_{\mathcal{L}} = 0$. Then $\mathcal{T}$ is compact.*

Theorem 2.5.49 *Let $\mathcal{X}$ be a normed space. Then the identity $\mathcal{I}: \mathcal{X} \to \mathcal{X}$ is compact, if and only if $\dim \mathcal{X} < +\infty$.*

Let us consider the following: assume that $\mathcal{T}: \mathcal{X} \to \mathcal{Y}$ is compact and invertible. If $\mathcal{T}^{-1}$ is continuous, then Theorem 2.5.45 tells us that $\mathcal{I} = \mathcal{T}^{-1} \circ \mathcal{T}: \mathcal{X} \to \mathcal{X}$ is compact. However, Theorem 2.5.49 allows this only if $\mathcal{X}$ has a finite dimension.

Theorem 2.5.50 *If $\mathcal{T} \in \mathcal{K}(\mathcal{X},\mathcal{Y})$ and $\dim \mathcal{X} = +\infty$ for normed spaces $(\mathcal{X}, \|\cdot\|_{\mathcal{X}})$ and $(\mathcal{Y}, \|\cdot\|_{\mathcal{Y}})$, then $\mathcal{T}$* cannot *have a continuous inverse.*

Talking about continuous inverses, we should add the Neumann series, which is a very abstract but highly useful generalization of the geometric series.

Theorem 2.5.51 (Neumann Series) *Let $\mathcal{X}$ be a Banach space and $\mathcal{T} \in \mathcal{L}(\mathcal{X})$ be an operator for which $\|\mathcal{T}\|_{\mathcal{L}} < 1$. Then $\mathcal{I} - \mathcal{T}$ has a continuous inverse $(\mathcal{I} - \mathcal{T})^{-1} \in \mathcal{L}(\mathcal{X})$ and $(\mathcal{I} - \mathcal{T})^{-1} = \sum_{n=0}^{\infty} \mathcal{T}^n$, where the convergence of the series is uniform, that is, in the $\|\cdot\|_{\mathcal{L}}$-sense.*

This is enough about continuous inverses for the moment. We have discussed some basic properties of continuous linear operators. Remember that functionals are special operators. So, what can we say if $\mathcal{F} \in \mathcal{L}(\mathcal{X},\mathbb{K})$? For example, Theorem 2.5.47 tells us that all continuous linear functionals are also compact. Moreover, due to Theorem 2.5.32, $(\mathcal{L}(\mathcal{X},\mathbb{K}), \|\cdot\|_{\mathcal{L}})$ is always a Banach space.

Definition 2.5.52 Let $\mathcal{X}$ be a normed vector space over the field $\mathbb{K}$. Then $\mathcal{X}^* := \mathcal{L}(\mathcal{X},\mathbb{K})$ is called the **dual space** of $\mathcal{X}$. We also write $\|\cdot\|_{\mathcal{X}^*} := \|\cdot\|_{\mathcal{L}(\mathcal{X},\mathbb{K})}$.

Probably one of the most important theorems about a dual space is the Riesz representation theorem.

Theorem 2.5.53 (Riesz Representation Theorem) *Let $(\mathcal{H}, \langle\cdot,\cdot\rangle)$ be a given Hilbert space. Then the following holds true for the dual space $\mathcal{H}^*$:*

(a) For every $\mathcal{F} \in \mathcal{H}^*$, there exists a uniquely determined vector $y_{\mathcal{F}} \in \mathcal{H}$ such that $\mathcal{F}x = \langle x, y_{\mathcal{F}}\rangle$ for all $x \in \mathcal{H}$. Moreover, $\|\mathcal{F}\|_{\mathcal{H}^*} = \|y_{\mathcal{F}}\|$.
(b) If $y \in \mathcal{H}$, then the mapping $\mathcal{F}_y: \mathcal{H} \to \mathbb{K}$ with $\mathcal{F}_y x := \langle x, y\rangle$, $x \in \mathcal{H}$, is bounded and linear and satisfies $\|\mathcal{F}_y\|_{\mathcal{H}^*} = \|y\|_{\mathcal{H}}$.

Part b is easy to prove: the Cauchy–Schwarz inequality yields $|\mathcal{F}_y x| = |\langle x, y\rangle| \leq \|x\| \cdot \|y\|$, where $|\mathcal{F}_y y| = |\langle y, y\rangle| = \|y\| \cdot \|y\|$. Hence, $|\mathcal{F}_y|_{\mathcal{H}^*} = \sup_{x\in\mathcal{H}^*,\, x\neq 0}(|\mathcal{F}_y x|/\|x\|) = \|y\|$.

Remark 2.5.54 Note that the Riesz representation theorem requires an underlying Hilbert space. This should be stressed, because many authors, in particular in the applied sciences, claim the existence of something called the (Dirac) delta *function* δ, a function which is supposed to have the property that $\int_{\mathbb{R}} \delta(x)F(x)\,\mathrm{d}x = F(0)$, or, based on a substitution, $\int_{\mathbb{R}} \delta(x-y)F(x)\,\mathrm{d}x = F(y)$. At first sight, this looks like an application of the Riesz representation theorem with $\mathcal{F}F := F(0)$. However, this is false! For the required property of δ, we need the $\mathrm{L}^2(\mathbb{R})$-inner product, which means that $\mathcal{F}$ has to be regarded on $\mathrm{L}^2(\mathbb{R})$ or a complete subspace of it. However, $F(0)$ is not well defined for $F \in \mathrm{L}^2(\mathbb{R})$, because the elements of $\mathrm{L}^2(\mathbb{R})$ are only well defined 'almost everywhere'; see Example 2.5.19. Moveover, even if we ignored this fact, $\mathcal{F}$ would not be continuous: let $F_n(t) := \sqrt{n(1-n|t|)}$ for $-\frac{1}{n} \le t \le \frac{1}{n}$ and $F_n(t) := 0$ elsewhere, where $n \in \mathbb{N}$. Then we can easily check that $\|F_n\|^2_{\mathrm{L}^2(\mathbb{R})} = 2\int_0^{1/n} n(1-nt)\,\mathrm{d}t = -(1-nt)^2\big|_{t=0}^{t=1/n} = 1$, but $\mathcal{F}F_n = F_n(0) = \sqrt{n} \to +\infty \quad \text{as } n \to +\infty$. Indeed, one can show that there is no such function δ with the required property; see, for example, Michel (2013, remark 3.28). There is a Dirac delta *distribution* δ, which is actually a functional *but* not a function. We will not go into further detail here but leave the reader with the advice not to follow the common mistake of treating δ like a function.

For some of our derivations, the distinction between strong and weak convergence is also important.

Definition 2.5.55 Let $(\mathcal{X}, \|\cdot\|_{\mathcal{X}})$ be a normed space. A sequence $(x_n)_n \subset \mathcal{X}$ is called

(a) **strongly convergent** to $\xi \in \mathcal{X}$, if it converges to ξ, that is, $\|x_n - \xi\| \to 0$ as $n \to \infty$.

(b) **weakly convergent** to $\xi \in \mathcal{X}$, if $\mathcal{F}x_n \to \mathcal{F}\xi$ as $n \to \infty$ for every element $\mathcal{F}$ of the dual space $\mathcal{X}^*$. In this case, the notation $x_n \rightharpoonup \xi$ is also common.

The Riesz representation theorem (Theorem 2.5.53) immediately yields an alternative equivalent characterization of weak convergence in Hilbert spaces.

Theorem 2.5.56 *Let $(\mathcal{H}, \langle\cdot,\cdot\rangle_{\mathcal{H}})$ be a Hilbert space. A sequence $(x_n) \subset \mathcal{H}$ converges weakly to $\xi \in \mathcal{H}$ if and only if $\langle x_n, y\rangle \to \langle \xi, y\rangle$ as $n \to \infty$ for all $y \in \mathcal{H}$.*

Eventually, it is a simple continuity argument which allows us to conclude as follows: $x_n \to \xi \Rightarrow \mathcal{F}(x_n) \to \mathcal{F}(\xi)$ for all $\mathcal{F} \in \mathcal{X}^* \Rightarrow x_n \rightharpoonup \xi$.

Theorem 2.5.57 *A strongly convergent sequence in a normed space is always weakly convergent to the same limit.*

Theorem 2.5.58 *Let $(\mathcal{X}, \langle\cdot,\cdot\rangle_{\mathcal{X}})$ be a Hilbert space and let (x_n) be a sequence in $\mathcal{X}$. Then the following holds true: (x_n) converges strongly to $\xi \in \mathcal{X}$ if and only if $x_n \rightharpoonup \xi$ and $\|x_n\|_{\mathcal{X}} \to \|\xi\|_{\mathcal{X}}$.*

The latter theorem is a consequence of the fact that

$$\|x_n - \xi\|^2_{\mathcal{X}} = \|x_n\|^2_{\mathcal{X}} - 2\langle x_n, \xi\rangle_{\mathcal{X}} + \|\xi\|^2_{\mathcal{X}}.$$

Let us now have a look at the following example: let $\mathcal{T} \in \mathcal{L}(\mathcal{H}_1, \mathcal{H}_2)$ be an operator between two Hilbert spaces $\mathcal{H}_1$ and $\mathcal{H}_2$. Then we consider the functional $\mathcal{F}_y \colon \mathcal{H}_1 \to \mathbb{K}$ with $\mathcal{F}_y x := \langle \mathcal{T}x, y\rangle_{\mathcal{H}_2}$, where $y \in \mathcal{H}_2$ is fixed. Is $\mathcal{F}_y$ in the dual space $\mathcal{H}_1^*$? It is obviously linear, because $\mathcal{T}$ is linear and every inner product is bilinear. Moreover, the Cauchy–Schwarz inequality and the continuity of $\mathcal{T}$ help us to prove the continuity of $\mathcal{F}_y$:

$$\left|\langle \mathcal{T}x, y\rangle_{\mathcal{H}_2}\right| \leq \|\mathcal{T}x\|_{\mathcal{H}_2}\|y\|_{\mathcal{H}_2} \leq \|\mathcal{T}\|_{\mathcal{L}}\,\|x\|_{\mathcal{H}_1}\|y\|_{\mathcal{H}_2}.$$

Hence, $\|\mathcal{F}_y\|_{\mathcal{H}_1^*} \leq \|\mathcal{T}\|_{\mathcal{L}}\,\|y\|_{\mathcal{H}_2}$. Thus, the Riesz representation theorem implies that there is a unique element $f_{\mathcal{F}_y} \in \mathcal{H}_1$, briefly f_y, such that

$$\mathcal{F}_y x = \langle \mathcal{T}x, y\rangle_{\mathcal{H}_2} = \langle x, f_y\rangle_{\mathcal{H}_1} \quad \text{for all } x \in \mathcal{H}_1.$$

Note that f depends on y. So, we have a well-defined mapping $\mathcal{T}^* \colon \mathcal{H}_2 \to \mathcal{H}_1$ mapping $y \in \mathcal{H}_2$ to $\mathcal{T}^*y := f_y$. It is not difficult to show that this mapping is in $\mathcal{L}(\mathcal{H}_2, \mathcal{H}_1)$. It is called the adjoint operator of $\mathcal{T}$.

Theorem 2.5.59 *Let $\left(\mathcal{H}_1, \langle\cdot,\cdot\rangle_{\mathcal{H}_1}\right)$ and $\left(\mathcal{H}_2, \langle\cdot,\cdot\rangle_{\mathcal{H}_2}\right)$ be two Hilbert spaces corresponding to the field $\mathbb{K}$, and let $\mathcal{T} \in \mathcal{L}(\mathcal{H}_1, \mathcal{H}_2)$ be a given operator. Then there is one and only one operator $\mathcal{T}^* \in \mathcal{L}(\mathcal{H}_2, \mathcal{H}_1)$ such that*

$$\langle \mathcal{T}x, y\rangle_{\mathcal{H}_2} = \langle x, \mathcal{T}^*y\rangle_{\mathcal{H}_1} \qquad \text{for all } x \in \mathcal{H}_1 \text{ and } y \in \mathcal{H}_2.$$

*Moreover, $\|\mathcal{T}\|_{\mathcal{L}(\mathcal{H}_1,\mathcal{H}_2)} = \|\mathcal{T}^*\|_{\mathcal{L}(\mathcal{H}_2,\mathcal{H}_1)}$. This operator $\mathcal{T}^*$ is called the **adjoint operator** corresponding to $\mathcal{T}$. If $\mathcal{H}_1 = \mathcal{H}_2$ and $\mathcal{T} = \mathcal{T}^*$, then $\mathcal{T}$ is called **self-adjoint**.*

Theorem 2.5.60 (Properties of Adjoint Operators) *Let $\mathcal{H}_1, \mathcal{H}_2$, and $\mathcal{H}_3$ be Hilbert spaces corresponding to the field $\mathbb{K}$, $\mathcal{T}, \mathcal{T}_1, \mathcal{T}_2 \in \mathcal{L}(\mathcal{H}_1, \mathcal{H}_2)$ and $\mathcal{S} \in \mathcal{L}(\mathcal{H}_2, \mathcal{H}_3)$ be given operators, and $r \in \mathbb{K}$ be a given scalar. Then the following holds true for the adjoint operators $\mathcal{T}^*, \mathcal{T}_1^*, \mathcal{T}_2^*$, and $\mathcal{S}^*$:*

(a) $(\mathcal{T}_1 \pm \mathcal{T}_2)^* = \mathcal{T}_1^* \pm \mathcal{T}_2^*$ and $(r\mathcal{T})^* = \bar{r}\,\mathcal{T}^*$, where $\bar{r}$ is the complex conjugate of r.
(b) $(\mathcal{S}\mathcal{T})^* = \mathcal{T}^*\mathcal{S}^*$.
(c) $\mathcal{T}^{**} = \mathcal{T}$.
(d) If $\mathcal{T}$ is bijective, then $\mathcal{T}^*$ is also bijective and $(\mathcal{T}^*)^{-1} = (\mathcal{T}^{-1})^*$.

Moreover, the identity operator $\mathcal{I}$ and the zero operator 0 are self-adjoint: $\mathcal{I}^* = \mathcal{I}$ and $0^* = 0$.

Note the following causalities: if $\mathcal{T}$ is bijective, then $\mathcal{T}^{-1}$ is continuous due to the inverse mapping theorem (Theorem 2.5.41). Since Theorem 2.5.60 says that the bijectivity of $\mathcal{T}^*$ is implied, $(\mathcal{T}^*)^{-1}$ is also continuous.

Theorem 2.5.61 *Let $\mathcal{H}_1$ and $\mathcal{H}_2$ be two Hilbert spaces and $\mathcal{T} \in \mathcal{L}(\mathcal{H}_1, \mathcal{H}_2)$ be a given operator. Then $\mathcal{T}$ is compact, if and only if $\mathcal{T}^*$ is compact.*

These are some more important properties of adjoint operators. However, before we can discuss them, we need to summarize the concept of an orthogonal complement, which is also important in some other contexts. Moreover, also convex sets are relevant for some parts of the theory in this book.

Definition 2.5.62 Let $\mathcal{X}$ be a vector space. A subset $\mathcal{C} \subset \mathcal{X}$ is called **convex** if $\lambda x + (1-\lambda)y \in \mathcal{C}$ for all $x, y \in \mathcal{C}$ and all $\lambda \in [0,1]$.

Theorem 2.5.63 *Let $(\mathcal{H}, \langle\cdot,\cdot\rangle_{\mathcal{H}})$ be a Hilbert space and $\mathcal{C} \subset \mathcal{H}$ be a non-empty, convex, and closed subset. Then there exists one and only one element $x \in \mathcal{C}$ such that $\|x\|_{\mathcal{H}} = \inf\{\|y\|_{\mathcal{H}} : y \in \mathcal{C}\}$.*

Since $\mathcal{C} - z := \{y - z \mid y \in \mathcal{C}\}$ is also convex, if $\mathcal{C}$ is convex, we also have the following **best-approximation property**.

Corollary 2.5.64 *Let $(\mathcal{H}, \langle\cdot,\cdot\rangle_{\mathcal{H}})$ be a Hilbert space and $\mathcal{C} \subset \mathcal{H}$ be a non-empty, convex, and closed subset. If $z \in \mathcal{H}$ is arbitrary, then there is one and only one element $x \in \mathcal{C}$ which is closest to z, that is,*

$$\|x - z\|_{\mathcal{H}} = \inf\{\|y - z\|_{\mathcal{H}} : y \in \mathcal{C}\}.$$

It is clear that non-convex sets need not have this property: the sphere $S_r(z)$ is non-convex, and each $y \in S_r(z)$ has the same distance r to z. Moreover, $]0,1] \subset \mathbb{R}$ is not closed, and there is no $x \in]0,1]$ which is closest to a given $z \in \mathbb{R}^-$: for every $x \in]0,1]$, the number $x/2$ is closer to z, while $0 \notin]0,1]$.

Convex sets should not be mixed up with convex functions. The latter have some useful properties in real analysis such as Jensen's inequality; see Heuser (2009, pp. 282–283), Iske (2018, p. 73), Jensen (1906), and Rudin (1987, p. 62) for further details.

Definition 2.5.65 Let $C \subset \mathbb{R}^n$ be a convex set and let: $f\colon C \to \mathbb{R}$ be a given function. The function f is called **convex** if, for all $x, y \in C$ and all $\lambda \in [0,1]$, we have $f(\lambda x + (1-\lambda)y) \le \lambda f(x) + (1-\lambda)f(y)$.

Theorem 2.5.66 *A function $f\colon C \to \mathbb{R}$ with $C \subset \mathbb{R}^n$ is convex if and only if its **epigraph** $\operatorname{epi} f := \{(x,y) \mid x \in C,\ y \in \mathbb{R}, \text{and } y \ge f(x)\}$ is a convex set in $\mathbb{R}^{n+1}$.*

Theorem 2.5.67 *A twice continuously differentiable function $f\colon C \to \mathbb{R}$ on a convex and open set $C \subset \mathbb{R}^n$ is convex if and only if its Hessian $\nabla \otimes \nabla f$ is positive semi-definite everywhere in C.*

Figure 2.8 shows an example of a convex function and its epigraph, which needs to be a convex set. Note that Theorem 2.5.67 says that the graph of the function has a curvature of the type of a 'left turn'.

As we mentioned previously, Jensen's inequality is a typical property of convex functions.

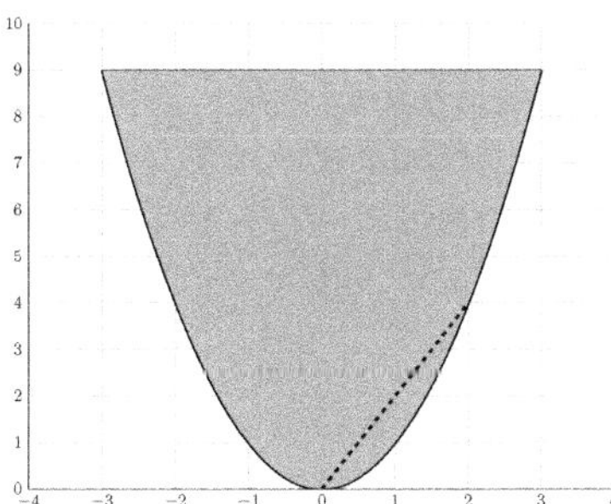

Figure 2.8 This figure shows the graph of $f: \mathbb{R} \to \mathbb{R}, x \mapsto x^2$ as an example of a convex function (note that $f'' \equiv 2 \geq 0$). In the case of a convex function, every secant between two points of the graph is located above the graph, as it is illustrated here for one example of a secant (dashed). An equivalent definition of convexity of a function is the convexity of the epigraph (grey).

Theorem 2.5.68 (Jensen's Inequality) *Let $f: C \to \mathbb{R}$ be a convex function on the convex set $C \subset \mathbb{R}^n$, and let $x^{(1)}, \ldots, x^{(k)} \in C$ be arbitrary points in its domain. Then $f\big(\sum_{j=1}^k \lambda_j x^{(j)}\big) \leq \sum_{j=1}^k \lambda_j f(x^{(j)})$ for all $\lambda_1, \ldots, \lambda_k \in [0,1]$ with $\sum_{j=1}^k \lambda_j = 1$.*

Corollary 2.5.69 *Let $x_1, \ldots, x_k \in \mathbb{R}$ be arbitrary numbers and let $p \geq 1$. Then $\sum_{j=1}^k |x_j| \leq k^{1-1/p}\big(\sum_{j=1}^k |x_j|^p\big)^{1/p}$. In (analogous) terms of the ℓ^p-norm (see Example 2.5.9, part d), this means for $x = (x_1, \ldots, x_k)^{\mathrm{T}} \in \mathbb{R}^k$ that $\|x\|_1 \leq k^{1-1/p}\|x\|_p$.*

Proof For $p > 1$, the function $f: \mathbb{R}_0^+ \to \mathbb{R}, x \mapsto x^p$ is convex, because $f''(x) = p(p-1)x^{p-2} > 0$ for all $x \in \mathbb{R}^+$. Hence, Jensen's inequality yields $\big(\sum_{j=1}^k k^{-1}|x_j|\big)^p \leq \sum_{j=1}^k k^{-1}|x_j|^p \Rightarrow k^{-1}\sum_{j=1}^k |x_j| \leq k^{-1/p}\big(\sum_{j=1}^k |x_j|^p\big)^{1/p}$ and Corollary 2.5.69 is proved (the case $p = 1$ is trivial). □

In the following, we summarize some basics on the decomposition of (particular) vector spaces.

Definition 2.5.70 Let V be a vector space and U, W be linear subspaces. We define $U + W := \{u + w \mid u \in U \text{ and } w \in W\}$. We say that V is the **direct sum** of U and W and write $V = U \oplus W$ if the following holds true: for every $v \in V$, there exist unique elements $u \in U$ and $w \in W$ such that $v = u + w$. U and W are called **algebraic complements** of each other.

Theorem 2.5.71 (Orthogonal Decomposition) *Let $(\mathcal{H}, \langle\cdot,\cdot\rangle)$ be an arbitrary* pre-Hilbert *space and $\mathcal{U}$ be a* complete *linear subspace. Then its* ***orthogonal complement***

$$\mathcal{U}^{\perp} := \{y \in \mathcal{H} \mid \langle y, x\rangle = 0 \text{ for all } x \in \mathcal{U}\} \tag{2.14}$$

is also an algebraic complement of $\mathcal{U}$*, that is,* $\mathcal{H} = \mathcal{U} \oplus \mathcal{U}^{\perp}$*. Moreover,* $\mathcal{U}^{\perp}$ *is a closed linear subspace of* $\mathcal{H}$*, where* $\mathcal{U} \cap \mathcal{U}^{\perp} = \{0\}$ *and* $(\mathcal{U}^{\perp})^{\perp} = \mathcal{U}$*. Furthermore, the operators* $\mathcal{P}_{\mathcal{U}}: \mathcal{H} \to \mathcal{U}$ *and* $\mathcal{P}_{\mathcal{U}^{\perp}}: \mathcal{H} \to \mathcal{U}^{\perp}$ *which are given by* $\mathcal{I} = \mathcal{P}_{\mathcal{U}} + \mathcal{P}_{\mathcal{U}^{\perp}}$*, where* $\mathcal{I}$ *is the identity on* $\mathcal{H}$*, are called* ***orthogonal projectors****. Each orthogonal projector is continuous and its operator norm equals 1 if it is not the zero operator.*

Due to Theorem 2.5.16, the conditions of Theorem 2.5.71 are also satisfied if $\mathcal{H}$ is a Hilbert space and $\mathcal{U}$ is a closed linear subspace.

Theorem 2.5.72 *Let* $\mathcal{H}_1$ *and* $\mathcal{H}_2$ *be two Hilbert spaces. Moreover, let* $\mathcal{T} \in \mathcal{L}(\mathcal{H}_1, \mathcal{H}_2)$ *be a given operator. Then the image* $\mathcal{T}(\mathcal{H}_1)$ *of* $\mathcal{T}$ *and the* ***null space*** $\ker \mathcal{T} := \{x \in \mathcal{H}_1 \mid \mathcal{T}x = 0\}$ *of* $\mathcal{T}$ *as well as their orthogonal complements satisfy the following identities:*

$$\mathcal{T}(\mathcal{H}_1)^{\perp} = \ker \mathcal{T}^*, \quad \overline{\mathcal{T}(\mathcal{H}_1)} = (\ker \mathcal{T}^*)^{\perp},$$
$$\mathcal{T}^*(\mathcal{H}_2)^{\perp} = \ker \mathcal{T}, \quad \overline{\mathcal{T}^*(\mathcal{H}_2)} = (\ker \mathcal{T})^{\perp}.$$

Note that the null space of a continuous linear operator $\mathcal{T}: \mathcal{X} \to \mathcal{Y}$ between two normed spaces is always a closed linear subspace of $\mathcal{X}$: if $x_1, x_2 \in \ker \mathcal{T}$ and $\lambda \in \mathbb{K}$, then $\mathcal{T}(\lambda x_1) = \lambda \mathcal{T} x_1 = 0$ and $\mathcal{T}(x_1 + x_2) = \mathcal{T}x_1 + \mathcal{T}x_2 = 0$. If $(x_n) \subset \ker \mathcal{T}$ and $x_n \to x \in \mathcal{X}$, then the continuity of $\mathcal{T}$ implies that $\mathcal{T}x_n \to \mathcal{T}x$, where $\mathcal{T}x_n = 0$ for all n, such that $x \in \ker \mathcal{T}$.

However, the image $\mathcal{T}(\mathcal{X})$ need not be closed, which is why Theorem 2.5.72 can only be true for the closures of the images. Nevertheless, for taking the orthogonal complement, we do not need the closure of the image: $\mathcal{U}^{\perp}$ in (2.14) is well defined also for non-closed linear subspaces $\mathcal{U}$ and, actually, $\overline{\mathcal{U}}^{\perp} = \mathcal{U}^{\perp}$ for the following reasons: the inclusion $\overline{\mathcal{U}}^{\perp} \subset \mathcal{U}^{\perp}$ is trivial. Let $x \in \mathcal{U}^{\perp}$, that is, $\langle x, y \rangle = 0$ for all $y \in \mathcal{U}$. If $y \in \overline{\mathcal{U}}$, then there is a sequence $(y_n) \subset \mathcal{U}$ with $\lim_{n\to\infty} \|y_n - y\| = 0$. Hence, the Cauchy–Schwarz inequality yields (the continuity of the inner product, that is) $|\langle x, y \rangle - 0| = |\langle x, y - y_n \rangle| \le \|x\| \cdot \|y - y_n\| \to 0$ such that $\langle x, y \rangle = 0$ and $x \in \overline{\mathcal{U}}^{\perp}$.

The fact that images of linear continuous operators need not be closed is connected to another aspect which we have mentioned earlier here.

Theorem 2.5.73 *Let* $\mathcal{X}$ *and* $\mathcal{Y}$ *be Banach spaces and* $\mathcal{T} \in \mathcal{L}(\mathcal{X}, \mathcal{Y})$ *be an injective operator. Then the inverse* $\mathcal{T}^{-1}: \mathcal{T}(\mathcal{X}) \to \mathcal{X}$ *is continuous, if and only if the image* $\mathcal{T}(\mathcal{X})$ *is closed.*

Another big topic in functional analysis is the spectral theory. We will only very briefly touch this here, since we need a few theorems. The spectral theory itself is a generalization of the theory of eigenvalues which is known for matrices, more precisely for linear operators between finite-dimensional spaces.

Definition 2.5.74 Let $\mathcal{X}$ be a Banach space corresponding to the field $\mathbb{K}$ and let $\mathcal{T} \in \mathcal{L}(\mathcal{X})$ be a given operator. Then the **spectrum** $\sigma(\mathcal{T})$ consists of all $\lambda \in \mathbb{K}$ for which $\lambda\mathcal{I} - \mathcal{T}$ *does not have* a continuous inverse $(\lambda\mathcal{I} - \mathcal{T})^{-1} \in \mathcal{L}(\mathcal{X})$. This spectrum is subdivided into the **point spectrum**

$$\sigma_{\mathrm{p}}(\mathcal{T}) := \{\lambda \in \mathbb{K} \mid \lambda\mathcal{I} - \mathcal{T} \text{ is } \textit{not} \text{ injective}\},$$

whose elements are called the **eigenvalues** of $\mathcal{T}$; the **continuous spectrum**

$$\sigma_{\mathrm{c}}(\mathcal{T}) := \{\lambda \in \sigma(\mathcal{T}) \setminus \sigma_{\mathrm{p}}(\mathcal{T}) \mid (\lambda\mathcal{I} - \mathcal{T})(\mathcal{X}) \text{ is dense in } \mathcal{X}\};$$

and the **residual spectrum**

$$\sigma_{\mathrm{r}}(\mathcal{T}) := \{\lambda \in \sigma(\mathcal{T}) \setminus \sigma_{\mathrm{p}}(\mathcal{T}) \mid (\lambda\mathcal{I} - \mathcal{T})(\mathcal{X}) \text{ is } \textit{not} \text{ dense in } \mathcal{X}\}.$$

Remember that the inverse mapping theorem (Theorem 2.5.41) tells us that, as soon as $\lambda\mathcal{I} - \mathcal{T}$ is bijective, the inverse $(\lambda\mathcal{I} - \mathcal{T})^{-1}$ is continuous, that is, $\lambda \notin \sigma(\mathcal{T})$.

Theorem 2.5.75 *Let $\mathcal{X}$ be a Banach space and $\mathcal{T} \in \mathcal{L}(\mathcal{X})$ be compact. Then $\ker(\mathcal{I} - \mathcal{T})$ is finite dimensional and $(\mathcal{I} - \mathcal{T})(\mathcal{X})$ is closed in $\mathcal{X}$.*

Note that, in the finite-dimensional case, $\sigma(\mathcal{T}) = \sigma_{\mathrm{p}}(\mathcal{T})$. Since compact operators are somehow 'closer' to operators on finite-dimensional spaces, it is not surprising that the point spectrum dominates the spectrum of compact operators.

Theorem 2.5.76 *Let $\mathcal{X}$ be a Banach space and $\mathcal{T} \in \mathcal{L}(\mathcal{X})$ be compact. Then $\sigma(\mathcal{T}) \setminus \{0\} = \sigma_{\mathrm{p}}(\mathcal{T}) \setminus \{0\}$.*

Let us collect some additionally useful theorems on spectra (see also Folland, 1976).

Theorem 2.5.77 *Let $\mathcal{H}$ be a Hilbert space and $\mathcal{T} \in \mathcal{L}(\mathcal{H})$. Then $\sigma(\mathcal{T}) = \sigma(\mathcal{T}^*)$.*

Theorem 2.5.78 (Fredholm's Theorem) *Let $\mathcal{H}$ be a Hilbert space and let $\mathcal{T} \in \mathcal{L}(\mathcal{H})$ be a compact operator. Then the following holds true:*

*(a) $\sigma(\mathcal{T})$ is finite or countable, where the only possible accumulation point is 0. Moreover, for $\lambda \in \sigma(\mathcal{T}) \setminus \{0\}$, the **eigenspace** $\mathrm{E}(\lambda; \mathcal{T}) := \{x \in \mathcal{H} \mid \mathcal{T}x = \lambda x\}$ is of finite dimension.*
(b) If $\lambda \neq 0$, then $\dim \mathrm{E}(\lambda; \mathcal{T}) = \dim \mathrm{E}(\bar{\lambda}; \mathcal{T}^*)$, where $\bar{\lambda}$ is the complex conjugate of λ.
(c) If $\lambda \neq 0$, then the image $(\lambda\mathcal{I} - \mathcal{T})(\mathcal{H})$ is closed in $\mathcal{H}$.

Corollary 2.5.79 *Let $\mathcal{H}$ be a Hilbert space and let $\mathcal{T} \in \mathcal{L}(\mathcal{H})$ be a compact operator. Moreover, let $\lambda \neq 0$. Then the following holds true:*

(a) The equation $(\lambda\mathcal{I} - \mathcal{T})x = y$ is solvable if and only if $y \in \mathrm{E}(\bar{\lambda}; \mathcal{T}^*)^{\perp}$.
(b) $\lambda\mathcal{I} - \mathcal{T}$ is surjective if and only if it is injective.

Note that $\mathrm{E}(\bar{\lambda}; \mathcal{T}^*) = \ker(\bar{\lambda}\mathcal{I} - \mathcal{T}^*) = \ker(\lambda\mathcal{I} - \mathcal{T})^*$ such that part a is a consequence of Theorem 2.5.72 and the fact that $(\lambda\mathcal{I} - \mathcal{T})(\mathcal{H})$ is closed (see Theorem 2.5.78).

There is something interesting about the latter results. Let us assume that $\mathcal{H}$ is an infinite-dimensional Hilbert space, like an $\mathrm{L}^2(D)$-space as it is often the case in applications. Let also $\mathcal{T} \in \mathcal{L}(\mathcal{H})$ be compact. Then $\mathcal{T}$ cannot have a continuous inverse (see Theorem 2.5.50). An inverse problem $\mathcal{T}F = G$ must be ill-posed in this case, as we will call it later (see Definition 3.3.2). However, as we learned now, this $\lambda = 0$ case is a very special case of $(\mathcal{T} - \lambda\mathcal{I})F = G$, because, for all other $\lambda \in \mathbb{K}$, the inverse $(\mathcal{T} - \lambda\mathcal{I})^{-1}$ *is continuous*, if it exists (see Theorems 2.5.73 and 2.5.78).

In the context of spectral theory, there are also some other useful theorems. They are connected to a decomposition of the domain into the null space (kernel) and its orthogonal complement as well as the decomposition of the co-domain into the range (image) and its orthogonal complement.

Theorem 2.5.80 *Let $(\mathcal{H}, \langle\cdot,\cdot\rangle)$ be a Hilbert space and let $\mathcal{T} \in \mathcal{L}(\mathcal{H})$ be a compact and self-adjoint operator. Then there exists a finite or countably infinite orthonormal system of eigenvectors $(u_n)_{n\in\mathcal{N}}$ in $\mathcal{H}$ (with $\mathcal{N} \subset \mathbb{N}_0$) and a corresponding system of eigenvalues $(\lambda_n)_{n\in\mathcal{N}} \subset \mathbb{R} \setminus \{0\}$ such that*

$$\mathcal{T}x = \sum_{n\in\mathcal{N}} \lambda_n \langle x, u_n\rangle u_n \tag{2.15}$$

for all $x \in \mathcal{H}$. If $\mathcal{N}$ is infinite, then $(\lambda_n)_{n\in\mathcal{N}}$ tends to zero.

The more general case where we have two Hilbert spaces $(\mathcal{X}, \langle\cdot,\cdot\rangle_{\mathcal{X}})$ and $(\mathcal{Y}, \langle\cdot,\cdot\rangle_{\mathcal{Y}})$ and a compact operator $\mathcal{T}: \mathcal{X} \to \mathcal{Y}$ can be easily deduced from Theorem 2.5.80 and is also well known. Clearly, $\mathcal{T}^*\mathcal{T}$ is self-adjoint. Hence, we find a representation $\mathcal{T}^*\mathcal{T}x = \sum_{n\in\mathcal{N}} \lambda_n \langle x, u_n\rangle_{\mathcal{X}} u_n$, $x \in \mathcal{X}$, as before. Then $\lambda_n = \lambda_n\langle u_n, u_n\rangle_{\mathcal{X}} = \langle \mathcal{T}^*\mathcal{T}u_n, u_n\rangle_{\mathcal{X}} = \|\mathcal{T}u_n\|_{\mathcal{Y}}^2 \geq 0$ such that $\lambda_n > 0$ for all $n \in \mathcal{N}$. Moreover, $(u_n)_{n\in\mathcal{N}}$ is obviously a basis of $(\ker \mathcal{T})^{\perp}$.

Furthermore, $\langle \mathcal{T}u_n, \mathcal{T}u_m\rangle_{\mathcal{Y}} = \langle \mathcal{T}^*\mathcal{T}u_n, u_m\rangle_{\mathcal{X}} = \lambda_n\delta_{nm}$ such that the system $(\mathcal{T}u_n)_{n\in\mathcal{N}}$ is an orthogonal basis of $\mathcal{T}(\mathcal{X})$. Correspondingly, we can expand $\mathcal{T}x$ for arbitrary $x \in \mathcal{X}$ in this basis:

$$\begin{aligned}\mathcal{T}x &= \sum_{n\in\mathcal{N}} \langle \mathcal{T}x, \mathcal{T}u_n\rangle_{\mathcal{Y}} \lambda_n^{-1} \mathcal{T}u_n = \sum_{n\in\mathcal{N}} \langle \mathcal{T}^*\mathcal{T}x, u_n\rangle_{\mathcal{X}} \lambda_n^{-1/2} \lambda_n^{-1/2} \mathcal{T}u_n \\ &= \sum_{n\in\mathcal{N}} \lambda_n^{1/2} \langle x, u_n\rangle_{\mathcal{X}} \lambda_n^{-1/2} \mathcal{T}u_n.\end{aligned}$$

The fact that $\|\lambda_n^{-1/2}\mathcal{T}u_n\|_{\mathcal{Y}} = 1$ leads us to the following fundamental theorem.

Theorem 2.5.81 (Singular-Value Decomposition) *Let two Hilbert spaces $(\mathcal{X}, \langle\cdot,\cdot\rangle_{\mathcal{X}})$ and $(\mathcal{Y}, \langle\cdot,\cdot\rangle_{\mathcal{Y}})$ and a compact linear operator $\mathcal{T}: \mathcal{X} \to \mathcal{Y}$ be given. Then there exist orthonormal systems $(u_n)_{n\in\mathcal{N}} \subset \mathcal{X}$ and $(v_n)_{n\in\mathcal{N}} \subset \mathcal{Y}$, where $\mathcal{N} \subset \mathbb{N}_0$, and real numbers $(\sigma_n)_{n\in\mathcal{N}} \subset \mathbb{R}^+$ such that*

$$\mathcal{T}x = \sum_{n\in\mathcal{N}} \sigma_n \langle x, u_n\rangle_{\mathcal{X}}\, v_n \tag{2.16}$$

*for all $x \in \mathcal{X}$. If $\mathcal{N}$ is infinite, then $(\sigma_n)_{n\in\mathcal{N}}$ tends to zero. The values σ_n are called the **singular values** of $\mathcal{T}$ and $\{(\sigma_n, u_n, v_n)\}_{n\in\mathcal{N}}$ is called a **singular system** of $\mathcal{T}$.*

The convergence in (2.15) and (2.16), which is relevant if $\mathcal{N}$ is infinite, is given in the sense of $\mathcal{H}$ and $\mathcal{Y}$, respectively.

Conversely, if an operator is given by (2.16) with, for all $x \in \mathcal{X}$, a converging series (or finite sum), then it is easy to see that $\|\mathcal{T}\|_{\mathcal{L}} = \sup_{n\in\mathcal{N}} \sigma_n$. Hence, $\mathcal{T}$ is bounded if and only if $(\sigma_n)_{n\in\mathcal{N}}$ is bounded. Moreover, if (σ_n) converges to zero (or is finite), then $\mathcal{T}$ is compact: the finite case is trivial due to Theorem 2.5.47. Otherwise, we define a sequence of operators $\mathcal{T}_k x := \sum_{n\in\mathcal{N},\, n\leq k} \sigma_n \langle x, u_n\rangle_{\mathcal{X}} v_n$, $x \in \mathcal{X}$, $k \in \mathbb{N}_0$. Then $\|\mathcal{T} - \mathcal{T}_k\|_{\mathcal{L}} = \sup_{n\in\mathcal{N},n>k} \sigma_n \to 0$ as $k \to \infty$. Since every $\mathcal{T}_k$ has a finite-dimensional image and is, therefore, compact, Theorem 2.5.46 yields the compactness of $\mathcal{T}$. Hence, in total, $\mathcal{T}$ is compact if and only if (σ_n) tends to zero.

Let us come to another useful topic of functional analysis. We know that, if we have a metric space $(\mathcal{X},d)$ and a subset $\mathcal{S} \subset \mathcal{X}$, then we can 'easily' clarify what the closure of $\mathcal{S}$ is, namely the 'smallest' closed set which contains $\mathcal{S}$ (see Definition 2.5.13):

$$\overline{\mathcal{S}} := \bigcap \{\mathcal{C} \subset \mathcal{X} \mid \mathcal{S} \subset \mathcal{C} \text{ and } \mathcal{C} \text{ is closed}\}.$$

Since arbitrary intersections of closed sets are closed (see Theorem 2.5.14), $\overline{\mathcal{S}}$ is closed, and obviously there is no proper subset of $\overline{\mathcal{S}}$ which is closed and still contains $\mathcal{S}$. The question is now: is there anything analogous for 'complete' instead of 'closed'? The answer is: yes, but the underlying theory is far more complicated. Therefore, we will also here only take note of the facts but omit the proofs.

Definition 2.5.82 Let $(\mathcal{X},d_{\mathcal{X}})$ and $(\mathcal{Y},d_{\mathcal{Y}})$ be metric spaces, where the latter is complete. If there exists a mapping $j\colon \mathcal{X} \to \mathcal{Y}$ which is **isometric**, that is, $d_{\mathcal{Y}}(j(x_1), j(x_2)) = d_{\mathcal{X}}(x_1,x_2)$ for all $x_1,x_2 \in \mathcal{X}$, and whose image $j(\mathcal{X}) \subset \mathcal{Y}$ is dense in $\mathcal{Y}$, then $\mathcal{Y}$ is called a **completion** of $\mathcal{X}$.

Note that an isometric mapping is always injective.

Theorem 2.5.83 *Every metric space $(\mathcal{X},d)$ has a completion $(\hat{\mathcal{X}},\hat{d})$. If $(\hat{\mathcal{X}}_1,\hat{d}_1)$ and $(\hat{\mathcal{X}}_2,\hat{d}_2)$ are two completions of $(\mathcal{X},d)$, then there exists a bijective isometric mapping $\hat{j}\colon \hat{\mathcal{X}}_1 \leftrightarrow \hat{\mathcal{X}}_2$. In other words, completions are unique up to isometry.*

Since a bijective isometry of two spaces means that they are basically the same spaces but only equipped with, maybe, different names and notations, one usually associates the metric d of $\mathcal{X}$ with the metric $\hat{d}$ of the completion $\hat{\mathcal{X}}$ as well as $\mathcal{X}$ with its embedding $j(\mathcal{X}) \subset \hat{\mathcal{X}}$. Briefly, $\mathcal{X}$ is treated as a subspace of $\hat{\mathcal{X}}$.

2.6 Curves and Surfaces

We will have to deal with surfaces (e.g., as models for the Earth's surface) and curves in this book as well as integrals across or along such geometrical objects. Only some basics on curves and surfaces are summarized here. For further details, please consult, for instance, Lang (2001, chapters XV and XX), Protter and Morrey (1977, chapter 16), Riley et al. (2008, chapter 11), Trim (1993, chapter 15), and Walter (1990, chapters 5 and 6). Let us first define what curves and surfaces are.

Definition 2.6.1 Let $a < b$ be real numbers such that $I := [a,b]$ is an interval in $\mathbb{R}$ and let $D \subset \mathbb{R}^2$ be a bounded and measurable set whose boundary ∂D has a vanishing area ($\lambda_2(\partial D) = 0$, if λ_2 is the Lebesgue measure in $\mathbb{R}^2$). Moreover, let $n \geq 2$.

(a) A continuously differentiable mapping $\gamma\colon I \to \mathbb{R}^n$ with $\gamma'(t) \neq 0$ for all $t \in]a,b[$ is called a **simple arc**, and its image $\gamma(I)$ is called a **curve**. We can also say that γ parameterizes the curve $\gamma(I)$. We also assume that γ is injective on $]a,b[$.

(b) A continuously differentiable and Lipschitz continuous mapping $\Phi\colon D \to \mathbb{R}^3$ which is injective on the interior $\operatorname{int} D$ of D and has a Jacobian with a full rank, that is, $\operatorname{rk} \Phi'(x) = 2$ for all $x \in \operatorname{int} D$, is called a **parameterization** of the **surface** $\Phi(D)$. The domain D is also called the **parameter range**.

We have already had a curve and a surface in Example 2.5.4: a circle, which can be parameterized by

$$\gamma(\varphi) := r \begin{pmatrix} \cos\varphi \\ \sin\varphi \end{pmatrix}, \quad \varphi \in [0, 2\pi], \tag{2.17}$$

is a curve. And, in the next dimension, a sphere is a surface, which can be parameterized by

$$\Phi(\varphi, t) := \begin{pmatrix} r\sqrt{1-t^2}\cos\varphi \\ r\sqrt{1-t^2}\sin\varphi \\ rt \end{pmatrix}, \quad \varphi \in [0, 2\pi], t \in [-1, 1]. \tag{2.18}$$

Certainly, there is more than one way to parameterize a given curve or surface. For example, $\tilde{\varphi}(t) := r(\sin\varphi, \cos\varphi)^{\mathrm{T}}$, $\varphi \in [-\pi, \pi]$, yields the same circle as γ.

We can use curves and surfaces as domains for integrations.

Definition 2.6.2 Let an interval $I \subset \mathbb{R}$ and a parameter range $D \subset \mathbb{R}^2$ be given as in Definition 2.6.1.

(a) Let $\gamma\colon I \to \mathbb{R}^n$ be a simple arc and $F\colon \gamma(I) \to \mathbb{R}$ be a continuous function. Then the **line integral** of F along $\gamma(I)$ is defined as

$$\int_{\gamma(I)} F(x)\, \mathrm{dl}(x) := \int_a^b F(\gamma(t))\, |\gamma'(t)|\, \mathrm{d}t.$$

(b) Let $\Phi\colon D \to \mathbb{R}^3$ be a given parameterization of the surface $\Phi(D)$ and let $G\colon \Phi(D) \to \mathbb{R}$ be a continuous function. Then the **surface integral** of G across $\Phi(D)$ is defined as

$$\int_{\Phi(D)} G(x)\, \mathrm{d}\omega(x) := \int_D G(\Phi(s,t)) \left| \left(\frac{\partial \Phi}{\partial s} \times \frac{\partial \Phi}{\partial t} \right)(s,t) \right| \mathrm{d}(s,t).$$

Note that F and G need not be continuous functions for the existence of the preceding integrals, this condition is only sufficient.

For the previous example of a circle, we have

$$|\gamma'(\varphi)| = \left| \begin{pmatrix} -r\sin\varphi \\ r\cos\varphi \end{pmatrix} \right| = r.$$

For a sphere with the parameterization Φ from earlier, we have

$$\left| \left(\frac{\partial \Phi}{\partial \varphi} \times \frac{\partial \Phi}{\partial t} \right)(\varphi, t) \right| = \left| \begin{pmatrix} -r\sqrt{1-t^2}\sin\varphi \\ r\sqrt{1-t^2}\cos\varphi \\ 0 \end{pmatrix} \times \begin{pmatrix} -\frac{rt}{\sqrt{1-t^2}}\cos\varphi \\ -\frac{rt}{\sqrt{1-t^2}}\sin\varphi \\ r \end{pmatrix} \right|$$

$$= \left| \begin{pmatrix} r^2\sqrt{1-t^2}\cos\varphi \\ r^2\sqrt{1-t^2}\sin\varphi \\ r^2 t(\sin^2\varphi + \cos^2\varphi) \end{pmatrix} \right| = r^2.$$

Accordingly, for volume integrals over a ball $B_R(0)$, the well-known substitution rule in the case of the use of polar coordinates requires the Jacobian of

$$\Psi(r, \varphi, t) := \begin{pmatrix} r\sqrt{1-t^2}\cos\varphi \\ r\sqrt{1-t^2}\sin\varphi \\ rt \end{pmatrix}$$

with $r \in [0, R], \varphi \in [0, 2\pi]$, and $t \in [-1, 1]$. This is given by

$$\det \Psi'(r, \varphi, t) = \begin{vmatrix} \sqrt{1-t^2}\cos\varphi & -r\sqrt{1-t^2}\sin\varphi & -\frac{rt}{\sqrt{1-t^2}}\cos\varphi \\ \sqrt{1-t^2}\sin\varphi & r\sqrt{1-t^2}\cos\varphi & -\frac{rt}{\sqrt{1-t^2}}\sin\varphi \\ t & 0 & r \end{vmatrix} = r^2$$

such that

$$\int_{B_R(0)} f(x)\,\mathrm{d}x = \int_0^R r^2 \int_\Omega f(r\xi)\,\mathrm{d}\omega(\xi)\,\mathrm{d}r$$

for every measurable and bounded function $f\colon B_R(0) \to \mathbb{R}$.

Let us come back to general curves and surfaces again. There are some other geometrical objects associated to them which are also important.

Definition 2.6.3 Let an interval $I \subset \mathbb{R}$ and a parameter range $D \subset \mathbb{R}^2$ be given as in Definition 2.6.1.

(a) If $\gamma\colon I \to \mathbb{R}^n$ is a simple arc, then $\gamma'(t)$ with $t \in]a, b[$ represents a **tangential vector** to the curve $\gamma(I)$ in the point $\gamma(t)$. The **arc length** of the section $\gamma|_{[a,c]}$ with $a \leq c \leq b$, is given by $s(c) := \int_a^c |\gamma'(t)|\,\mathrm{d}t$. We can use the arc length s for an alternative parameterization $\tilde{\gamma}(s) := \gamma(t(s))$. Then $\tilde{\gamma}'(s)$ is a unit tangential vector. Moreover, if $\tilde{\gamma} \in \mathrm{C}^{(2)}(I, \mathbb{R}^n)$, then $|\tilde{\gamma}''(s)|$ represents the **curvature** of $\gamma(I)$ in $\tilde{\gamma}(s)$, where $|\tilde{\gamma}''(s)|^{-1}$ is the **curvature radius** and $\tilde{\gamma}''(s)$ is a **normal vector** to the curve in $\tilde{\gamma}(s)$.

(b) If $\Phi\colon D \to \mathbb{R}^3$ parameterizes the surface $\Phi(D)$, then $\frac{\partial\Phi}{\partial u}(u,v)$ and $\frac{\partial\Phi}{\partial v}(u,v)$ span the **tangential plane**, and consequently $\left(\frac{\partial\Phi}{\partial u} \times \frac{\partial\Phi}{\partial v}\right)(u,v)$ is a **normal vector** to the surface in $\Phi(u,v$ for $(u,v) \in \operatorname{int} D$.

For a circle with parameterization (2.17), a tangential vector is given by

$$\gamma'(\varphi) = \begin{pmatrix} -r\sin\varphi \\ r\cos\varphi \end{pmatrix}, \quad \varphi \in]0,2\pi[.$$

This corresponds to the vector $\varepsilon^\varphi(\varphi)$ in part a of Example 2.5.4. The arc length is $s(\varphi) = \int_0^\varphi r\,\mathrm{d}t = r\varphi$ such that $\tilde{\gamma}(s) = (r\cos(s/r), r\sin(s/r))^{\mathrm{T}}$ is a parameterization with respect to the arc length. Hence, the curvature is everywhere $|(-r^{-1}\cos(sr^{-1}), -r^{-1}\sin(sr^{-1}))^{\mathrm{T}}| = r^{-1}$ such that the curvature radius equals r along the circle.

For a sphere with parameterization (2.18), the tangential plane in $\Phi(\varphi,t)$ is spanned by

$$\frac{\partial\Phi}{\partial\varphi}(\varphi,t) = \begin{pmatrix} -r\sqrt{1-t^2}\sin\varphi \\ r\sqrt{1-t^2}\cos\varphi \\ 0 \end{pmatrix} \text{ and } \frac{\partial\Phi}{\partial t}(\varphi,t) = \begin{pmatrix} -\frac{rt}{\sqrt{1-t^2}}\cos\varphi \\ -\frac{rt}{\sqrt{1-t^2}}\sin\varphi \\ r \end{pmatrix}$$

such that their vector product

$$\begin{pmatrix} r^2\sqrt{1-t^2}\cos\varphi \\ r^2\sqrt{1-t^2}\sin\varphi \\ r^2 t \end{pmatrix}$$

is a normal vector. By normalizing these vectors, we get the tangential vectors $\varepsilon^\varphi(\varphi)$ and $\varepsilon^t(\varphi,t)$ as well the outer unit normal vector $\varepsilon^r(\varphi,t)$ from part b of Example 2.5.4. Note that the requirement $\operatorname{rk}\Phi' = 2$ guarantees (in the interior of the parameter range) that the tangential vectors are linearly independent and indeed span the tangential plane, and that the normal vector is not the zero vector.

We will often assume that the Earth's surface is a regular surface. Let us define this here.

Definition 2.6.4 A surface $\Sigma \subset \mathbb{R}^3$ is called a **regular surface** if the following holds true:

(1) Σ subdivides the $\mathbb{R}^3$ into a bounded region (see Definition 2.1.2 for the definition of a region), which we call the **interior** Σ_{int}, and an unbounded region, which we call the **exterior** Σ_{ext}, such that the $\mathbb{R}^3$ is the disjoint union $\mathbb{R}^3 = \Sigma \,\dot{\cup}\, \Sigma_{\text{int}} \,\dot{\cup}\, \Sigma_{\text{ext}}$.
(2) Σ is a closed and compact surface which is free of double points.
(3) There exists a parameterization for Σ which is twice continuously differentiable, where the Jacobian matrix of this parameterization has maximal rank (i.e., rank 2) everywhere in the interior of the parameter range.

A few remarks on this definition are worth making:

- The term 'closed' is not meant in the topological sense (this requirement would be completely unnecessary in this case, because we already require that Σ is compact) but in the sense of 'there is no hole in it', that is, $\partial\Sigma = \emptyset$.
- Some publications require that $0 \in \Sigma_{\text{int}}$, which is only a matter of setting the coordinate system appropriately, because this simplifies some notations and proofs (see, e.g., Freeden, 1999, p. 56). However, we will need here some statements of the form 'for all regular surfaces', where we definitely also need surfaces with $0 \notin \Sigma_{\text{int}}$. For this reason, we skip this requirement here.
- Since we require a twice continuously differentiable parameterization of Σ, there exists a unit normal vector field on Σ which is (at least once) continuously differentiable.

For a similar definition of regular surfaces, see Canuto and Tabacco (2010, section 6.7).

3
On Gravitation, Harmonic Functions, and Related Topics

3.1 The Gravitational Potential

We start with the investigation of **Newton's gravitational potential**. Often, gravitation and gravity are distinguished in the sense that gravity represents the sum of the gravitational force and the centrifugal force. This distinction has, however, not consistently been performed in the literature. For this reason and due to the fact that the centrifugal force does not play an important rule in this book, we will use the word 'gravity' here as an equivalent expression for 'gravitation'.

The formulae of (classical) gravitation are well known. The most simple one is the following:

$$V(y) = GM\,\frac{1}{|x-y|}.$$

It represents the **gravitational potential** $V(y)$ at an arbitrary point $y \in \mathbb{R}^3 \setminus \{x\}$, while the gravitational field is produced by a point mass $M \in \mathbb{R}$ at the position $x \in \mathbb{R}^3$ and G is the gravitational constant. Correspondingly, another point mass m at point y is attracted by the gravitational force

$$m\nabla_y V(y) = GmM\,\frac{x-y}{|x-y|^3}.$$

In the presence of finitely many point masses $M_1, \dots, M_N \in \mathbb{R}$ at positions $x^{(1)}, \dots, x^{(N)} \in \mathbb{R}^3$, the gravitational potential can be summed up to

$$V(y) = G\sum_{k=1}^{N} \frac{M_k}{\left|x^{(k)} - y\right|}. \tag{3.1}$$

If we have a continuum, for example a planet, $D \subset \mathbb{R}^3$ (let us say a measurable and bounded set) and a mass density distribution $F\colon D \to \mathbb{R}$ (which can easily be assumed to be bounded and integrable), then we could first subdivide D into a disjoint union of 'many' 'small' subsets $D = \dot{\bigcup}_{k=1}^{N} D_k$ and assume that each D_k is sufficiently small such that F is approximately constant in each D_k (i.e., $F(x) \approx F(x^{(k)})$ for all $x \in D_k$, a fixed $x^{(k)} \in D_k$, and for each $k = 1, \dots, N$). Then we approximate the continuum D by N point masses and apply (3.1), which leads us to

$$V(y) \approx G \sum_{k=1}^{N} \frac{F\left(x^{(k)}\right) \operatorname{Vol}\left(D_k\right)}{\left|x^{(k)} - y\right|}, \tag{3.2}$$

where $\operatorname{Vol}(D_k)$ is the volume of D_k. This approximation (3.2) becomes more and more accurate (and, in the limit, exact) if we let the sizes of all D_k shrink appropriately (and at the same time $N \to \infty$). This approach leads to an integral and the celebrated formula (see, in chronological order, Newton, 1687; Koyré and Cohen, 1726; Wolfers, 1872; Tipler, 1982; Hofmann-Wellenhof and Moritz, 2005)

$$V(y) = G \int_D \frac{F(x)}{|x - y|} \, \mathrm{d}x, \quad y \in \mathbb{R}^3, \tag{3.3}$$

for the **gravitational potential** V of a planet $D \subset \mathbb{R}^3$ with the mass density distribution $F\colon D \to \mathbb{R}$. We will, therefore, start our discussions by investigating what we can find out about the function V in (3.3).

Note that F could also be the density distribution of an electric current. In this case, an appropriate constant for G yields the **Coulomb potential** V; see also Section 6.1.

Most of the proofs of this section are motivated by the derivations in Kellogg (1967, chapter VI) and Mikhlin (1970, pp. 162–164, 228–236).

We first have to make sure that V is, as we stated previously, defined on the whole space $\mathbb{R}^3$. This is not trivial, since we integrate a singularity in $|x - y|^{-1}$. Therefore, we have to impose an appropriate condition on F.

However, we first have to do some preliminary work.

Lemma 3.1.1 *Let $B_R(y)$ be a ball with arbitrary radius $R > 0$ and arbitrary centre $y \in \mathbb{R}^3$. Then $\int_{B_R(y)} |x - y|^{-p} \, \mathrm{d}x = \frac{4\pi}{3-p} R^{3-p}$ for all $p < 3$.*

Proof With the use of polar coordinates

$$x = y + \begin{pmatrix} r\sqrt{1-t^2} \cos\varphi \\ r\sqrt{1-t^2} \sin\varphi \\ rt \end{pmatrix},$$

we get

$$\int_{B_R(y)} \frac{1}{|x - y|^p} \, \mathrm{d}x = \int_0^R \int_{-1}^1 \int_0^{2\pi} \frac{r^2}{r^p} \, \mathrm{d}\varphi \, \mathrm{d}t \, \mathrm{d}r$$
$$= 4\pi \int_0^R r^{2-p} \, \mathrm{d}r = \frac{4\pi}{3 - p} R^{3-p}. \qquad \square$$

Theorem 3.1.2 *Let the function F in (3.3) be measurable and bounded in D, where $D = \Sigma_{\mathrm{int}}$ is the interior of a regular surface $\Sigma \subset \mathbb{R}^3$. Then the potential V is defined on the whole space $\mathbb{R}^3$, it is bounded, and it is continuous in every $y \in \mathbb{R}^3$.*

Proof Within this proof, we neglect the constant factor G, since it does not play a role concerning the discussed properties of V.

(1) V exists everywhere and is bounded:

For every $y \in \mathbb{R}^3$, we obtain

$$|V(y)| \leq \int_D \frac{|F(x)|}{|x-y|}\,\mathrm{d}x \leq \sup_{x\in D}|F(x)| \int_D \frac{1}{|x-y|}\,\mathrm{d}x.$$

Since F is a bounded function, $C := \sup_{x\in D}|F(x)|$ is finite. Let now $d \in \mathbb{R}^+$ be a fixed number. If $y \notin \overline{D}$, then $|x-y| \geq \operatorname{dist}(y, D) > 0$ for all $x \in D$. Hence, $V(y)$ exists and $|V(y)| \leq C[\operatorname{dist}(y, D)]^{-1}\lambda(D)$ for all $y \notin \overline{D}$, where λ is the Lebesgue measure (i.e., roughly speaking, $\lambda(D)$ is the volume of D). Thus, if $\operatorname{dist}(y, D) \geq d$, then $|V(y)| \leq Cd^{-1}\lambda(D)$.

Let now $y \in \overline{D}$ or $\operatorname{dist}(y, D) < d$. We set $R := \operatorname{diam}\overline{D} + d$. Since D is a bounded set, R is finite, and we have $D \subset B_R(y)$. As a consequence,

$$|V(y)| \leq C \int_{B_R(y)} \frac{1}{|x-y|}\,\mathrm{d}x = 2\pi C R^2$$

for all $y \in \mathbb{R}^3$, where we used Lemma 3.1.1. Consequently, V is bounded on the whole $\mathbb{R}^3$.

(2) Some constructions:

We have to make some preparations for the following proof of the continuity of V. First of all, let $y \in \mathbb{R}^3$ be an arbitrary point (where we want to prove the continuity), and let $\varepsilon > 0$ be an arbitrary real number. Corresponding to this choice, we define $\delta := \delta(\varepsilon) := \sqrt{\varepsilon/(52\pi C)}$.

Furthermore, let $z \in \mathbb{R}^3$ with $|z-y| < \delta$ (i.e., $z \in B_\delta(y)$) be given. The proof will be finished, once we have shown that $|V(z) - V(y)| < \varepsilon$.

(3) Subdivision of the integral:

One problem that we are facing here is the singularity of the integrand, at least if $y, z \in D$. For this reason, we distinguish the treatment of the integrand inside and outside a small ball $B_{2\delta}(y)$. We consider now the difference of the potential values, use the triangle inequality, subdivide the domain of integration, and use the boundedness of F. We get

$$\begin{aligned}
|V(z) - V(y)| &= \left| \int_D \frac{F(x)}{|x-z|}\,\mathrm{d}x - \int_D \frac{F(x)}{|x-y|}\,\mathrm{d}x \right| \\
&= \left| \int_D F(x) \left(\frac{1}{|x-z|} - \frac{1}{|x-y|} \right) \mathrm{d}x \right| \\
&\leq \int_D |F(x)| \left| \frac{1}{|x-z|} - \frac{1}{|x-y|} \right| \mathrm{d}x \\
&\leq \int_{B_{2\delta}(y)} |F(x)| \left| \frac{1}{|x-z|} - \frac{1}{|x-y|} \right| \mathrm{d}x \\
&\quad + \int_{D\setminus B_{2\delta}(y)} |F(x)| \left| \frac{1}{|x-z|} - \frac{1}{|x-y|} \right| \mathrm{d}x
\end{aligned}$$

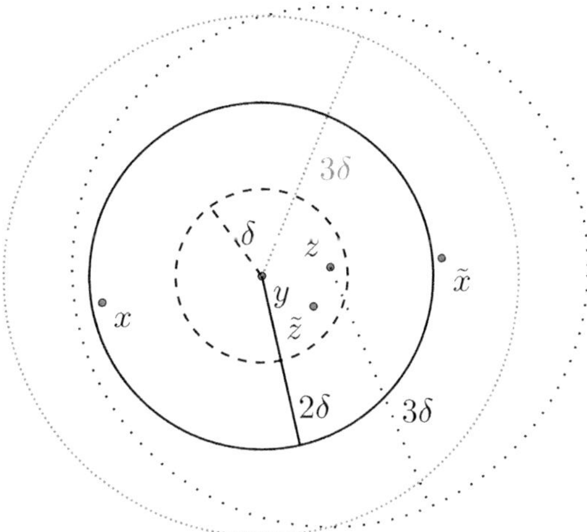

Figure 3.1 Illustration of the spheres used in the proof of Theorem 3.1.2, where this image is simplified to the two-dimensional case (here $z \in B_\delta(y)$, $\tilde{z} \in \overline{B_\delta(y)}$, $x \in B_{2\delta}(y)$, and $\tilde{x} \in \overline{D \setminus B_{2\delta}(y)}$; note that $x \in B_{3\delta}(z)$).

$$\begin{aligned}
&\le \int_{B_{2\delta}(y)} \frac{|F(x)|}{|x-z|}\,\mathrm{d}x + \int_{B_{2\delta}(y)} \frac{|F(x)|}{|x-y|}\,\mathrm{d}x \\
&\quad + \int_{D\setminus B_{2\delta}(y)} |F(x)| \left| \frac{1}{|x-z|} - \frac{1}{|x-y|} \right| \mathrm{d}x \\
&\le C \int_{B_{2\delta}(y)} \frac{1}{|x-z|}\,\mathrm{d}x + C \int_{B_{2\delta}(y)} \frac{1}{|x-y|}\,\mathrm{d}x \\
&\quad + C \int_{D\setminus B_{2\delta}(y)} \left| \frac{1}{|x-z|} - \frac{1}{|x-y|} \right| \mathrm{d}x. \qquad (3.4)
\end{aligned}$$

For the case that $B_{2\delta}(y) \not\subset D$, we extend F, for example, by setting $F(x) := 0$ for all $x \in \mathbb{R}^3 \setminus D$ (note that F only has to be measurable and bounded). This is compensated by the second inequality in (3.4).

(4) The integrals over the balls:

The first two integrals in our latter result are easy to handle. From Lemma 3.1.1, we obtain that $\int_{B_{2\delta}(y)} \frac{1}{|x-y|}\,\mathrm{d}x = 8\pi\delta^2$.

Furthermore, for $x \in B_{2\delta}(y)$, we get (see also Figure 3.1)

$$|x-z| \le |x-y| + |y-z| \le 2\delta + \delta = 3\delta.$$

As a consequence, $B_{2\delta}(y) \subset B_{3\delta}(z)$, and, therefore,

$$\int_{B_{2\delta}(y)} \frac{1}{|x-z|}\,\mathrm{d}x \le \int_{B_{3\delta}(z)} \frac{1}{|x-z|}\,\mathrm{d}x = 18\pi\delta^2.$$

Hence, the first two summands at the end of (3.4) can be estimated by $26\pi C\delta^2 = \varepsilon/2$.

(5) The remaining integral:

The use of the ball $B_{2\delta}(y)$ was to avoid the (possible) singularity of the integrand. For $\tilde{x} \in \overline{D \setminus B_{2\delta}(y)}$ and $\tilde{z} \in \overline{B_\delta(y)}$, we have (see also Figure 3.1) $2\delta \le |\tilde{x} - y| \le |\tilde{x} - \tilde{z}| + |\tilde{z} - y|$ and, therefore, $|\tilde{x} - \tilde{z}| \ge 2\delta - |\tilde{z} - y| \ge 2\delta - \delta = \delta$.

As a consequence, the function $\overline{D \setminus B_{2\delta}(y)} \times \overline{B_\delta(y)} \ni (\tilde{x}, \tilde{z}) \mapsto |\tilde{x} - \tilde{z}|^{-1}$ is continuous on a compact domain (note that D is bounded) and, consequently, uniformly continuous. Hence, corresponding to our previously chosen $\varepsilon > 0$, we find a $\delta_1 > 0$ such that

$$\left| \frac{1}{|\tilde{w} - \tilde{z}|} - \frac{1}{|\tilde{x} - \tilde{y}|} \right| < \frac{\varepsilon}{2C\lambda(D)}$$

for all $(\tilde{w}, \tilde{z}), (\tilde{x}, \tilde{y}) \in \overline{D \setminus B_{2\delta}(y)} \times \overline{B_\delta(y)}$ with $|(\tilde{w}, \tilde{z}) - (\tilde{x}, \tilde{y})| < \delta_1$. This is, in particular, the case if $\tilde{x} = \tilde{w}$ and $\tilde{y} = y$. In other words: if $|z - y| < \delta_1$ for $z \in \overline{B_\delta(y)}$, then

$$\left| \frac{1}{|x - z|} - \frac{1}{|x - y|} \right| < \frac{\varepsilon}{2C\lambda(D)} \quad \text{for all } x \in \overline{D \setminus B_{2\delta}(y)}.$$

Hence, the last summand in (3.4) can be estimated as follows: if we have $|z - y| < \min(\delta_1, \delta)$, then

$$C \int_{D \setminus B_{2\delta}(y)} \left| \frac{1}{|x - z|} - \frac{1}{|x - y|} \right| \mathrm{d}x \le C \frac{\varepsilon}{2C\lambda(D)} \int_{D \setminus B_{2\delta}(y)} 1 \,\mathrm{d}x \le \frac{\varepsilon}{2}.$$

(6) V is continuous:

We started with an arbitrary point $y \in \mathbb{R}^3$ and an arbitrary real number $\varepsilon > 0$. We obtained constants $\delta, \delta_1 \in \mathbb{R}^+$ such that we can set $\delta_2 := \min(\delta, \delta_1)$. Combining our previous considerations, we can now conclude that $|V(z) - V(y)| \le \varepsilon/2 + \varepsilon/2 = \varepsilon$, if $|z - y| < \delta_2$. □

In the following, we will show that the Newton potential satisfies two particular differential equations: more precisely, the Laplace equation outside the Earth and the Poisson equation inside the Earth (where actually the Poisson equation is an inhomogeneous Laplace equation). For this purpose, we first have to verify the differentiability of the potential.

Theorem 3.1.3 *Let the function F in (3.3) be measurable and bounded on D, where $D = \Sigma_{\text{int}}$ is the interior of a regular surface $\Sigma \subset \mathbb{R}^3$. Then the potential V is partially differentiable with respect to all coordinates on the whole space $\mathbb{R}^3$.*

Proof Note that Theorems 2.4.8 and 2.4.9 cannot be used here, because the integrand of (3.3) is singular, if $y \in D$.

(1) The suspected derivative:

If we knew that we could interchange the gradient operator ∇ and the integration, we would obtain

$$g(y) := \int_D F(x) \nabla_y \frac{1}{|x - y|} \,\mathrm{d}x = \int_D F(x) \frac{x - y}{|x - y|^3} \,\mathrm{d}x, \quad y \in \mathbb{R}^3,$$

as the suspected derivative (note that we set again $G = 1$ for reasons of simplicity of the formulae). Our task is, therefore, to show that

$$d_j(y,h) := \frac{V(y+h\varepsilon^j) - V(y)}{h} - g_j(y), \qquad y \in \mathbb{R}^3,\ h \in \mathbb{R} \setminus \{0\},$$

tends to 0 as $h \to 0$ for all $j = 1,2,3$ and all $y \in \mathbb{R}^3$. Remember that $\varepsilon^j = (\delta_{ij})_{i=1,2,3}$ is the jth canonical orthonormal basis vector in $\mathbb{R}^3$. The following considerations can be done now for every choice of $j \in \{1,2,3\}$.

(2) The function d_j exists everywhere:

By some simple calculations, we obtain

$$\begin{aligned}
d_j(y,h) &= \int_D F(x) \left(\frac{1}{h|x-y-h\varepsilon^j|} - \frac{1}{h|x-y|} - \frac{x_j - y_j}{|x-y|^3} \right) \mathrm{d}x \\
&= \int_D F(x) \left(\frac{|x-y| - |x-y-h\varepsilon^j|}{h|x-y-h\varepsilon^j|\,|x-y|} - \frac{x_j - y_j}{|x-y|^3} \right) \mathrm{d}x \\
&= \int_D F(x) \left(\frac{|x-y|^2 - |x-y-h\varepsilon^j|^2}{h|x-y-h\varepsilon^j|\,|x-y|(|x-y| + |x-y-h\varepsilon^j|)} \right. \\
&\qquad \left. - \frac{x_j - y_j}{|x-y|^3} \right) \mathrm{d}x \\
&= \int_D F(x) \left(\frac{|x|^2 - 2x\cdot y + |y|^2}{h|x-y-h\varepsilon^j|\,|x-y|(|x-y| + |x-y-h\varepsilon^j|)} \right. \\
&\qquad - \frac{|x|^2 + |y|^2 + h^2 - 2x\cdot y - 2hx_j + 2hy_j}{h|x-y-h\varepsilon^j|\,|x-y|(|x-y| + |x-y-h\varepsilon^j|)} \\
&\qquad \left. - \frac{x_j - y_j}{|x-y|^3} \right) \mathrm{d}x \\
&= \int_D F(x) \left(\frac{2hx_j - 2hy_j - h^2}{h|x-y-h\varepsilon^j|\,|x-y|(|x-y| + |x-y-h\varepsilon^j|)} \right. \\
&\qquad \left. - \frac{x_j - y_j}{|x-y|^3} \right) \mathrm{d}x. \qquad (3.5)
\end{aligned}$$

Furthermore, we have $|x_j - y_j| \leq |x-y|$ and

$$\begin{aligned}
\left|2hx_j - 2hy_j - h^2\right| &\leq |h|\big(|x_j - y_j| + |x_j - y_j - h|\big) \\
&\leq |h|\big(|x-y| + |x-y-h\varepsilon^j|\big).
\end{aligned}$$

Moreover, for $a,b \in \mathbb{R}$, we obtain $0 \leq (a-b)^2 = a^2 - 2ab + b^2$ such that $2ab \leq a^2 + b^2$. Hence, the integrand in (3.5) may be estimated by (again, we use the abbreviation $C := \sup_{x\in D} |F(x)|$)

$$\left| F(x)\left(\frac{2hx_j - 2hy_j - h^2}{h|x-y-h\varepsilon^j|\,|x-y|\big(|x-y|+|x-y-h\varepsilon^j|\big)} - \frac{x_j-y_j}{|x-y|^3}\right)\right|$$
$$\leq C\left(\frac{1}{|x-y-h\varepsilon^j|\,|x-y|} + \frac{1}{|x-y|^2}\right)$$
$$= C\,\frac{|x-y-h\varepsilon^j|\,|x-y| + |x-y-h\varepsilon^j|^2}{|x-y-h\varepsilon^j|^2|x-y|^2}$$
$$\leq C\,\frac{|x-y-h\varepsilon^j|^2 + |x-y|^2 + 2|x-y-h\varepsilon^j|^2}{2|x-y-h\varepsilon^j|^2\,|x-y|^2}$$
$$\leq C\left(\frac{3}{2|x-y|^2} + \frac{1}{2|x-y-h\varepsilon^j|^2}\right). \tag{3.6}$$

From Lemma 3.1.1, we get that the integral of the right-hand side of (3.6) (over a ball $B_R(y) \supset D$ resp. $B_R(y+h\varepsilon^j) \supset D$) is finite. Hence, $d_j(y,h)$ is also finite. This result is true for all $y \in \mathbb{R}^3$ and all $h \in \mathbb{R}\setminus\{0\}$.

(3) The value $d_j(y,0)$:

If we insert $h = 0$ in the last line of the equalities in (3.5) (after canceling h in the first fraction), we see that

$$d_j(y,0) := \int_D F(x)\left(\frac{2(x_j-y_j)}{2|x-y|^3} - \frac{x_j-y_j}{|x-y|^3}\right) \mathrm{d}x = 0 \tag{3.7}$$

for all $y \in \mathbb{R}^3$. It remains to show that $\lim_{h\to 0} d_j(y,h) = d_j(y,0)$. In other words, we have to show that d_j is continuous in $h = 0$.

(4) The continuity of $h \mapsto d_j(y,h)$ at 0:

Let $y \in \mathbb{R}^3$ and (a sufficiently small number) $h_0 > 0$ be fixed. Without loss of generality, we assume that $|h| < h_0$. We split up now the integral in (3.5) into two parts:

$$\begin{aligned} &d_j(y,h)\\ &= \int_{D\setminus B_{h_0}(y)} F(x)\left(\frac{2(x_j-y_j)-h}{|x-y-h\varepsilon^j|\,|x-y|(|x-y|+|x-y-h\varepsilon^j|)}\right.\\ &\qquad\left. - \frac{x_j-y_j}{|x-y|^3}\right)\mathrm{d}x\\ &+ \int_{D\cap B_{h_0}(y)} F(x)\left(\frac{2(x_j-y_j)-h}{|x-y-h\varepsilon^j|\,|x-y|(|x-y|+|x-y-h\varepsilon^j|)}\right.\\ &\qquad\left. - \frac{x_j-y_j}{|x-y|^3}\right)\mathrm{d}x. \end{aligned} \tag{3.8}$$

The first integral is continuous in h since $y, y+h\varepsilon^j \in B_{h_0}(y)$. In analogy to (3.7), we conclude that it vanishes for $h = 0$. Hence, the first integral can be made arbitrarily small for $h \to 0$. The second integral can be handled, in combination with (3.6), as follows (note that $B_{h_0}(y) \subset B_{2h_0}(y+h\varepsilon^j)$, because $|x-(y+h\varepsilon^j)| \leq |x-y|+|h| < h_0 + h_0 = 2h_0$ for all $x \in B_{h_0}(y)$):

$$\left| \int_{D\cap B_{h_0}(y)} F(x)\left(\frac{2(x_j-y_j)-h}{|x-y-h\varepsilon^j|\,|x-y|\big(|x-y|+|x-y-h\varepsilon^j|\big)} - \frac{x_j-y_j}{|x-y|^3}\right) \mathrm{d}x \right|$$

$$\begin{aligned}
&\leq C \int_{B_{h_0}(y)} \left(\frac{3}{2|x-y|^2} + \frac{1}{2|x-y-h\varepsilon^j|^2}\right) \mathrm{d}x \\
&\leq C \left(\int_{B_{h_0}(y)} \frac{3}{2|x-y|^2}\,\mathrm{d}x + \int_{B_{2h_0}(y+h\varepsilon^j)} \frac{1}{2|x-(y+h\varepsilon^j)|^2}\,\mathrm{d}x\right) \\
&= C\ (6\pi h_0 + 4\pi h_0) = 10\pi\, h_0\, C,
\end{aligned}$$

where we used Lemma 3.1.1.

We summarize: for a given $\varepsilon > 0$, we can find an $h_0 > 0$ (in particular, $h_0 < \varepsilon(20\pi C)^{-1}$) such that, for all $h \in \mathbb{R}$ with $|h| < h_0$, the last integral in (3.8) can be estimated by $\varepsilon/2$ *and* the first integral is also less than $\varepsilon/2$. This yields the continuity of $d_j(y,h)$ with respect to h in $h = 0$. Hence, $\lim_{h\to 0} d_j(y,h) = 0$ and the potential V is partially differentiable everywhere in $\mathbb{R}^3$. □

Corollary 3.1.4 *Let the function* F *in*

$$V(y) = G \int_D \frac{F(x)}{|x-y|}\,\mathrm{d}x, \quad y \in \mathbb{R}^3,$$

be measurable and bounded, where $D = \Sigma_{\mathrm{int}}$ *is the interior of a regular surface* $\Sigma \subset \mathbb{R}^3$. *Then*

$$\nabla V(y) = G \int_D F(x)\,\frac{x-y}{|x-y|^3}\,\mathrm{d}x \quad \textit{for all } y \in \mathbb{R}^3.$$

In physical geodesy, the gravitational potential represents an essential observable for the geosystem. For instance, equipotential surfaces of V are used to define the geoid, which is linked to the mean sea level. Moreover, mass transports can be detected by analyzing temporal variations of the gravitational field. In particular, the GRACE mission has provided us with an unprecedented quality of time-variable gravity models for such purposes (see, e.g., Schmidt et al., 2008; Tapley et al., 2004b). In general, satellite-based gravity data are, in the first place, not directly data of the potential V. For instance, satellite-to-satellite tracking (SST) can be used to get data of the first derivative ∇V and satellite gravity gradiometry (SGG) is appropriate for getting data of the second derivative, that is, the Hessian $\nabla \otimes \nabla V$, or at least some components of it (for further details, see also Freeden et al., 2002). For this reason, it is important to have a closer look at the first- and second-order derivatives of the gravitational potential V. In particular, we will see that V satisfies a particular second-order partial differential equation, which is an essential foundation for the modelling of the gravitational field.

Theorem 3.1.5 *Let the function* F *in*

$$V(y) = G \int_D \frac{F(x)}{|x-y|}\,\mathrm{d}x, \quad y \in \mathbb{R}^3,$$

be measurable and bounded, where $D = \overline{\Sigma_{\text{int}}}$ is the interior of a regular surface $\Sigma \subset \mathbb{R}^3$. Then V is infinitely differentiable on $\mathbb{R}^3 \setminus D$ and satisfies the (homogeneous) ***Laplace equation***

$$\Delta_y V(y) = 0 \quad \text{for all } y \in \mathbb{R}^3 \setminus D,$$

where Δ is the Laplace operator (see Definition 2.3.3).

Proof Let $\varepsilon > 0$ be arbitrary and $N_\varepsilon \subset \mathbb{R}^3 \setminus D$ be an arbitrary open subset such that $|x - y| \geq \varepsilon$ for all $x \in D$ and all $y \in N_\varepsilon$. Note that, since D is closed, we can find, for every $y \in \mathbb{R}^3 \setminus D$, an $\varepsilon > 0$ and an N_ε of this kind such that $y \in N_\varepsilon$. Hence, our result will be valid eventually for all $y \in \mathbb{R}^3 \setminus D$. We have $|x - y|^{-1} \leq \varepsilon^{-1}$ for all $x \in D$ and all $y \in N_\varepsilon$ such that

$$\left| \frac{\partial}{\partial y_j} \frac{1}{|x - y|} \right| = \left| \frac{x_j - y_j}{|x - y|^3} \right| \leq \frac{1}{|x - y|^2} \leq \frac{1}{\varepsilon^2}$$

for all $x \in D$ and all $y \in N_\varepsilon$ and (δ_{jk} is the Kronecker delta)

$$\begin{aligned} \left| \frac{\partial^2}{\partial y_k \partial y_j} \frac{1}{|x - y|} \right| &= \left| -\frac{\delta_{jk}}{|x - y|^3} + \frac{3(x_j - y_j)(x_k - y_k)}{|x - y|^5} \right| \\ &\leq \frac{4}{|x - y|^3} \leq \frac{4}{\varepsilon^3} \end{aligned}$$

for all $x \in D$ and all $y \in N_\varepsilon$. Due to the conditions on F, we are, consequently, allowed to apply Theorem 2.4.9 twice such that, first of all,

$$\frac{\partial}{\partial y_j} V(y) = G \int_D F(x) \frac{\partial}{\partial y_j} \frac{1}{|x - y|} \, \mathrm{d}x,$$

and then,

$$\frac{\partial^2}{\partial y_k \partial y_j} V(y) = G \int_D F(x) \frac{\partial^2}{\partial y_k \partial y_j} \frac{1}{|x - y|} \, \mathrm{d}x$$

for all $y \in N_\varepsilon$. In particular,

$$\Delta_y V(y) = G \int_D F(x) \, \Delta_y \frac{1}{|x - y|} \, \mathrm{d}x,$$

where

$$\begin{aligned} \Delta_y \frac{1}{|x - y|} &= \sum_{j=1}^{3} \left(\frac{-1}{|x - y|^3} + 3 \frac{(x_j - y_j)^2}{|x - y|^5} \right) \\ &= -\frac{3}{|x - y|^3} + 3 \frac{|x - y|^2}{|x - y|^5} = 0 \end{aligned} \tag{3.9}$$

for $x \neq y$. This yields, eventually, $\Delta_y V(y) = 0$ for all $y \in \mathbb{R}^3 \setminus D$. Furthermore, in analogy to the preceding considerations, we can easily deduce that all derivatives of $|x - y|^{-1}$ are bounded by a term of the form $c_n \, \varepsilon^{-n}$. As a consequence, all derivatives of V exist on $\mathbb{R}^3 \setminus D$, and the differentiation may be interchanged with the integration here. □

We get two other consequences. The first one has just been proved in (3.9). The second one needs some further considerations.

Corollary 3.1.6 (Fundamental Solution) *Let $x \in \mathbb{R}^3$ be arbitrary. Then the function $N(y) := |x-y|^{-1}$, $y \in \mathbb{R}^3 \setminus \{x\}$, is harmonic, that is, it satisfies the Laplace equation $\Delta_y |x-y|^{-1} = 0$ for all $y \neq x$.*

Corollary 3.1.7 *Under the conditions of Theorem 3.1.5, we have $|V(y)| = \mathcal{O}(|y|^{-1})$ and $|\nabla V(y)| = \mathcal{O}(|y|^{-2})$ as $|y| \to \infty$. For the Landau symbol $\mathcal{O}$, see Definition 2.1.7. A function with these properties is called* ***regular at infinity****.*

Proof Without loss of generality, let us assume that $0 \in D$ (otherwise enlarge D and define F as vanishing on the additional part of D). Let $d :=$ diam D be the diameter of D. Then, for all $y \in \mathbb{R}^3$ with $|y| > 2d$ and all $x \in D$, we get $|x| = |x - 0| \leq d < |y|/2$ such that $|y-x| \geq |y| - |x| > |y| - |y|/2 = |y|/2$. Hence, with $C := \sup_{x \in D} |F(x)|$, we obtain

$$|V(y)| \leq G \int_D \frac{|F(x)|}{|x-y|}\,\mathrm{d}x \leq GC \int_D \frac{2}{|y|}\,\mathrm{d}x = 2GC\,\lambda(D)\,\frac{1}{|y|},$$

where $\lambda(D) = \int_D 1\,\mathrm{d}x$ is the volume (i.e., the Lebesgue measure) of D. In a similar manner, we obtain, using Corollary 3.1.4,

$$\begin{aligned} |\nabla_y V(y)| &\leq G \int_D |F(x)| \left| \nabla_y \frac{1}{|x-y|} \right| \mathrm{d}x \leq GC \int_D \frac{1}{|x-y|^2}\,\mathrm{d}x \\ &\leq GC \int_D \frac{4}{|y|^2}\,\mathrm{d}x = 4GC\,\lambda\,(D)\,\frac{1}{|y|^2}. \end{aligned}$$

This completes the proof of the corollary. □

We now have a look at the interior of the Earth. From Theorem 3.1.3, we already know that V is partially differentiable in the whole space $\mathbb{R}^3$ if F is measurable and bounded. For obtaining a differentiability of the second order in the interior, we have to require a stronger condition on F.

Definition 3.1.8 Let $D \subset \mathbb{R}^n$ and $F : D \to \mathbb{R}$ be given. The function F is called **α-Hölder continuous** for an $\alpha \in \mathbb{R}^+$ if there exists a constant $C \in \mathbb{R}^+$ such that $|F(x) - F(y)| \leq C|x-y|^\alpha$ for all $x, y \in D$. The set of all α-Hölder continuous (scalar) functions on D is denoted by $\mathrm{C}^{(0,\alpha)}(D)$. The subset of $\mathrm{C}^{(k)}(D)$ where all derivatives of order k are α-Hölder continuous is denoted by $\mathrm{C}^{(k,\alpha)}(D)$. Furthermore, 1-Hölder continuous functions are also called **Lipschitz continuous**. If the value of α is not important, then also the term Hölder continuous can simply be used. For vectorial functions, the property is understood componentwise. Accordingly, we get the set $\mathrm{c}^{(k,\alpha)}(D)$ for mappings to $\mathbb{R}^3$.

Let us keep $x \in D$ arbitrary but fixed and assume that $F : D \to \mathbb{R}$ is α-Hölder continuous. Then, for every sequence $(y_n) \subset D$ with $y_n \overset{n\to\infty}{\longrightarrow} x$, we see that $|F(x) - F(y_n)| \leq C|x - y_n|^\alpha \overset{n\to\infty}{\longrightarrow} 0$ such that $F(y_n) \overset{n\to\infty}{\longrightarrow} F(x)$. This explains why we say that F is

Hölder *continuous*. More precisely, since $\delta(\varepsilon) := (\varepsilon C^{-1})^{1/\alpha}$ yields $|F(x) - F(y)| \le \varepsilon$ for all $x, y \in D$ with $|x - y| \le \delta$, F is also uniformly continuous. Furthermore, if $\alpha > 1$, then

$$\frac{|F(x) - F(y_n)|}{|x - y_n|} \le C|x - y_n|^{\alpha - 1} \xrightarrow[n\to\infty]{} 0$$

implies that F is differentiable in x *and* $F'(x) = 0$. Since we can argue this way for all $x \in D$, F is constant on connected subsets of D. For this reason, it does not make sense to further investigate functions which are α-Hölder continuous with $\alpha > 1$. Note that the converse implication (every constant function is α-Hölder continuous (for actually every $\alpha \in \mathbb{R}^+$)) is trivial.

Furthermore, if $F\colon D \to \mathbb{R}$ is continuously differentiable, then the mean-value theorem of differentiation says that, for every $x, y \in D$, there exist $\xi_1, \ldots, \xi_n \in \mathbb{R}$ such that

$$\begin{aligned}
|F(x) - F(y)| &\le |F(x) - F(y_1, x_2, \ldots, x_n)| + |F(y_1, x_2, \ldots, x_n) - F(y)| \\
&\le |F(x) - F(y_1, x_2, \ldots, x_n)| + |F(y_1, x_2, \ldots, x_n) - F(y_1, y_2, x_3, \ldots, x_n)| \\
&\quad + \cdots + |F(y_1, \ldots, y_{n-1}, x_n) - F(y)| \\
&= \left|\frac{\partial F}{\partial x_1}(\xi_1, x_2, \ldots, x_n)\right| |x_1 - y_1| + \cdots + \left|\frac{\partial F}{\partial x_n}(y_1, \ldots, y_{n-1}, \xi_n)\right| |x_n - y_n|.
\end{aligned}$$

For this conclusion, we need to guarantee that

$$(y_1, \ldots, y_{i-1}, t y_i + (1 - t)x_i, x_{i+1}, \ldots, x_n) \in D$$

for all $i = 1, \ldots, n$ and all $t \in [0, 1]$. We have then actually that there is $\tau_i \in [0, 1]$ such that $\xi_i = \tau_i y_i + (1 - \tau_i)x_i$.

Now, the Cauchy–Schwarz inequality in $\mathbb{R}^n$ yields

$$|F(x) - F(y)| \le \left|\left(\frac{\partial F}{\partial x_1}(\xi_1, x_2, \ldots, x_n), \ldots, \frac{\partial F}{\partial x_n}(y_1, \ldots, y_{n-1}, \xi_n)\right)\right| |x - y|.$$

If we additionally require that D is compact, then the continuity of ∇F implies that there exists a constant γ such that $|F(x) - F(y)| \le \gamma\, |x - y|$ for all $x, y \in D$. Thus, F is 1-Hölder continuous.

Let us summarize what we obtained so far.

Theorem 3.1.9 *Let $D \subset \mathbb{R}^n$ and $F\colon D \to \mathbb{R}$ be given.*

(a) If F is α-Hölder continuous for $\alpha > 0$, then F is uniformly continuous.
(b) F is α-Hölder continuous for $\alpha > 1$ if and only if F is constant on connected subsets of D.
(c) If F is continuously differentiable and its domain D is a compact orthotope $\prod_{i=1}^n [a_i, b_i]$, then F is α-Hölder continuous for $\alpha = 1$, that is, F is Lipschitz continuous.

Remark 3.1.10 The converse implications for parts (a) and (c) of Theorem 3.1.9 are false. For part (a), consider the function

$$F(x) := \begin{cases} \frac{1}{\log x}, & x \in \left]0, \frac{1}{2}\right], \\ 0, & x = 0 \end{cases}$$

on the compact domain $\left[0, \frac{1}{2}\right]$ (see also Atiyah et al., 2001a, p. 424). Since $\lim_{x\to 0+} F(x) = 0 = F(0)$, F is uniformly continuous (it is continuous on a compact domain). Moreover, $|F(x) - F(0)| = |\log x|^{-1} = -(\log x)^{-1}$ for all $x \in]0, 1/2]$.

If we found constants $\alpha, C \in \mathbb{R}^+$ such that $|F(x) - F(0)| \leq C|x - 0|^\alpha$, $x \in]0, 1/2]$, then we would get $-(\log x)^{-1} \leq Cx^\alpha$, $x \in]0, 1/2]$, such that $-C^{-1} \geq x^\alpha \log x$, $x \in]0, 1/2]$. However, the limit $x \to 0+$ yields the contradiction $-C^{-1} \geq 0$. The latter requires L'Hospital's rule: $\log x / x^{-\alpha}$ is a fraction where $\log x \to -\infty$ and $x^{-\alpha} \to +\infty$. Since $x^{-1}/(-\alpha x^{-\alpha-1}) = -(1/\alpha)x^\alpha \to 0$, we obtain $x^\alpha \log x \to 0$. Hence, F is not Hölder continuous.

For an example of a non-differentiable but 1-Hölder continuous function, see Hardy (1916).

We will now show that the gravitational potential V satisfies the well-known **Poisson equation**

$$\Delta V = -4\pi G F$$

if F is Hölder continuous. At this point, we are, for the first time, confronted with a gap between theory and practice. The real mass density distribution is not even continuous. For example, the core–mantle boundary is an interface between fluid and solid material which causes a notable jump in F. One could certainly argue that we could approximate F by a continuously differentiable function with a high slope instead of a jump. However, in such cases, we should not forget that such a modelling error can be a source of an inaccuracy.

To emphasize the need of a stronger condition on F than our previous conditions, we consider the function

$$F(x) := \begin{cases} 1, & x \in \mathbb{Q}^3 \cap D, \\ 0, & x \in D \setminus \mathbb{Q}^3, \end{cases}$$

which equals 1 for all $x \in D$ with rational coordinates and vanishes elsewhere. Obviously, F is measurable and bounded. Since $\mathbb{Q}^3 \cap D$ is a set of measure zero, we get

$$V(y) = G \int_D \frac{F(x)}{|x - y|} \, dx = 0$$

for all $y \in \mathbb{R}^3$ such that $\Delta V = 0$ in the whole space $\mathbb{R}^3$. Hence, $\Delta V \neq -4\pi G F$ on $\mathbb{Q}^3 \cap D$.

Eventually, we prove the Poisson equation for the gravitational potential (note that the Poisson equation is an inhomogeneous Laplace equation).

Theorem 3.1.11 *Let $D = \overline{\Sigma_{\text{int}}}$ be the interior of a regular surface $\Sigma \subset \mathbb{R}^3$ and $F\colon D \to \mathbb{R}$ be α-Hölder continuous for $\alpha > 0$. Then the gravitational potential V with*

$$V(y) := G \int_D \frac{F(x)}{|x-y|}\,\mathrm{d}x, \quad y \in \mathbb{R}^3,$$

is twice continuously differentiable in the interior of D and satisfies there the ***Poisson equation***

$$\Delta_y V(y) = -4\pi G\, F(y), \quad y \in \Sigma_{\text{int}}.$$

Proof Unfortunately, this proof is rather lengthy. One essential part consists of proving the differentiability of the second order. Since D is compact, F fulfills the conditions of Theorem 3.1.3 and Corollary 3.1.4 such that we already know that ∇V exists everywhere in $\mathbb{R}^3$ and

$$\frac{\partial}{\partial y_k} V(y) = G \int_D F(x) \frac{\partial}{\partial y_k} \frac{1}{|x-y|}\,\mathrm{d}x, \quad y \in \mathbb{R}^3, \tag{3.10}$$

for all $k \in \{1,2,3\}$. We now have to investigate what we can find out about

$$G \frac{\partial}{\partial y_j} \int_D F(x) \frac{\partial}{\partial y_k} \frac{1}{|x-y|}\,\mathrm{d}x, \quad j = 1,2,3, \tag{3.11}$$

for $y \in D$. Remember that the main difficulty connected to $y \in D$ is the singularity of the integrand. In the following, we set, for each $k \in \{1,2,3\}$,

$$W_k(x,y) := G\, F(x) \frac{\partial}{\partial y_k} \frac{1}{|x-y|}, \quad x \in D,\ y \in \mathbb{R}^3 \setminus \{x\}. \tag{3.12}$$

(1) Exclusion of the singularity:

Let $y \in \operatorname{int} D$ be an arbitrary point in the interior of D and $\varepsilon > 0$ be a sufficiently small radius such that the (open) ball $B_\varepsilon(y)$ is completely contained in $\operatorname{int} D$. From (3.10), we get

$$\frac{\partial}{\partial y_k} V(y) = \lim_{\varepsilon \to 0+} \int_{D \setminus B_\varepsilon(y)} W_k(x,y)\,\mathrm{d}x.$$

Let now $S \subset \operatorname{int} D$ be a closed subset with $y \in S$. Without loss of generality, we assume that $\varepsilon < \inf\{|z - \tilde{z}| : z \in \partial D, \tilde{z} \in \partial S\} =: \delta$. Note that δ is always positive due to the choice of S.

We will now have a look at (3.11) but for $D \setminus B_\varepsilon(y)$ as the integration domain, that is, we exclude the singularity. We restrict our attention here to the case $j = 1$. The considerations for the other coordinates are completely analogous. Setting $y^{(h)} := (y_1 + h, y_2, y_3)^{\mathrm{T}}$, we get

$$\begin{aligned} &\frac{\partial}{\partial y_1} \int_{D \setminus B_\varepsilon(y)} W_k(x,y)\,\mathrm{d}x \\ &= \lim_{h \to 0} \frac{1}{h} \left[\int_{D \setminus B_\varepsilon(y^{(h)})} W_k\left(x, y^{(h)}\right) \mathrm{d}x - \int_{D \setminus B_\varepsilon(y)} W_k(x,y)\,\mathrm{d}x \right] \end{aligned}$$

$$= \lim_{h\to 0} \frac{1}{h}\left[\int_{D\setminus B_\varepsilon(y^{(h)})} W_k\left(x, y^{(h)}\right) - W_k(x,y) + W_k(x,y)\,\mathrm{d}x \right.$$
$$\left. - \int_{D\setminus B_\varepsilon(y)} W_k(x,y)\,\mathrm{d}x\right]. \tag{3.13}$$

(2) The first half of the term:

We start by considering the first half of (3.13) as follows:

$$\int_{D\setminus B_\varepsilon\left(y^{(h)}\right)} \frac{1}{h}\left[W_k\left(x, y^{(h)}\right) - W_k(x,y)\right]\mathrm{d}x$$
$$= \int_D \chi_{D\setminus B_\varepsilon\left(y^{(h)}\right)}(x)\,\frac{1}{h}\left[W_k\left(x, y^{(h)}\right) - W_k(x,y)\right]\mathrm{d}x,$$

where χ is the characteristic function (see Definition 2.1.9). Obviously,

$$\chi_{D\setminus B_\varepsilon\left(y^{(h)}\right)}(x)\,\frac{1}{h}\left[W_k\left(x, y^{(h)}\right) - W_k(x,y)\right]$$
$$\xrightarrow[h\to 0]{} \chi_{D\setminus B_\varepsilon(y)}(x)\,\frac{\partial}{\partial y_1} W_k(x,y)$$

for almost every $x \in D$. From the mean value theorem of differentiation, we know that, for every h with $|h| \le h_0$ (where $h_0 < \varepsilon/2$ is a fixed and sufficiently small positive number), there exists a point $\eta_h \in B_{h_0}(y)$, more precisely $\eta_h = (y_1 + \vartheta h, y_2, y_3)$ with $\vartheta \in [0,1]$, such that

$$\frac{1}{h}\left[W_k\left(x, y^{(h)}\right) - W_k(x,y)\right] = \frac{\partial}{\partial y_1} W_k(x, \eta_h).$$

Note that η_h also depends on x. Since $D\setminus B_\varepsilon(y^{(h)}) \subset D\setminus B_{\varepsilon - h_0}(y)$ for all $|h| \in [0, h_0]$ and since

$$[D \setminus B_{\varepsilon-h_0}(y)] \times \overline{B_{\varepsilon/2}(y)} \ni (x,\eta) \mapsto \frac{\partial}{\partial \eta_1} W_k(x,\eta)$$

is continuous on a compact domain, it is bounded such that there is a constant C_1 where $|h^{-1}[W_k(x,y^{(h)}) - W_k(x,y)]| \le C_1$ and $\int_D C_1\,\mathrm{d}x$ is finite. As a consequence, we may apply Theorem 2.4.5 and get

$$\lim_{h\to 0} \frac{1}{h}\int_{D\setminus B_\varepsilon\left(y^{(h)}\right)} W_k\left(x, y^{(h)}\right) - W_k(x,y)\,\mathrm{d}x = \int_{D\setminus B_\varepsilon(y)} \frac{\partial}{\partial y_1} W_k(x,y)\,\mathrm{d}x. \tag{3.14}$$

(3) The second half of the term:

The remaining term in (3.13) is the limit of

$$\frac{1}{h}\left[\int_{D\setminus B_\varepsilon\left(y^{(h)}\right)} W_k(x,y)\,\mathrm{d}x - \int_{D\setminus B_\varepsilon(y)} W_k(x,y)\,\mathrm{d}x\right]$$
$$= \frac{1}{h}\left[\int_{D_1} W_k(x,y)\,\mathrm{d}x - \int_{D_2} W_k(x,y)\,\mathrm{d}x\right], \tag{3.15}$$

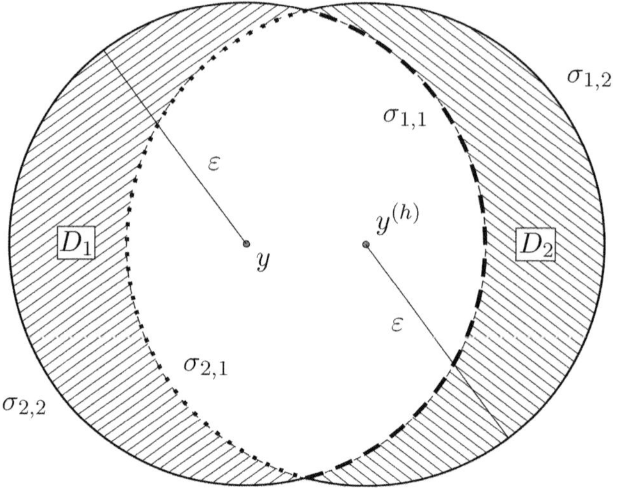

Figure 3.2 Illustration of the domains of integration in (3.15).

where the domains D_1 and D_2 are defined by

$$D_1 := \left(D \setminus B_\varepsilon\left(y^{(h)}\right)\right) \setminus (D \setminus B_\varepsilon(y)) = B_\varepsilon(y) \setminus B_\varepsilon\left(y^{(h)}\right),$$
$$D_2 := (D \setminus B_\varepsilon(y)) \setminus \left(D \setminus B_\varepsilon\left(y^{(h)}\right)\right) = B_\varepsilon\left(y^{(h)}\right) \setminus B_\varepsilon(y);$$

see also Figure 3.2. Here, we consider that the parts of the integration on $(D \setminus B_\varepsilon(y^{(h)})) \cap (D \setminus B_\varepsilon(y))$ cancel out themselves.

Note that both domains D_1 and D_2 have similar geometries. We need to parameterize them and start with D_2. Its boundaries consist of parts of the spheres $S_\varepsilon(y)$ and $S_\varepsilon\left(y^{(h)}\right)$. Let us denote these parts by $\sigma_{1,1}$ and $\sigma_{1,2}$, respectively; see again Figure 3.2. We can represent every $x \in D_2$ uniquely by $x = y + t(z - y)$ with $z \in \sigma_{1,1}$ and $t \in \mathbb{R}$ (with a still-to-investigate parameter range), that is, we use the rays from y through arbitrary $z \in \sigma_{1,1}$. While the lower bound for t (corresponding to $x \in \sigma_{1,1}$) is obviously 1, the upper bound $\bar{t}$ is not so easy to obtain. Since $\sigma_{1,2} \subset S_\varepsilon(y^{(h)})$, we get

$$(y_1 + h - [y_1 + \bar{t}(z_1 - y_1)])^2 + \sum_{j=2}^{3}(y_j - [y_j + \bar{t}(z_j - y_j)])^2 = \varepsilon^2,$$

that is, $[h - \bar{t}(z_1 - y_1)]^2 + \sum_{j=2}^3[\bar{t}(z_j - y_j)]^2 = \varepsilon^2$. This can be simplified to $h^2 - 2h\bar{t}(z_1 - y_1) + \bar{t}^2\sum_{j=1}^3(z_j - y_j)^2 = \varepsilon^2$, where $\sum_{j=1}^3(z_j - y_j)^2 = \varepsilon^2$, since $z \in S_\varepsilon(y)$. With $z = y + \varepsilon\xi$, where $\xi \in \Omega$ (remember that $\Omega = S_1(0)$), we arrive at the quadratic equation $(h^2 - \varepsilon^2)\varepsilon^{-2} - 2h\varepsilon^{-1}\xi_1\bar{t} + \bar{t}^2 = 0$ with the two solutions

$$\bar{t}_{1,2} = \frac{h}{\varepsilon}\,\xi_1 \pm \sqrt{\frac{h^2}{\varepsilon^2}\,\xi_1^2 - \frac{h^2 - \varepsilon^2}{\varepsilon^2}} = \frac{h}{\varepsilon}\,\xi_1 \pm \sqrt{1 + \frac{h^2}{\varepsilon^2}\left(\xi_1^2 - 1\right)}.$$

Since $-(h^2-\varepsilon^2)\varepsilon^{-2} = 1 - h^2\varepsilon^{-2}$ and h can be considered as arbitrarily small (at least, $|h| < \varepsilon$), the '$-$' version of $\bar{t}_{1,2}$ would be negative. Furthermore, with the Taylor expansion $\sqrt{1+s} = 1 + \mathcal{O}(s)$ with respect to $s \to 0$ (see Definition 2.1.7 for the Landau symbols), we get

$$\bar{t} - \frac{h}{\varepsilon}\xi_1 + \sqrt{1 + \frac{h^2}{\varepsilon^2}\left(\xi_1^2 - 1\right)} - \frac{h}{\varepsilon}\xi_1 + 1 + \mathcal{O}(h^2)$$

for $h \to 0$. Note that $\bar{t}$ depends on $\xi \in \Omega$. With this parameterization, we can represent the integral over D_2 as follows:

$$\frac{1}{h}\int_{D_2} W_k(x,y)\,\mathrm{d}x = \frac{1}{h}\int_{\Omega_{1,1}}\int_{\varepsilon}^{\varepsilon\bar{t}(\xi)} W_k(y + r\xi, y) r^2\,\mathrm{d}r\,\mathrm{d}\omega(\xi),$$

where $\Omega_{1,1} := \{\xi \in \Omega\colon y + \varepsilon\,\xi \in \sigma_{1,1}\}$.

The mean value theorem of integration yields the existence of a parameter $\tau(\xi)$ between 1 and $\bar{t}(\xi)$ such that

$$\begin{aligned}\frac{1}{h}\int_{D_2} W_k(x,y)\,\mathrm{d}x &= \frac{1}{h}\int_{\Omega_{1,1}} W_k(y + \varepsilon\tau(\xi)\xi, y)(\tau(\xi))^2\varepsilon^3(\bar{t}(\xi) - 1)\,\mathrm{d}\omega(\xi)\\ &= \int_{\Omega_{1,1}} W_k(y + \varepsilon\tau(\xi)\xi, y)(\tau(\xi))^2\left[\varepsilon^2\xi_1 + \mathcal{O}(h)\right]\mathrm{d}\omega(\xi).\end{aligned}$$

Analogous considerations are possible for D_1 such that we obtain a corresponding integral over $\Omega_{2,2} := \{\xi \in \Omega\colon y + \varepsilon\xi \in \sigma_{2,2}\}$. In detail, we parameterize D_1 by $x = y + t(z-y)$ with $z \in \sigma_{2,2} := \partial D_1 \cap S_\varepsilon(y)$. In this case, the upper bound for the parameter is easily given by $t = 1$. For the lower bound, $x \in \sigma_{2,1} := \partial D_1 \cap S_\varepsilon\left(y^{(h)}\right)$ requires that

$$(y_1 + h - [y_1 + \bar{t}(z_1 - y_1)])^2 + \sum_{j=2}^{3}(y_j - [y_j + \bar{t}(z_j - y_j)])^2 = \varepsilon^2,$$

which leads to $h^2 - 2h\bar{t}(z_1 - y_1) + \bar{t}^2\varepsilon^2 = \varepsilon^2$. With $z = y + \varepsilon\xi$, $\xi \in \Omega$, we obtain $\bar{t} = h\varepsilon^{-1}\xi_1 + 1 + \mathcal{O}(h^2)$ for $h \to 0$ as the lower bound.

It should be noted that the lower and upper bound may be interchanged here. We used this nomenclature in accordance with Figure 3.2, where D_2 corresponds to $\xi_1 \geq 0$ and D_1 has directions $\xi \in \Omega$ with $\xi_1 \leq 0$. It is, however, important that each D_l corresponds to a fixed sign of ξ_1 since y and $y^{(h)}$ have the same x_2- and x_3-coordinates.

The preceding considerations yield

$$\frac{1}{h}\int_{D_1} W_k(x,y)\,\mathrm{d}x = \frac{1}{h}\int_{\Omega_{2,2}}\int_{\varepsilon\bar{t}(\xi)}^{\varepsilon} W_k(y + r\xi, y) r^2\,\mathrm{d}r\,\mathrm{d}\omega(\xi).$$

The mean value theorem of integration now provides us with $\tau(\xi)$ between $\bar{t}(\xi)$ and 1 such that

$$\frac{1}{h}\int_{D_1} W_k(x,y)\,\mathrm{d}x = \frac{1}{h}\int_{\Omega_{2,2}} W_k(y+\varepsilon\tau(\xi)\xi,y)(\tau(\xi))^2\varepsilon^3(1-\bar{t}(\xi))\,\mathrm{d}\omega(\xi)$$
$$= -\int_{\Omega_{2,2}} W_k(y+\varepsilon\tau(\xi)\xi,y)(\tau(\xi))^2\left[\varepsilon^2\xi_1+\mathcal{O}(h)\right]\mathrm{d}\omega(\xi).$$

Note the '$-$' sign in front of the integral! In the limit $h \to 0$, the union $\Omega_{1,1} \cup \Omega_{2,2}$ (more precisely, $\Omega_{1,1}(h) \cup \Omega_{2,2}(h)$) becomes the unit sphere Ω. Moreover, $\bar{t}(\xi) \to 1$ and, thus, $\tau(\xi) \to 1$ for both integrals (over D_1 and over D_2). In combination with (3.15), the continuity of W_k, and Theorem 2.4.5, we obtain

$$\lim_{h\to 0}\frac{1}{h}\left[\int_{D\setminus B_\varepsilon\left(y^{(h)}\right)} W_k(x,y)\,\mathrm{d}x - \int_{D\setminus B_\varepsilon(y)} W_k(x,y)\,\mathrm{d}x\right]$$
$$= -\int_\Omega W_k(y+\varepsilon\xi,y)\,\varepsilon^2\xi_1\,\mathrm{d}\omega(\xi) = -\int_{S_\varepsilon(y)} W_k(x,y)\,\frac{x_1-y_1}{\varepsilon}\,\mathrm{d}\omega(x).$$

Altogether, we get for the differentiated integral in (3.13) the result (remember (3.14))

$$\frac{\partial}{\partial y_1}\int_{D\setminus B_\varepsilon(y)} W_k(x,y)\,\mathrm{d}x = \int_{D\setminus B_\varepsilon(y)} \frac{\partial}{\partial y_1} W_k(x,y)\,\mathrm{d}x - \int_{S_\varepsilon(y)} W_k(x,y)\,\frac{x_1-y_1}{\varepsilon}\,\mathrm{d}\omega(x). \tag{3.16}$$

(4) Simplifying the intermediate result:
With the definition of W_k in (3.12), our result in (3.16) becomes

$$G\frac{\partial}{\partial y_1}\int_{D\setminus B_\varepsilon(y)} F(x)\frac{\partial}{\partial y_k}\frac{1}{|x-y|}\,\mathrm{d}x$$
$$= G\int_{D\setminus B_\varepsilon(y)} F(x)\frac{\partial^2}{\partial y_1\partial y_k}\frac{1}{|x-y|}\,\mathrm{d}x$$
$$- G\int_{S_\varepsilon(y)} F(x)\frac{x_1-y_1}{\varepsilon}\frac{\partial}{\partial y_k}\frac{1}{|x-y|}\,\mathrm{d}\omega(x). \tag{3.17}$$

With $x = y+\varepsilon\xi$, $\xi\in\Omega$ and $\frac{\partial}{\partial y_k}|x-y|^{-1} = (x_k-y_k)|x-y|^{-3}$, we get

$$G\int_{S_\varepsilon(y)} F(x)\frac{x_1-y_1}{\varepsilon}\frac{\partial}{\partial y_k}\frac{1}{|x-y|}\,\mathrm{d}\omega(x) = G\int_\Omega F(y+\varepsilon\xi)\,\xi_1\xi_k\,\frac{\varepsilon}{\varepsilon^3}\,\varepsilon^2\,\mathrm{d}\omega(\xi). \tag{3.18}$$

(5) Starting with the limit $\varepsilon \to 0+$:
Since F is continuous and Ω is compact, the last integrand is bounded on all $\{y+\varepsilon\xi \mid \xi\in\Omega,\ \varepsilon\in[0,\varepsilon_0]\}$ for some $\varepsilon_0 > 0$ (note that $y \in \operatorname{int} D$) and we may apply Theorem 2.4.5 to conclude (again with the continuity of F) that

$$G\lim_{\varepsilon\to 0+}\int_\Omega F(y+\varepsilon\xi)\,\xi_1\xi_k\,\mathrm{d}\omega(\xi) = G\int_\Omega F(y)\,\xi_1\xi_k\,\mathrm{d}\omega(\xi). \tag{3.19}$$

Note that every $\xi \in \Omega$ is the outer unit normal vector at $\xi \in \Omega$, that is, $\nu(\xi) = \xi$ for all $\xi \in \Omega$, such that Gauß's law (see Theorem 2.4.2) yields

$$\int_{\Omega} \xi_1 \xi_k \,\mathrm{d}\omega(\xi) = \int_{\Omega} \xi_1 \nu(\xi) \cdot \varepsilon^k \,\mathrm{d}\omega(\xi) = \int_{B_1(0)} \mathrm{div}_x \left(x_1 \varepsilon^k \right) \mathrm{d}x = \frac{4\pi}{3} \delta_{1k}. \quad (3.20)$$

Hence, (3.18)–(3.20) yield for the last integral in (3.17) the limit relation

$$\lim_{\varepsilon \to 0+} \left(-G \int_{S_\varepsilon(y)} F(x) \frac{x_1 - y_1}{\varepsilon} \frac{\partial}{\partial y_k} \frac{1}{|x-y|} \,\mathrm{d}\omega(x) \right) = -G\, F(y) \frac{4\pi}{3} \delta_{1k}. \quad (3.21)$$

Note that, due to the construction of S earlier in this proof, we can argue in (3.19) that

$$\begin{aligned}
&\sup_{y \in S} \left| \int_{\Omega} F(y + \varepsilon\xi) \xi_1 \xi_k \,\mathrm{d}\omega(\xi) - \int_{\Omega} F(y) \xi_1 \xi_k \,\mathrm{d}\omega(\xi) \right| \\
&\quad = \sup_{y \in S} \left| \int_{\Omega} (F(y + \varepsilon\xi) - F(y))\, \xi_1 \xi_k \,\mathrm{d}\omega(\xi) \right| \\
&\quad \le \int_{\Omega} \sup_{y \in S} |F(y + \varepsilon\xi) - F(y)| \,\mathrm{d}\omega(\xi) \to \int_{\Omega} 0 \,\mathrm{d}\omega(\xi) = 0
\end{aligned}$$

as $\varepsilon \to 0+$ due to the uniform continuity of F on the compact set S. Consequently, the convergence in (3.21) is uniform with respect to $y \in S$.

(6) One more limit:

It remains to investigate the behaviour of the first integral on the right-hand side of (3.17) for $\varepsilon \to 0+$. Remember how we constructed S and D and defined δ rather at the beginning of this proof. We first subdivide the domain and also use the basic trick of adding and subtracting an additional term. Note that we now have $B_\delta(y)$ and $B_\varepsilon(y)$ and the spherical shell $\Omega_{\varepsilon,\delta}(y) := \{x \in \mathbb{R}^3 \,|\, \varepsilon < |x-y| < \delta\}$. We get

$$\begin{aligned}
G \int_{D \setminus B_\varepsilon(y)} F(x) \frac{\partial^2}{\partial y_1 \partial y_k} \frac{1}{|x-y|} \,\mathrm{d}x &= G \int_{D \setminus B_\delta(y)} F(x) \frac{\partial^2}{\partial y_1 \partial y_k} \frac{1}{|x-y|} \,\mathrm{d}x \\
&\quad + G \int_{\Omega_{\varepsilon,\delta}(y)} \left[F(x) - F(y) \right] \frac{\partial^2}{\partial y_1 \partial y_k} \frac{1}{|x-y|} \,\mathrm{d}x \\
&\quad + G\, F(y) \int_{\Omega_{\varepsilon,\delta}(y)} \frac{\partial^2}{\partial y_1 \partial y_k} \frac{1}{|x-y|} \,\mathrm{d}x. \quad (3.22)
\end{aligned}$$

Since we have

$$\frac{\partial^2}{\partial y_1 \partial y_k} \frac{1}{|x-y|} = \frac{\partial}{\partial y_1} \frac{x_k - y_k}{|x-y|^3} = 3 \frac{(x_1 - y_1)(x_k - y_k)}{|x-y|^5} - \frac{\delta_{1k}}{|x-y|^3},$$

we obtain for the latter integral with $x = y + r\xi$, $r \in]\varepsilon, \delta[$, $\xi \in \Omega$, the identity (note that we use (3.20) for the surface integral)

$$\int_{\Omega_{\varepsilon,\delta}(y)} \frac{\partial^2}{\partial y_1 \partial y_k} \frac{1}{|x-y|} \, \mathrm{d}x = \int_\Omega \int_\varepsilon^\delta \left(3\xi_1\xi_k r^{-3} - \delta_{1k} r^{-3}\right) r^2 \, \mathrm{d}r \, \mathrm{d}\omega(\xi)$$

$$= \int_\Omega (3\xi_1\xi_k - \delta_{1k}) \, \mathrm{d}\omega(\xi) \, \log r|_\varepsilon^\delta = (4\pi\,\delta_{1k} - 4\pi\,\delta_{1,k}) \log \frac{\delta}{\varepsilon} = 0. \quad (3.23)$$

Hence, the last integral in (3.22) vanishes. For the integrand of the penultimate integral in (3.22), we use the Hölder continuity (!) of F to derive the estimate

$$\left|[F(x) - F(y)] \frac{\partial^2}{\partial y_1 \partial y_k} \frac{1}{|x-y|}\right|$$
$$= |F(x) - F(y)| \cdot \left| 3\frac{(x_1 - y_1)(x_k - y_k)}{|x-y|^5} - \frac{\delta_{1,k}}{|x-y|^3} \right|$$
$$\leq C_2|x-y|^\alpha \left(3|x-y|^{-3} + \delta_{1,k}|x-y|^{-3}\right) \leq 4C_2|x-y|^{\alpha-3}.$$

Since

$$\int_{B_\varepsilon(y)} |x-y|^{\alpha-3} \, \mathrm{d}x = \int_\Omega \int_0^\varepsilon r^{\alpha-3+2} \, \mathrm{d}r \, \mathrm{d}\omega(\xi) = \frac{4\pi}{\alpha} \varepsilon^\alpha < +\infty,$$

we get the uniform convergence

$$\lim_{\varepsilon \to 0+} \sup_{y \in S} \left| G \int_{\Omega_{\varepsilon,\delta}(y)} [F(x) - F(y)] \frac{\partial^2}{\partial y_1 \partial y_k} \frac{1}{|x-y|} \, \mathrm{d}x \right.$$
$$\left. - G \int_{B_\delta(y)} [F(x) - F(y)] \frac{\partial^2}{\partial y_1 \partial y_k} \frac{1}{|x-y|} \, \mathrm{d}x \right|$$
$$= \lim_{\varepsilon \to 0+} \sup_{y \in S} \left| G \int_{B_\varepsilon(y)} \left([F(x) - F(y)] \frac{\partial^2}{\partial y_1 \partial y_k} \frac{1}{|x-y|} \right) \mathrm{d}x \right| = 0.$$

In combination with (3.22) and (3.23), we obtain the uniform convergence (with respect to $y \in S$) in

$$G \lim_{\varepsilon \to 0+} \int_{D \setminus B_\varepsilon(y)} F(x) \frac{\partial^2}{\partial y_1 \partial y_k} \frac{1}{|x-y|} \, \mathrm{d}x$$
$$= G \int_{D \setminus B_\delta(y)} F(x) \frac{\partial^2}{\partial y_1 \partial y_k} \frac{1}{|x-y|} \, \mathrm{d}x$$
$$+ G \int_{B_\delta(y)} [F(x) - F(y)] \frac{\partial^2}{\partial y_1 \partial y_k} \frac{1}{|x-y|} \, \mathrm{d}x. \quad (3.24)$$

(7) We are reaching the Q.E.D.:

Remember that our task was to determine the derivative in (3.11). We excluded a ball around the singularity at $y \in S$ and achieved the intermediate result in (3.17). For the right-hand side, we obtained the uniform limits in (3.21) and (3.24). Since both limits on the right-hand side are uniform, the left-hand side of (3.17) must also converge uniformly for $\varepsilon \to 0+$. This implies that $\int_D F(x) \frac{\partial}{\partial y_k} |x-y|^{-1} \, \mathrm{d}x$ is also differentiable with respect to y_1 in all $y \in \operatorname{int} D$. For analogy reasons, this holds for all components of y. This means that (note also (3.10))

$$\begin{aligned}\frac{\partial^2}{\partial y_j \partial y_k} V(y) &= G \frac{\partial}{\partial y_j} \int_D F(x) \frac{\partial}{\partial y_k} \frac{1}{|x-y|} \,\mathrm{d}x \\ &= G \lim_{\varepsilon \to 0+} \int_{D \setminus B_\varepsilon(y)} F(x) \frac{\partial^2}{\partial y_j \partial y_k} \frac{1}{|x-y|} \,\mathrm{d}x - G\, F(y) \frac{4\pi}{3} \delta_{jk}\end{aligned}$$

for all $y \in \operatorname{int} D$ and all $j,k \in \{1,2,3\}$. This means that V is twice continuously differentiable and

$$\begin{aligned}\Delta_y V(y) &= \sum_{j=1}^{3} \frac{\partial^2}{\partial y_j^2} V(y) \\ &= G \lim_{\varepsilon \to 0+} \int_{D \setminus B_\varepsilon(y)} F(x) \underbrace{\Delta_y \frac{1}{|x-y|}}_{=0} \,\mathrm{d}x - 3\,G\, F(y) \frac{4\pi}{3} \\ &= -4\pi\, G\, F(y)\end{aligned}$$

for all $y \in \operatorname{int} D = \Sigma_{\mathrm{int}}$. Q.E.D. □

From the theory of elliptic partial differential equations, it is known that the regularity of the mass density function F, that is, the regularity of the right-hand side of the Poisson equation, propagates to the solution. The proof of the next theorem, which yields such a property, is omitted; see, for example, Gilbarg and Trudinger (1977, Theorem 6.17).

Theorem 3.1.12 *Let $D = \Sigma_{\mathrm{int}}$ be the interior of a regular surface $\Sigma \subset \mathbb{R}^3$ and $F \in \mathrm{C}^{(k,\alpha)}(D)$ with $k \in \mathbb{N}_0$ and $\alpha > 0$. Then the corresponding gravitational potential*

$$V(y) := G \int_D \frac{F(x)}{|x-y|} \,\mathrm{d}x, \quad y \in \mathbb{R}^3,$$

satisfies $V \in \mathrm{C}^{(k+2,\alpha)}(D)$. In particular, if $F \in \mathrm{C}^{(\infty)}(D)$, then $V \in \mathrm{C}^{(\infty)}(D)$.

We summarize what we obtained so far. If $D = \overline{\Sigma_{\mathrm{int}}}$ is the (closed) interior of a regular surface Σ and $F\colon D \to \mathbb{R}$ is measurable and bounded, then the gravitational potential

$$V(y) = G \int_D \frac{F(x)}{|x-y|} \,\mathrm{d}x, \quad y \in \mathbb{R}^3,$$

has the following properties:

- V exists on the whole space $\mathbb{R}^3$ and is bounded, continuous, and partially differentiable everywhere (see Theorems 3.1.2 and 3.1.3). We may also exchange differentiation and integration here (see Corollary 3.1.4).
- V is infinitely differentiable in the outer space, that is, in $\mathbb{R}^3 \setminus D = \Sigma_{\mathrm{ext}}$, and satisfies the (homogeneous) Laplace equation $\Delta_y V(y) = 0$ there (see Theorem 3.1.5). Moreover, V is regular at infinity (see Corollary 3.1.7).

- If F is α-Hölder continuous for some $\alpha > 0$, then V is also twice continuously differentiable in the interior space, that is, in int $D = \Sigma_{\text{int}}$, and satisfies the Poisson equation $\Delta_y V(y) = -4\pi\, G\, F(y)$ there (see Theorem 3.1.11).
- Assume that F is α-Hölder continuous. Since we can expect ΔV to be discontinuous on the surface $\Sigma = \partial D$, V *cannot* (in general) be twice continuously differentiable on the whole space $\mathbb{R}^3$.

In the following, motivations from practical applications cause us to investigate, in particular, the following issues:

- We can determine V (or some related quantities) outside the Earth. It is, therefore, important to investigate the properties of the harmonic functions, that is, the solutions of the partial differential equation $\Delta V = 0$. We need basis systems, methods for numerically approximating the solution, existence and uniqueness results, etc.
- The potential V opens a door for us to determine the mass density F, which is important for identifying mass anomalies or mass transports. From the mathematical point of view, we have to solve an inverse problem, which is represented by a Fredholm integral equation of the first kind. Questions that arise are: is the equation solvable? Is the solution unique? Does noise on the data have an essential influence on the solution? And most of all: how can we get an (approximate) solution?

We will address these and other questions in the next sections and chapters. The example of the gravitational potential already gives us a glimpse at the variety of different mathematical areas which are involved in geomathematics.

3.2 Some Fundamental Properties of Harmonic Functions

We saw in Theorem 3.1.5 that the gravitational potential V satisfies the Laplace equation $\Delta V = 0$ outside the planet, that is, V is harmonic there. Harmonic functions play an important role in a series of applications, not only in gravitational field modelling. For this reason, this class of functions has already been intensively investigated. The corresponding subjects in mathematics are the theory of elliptic partial differential equations and, in particular, the potential theory. We will derive here some widely known and frequently used theorems on harmonic functions which are examples of starting points of these two mathematical theories. The considerations in this section are primarily based on Kellogg (1967) and Walter (1971).

We begin with the Kelvin transform.

Definition 3.2.1 Let $R \subset \mathbb{R}^3$ be a region where $0 \notin R$. Then the **Kelvin transform** maps functions $F\colon R \to \mathbb{R}$ to functions $F^*\colon R^* \to \mathbb{R}$ such that

$$R^* := \left\{ x \in \mathbb{R}^3 \setminus \{0\} \,\middle|\, \frac{x}{|x|^2} \in R \right\}; \qquad F^*(x) := \frac{1}{|x|} F\left(\frac{x}{|x|^2}\right), \quad x \in R^*.$$

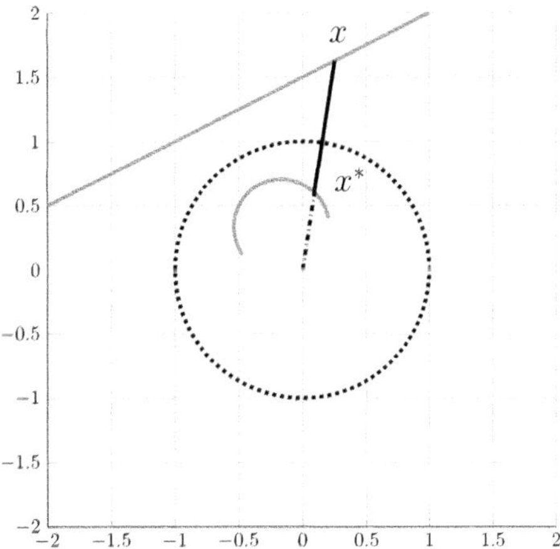

Figure 3.3 The grey sets are examples of domains of a function and its Kelvin transform, that is, they are mirrored at the unit sphere (dotted). Note that $|x^*|$ (length of the dash-dotted line) is the reciprocal of $|x|$ (lengths of dash-dotted and solid lines together).

We can imagine that R^* is obtained from R by mirroring the set at the unit sphere; see Figure 3.3. Some easy geometrical consequences are associated to the Kelvin transform:

- If $x^* = x/|x|^2$, then $|x^*| = |x|/|x|^2 = 1/|x|$.
- Hence, the sphere with centre 0 and radius $r > 0$ is mirrored to the sphere with centre 0 and radius r^{-1}.
- The origin $x = 0$ can be regarded as mirrored to infinity (in some three-dimensional sense of infinity).
- The Kelvin transform is the inverse of itself, that is,

$$(x^*)^* = \frac{x^*}{|x^*|^2} = \frac{x}{|x|^2} \Big/ \left|\frac{x}{|x|^2}\right|^2 = \frac{x}{|x|^2} \cdot |x|^2 = x,$$

$$F^{**}(x) = \frac{1}{|x|} F^*\left(\frac{x}{|x|^2}\right) = \frac{1}{|x|} \frac{1}{\left|\frac{x}{|x|^2}\right|} F\left(\frac{x}{|x|^2} \Big/ \left|\frac{x}{|x|^2}\right|^2\right) = F(x).$$

Theorem 3.2.2 *Let $R \subset \mathbb{R}^3$ be a region with $0 \notin R$ and $F \in \mathrm{C}^{(2)}(R)$ be a given function. Then F and its Kelvin transform F^* satisfy*

$$\Delta_x F^*(x) = \frac{1}{|x|^5} \left(\Delta_y F(y)\right) |_{y=x/|x|^2}$$

for all $x \in R^*$, where $F^* \in \mathrm{C}^{(2)}(R^*)$.

Proof It is obvious that $F^* \in \mathrm{C}^{(2)}(R^*)$, since F^* is a composition of a $\mathrm{C}^{(2)}$-function and $\mathrm{C}^{(\infty)}$-functions.

Let us now calculate what the application of the Laplace operator to F^* yields. With the product rule and Corollary 3.1.6, we get

$$\begin{aligned}
\Delta_x F^*(x) &= \sum_{j=1}^{3} \frac{\partial^2}{\partial x_j^2} \left[\frac{1}{|x|} F\left(\frac{x}{|x|^2} \right) \right] \\
&= \sum_{j=1}^{3} \left(\frac{\partial^2}{\partial x_j^2} \frac{1}{|x|} \right) F\left(\frac{x}{|x|^2} \right) + 2 \sum_{j=1}^{3} \left(\frac{\partial}{\partial x_j} \frac{1}{|x|} \right) \frac{\partial}{\partial x_j} F\left(\frac{x}{|x|^2} \right) \\
&\quad + \frac{1}{|x|} \sum_{j=1}^{3} \frac{\partial^2}{\partial x_j^2} F\left(\frac{x}{|x|^2} \right) \\
&= 0 + 2\left(\nabla_x \frac{1}{|x|} \right) \cdot \nabla_x F\left(\frac{x}{|x|^2} \right) + \frac{1}{|x|} \Delta_x F\left(\frac{x}{|x|^2} \right) \qquad (3.25)
\end{aligned}$$

for all $x \in R^*$. It is easy to verify that $\nabla_x |x|^{-1} = -x|x|^{-3}$ for all $x \neq 0$. Applying the chain rule, we obtain

$$\begin{aligned}
\frac{\partial}{\partial x_j} F\left(\frac{x}{|x|^2} \right) &= \left(\nabla_y F(y) \right)\Big|_{y=\frac{x}{|x|^2}} \cdot \left(-\frac{2x_j}{|x|^4} x + \frac{1}{|x|^2} \varepsilon^j \right) \qquad (3.26) \\
&= -\frac{2x_j}{|x|^4} x \cdot \left(\nabla_y F(y) \right)\Big|_{y=\frac{x}{|x|^2}} + \frac{1}{|x|^2} \left(\frac{\partial}{\partial y_j} F(y) \right)\Bigg|_{y=\frac{x}{|x|^2}},
\end{aligned}$$

$j = 1, 2, 3$, $x \in R^*$. As a consequence, we get for the second summand in (3.25) the result

$$\begin{aligned}
&2\left(\nabla_x \frac{1}{|x|} \right) \cdot \nabla_x F\left(\frac{x}{|x|^2} \right) \\
&= 4 \frac{|x|^2}{|x|^7} x \cdot \left(\nabla_y F(y) \right)\Big|_{y=\frac{x}{|x|^2}} - \frac{2}{|x|^5} x \cdot \left(\nabla_y F(y) \right)\Big|_{y=\frac{x}{|x|^2}} \\
&= \frac{2}{|x|^5} x \cdot \left(\nabla_y F(y) \right)\Big|_{y=\frac{x}{|x|^2}} \qquad (3.27)
\end{aligned}$$

for all $x \in R^*$. It remains to discuss the last term in (3.25). By differentiating (3.26) once again, we obtain, for $j = 1, 2, 3$,

$$\begin{aligned}
\frac{\partial^2}{\partial x_j^2} F\left(\frac{x}{|x|^2} \right) &= -\frac{2|x|^4 - 2x_j \cdot 2|x|^2 \cdot 2x_j}{|x|^8} x \cdot \left(\nabla_y F(y) \right)\Big|_{y=\frac{x}{|x|^2}} \\
&\quad - \frac{2x_j}{|x|^4} \left(\frac{\partial}{\partial y_j} F(y) \right)\Bigg|_{y=\frac{x}{|x|^2}} - \frac{2x_j}{|x|^4} x \cdot \left[\frac{\partial}{\partial x_j} \left(\nabla_y F(y) \right)\Big|_{y=\frac{x}{|x|^2}} \right] \\
&\quad - \frac{2x_j}{|x|^4} \left(\frac{\partial}{\partial y_j} F(y) \right)\Bigg|_{y=\frac{x}{|x|^2}} + \frac{1}{|x|^2} \frac{\partial}{\partial x_j} \left(\frac{\partial}{\partial y_j} F(y) \right)\Bigg|_{y=\frac{x}{|x|^2}}. \qquad (3.28)
\end{aligned}$$

By setting $G_k(y) := \frac{\partial}{\partial y_k} F(y)$, $k = 1, 2, 3$, $y \in R$, and $G = (G_1, G_2, G_3)^{\mathrm{T}}$, we see that

$$\frac{\partial}{\partial x_j} G_k\left(\frac{x}{|x|^2} \right) = \left(\nabla_y G_k(y) \right)\Big|_{y=\frac{x}{|x|^2}} \cdot \left(-\frac{2x_j}{|x|^4} x + \frac{1}{|x|^2} \varepsilon^j \right),$$

$j, k = 1, 2, 3$, such that

$$\frac{\partial}{\partial x_j} \left(\nabla_y F(y)\right)\Big|_{y=\frac{x}{|x|^2}} = \left[(\nabla_y \otimes \nabla_y) F(y)\right]\Big|_{y=\frac{x}{|x|^2}} \cdot \left(-\frac{2x_j}{|x|^4}\, x + \frac{1}{|x|^2}\, \varepsilon^j\right).$$

Hence, the previous line and (3.28) yield

$$\begin{aligned}\frac{\partial^2}{\partial x_j^2} F\left(\frac{x}{|x|^2}\right) &= \frac{8x_j^2 - 2|x|^2}{|x|^6}\, x \cdot G\left(\frac{x}{|x|^2}\right) - 4\frac{x_j}{|x|^4}\, G_j\left(\frac{x}{|x|^2}\right) \\ &\quad - \frac{2x_j}{|x|^4} x^{\mathrm{T}} \left[(\nabla_y \otimes \nabla_y) F(y)\right]\Big|_{y=\frac{x}{|x|^2}} \cdot \left(-2\frac{x_j}{|x|^4}\, x + \frac{1}{|x|^2}\, \varepsilon^j\right) \\ &\quad + \frac{1}{|x|^2} (\varepsilon^j)^{\mathrm{T}} \left[(\nabla_y \otimes \nabla_y) F(y)\right]\Big|_{y=\frac{x}{|x|^2}} \cdot \left(-2\frac{x_j}{|x|^4}\, x + \frac{1}{|x|^2}\, \varepsilon^j\right)\end{aligned}$$

for $j = 1, 2, 3$. By summing over all j, we arrive at $\left(\text{with } x = \sum_{j=1}^3 x_j \varepsilon^j\right)$

$$\begin{aligned}\Delta_x F\left(\frac{x}{|x|^2}\right) &= \left(\frac{8|x|^2 - 6|x|^2}{|x|^6} - \frac{4}{|x|^4}\right) x \cdot G\left(\frac{x}{|x|^2}\right) \\ &\quad + \left(\frac{4}{|x|^6} - \frac{2}{|x|^6} - \frac{2}{|x|^6}\right) x^{\mathrm{T}} \left[(\nabla_y \otimes \nabla_y) F(y)\right]\Big|_{y=\frac{x}{|x|^2}} \cdot x \\ &\quad + \frac{1}{|x|^4} (\Delta_y F(y))\Big|_{y=\frac{x}{|x|^2}}.\end{aligned} \tag{3.29}$$

We get the final result, eventually, by inserting (3.27) and (3.29) in (3.25):

$$\begin{aligned}\Delta_x F^*(x) &= \frac{2}{|x|^5}\, x \cdot G\left(\frac{x}{|x|^2}\right) - \frac{2}{|x|^5}\, x \cdot G\left(\frac{x}{|x|^2}\right) + \frac{1}{|x|^5} \left(\Delta_y F(y)\right)\Big|_{y=\frac{x}{|x|^2}} \\ &= \frac{1}{|x|^5} \left(\Delta_y F(y)\right)\Big|_{y=\frac{x}{|x|^2}}, \quad x \in R^*.\end{aligned}$$ □

There is an immediate consequence of Theorem 3.2.2.

Corollary 3.2.3 *Let $R \subset \mathbb{R}^3$ be a region where $0 \notin R$ and let $U \in \mathrm{C}^{(2)}(R)$ be a given function. Then the following holds true: the function U is harmonic on R if and only if its Kelvin transform U^* is harmonic on R^*.*

Example 3.2.4 From Corollary 3.1.6, we already know that the fundamental solution $F(x) := |x - y|^{-1}$, $x \in \mathbb{R}^3 \setminus \{y\}$, is harmonic. The Kelvin transform (formally, we restrict the domain of F here to $\mathbb{R}^3 \setminus \{0, y\}$ to be able to apply the Kelvin transform) yields

$$F^*(x) = \frac{1}{|x|} \frac{1}{\left|\frac{x}{|x|^2} - y\right|} = \frac{|x|}{|x - |x|^2 y|}, \quad x \in \mathbb{R}^3 \setminus \left\{0, \frac{y}{|y|^2}\right\},$$

as another harmonic function. In particular, for $y = 0$, we get $F(x) = \frac{1}{|x|}$ and $F^* \equiv 1$, which are obviously both harmonic. We will study further examples later in this book.

We continue with an integral theorem which is basically a consequence of Green's second identity (see Theorem 2.4.2). It is sometimes also called **Green's third identity**. It is not only valid for harmonic functions, but we will use it later for the particular case of harmonic functions.

Theorem 3.2.5 (Fundamental Theorem) *Let $D = \Sigma_{\mathrm{int}}$ be the interior of a regular surface Σ with outer unit normal ν and let $U \in \mathrm{C}^{(1)}(\overline{D}) \cap \mathrm{C}^{(2)}(D)$ be a function with $\Delta U \in \mathrm{L}^2(D)$. Then we have, for all $x_0 \in D$, the identity*

$$4\pi U(x_0) = \int_\Sigma \left(\frac{1}{|x - x_0|} \frac{\partial U}{\partial \nu}(x) - U(x) \frac{\partial}{\partial \nu(x)} \frac{1}{|x - x_0|} \right) \mathrm{d}\omega(x) - \int_D \frac{\Delta U(x)}{|x - x_0|} \,\mathrm{d}x.$$

Proof Note that Green's identities require that the considered functions are twice continuously differentiable. However, $V(x) := |x - x_0|^{-1}$ is not even defined in $x = x_0$. For this purpose, we apply Green's second identity to U and V on $D_\varepsilon := D \setminus B_\varepsilon(x_0)$ for a sufficiently small $\varepsilon > 0$. This is possible for all $x_0 \in D$, because D is open. Note that D_ε has two boundaries: Σ and $S_\varepsilon(x_0)$, where the outer unit normal to $S_\varepsilon(x_0)$ in $x \in S_\varepsilon(x_0)$ is $-\varepsilon^{-1}(x - x_0)$. We obtain

$$\begin{aligned} \int_{D_\varepsilon} U(x) \Delta_x \frac{1}{|x - x_0|} &- \frac{1}{|x - x_0|} \Delta_x U(x) \,\mathrm{d}x \\ = \int_\Sigma U(x) &\frac{\partial}{\partial \nu(x)} \frac{1}{|x - x_0|} - \frac{1}{|x - x_0|} \frac{\partial}{\partial \nu(x)} U(x) \,\mathrm{d}\omega(x) \\ + \int_{S_\varepsilon(x_0)} &U(x) \left(-\frac{x - x_0}{\varepsilon} \right) \cdot \nabla_x \frac{1}{|x - x_0|} \\ &- \frac{1}{|x - x_0|} \left(-\frac{x - x_0}{\varepsilon} \right) \cdot \nabla_x U(x) \,\mathrm{d}\omega(x). \end{aligned} \tag{3.30}$$

For the integral over $S_\varepsilon(x_0)$, we use the fact that ∇U is bounded, since $U \in \mathrm{C}^{(1)}(\overline{D})$ and $\overline{D}$ is compact. Moreover, $\nabla_x |x - x_0|^{-1} = -(x - x_0)|x - x_0|^{-3}$ for all $x \neq x_0$ such that the first part of the integral can be simplified to

$$-\frac{1}{\varepsilon} \int_{S_\varepsilon(x_0)} U(x)(x - x_0) \cdot \nabla_x \frac{1}{|x - x_0|} \,\mathrm{d}\omega(x) = \frac{1}{\varepsilon^2} \int_{S_\varepsilon(x_0)} U(x) \,\mathrm{d}\omega(x).$$

Since U is continuous, we can estimate this integral by

$$\begin{aligned} \frac{1}{\varepsilon^2} \int_{S_\varepsilon(x_0)} \min_{y \in \overline{B_\varepsilon(x_0)}} U(y) \,\mathrm{d}\omega(x) &\leq \frac{1}{\varepsilon^2} \int_{S_\varepsilon(x_0)} U(x) \,\mathrm{d}\omega(x) \\ &\leq \frac{1}{\varepsilon^2} \int_{S_\varepsilon(x_0)} \max_{y \in \overline{B_\varepsilon(x_0)}} U(y) \,\mathrm{d}\omega(x) \end{aligned}$$

such that

$$\frac{4\pi\varepsilon^2}{\varepsilon^2} \min_{y \in \overline{B_\varepsilon(x_0)}} U(y) \leq \frac{1}{\varepsilon^2} \int_{S_\varepsilon(x_0)} U(x) \,\mathrm{d}\omega(x) \leq \frac{4\pi\varepsilon^2}{\varepsilon^2} \max_{y \in \overline{B_\varepsilon(x_0)}} U(y).$$

Hence, the continuity of U yields

$$\lim_{\varepsilon\to 0+}\left(\frac{1}{\varepsilon^2}\int_{S_\varepsilon(x_0)} U(x)\,\mathrm{d}\omega(x)\right) = 4\pi\, U(x_0).$$

Second, we get

$$\begin{aligned}
&\left|\int_{S_\varepsilon(x_0)} \frac{1}{|x-x_0|}\frac{x-x_0}{\varepsilon}\cdot\nabla_x U(x)\,\mathrm{d}\omega(x)\right| \\
&\le \frac{1}{\varepsilon}\int_{S_\varepsilon(x_0)}\left|\frac{x-x_0}{\varepsilon}\right|\cdot|\nabla_x U(x)|\,\mathrm{d}\omega(x) \\
&\le \frac{1}{\varepsilon}\int_{S_\varepsilon(x_0)}\max_{y\in\overline{D}}|\nabla_y U(y)|\,\mathrm{d}\omega(x) = 4\pi\varepsilon\cdot\max_{y\in\overline{D}}|\nabla_y U(y)|.
\end{aligned}$$

Obviously, this integral vanishes in the limit $\varepsilon \to 0+$. For the second part of the volume integral in (3.30), we use the Cauchy–Schwarz inequality (here, we need that $\Delta U \in \mathrm{L}^2(D)$) and Lemma 3.1.1 to obtain

$$\begin{aligned}
&\left|\int_D \frac{1}{|x-x_0|}\Delta_x U(x)\,\mathrm{d}x - \int_{D_\varepsilon}\frac{1}{|x-x_0|}\Delta_x U(x)\,\mathrm{d}x\right| \\
&= \left|\int_{B_\varepsilon(x_0)}\frac{1}{|x-x_0|}\Delta_x U(x)\,\mathrm{d}x\right| \\
&\le \left(\int_{B_\varepsilon(x_0)}\frac{1}{|x-x_0|^2}\,\mathrm{d}x\right)^{1/2}\left(\int_{B_\varepsilon(x_0)}(\Delta_x U(x))^2\,\mathrm{d}x\right)^{1/2} \\
&\le \sqrt{4\pi\varepsilon}\,\|\Delta U\|_{\mathrm{L}^2(D)}, \qquad (3.31)
\end{aligned}$$

which obviously also tends to 0 as $\varepsilon \to 0+$.

Finally, we can apply the limit $\varepsilon \to 0+$ to (3.30) and get the following (note that $\Delta_x|x-x_0|^{-1} = 0$ for $x \ne x_0$; see Corollary 3.1.6):

$$-\int_D \frac{\Delta_x U(x)}{|x-x_0|}\,\mathrm{d}x = \int_\Sigma U(x)\frac{\partial}{\partial\nu(x)}\frac{1}{|x-x_0|} - \frac{1}{|x-x_0|}\frac{\partial}{\partial\nu(x)}U(x)\,\mathrm{d}\omega(x) + 4\pi\, U(x_0).$$

This is the desired result. □

We immediately get the following consequence for harmonic functions.

Corollary 3.2.6 *Let the conditions of Theorem 3.2.5 be satisfied, where U is additionally harmonic on D. Then*

$$4\pi\, U(x_0) = \int_\Sigma\left(\frac{1}{|x-x_0|}\frac{\partial U}{\partial\nu}(x) - U(x)\frac{\partial}{\partial\nu(x)}\frac{1}{|x-x_0|}\right)\mathrm{d}\omega(x)$$

for all $x_0 \in D$.

This already shows us that the knowledge of U and $\frac{\partial U}{\partial\nu}$ on the surface $\partial D = \Sigma$ only suffices to define a harmonic function on $D = \Sigma_{\mathrm{int}}$ uniquely. Corollary 3.2.6 additionally leads us to the following result.

Theorem 3.2.7 (Gauß's Mean Value Theorem) *Let the function U be harmonic on the ball $B_R(x_0)$ and continuous on $\overline{B_R(x_0)}$. Then*

$$U(x_0) = \frac{1}{4\pi R^2} \int_{S_R(x_0)} U(x)\,\mathrm{d}\omega(x) = \frac{3}{4\pi R^3} \int_{B_R(x_0)} U(x)\,\mathrm{d}x. \tag{3.32}$$

Proof Let $0 < \rho < R$. Then Corollary 3.2.6 yields the following identity (note that we have $U \in \mathrm{C}^{(2)}(\overline{B_\rho(x_0)})$):

$$\begin{aligned} U(x_0) &= \frac{1}{4\pi} \int_{S_\rho(x_0)} \frac{1}{|x-x_0|} \frac{\partial U}{\partial \nu}(x) - U(x) \frac{\partial}{\partial \nu(x)} \frac{1}{|x-x_0|}\,\mathrm{d}\omega(x) \\ &= \frac{1}{4\pi} \int_{S_\rho(x_0)} \frac{1}{\rho} \frac{\partial U}{\partial \nu}(x) - U(x) \underbrace{\frac{x-x_0}{\rho}}_{=\nu(x)} \cdot \underbrace{\frac{x_0-x}{|x-x_0|^3}}_{=\nabla_x |x-x_0|^{-1}}\,\mathrm{d}\omega(x) \\ &= \frac{1}{4\pi\rho} \int_{S_\rho(x_0)} \frac{\partial U}{\partial \nu}(x)\,\mathrm{d}\omega(x) + \frac{1}{4\pi\rho^2} \int_{S_\rho(x_0)} U(x)\,\mathrm{d}\omega(x). \end{aligned}$$

Furthermore, Green's second identity (see Theorem 2.4.2) yields

$$\int_{B_\rho(x_0)} U(x) \underbrace{\Delta_x 1}_{=0} - 1 \underbrace{\Delta_x U(x)}_{=0}\,\mathrm{d}x = \int_{S_\rho(x_0)} U(x) \underbrace{\frac{\partial}{\partial \nu(x)} 1}_{=0} - 1 \frac{\partial U}{\partial \nu}(x)\,\mathrm{d}\omega(x) \tag{3.33}$$

such that

$$U(x_0) = \frac{1}{4\pi\rho^2} \int_{S_\rho(x_0)} U(x)\,\mathrm{d}\omega(x). \tag{3.34}$$

In the limit $\rho \to R-$, we obtain

$$U(x_0) = \frac{1}{4\pi R^2} \int_{S_R(x_0)} U(x)\,\mathrm{d}\omega(x).$$

Furthermore, an additional radial integration of (3.34) leads us to

$$\int_0^R U(x_0)\rho^2\,\mathrm{d}\rho = \int_0^R \frac{1}{4\pi\rho^2} \int_{S_\rho(x_0)} U(x)\,\mathrm{d}\omega(x)\,\rho^2\,\mathrm{d}\rho$$

such that

$$\frac{R^3}{3} U(x_0) = \frac{1}{4\pi} \int_{B_R(x_0)} U(x)\,\mathrm{d}x. \qquad \square$$

Since $4\pi R^2$ is the surface area and $\frac{4}{3}\pi R^3$ is the volume of the ball $B_R(x_0)$, Gauß's mean value theorem, indeed, tells us that the value $U(x_0)$ of a harmonic function U is its mean value (one might also want to say its arithmetic mean; see Kellogg, 1967), over the neighbourhood around x_0 or the boundary of this neighbourhood.

Let us also remember what we proved in (3.33).

Corollary 3.2.8 *If U is harmonic on $B_R(x_0)$, then $\int_{S_\rho(x_0)} \frac{\partial U}{\partial \nu}(x)\, \mathrm{d}\omega(x) = 0$ for all $\rho \in]0, R[$, where ν in the outer unit normal on the sphere $S_\rho(x_0)$.*

This identity is also valid for more general domains.

Theorem 3.2.9 *Let $U \in \mathrm{C}^{(2)}(D) \cap \mathrm{C}^{(1)}(\overline{D})$ be a function which is harmonic on an open and bounded set D for which Green's identities are valid. Moreover, let ∂D be the boundary of D with outer unit normal ν. Then $\int_{\partial D} \frac{\partial U}{\partial \nu}(x)\, \mathrm{d}\omega(x) = 0$.*

Proof We apply Green's second identity from Theorem 2.4.2 and get

$$\int_D 1 \underbrace{\Delta U(x)}_{=0} - U(x) \underbrace{\Delta 1}_{=0}\, \mathrm{d}x = \int_{\partial D} \left(1 \frac{\partial U}{\partial \nu}(x) - U(x) \underbrace{\frac{\partial}{\partial \nu(x)} 1}_{=0} \right) \mathrm{d}\omega(x)$$

such that $0 = \int_{\partial D} \frac{\partial U}{\partial \nu}(x)\, \mathrm{d}\omega(x)$. □

To keep the statement of Theorem 3.2.9 as general as possible, the formulation 'an open and bounded set D for which Green's identities are valid' was chosen here. This includes interiors Σ_{int} of regular surfaces but also sets like $D = B_R(x_0) \setminus \overline{B_r(y_0)}$ with two boundaries such as $S_R(x_0) \cup S_r(y_0)$.

We come to the next important result on harmonic functions, the maximum principle. There exist two ways to formulate it.

Theorem 3.2.10 (Maximum Principle I) *Let $R \subset \mathbb{R}^3$ be a region and let U be a harmonic function on R. If U is not a constant, then U neither has a maximum nor a minimum in R.*

Proof We prove this theorem by contradiction. Let us assume that there is a point $x_0 \in R$ such that $U(x_0)$ is a maximum of U (in the case of a minimum, replace U by $-U$), that is, $U(x) \leq U(x_0)$ for all $x \in R$.

Since U is not constant, there exists a point $x_2 \in R$ where the inequality $U(x_2) < U(x_0) =: M$ holds true. Since R is connected (see Definition 2.1.2), there exists a polygonal chain, which we denote here by P, which starts in x_0, remains in R, and ends in x_2. Due to the continuity of U, there must exist a *last* point x_1 along P where

$$U(x_1) = M \tag{3.35}$$

holds. Note that x_0 and x_1 *might* coincide, whereas x_1 and x_2 *cannot* coincide.

Furthermore, R is open and, thus, there must exist a radius $\varepsilon > 0$ such that $\overline{B_\varepsilon(x_1)} \subset R$ but $x_2 \notin B_\varepsilon(x_1)$. Let x_3 be an intersection point of $S_\varepsilon(x_1)$ with that part of P which is between x_1 and x_2; see Figure 3.4 (note that x_3 might not be unique, but this does not matter).

Since x_1 is the *last* point on the way along P from x_0 to x_2 where U attains its maximum M, we must have $U(x_3) < M$. Due to the continuity of U and the fact that R is open, there exists, consequently, a $\delta > 0$ such that $B_\delta(x_3) \subset R$ and $U(x) < M$ for all $x \in B_\delta(x_3)$.

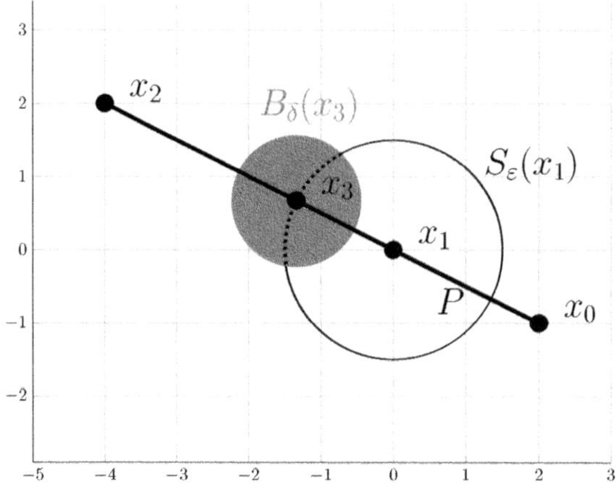

Figure 3.4 Illustration of the geometrical construction for the proof of maximum principle I (Theorem 3.2.10).

Now we apply Gauß's mean value theorem (Theorem 3.2.7). Because $S_\varepsilon(x_1) \cap B_\delta(x_3)$ has a positive surface measure, we get

$$\begin{aligned} U(x_1) &= \frac{1}{4\pi\varepsilon^2} \int_{S_\varepsilon(x_1)} U(x)\, d\omega(x) \\ &= \frac{1}{4\pi\varepsilon^2} \left(\int_{S_\varepsilon(x_1)\cap B_\delta(x_3)} \underbrace{U(x)}_{<M}\, d\omega(x) + \int_{S_\varepsilon(x_1)\setminus B_\delta(x_3)} \underbrace{U(x)}_{\leq M}\, d\omega(x) \right) \\ &< \frac{1}{4\pi\varepsilon^2} \cdot M \cdot \int_{S_\varepsilon(x_1)} 1\, d\omega(x) = M. \end{aligned}$$

This is a contradiction to (3.35). Hence, there can be neither a maximum nor a minimum of U in R. □

The second formulation of the maximum principle immediately follows from the first version.

Theorem 3.2.11 (Maximum Principle II) *Let $R \subset \mathbb{R}^3$ be a bounded region and $U \in \mathrm{C}^{(2)}(R) \cap \mathrm{C}(\overline{R})$ be a harmonic and non-constant function. Then U attains its maximal and minimal value at the boundary ∂R only.*

Proof U is continuous on the compact domain $\overline{R}$. Hence, it must have a maximum and a minimum on $\overline{R}$. Due to Theorem 3.2.10, these extrema cannot be in the interior of R. □

A consequence of the maximum principle is as follows: if we consider the gravitational potential, for example, between the Earth's surface and a satellite orbit, then the potential can achieve its maximal and minimal values only at the Earth's surface and at the orbit. In combination with Corollary 3.1.7, we can expect that the potential has its strongest anomalies at the Earth's surface. These structures become weaker and weaker, if the distance to the centre of the Earth increases.

What have we learned in this section?

- The Kelvin transform mirrors functions and their domains at the unit sphere such that harmonic functions remain harmonic after the mirroring (see Corollary 3.2.3). The formula can certainly be modified easily such that a sphere with a different radius serves as the mirror.
- Based on the fundamental theorem, harmonic functions on the interior Σ_{int} of a regular surface Σ can uniquely be recovered from their values and their normal derivatives at Σ (see Corollary 3.2.6).
- If we construct a ball $B_R(x_0)$ around a point x_0 in the interior of the domain of a harmonic function U, then $U(x_0)$ is the mean value (in a continuous sense, where an integral is used for calculating the mean value) of U on $B_R(x_0)$ and $S_R(x_0)$, respectively (see Theorem 3.2.7).
- On any sphere $S_R(x_0)$, the integral over the normal derivative of a harmonic function vanishes (see Corollary 3.2.8).
- The extrema of functions which are harmonic on a bounded region R, non-constant and continuous on $\overline{R}$ are located on the boundary ∂R (see Theorem 3.2.11).

3.3 Boundary-Value Problems, Green's Function, and Layer Potentials

In Section 3.2, we saw that harmonic functions are already determined by their values and their derivatives at the boundary (at least if the domain is bounded). Indeed, in practical applications, partial differential equations such as the Laplace equation are often combined with given boundary values. This corresponds to initial-value problems as they occur for ordinary differential equations. Indeed, there are some typical boundary-value problems associated to the Laplace equation. Parts of our discussions here are based on Freeden and Gerhards (2013), Kellogg (1967), and Walter (1971).

Problem 3.3.1 Let Σ be a regular surface with outer unit normal ν and let $F \in \mathrm{C}(\Sigma)$ be a given function. The following boundary-value problems are distinguished:

(a) **Interior Dirichlet Problem (IDP):** find $U \in \mathrm{C}^{(2)}(\Sigma_{\text{int}}) \cap \mathrm{C}(\overline{\Sigma_{\text{int}}})$ such that $\Delta U = 0$ in Σ_{int} and $U = F$ on Σ.
(b) **Interior Neumann Problem (INP):** find $U \in \mathrm{C}^{(2)}(\Sigma_{\text{int}}) \cap \mathrm{C}^{(1)}(\overline{\Sigma_{\text{int}}})$ such that $\Delta U = 0$ in Σ_{int} and $\frac{\partial}{\partial \nu} U = F$ on Σ.
(c) **Exterior Dirichlet Problem (EDP):** find $U \in \mathrm{C}^{(2)}(\Sigma_{\text{ext}}) \cap \mathrm{C}(\overline{\Sigma_{\text{ext}}})$ such that $\Delta U = 0$ in Σ_{ext}, $U = F$ on Σ, and U is regular at infinity (see Corollary 3.1.7).
(d) **Exterior Neumann Problem (ENP):** find $U \in \mathrm{C}^{(2)}(\Sigma_{\text{ext}}) \cap \mathrm{C}^{(1)}(\overline{\Sigma_{\text{ext}}})$ such that $\Delta U = 0$ in Σ_{ext}, $\frac{\partial}{\partial \nu} U = F$ on Σ, and U is regular at infinity.

There are a couple of questions which arise when one formulates such problems. These questions were used by Hadamard for a categorization of problems; see Hadamard (1902).

Definition 3.3.2 A problem is called a **well-posed problem in the sense of Hadamard** if the following three criteria are satisfied:

(a) A solution exists.
(b) There is not more than one solution.
(c) The solution continuously depends on the given data (such a solution is called **stable**).

Otherwise, the problem is called **ill-posed**.

We will take care of these criteria step by step.

Theorem 3.3.3 (Stability of the IDP Solution) *Let Σ be a regular surface and let $F, G \in \mathrm{C}(\Sigma)$ be two given functions. Moreover, let $U, V \in \mathrm{C}^{(2)}(\Sigma_{\mathrm{int}}) \cap \mathrm{C}(\overline{\Sigma_{\mathrm{int}}})$ satisfy*

$$\Delta U = 0 = \Delta V \quad \text{in } \Sigma_{\mathrm{int}}, \qquad U = F \text{ and } V = G \quad \text{on } \Sigma. \tag{3.36}$$

If there exists a constant $\varepsilon > 0$ such that $|F(x) - G(x)| \leq \varepsilon$ for all $x \in \Sigma$, then $|U(x) - V(x)| \leq \varepsilon$ for all $x \in \overline{\Sigma_{\mathrm{int}}}$. In other words, if $\|F - G\|_{\mathrm{C}(\Sigma)} \leq \varepsilon$, then $\|U - V\|_{\mathrm{C}(\overline{\Sigma_{\mathrm{int}}})} \leq \varepsilon$.

Proof Actually, not much work has to be done. Obviously, $U - V + \varepsilon$ is harmonic, since $\Delta(U - V + \varepsilon) = \Delta U - \Delta V + \Delta\varepsilon = 0$. Analogously, $U - V - \varepsilon$ is harmonic. Moreover, (3.36) and the required proximity of F and G on Σ imply that $U(x) - \varepsilon = F(x) - \varepsilon \leq G(x) = V(x) \leq F(x) + \varepsilon = U(x) + \varepsilon$ for all $x \in \Sigma$. This means that $U(x) - V(x) - \varepsilon \leq 0 \leq U(x) - V(x) + \varepsilon$ for all $x \in \Sigma$. Hence, maximum principle II (Theorem 3.2.11) tells us that we also have $U(x) - V(x) - \varepsilon \leq 0 \leq U(x) - V(x) + \varepsilon$ for all $x \in \overline{\Sigma_{\mathrm{int}}}$. This is equivalent to $|U(x) - V(x)| \leq \varepsilon$ for all $x \in \overline{\Sigma_{\mathrm{int}}}$. □

Corollary 3.3.4 (Uniqueness of the IDP Solution) *Let Σ be a regular surface and let $F \in \mathrm{C}(\Sigma)$ be an arbitrary function. Then there exists at most one solution to the IDP with boundary values F.*

Proof Let us assume that we have two solutions U and V. Then we have $|U(x) - V(x)| = |F(x) - F(x)| = 0 \leq \varepsilon$ for all $x \in \Sigma$ and (in particular) all $\varepsilon > 0$. Hence, $|U(x) - V(x)| \leq \varepsilon$ for all $x \in \overline{\Sigma_{\mathrm{int}}}$ and all $\varepsilon > 0$, that is, $U \equiv V$ on $\overline{\Sigma_{\mathrm{int}}}$. □

So far, we can tick the second and the third criterion by Hadamard in the case of the IDP. This is not the case for the INP, as we can easily deduce. With boundary values $F \equiv 0$, every constant function would solve the corresponding INP. Hence, the solution is not unique. Fortunately, constant summands to the solution are the only degree of freedom.

Theorem 3.3.5 (Non-uniqueness of the INP Solution) *Let a regular surface Σ with outer unit normal ν be given and let $F \in \mathrm{C}(\Sigma)$ be an arbitrary function. If U and V are solutions of the INP with boundary values F, then there exists a constant $c \in \mathbb{R}$ such that $U(x) = V(x) + c$ for all $x \in \overline{\Sigma_{\mathrm{int}}}$.*

Proof Obviously, $W := U - V$ solves the INP:

$$\Delta W = 0 \quad \text{in } \Sigma_{\text{int}}, \qquad\qquad \frac{\partial W}{\partial \nu} = 0 \quad \text{on } \Sigma.$$

Let us now apply Green's first identity (see Theorem 2.4.2) to W such that

$$\int_{\Sigma_{\text{int}}} W(x) \underbrace{\Delta W(x)}_{=0} + |\nabla W(x)|^2 \, \mathrm{d}x = \int_{\Sigma} W(x) \underbrace{\frac{\partial W}{\partial \nu}(x)}_{=0} \, \mathrm{d}\omega(x). \tag{3.37}$$

Hence, $\int_{\Sigma_{\text{int}}} |\nabla W(x)|^2 \, \mathrm{d}x = 0$. Since ∇W is continuous, this implies that $\nabla W \equiv 0$, that is, W is constant. □

Note that we could also use Green's first identity to show the uniqueness of the solution of the IDP. We would merely have to require that the solution is also in $\mathrm{C}^{(1)}(\overline{\Sigma_{\text{int}}})$ to be able to apply Green's identity. Then the surface integral in (3.37) would also vanish but this time because $W \equiv 0$ on Σ. We would conclude again that W is constant. Since $W \equiv 0$ on Σ, this constant is given by 0.

Let us now have a look at the exterior boundary-value problems. The proofs are, in parts, similar to the interior case.

Theorem 3.3.6 (Stability of the EDP Solution) *Let a regular surface Σ be given and let $F, G \in \mathrm{C}(\Sigma)$ be two given functions. Furthermore, let U, V be solutions of the EDP corresponding to the boundary values F and G, respectively. If there exists a constant $\varepsilon > 0$ such that* $\max_{x \in \Sigma} |F(x) - G(x)| \leq \varepsilon$, *then* $\sup_{x \in \overline{\Sigma_{\text{ext}}}} |U(x) - V(x)| \leq \varepsilon$.

Proof Note that the application of maximum principle II (Theorem 3.2.11) requires a bounded domain. For this reason, let $R > 0$ be a sufficiently large radius such that $\overline{\Sigma_{\text{int}}} \subset B_R(0)$; see Figure 3.5.

As in the proof of Theorem 3.3.3, we have that $U(x) - V(x) - \varepsilon \leq 0 \leq U(x) - V(x) + \varepsilon$ for all $x \in \Sigma$, where $U - V - \varepsilon$ and $U - V + \varepsilon$ both are harmonic functions. Since $\Sigma_{\text{ext}} \cap B_R(0)$ is a bounded region (at least for sufficiently large R), maximum principle II implies that

$$U(x) - V(x) - \varepsilon \leq \max \left[0, \max_{y \in S_R(0)} (U(y) - V(y) - \varepsilon) \right]$$

for all $x \in \overline{\Sigma_{\text{ext}} \cap B_R(0)}$. With analogous considerations for $U - V + \varepsilon$, we come to the conclusion that

$$U(x) - V(x) + \varepsilon \geq \min \left[0, \min_{y \in S_R(0)} (U(y) - V(y) + \varepsilon) \right]$$

and, hence,

$$\begin{aligned} U(x) - V(x) &\geq \min \left[-\varepsilon, \min_{y \in S_R(0)} (U(y) - V(y)) \right] \\ &= -\max \left[\varepsilon, -\min_{y \in S_R(0)} (U(y) - V(y)) \right] = -\max \left[\varepsilon, \max_{y \in S_R(0)} (V(y) - U(y)) \right] \end{aligned}$$

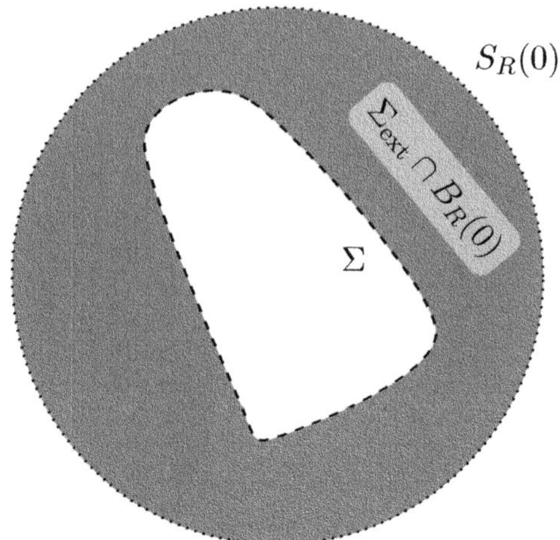

Figure 3.5 For proving the uniqueness of the solutions of the exterior boundary-value problems, we need bounded domains. For this purpose, we first consider the intersections $\Sigma_{\mathrm{ext}} \cap B_R(0)$ for sufficiently large $R > 0$ and finally let R tend to infinity.

for all $x \in \overline{\Sigma_{\mathrm{ext}} \cap B_R(0)}$. In total, we get

$$|U(x) - V(x)| \leq \max \left(\varepsilon, \max_{y \in S_R(0)} |U(y) - V(y)| \right)$$

for all $x \in \overline{\Sigma_{\mathrm{ext}} \cap B_R(0)}$. Since U and V have to be regular at infinity (see Corollary 3.1.7), $\lim_{R\to\infty} \max_{y \in S_R(0)} |U(y) - V(y)| = 0$ must hold. Hence, using the limit $R \to \infty$, we obtain that $|U(x) - V(x)| \leq \varepsilon$ for all $x \in \overline{\Sigma_{\mathrm{ext}}}$. □

In complete analogy to Corollary 3.3.4, we obtain the uniqueness of the EDP solution.

Corollary 3.3.7 (Uniqueness of the EDP Solution) *Let Σ be a given regular surface and let $F \in \mathrm{C}(\Sigma)$ be an arbitrary function. Then there exists at most one solution to the EDP with boundary values F.*

For the INP, we had found out that the solution is non-unique. In the case of the ENP, the additional requirement of a regularity at infinity helps us. We can show the uniqueness of the ENP solution similarly as we proved the non-uniqueness of the INP solution in Theorem 3.3.5, but we need again a bounded domain – this time because of Green's first identity.

Theorem 3.3.8 (Uniqueness of the ENP Solution) *Let Σ be a regular surface with outer unit normal ν and let $F \in \mathrm{C}(\Sigma)$ be an arbitrary function. Then there exists at most one solution to the ENP with boundary values F.*

Proof Let $R > 0$ be (again) sufficiently large such that $\overline{\Sigma_{\mathrm{int}}} \subset B_R(0)$; see Figure 3.5. If U and V solve the same ENP with boundary values F, then we apply Green's first identity to $W := U - V$ on $\Sigma_{\mathrm{ext}} \cap B_R(0)$ and get

$$\int_{\Sigma_{\text{ext}} \cap B_R(0)} W(x) \underbrace{\Delta W(x)}_{=0} + |\nabla W(x)|^2 \, \mathrm{d}x$$
$$= - \int_{\Sigma} W(x) \underbrace{\frac{\partial W}{\partial \nu}(x)}_{=0} \, \mathrm{d}\omega(x) + \int_{S_R(0)} W(x) \frac{\partial W}{\partial n}(x) \, \mathrm{d}\omega(x).$$

Here, n is the outer unit normal to $S_R(0)$, where 'outer' has to be understood with respect to $\Sigma_{\text{ext}} \cap B_R(0)$, that is, $n(x) = x/R$ (the 'outer' unit normal to Σ is, therefore, here $-\nu$. However, since $\frac{\partial W}{\partial \nu} = 0$ on Σ, this makes no difference). As a consequence, we arrive at

$$\int_{\Sigma_{\text{ext}} \cap B_R(0)} |\nabla W(x)|^2 \, \mathrm{d}x = \int_{S_R(0)} W(x) \frac{x}{R} \cdot \nabla W(x) \, \mathrm{d}\omega(x). \tag{3.38}$$

Since U and V (and, thus, also W) have to be regular at infinity, there exist constants $C_1, C_2 \in \mathbb{R}^+$ such that, for sufficiently large $R > 0$,

$$|W(x)| \le \frac{C_1}{|x|} \quad \text{and} \quad |\nabla W(x)| \le \frac{C_2}{|x|^2}$$

for all $|x| \ge R$. Hence,

$$\left| \int_{S_R(0)} W(x) \frac{x}{R} \cdot \nabla W(x) \, \mathrm{d}\omega(x) \right| \le \int_{S_R(0)} \frac{C_1}{R} \cdot 1 \cdot \frac{C_2}{R^2} \, \mathrm{d}\omega(x)$$
$$= C_1 C_2 \cdot 4\pi \cdot \frac{1}{R} \tag{3.39}$$

for sufficiently large R. Combining (3.38) and (3.39), we get

$$\int_{\Sigma_{\text{ext}}} |\nabla W(x)|^2 \, \mathrm{d}x = \lim_{R \to \infty} \int_{\Sigma_{\text{ext}} \cap B_R(0)} |\nabla W(x)|^2 \, \mathrm{d}x = 0.$$

Thus, W is constant on $\overline{\Sigma_{\text{ext}}}$. Since the regularity at infinity implies that $\lim_{|x| \to \infty} W(x) = 0$, only $W \equiv 0$ (and, consequently, $U \equiv V$) is possible. □

Note that also here the proof is almost analogous for the EDP, if we have $U, V \in \mathrm{C}^{(1)}(\overline{\Sigma_{\text{ext}}})$ as an additional requirement.

Example 3.3.9 Let us consider the case of a Dirichlet problem with constant boundary values, for example, $F \equiv 1$, with a spherical boundary $\Sigma = S_1(0)$. In the case of the IDP, the unique solution is $U \equiv 1$. On the other hand, $U \equiv 1$ also solves the Laplace equation on Σ_{ext} but it is not regular at infinity. In this case, the unique solution of the EDP would be $U(x) = |x|^{-1}$. From Corollary 3.1.6, we know that this function is harmonic in $\mathbb{R}^3 \setminus \{0\}$. Moreover, it is obviously regular at infinity and $U|_{S_1(0)} \equiv 1$.

This simple example of unique IDP and EDP solutions shows the following: there is, in general, more than one harmonic function corresponding to a given restriction to a surface Σ. We need another requirement to get a unique solution. In the case of the EDP, this is the regularity at infinity. Note that we indeed used the regularity at infinity in the proof of Theorem 3.3.6, which led us to the uniqueness result in Corollary 3.3.7.

In the case of the IDP, it is the implicit 'interior-space condition' which gives us the uniqueness, because $U(x) = |x|^{-1}$ would satisfy $U|_\Sigma \equiv 1$ with $\Sigma = S_1(0)$ but $\Delta U = 0$ *only* on $\Sigma_{\text{int}} \setminus \{0\}$. Hence, $U(x) = |x|^{-1}$ is excluded and $U \equiv 1$ is left as the solution.

We have not answered all questions about Hadamard's criteria for a well-posed problem (see Definition 3.3.2) here. In particular, the existence of the solution has not been addressed, yet. The answer, more precisely the proof, is, indeed, rather complicated. However, it gets a bit easier for the particular case of a spherical boundary. In geomathematics, we can often deal pretty well with this simplification, and there is a huge amount of knowledge on harmonic functions inside or (more important for us) outside a sphere. We will, therefore, have a closer look at the case of a sphere in the next section. However, we can start our considerations on the existence question with the general case of a domain $D = \Sigma_{\text{int}}$.

There is a (partially) constructive way to tackle the existence problem. This is given by Green's function.

Definition 3.3.10 Let $D = \Sigma_{\text{int}}$ be the interior of a regular surface Σ. A function $G\colon D_G \to \mathbb{R}$ with $D_G := \{(x,y) \in D \times \overline{D} \,|\, x \neq y\}$ and the form

$$G(x,y) = \frac{1}{4\pi} \frac{1}{|x-y|} + R(x,y), \quad (x,y) \in D_G,$$

is called **Green's function** for Δ and D if the following conditions are satisfied:

(a) For each $x \in D$, the function $R(x,\cdot)\colon y \mapsto R(x,y)$ is an element of $\mathrm{C}^{(2)}(D) \cap \mathrm{C}^{(1)}(\overline{D})$.
(b) $\Delta_y R(x,y) = 0$ for all $x, y \in D$.
(c) $G(x,y) = 0$ for all $x \in D$ and all $y \in \Sigma$.

Let us first clarify that a Green's function is well defined.

Theorem 3.3.11 *For a given $D = \Sigma_{\text{int}}$, the Green's function in Definition 3.3.10 is uniquely determined if it exists.*

Proof We assume that we found two Green's functions

$$G_j(x,y) = \frac{1}{4\pi} \frac{1}{|x-y|} + R_j(x,y), \quad j = 1,2,\ (x,y) \in D_G,$$

for Δ and D. From Definition 3.3.10, we conclude for the difference $H := G_1 - G_2$ that it fulfills

$$\Delta_y H(x,y) = 0 \quad \text{for all } y \in D, \qquad H(x,y) = 0 \quad \text{for all } y \in \Sigma$$

for each $x \in D$. This is an IDP, and we know that it cannot have more than one solution (see Corollary 3.3.4). Consequently, we have $H(x,\cdot) \equiv 0$ for each fixed $x \in D$ and, since x is arbitrary, $H \equiv 0$. □

The answer to our question on the existence of a solution is given by the following theorem, at least in parts.

Theorem 3.3.12 *Let $D = \Sigma_{\text{int}}$ be the interior of a regular surface Σ. If there exists a Green's function G for Δ and D, then there also exists a unique solution of the IDP, with $U \in C^{(2)}(D) \cap C^{(1)}(\overline{D})$,*

$$\Delta U = 0 \quad \text{in } D, \qquad\qquad U = F \quad \text{on } \Sigma,$$

corresponding to a given function $F \in C(\Sigma)$. This solution is represented by

$$U(x) = -\int_\Sigma F(y) \frac{\partial}{\partial \nu(y)} G(x,y)\, d\omega(y), \quad x \in D, \tag{3.40}$$

where ν is the outer unit normal to Σ.

We are not yet able to prove this theorem. We have to do some further work. Let us start as follows: from the fundamental theorem (Theorem 3.2.5), more precisely Corollary 3.2.6, we get, if U is harmonic in D, the identity

$$U(x) = \frac{1}{4\pi} \int_\Sigma \left(\frac{1}{|x-y|} \frac{\partial}{\partial \nu(y)} U(y) - U(y) \frac{\partial}{\partial \nu(y)} \frac{1}{|x-y|} \right) d\omega(y). \tag{3.41}$$

Moreover, with Green's second identity (Theorem 2.4.2) applied to U and R (the remainder term in Green's function), we obtain (for all $x \in D$)

$$\int_D R(x,y) \underbrace{\Delta_y U(y)}_{=0} - U(y) \underbrace{\Delta_y R(x,y)}_{=0}\, dy$$
$$= \int_\Sigma R(x,y) \frac{\partial}{\partial \nu(y)} U(y) - U(y) \frac{\partial}{\partial \nu(y)} R(x,y)\, d\omega(y).$$

Hence,

$$0 = \int_\Sigma R(x,y) \frac{\partial}{\partial \nu(y)} U(y) - U(y) \frac{\partial}{\partial \nu(y)} R(x,y)\, d\omega(y) \tag{3.42}$$

for all $x \in D$. We add (3.41) and (3.42) and get

$$U(x) = \int_\Sigma G(x,y) \frac{\partial}{\partial \nu(y)} U(y) - U(y) \frac{\partial}{\partial \nu(y)} G(x,y)\, d\omega(y)$$

for all $x \in D$, where $G(x,y) = 0$ for $y \in \Sigma$. Hence, we get (3.40). However, we argued here as follows: if U solves the IDP, then U is represented by (3.40). For this reason, we first have to prove that the IDP is solvable. This leads us to the investigation of so-called layer potentials.

Definition 3.3.13 Let $F\colon \Sigma \to \mathbb{R}$ be a measurable and bounded function on the regular surface Σ. Then the **single layer potential** P_s is defined by

$$P_s(x) := \int_\Sigma \frac{F(y)}{|x-y|}\, d\omega(y), \quad x \in \mathbb{R}^3.$$

Moreover, the **double layer potential** P_d is defined by

$$P_d(x) := \int_\Sigma F(y) \frac{\partial}{\partial \nu(y)} \frac{1}{|x-y|}\, d\omega(y), \quad x \in \mathbb{R}^3,$$

where ν is the outer unit normal to Σ.

Note that these layer potentials are often defined with a factor $\frac{1}{4\pi}$, which we skip here. Moreover, the single layer potential is also sometimes called the **simple layer potential**. The single layer potential strongly resembles Newton's gravitational potential, except that the mass distribution F is only given at a surface instead of a volumetric domain.

In order to understand the double layer potential, we consider two electric point charges $+Q$ and $-Q$ at the distinct points $y, z \in \mathbb{R}^3$. Then their potential is given by (up to constant factors)

$$P(x) = \frac{Q}{|x-y|} - \frac{Q}{|x-z|}, \quad x \in \mathbb{R}^3 \setminus \{y, z\}.$$

If we consider the straight line $y + t(z-y)$, $0 \le t \le 1$, then the mean value theorem of differentiation tells us that there exists $\tau \in]0,1[$ such that, in $w := y + \tau(z-y)$, the following holds true:

$$P(x) = -(z-y) \cdot \nabla_v \left. \frac{Q}{|x-v|} \right|_{v=w} = |y-z| \frac{y-z}{|y-z|} \cdot \nabla_v \left. \frac{Q}{|x-v|} \right|_{v=w}.$$

Note that the dot product on the right-hand side is a directional derivative. Now, we let z tend to y along the straight line $y + t(z-y)$. In this case, $\nu := \frac{y-z}{|y-z|}$ does not change and w tends to y. Moreover, simultaneously, we let Q grow such that $m := Q|y-z|$ also remains constant. Then we get

$$P(x) = m\, \nu \cdot \nabla_v \left. \frac{1}{|x-v|} \right|_{v=w} \longrightarrow m\, \nu \cdot \nabla_y \frac{1}{|x-y|}.$$

If we transfer this from point charges to a continuous distribution, we come to the double layer potential. Hence, the double layer potential can be regarded as a limit case where two layers tend to each other. For us, the double layer potential is important because it has already occurred in (3.40).

Let us see what we can find out about the layer potentials.

Theorem 3.3.14 *Let $F : \Sigma \to \mathbb{R}$ be measurable and bounded and let Σ be a regular surface. Then the corresponding single layer potential P_s is infinitely differentiable on $\mathbb{R}^3 \setminus \Sigma$ and the function is harmonic in $\mathbb{R}^3 \setminus \Sigma$: $\Delta_x P_s(x) = 0$ for all $x \in \mathbb{R}^3 \setminus \Sigma$. Moreover, P_s exists everywhere on $\mathbb{R}^3$, is bounded, and is continuous.*

Proof The proof is analogous to the proofs of Theorems 3.1.2 and 3.1.5. We have to take into account that the integrals in (3.4) are now surface integrals. For example, we get integrals of the form

$$\int_{B_r(y)\cap\Sigma} \frac{1}{|x-y|} \, \mathrm{d}\omega(x) \tag{3.43}$$

and we have to show that they tend to zero as $r \to 0+$. This can be explained as follows: from the properties of a regular surface (see Definition 2.6.4), we can conclude that, for sufficiently small $r > 0$, the surface part $B_r(y) \cap \Sigma$ (or, maybe, a superset of it) can be parameterized by a $C^{(1)}$-function $\Phi\colon D_r \to \mathbb{R}$, which represents one Cartesian coordinate in terms of the others. Without loss of generality, we assume that

$$x_3 = \Phi(x_1, x_2),\ (x_1, x_2) \in D_r \Leftrightarrow (x_1, x_2, x_3) \in B_r(y) \cap \Sigma.$$

Note that the domain of integration in (3.43) is empty for sufficiently small r, if $y \notin \Sigma$ (remember that Σ is compact). The case $y \notin \Sigma$ is only relevant for the boundedness. In this case, we choose $\varepsilon > 0$ and distinguish $\operatorname{dist}(y, \Sigma) \geq \varepsilon$, where $\int_\Sigma |x - y|^{-1}\,\mathrm{d}\omega(x) \leq \varepsilon^{-1} \int_\Sigma 1\,\mathrm{d}\omega(x)$.

We now proceed for the case $\operatorname{dist}(y, \Sigma) < \varepsilon$, which includes the case $y \in \Sigma$, and set $R := \varepsilon + \operatorname{diam} \Sigma$. Obviously,

$$|x - y|^2 = \sum_{j=1}^{3} (x_j - y_j)^2 \geq |(x_1, x_2) - (y_1, y_2)|^2 . \tag{3.44}$$

We use this now to estimate (3.43) by

$$\int_{B_r(y)\cap\Sigma} \frac{1}{|x - y|}\,\mathrm{d}\omega(x) \leq \int_{B_r(y)\cap\Sigma} \frac{1}{|(x_1, x_2) - (y_1, y_2)|}\,\mathrm{d}\omega(x)$$

$$\leq \int_{D_r} \frac{1}{|(x_1, x_2) - (y_1, y_2)|} \left| \begin{pmatrix} -\frac{\partial\Phi}{\partial x_1}(x_1, x_2) \\ -\frac{\partial\Phi}{\partial x_2}(x_1, x_2) \\ 1 \end{pmatrix} \right| \mathrm{d}(x_1, x_2).$$

Since Φ is a $C^{(1)}$-function, there must exist a constant C such that

$$\int_{B_r(y)\cap\Sigma} \frac{1}{|x - y|}\,\mathrm{d}\omega(x) \leq C \int_{D_r} \frac{1}{|(x_1, x_2) - (y_1, y_2)|}\,\mathrm{d}(x_1, x_2).$$

Moreover, the parameterization $x = (x_1, x_2, \Phi(x_1, x_2))$, $(x_1, x_2) \in D_r$, of $B_r(y) \cap \Sigma$ implies that D_r is a subset of the two-dimensional ball $B_r(y_1, y_2)$ around (y_1, y_2); see (3.44). Hence,

$$\int_{B_r(y)\cap\Sigma} \frac{1}{|x - y|}\,\mathrm{d}\omega(x) \leq C \int_{B_r(y_1, y_2)} \frac{1}{|(x_1, x_2) - (y_1, y_2)|}\,\mathrm{d}(x_1, x_2) = 2\pi C r.$$

Thus, in analogy to the proof of Theorem 3.1.2, we can choose again, for all $\varepsilon > 0$ and all $y \in \mathbb{R}^3$, an appropriate $\delta > 0$ such that $|y - z| < \delta$ implies $|P_{\mathrm{s}}(y) - P_{\mathrm{s}}(z)| < \varepsilon$.

The preceding considerations also yield that $\int_\Sigma |x - y|^{-1}\,\mathrm{d}\omega(x) \leq 2\pi C R$, because $B_R(y) \cap \Sigma = \Sigma$ for all y with $\operatorname{dist}(y, \Sigma) < \varepsilon$. Furthermore, for trivial reasons, the inequality $\int_{\Sigma\setminus B_r(y)} |x - y|^{-1}\,\mathrm{d}\omega(x) \leq r^{-1} \int_\Sigma 1\,\mathrm{d}\omega(x) < +\infty$ is true. This leads us to the existence and the boundedness of P_{s} everywhere on $\mathbb{R}^3$. The fact that P_{s} is bounded can also be shown in analogy to the proof of Theorem 3.1.2 by distinguishing points y with $\operatorname{dist}(y, \Sigma) \geq d$ from the rest (where $d > 0$ is fixed). □

Let us now take care of the double layer potential. Again, we address the questions on the existence, continuity, and harmonicity.

Theorem 3.3.15 *Let $F\colon \Sigma \to \mathbb{R}$ be measurable and bounded, where Σ is a regular surface. Then the corresponding double layer potential P_{d} is infinitely differentiable and harmonic on $\mathbb{R}^3 \setminus \Sigma$.*

Proof It appears to be intuitive to look for an analogous proof to the proof of Theorem 3.1.5. In fact, $\Delta_x(\nu(y) \cdot \nabla_y |x-y|^{-1}) = \nu(y) \cdot \nabla_y \Delta_x |x-y|^{-1} = 0$, if $x \neq y$. Again, all derivatives are bounded in an open set $N_\varepsilon \subset \mathbb{R}^3 \setminus \Sigma$ with $|x-y| \geq \varepsilon$ for all $x \in N_\varepsilon$ and $y \in \Sigma$. The rest is, indeed, analogous to the proof of Theorem 3.1.5. □

Corollary 3.3.16 *The function $x \mapsto \nu(y) \cdot \nabla_y |x-y|^{-1} = \frac{\partial}{\partial \nu(y)} |x-y|^{-1}$ is harmonic in $\mathbb{R}^3 \setminus \{y\}$ for all $y \in \mathbb{R}^3$.*

Theorem 3.3.17 *Let $F\colon \Sigma \to \mathbb{R}$ be a measurable and bounded function on the regular surface Σ. Then the corresponding double layer potential P_{d} exists on the whole space $\mathbb{R}^3$. Moreover, the restriction $P_{\mathrm{d}}|_\Sigma$ is continuous on Σ.*

Proof For $x \notin \Sigma$, it is easy to see that $P_{\mathrm{d}}(x)$ exists:

$$|P_{\mathrm{d}}(x)| \leq \int_\Sigma |F(y)|\,|\nu(y)| \left| \nabla_y \frac{1}{|x-y|} \right| \mathrm{d}\omega(y)$$
$$\leq \sup_{z \in \Sigma} |F(z)| \int_\Sigma \frac{1}{|x-y|^2} \,\mathrm{d}\omega(y). \tag{3.45}$$

If $x \notin \Sigma$, then $\Sigma \ni y \mapsto |x-y|^{-2}$ is bounded (note that Σ is compact) and the preceding integral is finite.

For the case $x \in \Sigma$, we first proceed like in the proof of Theorem 3.3.14 and assume that we have a parameterization $\Phi\colon D \to \mathbb{R}$ of $B_r(x) \cap \Sigma$ for a sufficiently small $r > 0$ such that

$$x_3 = \Phi(x_1,x_2),\ (x_1,x_2) \in D \Leftrightarrow (x_1,x_2,x_3) \in B_r(x) \cap \Sigma. \tag{3.46}$$

We first subdivide the integral of the double layer potential as

$$P_{\mathrm{d}}(x) = \int_{\Sigma \cap B_r(x)} F(y)\,\nu(y) \cdot \nabla_y \frac{1}{|x-y|} \,\mathrm{d}\omega(y)$$
$$+ \int_{\Sigma \setminus B_r(x)} F(y)\,\nu(y) \cdot \nabla_y \frac{1}{|x-y|} \,\mathrm{d}\omega(y). \tag{3.47}$$

The second integral in the latter equation is finite for the same reason that (3.45) is finite. Let us therefore concentrate on the first integral. With the parameterization (3.46), we get (with $\tilde{x} = (x_1,x_2)$, $\tilde{y} = (y_1,y_2)$)

$$\int_{\Sigma \cap B_r(x)} F(y)\,\nu(y) \cdot \nabla_y \frac{1}{|x-y|} \,\mathrm{d}\omega(y) \tag{3.48}$$
$$= \int_D F\left(\tilde{y}, \Phi(\tilde{y})\right)\ \nu\left(\tilde{y}, \Phi(\tilde{y})\right) \cdot \frac{(\tilde{x}, \Phi(\tilde{x})) - (\tilde{y}, \Phi(\tilde{y}))}{|(\tilde{x}, \Phi(\tilde{x})) - (\tilde{y}, \Phi(\tilde{y}))|^3}\, g(\tilde{y}) \,\mathrm{d}\tilde{y},$$

where $g(\tilde{y}) = \left|\left(-\frac{\partial\Phi}{\partial y_1}(\tilde{y}), -\frac{\partial\Phi}{\partial y_2}(\tilde{y}), 1\right)\right|$, $\tilde{y} \in D$. Moreover, due to the parameterization, the unit normal vector on $\Sigma \cap B_r(x)$ is given by

$$\nu\left(\tilde{y}, \Phi(\tilde{y})\right) = \frac{\left(-\frac{\partial\Phi}{\partial y_1}(\tilde{y}), -\frac{\partial\Phi}{\partial y_2}(\tilde{y}), 1\right)^{\mathrm{T}}}{\left|\left(-\frac{\partial\Phi}{\partial y_1}(\tilde{y}), -\frac{\partial\Phi}{\partial y_2}(\tilde{y}), 1\right)^{\mathrm{T}}\right|}, \quad \tilde{y} \in D.$$

We denote the integrand on the right-hand side of (3.48) by $I(\tilde{y})$. We have

$$I(\tilde{y}) = F\left(\tilde{y}, \Phi(\tilde{y})\right)\left[(\tilde{y} - \tilde{x}) \cdot \nabla_{\tilde{y}}\, \Phi(\tilde{y}) + \Phi(\tilde{x}) - \Phi(\tilde{y})\right] \\ \times \left[|\tilde{x} - \tilde{y}|^2 + (\Phi(\tilde{x}) - \Phi(\tilde{y}))^2\right]^{-3/2}. \tag{3.49}$$

Since Σ is a regular surface, we can assume that Φ is a $\mathrm{C}^{(2)}$-function. Hence, we can apply Taylor's theorem to the first square-bracket term in (3.49) such that (note also that $|x - y| \geq |\tilde{x} - \tilde{y}|$)

$$|I(\tilde{y})| \leq |F(\tilde{y}, \Phi(\tilde{y}))|\, |\tilde{x} - \tilde{y}|^2 \cdot C \cdot |\tilde{x} - \tilde{y}|^{-3},$$

where $C \in \mathbb{R}^+$ is a constant. Consequently,

$$\left|\int_{\Sigma \cap B_r(x)} F(y)\, \nu(y) \cdot \nabla_y \frac{1}{|x - y|}\, \mathrm{d}\omega(y)\right| \leq C \int_D \frac{|F(\tilde{y}, \Phi(\tilde{y}))|}{|\tilde{x} - \tilde{y}|}\, \mathrm{d}\tilde{y}. \tag{3.50}$$

Since F is bounded and $\int_D |\tilde{x} - \tilde{y}|^{-1}\, \mathrm{d}\tilde{y} \leq \int_{B_r(\tilde{x})} |\tilde{x} - \tilde{y}|^{-1}\, \mathrm{d}\tilde{y} = 2\pi r$, the right-hand side of (3.50) is finite. Hence, $P_{\mathrm{d}}(x)$ exists for all $x \in \mathbb{R}^3$.

For the proof of the continuity on Σ, let $x, z \in \Sigma$ be two distinct points. Then

$$P_{\mathrm{d}}(x) - P_{\mathrm{d}}(z) = \int_\Sigma F(y)\, \nu(y) \cdot \nabla_y \left(\frac{1}{|x - y|} - \frac{1}{|z - y|}\right) \mathrm{d}\omega(y).$$

We consider again a sufficiently small neighbourhood $B_r(x)$ and a parameterization Φ like above and let $z \in B_r(x) \cap \Sigma$. Also here, we want to use an analogy argument with respect to the proof of Theorem 3.1.2. The analogous identity to (3.4) would then contain terms which are of the form or can be estimated by

$$C \int_{\Sigma \cap B_r(x)} \left|\nu(y) \cdot \nabla_y \frac{1}{|w - y|}\right| \mathrm{d}y, \quad w \in \Sigma \cap B_r(x),$$

or

$$C \int_{\Sigma \setminus B_r(x)} \left|\nu(y) \cdot \nabla_y \left(\frac{1}{|x - y|} - \frac{1}{|z - y|}\right)\right| \mathrm{d}\omega(y).$$

The integrals of the first kind can be handled like in the proof of Theorem 3.1.2 and like earlier in this proof and therefore tend to 0 as $r \to 0+$. For the second integral, the uniform continuity argument from the proof of Theorem 3.1.2 can be applied in the same manner to the modified integrand. Hence, the continuity of V_{d} on Σ holds true. □

Theorem 3.3.18 *Let F be a measurable and bounded function on the regular surface Σ. Then the corresponding single and double layer potentials are regular at infinity.*

Proof Parts of this proof are analogous to the proof of Corollary 3.1.7. Let $\sigma := \max_{x\in\Sigma}|x|$. We now restrict our considerations to those $y \in \mathbb{R}^3$ which satisfy $|y| > 2\sigma$. Then all $x \in \Sigma$ satisfy $|x| < |y|/2$ and $|y-x| \geq |y|/2$. Thus, with $C := \|F\|_\infty \int_\Sigma 1\,\mathrm{d}\omega$, we have

$$|P_{\mathrm{s}}(y)| = \left|\int_\Sigma \frac{F(x)}{|y-x|}\,\mathrm{d}\omega(x)\right| \leq \|F\|_\infty \int_\Sigma \frac{2}{|y|}\,\mathrm{d}\omega(x) = \frac{2C}{|y|}.$$

Moreover, in view of Theorem 3.3.14 and its proof we can conclude that

$$\left|\nabla_y P_{\mathrm{s}}(y)\right| = \left|\int_\Sigma F(x)\nabla_y \frac{1}{|y-x|}\,\mathrm{d}\omega(x)\right| \leq \|F\|_\infty \int_\Sigma \frac{1}{|y-x|^2}\,\mathrm{d}\omega(x) \leq \frac{4C}{|y|^2}.$$

For the double layer potential, we get similar results:

$$|P_{\mathrm{d}}(y)| = \left|\int_\Sigma F(x)\frac{\partial}{\partial\nu(x)}\frac{1}{|y-x|}\,\mathrm{d}\omega(x)\right| \leq \|F\|_\infty \int_\Sigma \frac{1}{|y-x|^2}\,\mathrm{d}\omega(x) \leq \frac{4C}{|y|^2},$$

and (with Theorem 3.3.15 and its proof)

$$\begin{aligned}\left|\nabla_y P_{\mathrm{d}}(y)\right| &= \left|\int_\Sigma F(x)\nabla_y \frac{\partial}{\partial\nu(x)}\frac{1}{|y-x|}\,\mathrm{d}\omega(x)\right|\\ &= \left|\int_\Sigma F(x)\frac{\partial}{\partial\nu(x)}\nabla_y \frac{1}{|y-x|}\,\mathrm{d}\omega(x)\right|\\ &\leq \|F\|_\infty \int_\Sigma \left\|(\nabla_x\otimes\nabla_y)\frac{1}{|y-x|}\right\|\,\mathrm{d}\omega(x),\end{aligned}$$

where $\|\cdot\|$ is here the Euclidean norm of the occurring matrix.

A few basic analytic calculations yield

$$\begin{aligned}(\nabla_x\otimes\nabla_y)\frac{1}{|x-y|} &= \left(\frac{\partial}{\partial x_i}\frac{x_j-y_j}{|x-y|^3}\right)_{i,j=1,2,3}\\ &= |x-y|^{-6}\left(\delta_{ij}|x-y|^3 - (x_j-y_j)3\,|x-y|(x_i-y_i)\right)_{i,j=1,2,3}\\ &= \frac{1}{|x-y|^3}I_3 - 3\,\frac{(x-y)\otimes(x-y)}{|x-y|^5},\end{aligned}$$

where I_3 is the 3×3 identity matrix. Hence,

$$\begin{aligned}\left\|(\nabla_x\otimes\nabla_y)\frac{1}{|x-y|}\right\| &\leq \frac{\sqrt{3}}{|x-y|^3} + \frac{3}{|x-y|^5}\left(\sum_{i,j=1}^3 (x_i-y_i)^2(x_j-y_j)^2\right)^{1/2}\\ &\leq \frac{\sqrt{3}+3}{|x-y|^3},\end{aligned}$$

such that $|\nabla_y P_{\mathrm{d}}(y)| \leq 8(\sqrt{3}+3)C|y|^{-3}$. □

Note that we only showed the continuity of P_{d} *restricted* to Σ in Theorem 3.3.17. This means that, for convergent sequences $(x_n) \subset \Sigma$, the values $P_{\mathrm{d}}(x_n)$ tend to $P_{\mathrm{d}}(\lim x_n)$. Moreover, due to Theorem 3.3.15, P_d is continuous on $\mathbb{R}^3 \setminus \Sigma$. Hence, $P_{\mathrm{d}}(\lim x_n) = \lim P_{\mathrm{d}}(x_n)$ if (x_n) and $\lim x_n$ are both in $\mathbb{R}^3 \setminus \Sigma$. However, for convergent sequences (x_n) *outside* Σ with $\lim x_n \in \Sigma$ this is not true in general, as the following example shows.

Example 3.3.19 *The double layer potential is, in general, not continuous.* As an example, we consider the double layer potential corresponding to a constant 'density' $F \equiv 1$ and a spherical surface $\Sigma = S_R(x_0)$, that is,

$$P_{\mathrm{d}}(x) = \int_{S_R(x_0)} \frac{\partial}{\partial \nu(y)} \frac{1}{|x-y|} \,\mathrm{d}\omega(y), \quad x \in \mathbb{R}^3.$$

From Corollary 3.1.6, we know that $y \mapsto |x-y|^{-1}$ is harmonic for all $y \neq x$. Hence, Corollary 3.2.8 implies that $P_{\mathrm{d}}(x) = 0$, if $x \notin \overline{B_R(x_0)}$.

In the case $x \in B_R(x_0)$, we choose a sufficiently small $\varepsilon > 0$ such that $\overline{B_\varepsilon(x)} \subset B_R(x_0)$ and observe that $y \mapsto |x-y|^{-1}$ is still harmonic in the case that $y \in B_R(x_0) \setminus B_\varepsilon(x) =: D_\varepsilon$. Hence, Theorem 3.2.9 implies that

$$\begin{aligned} 0 &= \int_{\partial D_\varepsilon} \frac{\partial}{\partial \nu(y)} \frac{1}{|x-y|} \,\mathrm{d}\omega(y) \\ &= \underbrace{\int_{S_R(x_0)} \frac{\partial}{\partial \nu(y)} \frac{1}{|x-y|} \,\mathrm{d}\omega(y)}_{=P_{\mathrm{d}}(x)} + \int_{S_\varepsilon(x)} \frac{\partial}{\partial \nu(y)} \frac{1}{|x-y|} \,\mathrm{d}\omega(y). \end{aligned}$$

Note that ν on $S_\varepsilon(x)$ must be directed outside D_ε, that is, $\nu(y) = \varepsilon^{-1}(x-y)$ for $y \in S_\varepsilon(x)$. Thus, for $x \in B_R(x_0)$,

$$\begin{aligned} P_{\mathrm{d}}(x) &= -\int_{S_\varepsilon(x)} \frac{\partial}{\partial \nu(y)} \frac{1}{|x-y|} \,\mathrm{d}\omega(y) = -\int_{S_\varepsilon(x)} \frac{x-y}{\varepsilon} \cdot \frac{x-y}{|x-y|^3} \,\mathrm{d}\omega(y) \\ &= -\frac{1}{\varepsilon^2} \cdot 4\pi\varepsilon^2 = -4\pi. \end{aligned}$$

We already see here the discontinuity of P_{d} at points $x \in \Sigma$. When approaching x from the interior, we get the limit -4π. However, the limit from the exterior yields 0.

Let us nevertheless also calculate the potential at the surface. For $x \in \Sigma = S_R(x_0)$, we exclude again the singularity and first consider the domain $D_\varepsilon := B_R(x_0) \setminus B_\varepsilon(x)$. The boundary is given by (see also Figure 3.6)

$$\partial D_\varepsilon = \underbrace{(S_R(x_0) \setminus B_\varepsilon(x))}_{=:\alpha_\varepsilon} \cup \underbrace{(S_\varepsilon(x) \cap B_R(x_0))}_{=:\beta_\varepsilon}.$$

Moreover, let $\gamma_\varepsilon := S_R(x_0) \cap B_\varepsilon(x)$. The integral $\int_{\gamma_\varepsilon} \frac{\partial}{\partial \nu(y)} |x-y|^{-1} \,\mathrm{d}\omega(y) =: I_\varepsilon(x)$ looks like a double layer potential, except for the fact that the domain is not a closed surface, as we required it for regular surfaces. However, a closer look at the proof of

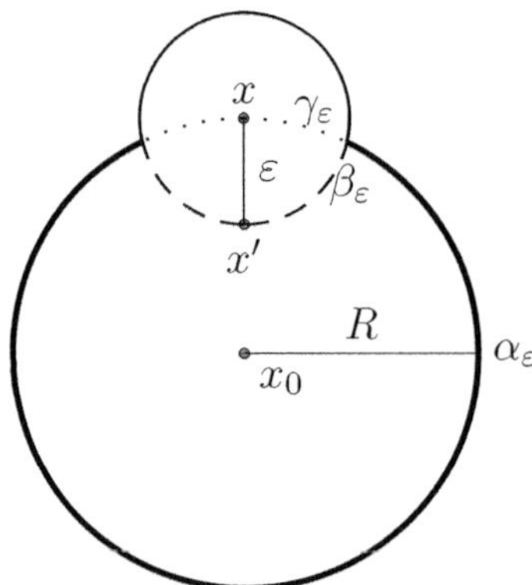

Figure 3.6 Illustration of the surfaces α_ε, β_ε, and γ_ε.

Theorem 3.3.17 (where only the first integral in (3.47) is relevant) reveals that also integrals of the type $I_\varepsilon(x)$ are finite and, in particular,

$$\lim_{\varepsilon\to 0+} I_\varepsilon(x) = 0. \tag{3.51}$$

We now get for the whole double layer potential $P_{\mathrm{d}}(x)$

$$\int_\Sigma \frac{\partial}{\partial\nu(y)}\frac{1}{|x-y|}\,\mathrm{d}\omega(y) = I_\varepsilon(x) + \int_{\alpha_\varepsilon} \frac{\partial}{\partial\nu(y)}\frac{1}{|x-y|}\,\mathrm{d}\omega(y). \tag{3.52}$$

Moreover, with Theorem 3.2.9, we conclude that $0 = \int_{\partial D_\varepsilon} \frac{\partial}{\partial\nu(y)}|x-y|^{-1}\,\mathrm{d}\omega(y)$ such that

$$\int_{\alpha_\varepsilon} \frac{\partial}{\partial\nu(y)}\frac{1}{|x-y|}\,\mathrm{d}\omega(y) = -\int_{\beta_\varepsilon} \frac{\partial}{\partial\nu(y)}\frac{1}{|x-y|}\,\mathrm{d}\omega(y), \tag{3.53}$$

where $\nu(y)$ must be oriented outside D_ε.

We use (3.53) in (3.52) and arrive at

$$\begin{aligned} P_{\mathrm{d}}(x) &= I_\varepsilon(x) - \int_{\beta_\varepsilon} \frac{\partial}{\partial\nu(y)}\frac{1}{|x-y|}\,\mathrm{d}\omega(y) \\ &= I_\varepsilon(x) - \int_{\beta_\varepsilon} \frac{x-y}{|x-y|}\cdot\frac{x-y}{|x-y|^3}\,\mathrm{d}\omega(y) = I_\varepsilon(x) - \frac{1}{\varepsilon^2}\int_{\beta_\varepsilon} 1\,\mathrm{d}\omega(y). \end{aligned} \tag{3.54}$$

It remains to calculate the limit $\varepsilon \to 0+$ for the latter integral. This, actually, leads to the concept of a solid angle, but we will not discuss this in detail here, since our special case of a sphere is easier to handle. Obviously, β_ε is a part of the sphere $S_\varepsilon(x)$, whose elements can be represented by $y = x + \varepsilon\,\xi, \xi \in \Omega$. Moreover, β_ε is symmetric with respect to $x' := x + \frac{\varepsilon}{R}(x_0 - x)$. We now move our coordinate system such that x is the origin and x' corresponds to $(0,0,\varepsilon)^{\mathrm{T}}$. In polar coordinates, this means that x' corresponds to $r = \varepsilon$ and $t = 1$ (with arbitrary φ).

Moreover, β_ε corresponds to those points $y = x + \varepsilon\,\xi \in S_\varepsilon(x)$ for which $|x + \varepsilon\,\xi - x_0| \leq R$ holds. This is equivalent to

$$\underbrace{|x - x_0|^2}_{=R^2} + 2\varepsilon(x - x_0) \cdot \xi + \varepsilon^2 \leq R^2,$$

that is, $a_\varepsilon := \varepsilon/(2R) \leq [(x_0 - x)/R] \cdot \xi$, or, in our new coordinate system, $r = \varepsilon$ and $t \geq a_\varepsilon$. Thus,

$$\int_{\beta_\varepsilon} 1 \,\mathrm{d}\omega(y) = \int_{a_\varepsilon}^{1} \int_0^{2\pi} \varepsilon^2 \,\mathrm{d}\varphi \,\mathrm{d}t = 2\pi\varepsilon^2 \left(1 - \frac{\varepsilon}{2R}\right). \tag{3.55}$$

Finally, inserting (3.51) and (3.55) in (3.54) and taking the limit $\varepsilon \to 0+$, we obtain $P_\mathrm{d}(x) = -2\pi$, $x \in \Sigma$.

Let us summarize what we found out.

Remark 3.3.20 If $\Sigma = S_R(x_0)$ and $F \equiv 1$, then the corresponding double layer potential P_d satisfies

- $P_\mathrm{d}(x) = 0$ for $x \notin \overline{B_R(x_0)}$ such that $P_\mathrm{d}(y) \to 0$ for $y \to x \in \Sigma$ with $y \notin \overline{B_R(x_0)}$.
- $P_\mathrm{d}(x) = -4\pi$ for $x \in B_R(x_0)$ such that $P_\mathrm{d}(y) \to -4\pi$ for $y \to x \in \Sigma$ with $y \in B_R(x_0)$.
- $P_\mathrm{d}(x) = -2\pi$ for $x \in \Sigma$.

This result can, indeed, be generalized to regular surfaces. We omit the proof here and refer to Folland (1976, Proposition 3.19).

Theorem 3.3.21 *Let Σ be a regular surface and $F \equiv 1$. Then the corresponding double layer potential satisfies*

$$P_\mathrm{d}(x) = \begin{cases} 0, & \text{if } x \in \Sigma_{\mathrm{ext}}, \\ -4\pi, & \text{if } x \in \Sigma_{\mathrm{int}}, \\ -2\pi, & \text{if } x \in \Sigma. \end{cases}$$

Unfortunately, there is still much work to do, before we are able to prove the existence of the solutions of the boundary-value problems. The following remaining steps are based on Folland (1976).

Definition 3.3.22 Let $D \subset \mathbb{R}^n$ and $E \subset \mathbb{R}^m$ be two measurable sets. Moreover, let a measurable function $K\colon E \times D \to \mathbb{R}$ and two scalar function spaces $\mathcal{X}(D)$ and $\mathcal{Y}(E)$ with domains D and E, respectively, be given such that the operator

$$\mathcal{T}\colon \mathcal{X}(D) \to \mathcal{Y}(E), \qquad F \mapsto \int_D K(\cdot, x) F(x) \,\mathrm{d}x$$

is well defined. Then $\mathcal{T}$ is called a **Fredholm integral operator of the first kind**. If $D = E$ and $\mathcal{X}(D) = \mathcal{Y}(E)$, then $\mathcal{T} + \mathcal{I}$, where $\mathcal{I}$ is the identity, is called a **Fredholm integral operator of the second kind**. Moreover, K is called the **integral**

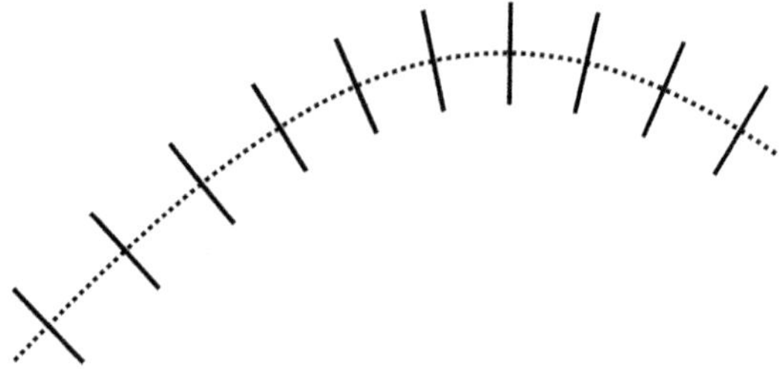

Figure 3.7 The dotted line corresponds to a segment of a (slice of) Σ. The solid lines show the graphs of the normal line functions $t \mapsto N(y,t)$ for selected points $y \in \Sigma$ and a small interval for t.

kernel. We say that K is of **order** α if $D = E$, $0 < \alpha < n-1$, and there exists a bounded function $A\colon D \times D \to \mathbb{R}$ such that $K(x,y) = A(x,y)|x-y|^{-\alpha}$ for all $x, y \in D$ with $x \neq y$. Eventually, equations of the type $\mathcal{T}F = G$ or $(\mathcal{T} + \mathcal{I})F = G$ are called **Fredholm integral equations** of the first or second kind, respectively.

Actually, Newton's gravitational potential, shown in (3.3), represents a Fredholm integral operator of the first kind with a kernel of order 1. We will discuss this aspect later in this book.

Before we come to an important result, we need some lemmata.

Lemma 3.3.23 *Let Σ be a regular surface with outer unit normal ν. Then there exists a constant $C \in \mathbb{R}^+$ such that $\int_\Sigma \left|\frac{\partial}{\partial\nu(y)}|x-y|^{-1}\right| \mathrm{d}\omega(y) \leq C$ for all $x \in \mathbb{R}^3 \setminus \Sigma$.*

Proof Let us construct a function $N\colon \Sigma \times \mathbb{R} \to \mathbb{R}^3$ by $N(y,t) := y + t\nu(y)$. Then $N(y,\cdot)$ parameterizes a straight line through y in the direction of the normal vector $\nu(y)$, as shown in Figure 3.7.

Let us assume that there is a $\delta > 0$ and an open set $S \supset \Sigma$ such that the restricted mapping $N\colon \Sigma \times]-\delta/2, \delta/2[\to S$ is bijective (this is, indeed, true; see Folland, 1976, pp. 8–9).

This implies the following: if $x \in \mathbb{R}^3$ has a distance $\operatorname{dist}(x,\Sigma)$ which is less than $\delta/2$, then there is one and only one $y \in \Sigma$ such that $y + t\nu(y) = x$ for exactly one $t \in]-\delta/2, \delta/2[$.

The other points (where $\operatorname{dist}(x,\Sigma) \geq \delta/2$) are easy to handle. For them,

$$\left|\frac{\partial}{\partial\nu(y)}\frac{1}{|x-y|}\right| \leq |\nu(y)|\,\frac{|x-y|}{|x-y|^3} \leq \frac{4}{\delta^2},$$

$$\int_\Sigma \left|\frac{\partial}{\partial\nu(y)}\frac{1}{|x-y|}\right| \mathrm{d}\omega(y) \leq \frac{4}{\delta^2}\int_\Sigma 1\,\mathrm{d}\omega(y) =: C_1. \tag{3.56}$$

Let now $\operatorname{dist}(x,\Sigma) < \delta/2$ and $x = z + t\nu(z)$, $z \in \Sigma$, $|t| < \delta/2$. We set $U_\delta := B_\delta(z) \cap \Sigma$. If $y \in \Sigma \setminus U_\delta$, then $|y-x| \geq |y-z| - |z-x| \geq \delta - \delta/2 = \delta/2$. As a consequence,

$$\int_{\Sigma\setminus U_\delta} \left|\frac{\partial}{\partial\nu(y)}\frac{1}{|x-y|}\right| \mathrm{d}\omega(y) \leq \int_{\Sigma\setminus U_\delta} \frac{1}{|x-y|^2}\,\mathrm{d}\omega(y) \leq \frac{4}{\delta^2}\int_\Sigma 1\,\mathrm{d}\omega(y) = C_1. \tag{3.57}$$

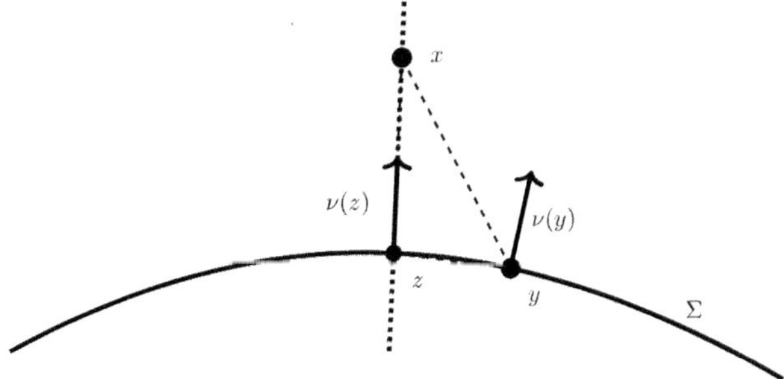

Figure 3.8 The point y is in a neighbourhood of the point $z \in \Sigma$ with $x = z + t\nu(z)$.

In the remaining case, where $y \in U_\delta$, we get

$$\left|\frac{\partial}{\partial\nu(y)} \frac{1}{|x-y|}\right| = \left|\nu(y) \cdot \frac{x-y}{|x-y|^3}\right| \leq \left|\nu(y) \cdot \frac{x-z}{|x-y|^3}\right| + \left|\nu(y) \cdot \frac{z-y}{|x-y|^3}\right|$$

$$\leq \frac{|x-z|}{|x-y|^3} + \frac{|z-y|}{|x-y|^3} \left|\cos \sphericalangle(\nu(y), z-y)\right|. \tag{3.58}$$

Since Σ is a regular surface, its smoothness enables us to start a low-order Taylor expansion of $y \mapsto \cos \sphericalangle(\nu(y),\ z-y)$ around the centre $y = z$, where $\sphericalangle(\nu(y),\ z-y) \to \pi/2$ as $y \to z$. Hence,

$$\cos \sphericalangle(\nu(y),\ z-y) = 0 + \mathcal{O}\left(|z-y|\right) \tag{3.59}$$

as $y \to z$; see also Figure 3.8.

Similarly, $\cos \sphericalangle(\nu(z), z-y) = \mathcal{O}\left(|z-y|\right)$ as $y \to z$. As a consequence, the law of cosines in the triangle given by x, y, and z yields

$$\begin{aligned}|x-y|^2 &= |x-z|^2 + |z-y|^2 - 2|x-z|\,|z-y| \cos \sphericalangle(\nu(z), y-z)\\ &= |x-z|^2 + |z-y|^2 + \mathcal{O}\left(|z-y|^2\right)\end{aligned}$$

for $y \to z$.

As a consequence, there exist constants $C_2, C_3 \in \mathbb{R}^+$, which are independent of x, such that

$$\begin{aligned}|\cos \sphericalangle(\nu(y), z-y)| &\leq C_2|z-y|,\\ |x-y|^2 &\geq |x-z|^2 + |z-y|^2 - C_3|z-y|^2,\\ |x-y|^2 &\geq |z-y|^2,\end{aligned}$$

if y is sufficiently close to z. Hence,

$$\frac{|z-y|}{|x-y|^3} \left|\cos \sphericalangle(\nu(y), z-y)\right| \leq \frac{|z-y|}{|z-y|^3} \left|\cos \sphericalangle(\nu(y), z-y)\right| \leq C_2|z-y|^{-1}, \tag{3.60}$$

if y is sufficiently close to z. Furthermore, the Taylor expansion of the function $\mathbb{R} \setminus \{-b\} \ni c \mapsto a(b+c)^{-3/2}$ around $c = 0$ (with $b \neq 0$) implies that there exists $\theta \in [0,1]$ such that

$$\frac{a}{(b+c)^{3/2}} = \frac{a}{b^{3/2}} - \frac{3}{2}\frac{a}{(b+\theta c)^{5/2}}\, c,$$

where c has to be close to 0. We set $a := |x-z|$, $b := |x-z|^2 + |z-y|^2$, and $c := -\,C_3|z-y|^2$. Hence, there is a constant $C_4 \in \mathbb{R}^+$ such that

$$\begin{aligned}\frac{|x-z|}{|x-y|^3} &\leq \frac{|x-z|}{(|x-z|^2+|z-y|^2 - C_3|z-y|^2)^{3/2}} \\ &\leq \frac{|x-z|}{(|x-z|^2+|z-y|^2)^{3/2}} + C_4\frac{|x-z|\,|z-y|^2}{(|x-z|^2+|z-y|^2)^{5/2}}\end{aligned} \tag{3.61}$$

for y sufficiently close to z. Here, we used that

$$\lim_{y\to z}\frac{|x-z|^2 + (1-\theta C_3)|z-y|^2}{|x-z|^2+|z-y|^2} = 1,$$

where x and y can be considered to be elements of appropriate compact sets (e.g. $x \in \overline{S}$), which implies that this convergence is uniform. Hence, there is indeed a $\delta > 0$ (independent of x) such that the error term in the Taylor expansion can be estimated, for $y \in U_\delta$, by

$$\frac{|x-z|\,|z-y|^2}{\left(|x-z|^2+(1-\theta C_3)|z-y|^2\right)^{5/2}} \leq 2\,\frac{|x-z|\,|z-y|^2}{\left(|x-z|^2+|z-y|^2\right)^{5/2}}.$$

For calculating the remaining integral over U_δ, we use polar coordinates with the radial coordinate $r := |y-z|$. This leads us in combination with (3.58), (3.60), and (3.61) to the following (we set $a := |x-z|$ and $s := ra^{-1}$ and $C_5 \in \mathbb{R}^+$ is another constant):

$$\begin{aligned}&\int_{U_\delta}\left|\frac{\partial}{\partial\nu(y)}\frac{1}{|x-y|}\right| d\omega(y) \\ &\leq C_5\int_0^\delta\left[\frac{a}{(a^2+r^2)^{3/2}} + C_4\frac{a\,r^2}{(a^2+r^2)^{5/2}}\right] r\,\mathrm{d}r + C_5C_2\int_0^\delta \frac{1}{r}\,r\,\mathrm{d}r \\ &= C_5\int_0^{\delta/a}\left[\frac{1}{a^2(1+s^2)^{3/2}} + C_4\frac{s^2}{a^2(1+s^2)^{5/2}}\right] s\,a^2\,\mathrm{d}s + C_5\delta C_2 \\ &\leq C_5\int_0^\infty \frac{s}{(1+s^2)^{3/2}} + C_4\frac{s^3}{(1+s^2)^{5/2}}\,\mathrm{d}s + C_5\delta C_2 \\ &= C_5\left(1+\frac{2}{3}\,C_4+\delta C_2\right),\end{aligned}$$

since $\int s(1+s^2)^{-3/2}\,\mathrm{d}s = -\sqrt{1+s^2}^{-1} + \text{const}$ and $\int s^3(1+s^2)^{-5/2}\,\mathrm{d}s = -(1/3)(3s^2+2)(s^2+1)^{-3/2} + \text{const}$. This completes the proof together with (3.56) and (3.57). □

Lemma 3.3.24 *The kernel* $K(x,y) := \frac{\partial}{\partial\nu(y)}\frac{1}{|x-y|}$ *with* $x, y \in \Sigma$ *and* $x \neq y$ *is an integral kernel of order* 1.

Proof We have

$$K(x,y) = \nu(y) \cdot \frac{x-y}{|x-y|^3} = \frac{|x-y| \cos \sphericalangle(\nu(y), x-y)}{|x-y|^3}.$$

In the proof of Lemma 3.3.23, we have already seen that $\cos \sphericalangle(\nu(y), x-y) = \mathcal{O}(|x-y|)$ as $x \to y$. As a consequence, the function

$$A(x,y) := \frac{\cos \sphericalangle(\nu(y), x-y)}{|x-y|}, \quad x, y \in \Sigma,\ x \neq y,$$

is bounded on $\Sigma \times \Sigma$. Hence, $K(x,y) = A(x,y)\,|x-y|^{-1}$ is an integral kernel of order 1. □

Lemma 3.3.25 *Let Σ be a regular surface, $F \in \mathrm{C}(\Sigma)$ be a given function, and $x_0 \in \Sigma$ be a given point. If $F(x_0) = 0$ and P_{d} is the double layer potential corresponding to F, then P_{d} is continuous in x_0.*

Proof We will use the ε-δ-criterion to prove the continuity. Let, therefore, $\varepsilon > 0$ be arbitrary. Moreover, we introduce the constants

$$C_1 := \sup_{x \in \mathbb{R}^3 \setminus \Sigma} \int_\Sigma \left| \frac{\partial}{\partial \nu(y)} \frac{1}{|x-y|} \right| \mathrm{d}\omega(y), \qquad C_2 := \sup_{x \in \Sigma} \int_\Sigma \left| \frac{\partial}{\partial \nu(y)} \frac{1}{|x-y|} \right| \mathrm{d}\omega(y).$$

The former is finite due to Lemma 3.3.23. For the latter, we find, in analogy to (3.59), a constant $C_3 \in \mathbb{R}^+$ such that

$$\begin{aligned}
\int_\Sigma \left| \nu(y) \cdot \frac{x-y}{|x-y|^3} \right| \mathrm{d}\omega(y) &= \int_\Sigma \frac{|\cos \sphericalangle(\nu(y), x-y)|}{|x-y|^2} \mathrm{d}\omega(y) \\
&\leq \int_\Sigma \frac{C_3\, |x-y|}{|x-y|^2} \mathrm{d}\omega(y) = C_3 \int_\Sigma \frac{1}{|x-y|} \mathrm{d}\omega(y).
\end{aligned}$$

This integral is finite, because the single layer potential exists everywhere and is bounded (see Theorem 3.3.14).

Since F is continuous, there exists an $\eta > 0$ such that, for all $y \in \Sigma \cap B_\eta(x_0) =: N_\eta$, we have $|F(y)| = |F(y) - F(x_0)| < \varepsilon[3 \max(C_1, C_2)]^{-1}$. Note that this η is actually uniform, because F is continuous on the compact domain Σ. As a consequence, the triangle inequality yields, if $x \in \overline{B_{\eta/2}(x_0)}$,

$$\begin{aligned}
|P_{\mathrm{d}}(x) - P_{\mathrm{d}}(x_0)| \leq &\int_{N_\eta} \left| \frac{\partial}{\partial \nu(y)} \frac{1}{|x-y|} \right| |F(y)|\, \mathrm{d}\omega(y) \\
&+ \int_{N_\eta} \left| \frac{\partial}{\partial \nu(y)} \frac{1}{|x_0-y|} \right| |F(y)|\, \mathrm{d}\omega(y) \\
&+ \int_{\Sigma \setminus N_\eta} \left| \frac{\partial}{\partial \nu(y)} \left(\frac{1}{|x-y|} - \frac{1}{|x_0-y|} \right) \right| |F(y)|\, \mathrm{d}\omega(y).
\end{aligned}$$

Due to our constructions, each of the first two integrals on the right-hand side is bounded from above by $\varepsilon/3$. Let us consider the latter as a function of $x \in \overline{B_{\eta/2}(x_0)}$. Since the integrand is non-singular, this function of x is continuous and, due to the compactness of the domain, also uniformly continuous. As a consequence, it converges uniformly to its value at

$x = x_0$ as $x \to x_0$, which is zero. Hence, there is a $\delta < \eta/2$ such that, for all $x \in B_\delta(x_0)$, the latter integral is less than $\varepsilon/3$. Hence, $|P_\mathrm{d}(x) - P_\mathrm{d}(x_0)| < 3\varepsilon/3 = \varepsilon$ for all $x \in B_\delta(x_0)$. □

Now we can prove an essential result about the solutions of the Dirichlet problems (see also Remark 3.3.20).

Theorem 3.3.26 *Let $F \in \mathrm{C}(\Sigma)$ be a given function (on the regular surface Σ with outer unit normal ν) and $U : \mathbb{R}^3 \to \mathbb{R}$ be the double layer potential P_d to F, that is,*

$$U(x) := \int_\Sigma F(y) \frac{\partial}{\partial \nu(y)} \frac{1}{|x-y|} \,\mathrm{d}\omega(y), \quad x \in \mathbb{R}^3.$$

Then the restrictions of U to Σ_{int} and Σ_{ext}, respectively, are continuously extendable to their closures $\overline{\Sigma_{\mathrm{int}}}$ and $\overline{\Sigma_{\mathrm{ext}}}$, respectively. Moreover, the limits

$$U_+(x) := \lim_{t\to 0+} U(x + t\nu(x)), \qquad U_-(x) := \lim_{t\to 0+} U(x - t\nu(x)), \qquad x \in \Sigma,$$

are uniform on Σ and yield continuous functions U_+ and U_-. They satisfy the Fredholm integral equations of the second kind:

$$U_+(x) = 2\pi F(x) + \int_\Sigma F(y) \frac{\partial}{\partial \nu(y)} \frac{1}{|x-y|} \,\mathrm{d}\omega(y), \quad x \in \Sigma, \tag{3.62}$$

$$U_-(x) = -2\pi F(x) + \int_\Sigma F(y) \frac{\partial}{\partial \nu(y)} \frac{1}{|x-y|} \,\mathrm{d}\omega(y), \quad x \in \Sigma. \tag{3.63}$$

Proof Let $x \in \Sigma$. Then, by using Theorem 3.3.21 and Lemma 3.3.25, we deduce that

$$\begin{aligned}
U_\pm(x) &= \lim_{t\to 0+} \left[\int_\Sigma (F(x) + F(y) - F(x)) \frac{\partial}{\partial \nu(y)} \frac{1}{|x \pm t\nu(x) - y|} \,\mathrm{d}\omega(y) \right] \\
&= \lim_{t\to 0+} \left[F(x) \int_\Sigma \frac{\partial}{\partial \nu(y)} \frac{1}{|x \pm t\nu(x) - y|} \,\mathrm{d}\omega(y) \right. \\
&\qquad \left. + \int_\Sigma (F(y) - F(x)) \frac{\partial}{\partial \nu(y)} \frac{1}{|x \pm t\nu(x) - y|} \,\mathrm{d}\omega(y) \right] \\
&= c_\pm F(x) + \int_\Sigma (F(y) - F(x)) \frac{\partial}{\partial \nu(y)} \frac{1}{|x-y|} \,\mathrm{d}\omega(y),
\end{aligned} \tag{3.64}$$

where

$$c_\pm = \begin{cases} 0 & \text{in the case ` + ',} \\ -4\pi & \text{in the case ` − '.} \end{cases}$$

Hence, using again Theorem 3.3.21, we get

$$U_\pm(x) = c_\pm F(x) + \int_\Sigma F(y) \frac{\partial}{\partial \nu(y)} \frac{1}{|x-y|} \,\mathrm{d}\omega(y) + 2\pi F(x).$$

This result corresponds to (3.63) and (3.62). Let us have a closer look at the convergence. For the first part in (3.64), the integrals are constant. This is obviously uniform. For the second integral, the convergence is a result of Lemma 3.3.25. A closer look at the proof reveals that this convergence is also uniform. This uniform convergence in combination with the continuity of U on $\mathbb{R}^3 \setminus \Sigma$ (see Theorem 3.3.15) implies the continuity of the extensions on $\overline{\Sigma_{\mathrm{int}}}$ and $\overline{\Sigma_{\mathrm{ext}}}$, respectively. □

Theorem 3.3.26 has brought up a Fredholm integral operator in the context of the Dirichlet problems. Though we still need some patience, we are not very far from the proof of the existence of the IDP and EDP solutions any more.

In view of Theorem 3.3.26, we study now the following integral equations (with $\widetilde{K} = -\frac{1}{2\pi}K$ or $\widetilde{K} = \frac{1}{2\pi}K$, we see that we have Fredholm integral equations of the second kind):

$$2\pi F(x) + \int_\Sigma K(x,y)F(y)\,\mathrm{d}\omega(y) = U(x), \quad x \in \Sigma,$$

$$-2\pi F(x) + \int_\Sigma K(x,y)F(y)\,\mathrm{d}\omega(y) = U(x), \quad x \in \Sigma,$$

where, in our particular case, $K(x,y) = \frac{\partial}{\partial\nu(y)}|x-y|^{-1}$, $x, y \in \Sigma$, $x \neq y$. Let us investigate what we can find out about such equations.

Lemma 3.3.27 *Let Σ be a regular surface and K be a continuous integral kernel of order $\alpha \in]0,2[$ which is defined for all $(x,y) \in \Sigma^2$ with $x \neq y$. Moreover, let $\mathcal{T}_K : \mathrm{L}^2(\Sigma) \to \mathrm{L}^2(\Sigma)$ be the corresponding Fredholm integral operator of the first kind. Then the following holds true:*

(a) $\mathcal{T}_K$ is a bounded operator.
(b) If $U : \Sigma \to \mathbb{R}$ is bounded, then $\mathcal{T}_K U$ is continuous on Σ.
(c) The Fredholm integral operator of the second kind $\mathcal{T}_K + \mathcal{I}$ has the following property: if $U \in \mathrm{L}^2(\Sigma)$ and $(\mathcal{T}_K + \mathcal{I})U \in \mathrm{C}(\Sigma)$, then $U \in \mathrm{C}(\Sigma)$.

Proof Part b needs to be proved separately.

(1) Part b:

Let $U : \Sigma \to \mathbb{R}$ be a bounded function, let $y_1 \in \Sigma$ be an arbitrary point, and let $\delta > 0$ be (for the moment) an arbitrary real number. Since K is a kernel of order $\alpha \in]0,2[$, we may write $K(y,x) = A(y,x)|x-y|^{-\alpha}$ for $x \neq y$, where A is bounded. Then

$$\begin{aligned}
&|(\mathcal{T}_K U)(y_1) - (\mathcal{T}_K U)(y_2)| \\
&= \left| \int_\Sigma (K(y_1,x) - K(y_2,x))\, U(x)\,\mathrm{d}\omega(x) \right| \\
&\leq \int_{\Sigma \cap B_\delta(y_1)} |K(y_1,x) - K(y_2,x)|\, |U(x)|\,\mathrm{d}\omega(x) \\
&\quad + \int_{\Sigma \setminus B_\delta(y_1)} |K(y_1,x) - K(y_2,x)|\, |U(x)|\,\mathrm{d}\omega(x) \\
&\leq \int_{\Sigma \cap B_\delta(y_1)} \left(\frac{|A(y_1,x)|}{|x-y_1|^\alpha} \,|\, U(x)| + \frac{|A(y_2,x)|}{|x-y_2|^\alpha} \,|\, U(x)| \right) \mathrm{d}\omega(x) \\
&\quad + \int_{\Sigma \setminus B_\delta(y_1)} |K(y_1,x) - K(y_2,x)|\, |U(x)|\,\mathrm{d}\omega(x),
\end{aligned}$$

where $y_2 \in B_\delta(y_1) \cap \Sigma$. Remember that A is bounded. Moreover, U is bounded such that there is a constant $C \in \mathbb{R}^+$ with

$$|(\mathcal{T}_K U)(y_1) - (\mathcal{T}_K U)(y_2)| \leq C \int_{\Sigma \cap B_\delta(y_1)} \left(\frac{1}{|x - y_1|^\alpha} + \frac{1}{|x - y_2|^\alpha} \right) \mathrm{d}\omega(x)$$
$$+ C \int_{\Sigma \setminus B_\delta(y_1)} |K(y_1, x) - K(y_2, x)| \, \mathrm{d}\omega(x). \quad (3.65)$$

The first integral can be estimated by using polar coordinates. It reveals that the whole term is of the order $\mathcal{O}(\delta^{2-\alpha})$. Since $\alpha < 2$, we can find, for given $\varepsilon > 0$, a sufficiently small $\delta > 0$ such that the first integral is bounded from above by $\frac{\varepsilon}{2C}$. For the second integral, we observe that, if (additionally) $y_2 \in \overline{B_{\delta/2}(y_1)}$ and $|x - y_1| \geq \delta$, then $|x - y_2| \geq |x - y_1| - |y_1 - y_2| \geq \delta - \frac{\delta}{2} = \frac{\delta}{2}$. Hence, the function

$$\overline{B_{\delta/2}(y_1) \cap \Sigma} \times (\Sigma \setminus B_\delta(y_1)) \ni (y, x) \mapsto K(y, x)$$

is continuous and, due to the compact domain, also uniformly continuous. Thus, for a given $\varepsilon > 0$, there exists $\delta_1 < \delta/2$ such that

$$|K(y_1, x) - K(y_2, x)| \leq \frac{\varepsilon}{2} \left(\int_\Sigma C \, \mathrm{d}\omega(x) \right)^{-1}$$

for all $x \in \Sigma \setminus B_\delta(y_1)$ and all $y_2 \in B_{\delta_1}(y_1) \cap \Sigma$. Consequently, for all $y_2 \in B_{\delta_1}(y_1) \cap \Sigma$, we have $|(\mathcal{T}_K U)(y_1) - (\mathcal{T}_K U)(y_2)| \leq \varepsilon$.

(2) Parts a and c:

Let $\varepsilon > 0$ be again an arbitrary real number. We construct a function $\phi \in \mathrm{C}(\Sigma \times \Sigma)$ with the following properties:

- $0 \leq \phi(x, y) \leq 1$ for all $x, y \in \Sigma$.
- $\phi(x, y) = 1$, if $|x - y| < \frac{\varepsilon}{2}$.
- $\phi(x, y) = 0$, if $|x - y| > \varepsilon$.

Such a function ϕ must exist. The 'transition zone' between $|x - y| = \frac{\varepsilon}{2}$ and $|x - y| = \varepsilon$ suffices to 'move' the values from 1 down to 0 in a continuous way.

We now set $K_0 := \phi K$ and $K_1 := (1 - \phi)K$, where $\mathcal{T}_{K_0}$ and $\mathcal{T}_{K_1}$ are the corresponding Fredholm integral operators of the first kind. The Cauchy–Schwarz inequality implies then

$$|(\mathcal{T}_{K_1} U)(y_1) - (\mathcal{T}_{K_1} U)(y_2)|^2 \leq \left(\int_\Sigma |K_1(y_1, x) - K_1(y_2, x)| \, |U(x)| \, \mathrm{d}\omega(x) \right)^2$$
$$\leq \int_\Sigma (K_1(y_1, x) - K_1(y_2, x))^2 \, \mathrm{d}\omega(x) \, \|U\|^2_{\mathrm{L}^2(\Sigma)}. \quad (3.66)$$

Similarly, as we argued earlier for the second integral in (3.65), the right-hand side of (3.66) converges to zero as $y_2 \to y_1$ such that $\mathcal{T}_{K_1} U$ is a continuous function. Hence, the function

$$F := (\mathcal{T}_K + \mathcal{I})U - \mathcal{T}_{K_1} U \quad (3.67)$$

is continuous if $(\mathcal{T}_K + \mathcal{I})U$ is continuous. Let us now look at the operator $\mathcal{T}_{K_0}$. With the Cauchy–Schwarz inequality, we obtain

$$\begin{aligned}
&\int_\Sigma \left| \phi(y,x) \frac{A(y,x)}{|y-x|^\alpha} \right| \,\mathrm{d}\omega(x) \\
&= \int_\Sigma \left(\frac{\phi(y,x)}{|y-x|^\alpha} \right)^{2/2} |A(y,x)| \,\mathrm{d}\omega(x) \\
&\le \left(\int_\Sigma \frac{\phi(y,x)}{|y-x|^\alpha} \,\mathrm{d}\omega(x) \right)^{1/2} \left(\int_\Sigma \frac{\phi(y,x)}{|y-x|^\alpha} (A(y,x))^2 \,\mathrm{d}\omega(x) \right)^{1/2} \\
&\le \|A\|_\infty \int_{\Sigma \cap B_\varepsilon(y)} \frac{1}{|y-x|^\alpha} \,\mathrm{d}\omega(x) \le C_1 \|A\|_\infty \,\varepsilon^{2-\alpha},
\end{aligned} \tag{3.68}$$

where $C_1 \in \mathbb{R}^+$ is a constant. By applying the Cauchy–Schwarz inequality in a similar manner again, we get

$$\begin{aligned}
|(\mathcal{T}_{K_0} U)(y)| &\le \int_\Sigma \left| \phi(y,x) \frac{A(y,x)}{|y-x|^\alpha} \right| |U(x)| \,\mathrm{d}\omega(x) \\
&\le \left(\int_\Sigma \left| \phi(y,x) \frac{A(y,x)}{|y-x|^\alpha} \right| \,\mathrm{d}\omega(x) \right)^{1/2} \\
&\quad \times \left(\int_\Sigma \left| \phi(y,x) \frac{A(y,x)}{|y-x|^\alpha} \right| (U(x))^2 \,\mathrm{d}\omega(x) \right)^{1/2}
\end{aligned}$$

such that Fubini's theorem and (3.68) (which also holds, if we integrate with respect to y) yield

$$\begin{aligned}
\left\| \mathcal{T}_{K_0} U \right\|^2_{\mathrm{L}^2(\Sigma)} &\le C_1 \|A\|_\infty \varepsilon^{2-\alpha} \int_\Sigma \int_\Sigma \left| \phi(y,x) \frac{A(y,x)}{|y-x|^\alpha} \right| (U(x))^2 \,\mathrm{d}\omega(x) \,\mathrm{d}\omega(y) \\
&\le \left(C_1 \|A\|_\infty \,\varepsilon^{2-\alpha} \right)^2 \int_\Sigma (U(x))^2 \,\mathrm{d}\omega(x).
\end{aligned}$$

This does not only show that $\mathcal{T}_{K_0}$ is bounded, it also shows that $\mathcal{T}_K$ is bounded, if we choose $\varepsilon > 2 \operatorname{diam} \Sigma$ such that $\phi \equiv 1$. Hence, there exists another constant $C_2 \in \mathbb{R}^+$ such that the operator norm of $\mathcal{T}_{K_0}$ can be estimated by $\|\mathcal{T}_{K_0}\|_{\mathcal{L}(\mathrm{L}^2(\Sigma), \mathrm{L}^2(\Sigma))} \le C_2 \,\varepsilon^{2-\alpha}$. As a consequence, we can choose $\varepsilon > 0$ sufficiently small such that the latter operator norm is less than 1. Hence, according to Theorem 2.5.51, $\mathcal{T}_{K_0} + \mathcal{I}$ is invertible and its inverse can be represented by the Neumann series $(\mathcal{T}_{K_0} + \mathcal{I})^{-1} = \sum_{n=0}^\infty (-\mathcal{T}_{K_0})^n$. Let now $(\mathcal{T}_K + \mathcal{I})U$ be continuous.

Due to part b of this theorem, we know that every $\left(-\mathcal{T}_{K_0}\right)^n F$ is continuous (remember that F is continuous, and continuous functions on the compact surface Σ are always bounded). Moreover, (3.67) and the construction of K_0 and K_1 imply that $F = (\mathcal{I} + \mathcal{T}_{K_0})U$ such that

$$U = (\mathcal{I} + \mathcal{T}_{K_0})^{-1} F = \sum_{n=0}^\infty \left(-\mathcal{T}_{K_0}\right)^n F. \tag{3.69}$$

For concluding that U is continuous, we need to know that the preceding series converges uniformly. For this purpose, we consider $\mathcal{T}_{K_0}$ as an operator from $\mathrm{L}^\infty(\Sigma)$ to $\mathrm{L}^\infty(\Sigma)$. With (3.68), we get that

$$\left|(\mathcal{T}_{K_0}G)(y)\right| = \left|\int_\Sigma K_0(y,x)G(x)\,\mathrm{d}\omega(x)\right| \leq \|G\|_\infty C_1 \|A\|_\infty \varepsilon^{2-\alpha}$$

for all $G \in \mathrm{L}^\infty(\Sigma)$. Hence, we might have to further decrease $\varepsilon > 0$ in order to eventually obtain that $\|\mathcal{T}_{K_0}\|_{\mathcal{L}(\mathrm{L}^\infty(\Sigma),\mathrm{L}^\infty(\Sigma))} < 1$. Thus, the series in (3.69) also converges in the $\|\cdot\|_\infty$-norm, that is, uniformly. □

Note that we can replace the kernel K by $\lambda^{-1}K$ for any $\lambda \in \mathbb{R}\setminus\{0\}$. Hence, part c of Lemma 3.3.27 is also valid for $\mathcal{T}_K + \lambda\mathcal{I}$ for any $\lambda \in \mathbb{R}\setminus\{0\}$.

Lemma 3.3.28 *Let Σ be a regular surface and K be an integral kernel on $\Sigma \times \Sigma \setminus \{(x,x) \mid x \in \Sigma\}$ of order $\alpha \in]0,2[$. Then the corresponding Fredholm integral operator of the first kind $\mathcal{T}_K : \mathrm{L}^2(\Sigma) \to \mathrm{L}^2(\Sigma)$ is compact.*

The proof of the compactness of $\mathcal{T}_K$ requires further functional analytic concepts. We omit it for this reason see Folland (1976, Proposition 3.12).

For the single layer potential, a result can be proved which is the counterpart to Theorem 3.3.26. We omit the proof here and refer to Folland (1976, pp. 173–177) and Walter (1971, pp. 89–93).

Theorem 3.3.29 *Let $F \in \mathrm{C}(\Sigma)$ be a given function (on the regular surface Σ with outer unit normal ν) and $U : \mathbb{R}^3 \to \mathbb{R}$ be (the single layer potential) given by $U(x) := \int_\Sigma F(y)\,|x-y|^{-1}\,\mathrm{d}\omega(y)$, $x \in \mathbb{R}^3$. Then the restrictions of U to $\overline{\Sigma_{\mathrm{int}}}$ and $\overline{\Sigma_{\mathrm{ext}}}$, respectively, are continuous there and continuously differentiable in Σ_{int} and Σ_{ext}, respectively. Moreover, the limits*

$$\partial_{\nu+}U(x) := \lim_{t\to 0+}\left[\nu(x)\cdot\nabla U(x+t\nu(x))\right], \quad x \in \Sigma,$$

$$\partial_{\nu-}U(x) := \lim_{t\to 0+}\left[\nu(x)\cdot\nabla U(x-t\nu(x))\right], \quad x \in \Sigma,$$

exist and are uniform such that $\partial_{\nu+}U$ and $\partial_{\nu-}U$ are continuous on Σ. Furthermore, they satisfy the Fredholm integral equations of the second kind:

$$\partial_{\nu+}U(x) = -2\pi F(x) + \int_\Sigma F(y)\frac{\partial}{\partial\nu(x)}\frac{1}{|x-y|}\,\mathrm{d}\omega(y), \quad x \in \Sigma, \tag{3.70}$$

$$\partial_{\nu-}U(x) = +2\pi F(x) + \int_\Sigma F(y)\frac{\partial}{\partial\nu(x)}\frac{1}{|x-y|}\,\mathrm{d}\omega(y), \quad x \in \Sigma. \tag{3.71}$$

Why is Lemma 3.3.27 helpful for our purposes? If we define the abbreviation $K(x,y) := \frac{\partial}{\partial\nu(y)}|x-y|^{-1}$ for $x, y \in \mathbb{R}^3$ with $x \neq y$, then the double layer potential $G := \mathcal{T}_K F$, $F \in \mathrm{C}(\Sigma)$, satisfies the integral equations

$$G_- = -2\pi F + \mathcal{T}_K F, \qquad G_+ = +2\pi F + \mathcal{T}_K F$$

according to Theorem 3.3.26. Moreover, G_- and G_+ are continuous on Σ. Lemma 3.3.27 now tells us that the continuity of G_- and G_+, respectively, implies that a solution F of each integral equation is continuous.

Let us now set $K^*(y,x) := K(x,y) = \frac{\partial}{\partial\nu(y)}|x-y|^{-1}$ for $x,y \in \mathbb{R}^3$ with $x \neq y$. Then we have, for $F_1, F_2 \in \mathrm{L}^2(\Sigma)$, due to Fubini's theorem,

$$\begin{aligned}\langle F_1, \mathcal{T}_K F_2\rangle_{\mathrm{L}^2(\Sigma)} &= \int_\Sigma F_1(x) \int_\Sigma F_2(y) \frac{\partial}{\partial\nu(y)} \frac{1}{|x-y|} \, \mathrm{d}\omega(y)\, \mathrm{d}\omega(x) \\ &= \int_\Sigma F_2(y) \int_\Sigma F_1(x) \frac{\partial}{\partial\nu(y)} \frac{1}{|x-y|} \, \mathrm{d}\omega(x)\, \mathrm{d}\omega(y) \\ &= \langle F_2, \mathcal{T}_{K^*} F_1\rangle_{\mathrm{L}^2(\Sigma)}.\end{aligned}$$

In other words, the adjoint operator of $\mathcal{T}_K$ is given by $\mathcal{T}_K^* = \mathcal{T}_{K^*}$. In this terminology and with $N(x,y) := |x-y|^{-1}$ for $x,y \in \mathbb{R}^3$ with $x \neq y$, the single layer potential $G := \mathcal{T}_N F$, $F \in \mathrm{C}(\Sigma)$, satisfies the integral equations

$$\partial_{\nu-} G = +2\pi F + \mathcal{T}_{K^*} F, \qquad \partial_{\nu+} G = -2\pi F + \mathcal{T}_{K^*} F,$$

according to Theorem 3.3.29. The consequences due to Lemma 3.3.27 are analogous to the previously discussed case of $\mathcal{T}_K$.

Definition 3.3.30 Let $K(x,y) := \frac{\partial}{\partial\nu(y)}|x-y|^{-1}$ and $K^*(x,y) := K(y,x)$ for $x,y \in \Sigma$ with $x \neq y$. For the following discussions, we define the spaces

$$\begin{aligned}\mathcal{V}_- &:= \left\{G \in \mathrm{L}^2(\Sigma) \,|\, (-2\pi\, \mathcal{I} + \mathcal{T}_K)G = 0\right\}, \\ \mathcal{V}_+ &:= \left\{G \in \mathrm{L}^2(\Sigma) \,|\, (+2\pi\, \mathcal{I} + \mathcal{T}_K)G = 0\right\}, \\ \mathcal{W}_- &:= \left\{G \in \mathrm{L}^2(\Sigma) \,|\, (-2\pi\, \mathcal{I} + \mathcal{T}_K^*)G = 0\right\}, \\ \mathcal{W}_+ &:= \left\{G \in \mathrm{L}^2(\Sigma) \,|\, (+2\pi\, \mathcal{I} + \mathcal{T}_K^*)G = 0\right\}\end{aligned}$$

as the null spaces of the operators which are obviously linked to the Dirichlet and the Neumann problems.

We have to further investigate these spaces here. Note that Lemma 3.3.27 implies that they are subspaces of $\mathrm{C}(\Sigma)$.

Lemma 3.3.31 *The spaces $\mathcal{V}_+$ and $\mathcal{W}_+$ are one-dimensional. Moreover, $\mathcal{V}_- = \mathcal{W}_- = \{0\}$.*

Proof If $G \equiv 1$, then Theorem 3.3.21 tells us that $\mathcal{T}_K G \equiv -2\pi$ such that $G \in \mathcal{V}_+$ (and therefore also every constant function). Hence, in combination with Theorem 2.5.78 (remember that $\mathcal{T}_K$ and $\mathcal{T}_K^*$ are compact; see Lemmata 3.3.24 and 3.3.28), we have $1 \leq \dim \mathcal{V}_+ = \dim \mathcal{W}_+$.

Let now P_s be the single layer potential to an arbitrary function $H \in \mathcal{W}_+$. According to Theorem 3.3.29, we have $\partial_{\nu-} P_\mathrm{s} = (2\pi\, \mathcal{I} + \mathcal{T}_K^*)H = 0$. Since a single layer potential is harmonic in $\mathbb{R}^3 \setminus \Sigma$ (see Theorem 3.3.14), our P_s solves the INP with vanishing boundary values. Hence, P_s is constant on Σ_int (see Theorem 3.3.5). Note that P_s linearly depends on H. Moreover, with (3.70) and (3.71), we see that $\partial_{\nu+} P_\mathrm{s} = \partial_{\nu+} P_\mathrm{s} - \partial_{\nu-} P_\mathrm{s} = -4\pi H$ such that $P_\mathrm{s} \equiv 0$, if and only if $H \equiv 0$. Thus, we have a one-to-one mapping between $H \in \mathcal{W}_+$ and the space of all constant functions. Consequently, $\dim \mathcal{W}_+ = \dim \mathcal{V}_+ = 1$.

From Theorem 2.5.78, we also have that $\dim \mathcal{V}_- = \dim \mathcal{W}_-$. Let now P_s be the single layer potential to a function $H \in \mathcal{W}_-$. Now, Theorem 3.3.29 tells us that $\partial_{\nu+} P_\mathrm{s} = (-2\pi \mathcal{I} + \mathcal{T}_{K^*}) H = 0$. However, now we refer to the uniqueness theorem for the ENP (P_s is regular at infinity; see Theorem 3.3.18), as shown in Theorem 3.3.8, and conclude that $P_\mathrm{s} = 0$ on Σ_ext. Note that (see Theorem 3.3.14) P_s is continuous on the entire $\mathbb{R}^3$ such that $P_\mathrm{s} = 0$ also on Σ. Moreover, P_s is harmonic in Σ_int, which implies that $P_\mathrm{s} = 0$ also on $\overline{\Sigma_\mathrm{int}}$. In analogy to the first part of this proof, we now have $H = 0$. □

Lemma 3.3.32 *The space* $\mathrm{L}^2(\Sigma)$ *can be decomposed and represented as follows:* $\mathrm{L}^2(\Sigma) = \mathcal{V}_+^\perp \oplus \mathcal{W}_+ = \mathcal{V}_-^\perp = \mathcal{W}_-^\perp$.

Proof The '−'-case is trivial. We only show the '+'-case. Let $G \in \mathcal{V}_+^\perp \cap \mathcal{W}_+$. Then $(2\pi\mathcal{I} + \mathcal{T}_K^*)G = 0$ and, due to Corollary 2.5.79, there is a solution H of $(2\pi\mathcal{I} + \mathcal{T}_K^*)H = -4\pi G$. Now we consult part c of Lemma 3.3.27, which tells us that G is continuous and, consequently, also H must be continuous.

Let further $P_{\mathrm{s},G}$ and $P_{\mathrm{s},H}$ be the single layer potentials corresponding to G and H, respectively. Hence, Theorem 3.3.29 implies that

$$\begin{aligned}
\partial_{\nu-} P_{\mathrm{s},G}(x) &= 2\pi G(x) + (\mathcal{T}_K^* G)(x) = 0,\\
\partial_{\nu-} P_{\mathrm{s},H}(x) &= 2\pi H(x) + (\mathcal{T}_K^* H)(x) = -4\pi G(x),\\
\partial_{\nu+} P_{\mathrm{s},G}(x) &= -2\pi G(x) + (\mathcal{T}_K^* G)(x) = -4\pi G(x) + 2\pi G(x) + (\mathcal{T}_K^* G)(x)\\
&= -4\pi G(x) = \partial_{\nu-} P_{\mathrm{s},H}(x).
\end{aligned}$$

Furthermore, by using the preceding result, Green's first and second identities (see Theorem 2.4.2), and the harmonicity of the single layer potential (see Theorem 3.3.14), we obtain

$$\begin{aligned}
\int_{\Sigma_\mathrm{ext}} |\nabla P_{\mathrm{s},G}(x)|^2 \,\mathrm{d}x &= \int_{\Sigma_\mathrm{ext}} P_{\mathrm{s},G}(x) \underbrace{\Delta P_{\mathrm{s},G}(x)}_{=0} + |\nabla P_{\mathrm{s},G}(x)|^2 \,\mathrm{d}x\\
&= -\int_\Sigma P_{\mathrm{s},G}(x) \partial_{\nu+} P_{\mathrm{s},G}(x) \,\mathrm{d}\omega(x)\\
&= -\int_\Sigma P_{\mathrm{s},G}(x) \underbrace{\partial_{\nu-} P_{\mathrm{s},H}(x)}_{=\partial_{\nu+} P_{\mathrm{s},G}(x)} - P_{\mathrm{s},H}(x) \underbrace{\partial_{\nu-} P_{\mathrm{s},G}(x)}_{=0} \,\mathrm{d}\omega(x)\\
&= -\int_{\Sigma_\mathrm{int}} P_{\mathrm{s},G}(x) \underbrace{\Delta P_{\mathrm{s},H}(x)}_{=0} - P_{\mathrm{s},H}(x) \underbrace{\Delta P_{\mathrm{s},G}(x)}_{=0} \,\mathrm{d}x = 0.
\end{aligned}$$

Note the similarity to the proof of Theorem 3.3.8, where we borrow here the justification for the use of Green's identities on Σ_ext. Finally, $P_{\mathrm{s},G}$ is constant in Σ_ext such that $G = -(4\pi)^{-1} \partial_{\nu+} P_{\mathrm{s},G} = 0$. Clearly (see Theorem 2.5.71), $\mathrm{L}^2(\Sigma) = \mathcal{V}_+^\perp \oplus \mathcal{V}_+$, where $\dim \mathcal{V}_+ = 1 = \dim \mathcal{W}_+$. Due to the trivial intersection, we get $\mathrm{L}^2(\Sigma) = \mathcal{V}_+^\perp \oplus \mathcal{W}_+$. □

Lemma 3.3.33 *The space* $\mathrm{L}^2(\Sigma)$ *can also be decomposed and represented as* $\mathrm{L}^2(\Sigma) = \mathcal{V}_+ \oplus (2\pi\mathcal{I} + \mathcal{T}_K)(\mathrm{L}^2(\Sigma)) = (-2\pi\mathcal{I} + \mathcal{T}_K)(\mathrm{L}^2(\Sigma))$.

Proof Let $G \in \mathcal{V}_+ \cap (2\pi\mathcal{I} + \mathcal{T}_K)(\mathrm{L}^2(\Sigma))$. Then, due to Lemma 3.3.32, we have unique functions $H_1 \in \mathcal{V}_+^\perp$ and $H_2 \in \mathcal{W}_+$ such that $G = H_1 + H_2$. Due to Corollary 2.5.79, the image of $2\pi\mathcal{I} + \mathcal{T}_K$ is the orthogonal complement of the null space $\mathcal{W}_+$ of $2\pi\mathcal{I} + \mathcal{T}_K^*$, that is, $\langle G, H_2\rangle_{\mathrm{L}^2(\Sigma)} = 0$.

Furthermore, obviously, $\langle G, H_1\rangle_{\mathrm{L}^2(\Sigma)} = 0$. This is only possible if $G = 0$. As in the previous proof, we can argue that $\mathrm{L}^2(\Sigma) = \mathcal{W}_+ \oplus \mathcal{W}_+^\perp$, where $\dim \mathcal{W}_+ = 1 = \dim \mathcal{V}_+$ and $\mathcal{W}_+^\perp$ is the image of $2\pi\mathcal{I} + \mathcal{T}_K$. Since $\mathcal{V}_+ \cap \mathcal{W}_+^\perp = \{0\}$, the proof is completed. The '$-$'-case is again trivial. □

After a lot of hard work, we are now able to prove the existence of the solutions of the boundary-value problems (bvps).

Theorem 3.3.34 (Existence of the BVP Solutions) *The boundary-value problems IDP, EDP, and ENP are all solvable. The INP is solvable if and only if the boundary-value function* $F \in \mathrm{C}(\Sigma)$ *satisfies* $\int_\Sigma F(x)\,\mathrm{d}\omega(x) = 0$.

Proof We consider the boundary-value problems one after the other.

(1) Let us start with the IDP:

Let, also for the other problems, $F \in \mathrm{C}(\Sigma)$ be given as the boundary value. We know from Lemma 3.3.33 that there exists $H \in \mathrm{L}^2(\Sigma)$ such that $F = -2\pi H + \mathcal{T}_K H$. Let us now look at the function

$$U(x) := \int_\Sigma K(x,y)H(y)\,\mathrm{d}\omega(y), \quad x \in \overline{\Sigma_{\mathrm{int}}}.$$

It is harmonic (as a double layer potential) in Σ_{int} (see Theorem 3.3.15) and its boundary values (see Theorem 3.3.26) are $-2\pi H + \mathcal{T}_K H = F$, where, due to Lemma 3.3.27, H is continuous.

The function U is the solution we were looking for. Note that U is also in $\mathrm{C}^{(2)}(\Sigma_{\mathrm{int}}) \cap \mathrm{C}(\overline{\Sigma_{\mathrm{int}}})$, as we have seen before (we skip this point for the other problems, because it can be seen analogously).

(2) Now it is the turn of the EDP:

Lemma 3.3.33 now yields the existence of $H_1, H_2 \in \mathrm{L}^2(\Sigma)$ such that $H_1 \in \mathcal{V}_+$ and $F = H_1 + 2\pi H_2 + \mathcal{T}_K H_2$. We now consider the double layer potential $P_{\mathrm{d}}(x) := \int_\Sigma K(x,y)H_2(y)\,\mathrm{d}\omega(y)$, $x \in \overline{\Sigma_{\mathrm{ext}}}$. Again, P_{d} is harmonic (see Theorem 3.3.15) and its boundary values (see Theorem 3.3.26) are $2\pi H_2 + \mathcal{T}_K H_2$, where H_1 is continuous such that $F - H_1$ is continuous and then H_2 is continuous (see Lemma 3.3.27). Moreover, due to Theorem 3.3.18, P_{d} is regular at infinity.

Concerning H_1, Lemma 3.3.31 and its proof tell us that H_1 is constant and, for every constant such as H_1, there is $H_3 \in \mathcal{W}_+$ such that the single layer potential P_{s} associated to H_3 equals H_1 on Σ_{int}. Furthermore, due to Theorems 3.3.14 and 3.3.18, P_{s} is harmonic in Σ_{ext}, it is regular at infinity, and P_{s} is continuous on the whole $\mathbb{R}^3$, that is, it equals H_1 on $\overline{\Sigma_{\mathrm{int}}}$ and, in particular, on Σ.

Let now $U := P_s + P_d$ on $\overline{\Sigma_{\text{ext}}}$. Then U is harmonic in Σ_{ext}, is regular at infinity, and satisfies $U_+ = H_1 + 2\pi H_2 + \mathcal{T}_K H_2 = F$ on Σ.

(3) It is time for the INP:

The vanishing surface integral of F is equivalent to $F \in \mathcal{V}_+^{\perp}$. Since the orthogonal complement of the null space $\mathcal{V}_+$ of $2\pi\mathcal{I} + \mathcal{T}_K$ is the image of $2\pi\mathcal{I} + \mathcal{T}_K^*$, as a consequence of Theorems 2.5.72 and 2.5.78, there is an $H \in \mathrm{L}^2(\Sigma)$ such that $F = 2\pi H + \mathcal{T}_K^* H$ if and only if $F \in \mathcal{V}_+^{\perp}$, where H is again continuous because of Lemma 3.3.27. Now we consider the single layer potential $U(x) := \int_\Sigma H(y)\,|x-y|^{-1}\,\mathrm{d}\omega(y)$, $x \in \overline{\Sigma_{\text{int}}}$, which is harmonic on Σ_{int} (Theorem 3.3.14). Moreover, $\partial_{\nu-} U = F \Leftrightarrow 2\pi H + \mathcal{T}_K^* H = F$ due to Theorem 3.3.29.

(4) Finally, the ENP:

We know that $\mathcal{V}_- = \{0\}$ (see Lemma 3.3.31). Hence (see again Theorems 2.5.72 and 2.5.78), we have $\mathrm{L}^2(\Sigma) = (-2\pi\mathcal{I} + \mathcal{T}_K^*)(\mathrm{L}^2(\Sigma))$. Let now $H \in \mathrm{L}^2(\Sigma)$ solve $F = -2\pi H + \mathcal{T}_K^* H$ (which implies again that $H \in \mathrm{C}(\Sigma)$). Then the single layer potential $U(x) := \int_\Sigma H(y)\,|x-y|^{-1}\,\mathrm{d}\omega(y)$, $x \in \overline{\Sigma_{\text{ext}}}$, is harmonic in Σ_{ext}, regular at infinity, and has the property $\partial_{\nu+} U = -2\pi H + \mathcal{T}_K^* H = F$ on Σ. We are done! □

And now let us remember our discussions after Theorem 3.3.12. This means that we have finally also proved Theorem 3.3.12.

For reasons of completeness, let us also mention that Green's function is symmetric (and, consequently, also R), as we will see now.

Let us take two points $x, z \in D$ with $x \neq z$ and let us choose $\varepsilon > 0$ sufficiently small such that $B_\varepsilon(x) \cap B_\varepsilon(z) = \emptyset$, $B_\varepsilon(x) \subset D$, and $B_\varepsilon(z) \subset D$. We apply now Green's second identity (see Theorem 2.4.2) to the functions $y \mapsto G(x, y)$ and $y \mapsto G(z, y)$ and set $D_\varepsilon := D \setminus (B_\varepsilon(x) \cup B_\varepsilon(z))$. This yields

$$
\begin{aligned}
&\int_{D_\varepsilon} G(x,y)\,\underbrace{\Delta_y G(z,y)}_{=0} - G(z,y)\,\underbrace{\Delta_y G(x,y)}_{=0}\,\mathrm{d}y \qquad (3.72)\\
&= \int_\Sigma \underbrace{G(x,y)}_{=0}\,\frac{\partial}{\partial\nu(y)}\,G(z,y) - \underbrace{G(z,y)}_{=0}\,\frac{\partial}{\partial\nu(y)}\,G(x,y)\,\mathrm{d}\omega(y)\\
&\quad + \int_{S_\varepsilon(x)\cup S_\varepsilon(z)} G(x,y)\frac{\partial}{\partial n(y)}\,G(z,y) - G(z,y)\frac{\partial}{\partial n(y)}\,G(x,y)\,\mathrm{d}\omega(y),
\end{aligned}
$$

where ν is the outer unit normal to Σ and n is the outer unit normal to the involved spheres. For symmetry reasons, it suffices to consider here only the integral over $S_\varepsilon(x)$ and then conclude on the other integral by analogy.

On $S_\varepsilon(x)$, we have $n(y) = \varepsilon^{-1}(x - y)$. Hence, Remark 3.3.20 yields

$$
\begin{aligned}
&\int_{S_\varepsilon(x)} G(x,y)\,\frac{\partial}{\partial n(y)}\,G(z,y) - G(z,y)\,\frac{\partial}{\partial n(y)}\,G(x,y)\,\mathrm{d}\omega(y)\\
&= \int_{S_\varepsilon(x)} \left(\frac{1}{4\pi\varepsilon} + R(x,y)\right)\frac{\partial}{\partial n(y)}\left(\frac{1}{4\pi|z-y|} + R(z,y)\right)\,\mathrm{d}\omega(y)
\end{aligned}
$$

$$
\begin{aligned}
&-\int_{S_\varepsilon(x)} \left(\frac{1}{4\pi |z-y|} + R(z,y)\right) \frac{\partial}{\partial n(y)} \left(\frac{1}{4\pi |x-y|} + R(x,y)\right) \mathrm{d}\omega(y) \\
&= \int_{S_\varepsilon(x)} R(x,y) \frac{\partial}{\partial n(y)} \left(\frac{1}{4\pi |z-y|} + R(z,y)\right) \mathrm{d}\omega(y) \\
&\quad + \frac{1}{4\pi\varepsilon} \int_{S_\varepsilon(x)} \frac{\partial}{\partial n(y)} R(z,y)\, \mathrm{d}\omega(y) - \frac{1}{4\pi\varepsilon^2} \int_{S_\varepsilon(x)} \underbrace{\frac{1}{4\pi |z-y|} + R(z,y)}_{=G(z,y)} \mathrm{d}\omega(y) \\
&\quad - \int_{S_\varepsilon(x)} \left(\frac{1}{4\pi |z-y|} + R(z,y)\right) \frac{\partial}{\partial n(y)} R(x,y)\, \mathrm{d}\omega(y),
\end{aligned}
\tag{3.73}
$$

since

$$
\frac{\partial}{\partial n(y)} \frac{1}{|x-y|} = \frac{x-y}{\varepsilon} \cdot \frac{x-y}{|x-y|^3} = \frac{1}{\varepsilon^2}
$$

for $y \in S_\varepsilon(x)$. Let us have a look at the integrals in (3.73). All integrands are continuous on $\overline{B_\varepsilon(x)}$. Hence, each integral can be estimated by the maximum of its integrand times $4\pi\varepsilon^2$. Hence, in the limit $\varepsilon \to 0+$, only

$$
-\lim_{\varepsilon\to 0+} \left(\frac{1}{4\pi\varepsilon^2} \int_{S_\varepsilon(x)} G(z,y)\, \mathrm{d}\omega(y)\right) = -G(z,x)
$$

is left, where the limit is a consequence of the continuity of the integrand. If we proceed in an analogous manner with the integral over $S_\varepsilon(z)$, then we arrive at the limit

$$
-\lim_{\varepsilon\to 0+} \left(\frac{1}{4\pi\varepsilon^2} \int_{S_\varepsilon(z)} G(x,y)\, \mathrm{d}\omega(y)\right) = -G(x,z).
$$

Hence, the limit $\varepsilon \to 0+$ in (3.72) yields $0 = -G(z,x) + G(x,z)$.

Theorem 3.3.35 (Symmetry of Green's function) *If Σ is a regular surface and G is Green's function for Δ and $D := \Sigma_{\mathrm{int}}$, then $G(x,y) = G(y,x)$ for all $x,y \in D$ with $x \neq y$.*

Hence, $\Delta_x R(x,y) = 0$ for all $x,y \in D$.

The 'catch' of Theorem 3.3.12 is the existence of Green's function. Fortunately, there is a helpful but theoretical result, which will not be proved here (see, e.g., Folland, 1976, pp. 110, 343).

Theorem 3.3.36 (Existence of Green's Function) *Let $D = \Sigma_{\mathrm{int}}$ be the interior of a regular surface Σ with $\mathrm{C}^{(\infty)}$-parameterization. Then Green's function for Δ and D exists. Moreover, for every $x \in D$, the function $G(x,\cdot)$ is infinitely differentiable on $\overline{D} \setminus \{x\}$.*

The more tricky 'catch' is the question: how do we get G? As an example, we consider a very simple and geomathematically interesting $\mathrm{C}^{(\infty)}$-surface: the sphere $\Sigma = S_R(0)$; see Walter (1971, pp. 41–45).

Theorem 3.3.37 *The Green's function for the sphere* $\Sigma = S_R(0)$ *equals*

$$G(x,y) = \begin{cases} \frac{1}{4\pi\,|x-y|} - \frac{R}{4\pi\,|x|}\left|\frac{R^2}{|x|^2}x - y\right|^{-1}, & \text{if } x \neq 0, \\ \frac{1}{4\pi\,|y|} - \frac{1}{4\pi R}, & \text{if } x = 0, \end{cases} \tag{3.74}$$

$x \in B_R(0)$, $y \in \overline{B_R(0)}$, $x \neq y$.

Proof Let $D = \Sigma_{\mathrm{int}}$. We have

$$\left|\frac{R^2}{|x|^2}x\right| \cdot \frac{1}{R} = \frac{R}{|x|} \tag{3.75}$$

such that $\frac{R^2}{|x|^2}x$ is the mirror point of x with respect to $S_R(0)$. Hence, $\tilde{R}(x,\cdot) := G(x,\cdot) - (4\pi\,|x-\cdot|)^{-1}$ is in $\mathrm{C}^{(1)}(\overline{D}) \cap \mathrm{C}^{(2)}(D)$ (even in $\mathrm{C}^{(\infty)}(D)$, cf. Theorem 3.3.36).

The mirroring corresponds to the Kelvin transform (see Definition 3.2.1 and Theorem 3.2.2). Since a function $F(\cdot)$ is harmonic if and only if $F(R\cdot)$ is harmonic, it does not matter if we mirror at $\Omega = S_1(0)$ or at $S_R(0)$. Hence, $\Delta_y \tilde{R}(x,y) = 0$, if $x, y \in D$, because $y \mapsto |x-y|^{-1}$ is harmonic for $x \neq y$ (see Corollary 3.1.6).

Moreover, if $x \in D$ and $y \in \Sigma$ ($\Leftrightarrow |y| = R$), then

$$\begin{aligned} \left(\frac{R}{|x|}\left|\frac{R^2}{|x|^2}x - y\right|^{-1}\right)^2 &= \frac{R^2}{|x|^2}\left(\frac{R^4}{|x|^4}|x|^2 - 2\frac{R^2}{|x|^2}x\cdot y + |y|^2\right)^{-1} \\ &= \left(R^2 - 2x\cdot y + |x|^2\right)^{-1} = \frac{1}{|x-y|^2} \end{aligned} \tag{3.76}$$

and consequently $G(x,y) = 0$. Hence, all conditions of Definition 3.3.10 are satisfied (the case $x = 0$ is easy). □

Let us have a look at Theorem 3.3.12: if $F \in \mathrm{C}(S_R(0))$ is given, then the unique solution U of the corresponding IDP is given by

$$U(x) = -\int_{S_R(0)} F(y)\,\frac{\partial}{\partial\nu(y)}\,G(x,y)\,\mathrm{d}\omega(y), \quad |x| < R.$$

Let us now calculate the normal derivative of G. Since $\nu(y) = y/R$ for all $y \in S_R(0)$, we have (with (3.74) and (3.76))

$$\begin{aligned} \frac{\partial}{\partial\nu(y)}G(x,y) &= \frac{1}{4\pi}\frac{y}{R}\cdot\left(\frac{x-y}{|x-y|^3} - \frac{R}{|x|}\frac{\frac{R^2}{|x|^2}x - y}{\left|\frac{R^2}{|x|^2}x - y\right|^3}\right) \\ &= \frac{1}{4\pi}\frac{y}{R}\cdot\left(\frac{x-y}{|x-y|^3} - \frac{|x|^2}{R^2}\frac{\frac{R^2}{|x|^2}x - y}{|x-y|^3}\right) = \frac{1}{4\pi}\frac{|x|^2 - R^2}{R|x-y|^3} \end{aligned}$$

for $x \neq 0$ and, for $x = 0$,

$$\frac{\partial}{\partial\nu(y)}G(x,y) = \frac{1}{4\pi}\frac{y}{R}\cdot\left(-\frac{y}{|y|^3}\right) = -\frac{1}{4\pi R^2} = \frac{1}{4\pi}\frac{|x|^2 - R^2}{R|x-y|^3}.$$

If we write now $x = r\xi$ and $y = R\eta$ with $r = |x|$ and $\xi, \eta \in \Omega$, then

$$\frac{\partial}{\partial \nu(y)} G(x, y) = \frac{1}{4\pi} \frac{r^2 - R^2}{R(r^2 + R^2 - 2rR\xi \cdot \eta)^{3/2}}.$$

Let us summarize this in a theorem.

Theorem 3.3.38 (Poisson Integral Formula [IDP]) *Let a function $F \in \mathrm{C}(S_R(0))$ be given, where $R > 0$. Then the unique solution of the corresponding IDP is given, for all $x \in B_R(0)$, where $x = r\xi$ and $\xi \in \Omega$, by*

$$\begin{aligned} U(x) &= \int_{S_R(0)} F(y) \frac{R^2 - |x|^2}{4\pi R |x - y|^3} \, \mathrm{d}\omega(y) \\ &= \int_{\Omega} F(R\eta) \frac{R^2 - r^2}{4\pi (r^2 + R^2 - 2rR\xi \cdot \eta)^{3/2}} R \, \mathrm{d}\omega(\eta). \end{aligned}$$

Definition 3.3.39 The kernel P with

$$P(x, y) := \frac{R^2 - |x|^2}{4\pi R |x - y|^3}$$

for $x \in B_R(0)$, $y \in \overline{B_R(0)}$, and $x \neq y$ is called the **Poisson kernel** or the **Abel–Poisson kernel**.

Can we get such a formula also for the EDP? Indeed, the Kelvin transform (see Definition 3.2.1 and Theorem 3.2.2) shows us how to get it: let $x \in \mathbb{R}^3 \setminus \overline{B_R(0)}$ and remember (3.75) for the more general mirroring (the substitution $x = R\tilde{x}$ for $\tilde{x} \in B_1(0)$ for transferring the Kelvin transform from $S_1(0)$ to $S_R(0)$ leads to $F^*(x) = R|x|^{-1} F(R^2|x|^{-2}x)$ as the transformed function with respect to $S_R(0)$; see also Kellogg, 1967, pp. 231–232) at $S_R(0)$. Then Theorem 3.3.38 and (3.76) yield

$$\begin{aligned} U^*(x) &= \frac{R}{|x|} \int_{S_R(0)} F(y) \frac{R^2 - \left| \frac{R^2}{|x|^2} x \right|^2}{4\pi R \left| \frac{R^2}{|x|^2} x - y \right|^3} \, \mathrm{d}\omega(y) \\ &= \frac{R}{|x|} \int_{S_R(0)} F(y) \frac{R^2 - \frac{R^4}{|x|^2}}{4\pi R |x - y|^3} \frac{|x|^3}{R^3} \, \mathrm{d}\omega(y) \\ &= \int_{S_R(0)} F(y) \frac{|x|^2 - R^2}{4\pi R |x - y|^3} \, \mathrm{d}\omega(y). \end{aligned}$$

Moreover, if (x_n) with $|x_n| > R$ for all n converges to $x \in S_R(0)$, then $x_n^* := (R/|x_n|)^2 x_n$ converges to x as well. Hence,

$$\lim_{n \to \infty} U^*(x_n) = \lim_{n \to \infty} \left(\frac{R}{|x_n|} U(x_n^*) \right) = F(x).$$

We omit here to verify that U^* is regular at infinity.

Theorem 3.3.40 (Poisson Integral Formula [EDP]) *Let a function $F \in \mathrm{C}(S_R(0))$ be given with $R > 0$. Then the unique solution of the corresponding EDP is given, for all $x \in \mathbb{R}^3 \setminus \overline{B_R(0)}$, where $x = r\xi$ and $\xi \in \Omega$, by*

$$U(x) = \int_{S_R(0)} F(y) \frac{|x|^2 - R^2}{4\pi R|x-y|^3} \, \mathrm{d}\omega(y)$$
$$= \int_\Omega F(R\eta) \frac{r^2 - R^2}{4\pi (r^2 + R^2 - 2rR\xi \cdot \eta)^{3/2}} R \, \mathrm{d}\omega(\eta).$$

Note the nice symmetry between the Poisson integral formulae for the IDP and the EDP.

The Poisson integral formula is not only helpful for solving the IDP or the EDP, if Σ is a sphere, it can also be used to prove some very fundamental properties of harmonic functions (see also Walter, 1971, pp. 45–47), as we will do now.

Theorem 3.3.41 *Every harmonic function on an open domain is infinitely differentiable.*

Proof Let $D \subset \mathbb{R}^3$ be open and let $U \in \mathrm{C}^{(2)}(D)$ satisfy $\Delta U = 0$ in D. Moreover, let $x_0 \in D$ be an arbitrary point. Since D is open, there exists a ball $B_{2R}(x_0) \subset D$. Let $B := B_R(x_0)$, then $U \in \mathrm{C}^{(2)}(\overline{B})$ and $\Delta U = 0$ in B. Hence, the Poisson integral formula for the IDP (see Theorem 3.3.38) yields (for the equally harmonic function $x \mapsto U(x_0 + x)$ on $B_R(0)$)

$$U(x_0 + x) = \frac{R^2 - |x|^2}{4\pi R} \int_{S_R(0)} \frac{U(x_0 + y)}{|x-y|^3} \, \mathrm{d}\omega(y), \quad x \in B_R(0).$$

This integrand is infinitely differentiable with respect to $x \in B_R(0)$, since $|y| = R$. The rest of the proof is analogous to the proof that Newton's gravitational potential is infinitely differentiable outside the domain of integration (see Theorem 3.1.5). □

Lemma 3.3.42 *For all $s \in]-1, 1[$ and all $n \in \mathbb{N}_0$, we have*

$$(1+s)^{-n/2} = \sum_{k=0}^{\infty} \binom{-n/2}{k} s^k, \tag{3.77}$$

where, here, $\binom{-n/2}{0} := 1$ and, else, $\binom{-n/2}{k} := (k!)^{-1} \prod_{j=0}^{k-1}(-n/2 - j)$.

Proof The series in (3.77) is actually a Taylor series. By setting $y(s) := (1+s)^{-n/2}$, we get $y(0) = 1$ and

$$y^{(k)}(s) = \left[\left(-\frac{n}{2}\right) \cdots \left(-\frac{n}{2} - k + 1\right)\right] (1+s)^{-k-n/2} \quad \text{for } k \geq 1,$$

which can be easily verified by induction. Hence,

$$T(s) := 1 + \sum_{k=1}^{\infty} \left[\left(-\frac{n}{2}\right) \cdots \left(-\frac{n}{2} - k + 1\right)\right] \frac{1}{k!} s^k = \sum_{k=0}^{\infty} \binom{-n/2}{k} s^k$$

is the corresponding Taylor series, which converges for $|s| < 1$, because

$$\left|\binom{-n/2}{k}\right| = \frac{1}{k!}\prod_{j=0}^{k-1}\frac{n+2j}{2} = \prod_{j=0}^{k-1}\frac{n+2j}{2(j+1)} \qquad \text{for } k \geq 1$$

such that $\left|\binom{-n/2}{k}\binom{-n/2}{k+1}^{-1}\right| = 2(k+1)(n+2k)^{-1} \to 1$ as $k \to \infty$. Furthermore, the function y satisfies the ordinary differential equation

$$y'(s) = -\frac{n}{2(1+s)}\,y(s), \qquad s \in]-1,1[, \tag{3.78}$$

with $y(0) = 1$. We will verify now that T also satisfies this first-order equation. By differentiating the Taylor series, we obtain

$$T'(s) = \sum_{k=1}^{\infty}\binom{-n/2}{k}ks^{k-1} = \sum_{k=0}^{\infty}\binom{-n/2}{k+1}(k+1)s^k.$$

We now use both of the preceding representations and, additionally, separate the summand with the smallest k. This yields

$$\begin{aligned}
-\frac{2}{n}(1+s)T'(s) &= 1 + s + \sum_{k=1}^{\infty}\left[\left(-\frac{n}{2}-1\right)\cdots\left(-\frac{n}{2}-k\right)\right]\frac{1}{k!}s^k \\
&\quad + \sum_{k=2}^{\infty}\left[\left(-\frac{n}{2}-1\right)\cdots\left(-\frac{n}{2}-k+1\right)\right]\frac{1}{(k-1)!}s^k \\
&= 1 + s + \left(-\frac{n}{2}-1\right)s \\
&\quad + \sum_{k=2}^{\infty}\left[\left(-\frac{n}{2}-1\right)\cdots\left(-\frac{n}{2}-k+1\right)\right]\frac{-n-2k+2k}{2k!}s^k \\
&= T(s).
\end{aligned}$$

Furthermore, $T(0) = 1$ obviously holds true. Let us now choose a fixed $\sigma \in]0,1[$ and denote the right-hand side of (3.78) by

$$f(s,\eta) := -\frac{n}{2(1+s)}\eta, \quad s \in [-\sigma,\sigma],\ \eta \in \mathbb{R}.$$

Then f is uniformly Lipschitz continuous in η in the sense that

$$|f(s,\eta_1) - f(s,\eta_2)| = \left|\frac{n}{2(1+s)}(\eta_1 - \eta_2)\right| \leq \frac{n}{2(1-\sigma)}|\eta_1 - \eta_2|$$

for all $s \in [-\sigma,\sigma]$ and all $\eta_1, \eta_2 \in \mathbb{R}$. Hence, the Picard–Lindelöf theorem, (see, for example, Heuser, 1991, pp. 139–141, or Roberts, 2010, pp. 46–47), implies that there is only one solution $y = T$ of (3.78) with $y(0) = 1$. Since σ was arbitrary in $]0,1[$, we have $y(s) = T(s)$ for all $s \in]-1,1[$. □

Theorem 3.3.43 *Every harmonic function U on an open domain D is analytic, that is, for every $x_0 \in D$, there is a ball $B_R(x_0) \subset D$, where U can be expanded into a (unique) power series with centre x_0.*

Proof Note that the uniqueness of a power series (with fixed centre x_0) is a basic result of analysis. Furthermore, we choose now an arbitrary point $x_0 \in D$ and a ball $B_R(x_0) \subset D$, which exists because D is open.

Let us first observe that

$$\frac{r^2 + R^2 - 2rR\xi \cdot \eta}{R^2} = 1 + \frac{r^2}{R^2} - 2\frac{r}{R}\xi \cdot \eta,$$

where $\xi, \eta \in \Omega$. If $0 \le r \le R/3$, then $|r^2/R^2 - 2(r/R)\xi \cdot \eta| \le 1/9 + 2/3 = 7/9$. Hence, Lemma 3.3.42 yields, for $|x| = r \le R/3$, $x = r\xi$, $y = R\eta$, and $\xi, \eta \in \Omega$, that the Abel–Poisson kernel can be represented as

$$\begin{aligned} P(x,y) &= \frac{R^2 - r^2}{4\pi R^4}\left(\frac{R^2}{r^2 + R^2 - 2rR\xi \cdot \eta}\right)^{3/2} \\ &= \frac{R^2 - r^2}{4\pi R^4}\sum_{k=0}^{\infty}\binom{-3/2}{k}\left(\frac{r^2 - 2rR\xi \cdot \eta}{R^2}\right)^k, \end{aligned}$$

where (as is usual for a power series) the convergence is uniform and absolute. Hence, the Poisson integral formula (Theorem 3.3.38) and the binomial theorem imply that

$$\begin{aligned} U(x_0 + x) &= \frac{R^2 - r^2}{4\pi R^2}\int_\Omega U(x_0 + R\eta)\sum_{k=0}^{\infty}\binom{-3/2}{k}\left(\frac{r^2 - 2rR\xi \cdot \eta}{R^2}\right)^k \mathrm{d}\omega(\eta) \\ &= \frac{R^2 - r^2}{4\pi R^2}\sum_{k=0}^{\infty}\binom{-3/2}{k}R^{-2k} \\ &\quad \times \int_\Omega U(x_0 + R\eta)\sum_{j=0}^{k}\binom{k}{j}r^{2j}(-2rR\xi \cdot \eta)^{k-j}\,\mathrm{d}\omega(\eta). \end{aligned}$$

We should observe here that r^2, r^{2j}, and $(r\xi \cdot \eta)^{k-j}$ are all polynomials in x_1, x_2, x_3. Hence, a rearrangement of the preceding series (which is possible due to the absolute convergence, as discussed previously) turns the latter result into a power series of $U(z)$ with $z = x_0 + x \in B_{R/3}(x_0)$. □

Note that the analytic functions constitute a proper subspace of the space of all infinitely differentiable functions.

There is just one more thing. A power series in $\mathbb{R}^3$ has the form

$$U(z) = \sum_{n=0}^{\infty}\overbrace{\sum_{\substack{\alpha \in \mathbb{N}_0^3 \\ |\alpha| = n}} C_\alpha (z - y)^\alpha}^{=:U_n(z-y)},$$

where $\alpha \in \mathbb{N}_0^3$ is a multi-index with $|\alpha| := \sum_{i=1}^3 \alpha_i$ and $x^\alpha := x_1^{\alpha_1}x_2^{\alpha_2}x_3^{\alpha_3}$ for $x \in \mathbb{R}^3$. Let us have a look at the inner sum. Obviously,

$$U_n(\lambda x) = \sum_{\substack{\alpha \in \mathbb{N}_0^3 \\ |\alpha| = n}} C_\alpha \lambda^{\alpha_1 + \alpha_2 + \alpha_3} x_1^{\alpha_1}x_2^{\alpha_2}x_3^{\alpha_3} = \lambda^n U_n(x)$$

for all $x \in \mathbb{R}^3$ and all $\lambda \in \mathbb{R}$.

Definition 3.3.44 A polynomial $P\colon \mathbb{R}^d \to \mathbb{R}$ is called **homogeneous** of degree $n \in \mathbb{N}_0$, if $P(\lambda x) = \lambda^n P(x)$ for all $x \in \mathbb{R}^d$ and all $\lambda \in \mathbb{R}$. The space of all such polynomials is denoted by $\mathrm{Hom}_n(\mathbb{R}^d)$.

The following theorem is easy to verify (we have already done a part of the work).

Theorem 3.3.45 *Let $d \in \mathbb{N}$ and $n \in \mathbb{N}_0$ be arbitrary. Then the following holds true:*

(a) $\mathrm{Hom}_n(\mathbb{R}^d)$ *is a vector space.*
(b) A polynomial $P\colon \mathbb{R}^d \to \mathbb{R}$ satisfies $P \in \mathrm{Hom}_n(\mathbb{R}^d)$ if and only if there exist coefficients $C_\alpha \in \mathbb{R}$ such that $P(x) = \sum_{\alpha\in\mathbb{N}_0^d,\, |\alpha|=n} C_\alpha x^\alpha$ for all $x \in \mathbb{R}^d$. The C_α in this representation are uniquely given by P.

So, we learned that every harmonic function U can be locally represented as $U(z) = \sum_{n=0}^\infty U_n(z-y)$, where U_n is a homogeneous polynomial of degree n. Hence, if we move the coordinate system such that $y = 0$, then $U(\lambda z) = \sum_{n=0}^\infty U_n(\lambda z) = \sum_{n=0}^\infty \lambda^n\, U_n(z)$ for all z in a neighourhood of 0 and all $\lambda \in [-1,1]$. Since we still have a power series here, we are allowed to do the following: $0 = \Delta_z\, U(\lambda z) = \sum_{n=0}^\infty \lambda^n \Delta_z\, U_n(z)$. Since a power series is unique (for a given centre), this power series of the zero function (with λ as the variable) is only possible if $\Delta_z\, U_n(z) = 0$ for all z and all n. Remember that U_n is a polynomial. If $\Delta\, U_n$ vanishes inside a ball, then $\Delta\, U_n = 0$ everywhere in $\mathbb{R}^3$.

Theorem 3.3.46 *If U is harmonic in an open set $D \subset \mathbb{R}^3$ and $x_0 \in D$ is arbitrary, then U can be represented, in a neighbourhood of x_0, as a series of homogeneous harmonic polynomials.*

Do not forget this last result of this section. It will be very helpful later. Let us already introduce a notation for this purpose.

Definition 3.3.47 The space

$$\mathrm{Harm}_n\left(\mathbb{R}^d\right) := \left\{P \in \mathrm{Hom}_n\left(\mathbb{R}^d\right) \,\middle|\, \Delta P = 0 \text{ in } \mathbb{R}^d\right\}$$

collects all **homogeneous harmonic polynomials** of degree n on $\mathbb{R}^d$.

Let us summarize what the main results of this section are.

- We considered two kinds of boundary values: either the function $U = F$ itself is given at the surface Σ (Dirichlet boundary values) or its normal derivative $\frac{\partial U}{\partial \nu} = F$ is given there (Neumann boundary values). Moreover, for both types of boundary values, we distinguished the interior problems IDP and INP, where the function is continued from Σ to a harmonic function in Σ_{int}, from the exterior problems EDP and ENP, where we attempt to continue the function U to a harmonic function in Σ_{ext}. In the latter case, the additional requirement of regularity at infinity is imposed (see Problem 3.3.1).
- The IDP, EDP, and ENP are solvable for every boundary-value function $F \in \mathrm{C}(\Sigma)$; see Theorem 3.3.34. Moreover, the solution is unique in all these cases (see Corollaries 3.3.4 and 3.3.7 and Theorem 3.3.8). For the IDP and the EDP, we also showed that the solution is stable (see Theorems 3.3.3 and 3.3.6). This means that these problems are well-posed in the sense of Hadamard (see Definition 3.3.2).

- The INP is ill-posed in the sense of Hadamard. It is solvable for boundary values $F \in \mathrm{C}(\Sigma)$ if and only if $\int_\Sigma F(x)\,\mathrm{d}\omega(x) = 0$ (see Theorem 3.3.34). Moreover, its solution is only unique up to constant summands (see Theorem 3.3.5).
- The uniqueness proofs were rather easy and used the maximum principle or Green's first identity. For the exterior problems, the regularity at infinity is also necessary to prove the uniqueness of the solution.
- The concept of Green's function is useful to represent the solution in terms of the boundary values (see Theorem 3.3.12). However, the 'catch' is that it can be very hard to find Green's function in the case of more complicated surfaces. What we know is:
  - If Green's function exists, then it is unique for the particular surface (see Theorem 3.3.11) and it is symmetric (see Theorem 3.3.35).
  - If the surface Σ has a $\mathrm{C}^{(\infty)}$-parameterization, then Green's function for Σ exists (see Theorem 3.3.36).
- The proof of the existence of the solutions was much more complicated than the proof of the uniqueness. We needed an excursion to potential theory concerning the properties of single and double layer potentials and an excursion to some parts of functional analysis.
- The single layer potential corresponding to a measurable and bounded density function F on Σ is infinitely differentiable and harmonic in $\mathbb{R}^3 \setminus \Sigma$. Moreover, it exists everywhere on $\mathbb{R}^3$, and is bounded and continuous (see Theorem 3.3.14). However, its normal derivative is, in general, not continuous on Σ. If F is continuous, then the single layer potential has a normal derivative which jumps (absolutely) by $4\pi F$ at Σ (see Theorem 3.3.29.)
- The double layer potential corresponding to a measurable and bounded function F on Σ is infinitely differentiable and harmonic in $\mathbb{R}^3 \setminus \Sigma$ (see Theorem 3.3.15) and is, therefore, also continuous on $\mathbb{R}^3 \setminus \Sigma$. Moreover, it exists everywhere in $\mathbb{R}^3$, and its restriction to Σ is continuous on Σ (see Theorem 3.3.17). However, the double layer potential is, in general, not continuous as a function on the whole space $\mathbb{R}^3$ (see Example 3.3.19 and Remark 3.3.20). If F is continuous on Σ, then the double layer potential jumps (absolutely) by $4\pi F$ at Σ (see Theorem 3.3.26).
- If Σ is a sphere, then a Green's function can be found relatively easily. The corresponding representation of the solutions of the IDP and the EDP, respectively, is called the Poisson integral formula (Theorems 3.3.38 and 3.3.40).
- Every harmonic function on an open domain is not only infinitely differentiable (see Theorem 3.3.41) but also analytic (see Theorem 3.3.43). Moreover, in the neighbourhood of every point of its domain, it is representable as a series of homogeneous harmonic polynomials (see Theorem 3.3.46).

We have seen that we can find out more about harmonic functions if their domain is the interior or the exterior of a sphere. In the case of geomathematics, the latter situation is very appropriate, because Newton's gravitational potential is harmonic outside the Earth (see Theorem 3.1.5) and the surface of the Earth is approximately a sphere. For this reason, the next section is dedicated to the particular case of a spherical boundary.

3.4 The Sphere as a Particular Boundary

What can we find out about functions U which satisfy the Laplace equation $\Delta U = 0$ on Σ_{int} or Σ_{ext} if the boundary is a sphere $\Sigma = S_R(0)$ as a particular case? One useful approach could be a separation ansatz for U in the sense that $U(x) = U(|x|(x/|x|)) = F(|x|)Y(x/|x|)$. We will write here $U(r\xi) = F(r)Y(\xi)$, where $\xi \in \Omega$ (remember that Ω is the unit sphere) and r is either in $[0, R]$ (interior problems) or in $[R, +\infty[$ (exterior problems). For the interior problems, we should observe that the separation only works for $x \neq 0$, since $0 = r\xi$ with $r = 0$ leaves ξ as an arbitrary unit vector. Since we look for functions U which are particularly continuous, this is, however, not a problem. Nevertheless, we should take care that solutions to the IDP or the INP which we might derive are indeed continuously extendable to $x = 0$.

For applying the separation ansatz, we also have to separate the Laplace operator. Let us start with the first-order differentiation, which is represented by a gradient.

From the chain rule, we have for a function in Cartesian and polar coordinates, respectively, the following identities (see also Example 2.5.4 for the used nomenclature):

$$\begin{aligned}
\frac{\partial}{\partial r} F(x(r,\varphi,t)) &= \nabla_x F(x)|_{x=x(r,\varphi,t)} \cdot \frac{\partial}{\partial r} x(r,\varphi,t), \\
\frac{\partial}{\partial \varphi} F(x(r,\varphi,t)) &= \nabla_x F(x)|_{x=x(r,\varphi,t)} \cdot \frac{\partial}{\partial \varphi} x(r,\varphi,t), \\
\frac{\partial}{\partial t} F(x(r,\varphi,t)) &= \nabla_x F(x)|_{x=x(r,\varphi,t)} \cdot \frac{\partial}{\partial t} x(r,\varphi,t).
\end{aligned}$$

From Example 2.5.4, we already know that

$$\frac{\partial x}{\partial r} = \varepsilon^r, \qquad \frac{\partial x}{\partial \varphi} = r\sqrt{1-t^2}\,\varepsilon^\varphi, \qquad \frac{\partial x}{\partial t} = \frac{r}{\sqrt{1-t^2}}\,\varepsilon^t,$$

where we have to exclude the poles ($|t| = 1$) for the latter formula. We can now write the chain rule and its result by a matrix-vector product

$$\begin{pmatrix} \frac{\partial F}{\partial r} \\ \frac{\partial F}{\partial \varphi} \\ \frac{\partial F}{\partial t} \end{pmatrix} = \begin{pmatrix} \frac{\partial x_1}{\partial r} & \frac{\partial x_2}{\partial r} & \frac{\partial x_3}{\partial r} \\ \frac{\partial x_1}{\partial \varphi} & \frac{\partial x_2}{\partial \varphi} & \frac{\partial x_3}{\partial \varphi} \\ \frac{\partial x_1}{\partial t} & \frac{\partial x_2}{\partial t} & \frac{\partial x_3}{\partial t} \end{pmatrix} \begin{pmatrix} \frac{\partial F}{\partial x_1} \\ \frac{\partial F}{\partial x_2} \\ \frac{\partial F}{\partial x_3} \end{pmatrix} = \underbrace{\begin{pmatrix} (\varepsilon^r)^{\mathrm{T}} \\ r\sqrt{1-t^2}\,(\varepsilon^\varphi)^{\mathrm{T}} \\ \frac{r}{\sqrt{1-t^2}}\,(\varepsilon^t)^{\mathrm{T}} \end{pmatrix}}_{=:M} \begin{pmatrix} \frac{\partial F}{\partial x_1} \\ \frac{\partial F}{\partial x_2} \\ \frac{\partial F}{\partial x_3} \end{pmatrix}.$$

We have already observed that $\varepsilon^r, \varepsilon^\varphi$, and ε^t are orthonormal. Hence,

$$M^{-1} = \left(\varepsilon^r, \frac{1}{r\sqrt{1-t^2}}\,\varepsilon^\varphi, \frac{\sqrt{1-t^2}}{r}\,\varepsilon^t\right).$$

As a consequence,

$$\nabla_x F = \left(\varepsilon^r, \ \frac{1}{r\sqrt{1-t^2}} \varepsilon^\varphi, \ \frac{\sqrt{1-t^2}}{r} \varepsilon^t \right) \begin{pmatrix} \frac{\partial F}{\partial r} \\ \frac{\partial F}{\partial \varphi} \\ \frac{\partial F}{\partial t} \end{pmatrix}$$

$$= \varepsilon^r \frac{\partial F}{\partial r} + \varepsilon^\varphi \frac{1}{r\sqrt{1-t^2}} \frac{\partial F}{\partial \varphi} + \varepsilon^t \frac{\sqrt{1-t^2}}{r} \frac{\partial F}{\partial t}.$$

Let us summarize this.

Theorem 3.4.1 *If F has all first-order partial derivatives on D, where $D \subset \mathbb{R}^3$ is some 'harmless' set, then the gradient of F can be written in polar coordinates as follows:*

$$(\nabla_x F(x))|_{x=x(r,\varphi,t)}$$

$$= \frac{\partial F\,(x(r,\varphi,t))}{\partial r}\, \varepsilon^r(\varphi,t)$$

$$+ \frac{1}{r}\left[\frac{1}{\sqrt{1-t^2}} \frac{\partial F\,(x(r,\varphi,t))}{\partial \varphi}\, \varepsilon^\varphi(\varphi) + \sqrt{1-t^2}\, \frac{\partial F\,(x(r,\varphi,t))}{\partial t}\, \varepsilon^t(\varphi,t) \right].$$

Definition 3.4.2 The **surface gradient** is defined as the following differential operator:

$$\nabla^* := \varepsilon^\varphi \frac{1}{\sqrt{1-t^2}} \frac{\partial}{\partial \varphi} + \varepsilon^t \sqrt{1-t^2}\, \frac{\partial}{\partial t}.$$

With this short-hand notation, we can write

$$\nabla_x F(x) = \xi\, \frac{\partial}{\partial r}\, F(r\xi) + \frac{1}{r}\, \nabla^*_\xi F(r\xi), \quad x = r\xi, \tag{3.79}$$

where $r = |x|$ and $\xi \in \Omega$. Here, ξ represents the unit normal vector to Ω (i.e., ε^r), which is the same as the normalized position vector $\xi = x/|x|$. In the latter sense, ∇^*_ξ stands for the differentiation with respect to the angular dependence, but the notation ∇^*_ξ itself remains independent of particular choices of coordinate systems on Ω, which is an advantage. In other words, we can define ∇^*_ξ also via (3.79).

What we actually need is a formula for the decomposition of Δ. Hence, we have to proceed. Let us start with already stating the result.

Theorem 3.4.3 *If F has all partial derivatives up to order 2 on D, where $D \subset \mathbb{R}^3$ is some 'harmless' set, then the **Laplace operator** can be decomposed into an angular part and a radial part as follows:*

$$(\Delta_x F(x))|_{x=x(r,\varphi,t)}$$

$$= \frac{\partial^2}{\partial r^2}\, F\,(x(r,\varphi,t)) + \frac{2}{r}\frac{\partial}{\partial r}\, F\,(x(r,\varphi,t))$$

$$+ \frac{1}{r^2}\left\{ \frac{\partial}{\partial t}\left[(1-t^2)\frac{\partial}{\partial t}\, F\,(x(r,\varphi,t)) \right] + \frac{1}{1-t^2}\frac{\partial^2}{\partial \varphi^2}\, F\,(x(r,\varphi,t)) \right\}.$$

Briefly,

$$\Delta_x F(x) = \frac{\partial^2}{\partial r^2} F(r\xi) + \frac{2}{r}\frac{\partial}{\partial r} F(r\xi) + \frac{1}{r^2}\Delta^*_\xi F(r\xi),$$

where Δ^* is called the ***Beltrami operator*** *or the* ***Laplace–Beltrami operator***.

Proof From Theorem 2.3.4, we know that $\Delta F(x) = \operatorname{div}\operatorname{grad} F(x) = \nabla\cdot(\nabla F)(x)$. For this reason, we can use Theorem 3.4.1 and get

$$\Delta F = \left(\varepsilon^r \frac{\partial}{\partial r} + \varepsilon^\varphi \frac{1}{r\sqrt{1-t^2}}\frac{\partial}{\partial\varphi} + \varepsilon^t \frac{1}{r}\sqrt{1-t^2}\frac{\partial}{\partial t}\right) \cdot \left(\varepsilon^r \frac{\partial}{\partial r} + \varepsilon^\varphi \frac{1}{r\sqrt{1-t^2}}\frac{\partial}{\partial\varphi} + \varepsilon^t \frac{1}{r}\sqrt{1-t^2}\frac{\partial}{\partial t}\right) F.$$

Actually, we would have to write here $\varepsilon^r(\varphi,t)$, $\varepsilon^\varphi(\varphi)$, and $\varepsilon^t(\varphi,t)$, which would have made the preceding formula less readable. However, we have to be aware of this dependence, because we need the product rule here:

$$\begin{aligned}
\Delta F &= \underbrace{\varepsilon^r\cdot\varepsilon^r}_{=1}\frac{\partial^2}{\partial r^2}F + \underbrace{\varepsilon^r\cdot\varepsilon^\varphi}_{=0}\frac{\partial}{\partial r}\left(\frac{1}{r\sqrt{1-t^2}}\frac{\partial}{\partial\varphi}F\right)\\
&+ \underbrace{\varepsilon^r\cdot\varepsilon^t}_{=0}\frac{\partial}{\partial r}\left(\frac{1}{r}\sqrt{1-t^2}\frac{\partial}{\partial t}F\right)\\
&+ \varepsilon^\varphi\cdot\left(\frac{\partial}{\partial\varphi}\varepsilon^r\right)\frac{1}{r\sqrt{1-t^2}}\frac{\partial}{\partial r}F + \underbrace{\varepsilon^\varphi\cdot\varepsilon^r}_{=0}\frac{1}{r\sqrt{1-t^2}}\frac{\partial^2}{\partial\varphi\partial r}F\\
&+ \varepsilon^\varphi\cdot\left(\frac{\partial}{\partial\varphi}\varepsilon^\varphi\right)\frac{1}{r^2(1-t^2)}\frac{\partial}{\partial\varphi}F + \underbrace{\varepsilon^\varphi\cdot\varepsilon^\varphi}_{=1}\frac{1}{r^2(1-t^2)}\frac{\partial^2}{\partial\varphi^2}F\\
&+ \varepsilon^\varphi\cdot\left(\frac{\partial}{\partial\varphi}\varepsilon^t\right)\frac{1}{r^2}\frac{\partial}{\partial t}F + \underbrace{\varepsilon^\varphi\cdot\varepsilon^t}_{=0}\frac{1}{r^2}\frac{\partial^2}{\partial\varphi\partial t}F\\
&+ \varepsilon^t\cdot\left(\frac{\partial}{\partial t}\varepsilon^r\right)\frac{1}{r}\sqrt{1-t^2}\frac{\partial}{\partial r}F + \underbrace{\varepsilon^t\cdot\varepsilon^r}_{=0}\frac{1}{r}\sqrt{1-t^2}\frac{\partial^2}{\partial t\partial r}F\\
&+ \underbrace{\varepsilon^t\cdot\varepsilon^\varphi}_{=0}\frac{1}{r^2}\sqrt{1-t^2}\frac{\partial}{\partial t}\left(\frac{1}{\sqrt{1-t^2}}\frac{\partial}{\partial\varphi}F\right) + \varepsilon^t\cdot\left(\frac{\partial}{\partial t}\varepsilon^t\right)\frac{1-t^2}{r^2}\frac{\partial}{\partial t}F\\
&+ \underbrace{\varepsilon^t\cdot\varepsilon^t}_{=1}\frac{1}{r^2}\sqrt{1-t^2}\left(-\frac{t}{\sqrt{1-t^2}}\frac{\partial}{\partial t}F + \sqrt{1-t^2}\frac{\partial^2}{\partial t^2}F\right).
\end{aligned}$$

The derivatives of the vectors $\varepsilon^r, \varepsilon^\varphi$, and ε^t are easy to obtain:

$$\frac{\partial}{\partial\varphi}\varepsilon^r = \begin{pmatrix} -\sqrt{1-t^2}\sin\varphi \\ \sqrt{1-t^2}\cos\varphi \\ 0 \end{pmatrix} = \sqrt{1-t^2}\,\varepsilon^\varphi,$$

$$\frac{\partial}{\partial t}\,\varepsilon^r = \begin{pmatrix} -\frac{t}{\sqrt{1-t^2}}\cos\varphi \\ -\frac{t}{\sqrt{1-t^2}}\sin\varphi \\ 1 \end{pmatrix} = \frac{1}{\sqrt{1-t^2}}\,\varepsilon^t,$$

$$\frac{\partial}{\partial\varphi}\,\varepsilon^\varphi = \begin{pmatrix} -\cos\varphi \\ -\sin\varphi \\ 0 \end{pmatrix},$$

$$\frac{\partial}{\partial\varphi}\,\varepsilon^t = \begin{pmatrix} t\sin\varphi \\ -t\cos\varphi \\ 0 \end{pmatrix} = -t\varepsilon^\varphi,$$

$$\frac{\partial}{\partial t}\,\varepsilon^t = \begin{pmatrix} -\cos\varphi \\ -\sin\varphi \\ \frac{-t}{\sqrt{1-t^2}} \end{pmatrix} = -\frac{1}{\sqrt{1-t^2}}\,\varepsilon^r.$$

We now only have to collect what we have done in this proof so far:

$$\begin{aligned}
\Delta F &= \frac{\partial^2}{\partial r^2}F + \frac{1}{r}\frac{\partial}{\partial r}F + \frac{1}{r^2(1-t^2)}\frac{\partial^2}{\partial\varphi^2}F - \frac{t}{r^2}\frac{\partial}{\partial t}F \\
&\quad + \frac{1}{r}\frac{\partial}{\partial r}F - \frac{t}{r^2}\frac{\partial}{\partial t}F + \frac{1-t^2}{r^2}\frac{\partial^2}{\partial t^2}F \\
&= \frac{\partial^2}{\partial r^2}F + \frac{2}{r}\frac{\partial}{\partial r}F + \frac{1}{r^2}\left(\frac{1}{1-t^2}\frac{\partial^2}{\partial\varphi^2}F - 2t\frac{\partial}{\partial t}F + (1-t^2)\frac{\partial^2}{\partial t^2}F\right) \\
&= \frac{\partial^2}{\partial r^2}F + \frac{2}{r}\frac{\partial}{\partial r}F + \frac{1}{r^2}\left[\frac{1}{1-t^2}\frac{\partial^2}{\partial\varphi^2}F + \frac{\partial}{\partial t}\left((1-t^2)\frac{\partial}{\partial t}F\right)\right].
\end{aligned}$$

□

Corollary 3.4.4 *The local orthonormal vectors $\varepsilon^r, \varepsilon^\varphi$, and ε^t satisfy:*

$$\frac{\partial}{\partial\varphi}\,\varepsilon^r = \sqrt{1-t^2}\,\varepsilon^\varphi, \qquad \frac{\partial}{\partial t}\,\varepsilon^r = \frac{1}{\sqrt{1-t^2}}\,\varepsilon^t,$$

$$\frac{\partial}{\partial\varphi}\,\varepsilon^\varphi = \begin{pmatrix} -\cos\varphi \\ -\sin\varphi \\ 0 \end{pmatrix}, \qquad \frac{\partial}{\partial t}\,\varepsilon^\varphi = 0,$$

$$\frac{\partial}{\partial\varphi}\,\varepsilon^t = -t\varepsilon^\varphi, \qquad \frac{\partial}{\partial t}\,\varepsilon^t = -\frac{1}{\sqrt{1-t^2}}\,\varepsilon^r$$

such that

$$\varepsilon^r\cdot\frac{\partial}{\partial\varphi}\,\varepsilon^r = 0, \qquad \varepsilon^r\cdot\frac{\partial}{\partial\varphi}\,\varepsilon^\varphi = -\sqrt{1-t^2}, \qquad \varepsilon^r\cdot\frac{\partial}{\partial\varphi}\,\varepsilon^t = 0,$$

$$\varepsilon^\varphi\cdot\frac{\partial}{\partial\varphi}\,\varepsilon^r = \sqrt{1-t^2}, \qquad \varepsilon^\varphi\cdot\frac{\partial}{\partial\varphi}\,\varepsilon^\varphi = 0, \qquad \varepsilon^\varphi\cdot\frac{\partial}{\partial\varphi}\,\varepsilon^t = -t,$$

$$\varepsilon^t \cdot \frac{\partial}{\partial \varphi} \varepsilon^r = 0, \qquad \varepsilon^t \cdot \frac{\partial}{\partial \varphi} \varepsilon^\varphi = t, \qquad \varepsilon^t \cdot \frac{\partial}{\partial \varphi} \varepsilon^t = 0,$$

$$\varepsilon^r \cdot \frac{\partial}{\partial t} \varepsilon^r = 0, \qquad \varepsilon^r \cdot \frac{\partial}{\partial t} \varepsilon^\varphi = 0, \qquad \varepsilon^r \cdot \frac{\partial}{\partial t} \varepsilon^t = \frac{-1}{\sqrt{1-t^2}},$$

$$\varepsilon^\varphi \cdot \frac{\partial}{\partial t} \varepsilon^r = 0, \qquad \varepsilon^\varphi \cdot \frac{\partial}{\partial t} \varepsilon^\varphi = 0, \qquad \varepsilon^\varphi \cdot \frac{\partial}{\partial t} \varepsilon^t = 0,$$

$$\varepsilon^t \cdot \frac{\partial}{\partial t} \varepsilon^r = \frac{1}{\sqrt{1-t^2}}, \qquad \varepsilon^t \cdot \frac{\partial}{\partial t} \varepsilon^\varphi = 0, \qquad \varepsilon^t \cdot \frac{\partial}{\partial t} \varepsilon^t = 0.$$

Note that some of the preceding orthogonalities are obvious without the need of an explicit calculation: for example, since $|\varepsilon^r|^2 = 1$ everywhere, we have

$$0 = \frac{\partial}{\partial \varphi} \left|\varepsilon^r(\varphi,t)\right|^2 = \frac{\partial}{\partial \varphi} \left(\varepsilon^r(\varphi,t) \cdot \varepsilon^r(\varphi,t)\right) = 2\varepsilon^r(\varphi,t) \cdot \frac{\partial}{\partial \varphi} \varepsilon^r(\varphi,t).$$

Analogous identities are valid for the derivative with respect to t and for the vectors ε^φ and ε^t.

Definition 3.4.5 The **Legendre operator** $\mathcal{L} : \mathrm{C}^{(2)}[-1,1] \to \mathrm{C}[-1,1]$ is defined by

$$\mathcal{L} F := -2t \frac{\mathrm{d}F}{\mathrm{d}t} + (1-t^2) \frac{\mathrm{d}^2 F}{\mathrm{d}t^2}, \qquad F \in \mathrm{C}^{(2)}[-1,1].$$

The Legendre operator represents the part of Δ^* which corresponds to the polar distance t.

Let us come back to the original question: what can we find out about the solution of one of our boundary-value problems if $\Sigma = S_R(0)$. The Laplace equation for the solution U and Theorem 3.4.3 yield (remember that $U(x) = F(r)Y(\xi)$ with $x = r\xi$, $\xi \in \Omega$)

$$\Delta_x U(x) = 0 \Leftrightarrow \left(F''(r) + \frac{2}{r} F'(r)\right) Y(\xi) + \frac{F(r)}{r^2} \Delta_\xi^* Y(\xi) = 0.$$

We set $D_r^{\mathrm{int}} := [0, R]$ and $D_r^{\mathrm{ext}} := [R, +\infty[$ such that $D_r \in \{D_r^{\mathrm{int}}, D_r^{\mathrm{ext}}\}$ stands for the domain corresponding to the considered interior or exterior boundary-value problem.

Now, outside the zeros of F and Y, we have

$$\frac{r^2 F''(r) + 2r F'(r)}{F(r)} = -\frac{\Delta_\xi^* Y(\xi)}{Y(\xi)}. \tag{3.80}$$

The exception of the zeros of F and Y is not a serious problem here. Unless one of these functions turns out to be vanishing on a neighbourhood of a point instead of just isolated points, F and Y are still unique by using the continuity of these functions.

In (3.80), the left-hand side only depends on r and the right-hand side only depends on ξ. This allows us to conclude as follows: by considering the solution at an arbitrary but fixed sphere $S_r(0)$, we get a *constant* left-hand side, which must coincide with the right-hand side *for all* $\xi \in \Omega$. Vice versa, an arbitrary but fixed direction in space $\xi \in \Omega$ turns the right-hand side into a constant, which equals the left-hand side for all r. Thus, both sides in (3.80) must be constant. That is, there is a constant $C_1 \in \mathbb{R}$ such that

$$r^2 F''(r) + 2r F'(r) = C_1 F(r), \tag{3.81}$$

$$\Delta_\xi^* Y(\xi) = -C_1 Y(\xi) \tag{3.82}$$

for all r and ξ, where F and Y, respectively, do not vanish. For F, it suffices to have an arbitrarily small interval where F does not vanish to conclude that all solutions of the Cauchy–Euler type equation in (3.81) can be described as linear combinations of two linearly independent solutions (see Roberts, 2010, p. 185). With the ansatz $F(r) = r^k$, we get from (3.81) the algebraic equation $k(k-1) + 2k = C_1$, that is, $k^2 + k - C_1 = 0$. The solutions are

$$k_{1,2} = -\frac{1}{2} \pm \sqrt{\frac{1}{4} + C_1} = \frac{1}{2}\left(-1 \pm \sqrt{1 + 4C_1}\right). \tag{3.83}$$

We will come back to this. Let us first look at (3.82) and use Theorem 3.4.3. We get

$$\mathcal{L}_t Y(\xi(\varphi,t)) + \frac{1}{1-t^2}\frac{\partial^2}{\partial\varphi^2} Y(\xi(\varphi,t)) = -C_1 Y(\xi(\varphi,t)), \tag{3.84}$$

where $\xi(\varphi,t)$ stands for the representation of ξ in the polar coordinates φ and t. We use now the same technique which we have used previously and transfer (3.84) with the separation ansatz $Y(\xi(\varphi,t)) = G(\varphi)H(t)$ into

$$(1-t^2)\frac{\mathcal{L}_t H(t)}{H(t)} + C_1(1-t^2) = -\frac{G''(\varphi)}{G(\varphi)},$$

which is again valid outside the zeros of the denominators. Again, the separation of the variables to different sides implies that both sides are constant. Hence, we have (another) constant $C_2 \in \mathbb{R}$ such that

$$(1-t^2)\mathcal{L}_t H(t) = \left[C_2 - C_1(1-t^2)\right] H(t), \tag{3.85}$$

$$G''(\varphi) = -C_2\, G(\varphi). \tag{3.86}$$

There is something we know about φ: as the polar coordinate of ξ, it represents the longitude such that $G^{(k)}(0) = G^{(k)}(2\pi)$ must hold at least for $k \le 2$. On the other hand, (3.86) is a second-order linear and homogeneous differential equation with constant coefficients, which is well known from the theory of oscillations. Its general solution has the form

$$G(\varphi) = \begin{cases} a\, e^{\sqrt{-C_2}\,\varphi} + b\, e^{-\sqrt{-C_2}\,\varphi}, & \text{if } C_2 < 0, \\ a + b\,\varphi, & \text{if } C_2 = 0, \\ a \sin\left(\sqrt{C_2}\,\varphi\right) + b\cos\left(\sqrt{C_2}\,\varphi\right), & \text{if } C_2 > 0. \end{cases}$$

Besides the trivial case where $a = b = 0$ and the constant case ($C_2 = 0$ and $b = 0$), the 'periodicity' $G^{(k)}(0) = G^{(k)}(2\pi)$ can only be achieved in the case of $C_2 > 0$ and only if

$$b = a\sin\left(2\pi\sqrt{C_2}\right) + b\cos\left(2\pi\sqrt{C_2}\right),$$

$$a\sqrt{C_2} = a\sqrt{C_2}\cos\left(2\pi\sqrt{C_2}\right) - b\sqrt{C_2}\sin\left(2\pi\sqrt{C_2}\right)$$

(for $k \ge 2$, the equation $G'' = -C_2 G$ implies that $G^{(k)}(0) = G^{(k)}(2\pi)$ does not cause any further constraints).

Again, we do not want the trivial solution where $a = b = 0$. Hence, the previous system of linear equations, which is equivalent to (with $c := 2\pi\sqrt{C_2}$)

$$0 = a \sin c + b(\cos c - 1), \qquad 0 = a(\cos c - 1) - b \sin c,$$

must have more than one solution. Consequently, the determinant

$$\begin{vmatrix} \sin c & \cos c - 1 \\ \cos c - 1 & -\sin c \end{vmatrix} = -\sin^2 c - \cos^2 c + 2\cos c - 1 = 2\cos c - 2$$

must vanish. In other words, $c = 2\pi j$, $j \in \mathbb{Z}$, that is, $C_2 = j^2$, $j \in \mathbb{N}_0$. Note that $j = 0$ ($\Leftrightarrow C_2 = 0$) corresponds to the constant solutions. Let us now have a look at (3.85), which is written out as

$$-2(1-t^2)t\, H'(t) + (1-t^2)^2 H''(t) = \left[j^2 - C_1(1-t^2)\right] H(t). \tag{3.87}$$

This equation also has several solutions depending on the values of C_1 and $C_2 = j^2$.

We see that there is certainly not only one particular choice of a basis system for harmonic functions inside or outside a ball. We will, therefore, later study a common generalized approach to this task.

Let us now come back to the equations we derived earlier. In (3.82), we saw that the angular part Y has to be an eigenfunction of the Beltrami operator, which resulted, for example, in (3.87).

In the following, we use the assumption that the solution H of (3.87) is expandable into a power series of the form

$$H(t) = \sum_{n=0}^{\infty} a_n t^n \tag{3.88}$$

such that $H'(t) = \sum_{n=1}^{\infty} a_n n t^{n-1}$ and $H''(t) = \sum_{n=2}^{\infty} a_n n(n-1) t^{n-2}$. We insert these series into (3.87) and get

$$\begin{aligned} &-2\sum_{n=1}^{\infty} a_n n t^n + 2\sum_{n=1}^{\infty} a_n n t^{n+2} \\ &+\sum_{n=2}^{\infty} a_n n(n-1)t^{n-2} - 2\sum_{n=2}^{\infty} a_n n(n-1)t^n + \sum_{n=2}^{\infty} a_n n(n-1)t^{n+2} \\ &\quad = (j^2 - C_1)\sum_{n=0}^{\infty} a_n t^n + C_1 \sum_{n=0}^{\infty} a_n t^{n+2}. \end{aligned}$$

We perform a few index shifts and observe that summands including factors of the type $n - k$ can be added for $n = k$ without changing anything. This leads us to

$$\begin{aligned} &\sum_{n=0}^{\infty}\left[-2a_n n + a_{n+2}(n+2)(n+1) - 2a_n n(n-1) + (C_1 - j^2)a_n\right]t^n \\ &\quad + \sum_{n=2}^{\infty} a_{n-2}\left[2(n-2) + (n-2)(n-3) - C_1\right] t^n = 0. \end{aligned}$$

Since a power series is uniquely representable, this implies that the sum of all coefficients of t^n must vanish. Hence,

$$a_n\left(-2n^2 + C_1 - j^2\right) + a_{n+2}(n+2)(n+1) + a_{n-2}\left(\underbrace{n^2 - 3n + 2}_{=(n-2)(n-1)} - C_1\right) = 0 \tag{3.89}$$

for $n \geq 2$ and

$$a_n\left(-2n^2 + C_1 - j^2\right) + a_{n+2}(n+2)(n+1) = 0 \tag{3.90}$$

for $n < 2$. This is a recursion where a_0 and a_1 are free to choose. Then (3.90) tells us that

$$a_{n+2} = \frac{2n^2 + j^2 - C_1}{(n+2)(n+1)} a_n \quad \text{for } n < 2. \tag{3.91}$$

Further, (3.89) leads us to

$$a_{n+2} = \frac{2n^2 + j^2 - C_1}{(n+2)(n+1)} a_n + \frac{-(n-2)(n-1) + C_1}{(n+2)(n+1)} a_{n-2} \quad \text{for } n \geq 2. \tag{3.92}$$

Obviously, the denominators never vanish such that these formulae are always valid.

We obtained here an interesting recursion. It tells us a lot about the solutions of (3.87), namely:

- The odd and the even degrees are completely decoupled, because the recurrence only incorporates '+2-steps'. Hence, we can separate the solutions of the differential equation into a sum of an odd function and an even function (see Definition 3.4.6), where both summands also solve the equation.
- The coefficients a_0 and a_1 are completely free to choose and all the others are uniquely determined, because (3.91) and (3.92) together are equivalent to the differential equation (3.87) with the power series ansatz (3.88).
- As a consequence, the initial assumption $a_0 \in \mathbb{R}$ (arbitrary) and $a_1 = 0$ generates all even solutions (and, analogously, $a_0 = 0$ and $a_1 \in \mathbb{R}$ [arbitrary] yields all odd solutions).
- In the case $j = 0$, the recursion is finite for particular values of C_1, that is, we get polynomial solutions. This is not yet obvious, but we will soon derive this.

Definition 3.4.6 A function $F: D \to \mathbb{R}$, where $D \subset \mathbb{R}$ and $\{-x \mid x \in D\} = D$, is called

(a) **Odd**, if $F(-x) = -F(x)$ for all $x \in D$.
(b) **Even**, if $F(-x) = F(x)$ for all $x \in D$.

Let us now have a look at the case $j = 0$, for which we move a bit backwards in our derivations. Namely, (3.87) simplifies for $j = 0$ to

$$-2tH'(t) + (1 - t^2)H''(t) = -C_1 H(t).$$

The power series ansatz now leads us to

$$-2\sum_{n=1}^{\infty} a_n n t^n + \sum_{n=2}^{\infty} a_n n(n-1)t^{n-2} - \sum_{n=2}^{\infty} a_n n(n-1)t^n = -C_1 \sum_{n=0}^{\infty} a_n t^n.$$

We perform again some index shifts and obtain (while adding also here some vanishing summands)

$$\sum_{n=0}^{\infty}\big[-2na_n + a_{n+2}(n+2)(n+1) - a_n n(n-1)\big]t^n = \sum_{n=0}^{\infty}(-C_1 a_n)t^n.$$

This is valid if and only if

$$a_n\big(-n^2 - n + C_1\big) + a_{n+2}(n+2)(n+1) = 0 \quad \text{for all } n \geq 0.$$

The recursion is, indeed, simpler here:

$$a_{n+2} = \frac{n^2 + n - C_1}{(n+2)(n+1)}\, a_n \qquad \text{for all } n \geq 0. \tag{3.93}$$

As a consequence, if $C_1 = k(k+1)$ for one $k \in \mathbb{N}_0$, then the choice

- $a_0 \neq 0,\ a_1 = 0$, if k is even,
- $a_0 = 0,\ a_1 \neq 0$, if k is odd,

yields an even/odd polynomial of degree k, because the preceding recursion stops for $n = k$.

Definition 3.4.7 The polynomials $P_k(t) = \sum_{n=0}^{k} a_n t^n$, $k \in \mathbb{N}_0$, which are given by the recurrence relation (3.93) with $C_1 = k(k+1)$ and

- $a_0 = (-1)^{k/2} k!\ 2^{-k}[(k/2)!\,]^{-2}$, $a_1 = 0$, if k is even,
- $a_0 = 0$, $a_1 = (-1)^{(k-1)/2}(k+1)!\ 2^{-k}\{[(k+1)/2]!\ [(k-1)/2]!\,\}^{-1}$, if k is odd,

are called the **Legendre polynomials**.

The choice of the initial coefficients looks weird here, but we will see later on that this results in $P_k(1) = 1$ for all $k \in \mathbb{N}_0$.

From what we learned so far, we can already state several properties of the Legendre polynomials.

Theorem 3.4.8 *All Legendre polynomials of even degrees are even functions. All Legendre polynomials of odd degrees are odd functions.*

Theorem 3.4.9 *Each Legendre polynomial P_n, $n \in \mathbb{N}_0$, is an eigenfunction of the Legendre operator to the eigenvalue $-n(n+1)$. In other words, each polynomial P_n solves the* ***Legendre differential equation***

$$(1-t^2)P_n''(t) - 2tP_n'(t) = -n(n+1)P_n(t), \quad t \in \mathbb{R}. \tag{3.94}$$

Let us come back to (3.83) now. With $C_1 = n(n+1)$, $n \in \mathbb{N}_0$, we get

$$k_{1,2} = \frac{1}{2}\left(-1 \pm \sqrt{1+4n(n+1)}\right) = \frac{1}{2}[-1 \pm (2n+1)]$$

such that $k_1 = n$, $k_2 = -n-1$. Hence, the general solution of (3.81) is given (for $C_1 = n(n+1)$) by

$$F(r) = ar^n + br^{-n-1}. \tag{3.95}$$

For the interior problems, r^{-n-1} would not be defined in $r = 0$ such that $F(r) = ar^n$ has to be used. For the exterior problems, only $F(r) = br^{-n-1}$ fulfils the regularity at infinity (see Corollary 3.1.7).

The recursion which we used to define the Legendre polynomials in Definition 3.4.7 can be resolved, as we will show in the following theorem. The corresponding formula is well known (see, e.g., Agarwal and O'Regan, 2009, Equation (7.8)).

Theorem 3.4.10 *The Legendre polynomials satisfy*

$$P_k(t) = \sum_{m=0}^{\left\lfloor \frac{k}{2} \right\rfloor} (-1)^m \frac{(2k-2m)!}{2^k m!\,(k-m)!\,(k-2m)!} t^{k-2m},$$

where $\lfloor \cdot \rfloor$ represents the Gauß bracket for rounding down (see Definition 2.1.10).

Proof We prove this theorem for the coefficients a_{k-2m} of

$$P_k(t) = \sum_{m=0}^{\left\lfloor \frac{k}{2} \right\rfloor} a_{k-2m} t^{k-2m}. \tag{3.96}$$

Note that we already know from Theorem 3.4.8 that P_k must obey (3.96). We now perform an induction with respect to m from $\lfloor k/2 \rfloor$ downward to 0 and show that

$$a_{k-2m} = (-1)^m \frac{(2k-2m)!}{2^k m!\,(k-m)!\,(k-2m)!}. \tag{3.97}$$

For the initial value $m = \lfloor k/2 \rfloor$, we have to distinguish two cases: if k is even, then (3.97) with $m = k/2$ yields

$$a_{k-k} = (-1)^{k/2} \frac{(2k-k)!}{2^k \left(\frac{k}{2}\right)!\left(\frac{k}{2}\right)!\,0!} = (-1)^{k/2} \frac{k!}{2^k \left[\left(\frac{k}{2}\right)!\right]^2},$$

which coincides with the value of a_0 in Definition 3.4.7. If k is odd, then (3.97) with $m = \frac{k-1}{2}$ yields

$$a_{k-(k-1)} = (-1)^{(k-1)/2} \frac{(2k-k+1)!}{2^k \left(\frac{k-1}{2}\right)!\left(\frac{k+1}{2}\right)!\,1!} = (-1)^{(k-1)/2} \frac{(k+1)!}{2^k \left(\frac{k-1}{2}\right)!\left(\frac{k+1}{2}\right)!},$$

which coincides with a_1 in Definition 3.4.7.

Let us now assume that (3.97) holds for one $m > 0$. Then this assumption and the recurrence formula (3.93) yield

$$\begin{aligned} a_{k-2(m-1)} &= \frac{(k-2m)^2 + k - 2m - k(k+1)}{(k-2m+2)(k-2m+1)} a_{k-2m} \\ &= \frac{-4km + 4m^2 - 2m}{(k-2m+2)(k-2m+1)} \cdot (-1)^m \cdot \frac{(2k-2m)!}{2^k m!\,(k-m)!\,(k-2m)!} \\ &= (-1)^{m-1} \frac{2m(2k-2m+1)(2k-2m)!}{[k-2(m-1)]!\; 2^k m!\,(k-m)!} \cdot \frac{k-m+1}{k-m+1} \\ &= (-1)^{m-1} \frac{[2k-2(m-1)]!}{[k-2(m-1)]!\; 2^k (m-1)!\,[k-(m-1)]!}, \end{aligned}$$

which completes the proof. □

Corollary 3.4.11 *The first Legendre polynomials are given, for all $t \in \mathbb{R}$, by $P_0(t) = 1$, $P_1(t) = t$, $P_2(t) = (3t^2 - 1)/2$.*

The validity of these formulae for P_0, P_1, and P_2 can be easily deduced from Theorem 3.4.10.

Let us remember why we started to investigate Legendre polynomials. We still have to take care of the more general equation (3.87), which we have only solved for $j = 0$ so far.

Theorem 3.4.12 *Equation (3.87) with $C_1 = n(n+1)$, $n \in \mathbb{N}_0$, and $j \in \mathbb{N}_0$ is solved by*

$$H(t) = (1-t^2)^{j/2} \frac{\mathrm{d}^j}{\mathrm{d}t^j} P_n(t), \quad t \in]-1, 1[, \tag{3.98}$$

where P_n is the Legendre polynomial of degree n.

Proof We use the ansatz $H(t) = K(t) P_n^{(j)}(t)$ for (3.87). By inserting this, we get

$$\begin{aligned} &-2(1-t^2)t\left(K'(t)P_n^{(j)}(t) + K(t)P_n^{(j+1)}(t)\right) \\ &+ (1-t^2)^2\left(K''(t)P_n^{(j)}(t) + 2K'(t)P_n^{(j+1)}(t) + K(t)P_n^{(j+2)}(t)\right) \\ &\quad = \left[j^2 - n(n+1)(1-t^2)\right]K(t)P_n^{(j)}(t), \end{aligned}$$

which is equivalent to

$$\begin{aligned} &(1-t^2)K(t)\left[-2tP_n^{(j+1)}(t) + (1-t^2)P_n^{(j+2)}(t) + n(n+1)P_n^{(j)}(t)\right] \\ &+ (1-t^2)\left[-2tK'(t)P_n^{(j)}(t) + (1-t^2)\left(K''(t)P_n^{(j)}(t) + 2K'(t)P_n^{(j+1)}(t)\right)\right] \\ &\quad = j^2 K(t)P_n^{(j)}(t). \end{aligned} \tag{3.99}$$

We differentiate (3.94) j times by using the Leibniz rule (Theorem 2.3.6):

$$\begin{aligned}(1-t^2)P_n^{(j+2)}(t)+j(-2t)P_n^{(j+1)}(t)+\frac{j(j-1)}{2}(-2)P_n^{(j)}(t)\\ -2t\,P_n^{(j+1)}(t)-2j\,P_n^{(j)}(t)=-n(n+1)P_n^{(j)}(t).\end{aligned}\tag{3.100}$$

In the first square-bracket term of (3.99), we can discover parts of (3.100). Hence, (3.99) holds if and only if

$$\begin{aligned}&(1-t^2)K(t)\left[2jt\,P_n^{(j+1)}(t)+j(j+1)P_n^{(j)}(t)\right]\\ &+(1-t^2)\left[-2tK'(t)P_n^{(j)}(t)+(1-t^2)\left(K''(t)P_n^{(j)}(t)+2K'(t)P_n^{(j+1)}(t)\right)\right]\\ &=j^2K(t)P_n^{(j)}(t).\end{aligned}\tag{3.101}$$

Sufficient for the validity of (3.101) is the following system of equations:

$$jtK(t)+(1-t^2)K'(t)=0,\tag{3.102}$$

$$(1-t^2)\big[j(j+1)K(t)-2tK'(t)+(1-t^2)K''(t)\big]=j^2K(t).\tag{3.103}$$

If we set $K(t):=(1-t^2)^{j/2}$, then

$$\begin{aligned}K'(t)&=\frac{j}{2}(1-t^2)^{(j-2)/2}\cdot(-2t)=-jt(1-t^2)^{(j-2)/2},\\ K''(t)&=-j(1-t^2)^{(j-2)/2}-jt\,\frac{j-2}{2}(1-t^2)^{(j-4)/2}\cdot(-2t)\\ &=-j(1-t^2)^{(j-2)/2}+j(j-2)t^2(1-t^2)^{(j-4)/2}.\end{aligned}$$

We immediately see that (3.102) is fulfilled. For (3.103), we need a few more calculations:

$$\begin{aligned}&(1-t^2)\big[j(j+1)K(t)-2tK'(t)+(1-t^2)K''(t)\big]\\ &=j(j+1)(1-t^2)^{(j+2)/2}+2jt^2(1-t^2)^{j/2}\\ &\quad-j(1-t^2)^{(j+2)/2}+j(j-2)t^2(1-t^2)^{j/2}\\ &=(1-t^2)^{j/2}\big[(1-t^2)j^2+j^2t^2\big]=j^2(1-t^2)^{j/2}.\end{aligned}$$

Hence, H from (3.98) solves (3.87). □

Since P_n is a polynomial of degree n, the preceding solutions only make sense for $j\le n$ (the rest is the zero function). Let us summarize our results. We used the separation ansatz $U(x)=F(r)G(\varphi)H(t)$, where (r,φ,t) are the polar coordinates of x. This has led us to

$$F(r)=r^\alpha,\quad \alpha\in\{n,-n-1\},$$

$$G(\varphi)=a\sin(j\varphi)+b\cos(j\varphi),\qquad H(t)=(1-t^2)^{j/2}\frac{\mathrm{d}^j}{\mathrm{d}t^j}P_n(t),$$

where $j=0,\dots,n$ and $n\in\mathbb{N}_0$. Moreover, $\alpha=n$ corresponds to interior boundary-value problems and $\alpha=-n-1$ belongs to the exterior counterparts. This result motivates us to set

$$\tilde{Y}_{n,j}(\xi(\varphi,t)) := (1-t^2)^{|j|/2}\frac{\mathrm{d}^{|j|}}{\mathrm{d}t^{|j|}}P_n(t)\cdot\begin{cases}\sin(j\varphi), & j=1,\ldots,n,\\ \cos(j\varphi), & j=-n,\ldots,0,\end{cases} \tag{3.104}$$

where $n \in \mathbb{N}_0$ and $\xi(\varphi,t)$ is the representation of $\xi \in \Omega$ in the polar coordinates φ and t. Let us have a closer look at these functions. Since, for an integrable function $L\colon \Omega \to \mathbb{R}$ with $L(\xi(\varphi,t)) = L_1(\varphi)L_2(t)$, we have

$$\int_\Omega L(\xi)\,\mathrm{d}\omega(\xi) = \int_{-1}^{1} L_2(t)\,\mathrm{d}t \int_0^{2\pi} L_1(\varphi)\,\mathrm{d}\varphi,$$

and due to the well-known fact that G_j with $G_j(\varphi) := \sin(j\varphi)$ for $j = 1,\ldots,n$ and $G_j(\varphi) := \cos(j\varphi)$ for $j = -n,\ldots,0$ satisfy

$$\int_0^{2\pi} G_j(\varphi)G_k(\varphi)\,\mathrm{d}\varphi = \delta_{jk}\frac{2\pi}{2-\delta_{j0}}, \tag{3.105}$$

where δ_{jk} is the Kronecker delta, we have that

$$\left\langle \tilde{Y}_{n,j}, \tilde{Y}_{m,k}\right\rangle_{\mathrm{L}^2(\Omega)} = 0 \text{ for } j \neq k$$

(with $|j|,|k| \leq \min(n,m)$). What happens if $j = k$ but $n \neq m$? In this particular case, we should recall that the functions $\tilde{Y}_{n,j}$ were derived as solutions of (3.82). There is namely something interesting about this equation (which is actually well known).

Lemma 3.4.13 *Let Y_n, $Y_m \in \mathrm{C}^{(2)}(\Omega)$ be eigenfunctions of the Beltrami operator to the eigenvalues $-n(n+1)$ and $-m(m+1)$, respectively, where $n,m \in \mathbb{N}_0$. If $n \neq m$, then $\langle Y_n, Y_m\rangle_{\mathrm{L}^2(\Omega)} = 0$.*

Proof Let H_n and H_m be the IDP solutions to the boundary-values Y_n and Y_m, respectively. Then Green's second identity (see Theorem 2.4.2) yields

$$\begin{aligned} 0 &= \int_{B_1(0)} H_n(x)\underbrace{\Delta H_m(x)}_{=0} - H_m(x)\underbrace{\Delta H_n(x)}_{=0}\,\mathrm{d}x \\ &= \int_\Omega Y_n(\xi)\frac{\partial H_m}{\partial \nu}(\xi) - Y_m(\xi)\frac{\partial H_n}{\partial \nu}(\xi)\,\mathrm{d}\omega(\xi), \end{aligned} \tag{3.106}$$

where ν is the outer unit normal $\nu(\xi) = \xi$ to Ω. From our derivations, we can deduce what H_n and H_m are, because they are unique:

$$H_n(r\xi) = r^n Y_n(\xi), \qquad H_m(r\xi) = r^m Y_m(\xi),$$

where $r \in]0,1]$ and $\xi \in \Omega$. Hence, with $p \in \{n,m\}$, we see that

$$\frac{\partial H_p}{\partial \nu}(\xi) = \xi\cdot\left[\left(\xi\frac{\partial}{\partial r} + \frac{1}{r}\nabla^*_\xi\right)H_p(r\xi)\right]\Bigg|_{r=1} = \left(\frac{\partial}{\partial r}H_p(r\xi)\right)\Bigg|_{r=1},$$

where we used Theorem 3.4.1 and the facts that $\xi = \varepsilon^r$ and $\varepsilon^r \cdot \nabla^* = 0$ (see Example 2.5.4). Thus, $\frac{\partial H_p}{\partial \nu}(\xi) = \left(p r^{p-1}Y_p(\xi)\right)\big|_{r=1} = p\,Y_p(\xi)$. If we insert this in (3.106), we obtain $0 = (m-n)\langle Y_n, Y_m\rangle_{\mathrm{L}^2(\Omega)}$. Hence, the inner product must vanish if $n \neq m$. □

Theorem 3.4.14 *For all integers $n, m \in \mathbb{N}_0$, $j \in \{-n, \ldots, n\}$ and $k \in \{-m, \ldots, m\}$, we have $\int_\Omega \tilde{Y}_{n,j}(\xi)\, \tilde{Y}_{m,k}(\xi)\, \mathrm{d}\omega(\xi) = 0$ if $n \neq m$ or $j \neq k$.*

Proof This is an immediate result of the previous considerations. □

This means that the functions $\tilde{Y}_{n,j}$ are orthogonal in the $\mathrm{L}^2(\Omega)$-sense. They are, however, not orthonormal. Before we can calculate their norm, we need to do some further investigations of the functions $\tilde{Y}_{n,j}$ and the Legendre polynomials. These functions have already been intensively investigated; see, for example, Abramowitz and Stegun (1972), Atkinson and Han (2012), Chihara (1978), Efthimiou and Frye (2014), Ferrers (1877), Freeden and Gutting (2013), Freeden and Schreiner (2009), Freeden et al. (1998), Heiskanen and Moritz (1981), Hobson (1965), MacRobert (1927), Magnus et al. (1966), Michel (2013), Müller (1966), Robin (1958), Robin (1959), and Szegö (1975). For parts of the remaining pages of this section, we will use some derivations from Agarwal and O'Regan (2009); Freeden et al. (1998), and Robin (1957).

Theorem 3.4.15 (Rodriguez's Formula) *The Legendre polynomials satisfy, for each degree $n \in \mathbb{N}_0$, the identity*

$$P_n(t) = \frac{1}{2^n n!} \frac{\mathrm{d}^n}{\mathrm{d}t^n} (t^2 - 1)^n, \quad t \in [-1, 1].$$

Proof We set $Q_n(t) := (t^2 - 1)^n$ and observe that

$$(t^2 - 1) Q_n'(t) = (t^2 - 1) n (t^2 - 1)^{n-1} \cdot 2t = 2nt\, Q_n(t).$$

Now we differentiate the latter result $(n+1)$ times and use the Leibniz rule (Theorem 2.3.6):

$$(t^2 - 1) Q_n^{(n+2)}(t) + (n+1) \cdot 2t \cdot Q_n^{(n+1)}(t) + \frac{(n+1) \cdot n}{2} \cdot 2 \cdot Q_n^{(n)}(t)$$
$$= 2nt\, Q_n^{(n+1)}(t) + 2n(n+1) Q_n^{(n)}(t).$$

We can simplify this to

$$(t^2 - 1) Q_n^{(n+2)}(t) + 2t\, Q_n^{(n+1)}(t) - n(n+1) Q_n^{(n)}(t) = 0.$$

Hence, $Q_n^{(n)}$ solves the differential equation (3.94) of the Legendre polynomials. In our preceding considerations, we saw that the Legendre polynomials and their multiples are the only polynomials which solve this equation. Thus, there exist constants $c_n \in \mathbb{R}$ such that $Q_n^{(n)} = c_n P_n$. Moreover, we have $Q_n(t) = \sum_{j=0}^{n} \binom{n}{j} t^{2j} (-1)^{n-j}$ such that

$$Q_n^{(k)}(t) = \sum_{j=\lceil k/2 \rceil}^{n} \binom{n}{j} \cdot 2j \cdot (2j-1) \cdots (2j-k+1) t^{2j-k} (-1)^{n-j}$$

and, consequently,

$$Q_n^{(n)}(0) = \binom{n}{n/2} n! (-1)^{n/2}, \text{ if } n \text{ is even, and}$$

$$Q_n^{(n)}(0) = 0 \text{ and } Q_n^{(n+1)}(0) = \binom{n}{(n+1)/2} (n+1)! (-1)^{(n-1)/2}, \text{ if } n \text{ is odd.}$$

On the other hand, Definition 3.4.7 tells us that

$$P_n(0) = (-1)^{n/2} \frac{n!}{2^n} \left[\left(\frac{n}{2} \right)! \right]^{-2}, \qquad \text{if } n \text{ is even, and}$$

$$P_n(0) = 0 \text{ and } P_n'(0) = (-1)^{(n-1)/2} \frac{(n+1)!}{2^n} \left[\left(\frac{n+1}{2} \right)! \left(\frac{n-1}{2} \right)! \right]^{-1},$$

if n is odd. Since

$$\binom{n}{n/2} = \frac{n!}{\left[\left(\frac{n}{2} \right)! \right]^2}, \qquad \text{if } n \text{ is even, and}$$

$$\binom{n}{(n+1)/2} = \frac{n!}{\left(\frac{n+1}{2} \right)! \left(\frac{n-1}{2} \right)!}, \qquad \text{if } n \text{ is odd,}$$

we are able to conclude that $Q_n^{(n)} = n!\, 2^n P_n$ for all $n \in \mathbb{N}_0$. □

Corollary 3.4.16 *The* ***associated Legendre functions***

$$P_{n,j}(t) := (1-t^2)^{j/2} \frac{\mathrm{d}^j}{\mathrm{d}t^j} P_n(t), \quad t \in [-1,1], \tag{3.107}$$

where $n \in \mathbb{N}_0$ and $j = 0, \ldots, n$ satisfy

$$P_{n,j}(t) = \frac{1}{2^n n!} (1-t^2)^{j/2} \frac{\mathrm{d}^{n+j}}{\mathrm{d}t^{n+j}} (t^2-1)^n, \quad t \in [-1,1].$$

Obviously, $P_{n,j}(\pm 1) = 0$ for $j = 1, \ldots, n$.

With this abbreviation, we can write

$$\tilde{Y}_{n,j}(\xi(\varphi,t)) = P_{n,|j|}(t) \cdot \begin{cases} \sin(j\varphi), & j = 1, \ldots, n, \\ \cos(j\varphi), & j = -n, \ldots, 0. \end{cases}$$

There is still work to do before we can calculate $\|\tilde{Y}_{n,j}\|_{\mathrm{L}^2(\Omega)}$. Besides, most of the intermediate steps are valuable results by themselves.

Theorem 3.4.17 *The so-called* ***generating function*** *of the Legendre polynomials, which is given by $F(x,t) := \sum_{n=0}^{\infty} P_n(x) t^n$, satisfies $F(x,t) = (1-2xt+t^2)^{-1/2}$, where the series converges pointwise for all $x,t \in \mathbb{R}$ with $|t| < \sqrt{1+x^2} - |x|$. Moreover, if $r \in \mathbb{R}^+$ is arbitrary but fixed, then the series converges absolutely and uniformly for all $x \in [-r,r]$ and all $t \in \mathbb{R}$ with $|t| < \sqrt{1+r^2} - r$.*

Proof Let $r \in \mathbb{R}^+$ be arbitrary. We consider now arbitrary $x \in [-r,r]$ and $t \in \mathbb{R}$ with $|t| < \sqrt{1+r^2} - r$. Then

$$\begin{aligned} |2xt - t^2| &\le 2|x||t| + |t|^2 \\ &< 2r\sqrt{1+r^2} - 2r^2 + 1 + r^2 - 2r\sqrt{1+r^2} + r^2 \end{aligned}$$

such that $|2xt - t^2| < 1$. Hence, we may use the well-known power series (see, e.g., Zeidler, 1996, p. 119)

$$(1+y)^{-1/2} = 1 + \sum_{k=1}^{\infty} \frac{(-1/2)\cdots(-k+1/2)}{k!} y^k, \quad |y| < 1,$$

with $y = -(2xt - t^2)$. This leads us to

$$\begin{aligned}
&(1 - 2xt + t^2)^{-1/2} \\
&= 1 + \sum_{k=1}^{\infty} \frac{(-1/2)\cdots(-k+1/2)}{k!} (2xt - t^2)^k (-1)^k \\
&= 1 + \sum_{k=1}^{\infty} \frac{(-1/2)\cdots(-k+1/2)}{k!} (-1)^k t^k \sum_{j=0}^{k} \binom{k}{j} (2x)^{k-j} (-t)^j \\
&=: \sum_{n=0}^{\infty} a_n(x) t^n
\end{aligned}$$

with the following ($n = k + j$ implies that $1 \le k \le n$, $k = n - j$, $0 \le j \le n - 1$, but $j \le k = n - j$ such that $2j \le n$):

$$\begin{aligned}
a_0(x) &= 1, \\
a_n(x) &= \sum_{j=0}^{\lfloor n/2 \rfloor} \frac{(-1/2)\cdots(-n+j+1/2)}{(n-j)!} \binom{n-j}{j} (2x)^{n-2j} (-1)^{n-j+j} \\
&= \sum_{j=0}^{\lfloor n/2 \rfloor} \frac{(-1/2)\cdots(-n+j+1/2)}{(n-2j)!\, j!} (2x)^{n-2j} (-1)^n \\
&= \sum_{j=0}^{\lfloor n/2 \rfloor} (-2)^{j-n} \frac{1 \cdot 3 \cdots (2n-2j-1)}{(n-2j)!\, j!} 2^{n-2j} (-1)^n x^{n-2j}.
\end{aligned} \tag{3.108}$$

Moreover, the following holds true:

$$\begin{aligned}
\frac{(2n-2j)!}{(n-j)!} &= \frac{(2n-2j)(2n-2j-1)\cdots 3 \cdot 2 \cdot 1}{(n-j)(n-j-1)\cdots 2 \cdot 1} \\
&= 2^{n-j}(2n-2j-1)(2n-2j-3)\cdots 3 \cdot 1.
\end{aligned} \tag{3.109}$$

If we insert (3.109) in (3.108), then we obtain

$$a_n(x) = \sum_{j=0}^{\lfloor n/2 \rfloor} (-1)^j 2^{-n} \frac{(2n-2j)!}{(n-j)!\,(n-2j)!\, j!} x^{n-2j} = P_n(x)$$

due to Theorem 3.4.10, where the case $n = 0$ is trivial. Furthermore, remember that a power series – such as the series we used in this proof – always converges absolutely and uniformly in the interior of its circle of convergence. □

Moreover, the series in Theorem 3.4.17 is known to converge pointwise for $x \in [-1,1]$ and $t \in]-1,1[$ and uniformly for $x \in [-1,1]$ and $t \in [-h,h]$ for an arbitrary but fixed $h \in]0,1[$.

Corollary 3.4.18 *For all $x, y \in \mathbb{R}^3$ with $|x| < |y|$, the fundamental solution can be expanded in Legendre polynomials as follows:*

$$\frac{1}{|x-y|} = \sum_{n=0}^{\infty} \frac{|x|^n}{|y|^{n+1}} P_n\left(\frac{x}{|x|} \cdot \frac{y}{|y|}\right).$$

Proof We set $t := |x|/|y|$ and $s := (x/|x|) \cdot (y/|y|)$. Then Theorem 3.4.17 yields

$$\begin{aligned}\frac{1}{|y|}\sum_{n=0}^{\infty}\left(\frac{|x|}{|y|}\right)^n P_n\left(\frac{x}{|x|}\cdot\frac{y}{|y|}\right) &= \frac{1}{|y|}\left(1-2\frac{x}{|x|}\cdot\frac{y}{|y|}\frac{|x|}{|y|}+\frac{|x|^2}{|y|^2}\right)^{-1/2}\\ &= \left(|y|^2 - 2x\cdot y + |x|^2\right)^{-1/2} = |x-y|^{-1}.\end{aligned}$$

□

The following theorem is very useful, also for practical numerical purposes, because it allows us to calculate the Legendre polynomials much more easily than Definition 3.4.7 or Theorems 3.4.10 and 3.4.15 allow it.

Theorem 3.4.19 (Recurrence Formula for Legendre Polynomials)
The Legendre polynomials obey the following recurrence relation

$$\begin{aligned} P_0(x) &= 1, \qquad\qquad P_1(x) = x,\\ (n+1)P_{n+1}(x) &= (2n+1)xP_n(x) - nP_{n-1}(x) \end{aligned} \tag{3.110}$$

for all $x \in \mathbb{R}$ and all $n \in \mathbb{N}$.

Proof We differentiate the generating function (see Theorem 3.4.17) with respect to t and get (within the admissible domain) the identity

$$-\frac{1}{2}(1-2xt+t^2)^{-3/2}(-2x+2t) = \sum_{n=1}^{\infty} P_n(x)nt^{n-1}. \tag{3.111}$$

Note that this is valid, because we have here, for every fixed $x \in \mathbb{R}$, a power series in t, which may be infinitely differentiated summandwise.

By multiplying the preceding result with $(1-2xt+t^2)$, we get

$$(1-2xt+t^2)^{-1/2}(x-t) = (1-2xt+t^2)\sum_{n=1}^{\infty} P_n(x)nt^{n-1}$$

such that

$$(x-t)\sum_{n=0}^{\infty} P_n(x)t^n = \sum_{n=1}^{\infty} P_n(x)nt^{n-1} + \sum_{n=1}^{\infty}(-2x)P_n(x)nt^n + \sum_{n=1}^{\infty} P_n(x)nt^{n+1}.$$

With minor index shifts, we obtain

$$\sum_{n=0}^{\infty} x P_n(x) t^n + \sum_{n=1}^{\infty} (-P_{n-1}(x)) t^n$$
$$= \sum_{n=0}^{\infty} P_{n+1}(x)(n+1)t^n + \sum_{n=1}^{\infty} (-2xn) P_n(x) t^n + \sum_{n=1}^{\infty} P_{n-1}(x)(n-1)t^n.$$

In the last series, we added the *vanishing* summand for $n = 1$.

Since a power series is unique for a fixed centre, the preceding identity is only valid if $xP_n(x) - P_{n-1}(x) = (n+1)P_{n+1}(x) - 2xnP_n(x) + (n-1)P_{n-1}(x)$, that is, $(2n+1)xP_n(x) - nP_{n-1}(x) = (n+1)P_{n+1}(x)$ for all $n \geq 1$ and all $x \in \mathbb{R}$. □

Corollary 3.4.20 *The Abel–Poisson kernel can be expanded as follows:*

$$\frac{1}{4\pi} \frac{R^2 - r^2}{(r^2 + R^2 - 2rRx)^{3/2}} = \sum_{n=0}^{\infty} \frac{2n+1}{4\pi R} \left(\frac{r}{R}\right)^n P_n(x), \ x \in [-1,1], 0 \leq r < R.$$

Proof By combining (3.111), where the vanishing summand for $n = 0$ can be added, and the generating function from Theorem 3.4.17, we get

$$\sum_{n=0}^{\infty} (2n+1) t^n P_n(x) = \frac{2xt - 2t^2}{(1 - 2xt + t^2)^{3/2}} + \frac{1 - 2xt + t^2}{(1 - 2xt + t^2)^{3/2}}$$
$$= \frac{1 - t^2}{(1 - 2xt + t^2)^{3/2}}.$$

With $t = r/R$, we obtain

$$\sum_{n=0}^{\infty} \frac{2n+1}{4\pi R} \left(\frac{r}{R}\right)^n P_n(x) = \frac{1 - r^2 R^{-2}}{4\pi R (1 - 2xrR^{-1} + r^2 R^{-2})^{3/2}}$$
$$= \frac{R^2 - r^2}{4\pi (R^2 - 2rRx + r^2)^{3/2}}.$$ □

Corollary 3.4.21 (More Recurrence Formulae for Legendre Polynomials)
For all $x \in \mathbb{R}$ and all $n \in \mathbb{N}$, we have

$$nP_n(x) = xP_n'(x) - P_{n-1}'(x), \tag{3.112}$$
$$(2n+1)P_n(x) = P_{n+1}'(x) - P_{n-1}'(x), \tag{3.113}$$
$$(n+1)P_n(x) = P_{n+1}'(x) - xP_n'(x), \tag{3.114}$$
$$(x^2 - 1)P_n'(x) = n(xP_n(x) - P_{n-1}(x)), \tag{3.115}$$
$$(x^2 - 1)P_n'(x) = (n+1)(P_{n+1}(x) - xP_n(x)), \tag{3.116}$$
$$(2n+1)(x^2 - 1)P_n'(x) = n(n+1)(P_{n+1}(x) - P_{n-1}(x)), \tag{3.117}$$
$$(2n+1)xP_n'(x) = nP_{n+1}'(x) + (n+1)P_{n-1}'(x). \tag{3.118}$$

Proof We only elaborate here the proofs of the first three formulae.

(1) The first formula:
We differentiate now the generating function from Theorem 3.4.17 with respect to x. This yields $t(1-2xt+t^2)^{-3/2} = \sum_{n=1}^{\infty} P_n'(x)t^n$. Together with (3.111), this implies that $(x-t)\sum_{n=1}^{\infty} P_n'(x)t^n = t\sum_{n=1}^{\infty} P_n(x)nt^{n-1}$. Hence, again with a little index shift, we arrive at

$$\sum_{n=1}^{\infty} xP_n'(x)t^n + \sum_{n=2}^{\infty}\left(-P_{n-1}'(x)\right)t^n = \sum_{n=1}^{\infty} P_n(x)nt^n,$$

which can only be true if (3.112) holds, where we should observe that $P_{n-1}' \equiv 0$ for $n=1$.

(2) The second formula:
By differentiating (3.110), we obtain

$$(n+1)P_{n+1}'(x) = (2n+1)P_n(x) + (2n+1)xP_n'(x) - nP_{n-1}'(x).$$

We use now (3.112) for $xP_n'(x)$ such that

$$\begin{aligned}(n+1)P_{n+1}'(x) = {} & (2n+1)P_n(x) + (2n+1)nP_n(x)\\ & + (2n+1)P_{n-1}'(x) - nP_{n-1}'(x).\end{aligned}$$

Eventually, we divide by $n+1$ and get $P_{n+1}'(x) = (2n+1)P_n(x) + P_{n-1}'(x)$, which proves (3.113).

(3) The third formula:
We only have to subtract (3.112) from (3.113) and immediately get (3.114).

The proofs of the remaining formulae (3.115)–(3.118) are left to the reader as an (easy) exercise. □

Let us now have a closer look at the integrals $\int_{-1}^{1} P_n(t)P_m(t)\,\mathrm{d}t$. Due to (3.104), we see that $P_n(t) = \tilde{Y}_{n,0}(\xi(\varphi,t))$ for all φ and t (i.e., $\tilde{Y}_{n,0}$ is independent of φ). Hence, Theorem 3.4.14 yields that

$$\int_{-1}^{1} P_n(t)P_m(t)\,\mathrm{d}t = \frac{1}{2\pi}\int_{\Omega} \tilde{Y}_{n,0}(\xi)\tilde{Y}_{m,0}(\xi)\,\mathrm{d}\omega(\xi) = 0,$$

if $n \neq m$. For the case $n = m$, we use Rodriguez's formula (Theorem 3.4.15):

$$\int_{-1}^{1} (P_n(t))^2\,\mathrm{d}t = \frac{1}{2^{2n}(n!)^2}\int_{-1}^{1}\left[\frac{\mathrm{d}^n}{\mathrm{d}t^n}(t^2-1)^n\right]\frac{\mathrm{d}^n}{\mathrm{d}t^n}(t^2-1)^n\,\mathrm{d}t.$$

We now integrate by parts n times if $n > 0$. On the way to the result, we have to evaluate terms of the kind

$$\left\{\left[\frac{\mathrm{d}^{n-k-1}}{\mathrm{d}t^{n-k-1}}(t^2-1)^n\right]\frac{\mathrm{d}^{n+k}}{\mathrm{d}t^{n+k}}(t^2-1)^n\right\}\Bigg|_{t=-1}^{t=1} = 0$$

for $k = 0, \ldots, n-1$. Finally, we get (using $t = \cos x$)

$$\int_{-1}^{1} (P_n(t))^2 \, \mathrm{d}t = \frac{(-1)^n}{2^{2n}(n!)^2} \int_{-1}^{1} (t^2-1)^n \frac{\mathrm{d}^{2n}}{\mathrm{d}t^{2n}} (t^2-1)^n \, \mathrm{d}t$$

$$= \frac{(2n)!}{2^{2n}(n!)^2} \int_{-1}^{1} (1-t^2)^n \, \mathrm{d}t = -\frac{(2n)!}{2^{2n}(n!)^2} \int_{0}^{\pi} \sin^{2n} x \cdot (-\sin x) \, \mathrm{d}x.$$

By induction and integration by parts, we can easily deduce that

$$\int_0^{\pi} \sin^{2n+1} x \, \mathrm{d}x = \left(-\cos x \cdot \sin^{2n} x\right)\Big|_0^{\pi} + 2n \int_0^{\pi} \cos^2 x \, \sin^{2n-1} x \, \mathrm{d}x$$

$$= 2n \int_0^{\pi} \sin^{2n-1} x - \sin^{2n+1} x \, \mathrm{d}x$$

and, consequently,

$$\int_0^{\pi} \sin^{2n+1} x \, \mathrm{d}x = \frac{2n}{2n+1} \int_0^{\pi} \sin^{2n-1} x \, \mathrm{d}x$$

$$= \frac{(2n)(2n-2)\cdots 2}{(2n+1)(2n-1)\cdots 3} \underbrace{\int_0^{\pi} \sin x \, \mathrm{d}x}_{=2}.$$

Hence, we obtain the following squared $\mathrm{L}^2[-1,1]$-norm:

$$\int_{-1}^{1} (P_n(t))^2 \, \mathrm{d}t = \frac{(2n)(2n-1)\cdots 2 \cdot 1}{2^{2n}[n(n-1)\cdots 2 \cdot 1]^2} \cdot \frac{(2n)(2n-2)\cdots 2}{(2n+1)(2n-1)\cdots 3} \cdot 2 = \frac{2}{2n+1}.$$

This is also true for $n = 0$, since $P_0 \equiv 1$. We have proved the following theorem.

Theorem 3.4.22 (Orthogonality and Norm of the Legendre Polynomials) *For all $n, m \in \mathbb{N}_0$, we have*

$$\int_{-1}^{1} P_n(t) P_m(t) \, \mathrm{d}t = \frac{2}{2n+1} \delta_{nm},$$

where δ_{nm} is the Kronecker delta.

We will now show the generalization to the associated Legendre functions. Remember that $P_n = P_{n,0}$.

Theorem 3.4.23 (Norm of the Associated Legendre Functions) *We get, for all $n \in \mathbb{N}_0$ and all $j = 0, \ldots, n$, the identity*

$$\int_{-1}^{1} (P_{n,j}(t))^2 \, \mathrm{d}t = \frac{2}{2n+1} \frac{(n+j)!}{(n-j)!}.$$

Proof We may assume that $j > 0$. From the definition of these functions in (3.107), namely, $P_{n,j}(t) = (1-t^2)^{j/2} P_n^{(j)}(t)$, we obtain that

$$\frac{\mathrm{d}}{\mathrm{d}t} P_{n,j}(t) = \frac{j}{2}(1-t^2)^{(j-2)/2} \cdot (-2t) \cdot P_n^{(j)}(t) + (1-t^2)^{j/2} P_n^{(j+1)}(t),$$

where $P_n^{(j)}$ is the jth derivative of $P_n = P_{n,0}$. We set now $P_{n,n+1}(t) := 0$ for all $t \in [-1, 1]$. Hence,

$$(1-t^2)\frac{\mathrm{d}}{\mathrm{d}t}P_{n,j}(t) = -jtP_{n,j}(t) + \sqrt{1-t^2}\, P_{n,j+1}(t).$$

With a rearrangement and a little index shift ($j+1 \to j$), we get

$$P_{n,j}(t) = \sqrt{1-t^2}\,\frac{\mathrm{d}}{\mathrm{d}t}P_{n,j-1}(t) + (j-1)\frac{t}{\sqrt{1-t^2}}\,P_{n,j-1}(t) \tag{3.119}$$

for all $t \in]-1, 1[$. Let us now integrate the square of (3.119)

$$\begin{aligned}\int_{-1}^{1}\left(P_{n,j}(t)\right)^2\,\mathrm{d}t &= \int_{-1}^{1}(1-t^2)\left(\frac{\mathrm{d}}{\mathrm{d}t}P_{n,j-1}(t)\right)^2\,\mathrm{d}t\\ &\quad + 2(j-1)\int_{-1}^{1} tP_{n,j-1}(t)\,\frac{\mathrm{d}}{\mathrm{d}t}P_{n,j-1}(t)\,\mathrm{d}t\\ &\quad + (j-1)^2\int_{-1}^{1}\frac{t^2}{1-t^2}\left(P_{n,j-1}(t)\right)^2\,\mathrm{d}t.\end{aligned} \tag{3.120}$$

The first integral on the right-hand side will now be handled via integration by parts, where the obtained primitive functions term vanishes for $t = \pm 1$. Then we use

$$\frac{\mathrm{d}}{\mathrm{d}t}\left[t(P_{n,j-1}(t))^2\right] = \left(P_{n,j-1}(t)\right)^2 + 2tP_{n,j-1}(t)\,\frac{\mathrm{d}}{\mathrm{d}t}P_{n,j-1}(t)$$

in the second integral of the right-hand side. Moreover, the last integral in (3.120) will be rearranged slightly by decomposing the fraction. We get

$$\begin{aligned}\int_{-1}^{1}\left(P_{n,j}(t)\right)^2\,\mathrm{d}t &= -\int_{-1}^{1} P_{n,j-1}(t)\,\frac{\mathrm{d}}{\mathrm{d}t}\left[(1-t^2)\frac{\mathrm{d}}{\mathrm{d}t}P_{n,j-1}(t)\right]\,\mathrm{d}t\\ &\quad - (j-1)\int_{-1}^{1}\left(P_{n,j-1}(t)\right)^2\,\mathrm{d}t + (j-1)\left[t\left(P_{n,j-1}(t)\right)^2\right]\Big|_{t=-1}^{t=1}\\ &\quad - (j-1)^2\int_{-1}^{1}\left(P_{n,j-1}(t)\right)^2\,\mathrm{d}t + (j-1)^2\int_{-1}^{1}\frac{1}{1-t^2}\left(P_{n,j-1}(t)\right)^2\,\mathrm{d}t.\end{aligned}$$

Either $j-1=0$ or $P_{n,j-1}(\pm 1) = 0$ (see Corollary 3.4.16) such that the third term on the right-hand side vanishes. Moreover, Theorem 3.4.12 tells us that $P_{n,j-1}$ solves the differential equation

$$(1-t^2)\frac{\mathrm{d}}{\mathrm{d}t}\left[(1-t^2)\frac{\mathrm{d}}{\mathrm{d}t}P_{n,j-1}(t)\right] = \left[(j-1)^2 - n(n+1)(1-t^2)\right]P_{n,j-1}(t).$$

Thus,

$$\int_{-1}^{1} \left(P_{n,j}(t)\right)^2 \, \mathrm{d}t = \left[n(n+1) - (j-1) - (j-1)^2\right] \int_{-1}^{1} \left(P_{n,j-1}(t)\right)^2 \, \mathrm{d}t$$
$$= \left(n^2 + n - j^2 + j\right) \int_{-1}^{1} \left(P_{n,j-1}(t)\right)^2 \, \mathrm{d}t$$
$$= (n+j)(n-j+1) \int_{-1}^{1} \left(P_{n,j-1}(t)\right)^2 \, \mathrm{d}t.$$

With a simple induction and Theorem 3.4.22, we can now conclude that

$$\int_{-1}^{1} \left(P_{n,j}(t)\right)^2 \, \mathrm{d}t$$
$$= \left[(n+j)(n+j-1)\cdots(n+1)\right]\left[(n-j+1)(n-j+2)\cdots n\right] \frac{2}{2n+1}$$
$$= \frac{(n+j)!}{(n-j)!} \frac{2}{2n+1}.$$

□

Let us summarize what we have done in this section.

Definition 3.4.24 The **fully normalized spherical harmonics** are defined by

$$Y_{n,j}(\xi(\varphi,t)) := \sqrt{\frac{(2n+1)(n-|j|)!\,(2-\delta_{j0})}{4\pi(n+|j|)!}}\, P_{n,|j|}(t) \begin{cases} \sin(j\varphi), & j = 1,\ldots,n, \\ \cos(j\varphi), & j = -n,\ldots,0, \end{cases}$$

where $\xi(\varphi,t)$ is the representation of $\xi \in \Omega$ in polar coordinates.

Theorems 3.4.14 and 3.4.23 together with (3.105) yield now the following theorem.

Theorem 3.4.25 *The fully normalized spherical harmonics represent an orthonormal system in* $(\mathrm{L}^2(\Omega), \langle\cdot,\cdot\rangle_{\mathrm{L}^2(\Omega)})$.

We have not proved yet that the fully normalized spherical harmonics are also complete in $\mathrm{L}^2(\Omega)$. We will see later that this is, indeed, the case. With this 'spoiler' in mind, we can say now that the unique IDP solution U to boundary values $F\colon \Omega \to \mathbb{R}$ can be represented by

$$U(r\xi) = \sum_{n=0}^{\infty} \sum_{j=-n}^{n} \langle F, Y_{n,j}\rangle_{\mathrm{L}^2(\Omega)} r^n Y_{n,j}(\xi), \quad r \in [0,1], \xi \in \Omega,$$

provided that the series has an appropriate convergence (e.g., we would have to interchange Δ with the limit of the series $\sum_{n=0}^{\infty}$). Correspondingly, the unique EDP solution U to boundary values $F\colon \Omega \to \mathbb{R}$ can (again under the assumption of an appropriate convergence) be represented by

$$U(r\xi) = \sum_{n=0}^{\infty} \sum_{j=-n}^{n} \langle F, Y_{n,j}\rangle_{\mathrm{L}^2(\Omega)} r^{-n-1} Y_{n,j}(\xi), \quad r \geq 1, \xi \in \Omega.$$

Just one more thing which fits quite well into our previous work: the functions $\Omega_{\mathrm{int}} \ni r\xi \mapsto r^n Y_{n,j}(\xi)$ and $\Omega_{\mathrm{ext}} \ni r\xi \mapsto r^{-n-1} Y_{n,j}(\xi)$ can be obtained from each other by the Kelvin transform (see Definition 3.2.1 and note Corollary 3.2.3).

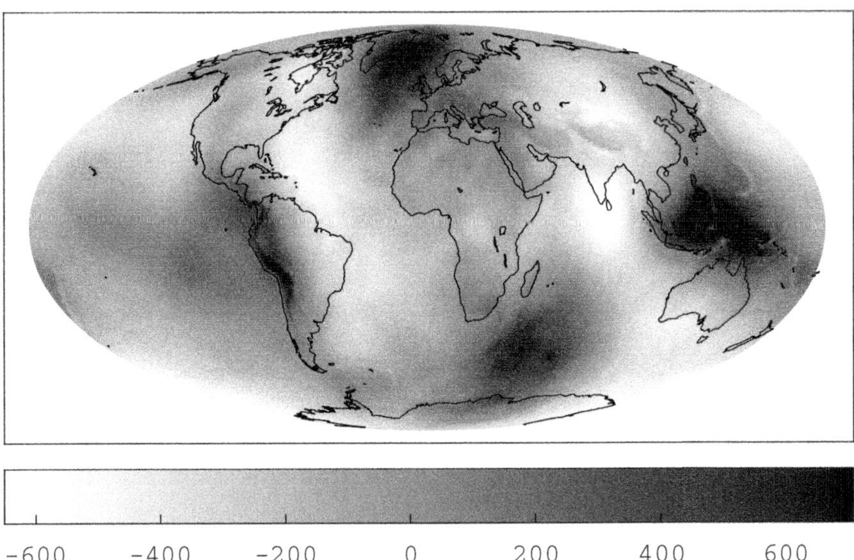

Figure 3.9 The Earth Gravitational Model 2008 (EGM2008) in $\mathrm{m}^2\mathrm{s}^{-2}$ due to Pavlis et al. (2012) from degrees 3 to 2190 and orders 0 to 2159.

Definition 3.4.26 The functions

$$H^{\mathrm{int}}_{n,j}(x) := |x|^n\, Y_{n,j}\left(\frac{x}{|x|}\right), \quad x \in \Omega_{\mathrm{int}},$$

are called **inner harmonics**. Moreover, the functions

$$H^{\mathrm{ext}}_{-n-1,j}(x) := |x|^{-n-1} Y_{n,j}\left(\frac{x}{|x|}\right), \quad x \in \Omega_{\mathrm{ext}},$$

are called **outer harmonics**.

Fully normalized spherical harmonics are a common tool for representing approximations to the Earth's gravitational potential. An example is the Earth Gravitational Model 2008 (EGM2008) due to Pavlis et al. (2012); see Figure 3.9.

3.5 A Brief Excursion to Other Dimensions

Within this chapter, we have concentrated on the three-dimensional case, which is reasonable due to the focus on geomathematics. Nevertheless, it can be interesting to have a brief look at some propositions and their counterparts in arbitrary dimensions. Some examples are summarized here but without proofs. Further details can be found in Folland (1976), Kellogg (1967), Mikhlin (1970), and Walter (1971).

Theorem 3.5.1 *Let $n \geq 2$ and $y \in \mathbb{R}^n$. Then the function $F\colon \mathbb{R}^n \setminus \{y\} \ni x \mapsto \gamma_n(x-y)$ with*

$$\gamma_n(x) := \frac{1}{(n-2)|x|^{n-2}}, \quad x \in \mathbb{R}^n \setminus \{0\}, \quad \text{if } n \geq 3,$$

and

$$\gamma_2(x) := -\log|x|, \qquad x \in \mathbb{R}^2 \setminus \{0\}, \quad \text{if } n = 2,$$

is harmonic. The functions F are called ***fundamental solutions*** *of the Laplace equation.*

Sometimes, the fundamental solutions are also defined with an additional factor such as ω_n^{-1}, where ω_n is the area of the unit sphere in $\mathbb{R}^n$.

With these fundamental solutions, one can define **volume potentials** as generalizations of Newton's gravitational potential by

$$V(x) := \int_D F(y) \frac{1}{|x-y|^{n-2}} \, \mathrm{d}y, \quad x \in \mathbb{R}^n, \quad \text{if } n \geq 3, \tag{3.121}$$

$$V(x) := \int_D F(y) \log|x-y| \, \mathrm{d}y, \quad x \in \mathbb{R}^2, \quad \text{if } n = 2, \tag{3.122}$$

where $D \subset \mathbb{R}^n$ is a bounded and open set and $F\colon D \to \mathbb{R}$ is a bounded and Lebesgue-integrable function. Indeed, one can also show the following in this generalized case:

- V is continuous and partially differentiable with respect to all variables in the whole $\mathbb{R}^n$.
- V is infinitely differentiable and harmonic in $\mathbb{R}^n \setminus \overline{D}$, that is, V satisfies the **Laplace equation** outside $\overline{D}$.
- If F is additionally Hölder continuous, then V is twice continuously differentiable in D and satisfies the **Poisson equation**:

$$\Delta V(x) = (2-n)\omega_n F(x), \quad x \in D, \quad \text{if } n \geq 3,$$
$$\Delta V(x) = -\omega_2 F(x), \qquad x \in D, \quad \text{if } n = 2.$$

Also for the main properties of harmonic functions which we discussed here, analogues in arbitrary dimensions are valid.

Definition 3.5.2 Let $R \subset \mathbb{R}^n, n \geq 2$, be a region where $0 \notin R$. Then the **Kelvin transform** maps functions $F\colon R \to \mathbb{R}$ to functions

$$F^*(x) := \frac{1}{|x|^{n-2}} F\left(\frac{x}{|x|^2}\right), \quad x \in R^*,$$

where $R^* := \{x \in \mathbb{R}^n \setminus \{0\} \mid x/|x|^2 \in R\}$.

Theorem 3.5.3 *Let $R \subset \mathbb{R}^n, n \geq 2$, be a region with $0 \notin R$ and $F \in \mathrm{C}^{(2)}(R)$ be a given function. Then its Kelvin transform F^* is an element of $\mathrm{C}^{(2)}(R^*)$ and*

$$\Delta_x F^*(x) = \frac{1}{|x|^{n+2}} \left(\Delta_y F(y)\right)\Big|_{y=\frac{x}{|x|^2}}, \quad x \in R^*.$$

Under appropriate conditions on the bounded region R and its boundary ∂R with outer unit normal ν, one also gets a **fundamental theorem** in $\mathbb{R}^n, n \geq 2$, of the following form:

$$\omega_n U(x_0) = \int_{\partial D} \gamma(x - x_0) \frac{\partial}{\partial \nu} U(x) - U(x) \frac{\partial}{\partial \nu(x)} \gamma(x - x_0) \, \mathrm{d}\omega(x)$$
$$- \int_D \gamma(x - x_0) \Delta U(x) \, \mathrm{d}x,$$

if $U \in \mathrm{C}^{(1)}(\overline{D}) \cap \mathrm{C}^{(2)}(D)$ with bounded second-order derivatives on D (note the slightly different requirement, since the argumentation as in (3.31) is not analogous for $n \geq 4$).

This results again in **Gauß's mean value theorem**. If U is harmonic in $B_R(x_0) \subset \mathbb{R}^n, n \geq 2$, and continuous on $\overline{B_R(x_0)}$, then

$$U(x_0) = \frac{1}{\omega_n R^{n-1}} \int_{S_R(x_0)} U(x) \, \mathrm{d}\omega(x) = \frac{n}{\omega_n R^n} \int_{B_R(x_0)} U(x) \, \mathrm{d}x.$$

This leads to completely analogous versions of **maximum principles I and II**.

Last but not least, also the boundary-value problems can be formulated and discussed in $\mathbb{R}^n$ for $n \geq 3$ in a similar way as we did it here for the case $n = 3$. Only regarding the regularity at infinity, the condition depends on the dimension. This results in analogous statements regarding existence, uniqueness, and stability for $n \geq 3$ as far as they were proved here for $n = 3$. For $n = 2$, there are some differences resulting in a modified theorem for the ENP but analogous statements for the other three boundary-value problems (for details, see, e.g., Folland, 1976, Theorem 3.41).

Exercises

3.1 Prove the harmonicity of the fundamental solutions in Theorem 3.5.1.

3.2 Verify that the generalization of the Newton potential to other dimensions in (3.121) and (3.122) is harmonic outside $\overline{D}$.

3.3 Suggest a definition for 'regular at infinity' in $\mathbb{R}^n, n \geq 3$, where your definition should satisfy the following two conditions (which you need to prove): first, it should coincide with the existing definition for $n = 3$. Second, the fundamental solution should satisfy it (you may simply use $y \mapsto |y|^{-n+2}$ here).

3.4 Prove Theorem 3.5.3 for the Kelvin transform in $\mathbb{R}^n$.

3.5 Gauß's mean value theorem is also valid in $\mathbb{R}^1$. Suggest and prove a corresponding analogous theorem. Moreover, derive an explicit formula for all harmonic functions on a region in $\mathbb{R}^1$.

3.6 Let $U \in \mathrm{C}^{(2)}(\overline{B_R(x_0)})$ with $R > 0$ and $x_0 \in \mathbb{R}^3$. Prove the following stronger version of Gauß's mean value theorem:

$$U(x_0) = \frac{1}{4\pi R^2} \int_{S_R(x_0)} U(x)\,\mathrm{d}\omega(x) - \frac{1}{4\pi} \int_{B_R(x_0)} \Delta U(x) \left(\frac{1}{|x - x_0|} - \frac{1}{R} \right) \mathrm{d}x.$$

3.7 The plumb line is given by the (opposite) direction of the gravitational field. More precisely, we define $n(x) := -|\nabla V(x)|^{-1} \nabla V(x)$ for all $x \in \mathbb{R}^3$ in which $|\nabla V(x)| \neq 0$, where V is the usual gravitational potential. If we choose now an arbitrary but fixed point $x_0 \in \mathbb{R}^3$ with $|\nabla V(x_0)| \neq 0$, then the plumb line through x_0 is given by the arc $\varphi\colon I \to \mathbb{R}^3$ ($I \subset \mathbb{R}$ is an interval, and the parameterization is done with respect to the arc length s) which satisfies $x_0 \in \varphi(I)$ and $\varphi'(s) = n(\varphi(s))$ in each $s \in I$. Prove that the curvature of the plumb line satisfies

$$\varphi''(s) = \left(|\nabla V|^{-1} \nabla^* |\nabla V| \right)\Big|_{\varphi(s)},$$

where ∇^* is the surface gradient with respect to the equipotential surface $\{y \in \mathbb{R}^3 \mid V(y) = c\}$ for a fixed $c = V(\varphi(s))$, that is, $\nabla^* F = \nabla F - (n \cdot \nabla F)n$. (Hint: to avoid the square root in the Euclidean norm $|\nabla V|$, investigate $\nabla^* |\nabla V|^2$ instead.)

3.8 Let $R \subset \mathbb{R}^n$ be a region and let $U\colon R \to \mathbb{R}$ be a continuous function. If U satisfies

$$U(x_0) \leq \frac{1}{\Omega_n R^n} \int_{B_R(x_0)} U(x)\,\mathrm{d}x \tag{3.123}$$

in every ball $B_R(x_0)$ with $\overline{B_R(x_0)} \subset R$, where $\Omega_n := \int_{B_1(0)} 1\,\mathrm{d}x$ is the volume of the unit ball in $\mathbb{R}^n$, then U is called subharmonic. If always '$\geq$' holds in (3.123) instead, then U is called superharmonic.

(a) Prove (for $n = 3$ and $U \in \mathrm{C}^{(2)}(\overline{R})$) that $\Delta U \geq 0$ in $R \Leftrightarrow U$ is subharmonic.

(b) Which well-known property from analysis is equivalent to subharmonicity in $\mathbb{R}^1$? (No proof is required.)

3.9 Let $D \subset \mathbb{R}^n$ and $E \subset \mathbb{R}^m$ be compact and non-void sets and let $K\colon E \times D \to \mathbb{R}$ be a continuous function. The corresponding Fredholm integral operator of the first kind $\mathcal{T}\colon \mathrm{C}(D) \to \mathrm{C}(E)$ is defined as in Definition 3.3.22. First, show that $\mathcal{T}$ is bounded, then show that $\mathcal{T}$ is also compact. (Hint: Use the Ascoli–Arzelà theorem for the latter.)

3.10 Let $D \subset \mathbb{R}^n$ and $E \subset \mathbb{R}^m$ be measurable and non-void sets and let $K \in \mathrm{L}^2(E \times D)$, where $\mathcal{T}\colon \mathrm{L}^2(D) \to \mathrm{L}^2(E)$ is again the corresponding Fredholm integral operator of the first kind. Prove the following propositions: $\mathcal{T}F \in \mathrm{L}^2(E)$ for all $F \in \mathrm{L}^2(D)$. $\mathcal{T}$ is a bounded operator. $\mathcal{T}$ is a compact operator. (Hint to the latter: you may use that bounded sequences in $\mathrm{L}^2(D)$ always have a weakly convergent subsequence.)

3.11 The unit disc $D := \left\{x \in \mathbb{R}^3 \,\middle|\, x_1^2 + x_2^2 \leq 1,\, x_3 = 0\right\}$ is not a closed surface, but the layer potentials can be defined completely analogously by replacing Σ with D. For reasons of simplicity, let the distribution function in the integrand be given by $F \equiv 1$.

(a) Calculate P_s along the x_3-axis, that is, $P_s(0,0,x_3)$, and verify that the jump $\partial_{\nu+}P_s - \partial_{\nu-}P_s$ in $(0,0,0)^T$ for $\nu = (0,0,1)^T$ is the same as in the known case for a regular surface.

(b) In analogy to part (a), investigate $(P_d)_+$ and $(P_d)_-$.

3.12 Prove Harnack's inequality: If $U \in C^{(2)}(B_R(0)) \cap C(\overline{B_R(0)})$ is harmonic and non-negative on $B_R(0)$, then

$$R\,\frac{R-|x|}{(R+|x|)^2}\,U(0) \le U(x) \le R\,\frac{R+|x|}{(R-|x|)^2}\,U(0) \quad \text{for all } x \in B_R(0).$$

3.13 Prove that the only functions which are non-negative and harmonic on the entire $\mathbb{R}^3$ are the constant functions (with non-negative constant).

3.14 Let $R \subset \mathbb{R}^3$ be a region and let $U \in C(R)$. Prove the following: if U satisfies Gauß's mean value property on every ball $\overline{B_r(x_0)} \subset R$, that is, (3.32) holds true on every such ball, then U is harmonic on R. (Hint: check what the essential requirements for the proof of maximum principle I were.)

3.15 Show that every uniformly convergent sequence of functions which are harmonic on a region R also has a harmonic function as its limit.

3.16 Download data (ideally, level 2 data) from a current gravitational model such as EGM-type models and plot the gravitational potential.

3.17 Verify that the Laplace operator on $\mathbb{R}^2$ is representable in two-dimensional polar coordinates $x = r\cos\varphi$, $y = r\sin\varphi$ for twice continuously differentiable functions U by

$$\Delta U = \frac{\partial^2 U}{\partial r^2} + \frac{1}{r}\frac{\partial U}{\partial r} + \frac{1}{r^2}\frac{\partial^2 U}{\partial \varphi^2}.$$

3.18 Use the separation ansatz $U(x,y) = F(r)Y(\varphi)$ to derive a representation for the general solution of the 2D Laplace equation $\Delta U = 0$ on the disc $\{(x,y) \in \mathbb{R}^2 \mid x^2 + y^2 \le R^2\}$.

4
Basis Functions

We have already seen that fully normalized spherical harmonics provide us with one way of expanding harmonic functions with respect to their angular dependence. There are many other options of choosing such basis functions and numerous other modelling problems for which different basis systems are required. In this chapter, we will discuss a selection of such basis systems which have turned out to be useful in geomathematics.

4.1 Spherical Analysis

When we tried to solve the Laplace equation, we proved a couple of propositions from spherical analysis. This includes the decompositions of the gradient ∇ and the Laplace operator Δ (see Theorems 3.4.1 and 3.4.3 as well as Definition 3.4.2), which we briefly recapitulate here:

$$\nabla_x = \xi \frac{\partial}{\partial r} + \frac{1}{r} \nabla^*_{\xi}, \tag{4.1}$$

$$\Delta_x = \frac{\partial^2}{\partial r^2} + \frac{2}{r}\frac{\partial}{\partial r} + \frac{1}{r^2} \Delta^*_{\xi}, \tag{4.2}$$

where $x = r\xi$, $r = |x|$, $\xi \in \Omega$. We will summarize here some further useful propositions in the context of differentiation and integration on the sphere. We will not prove everything and refer to Freeden et al. (1998), Freeden and Gutting (2013), Freeden and Schreiner (2009), and Michel (2013) for further details.

We start by defining another differential operator, the surface curl gradient (note that we defined $\mathrm{c}(\Omega) := \mathrm{C}(\Omega, \mathbb{R}^3)$).

Definition 4.1.1 The **surface curl gradient** $\mathrm{L}^*\colon \mathrm{C}^{(1)}(\Omega) \to \mathrm{c}(\Omega)$ is defined by $\mathrm{L}^*_{\xi} F(\xi) := \xi \times \nabla^*_{\xi} F(\xi)$, $\xi \in \Omega$, for all $F \in \mathrm{C}^{(1)}(\Omega)$.

Remember the local orthonormal basis system which is given in $\xi \in \Omega$ with the polar coordinates φ and t by the normal vector $\varepsilon^r(\varphi, t) = \xi$ and the tangential vectors $\varepsilon^{\varphi}(\varphi)$ and $\varepsilon^t(\varphi, t)$; see Example 2.5.4. We know that $\varepsilon^r \times \varepsilon^{\varphi} = \varepsilon^t$. Moreover, we have the following (see Definition 3.4.2):

$$\nabla^* = \varepsilon^{\varphi} \frac{1}{\sqrt{1-t^2}} \frac{\partial}{\partial \varphi} + \varepsilon^t \sqrt{1-t^2} \frac{\partial}{\partial t}. \tag{4.3}$$

Hence,

$$\mathrm{L}_\xi^* = \varepsilon^t \frac{1}{\sqrt{1-t^2}} \frac{\partial}{\partial \varphi} - \varepsilon^\varphi \sqrt{1-t^2} \frac{\partial}{\partial t}. \tag{4.4}$$

In analogy to the notations that we have for the (Euclidean) divergence, Jacobian matrix, and Hessian (see Definition 2.3.1), we can also define corresponding operators on the sphere.

Definition 4.1.2 Let $f \in \mathrm{c}^{(1)}(\Omega)$ and $G \in \mathrm{C}^{(2)}(\Omega)$ be given functions. Moreover, let $D^* \colon \mathrm{C}^{(1)}(\Omega) \to \mathrm{c}(\Omega)$ be a first-order differential operator on the sphere (such as ∇^* and L^*). Then we define

$$\left(D^* \cdot f\right)(\xi) := \sum_{j=1}^{3} D_j^* f_j(\xi),$$

$$\left(D^* \otimes f\right)(\xi) := \left(D_j^* f_k(\xi)\right)_{j,k=1,2,3},$$

$$\left(D^* \times f\right)(\xi) := \left(\sum_{k,l=1}^{3} \varepsilon_{jkl} D_k^* f_l(\xi)\right)_{j=1,2,3},$$

$$\left(D^* \otimes D^*\right) G(\xi) := \left(D_j^* D_k^* G(\xi)\right)_{j,k=1,2,3},$$

where D_j^* is the jth component of the operator D^* (i.e., $D_j^* F(\xi) = \varepsilon^j \cdot D^* F(\xi)$ for $F \in \mathrm{C}^{(1)}(\Omega)$), f_k is the kth component of the vectorial function f, and ε_{jkl} is the Levi-Cività alternating symbol (see Theorem 2.1.4).

Theorem 4.1.3 *The following formulae hold true:*

$$\nabla_\xi^* \cdot \xi = 2, \qquad \nabla_\xi^*(\xi \cdot \eta) = \eta - (\xi \cdot \eta)\xi,$$

$$\mathrm{L}_\xi^* \cdot \xi = 0, \qquad \mathrm{L}_\xi^*(\xi \cdot \eta) = \xi \times \eta,$$

$$\nabla_\xi^* \otimes \xi = \mathbf{i} - \xi \otimes \xi, \qquad \Delta_\xi^*(\xi \cdot \eta) = -2\xi \cdot \eta$$

as well as

$$\mathrm{L}_\xi^* \otimes \xi = \sum_{j=1}^{3} (\xi \times \varepsilon^j) \otimes \varepsilon^j,$$

$$\nabla_\xi^* \otimes (\xi \times \eta) = -\sum_{i=1}^{3} \varepsilon^i \otimes (\eta \times \varepsilon^i) - \xi \otimes (\xi \times \eta),$$

$$\mathrm{L}_\xi^* \otimes (\xi \times \eta) = \eta \otimes \xi - (\xi \cdot \eta)\mathbf{i},$$

for all $\xi, \eta \in \Omega$, where $\mathbf{i}$ is the 3×3 identity matrix (which is also denoted by I elsewhere in this book).

Proof For $x \in \mathbb{R}^3$, we have $\nabla_x \cdot x = 3$ such that (see (4.1))

$$3 = \left(\xi \frac{\partial}{\partial r} + \frac{1}{r} \nabla_\xi^*\right) \cdot (r\xi) = 1 + \nabla_\xi^* \cdot \xi, \quad r > 0, \xi \in \Omega,$$

and $\nabla_\xi^* \cdot \xi = 2$. Moreover, with the Levi-Cività symbol (see Theorem 2.1.4), we get $\mathrm{L}_\xi^* \cdot \xi = \sum_{i=1}^3 \left(\mathrm{L}_\xi^*\right)_i \xi_i = \sum_{i,j,k=1}^3 \varepsilon_{ijk}\, \xi_j \left(\nabla_\xi^*\right)_k \xi_i$. Furthermore, using $x = r\xi$, $r = |x|$, $\xi \in \Omega$, we obtain

$$\nabla_x \otimes x = \left(\xi \frac{\partial}{\partial r} + \frac{1}{r}\nabla_\xi^*\right) \otimes (r\xi) = \xi \otimes \xi + \nabla_\xi^* \otimes \xi,$$

where $\mathbf{i} = \nabla_x \otimes x$. Hence, $\mathrm{L}_\xi^* \cdot \xi = \sum_{i,j,k=1}^3 \varepsilon_{ijk}\, \xi_j (\delta_{ki} - \xi_k \xi_i) = -\xi \cdot (\xi \times \xi) = 0$. Moreover,

$$\eta = \nabla_x (x \cdot \eta) = \left(\xi \frac{\partial}{\partial r} + \frac{1}{r}\nabla_\xi^*\right)(r\xi \cdot \eta) = \xi(\xi \cdot \eta) + \nabla_\xi^* (\xi \cdot \eta)$$

such that $\nabla_\xi^*(\xi \cdot \eta) = \eta - (\xi \cdot \eta)\xi$. Hence, $\mathrm{L}_\xi^*(\xi \cdot \eta) = \xi \times \nabla_\xi^*(\xi \cdot \eta) = \xi \times \eta$.

Furthermore, (4.2) yields

$$0 = \Delta_x (x \cdot \eta) = \left(\frac{\partial^2}{\partial r^2} + \frac{2}{r}\frac{\partial}{\partial r} + \frac{1}{r^2}\Delta_\xi^*\right)(r\xi \cdot \eta)$$

such that $0 = 2r^{-1}\xi \cdot \eta + r^{-1}\Delta_\xi^*(\xi \cdot \eta)$ and, consequently, $\Delta_\xi^*(\xi \cdot \eta) = -2\xi \cdot \eta$.

For $\mathrm{L}_\xi^* \otimes \xi$, we consider an arbitrary component of this tensor:

$$\begin{aligned}\left(\mathrm{L}_\xi^* \otimes \xi\right)_{i,j} &= \sum_{k,l=1}^3 \varepsilon_{ikl}\xi_k \left(\nabla_\xi^*\right)_l \xi_j = \sum_{k,l=1}^3 \varepsilon_{ikl}\xi_k \left(\delta_{lj} - \xi_l \xi_j\right) \\ &= \left(\xi \times \varepsilon^j\right)_i - \underbrace{(\xi \times \xi)_i}_{=0} \xi_j.\end{aligned}$$

Hence, $\mathrm{L}_\xi^* \otimes \xi = \sum_{i,j=1}^3 \left(\mathrm{L}_\xi^* \otimes \xi\right)_{i,j} \varepsilon^i \otimes \varepsilon^j = \sum_{j=1}^3 (\xi \times \varepsilon^j) \otimes \varepsilon^j$. Finally, we consider the application of L_ξ^* and ∇_ξ^* to the vector product $\xi \times \eta$. Again, we have a look at an arbitrary component of the tensor and obtain

$$\begin{aligned}\left[\nabla_\xi^* \otimes (\xi \times \eta)\right]_{i,j} &= \left(\nabla_\xi^*\right)_i \sum_{k,l=1}^3 \varepsilon_{jkl}\xi_k \eta_l = \sum_{k,l=1}^3 \varepsilon_{jkl} \left(\delta_{ik} - \xi_i \xi_k\right) \eta_l \\ &= \left(\varepsilon^i \times \eta\right)_j - \xi_i (\xi \times \eta)_j\end{aligned}$$

such that $\nabla_\xi^* \otimes (\xi \times \eta) = -\sum_{i=1}^3 \varepsilon^i \otimes (\eta \times \varepsilon^i) - \xi \otimes (\xi \times \eta)$. Analogously, we get the following (using Theorem 2.1.4):

$$\begin{aligned}\left[\mathrm{L}_\xi^* \otimes (\xi \times \eta)\right]_{i,j} &= \sum_{k,l=1}^3 \varepsilon_{jkl} \sum_{m,n=1}^3 \varepsilon_{imn}\xi_m \delta_{nk} \eta_l = \sum_{k,l,m=1}^3 \varepsilon_{jkl}\varepsilon_{imk}\xi_m \eta_l \\ &= \sum_{k,l,m=1}^3 \varepsilon_{kjl}\varepsilon_{kmi}\xi_m \eta_l = \sum_{l,m=1}^3 \left(\delta_{jm}\delta_{li} - \delta_{ji}\delta_{lm}\right)\xi_m \eta_l = \xi_j \eta_i - \delta_{ij}\xi \cdot \eta\end{aligned}$$

such that $\mathrm{L}_\xi^* \otimes (\xi \times \eta) = \eta \otimes \xi - (\xi \cdot \eta)\mathbf{i}$, which completes the proof. □

For two of the obtained tensors in Theorem 4.1.3, an abbreviation is known (see Freeden et al., 1998).

Definition 4.1.4 We define the tensors $\mathbf{i}_{\tan}(\xi) := \mathbf{i} - \xi \otimes \xi$ and $\mathbf{j}_{\tan}(\xi) := \sum_{j=1}^{3}(\xi \times \varepsilon^j) \otimes \varepsilon^j$ for $\xi \in \Omega$.

By applying the chain rule, we immediately obtain the following corollary.

Corollary 4.1.5 *If $F \in \mathrm{C}^{(1)}[-1,1]$, then we have, for all $\xi, \eta \in \Omega$,*

$$\nabla_\xi^* F(\xi \cdot \eta) = F'(\xi \cdot \eta)\,[\eta - (\xi \cdot \eta)\xi], \qquad \mathrm{L}_\xi^* F(\xi \cdot \eta) = F'(\xi \cdot \eta)\xi \times \eta.$$

Concerning derivatives in Euclidean spaces, it is clear that $\nabla \cdot \nabla = \Delta$. Indeed, corresponding properties are valid for ∇^* and L^* as well. Moreover, ∇^* and L^* always produce tangential vector fields. We will address these issues in the next theorem.

Theorem 4.1.6 *Let $F \in \mathrm{C}^{(2)}(\Omega)$ and $G \in \mathrm{C}^{(1)}(\Omega)$. Then the following identities hold true for all $\xi \in \Omega$:*

$$\begin{aligned}
\nabla_\xi^* \cdot \nabla_\xi^* F(\xi) &= \Delta_\xi^* F(\xi), & \mathrm{L}_\xi^* \cdot \mathrm{L}_\xi^* F(\xi) &= \Delta_\xi^* F(\xi),\\
\mathrm{L}_\xi^* \cdot \nabla_\xi^* F(\xi) &= 0, & \nabla_\xi^* \cdot \mathrm{L}_\xi^* F(\xi) &= 0,\\
\xi \cdot \nabla_\xi^* G(\xi) &= 0, & \xi \cdot \mathrm{L}_\xi^* G(\xi) &= 0,\\
\left(\nabla_\xi^* G(\xi)\right) \cdot \left(\mathrm{L}_\xi^* G(\xi)\right) &= 0.
\end{aligned}$$

Proof We will not prove all identities for reasons of brevity. As an example of the first four equations, we show $\mathrm{L}^* \cdot \nabla^* = 0$. The fifth and sixth identities, which mean that $\nabla^* G$ and $\mathrm{L}^* G$ are tangential on Ω, immediately result from the fact that $\varepsilon^r = \xi$ and the representations of ∇^* and L^* in terms of ε^φ and ε^t (see (4.3) and (4.4)). Moreover, $\mathrm{L}_\xi^* G(\xi) = \xi \times \nabla_\xi^* G(\xi)$ is, as a vector product, orthogonal to $\nabla_\xi^* G(\xi)$. Now, let us take care of $\mathrm{L}^* \cdot \nabla^*$:

$$\mathrm{L}_\xi^* \cdot \nabla_\xi^* F(\xi) = \sum_{j=1}^{3} \left(\mathrm{L}_\xi^*\right)_j \left[\nabla_\xi^* F(\xi)\right]_j = \sum_{j,k,l=1}^{3} \varepsilon_{jkl}\xi_k \left(\nabla_\xi^*\right)_l \left[\nabla_\xi^* F(\xi)\right]_j,$$

$\xi \in \Omega$. Here, we used the Levi-Cività alternating symbol (see Theorem 2.1.4). We can now deduce that

$$\begin{aligned}
\mathrm{L}_\xi^* \cdot \nabla_\xi^* F(\xi) &= \sum_{j,k,l=1}^{3} \varepsilon_{klj}\xi_k \left(\nabla_\xi^*\right)_l \left[\nabla_\xi^* F(\xi)\right]_j\\
&= \sum_{k=1}^{3} \xi_k \sum_{j,l=1}^{3} \varepsilon_{klj} \left(\nabla_\xi^*\right)_l \left(\nabla_\xi^*\right)_j F(\xi).
\end{aligned} \tag{4.5}$$

Using Definition 3.4.2 and Corollary 3.4.4, we can calculate that (where $v := (-\cos\varphi, -\sin\varphi, 0)^{\mathrm{T}}$)

$$\nabla_\xi^* \otimes \nabla_\xi^* F(\xi) = \left(\varepsilon^\varphi \frac{1}{\sqrt{1-t^2}} \frac{\partial}{\partial \varphi} + \varepsilon^t \sqrt{1-t^2} \frac{\partial}{\partial t} \right)$$
$$\otimes \left(\varepsilon^\varphi \frac{1}{\sqrt{1-t^2}} \frac{\partial}{\partial \varphi} + \varepsilon^t \sqrt{1-t^2} \frac{\partial}{\partial t} \right) F$$
$$= \frac{1}{1-t^2} \frac{\partial F}{\partial \varphi} \varepsilon^\varphi \otimes v + \frac{1}{1-t^2} \frac{\partial^2 F}{\partial \varphi^2} \varepsilon^\varphi \otimes \varepsilon^\varphi - t \frac{\partial F}{\partial t} \varepsilon^\varphi \otimes \varepsilon^\varphi$$
$$+ \frac{\partial^2 F}{\partial \varphi \partial t} \varepsilon^\varphi \otimes \varepsilon^t + \frac{t}{1-t^2} \frac{\partial F}{\partial \varphi} \varepsilon^t \otimes \varepsilon^\varphi + \frac{\partial^2 F}{\partial t \partial \varphi} \varepsilon^t \otimes \varepsilon^\varphi$$
$$- \sqrt{1-t^2} \frac{\partial F}{\partial t} \varepsilon^t \otimes \varepsilon^r - t \frac{\partial F}{\partial t} \varepsilon^t \otimes \varepsilon^t + (1-t^2) \frac{\partial^2 F}{\partial t^2} \varepsilon^t \otimes \varepsilon^t,$$

where φ and t stand for the polar coordinates of $\xi \in \Omega$. We insert this result into (4.5) and obtain

$$\mathrm{L}_\xi^* \cdot \nabla_\xi^* F(\xi) = \frac{1}{1-t^2} \frac{\partial F}{\partial \varphi} \xi \cdot \left(\varepsilon^\varphi \times v \right) + \frac{\partial^2 F}{\partial \varphi \partial t} \xi \cdot \left(\varepsilon^\varphi \times \varepsilon^t \right)$$
$$+ \frac{t}{1-t^2} \frac{\partial F}{\partial \varphi} \xi \cdot \left(\varepsilon^t \times \varepsilon^\varphi \right) + \frac{\partial^2 F}{\partial t \partial \varphi} \xi \cdot \left(\varepsilon^t \times \varepsilon^\varphi \right)$$
$$- \sqrt{1-t^2} \frac{\partial F}{\partial t} \xi \cdot \left(\varepsilon^t \times \varepsilon^r \right).$$

Since $\varepsilon^\varphi \times v = (0,0,1)^{\mathrm{T}} = \varepsilon^3$ and $\varepsilon^\varphi \times \varepsilon^t = \varepsilon^r = \xi$ as well as $\varepsilon^t \times \varepsilon^r = \varepsilon^\varphi$, we can simplify the latter result to the following (note that $\xi \cdot \varepsilon^3 = t$):

$$\mathrm{L}_\xi^* \cdot \nabla_\xi^* F(\xi) = \frac{t}{1-t^2} \frac{\partial F}{\partial \varphi} + \frac{\partial^2 F}{\partial \varphi \partial t} - \frac{\partial^2 F}{\partial t \partial \varphi} - \frac{t}{1-t^2} \frac{\partial F}{\partial \varphi} = 0,$$

where we used Schwarz's theorem (remember that $F \in \mathrm{C}^{(2)}(\Omega)$). □

By combining Theorems 4.1.3 and 4.1.6, we obtain the following result.

Corollary 4.1.7 *If $G \in \mathrm{C}^{(2)}[-1,1]$, then we have, for all $\xi, \eta \in \Omega$,*

$$\Delta_\xi^* G(\xi \cdot \eta) = -2(\xi \cdot \eta) G'(\xi \cdot \eta) + \left[1 - (\xi \cdot \eta)^2 \right] G''(\xi \cdot \eta).$$

Proof From the previous results, we obtain

$$\Delta_\xi^* G(\xi \cdot \eta) = \nabla_\xi^* \cdot \nabla_\xi^* G(\xi \cdot \eta) = \nabla_\xi^* \cdot \left[G'(\xi \cdot \eta)(\eta - (\xi \cdot \eta)\xi) \right]$$
$$= \left[\nabla_\xi^* G'(\xi \cdot \eta) \right] \cdot \left[\eta - (\xi \cdot \eta)\xi \right] + G'(\xi \cdot \eta) \nabla_\xi^* \cdot \left[\eta - (\xi \cdot \eta)\xi \right]$$
$$= G''(\xi \cdot \eta) \left[\eta - (\xi \cdot \eta)\xi \right] \cdot \left[\eta - (\xi \cdot \eta)\xi \right] - G'(\xi \cdot \eta) \left[\xi \cdot \nabla_\xi^* (\xi \cdot \eta) + 2\xi \cdot \eta \right]$$
$$= G''(\xi \cdot \eta) \left[1 - (\xi \cdot \eta)^2 \right] - 2G'(\xi \cdot \eta) \xi \cdot \eta$$

for all $\xi, \eta \in \Omega$. □

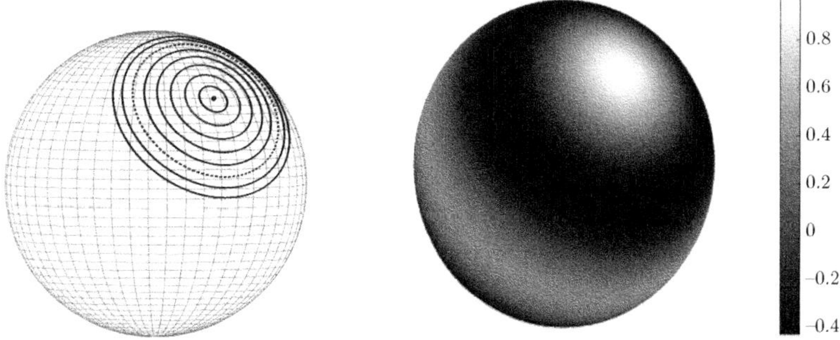

Figure 4.1 A zonal function $\Omega \ni \eta \mapsto G(\xi \cdot \eta)$ only depends on the Euclidean distance $|\xi - \eta|$. In other words, the function is constant on each circle around $\xi \in \Omega$ (see the left picture, where the dot stands for ξ). The right image shows the values of $P_4(\xi \cdot \eta)$ for the Legendre polynomial of degree 4.

Functions $\Omega \ni \xi \mapsto G(\xi \cdot \eta)$, as we encountered them here, play a particular role in the construction of approximations to spherical functions, as we will see later in this book. We dedicate here some additional considerations to these functions.

Definition 4.1.8 A function $F\colon \Omega \to \mathbb{R}$ is called a **zonal function** if there exists a point $\xi \in \Omega$ and a function $G\colon [-1,1] \to \mathbb{R}$ such that the identity $F(\eta) = G(\xi \cdot \eta)$ holds for all $\eta \in \Omega$. In this case, we also say that G is a ξ-zonal function.

By using a standard parameterization in polar coordinates and an appropriate coordinate transformation (see, e.g., Michel, 2013, Theorem 4.16), one can prove the following theorem.

Theorem 4.1.9 *Let $G\colon [-1,1] \to \mathbb{R}$ be an integrable function. Then we have*

$$\int_\Omega G(\xi \cdot \eta)\, \mathrm{d}\omega(\eta) = \int_\Omega G(\varepsilon^3 \cdot \eta)\, \mathrm{d}\omega(\eta) = 2\pi \int_{-1}^{1} G(t)\, \mathrm{d}t$$

for all $\xi \in \Omega$, that is, the integral is independent of ξ.

Zonal functions $\eta \mapsto G(\xi \cdot \eta)$ have the outstanding property that they only depend on the Euclidean distance $|\xi - \eta|$, because

$$|\xi - \eta|^2 = |\xi|^2 - 2\xi \cdot \eta + |\eta|^2 = 2(1 - \xi \cdot \eta)$$

for all $\xi, \eta \in \Omega$. Since, for fixed $\xi \in \Omega$, the condition $|\xi - \eta| = c$ for $\eta \in \mathbb{R}^3$ defines a sphere, zonal functions are constant on the intersection of these spheres with Ω. In other words, ξ-zonal functions are constant on each circle on Ω which has ξ as a centre; see also Figure 4.1.

For particular proofs and parameterizations, great circles on the sphere are of importance.

Definition 4.1.10 A **great circle** on the sphere $S_R(0)$ is an intersection of $S_R(0)$ and a plane containing the point 0; see also Figure 4.2.

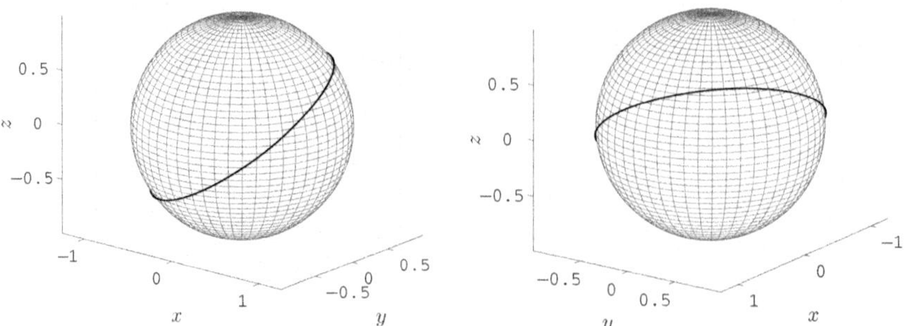

Figure 4.2 A great circle on the sphere is shown from two different perspectives.

For a given sphere – without loss of generality, let us take Ω – a great circle G is uniquely determined by two of its points $\xi, \eta \in \Omega$, if these are not collinear, that is, if $\xi \times \eta \neq 0$ (equivalently, $-1 < \xi \cdot \eta < 1$). In this case, ξ and η span a plane – the plane whose intersection with Ω yields the great circle G. Clearly, $\xi - (\xi \cdot \eta)\eta$ is located in the same plane and is orthogonal to η. Since $|\xi - (\xi \cdot \eta)\eta|^2 = |\xi|^2 - 2(\xi \cdot \eta)^2 + (\xi \cdot \eta)^2 |\eta|^2 = 1 - (\xi \cdot \eta)^2$, we can easily normalize this vector such that G is parameterized by

$$\phi(t) := (\cos t)\eta + (\sin t)\frac{\xi - (\xi \cdot \eta)\eta}{\sqrt{1 - (\xi \cdot \eta)^2}}, \quad t \in [-\pi, \pi].$$

We have $\phi(0) = \eta$ and

$$\phi(\arccos \xi \cdot \eta) = (\xi \cdot \eta)\eta + \sqrt{1 - (\xi \cdot \eta)^2}\, \frac{\xi - (\xi \cdot \eta)\eta}{\sqrt{1 - (\xi \cdot \eta)^2}} = \xi.$$

Obviously, $\phi \in \mathrm{C}^{(\infty)}[-\pi, \pi]$.

Let us now address some properties of integrals on the sphere.

We omit the proof of the following two theorems, which are somehow the surface counterparts of Theorem 2.4.2; see Amann and Escher (2008, pp. 459–461), Freeden et al. (1998, p. 16), and Freeden and Schreiner (2009, p. 40).

Theorem 4.1.11 *Let $F, G \in \mathrm{C}^{(2)}(\overline{\Gamma})$ be given functions and let $\Gamma \subset \Omega$ be a subset with sufficiently smooth boundary $\partial\Gamma$ (see Figure 4.3). Moreover, let the vector ν denote the outward unit normal surface vector field to $\partial\Gamma$. Then the following identities are valid.*

(a) Green's first surface identity:

$$\int_\Gamma \nabla_\xi^* G(\xi) \cdot \nabla_\xi^* F(\xi)\, \mathrm{d}\omega(\xi) + \int_\Gamma F(\xi) \Delta_\xi^* G(\xi)\, \mathrm{d}\omega(\xi)$$
$$= \int_{\partial\Gamma} F(\xi) \frac{\partial}{\partial\nu(\xi)} G(\xi)\, \mathrm{dl}(\xi).$$

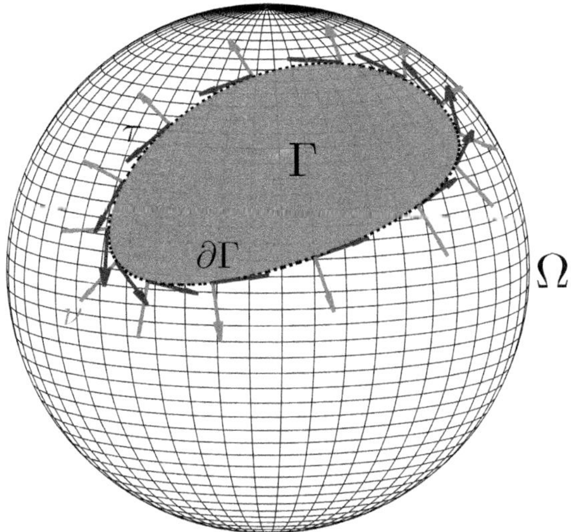

Figure 4.3 Subset Γ (grey) of the unit sphere Ω with boundary $\partial\Gamma$ (dotted), unit surface tangential vectors τ (dark grey), and unit surface normal vectors ν (light grey).

(b) Green's second surface identity:

$$\int_\Gamma \left(F(\xi)\Delta^*_\xi G(\xi) - G(\xi)\Delta^*_\xi F(\xi)\right) d\omega(\xi)$$
$$= \int_{\partial\Gamma} \left(F(\xi)\frac{\partial}{\partial\nu(\xi)}G(\xi) - G(\xi)\frac{\partial}{\partial\nu(\xi)}F(\xi)\right) dl(\xi).$$

If $\Gamma = \Omega$, then $\partial\Gamma = \emptyset$ as a limit case such that the line integrals vanish.

Theorem 4.1.12 (Surface Theorems of Gauß and Stokes) *Let $\Gamma \subset \Omega$ be a subset with sufficiently smooth boundary $\partial\Gamma$ with unit outer normal and positively oriented unit tangential surface vector fields ν and τ to $\partial\Gamma$ (see Figure 4.3). If the vector field $f \in c^{(1)}(\overline{\Gamma})$ is tangential (i.e., $f(\xi)\cdot\xi = 0$ for all $\xi \in \Omega$), then* ***Gauß's surface law***

$$\int_\Gamma \nabla^*_\xi \cdot f(\xi)\, d\omega(\xi) = \int_{\partial\Gamma} \nu(\xi)\cdot f(\xi)\, dl(\xi)$$

and the ***surface theorem of Stokes***

$$\int_\Gamma L^*_\xi \cdot f(\xi)\, d\omega(\xi) = \int_{\partial\Gamma} \tau(\xi)\cdot f(\xi)\, dl(\xi)$$

hold true.

Since, for all $F \in C^{(1)}(\Omega)$ and $f \in c^{(1)}(\Omega)$, we have

$$\nabla^*\cdot(F(\xi)f(\xi)) = \left(\nabla^* F(\xi)\right)\cdot f(\xi) + F(\xi)\nabla^*\cdot f(\xi),$$
$$L^*\cdot(F(\xi)f(\xi)) = \left(L^* F(\xi)\right)\cdot f(\xi) + F(\xi)L^*\cdot f(\xi),$$

$\xi \in \Omega$, Theorem 4.1.12 obviously also yields the following integral identities.

Corollary 4.1.13 *For all $F \in \mathrm{C}^{(1)}(\Omega)$ and all tangential vector fields $f \in \mathrm{c}^{(1)}(\Omega)$, the following identities hold true:*

$$\int_{\Omega} \left(\nabla^* F(\xi)\right) \cdot f(\xi)\, d\omega(\xi) = -\int_{\Omega} F(\xi)\nabla^* \cdot f(\xi)\, d\omega(\xi),$$

$$\int_{\Omega} \left(\mathrm{L}^* F(\xi)\right) \cdot f(\xi)\, d\omega(\xi) = -\int_{\Omega} F(\xi)\mathrm{L}^* \cdot f(\xi)\, d\omega(\xi),$$

$$\int_{\Omega} \nabla^* \cdot f(\xi)\, d\omega(\xi) = 0,$$

$$\int_{\Omega} \mathrm{L}^* \cdot f(\xi)\, d\omega(\xi) = 0.$$

An immediate consequence of Theorem 4.1.12 is that

$$\int_{\gamma} \tau(\xi) \cdot f(\xi)\, \mathrm{dl}(\xi) = 0 \tag{4.6}$$

for every sufficiently smooth, closed, and non-self-intersecting curve γ on Ω, if $\mathrm{L}^* \cdot f = 0$ on Ω. On the other hand, Theorem 4.1.6 tells us that $\mathrm{L}^* \cdot f = 0$ holds if there is a function $G \in \mathrm{C}^{(2)}(\Omega)$ with $f = \nabla^* G$. Indeed, there is a connection (see also Backus, 1986): let $0 < \varepsilon < 1$, $\eta \in \Omega$ and choose, for each $\xi \in \Omega$, a curve $\gamma_\xi \subset \Omega$ which starts in η and ends in ξ. Moreover, let $G \in \mathrm{C}^{(1)}(\Omega)$ be extended to all $x \in \mathbb{R}^3$ with $1 - \varepsilon < |x| < 1 + \varepsilon$ by $\tilde{G}(x) := G(x/|x|)$. Then $\nabla_x \tilde{G}(x) = |x|^{-1}\nabla^*_\zeta G(\zeta)$ with $x = |x|\zeta$. Moreover, it is a fundamental result of real analysis that

$$\int_{\gamma_\xi} \tau(\zeta) \cdot \nabla^*_\zeta G(\zeta)\, \mathrm{dl}(\zeta) = \int_{\gamma_\xi} \tau(\zeta) \cdot \left(\nabla_{r\zeta} \tilde{G}(r\zeta)\right)\big|_{r=1} \mathrm{dl}(\zeta) = \tilde{G}(\xi) - \tilde{G}(\eta)$$

is independent of the chosen curve γ_ξ from η to ξ. Vice versa, if $\mathrm{L}^* \cdot f = 0$ on Ω and $f \in \mathrm{c}^{(1)}(\Omega)$, then we set $G(\xi) := \int_{\gamma_\xi} \tau(\zeta) \cdot f(\zeta)\, \mathrm{dl}(\zeta) + G(\eta)$, $\xi \in \Omega$, with an arbitrarily chosen value $G(\eta)$ at a fixed point $\eta \in \Omega$.

If we choose the parameterization of γ_ξ sufficiently smooth and extend f to a volumetric domain (like we did with G), then we can use again standard arguments from real analysis to conclude that $G \in \mathrm{C}^{(2)}(\Omega)$, because f is continuously differentiable. Hence, our previous derivations reveal that $\nabla^* G$ satisfies $\int_{\gamma_\xi} \tau(\zeta) \cdot \nabla^*_\zeta G(\zeta)\, \mathrm{dl}(\zeta) = G(\xi) - G(\eta)$ such that

$$\int_{\gamma_\xi} \tau(\zeta) \cdot \left(f(\zeta) - \nabla^*_\zeta G(\zeta)\right) \mathrm{dl}(\zeta) = 0 \tag{4.7}$$

for all $\xi \in \Omega$. Since $\mathrm{L}^* \cdot (f - \nabla^* G) = 0$ (see Theorem 4.1.6), all closed curve integrals satisfy, due to (4.6), the identity

$$\int_{\gamma} \tau(\zeta) \cdot \left(f(\zeta) - \nabla^*_\zeta G(\zeta)\right) \mathrm{dl}(\zeta) = 0. \tag{4.8}$$

Hence, by combining closed curves with γ_ξ and using (4.7), we may conclude that (4.8) is true for any arbitrary (sufficiently smooth) curve γ on Ω (see also Figure 4.4). By shrinking a curve γ to an arbitrary point ξ and using the continuity of the involved functions, we get that $f(\xi) = \nabla^*_\xi G(\xi)$. We proved the following theorem.

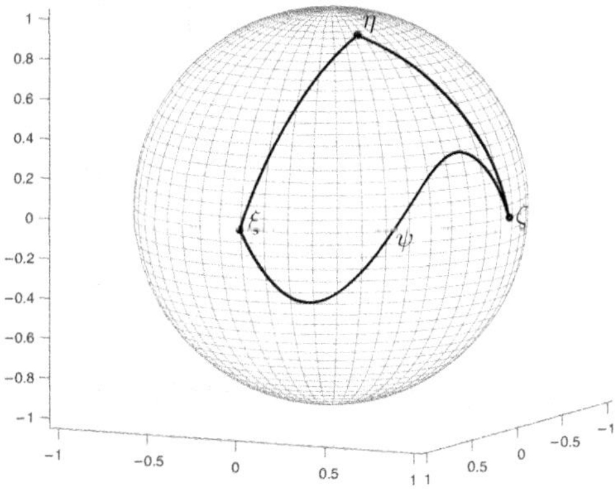

Figure 4.4 The integral along the arbitrary curve ψ from ξ to ζ can be added to the integrals along the curves from ζ to η and from η to ξ. This yields a closed curve, along which the integral of $\tau \cdot (f - \nabla^* G)$ vanishes. Since the added integrals along curves from and to η also vanish, also the integral of $\tau \cdot (f - \nabla^* G)$ along the arbitrary curve ψ must vanish.

Theorem 4.1.14 *If $f \in \mathrm{c}^{(1)}(\Omega)$ satisfies $\mathrm{L}^* \cdot f = 0$ on Ω, then there is a function $G \in \mathrm{C}^{(2)}(\Omega)$ such that $f = \nabla^* G$. Moreover,*

$$G(\xi) - G(\eta) = \int_\gamma \tau(\zeta) \cdot f(\zeta) \, \mathrm{dl}(\zeta)$$

for all $\xi, \eta \in \Omega$ and every (sufficiently smooth and non-self-intersecting) curve from η to ξ with unit tangential vector τ.

4.2 Spherical Harmonics and Legendre Polynomials

In Section 3.4, we derived a particular system of functions on the sphere, the fully normalized spherical harmonics $\{Y_{n,j}\}_{n\in\mathbb{N}_0;\ j=-n,\ldots,n}$, which is useful for representing harmonic functions (see Definition 3.4.24). Indeed, these functions have manifold applications in Earth sciences and beyond. This is the reason why we study them in further detail here. We will, however, follow a general ansatz of such functions for which the previously introduced system is only a particular case (see also Freeden et al., 1998; Michel, 2013; Müller, 1966). Let us also here first summarize a few propositions which we have already proved:

- The fully normalized spherical harmonic $Y_{n,j}$ is an eigenfunction of the Beltrami operator to the eigenvalue $-n(n+1)$. This was a consequence of their derivation; see (3.82) and the subsequent derivations.
- The system $\{Y_{n,j}\}_{n\in\mathbb{N}_0;\ j=-n,\ldots,n}$ is orthonormal in $\left(\mathrm{L}^2(\Omega), \langle\cdot,\cdot\rangle_{\mathrm{L}^2(\Omega)}\right)$; see Theorem 3.4.25.
- The functions $F_{n,j}(r\xi) := r^n Y_{n,j}(\xi)$, $r \in \mathbb{R}_0^+$, $\xi \in \Omega$, and $G_{n,j}(r\xi) := r^{-n-1} Y_{n,j}(\xi)$, $r \in \mathbb{R}^+$, $\xi \in \Omega$, are harmonic. This is implied by the separation ansatz at the beginning of Section 3.4 together with (3.81) and (3.95).

- A harmonic function on an open set $D \subset \mathbb{R}^3$ can be represented, in a neighbourhood of an arbitrary $x_0 \in D$, in a series of homogeneous harmonic polynomials. We denoted the set of all homogeneous harmonic polynomials of degree n on $\mathbb{R}^d$ by $\mathrm{Harm}_n(\mathbb{R}^d)$; see Theorem 3.3.46 and Definition 3.3.47.
- The functions $Y_{n,j}$ have the form (see Corollary 3.4.16 and Definition 3.4.24)

$$Y_{n,j}\,(\xi(\varphi,t)) = c_{n,j}(1-t^2)^{j/2}P_n^{(j)}(t)\sin(j\varphi), \quad \text{if } j>0,$$

$$Y_{n,j}\,(\xi(\varphi,t)) = c_{n,j}(1-t^2)^{|j|/2}P_n^{(|j|)}(t)\cos(j\varphi), \quad \text{if } j\le 0,$$

where $c_{n,j}$ is a constant. Using the polar coordinate representation $\xi(\varphi,t)$ of $\xi \in \Omega$ (see also Example 2.5.4) and the common formulae

$$\sin(j\varphi) = \sum_{\substack{k=0\\ k \text{ odd}}}^{j} (-1)^{(k-1)/2}\binom{j}{k} \cos^{j-k}\varphi\,\sin^k\varphi,$$

$$\cos(j\varphi) = \sum_{\substack{k=0\\ k \text{ even}}}^{j} (-1)^{k/2}\binom{j}{k} \cos^{j-k}\varphi\,\sin^k\varphi,$$

we see that all $Y_{n,j}$ are polynomials in ξ_1, ξ_2, and ξ_3. Indeed, one can show that the inner harmonic $F_{n,j}(r\xi) = r^n Y_{n,j}(\xi)$ is an element of $\mathrm{Harm}_n(\mathbb{R}^3)$. We will get this as a byproduct later.

The fully normalized spherical harmonics $Y_{n,j}$ of degrees $n = 1,2,3$ are shown in Figures 4.5–4.7. Note the typical polynomial character of these functions.

We will use these common features of the fully normalized spherical harmonics as a starting point for the generalization.

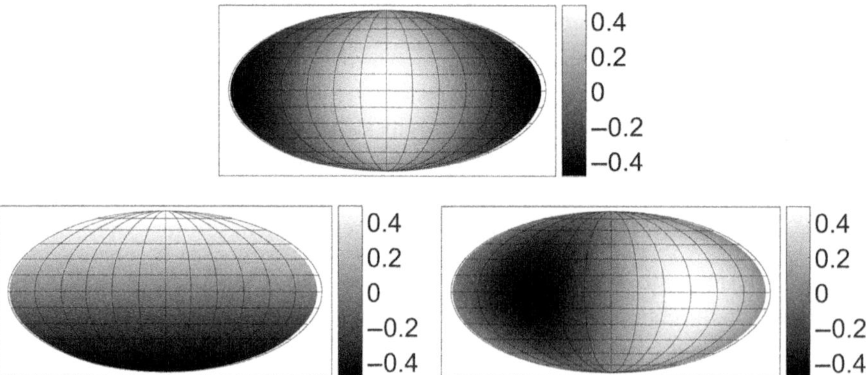

Figure 4.5 Scalar fully normalized spherical harmonics $Y_{n,j}$ of degree $n = 1$ and orders $j = -1,0,1$ (sorted in reading direction).

Definition 4.2.1 Let $D \subset \mathbb{R}^d$ be an arbitrary non-empty subset. Then we define $\mathrm{Harm}_n(D)$ as the set of all restrictions of all functions in $\mathrm{Harm}_n(\mathbb{R}^d)$ to D. Furthermore, we set

$$\mathrm{Harm}_{0\ldots n}(\mathbb{R}^3) := \bigoplus_{i=0}^{n} \mathrm{Harm}_i(\mathbb{R}^3), \quad \mathrm{Harm}_{0\ldots\infty}(\mathbb{R}^3) := \bigcup_{i=0}^{\infty} \mathrm{Harm}_{0\ldots i}(\mathbb{R}^3).$$

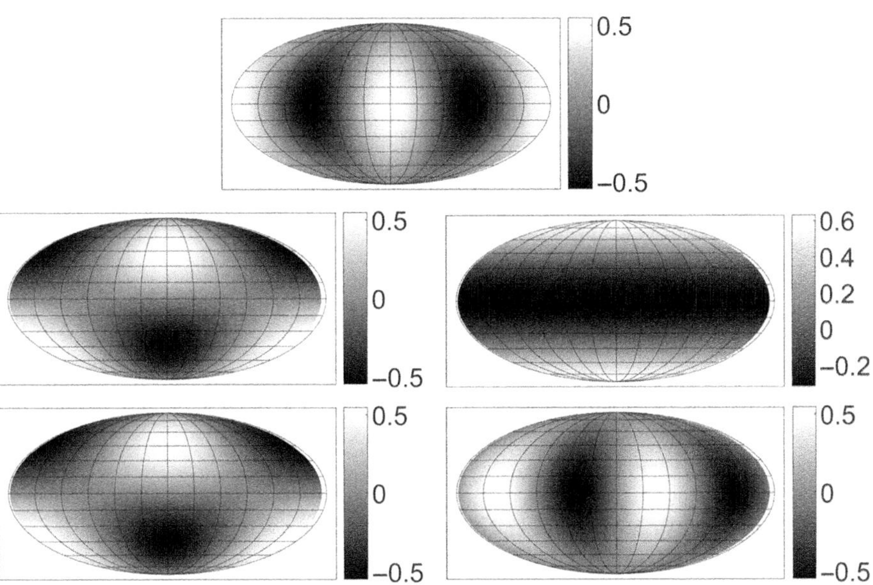

Figure 4.6 Scalar fully normalized spherical harmonics $Y_{n,j}$ of degree $n = 2$ and orders $j = -2, \ldots, 2$ (sorted in reading direction).

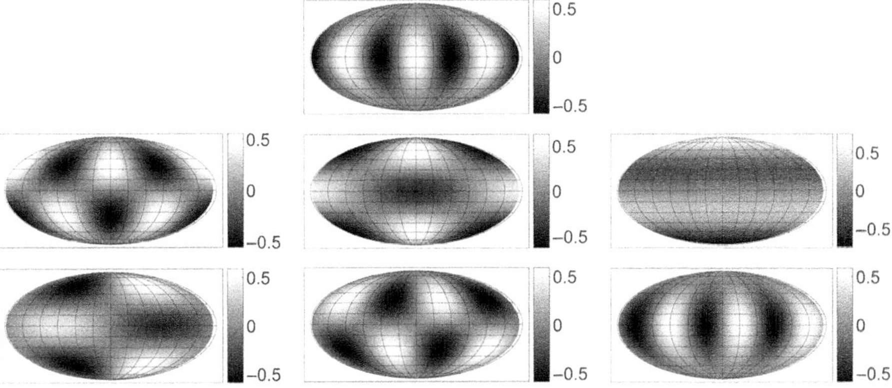

Figure 4.7 Scalar fully normalized spherical harmonics $Y_{n,j}$ of degree $n = 3$ and orders $j = -3, \ldots, 3$ (sorted in reading direction).

The restrictions of such functions to D are collected, correspondingly, in the sets $\mathrm{Harm}_{0\ldots n}(D)$ and $\mathrm{Harm}_{0\ldots\infty}(D)$, respectively.

Theorem 4.2.2 *The following dimension formulae are valid:*

$$\dim \mathrm{Hom}_n(\mathbb{R}^2) = n+1, \qquad \dim \mathrm{Hom}_n(\mathbb{R}^3) = \frac{n(n+1)}{2},$$

$$\dim \mathrm{Harm}_n(\mathbb{R}^3) = 2n+1.$$

Proof We start with the general homogeneous polynomials and then consider the harmonic homogeneous polynomials.

(1) The homogeneous polynomials:
For the homogeneous polynomials, we have already seen in Theorem 3.3.45 that $P \in \mathrm{Hom}_n(\mathbb{R}^d)$ is uniquely representable as

$$P(x) = \sum_{\substack{\alpha \in \mathbb{N}_0^d \\ \alpha_1+\cdots+\alpha_d=n}} C_\alpha\, x_1^{\alpha_1} \cdots x_d^{\alpha_d}, \quad x \in \mathbb{R}^d.$$

The rest is an easy combinatorial task: if $d = 2$, then all particular pairs $(\alpha_1,\alpha_2) \in \mathbb{N}_0^2$ with $\alpha_1 + \alpha_2 = n$ are given by $(0,n),(1,n-1),\ldots,(n,1)$, which are $n+1$ pairs. For $d = 3$, we can, for example, choose $\alpha_1 = 0,\ldots,n$ arbitrarily. Then we can choose $\alpha_2 = 0,\ldots,n-\alpha_1$ and $\alpha_3 = n-\alpha_1-\alpha_2$ is fixed. This yields $\sum_{\alpha_1=0}^{n}(n+1-\alpha_1) = \sum_{j=1}^{n} j = n(n+1)/2$ triples $\alpha \in \mathbb{N}_0^3$ with $\alpha_1+\alpha_2+\alpha_3 = n$.

(2) The harmonic homogeneous polynomials:
Let $H_n \in \mathrm{Harm}_n(\mathbb{R}^3)$. Using again Theorem 3.3.45, we can represent it as $H_n(x) = \sum_{j=0}^{n} x_1^j A_{n-j}(x_2,x_3)$, $x \in \mathbb{R}^3$, where $A_{n-j} \in \mathrm{Hom}_{n-j}(\mathbb{R}^2)$ for all $j = 0,\ldots,n$. Since H_n is also harmonic, this leads us to

$$\begin{aligned} 0 &= \sum_{j=0}^{n} \left(\frac{\partial^2}{\partial x_1^2} + \frac{\partial^2}{\partial x_2^2} + \frac{\partial^2}{\partial x_3^2} \right) \left[x_1^j A_{n-j}(x_2,x_3) \right] \\ &= \sum_{j=2}^{n} j(j-1) x_1^{j-2} A_{n-j}(x_2,x_3) + \sum_{j=0}^{n-2} x_1^j \left(\frac{\partial^2}{\partial x_2^2} + \frac{\partial^2}{\partial x_3^2} \right) A_{n-j}(x_2,x_3) \\ &= \sum_{j=0}^{n-2} x_1^j \left[(j+2)(j+1) A_{n-j-2}(x_2,x_3) + \left(\frac{\partial^2}{\partial x_2^2} + \frac{\partial^2}{\partial x_3^2} \right) A_{n-j}(x_2,x_3) \right]. \end{aligned}$$

Hence, H_n is harmonic if and only if the A_{n-j} satisfy the recurrence relation

$$A_{n-j-2}(x_2,x_3) = -\frac{1}{(j+2)(j+1)} \left(\frac{\partial^2}{\partial x_2^2} + \frac{\partial^2}{\partial x_3^2} \right) A_{n-j}(x_2,x_3)$$

for all $j = 0,\ldots,n-2$. In other words, A_n and A_{n-1} are completely free to choose whereas all the other A_{n-j} are then uniquely determined. Thus, the following must hold true:

$$\dim \mathrm{Harm}_n(\mathbb{R}^3) = \dim \mathrm{Hom}_n(\mathbb{R}^2) + \dim \mathrm{Hom}_{n-1}(\mathbb{R}^2) = 2n+1. \qquad \square$$

Theorem 4.2.3 *The restriction to the unit sphere does not change the dimension:* $\dim \mathrm{Harm}_n(\Omega) = 2n + 1$.

Proof Obviously, $\dim \mathrm{Harm}_n(\Omega) \le \dim \mathrm{Harm}_n(\mathbb{R}^3)$. Let us assume that there exist functions $H_1, \ldots, H_{2n+1} \in \mathrm{Harm}_n(\mathbb{R}^3)$ which are linearly independent while their restrictions to Ω become linearly dependent. If now

$$\sum_{j=1}^{2n+1} \alpha_j \, H_j\big|_\Omega = 0 \quad \text{with} \quad (\alpha_1, \ldots, \alpha_{2n+1}) \ne (0, \ldots, 0),$$

then $\sum_{j=1}^{2n+1} \alpha_j H_j$ solves the IDP on Ω_{int} with vanishing boundary values. However, the zero function on $\overline{\Omega_{\mathrm{int}}}$ is the only solution of this IDP (see Corollary 3.3.4). Hence, $\sum_{j=1}^{2n+1} \alpha_j H_j = 0$ on $\overline{\Omega_{\mathrm{int}}}$ and, since these are polynomials, also on the entire $\mathbb{R}^3$. This is a contradiction to the linear independence on $\mathbb{R}^3$. Consequently, $\dim \mathrm{Harm}_n(\Omega) = 2n+1$. □

Corollary 4.2.4 *For all $n \in \mathbb{N}_0$, we have* $\dim \mathrm{Harm}_{0\ldots n}(\Omega) = (n + 1)^2$.

Proof The dimension of $\mathrm{Harm}_{0\ldots n}(\Omega)$ is obtained via an easy calculation:

$$\sum_{k=0}^{n} (2k + 1) = 2\,\frac{n(n + 1)}{2} + n + 1 = n(n + 1) + n + 1 = (n + 1)^2. \qquad \square$$

We continue with a generalization of the fully normalized spherical harmonics, as it is, for instance, also done in Freeden et al. (1998), Michel (2013), and Müller (1966).

Definition 4.2.5 For all $n \in \mathbb{N}_0$, the finite sequence $\{Y_{n,j}\}_{j=-n,\ldots,n}$ represents an arbitrary choice of an orthonormal basis of $\left(\mathrm{Harm}_n(\Omega), \langle\cdot,\cdot\rangle_{\mathrm{L}^2(\Omega)}\right)$.

Note that this convention is always and only valid for Y with a double index.

Definition 4.2.6 The elements of $\mathrm{Harm}_n(\Omega)$ are called **spherical harmonics**.

Accordingly, Definition 3.4.26 of **inner and outer harmonics** can be generalized to arbitrary choices of functions $Y_{n,j}$.

Theorem 4.2.7 *The elements of* $\mathrm{Harm}_n(\Omega)$ *are eigenfunctions of the Beltrami operator to the eigenvalue* $-n(n + 1)$.

Proof With Theorems 3.3.45 and 3.4.3 and with $x = r\xi$, $\xi \in \Omega$, we obtain for every $H \in \mathrm{Harm}_n(\mathbb{R}^3)$, $n \in \mathbb{N}_0$ fixed, with its corresponding C_α, the identity

$$0 = \Delta_x H(x) = \Delta_x \sum_{\substack{\alpha\in\mathbb{N}_0^3\\|\alpha|=n}} C_\alpha x^\alpha = \left(\frac{\partial^2}{\partial r^2} + \frac{2}{r}\frac{\partial}{\partial r} + \frac{1}{r^2}\Delta_\xi^*\right) \sum_{\substack{\alpha\in\mathbb{N}_0^3\\|\alpha|=n}} C_\alpha (r\xi)^\alpha$$

$$= \left(\frac{\partial^2}{\partial r^2} + \frac{2}{r}\frac{\partial}{\partial r} + \frac{1}{r^2}\Delta_\xi^*\right) \sum_{\substack{\alpha\in\mathbb{N}_0^3\\|\alpha|=n}} C_\alpha\, (r\xi_1)^{\alpha_1}\, (r\xi_2)^{\alpha_2}\, (r\xi_3)^{\alpha_3}$$

$$= \left(\frac{\partial^2}{\partial r^2} + \frac{2}{r}\frac{\partial}{\partial r} + \frac{1}{r^2}\Delta_\xi^*\right)\left(r^n H(\xi)\right)$$

$$= [n(n-1)+2n]r^{n-2}H(\xi) + r^{n-2}\Delta_\xi^* H(\xi).$$

Hence, $\Delta_\xi^* H(\xi) = -n(n+1)H(\xi)$ for all $\xi \in \Omega$. □

We do not know yet if these are the only eigenfunctions and eigenvalues. We will come back to this question later on.

The combination of Lemma 3.4.13 and Theorem 4.2.7 yields further insight into the specific properties of spherical harmonics.

Theorem 4.2.8 *Let $n, m \in \mathbb{N}_0$ with $n \neq m$ and let $Y_n \in \mathrm{Harm}_n(\Omega)$ and $Y_m \in \mathrm{Harm}_m(\Omega)$ be two arbitrary spherical harmonics (of, consequently, different degrees). Then $\langle Y_n, Y_m \rangle_{\mathrm{L}^2(\Omega)} = 0$.*

A fundamental property of orthonormal basis systems of spherical harmonics is the addition theorem. There exist several ways of proving it, where each of them is rather lengthy. For this reason, we omit the proof here and refer to Freeden et al. (1998), Michel (2013), and Müller (1966) for examples of proofs.

Theorem 4.2.9 (Addition Theorem for Spherical Harmonics) *Let an arbitrary orthonormal basis $\{Y_{n,j}\}_{j=-n,\ldots,n}$ of $\left(\mathrm{Harm}_n(\Omega), \langle\cdot,\cdot\rangle_{\mathrm{L}^2(\Omega)}\right)$ be given. Then*

$$\sum_{j=-n}^{n} Y_{n,j}(\xi)Y_{n,j}(\eta) = \frac{2n+1}{4\pi} P_n(\xi \cdot \eta)$$

for all $\xi, \eta \in \Omega$, where P_n is the Legendre polynomial of degree n.

This is the right moment to remember what we have already found out about the Legendre polynomials $\{P_n\}_{n\in\mathbb{N}_0}$.

- The function $y = P_n$ is the only polynomial which solves the differential equation $(1-t^2)y''(t) - 2ty'(t) = -n(n+1)y(t)$, $t \in [-1,1]$, and satisfies

$$P_n(0) = (-1)^{n/2}\frac{n!}{2^n}\left[\left(\frac{n}{2}\right)!\right]^{-2} \quad \text{if } n \text{ is even,}$$

  and

$$P_n'(0) = (-1)^{(n-1)/2}\frac{(n+1)!}{2^n}\left[\left(\frac{n+1}{2}\right)!\left(\frac{n-1}{2}\right)!\right]^{-1} \quad \text{if } n \text{ is odd,}$$

  respectively; see Definition 3.4.7 and Theorem 3.4.9.
- P_n is an even function if n is even, and it is an odd function if n is odd; see Theorem 3.4.8.

- The Legendre polynomials have the following explicit representation (see Theorem 3.4.10):

$$P_n(t) = \sum_{k=0}^{\lfloor \frac{n}{2} \rfloor} (-1)^k \frac{(2n-2k)!}{2^n k!\,(n-k)!\,(n-2k)!} t^{n-2k}, \quad t \in [-1,1].$$

- The Rodriguez formula (see Theorem 3.4.15) says that

$$P_n(t) = \frac{1}{2^n n!} \frac{\mathrm{d}^n}{\mathrm{d}t^n} (t^2-1)^n, \quad t \in [-1,1].$$

- The following expansions in Legendre polynomials are valid (at least in the given convergence range):

$$(1-2th+h^2)^{-1/2} = \sum_{n=0}^{\infty} P_n(t) h^n, \quad t \in [-1,1],\, h \in]-1,1[,$$

$$\frac{1}{|x-y|} = \sum_{n=0}^{\infty} \frac{|x|^n}{|y|^{n+1}} P_n\left(\frac{x}{|x|} \cdot \frac{y}{|y|}\right), \quad x,y \in \mathbb{R}^3,\, |x| < |y|,$$

$$\frac{R^2-r^2}{(r^2+R^2-2rRt)^{3/2}} = \sum_{n=0}^{\infty} \frac{2n+1}{R} \left(\frac{r}{R}\right)^n P_n(t), \quad t \in [-1,1], 0 \le r < R.$$

  See Theorem 3.4.17 as well as Corollaries 3.4.18 and 3.4.20. Note that the convergence ranges for t and h are a bit different here than in Theorem 3.4.17, but it can easily be proved that both sides of the first identity here solve the same initial-value problem.
- With $P_0 \equiv 1$, $P_1(t) = t$, and the recurrence formula

$$(n+1)P_{n+1}(t) = (2n+1)tP_n(t) - nP_{n-1}(t), \quad n \ge 1, \tag{4.9}$$

  the Legendre polynomials can be easily calculated; see Theorem 3.4.19. There are numerous other recurrence relations, see Corollary 3.4.21.
- The Legendre polynomials are $\mathrm{L}^2[-1,1]$-orthogonal. More precisely, the identity $\int_{-1}^{1} P_n(t)P_m(t)\,\mathrm{d}t = \delta_{nm}\, 2/(2n+1)$ holds for all $n,m \in \mathbb{N}_0$; see Theorem 3.4.22.

The latter property together with the fact that all polynomials on an interval $[a,b]$ are dense in $(\mathrm{C}[a,b]), \|\cdot\|_\infty)$ due to the Stone–Weierstraß approximation theorem (see Theorem 2.5.23) immediately yields another important property of Legendre polynomials.

Theorem 4.2.10 *The Legendre polynomials P_n constitute a complete orthogonal system in $\left(\mathrm{L}^2[-1,1], \langle\cdot,\cdot\rangle_{\mathrm{L}^2[-1,1]}\right)$. Moreover, $\left\{\sqrt{(2n+1)/2}\, P_n\right\}_{n\in\mathbb{N}_0}$ is a complete orthonormal system in $\left(\mathrm{L}^2[-1,1], \langle\cdot,\cdot\rangle_{\mathrm{L}^2[-1,1]}\right)$.*

For an alternative proof, see the considerations after Theorem 4.2.21.

We will be able to prove further properties of the Legendre polynomials in this section. One of them is an easy consequence of the already derived properties.

Theorem 4.2.11 *The Legendre polynomials satisfy*

$$P_n(-1) = (-1)^n, \qquad P_n(1) = 1,$$

$$P_n'(-1) = (-1)^{n+1}\,\frac{n(n+1)}{2}, \qquad P_n'(1) = \frac{n(n+1)}{2},$$

$$P_n''(-1) = (-1)^n\,\frac{n(n+1)(n^2+n-2)}{8}, \qquad P_n''(1) = \frac{n(n+1)(n^2+n-2)}{8}.$$

Proof Clearly, P_0 and P_1 satisfy all properties that we want to prove here. Using (4.9), we get (with $n \geq 1$) that

$$(n+1)P_{n+1}(\pm 1) = (2n+1)\cdot(\pm 1)\cdot P_n(\pm 1) - nP_{n-1}(\pm 1)$$

such that (by induction) $P_{n+1}(1) = (2n+1-n)/(n+1) = 1$ and

$$P_{n+1}(-1) = \frac{(-2n-1)(-1)^n - n(-1)^{n-1}}{n+1} = (-1)^{n+1}\frac{2n+1-n}{n+1} = (-1)^{n+1},$$

where the latter is also implied by $P_{n+1}(1) = 1$ and the fact that $P_{n+1}(-t) = (-1)^{n+1}P_{n+1}(t)$, which is also applicable to the remaining proofs for values at $t = -1$ (note that $P_{n+1}'(-t) = (-1)^{n+2}P_{n+1}'(t)$ and so on). With (3.112) from Corollary 3.4.21, we get $nP_n(1) = P_n'(1) - P_{n-1}'(1)$ for $n \geq 1$ such that (again by induction) $P_n'(1) = n + (n-1)n/2 = n(n+1)/2$.

Let us now differentiate (3.112). The result implies that $nP_n'(1) = P_n'(1) + P_n''(1) - P_{n-1}''(1)$ for $n \geq 1$. Hence, another simple induction yields

$$\begin{aligned}
P_n''(1) &= (n-1)\frac{n(n+1)}{2} + \frac{(n-1)n\left[(n-1)^2 + n - 1 - 2\right]}{8} \\
&= \frac{n(n+1)}{8}\left[4n - 4 + (n-1)\frac{n^2 - 2n + 1 + n - 3}{n+1}\right] \\
&= \frac{n(n+1)}{8}\left[4n - 4 + (n-1)(n-2)\right] = \frac{n(n+1)(n^2+n-2)}{8}.
\end{aligned}$$

□

The addition theorem (see Theorem 4.2.9) can be used to prove further useful propositions such as the following two norm estimates.

Theorem 4.2.12 *Every spherical harmonic $Y_n \in \mathrm{Harm}_n(\Omega)$ satisfies*

$$\|Y_n\|_{\mathrm{C}(\Omega)} = \max_{\xi\in\Omega}|Y_n(\xi)| \leq \sqrt{\frac{2n+1}{4\pi}}\,\|Y_n\|_{\mathrm{L}^2(\Omega)}.$$

In particular,

$$\|Y_{n,j}\|_{\mathrm{C}(\Omega)} \leq \sqrt{\frac{2n+1}{4\pi}}.$$

Proof With the arbitrary orthonormal basis of $\mathrm{Harm}_n(\Omega)$, as introduced in Definition 4.2.5, we can write $|Y_n(\xi)| = \left|\sum_{j=-n}^{n}\langle Y_n, Y_{n,j}\rangle_{\mathrm{L}^2(\Omega)}\, Y_{n,j}(\xi)\right|$ for all $\xi \in \Omega$. Hence, the Cauchy–Schwarz inequality in $\mathbb{R}^{2n+1}$ (see Theorem 2.5.8) implies together with the addition theorem (Theorem 4.2.9), the Parseval identity (see Theorem 2.5.25), and Theorem 4.2.11 that

$$|Y_n(\xi)| \le \left(\sum_{j=-n}^{n} \langle Y_n, Y_{n,j}\rangle_{\mathrm{L}^2(\Omega)}^2\right)^{1/2} \left(\sum_{j=-n}^{n} Y_{n,j}(\xi)^2\right)^{1/2}$$
$$= \|Y_n\|_{\mathrm{L}^2(\Omega)} \left(\frac{2n+1}{4\pi} P_n(\xi\cdot\xi)\right)^{1/2} = \|Y_n\|_{\mathrm{L}^2(\Omega)} \sqrt{\frac{2n+1}{4\pi}}$$

for all $\xi \in \Omega$. This inequality immediately yields the results of Theorem 4.2.12. □

A corresponding result can also be proved for the Legendre polynomials out of the addition theorem.

Theorem 4.2.13 *The Legendre polynomials satisfy*

$$\|P_n\|_{\mathrm{C}[-1,1]} = \max_{t\in[-1,1]} |P_n(t)| = P_n(1) = 1.$$

Proof Let $t \in [-1,1]$ be arbitrary. Obviously, there exist $\xi, \eta \in \Omega$ such that $t = \xi \cdot \eta$. Then the addition theorem (Theorem 4.2.9) again in combination with the Cauchy–Schwarz inequality in $\mathbb{R}^{2n+1}$ (Theorem 2.5.8) and Theorem 4.2.11 yields

$$|P_n(\xi\cdot\eta)| = \frac{4\pi}{2n+1}\left|\sum_{j=-n}^{n} Y_{n,j}(\xi)Y_{n,j}(\eta)\right|$$
$$\le \frac{4\pi}{2n+1}\left(\sum_{j=-n}^{n} Y_{n,j}(\xi)^2\right)^{1/2}\left(\sum_{j=-n}^{n} Y_{n,j}(\eta)^2\right)^{1/2}$$
$$= \frac{4\pi}{2n+1}\left(\frac{2n+1}{4\pi}P_n(\xi\cdot\xi)\right)^{1/2}\left(\frac{2n+1}{4\pi}P_n(\eta\cdot\eta)\right)^{1/2} = 1,$$

where we already know that $1 = P_n(1)$. □

We need another fundamental property of spherical harmonics: all choices of function systems $\{Y_{n,j}\}_{n\in\mathbb{N}_0;\, j=-n,\ldots,n}$ are complete in $\left(\mathrm{L}^2(\Omega), \langle\cdot,\cdot\rangle_{\mathrm{L}^2(\Omega)}\right)$. This means that every $F \in \mathrm{L}^2(\Omega)$ can be expanded into a (Fourier) series in this orthonormal basis. For proving this, we first have to do some preparations. They start with remembering an important result from our considerations about boundary-value problems of the Laplace equation: the Poisson integral formula. We will call the following variant the **Poisson integral formula of constructive approximation on the sphere** (see, e.g., also Freeden et al., 1998, Theorem 3.4.1; and Michel, 2013, Theorem 5.19).

Theorem 4.2.14 (Poisson Integral Formula) *Let $F \in \mathrm{C}(\Omega)$. Then the following uniform convergence holds true:*

$$\lim_{r\to 1-} \sup_{\xi\in\Omega} \left|\frac{1}{4\pi}\int_\Omega \frac{(1-r^2)F(\eta)}{(1+r^2-2r\xi\cdot\eta)^{3/2}}\, \mathrm{d}\omega(\eta) - F(\xi)\right| = 0. \tag{4.10}$$

Proof We consider the interior Dirichlet problem (IDP): find a function $U \in \mathrm{C}^{(2)}(\Omega_{\mathrm{int}}) \cap \mathrm{C}(\overline{\Omega_{\mathrm{int}}})$ such that $\Delta U = 0$ in Ω_{int} and $U = F$ on Ω. The Poisson integral formula which we have already proved, namely Theorem 3.3.38, tells us that

$$U(r\xi) = \int_\Omega F(\eta) \frac{1 - r^2}{4\pi(1 + r^2 - 2r\xi \cdot \eta)^{3/2}} \, \mathrm{d}\omega(\eta), \quad 0 \leq r < 1, \xi \in \Omega,$$

solves this IDP. Since $U \in \mathrm{C}(\overline{\Omega_{\mathrm{int}}})$, it is uniformly continuous: for each $\varepsilon > 0$, there exists $\delta > 0$ (without loss of generality, $\delta < 1$) such that, for all $x, y \in \overline{\Omega_{\mathrm{int}}}$ with $|x - y| \leq \delta$, we have $|U(x) - U(y)| < \varepsilon$. In particular, if $y = \xi \in \Omega$ and $x = r\xi$ with $0 \leq r < 1$, then we have $|U(r\xi) - U(\xi)| < \varepsilon$ for all $r \in [1 - \delta, 1[$. In other words,

$$\left| \frac{1}{4\pi} \int_\Omega \frac{(1 - r^2) F(\eta)}{(1 + r^2 - 2r\xi \cdot \eta)^{3/2}} \, \mathrm{d}\omega(\eta) - F(\xi) \right| < \varepsilon$$

for all $\xi \in \Omega$ and all $r \in [1 - \delta, 1[$. This is the required result. □

Let us have a look at the Poisson integral formula in this version and the Abel–Poisson kernel. From Corollary 3.4.20, we know that it can be represented as follows (here, with $R = 1$ again):

$$\frac{1}{4\pi} \frac{1 - r^2}{(1 + r^2 - 2r\xi \cdot \eta)^{3/2}} = \sum_{n=0}^{\infty} \frac{2n + 1}{4\pi} r^n P_n(\xi \cdot \eta)$$

for $0 \leq r < 1$ and $\xi, \eta \in \Omega$. Obviously, for each fixed r, this series is uniformly convergent in ξ and η, because $|P_n(\xi \cdot \eta)| \leq 1$ for all $\xi, \eta \in \Omega$ (see Theorem 4.2.13) and $\sum_{n=0}^{\infty} (2n + 1) r^n < +\infty$. Hence, Theorem 2.4.5 allows us to interchange the series with the integral in (4.10) such that

$$\begin{aligned} \frac{1}{4\pi} \int_\Omega \frac{(1 - r^2) F(\eta)}{(1 + r^2 - 2r\xi \cdot \eta)^{3/2}} \, \mathrm{d}\omega(\eta) &= \sum_{n=0}^{\infty} \frac{2n + 1}{4\pi} r^n \int_\Omega F(\eta) P_n(\xi \cdot \eta) \, \mathrm{d}\omega(\eta) \\ &= \sum_{n=0}^{\infty} \sum_{j=-n}^{n} r^n \int_\Omega F(\eta) Y_{n,j}(\eta) \, \mathrm{d}\omega(\eta) \, Y_{n,j}(\xi) \end{aligned} \tag{4.11}$$

for all $\xi \in \Omega$ and all $r \in [0, 1[$, where we used the addition theorem (Theorem 4.2.9) here. Note that $\int_\Omega F(\eta) Y_{n,j}(\eta) \, \mathrm{d}\omega(\eta) = \langle F, Y_{n,j} \rangle_{\mathrm{L}^2(\Omega)}$ and we know that the functions $Y_{n,j}$ are orthonormal in $\mathrm{L}^2(\Omega)$. Hence, the Bessel inequality (see Theorem 2.5.25) implies that there is a constant $C > 0$ such that $\left| \langle F, Y_{n,j} \rangle_{\mathrm{L}^2(\Omega)} \right| \leq C$ for all $n \in \mathbb{N}_0$ and all $j = -n, \ldots, n$. In combination with the estimate $|Y_{n,j}(\xi)| \leq \sqrt{(2n + 1)/(4\pi)}$ (see Theorem 4.2.12), we are now in the position to conclude that the series in (4.11) also uniformly converges with respect to $\xi \in \Omega$ and for each *fixed* $r \in [0, 1[$.

Now let $\varepsilon > 0$. We consequently find an $N = N(r, \varepsilon) \in \mathbb{N}$ such that the remainder satisfies

$$\left\| \sum_{n=N+1}^{\infty} \sum_{j=-n}^{n} r^n \langle F, Y_{n,j} \rangle_{\mathrm{L}^2(\Omega)} Y_{n,j} \right\|_{\mathrm{C}(\Omega)} \leq \frac{\varepsilon}{2}.$$

Here is now the link to the Poisson integral formula from Theorem 4.2.14: to the very same $\varepsilon > 0$, we also find a $r_0 \in [0, 1[$ such that (remember (4.11))

$$\left\| \sum_{n=0}^{\infty} \sum_{j=-n}^{n} r_0^n \langle F, Y_{n,j} \rangle_{\mathrm{L}^2(\Omega)} Y_{n,j} - F \right\|_{\mathrm{C}(\Omega)} \leq \frac{\varepsilon}{2}.$$

The triangle inequality eventually yields that for each $\varepsilon > 0$, there exist $r_0 \in [0, 1[$ and $N = N(r_0, \varepsilon) \in \mathbb{N}$ such that

$$\left\| F - \sum_{n=0}^{N} \sum_{j=-n}^{n} r_0^n \langle F, Y_{n,j} \rangle_{\mathrm{L}^2(\Omega)} Y_{n,j} \right\|_{\mathrm{C}(\Omega)} \leq \varepsilon.$$

In other words, no matter how small the chosen error tolerance ε might be, for every continuous function $F : \Omega \to \mathbb{R}$, there exists a finite linear combination of $Y_{n,j}$-functions which uniformly approximates F with this error bound ε. We have already learned that there is a name for such a property (see Definition 2.5.26).

Theorem 4.2.15 *Each system $\{Y_{n,j}\}_{n\in\mathbb{N}_0\,;\, j=-n,\dots,n}$ is closed (in the sense of the approximation theory) in $\left(\mathrm{C}(\Omega), \|\cdot\|_{\mathrm{C}(\Omega)}\right)$.*

We now first use the fact that $\|\cdot\|_{\mathrm{C}(\Omega)}$ is 'stronger' than $\|\cdot\|_{\mathrm{L}^2(\Omega)}$ (see Theorem 2.5.20, part b) and then the property of $\mathrm{C}(\Omega)$ as a dense subset of $\mathrm{L}^2(\Omega)$ (see Theorem 2.5.21) to make the following conclusions.

Theorem 4.2.16 *Each system $\{Y_{n,j}\}_{n\in\mathbb{N}_0;\, j=-n,\dots,n}$ is closed (in the sense of the approximation theory) in $\left(\mathrm{C}(\Omega), \|\cdot\|_{\mathrm{L}^2(\Omega)}\right)$.*

Theorem 4.2.17 *Each system $\{Y_{n,j}\}_{n\in\mathbb{N}_0;\, j=-n,\dots,n}$ is closed (in the sense of the approximation theory) in $\left(\mathrm{L}^2(\Omega), \|\cdot\|_{\mathrm{L}^2(\Omega)}\right)$.*

The latter result is a major property of the $Y_{n,j}$-functions, because they are constructed to be $\langle\cdot,\cdot\rangle_{\mathrm{L}^2(\Omega)}$-orthonormal, and now Theorem 2.5.25 on complete orthonormal systems comes into play.

Theorem 4.2.18 *Every system $\{Y_{n,j}\}_{n\in\mathbb{N}_0;\, j=-n,\dots,n}$ is a complete orthonormal system in $\left(\mathrm{L}^2(\Omega), \langle\cdot,\cdot\rangle_{\mathrm{L}^2(\Omega)}\right)$.*

The fact that the $Y_{n,j}$ constitute a basis is a reason to come back to the identity $\Delta^* Y_{n,j} = -n(n+1)Y_{n,j}$ because the combination of both yields an alternative way of defining spherical harmonics.

Theorem 4.2.19 *All eigenvalues of the Beltrami operator $\Delta^* : \mathrm{C}^{(2)}(\Omega) \to \mathrm{C}(\Omega)$ are given by the sequence $(-n(n+1))_{n\in\mathbb{N}_0}$. The eigenspace corresponding to the eigenvalue $-n(n+1)$ for $n \in \mathbb{N}_0$ is* $\mathrm{Harm}_n(\Omega)$.

In other words, the spherical harmonics are *the* eigenfunctions of the Beltrami operator.

Proof We first determine the eigenvalues and then derive the associated eigenfunctions.

(1) Eigenvalues:

With Green's second surface identity (Theorem 4.1.11), we can deduce that $\int_\Omega \left(F(\xi)\Delta^*_\xi Y_{n,j}(\xi) - Y_{n,j}(\xi)\Delta^*_\xi F(\xi)\right) \mathrm{d}\omega(\xi) = 0$ for all $F \in \mathrm{C}^{(2)}(\Omega), n \in \mathbb{N}_0$, and $j \in \{-n,\ldots,n\}$. If we now assume that $\Delta^* F = \lambda F$ for a constant $\lambda \in \mathbb{R}$ and use that $\Delta^* Y_{n,j} = -n(n+1)Y_{n,j}$, then we get

$$(-n(n+1) - \lambda)\langle F, Y_{n,j}\rangle_{\mathrm{L}^2(\Omega)} = 0, \tag{4.12}$$

which holds again for all $n \in \mathbb{N}_0$ and all $j \in \{-n,\ldots,n\}$. If $\lambda \neq -n(n+1)$ for all $n \in \mathbb{N}_0$, then (4.12) can only be true if $\langle F, Y_{n,j}\rangle_{\mathrm{L}^2(\Omega)} = 0$ for all $n \in \mathbb{N}_0$ and all $j \in \{-n,\ldots,n\}$. Due to the completeness of the $Y_{n,j}$ (Theorem 4.2.18), this means that $F = 0$ and, hence, F is not an eigenfunction (remember that the zero vector is, by definition, never an eigenvector). Thus, if $F \neq 0$, then there must exist an $n \in \mathbb{N}_0$ such that $\lambda = -n(n+1)$. With our knowledge about the $Y_{n,j}$, we have already observed that these eigenvalues, indeed, all occur.

(2) Eigenfunctions:

If $\lambda = -k(k+1)$ for a fixed $k \in \mathbb{N}_0$, then (4.12) implies $\langle F, Y_{n,j}\rangle_{\mathrm{L}^2(\Omega)} = 0$ for all $n \in \mathbb{N}_0 \setminus \{k\}$ and all $j \in \{-n,\ldots,n\}$. On the other hand, the basis property of the $Y_{n,j}$ (see Theorem 4.2.18) implies that F has the expansion $F = \sum_{n=0}^\infty \sum_{j=-n}^n \langle F, Y_{n,j}\rangle_{\mathrm{L}^2(\Omega)} Y_{n,j}$ in the sense of $\mathrm{L}^2(\Omega)$. Hence, F must be representable as $F = \sum_{j=-k}^k \langle F, Y_{k,j}\rangle_{\mathrm{L}^2(\Omega)} Y_{k,j}$ and consequently $F \in \mathrm{Harm}_k(\Omega)$; see Definition 4.2.5. □

Since we constructed the fully normalized spherical harmonics as orthonormal eigenfunctions of Δ^*, we see now that, indeed, they are a particular case of the more general Definition 4.2.5.

Corollary 4.2.20 *The system of fully normalized spherical harmonics $Y_{n,j}$, $j = -n,\ldots,n$, constitutes an orthonormal basis of* $(\mathrm{Harm}_n(\Omega), \langle\cdot,\cdot\rangle_{\mathrm{L}^2(\Omega)})$.

In view of what we learned about the expansion of harmonic functions in homogeneous harmonic polynomials, we can now indeed conclude, for example, that the solution of the IDP ($\Delta U = 0$ in Ω_{int}, $U = F$ on Σ) is representable by the following formula $U(r\xi) = \sum_{n=0}^\infty \sum_{j=-n}^n \langle F, Y_{n,j}\rangle_{\mathrm{L}^2(\Omega)} r^n Y_{n,j}(\xi)$ with $r \in [0,1]$ and $\xi \in \Omega$ (see also the discussions after Theorem 3.4.25).

A similar result as in Theorem 4.2.19 can be shown for the Legendre polynomials P_n. Remember that they were introduced as solutions $y = P_n$ of the differential equation

$$(1-t^2)y''(t) - 2ty'(t) = -n(n+1)y(t).$$

The left-hand side of this equation corresponds to a well-known differential operator, the Legendre operator; see Definition 3.4.5.

Hence, $\mathcal{L}\, P_n = -n(n+1)P_n$. Moreover, $\mathcal{L}$ is a part of the Beltrami operator (see Theorem 3.4.3) such that $\Delta^*_\xi = \mathcal{L}_t + (1-t^2)^{-1} \frac{\partial^2}{\partial \varphi^2}$. Consequently, if $F \in \mathrm{C}^{(2)}[-1,1]$ and $\lambda \in \mathbb{R}$ such that $\mathcal{L}\, F = \lambda F$, then this is equivalent to

$$\left(\Delta^*_\xi - \frac{1}{1-t^2} \frac{\partial^2}{\partial \varphi^2} \right) F(\varepsilon^3 \cdot \xi) = \lambda F(\varepsilon^3 \cdot \xi) \quad \text{for all } \xi \in \Omega.$$

Note that $\xi \cdot \varepsilon^3$ yields the polar distance t of ξ (see Example 2.5.4). Hence, $\mathcal{L}\, F = \lambda F \Leftrightarrow \Delta^* F(\varepsilon^3 \cdot) = \lambda F(\varepsilon^3 \cdot)$. With Theorem 4.2.19, this means that $\mathcal{L}\, F = \lambda F$ if and only if there exists $n \in \mathbb{N}_0$ such that $\lambda = -n(n+1)$ and $F(\varepsilon^3 \cdot) \in \mathrm{Harm}_n(\Omega)$. Furthermore, Theorems 4.1.9 and 4.2.9 and Lemma 3.4.13 allow us to conclude that

$$\begin{aligned} \int_{-1}^{1} F(t) P_k(t)\, \mathrm{d}t &= \frac{1}{2\pi} \int_\Omega F(\varepsilon^3 \cdot \xi) P_k(\varepsilon^3 \cdot \xi)\, \mathrm{d}\omega(\xi) \\ &= \frac{2}{2k+1} \sum_{j=-k}^{k} \int_\Omega F(\varepsilon^3 \cdot \xi) Y_{k,j}(\xi)\, \mathrm{d}\omega(\xi)\, Y_{k,j}(\varepsilon^3) = 0 \end{aligned} \tag{4.13}$$

for all $k \in \mathbb{N}_0 \setminus \{n\}$ and all $j \in \{-k, \ldots, k\}$. In other words, $\mathcal{L}\, F = \lambda F$ is valid if and only if there exists $n \in \mathbb{N}_0$ with $\lambda = -n(n+1)$ and $\langle F, P_k \rangle_{\mathrm{L}^2[-1,1]} = 0$ for all $k \in \mathbb{N}_0 \setminus \{n\}$. By consulting Theorem 4.2.10, we arrive at the following result.

Theorem 4.2.21 *The eigenvalues of the Legendre operator $\mathcal{L}$ have the form $-n(n+1)$, $n \in \mathbb{N}_0$. The eigenfunctions corresponding to the eigenvalue $-n(n+1)$ have the form $c P_n$, where $c \in \mathbb{R} \setminus \{0\}$ is a constant and P_n is the Legendre polynomial of degree n.*

In this book, Theorem 4.2.10 on the completeness of the Legendre polynomials followed in parts from the Stone–Weierstraß approximation theorem. With the knowledge of the completeness of the $Y_{n,j}$, we also could have proved it now with an alternative argument similar to (4.13): for all $\eta \in \Omega$ and all $F \in \mathrm{L}^2[-1,1]$, we have

$$\int_{-1}^{1} F(t) P_k(t)\, \mathrm{d}t = \frac{2}{2k+1} \sum_{j=-k}^{k} \int_\Omega F(\eta \cdot \xi)\, Y_{k,j}(\xi)\, \mathrm{d}\omega(\xi)\, Y_{k,j}(\eta)$$

for each $k \in \mathbb{N}_0$. Hence, if $\langle F, P_k \rangle_{\mathrm{L}^2[-1,1]} = 0$ for a given $F \in \mathrm{L}^2[-1,1]$ and all $k \in \mathbb{N}_0$, then the right-hand side must also vanish accordingly. Since the zero function is also uniquely representable in the $Y_{k,j}$, the vanishing integrals imply that the function $\Omega \ni \xi \mapsto F(\eta \cdot \xi)$ must vanish almost everywhere. Hence, $F = 0$ also in $\mathrm{L}^2[-1,1]$.

Eventually, we briefly mention how fully normalized spherical harmonics can be calculated numerically. The following algorithm is based on Fengler (2005). Note that the fully normalized spherical harmonics include the factor

$$\tilde{P}_{n,j}(t) := \sqrt{\frac{2n+1}{4\pi} \frac{(n-j)!}{(n+j)!}}\, P_{n,j}(t), \quad t \in [-1,1]. \tag{4.14}$$

Since factorials quickly become extremely large, which can result in numerical instabilities, it should be utilized that the fraction $(n-j)!/(n+j)!$ is not large. Therefore, we use a property from Robin (1957, p. 100, (47)).

Theorem 4.2.22 *The Legendre functions $P_{n,j}$ satisfy*

$$(n-j)P_{n,j}(t) - (2n-1)tP_{n-1,j}(t) + (n+j-1)P_{n-2,j}(t) = 0$$

for all $t \in [-1,1], n \in \mathbb{N} \setminus \{1\}$, and $j \in \{0,\dots,n-1\}$, where $P_{n-2,n-1}(t) := 0$ for all t.

Proof We recall the results from (3.110), (3.112), and (3.114), which read

$$\begin{aligned} nP_n(t) &= (2n-1)tP_{n-1}(t) - (n-1)P_{n-2}(t), \qquad (4.15)\\ (n-1)P_{n-1}(t) &= tP'_{n-1}(t) - P'_{n-2}(t),\\ nP_{n-1}(t) &= P'_n(t) - tP'_{n-1}(t). \end{aligned}$$

By adding the latter two identities, we get

$$(2n-1)P_{n-1}(t) = P'_n(t) - P'_{n-2}(t). \qquad (4.16)$$

We now derive (4.15) j times and (4.16) $j-1$ times. This results (with the Leibniz rule; see Theorem 2.3.6) in

$$nP_n^{(j)}(t) = (2n-1)tP_{n-1}^{(j)}(t) + (2n-1)jP_{n-1}^{(j-1)}(t) - (n-1)P_{n-2}^{(j)}(t) \qquad (4.17)$$

and

$$(2n-1)P_{n-1}^{(j-1)}(t) = P_n^{(j)}(t) - P_{n-2}^{(j)}(t). \qquad (4.18)$$

We insert (4.18) into (4.17) and obtain

$$(n-j)P_n^{(j)}(t) = (2n-1)tP_{n-1}^{(j)}(t) - (n+j-1)P_{n-2}^{(j)}(t).$$

The multiplication with $(1-t^2)^{j/2}$ and a consultation of Corollary 3.4.16 yield the desired result. Note that $P_{n-2}^{(n-1)} \equiv 0$ such that the theorem is also valid in the case $j = n-1$. □

With this result in mind, we can now derive that

$$\begin{aligned} (n-j)\tilde{P}_{n,j}(t) = {}& \sqrt{(2n+1)(2n-1)}\,t\sqrt{\frac{n-j}{n+j}}\,\tilde{P}_{n-1,j}(t)\\ & - (n+j-1)\sqrt{\frac{2n+1}{2n-3}\frac{(n-j)(n-1-j)}{(n+j)(n-1+j)}}\,\tilde{P}_{n-2,j}(t) \end{aligned}$$

such that

$$\tilde{P}_{n,j}(t) = \sqrt{\frac{(2n+1)(2n-1)}{(n-j)(n+j)}}\, t\, \tilde{P}_{n-1,j}(t)$$
$$- \sqrt{\frac{(2n+1)(n-1-j)(n-1+j)}{(2n-3)(n-j)(n+j)}}\, \tilde{P}_{n-2,j}(t). \tag{4.19}$$

To have a complete recursion, we also need formulae for $\tilde{P}_{n,j}$ with $j > n-2$. We have that (remember Corollary 3.4.16)

$$P_{n,n}(t) = \frac{1}{2^n n!}(1-t^2)^{n/2}\frac{\mathrm{d}^{2n}}{\mathrm{d}t^{2n}}(t^2-1)^n = \frac{1}{2^n n!}(1-t^2)^{n/2}(2n)!$$

and, consequently, for $n \geq 1$,

$$P_{n,n}(t) = \frac{1}{2n}\sqrt{1-t^2}\cdot 2n\cdot(2n-1)P_{n-1,n-1}(t) = (2n-1)\sqrt{1-t^2}\, P_{n-1,n-1}(t).$$

Hence, again for $n \geq 1$, we obtain

$$\tilde{P}_{n,n}(t) = \sqrt{\frac{2n+1}{2n-1}\cdot\frac{1}{2n(2n-1)}}\,(2n-1)\sqrt{1-t^2}\,\tilde{P}_{n-1,n-1}(t)$$
$$= \sqrt{\frac{2n+1}{2n}}\sqrt{1-t^2}\,\tilde{P}_{n-1,n-1}(t).$$

Moreover, still for $n \geq 1$, we can derive that

$$P_{n,n-1}(t) = \frac{1}{2^n n!}(1-t^2)^{(n-1)/2}\frac{\mathrm{d}^{2n-1}}{\mathrm{d}t^{2n-1}}(t^2-1)^n$$
$$= \frac{1}{2^n n!}(1-t^2)^{(n-1)/2}(2n)!\, t$$
$$= \frac{1}{2^{n-1}(n-1)!}(1-t^2)^{(n-1)/2}(2n-1)!\, t = (2n-1)t\, P_{n-1,n-1}(t),$$

where always $t \in [-1,1]$ in the considerations here. Hence,

$$\tilde{P}_{n,n-1}(t) = \sqrt{\frac{2n+1}{2n-1}\cdot\frac{1}{2n-1}}\,(2n-1)t\,\tilde{P}_{n-1,n-1}(t) = \sqrt{2n+1}\, t\, \tilde{P}_{n-1,n-1}(t). \tag{4.20}$$

We also need to calculate the trigonometric part of the fully normalized spherical harmonics. First, we recall that $\xi \in \Omega$ satisfies in polar coordinates $\xi_1 = \sqrt{1-t^2}\cos\varphi$ and $\xi_2 = \sqrt{1-t^2}\sin\varphi$. If we now set $z := \xi_1 + \mathrm{i}\xi_2$ with the complex imaginary constant i ($\mathrm{i}^2 = -1$), then

$$z^j = (1-t^2)^{j/2}(\cos\varphi + \mathrm{i}\sin\varphi)^j = (1-t^2)^{j/2}e^{\mathrm{i}j\varphi}$$
$$= (1-t^2)^{j/2}[\cos(j\varphi) + \mathrm{i}\sin(j\varphi)].$$

This shows us that we obtain the factor $(1-t^2)^{j/2}$ of the associated Legendre functions from the calculation of the trigonometric part of the fully normalized spherical harmonics. We have the following (see Definition 3.4.24 and Corollary 3.4.16):

$$Y_{n,j}(\xi(\varphi,t)) = \sqrt{\frac{2n+1}{4\pi}\frac{(n-|j|)!}{(n+|j|)!}(2-\delta_{j0})}\,\frac{1}{2^n n!}$$
$$\times \frac{\mathrm{d}^{n+|j|}}{\mathrm{d}t^{n+|j|}}(t^2-1)^n \begin{cases} \Im(\xi_1+\mathrm{i}\xi_2)^j, & j=1,\ldots,n, \\ \Re(\xi_1+\mathrm{i}\xi_2)^{-j}, & j=-n,\ldots,0, \end{cases}$$

where $\Im$ and $\Re$ stand for the imaginary and the real part of a complex number. Together with (4.19), we now have found a way to calculate the fully normalized spherical harmonics numerically. Note that the previous considerations also yield that

$$(1-t^2)^{-n/2}\tilde{P}_{n,n}(t) = \sqrt{\frac{2n+1}{2n}}(1-t^2)^{-(n-1)/2}\tilde{P}_{n-1,n-1}(t),$$
$$(1-t^2)^{-(n-1)/2}\tilde{P}_{n,n-1}(t) = \sqrt{2n+1}\,t(1-t^2)^{-(n-1)/2}\tilde{P}_{n-1,n-1}(t).$$

Algorithm 4.2.23 *The following algorithm yields $Y_{n,j}(\xi)$ at a given point $\xi \in \Omega$ for all degrees $n \leq N$ and all orders $j=-n,\ldots,n$.*
Given: *$\xi \in \Omega$, maximal degree $N \in \mathbb{N}$.*

(1) *Let $z := \xi_1 + \mathrm{i}\xi_2$ and $P_0 := 1$.*
(2) *for $j = 1$ to N,*

$P_{j+1} := P_j \cdot z$

end
(3) *Let $X_{0,0} := \sqrt{\frac{1}{4\pi}}$.*
(4) *for $n = 1$ to N,*

$X_{n,n} := \sqrt{\frac{2n+1}{2n}}\,X_{n-1,n-1}$

end
(5) *for $j = 0$ to $N-1$,*

$X_{j+1,j} := \sqrt{2j+3}\,t X_{j,j}$
for $n = j+2$ to N,

$$X_{n,j} := \sqrt{\frac{(2n+1)(2n-1)}{(n-j)(n+j)}}\,t X_{n-1,j} - \sqrt{\frac{(2n+1)(n-1-j)(n-1+j)}{(2n-3)(n-j)(n+j)}}\,X_{n-2,j}$$

end

end
(6) *for $n = 1$ to N,*

for $j = 1$ to N,

$X_{n,j} := X_{n,j} \cdot \sqrt{2}$

end

end

(7) *for* $n = 0$ *to* N,

for $j = -n$ *to* 0,

$$Y_{n,j} := X_{n,|j|} \cdot \Re P_{|j|}$$

end

for $j = 1$ *to* n,

$$Y_{n,j} := X_{n,j} \cdot \Im P_j$$

end

end

The array $Y_{n,j}$ *contains the values of the fully normalized spherical harmonics* $Y_{n,j}(\xi)$.

Note that for large degrees (where the maximal degree of the gravitational model EGM2008 is already large in this sense), the calculation of the fully normalized spherical harmonics becomes unstable due to the common machine accuracy. This problem can be overcome by scaling the occurring terms appropriately; see Holmes and Featherstone (2002). More advanced techniques are also available for extremely large degrees in Fukushima (2012).

The recurrence relation in Theorem 4.2.22 can also be used to prove a well-known theorem on spherical harmonics; see also Seibert (2018, Theorem 3.3.5) for a generalization of this formula.

Theorem 4.2.24 (Christoffel–Darboux Formula) *In the case of the fully normalized spherical harmonics, we have*

$$(t_\xi - t_\eta) \sum_{n=|j|}^{L-1} Y_{n,j}(\xi) Y_{n,j}(\eta) = c_{L,j} \left(Y_{L,j}(\xi) Y_{L-1,j}(\eta) - Y_{L-1,j}(\xi) Y_{L,j}(\eta) \right) \tag{4.21}$$

for all $L \in \mathbb{N}$, *all* $\xi, \eta \in \Omega$, *and all* $j \in \{-L+1, \ldots, L-1\}$, *where* t_ξ *and* t_η *are the polar distances of* ξ *and* η, *respectively, and*

$$c_{L,j} := \sqrt{(L^2 - j^2)(4L^2 - 1)^{-1}} = \sqrt{(L-j)(L+j)(2L+1)^{-1}(2L-1)^{-1}}.$$

Proof With (4.19) and the abbreviations in (4.14) and Definition 3.4.24, we obtain $Y_{n,j}(\xi) = c_{n,j}^{-1} t_\xi Y_{n-1,j}(\xi) - c_{n-1,j} c_{n,j}^{-1} Y_{n-2,j}(\xi)$, $\xi \in \Omega$, for all $n \geq 2$ and all $|j| \leq n-2$. Hence, $t_\xi Y_{n-1,j}(\xi) = c_{n,j} Y_{n,j}(\xi) + c_{n-1,j} Y_{n-2,j}(\xi)$, $\xi \in \Omega$. Certainly, this is also analogously valid at the point $\eta \in \Omega$. With a little index shift ($n-1 \to n$), we get

$$t_\xi Y_{n,j}(\xi) = c_{n+1,j} Y_{n+1,j}(\xi) + c_{n,j} Y_{n-1,j}(\xi), \quad \xi \in \Omega,$$

$$t_\eta Y_{n,j}(\eta) = c_{n+1,j} Y_{n+1,j}(\eta) + c_{n,j} Y_{n-1,j}(\eta), \quad \eta \in \Omega,$$

for all $n \geq 1$ and all $j \in \mathbb{Z}$ with $|j| \leq n-1$. Moreover, for $|j| = n$, we have $c_{n,j} = c_{n,n} = 0$ and $c_{n+1,j} = c_{n+1,n} = \sqrt{(2n+1)(2n+1)^{-1}(2n+3)^{-1}} = (2n+3)^{-1/2}$. Hence, (4.20) shows us that the preceding identities are also valid for $|j| = n$. With this knowledge in mind, we observe that

$$t_\xi\, Y_{n,j}(\xi)Y_{n,j}(\eta) = c_{n+1,j}\, Y_{n+1,j}(\xi)Y_{n,j}(\eta) + c_{n,j}\, Y_{n-1,j}(\xi)Y_{n,j}(\eta)$$
$$t_\eta\, Y_{n,j}(\eta)Y_{n,j}(\xi) = c_{n+1,j}\, Y_{n+1,j}(\eta)Y_{n,j}(\xi) + c_{n,j}\, Y_{n-1,j}(\eta)Y_{n,j}(\xi)$$

and, consequently, again using that $c_{n,j} = 0$ for $n = |j|$, we get

$$\begin{aligned}
&(t_\xi - t_\eta) \sum_{n=|j|}^{L-1} Y_{n,j}(\xi)\, Y_{n,j}(\eta) \\
&= \sum_{n=|j|}^{L-1} c_{n+1,j}\, Y_{n+1,j}(\xi)\, Y_{n,j}(\eta) + \sum_{n=|j|}^{L-1} c_{n,j}\, Y_{n-1,j}(\xi)\, Y_{n,j}(\eta) \\
&\quad - \sum_{n=|j|}^{L-1} c_{n+1,j}\, Y_{n+1,j}(\eta)\, Y_{n,j}(\xi) - \sum_{n=|j|}^{L-1} c_{n,j}\, Y_{n-1,j}(\eta)\, Y_{n,j}(\xi) \\
&= \sum_{n=|j|+1}^{L} c_{n,j}\, Y_{n,j}(\xi)\, Y_{n-1,j}(\eta) + \sum_{n=|j|}^{L-1} c_{n,j}\, Y_{n-1,j}(\xi)\, Y_{n,j}(\eta) \\
&\quad - \sum_{n=|j|+1}^{L} c_{n,j}\, Y_{n,j}(\eta)\, Y_{n-1,j}(\xi) - \sum_{n=|j|}^{L-1} c_{n,j}\, Y_{n-1,j}(\eta)\, Y_{n,j}(\xi) \\
&= c_{L,j} Y_{L,j}(\xi) Y_{L-1,j}(\eta) - c_{L,j}\, Y_{L,j}(\eta)\, Y_{L-1,j}(\xi)
\end{aligned}$$

It is easy to see that this is (4.21). □

4.3 Vector Spherical Harmonics

In this section, the well-known theory of vector spherical harmonics is summarized. The considerations are essentially based on Edmonds (1957), Freeden et al. (1998), Leweke (2018), and Morse and Feshbach (1953a,b). Not all properties of vector spherical harmonics will be discussed here. For further details and applications, see these references as well as, for example: Balandin et al. (2012), Barrera et al. (1985), Carrascal et al. (1991), Clapp and Li (1970), Dahlen and Tromp (1998), Freeden and Gutting (2008), Freeden and Gutting (2013), Freeden and Schreiner (2009), Gubbins et al. (2011), Hill (1954), Sanna (2000), von Brecht (2016), and Weinberg (1994).

Intuitively, one would guess that each scalar spherical harmonic $Y_{n,j}$, which we have already discussed in detail, should be transformed into three vectorial functions in an appropriate way in order to get an orthonormal basis for $\mathrm{L}^2(\Omega, \mathbb{R}^3) = \mathrm{l}^2(\Omega)$. Indeed, this is a viable strategy and, certainly, there is more than one way of constructing such systems. Two very easy possibilities are represented by multiplying the $Y_{n,j}$ with vectors which are orthonormal in the Euclidean (i.e., Cartesian) sense. In other words, the systems $\{\varepsilon^i Y_{n,j}\}_{i=1,2,3;\, n\in\mathbb{N}_0;\, j=-n,\dots,n}$ and

$$\left\{\varepsilon^r Y_{n,j}\right\}_{\substack{n\in\mathbb{N}_0\\ j=-n,\dots,n}} \cup \left\{\varepsilon^\varphi Y_{n,j}\right\}_{\substack{n\in\mathbb{N}_0\\ j=-n,\dots,n}} \cup \left\{\varepsilon^t Y_{n,j}\right\}_{\substack{n\in\mathbb{N}_0\\ j=-n,\dots,n}}$$

are both complete orthonormal systems in $\mathrm{l}^2(\Omega)$, which is rather obvious. The latter system is definitely preferable in Earth sciences, because it allows one to separate fields $f \in \mathrm{l}^2(\Omega)$ into their normal and tangential components.

However, these simply constructed bases do not suffice for all applications or are, at least, not as practicable as some other available options. For example, when talking about gravitation, then the gradient ∇V of the gravitational potential is an important physical quantity, because it is proportional to the gravitational force and it also becomes available from satellite-to-satellite tracking techniques; see, for example, Freeden et al. (2002).

From Theorem 3.4.1, we know that

$$\nabla_x F(x) = \xi \frac{\partial}{\partial r} F(r\xi) + \frac{1}{r} \nabla^*_\xi F(r\xi), \quad x = r\xi,\ \xi \in \Omega, \tag{4.22}$$

for (continuously) differentiable functions F. In this respect, it makes sense to introduce the following established alternative approach (see also Theorem 6.3.8).

Definition 4.3.1 For functions $F \in \mathrm{C}(\Omega)$ and $G \in \mathrm{C}^{(1)}(\Omega)$, we define the operators $o^{(1)}_\xi F(\xi) := \xi F(\xi)$, $o^{(2)}_\xi G(\xi) := \nabla^*_\xi G(\xi)$, and $o^{(3)}_\xi G(\xi) := \mathrm{L}^*_\xi G(\xi)$, where the surface curl gradient L^* is defined in Definition 4.1.1. Moreover, with $0_i := 0$ for $i = 1$ and $0_i := 1$ for $i \in \{2,3\}$, we define the functions $y^{(i)}_{n,j} := \|o^{(i)} Y_{n,j}\|^{-1}_{\mathrm{l}^2(\Omega)} o^{(i)} Y_{n,j}$ and call $i \in \{1,2,3\}$ the type, $n \geq 0_i$ the degree, and $j \in \{-n,\dots,n\}$ the order of $y^{(i)}_{n,j}$.

These functions are called the **Morse–Feshbach vector spherical harmonics** (see Morse and Feshbach, 1953a,b). These functions are normalized by definition and their norms are easy to determine: for type $i = 1$, we get $\int_\Omega \left|o^{(1)}_\xi Y_{n,j}(\xi)\right|^2 \mathrm{d}\omega(\xi) = \int_\Omega (Y_{n,j}(\xi))^2 \mathrm{d}\omega(\xi) = 1$ and, for type $i = 2$, by using Green's first surface identity (Theorem 4.1.11) and Theorem 4.2.19,

$$\begin{aligned}\int_\Omega \left|o^{(2)}_\xi Y_{n,j}(\xi)\right|^2 \mathrm{d}\omega(\xi) &= \int_\Omega \left|\nabla^*_\xi Y_{n,j}(\xi)\right|^2 \mathrm{d}\omega(\xi)\\ &= -\int_\Omega Y_{n,j}(\xi)\Delta^*_\xi Y_{n,j}(\xi)\,\mathrm{d}\omega(\xi) = n(n+1)\underbrace{\|Y_{n,j}\|^2_{\mathrm{L}^2(\Omega)}}_{=1}.\end{aligned}$$

Moreover, with the Lagrange identity (see Theorem 2.1.5) and Theorem 4.1.6, we obtain

$$\begin{aligned}\int_\Omega \left|o^{(3)}_\xi Y_{n,j}(\xi)\right|^2 \mathrm{d}\omega(\xi) &= \int_\Omega \left|\xi \times \nabla^*_\xi Y_{n,j}(\xi)\right|^2 \mathrm{d}\omega(\xi)\\ &= \int_\Omega \underbrace{|\xi|^2}_{=1} \left|\nabla^*_\xi Y_{n,j}(\xi)\right|^2 - \underbrace{\left(\xi \cdot \nabla^*_\xi Y_{n,j}(\xi)\right)^2}_{=0} \mathrm{d}\omega(\xi) = n(n+1).\end{aligned}$$

Hence, the Morse–Feshbach vector spherical harmonics have the form

$$y^{(1)}_{n,j}(\xi) = \xi Y_{n,j}(\xi), \qquad y^{(2)}_{n,j}(\xi) = \frac{1}{\sqrt{n(n+1)}} \nabla^*_\xi Y_{n,j}(\xi),$$
$$y^{(3)}_{n,j}(\xi) = \frac{1}{\sqrt{n(n+1)}} \mathrm{L}^*_\xi Y_{n,j}(\xi).$$

Next, we verify the orthogonality of these functions. For equal degree-order pairs (n, j), Theorem 4.1.6 yields, if $n \geq 1$, that

$$y^{(1)}_{n,j}(\xi) \cdot y^{(2)}_{n,j}(\xi) = Y_{n,j}(\xi)\xi \cdot \nabla^* Y_{n,j}(\xi)[n(n+1)]^{-1/2} = 0,$$
$$y^{(1)}_{n,j}(\xi) \cdot y^{(3)}_{n,j}(\xi) = Y_{n,j}(\xi)\xi \cdot \mathrm{L}^* Y_{n,j}(\xi)[n(n+1)]^{-1/2} = 0,$$
$$y^{(2)}_{n,j}(\xi) \cdot y^{(3)}_{n,j}(\xi) = \left[\nabla^* Y_{n,j}(\xi)\right] \cdot \left[\mathrm{L}^* Y_{n,j}(\xi)\right] [n(n+1)]^{-1} = 0.$$

In other words, these three functions are pairwise orthogonal in the Euclidean (or Cartesian) sense, that is, pointwise orthogonal. For arbitrary pairs (n, j) and (m,k), it requires some more work to prove the orthogonality. However, the preceding arguments remain valid for

$$y^{(1)}_{n,j}(\xi) \cdot y^{(2)}_{m,k}(\xi) = 0, \qquad y^{(1)}_{n,j}(\xi) \cdot y^{(3)}_{m,k}(\xi) = 0. \tag{4.23}$$

Note that, in an arbitrary point $\xi \in \Omega$, the vector ξ is also the outer unit normal to Ω in this point ξ. In this respect, (4.23) simply reflects the fact that Morse–Feshbach vector spherical harmonics of type $i = 1$ are normal vector fields to the sphere and those of type $i = 2$ or $i = 3$ are tangential. Merely the combination of $y^{(2)}_{n,j}$ and $y^{(3)}_{m,k}$ or of $y^{(i)}_{n,j}$ and $y^{(i)}_{m,k}$, $i = 1,2,3$, does not necessarily yield a pair which is pointwise orthogonal if $(n, j) \neq (m,k)$. Nevertheless, we have the $\mathrm{l}^2(\Omega)$-orthogonality. Corollary 4.1.13 yields together with Theorem 4.1.6 that

$$\int_\Omega \nabla^* Y_{n,j}(\xi) \cdot \mathrm{L}^* Y_{m,k}(\xi)\, d\omega(\xi) = -\int_\Omega Y_{n,j}(\xi) \underbrace{\nabla^*_\xi \cdot \mathrm{L}^*_\xi Y_{m,k}(\xi)}_{=0}\, d\omega(\xi) = 0$$

For $(n, j) \neq (m,k)$, we get additionally

$$\int_\Omega \left(\xi Y_{n,j}(\xi)\right) \cdot \left(\xi Y_{m,k}(\xi)\right) d\omega(\xi) = \int_\Omega Y_{n,j}(\xi)\, Y_{m,k}(\xi) d\omega(\xi) = 0,$$
$$\int_\Omega \left(\nabla^* Y_{n,j}(\xi)\right) \cdot \left(\nabla^* Y_{m,k}(\xi)\right) d\omega(\xi) = -\int_\Omega Y_{n,j}(\xi) \underbrace{\Delta^* Y_{m,k}(\xi)}_{=-m(m+1)Y_{m,k}(\xi)} d\omega(\xi) = 0,$$
$$\int_\Omega \left(\mathrm{L}^* Y_{n,j}(\xi)\right) \cdot \left(\mathrm{L}^* Y_{m,k}(\xi)\right) d\omega(\xi) = \int_\Omega \left(\nabla^* Y_{n,j}(\xi)\right) \cdot \left(\nabla^* Y_{m,k}(\xi)\right) d\omega(\xi) = 0$$

in analogy to the calculation of the preceding norms.

So, we have the following intermediate result.

Theorem 4.3.2 *The system* $\left\{y^{(i)}_{n,j}\right\}_{i\in\{1,2,3\},n\geq 0_i,\, j\in\{-n,\ldots,n\}}$ *is orthonormal in* $\mathrm{l}^2(\Omega)$.

Plots of some Morse–Feshbach vector spherical harmonics can be seen in Figures 4.8–4.13. The functions of type 1 are normal to the sphere whereas the others are tangential. Moreover, due to Theorem 4.1.6, we have $\mathrm{L}^* \cdot y_{n,j}^{(2)} = 0$ and $\nabla^* \cdot y_{n,j}^{(3)} = 0$, that is, the functions of type 2 are surface-curl-free and those of type 3 are surface-divergence-free.

The next step is to show that the Morse–Feshbach vector spherical harmonics also constitute a basis in $\mathrm{l}^2(\Omega)$.

Theorem 4.3.3 *The orthonormal system* $\left\{y_{n,j}^{(i)}\right\}_{i\in\{1,2,3\},n\geq 0_i,j\in\{-n,\ldots,n\}}$ *is complete in* $\mathrm{l}^2(\Omega)$.

Proof We first consider functions $f \in \mathrm{c}^{(1)}(\Omega)$ only. Let us assume that

$$\int_\Omega y_{n,j}^{(i)}(\xi) \cdot f(\xi)\, \mathrm{d}\omega(\xi) = 0 \tag{4.24}$$

for all $i \in \{1,2,3\}, n \geq 0_i$, and $j \in \{-n,\ldots,n\}$. For type $i = 1$, this means that $\int_\Omega Y_{n,j}(\xi)\xi \cdot f(\xi)\, \mathrm{d}\omega(\xi) = 0$ for all $n \in \mathbb{N}_0$ and all $j \in \{-n,\ldots,n\}$. Due to the completeness of the scalar spherical harmonics (see Theorem 4.2.18), this means that $\xi \cdot f(\xi) = 0$ for all $\xi \in \Omega$ (since f is continuous, this really holds true pointwise). Moreover, types $i = 2$ and $i = 3$ in (4.24) in combination with all identities in Corollary 4.1.13 lead us to

$$\int_\Omega Y_{n,j}(\xi)\, \nabla_\xi^* \cdot f(\xi)\, \mathrm{d}\omega(\xi) = 0 = \int_\Omega Y_{n,j}(\xi) \mathrm{L}_\xi^* \cdot f(\xi)\, \mathrm{d}\omega(\xi)$$

for all $n \in \mathbb{N}_0$ and all $j \in \{-n,\ldots,n\}$. Note that $n = 0$ is included here! Hence, $\nabla^* \cdot f = \mathrm{L}^* \cdot f = 0$. Now, we can use Theorem 4.1.14, which tells us that there exists a function $G \in \mathrm{C}^{(2)}(\Omega)$ such that $f = \nabla^* G$. Hence, $0 = \nabla^* \cdot f = \Delta^* G$ and consequently G must be a constant function such that $f = 0$. This means that the functions $y_{n,j}^{(i)}$ are complete in $(\mathrm{c}^{(1)}(\Omega), \langle\cdot,\cdot\rangle_{\mathrm{l}^2(\Omega)})$ and consequently also closed; see Definition 2.5.26 and Theorem 2.5.27. Finally, Theorem 2.5.22 shows us that these properties can be extended (for each component of the vectorial functions) to all $f \in \mathrm{l}^2(\Omega)$. □

Note that this system is also closed in $(\mathrm{c}(\Omega), \|\cdot\|_{\mathrm{c}(\Omega)})$; see, for example, (Freeden et al., 1998, Theorem 12.3.5).

Let us have a look at further properties which can be proved for vector spherical harmonics. For instance, we have an addition theorem for scalar spherical harmonics (see Theorem 4.2.9) which says that

$$\sum_{j=-n}^{n} Y_{n,j}(\xi) Y_{n,j}(\eta) = \frac{2n+1}{4\pi} P_n(\xi \cdot \eta)$$

for all $\xi, \eta \in \Omega$, where P_n is the Legendre polynomial of degree n. Analogously, we can state the well-known vectorial version of the addition theorem: let $n \in \mathbb{N}_0$ and $i_1, i_2 \in \{1,2,3\}$, where $n \geq 1$, if $i_1 \neq 1$ or $i_2 \neq 1$. Then

$$\sum_{j=-n}^{n} y_{n,j}^{(i_1)}(\xi) \otimes y_{n,j}^{(i_2)}(\eta) = \sum_{j=-n}^{n} \left(\mu_n^{(i_1)} \mu_n^{(i_2)}\right)^{-1/2} \left[o_\xi^{(i_1)} Y_{n,j}(\xi)\right] \otimes \left[o_\eta^{(i_2)} Y_{n,j}(\eta)\right],$$

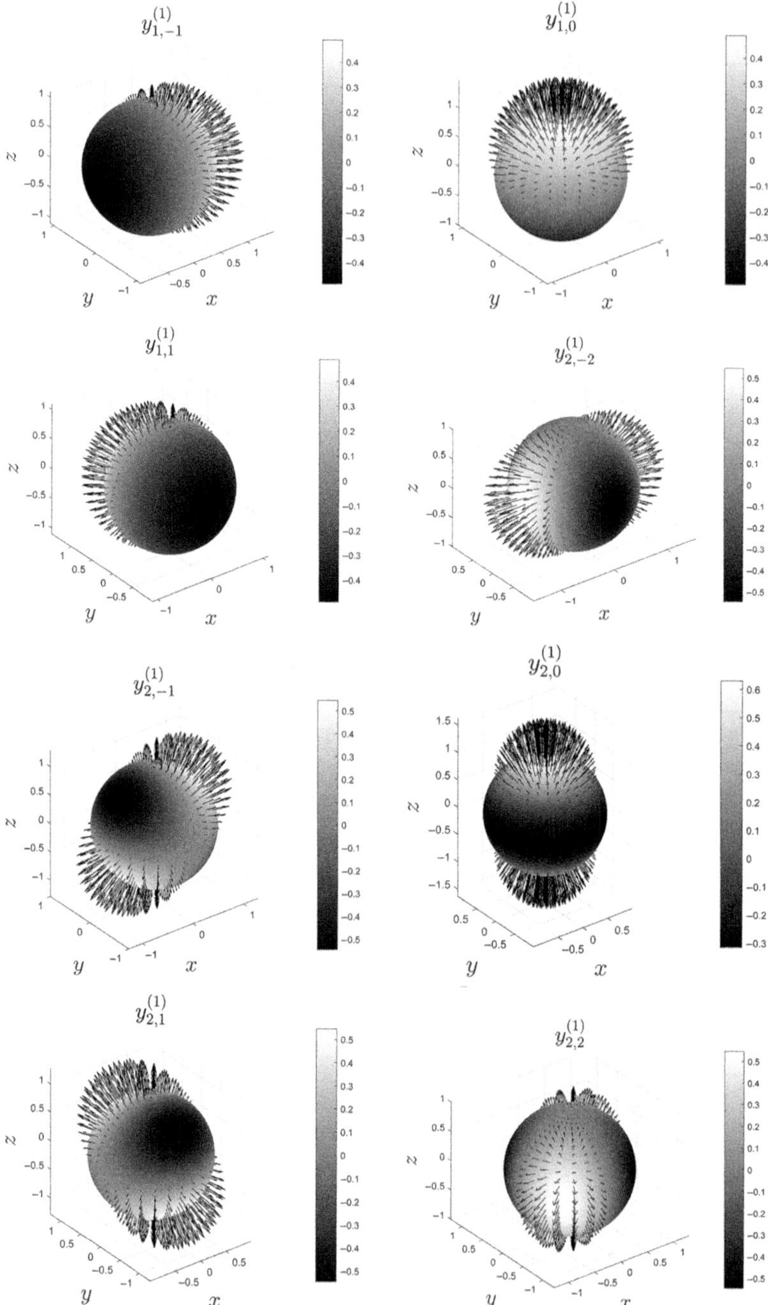

Figure 4.8 Vectorial Morse–Feshbach spherical harmonics $y^{(i)}_{n,j}$ of type $i = 1$, degrees $n \in \{1,2\}$, and orders $j \in \{-n,\dots,n\}$ (based on the scalar fully normalized spherical harmonics, which are represented by the shading): it can be seen that the functions of type $i = 1$ are normal fields to the sphere.

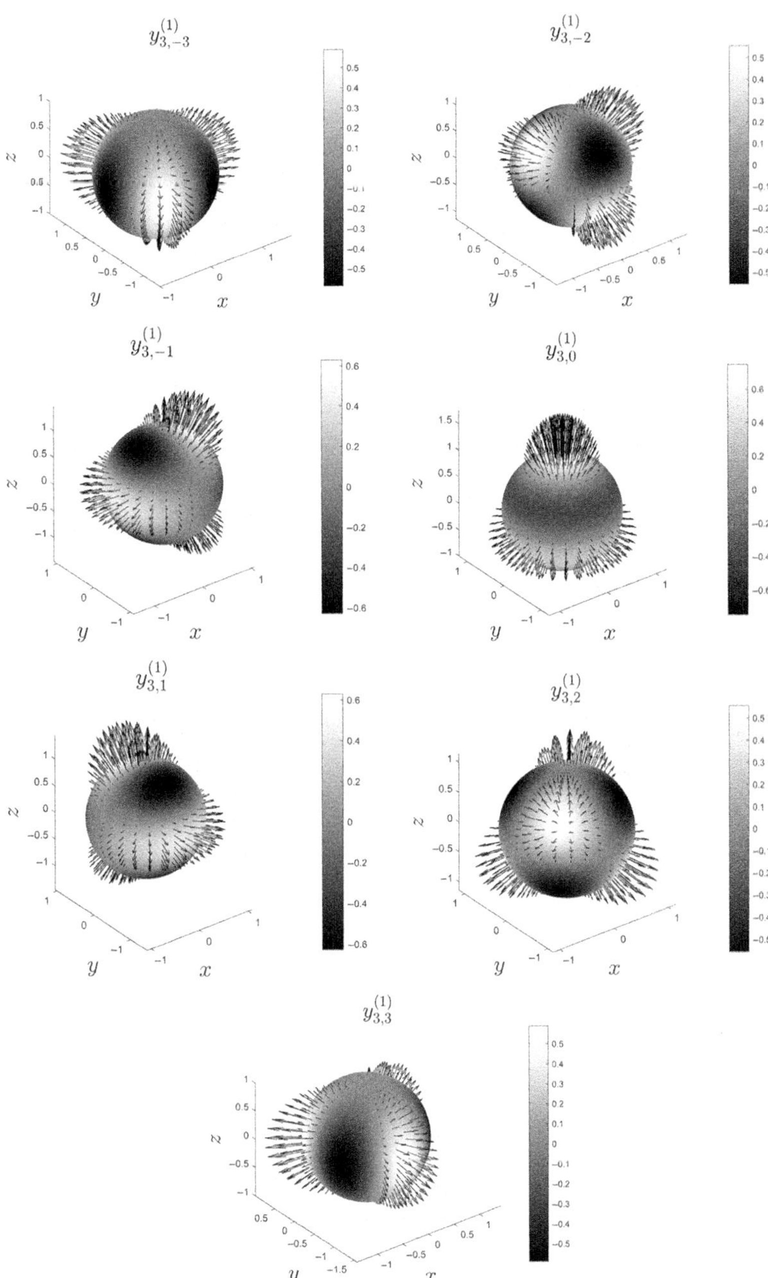

Figure 4.9 Vectorial Morse–Feshbach spherical harmonics $y^{(i)}_{n,j}$ of type $i = 1$, degree $n = 3$, and orders $j \in \{-n, \ldots, n\}$ (based on the scalar fully normalized spherical harmonics, which are represented by the shading): it can be seen that the functions of type $i = 1$ are normal fields to the sphere.

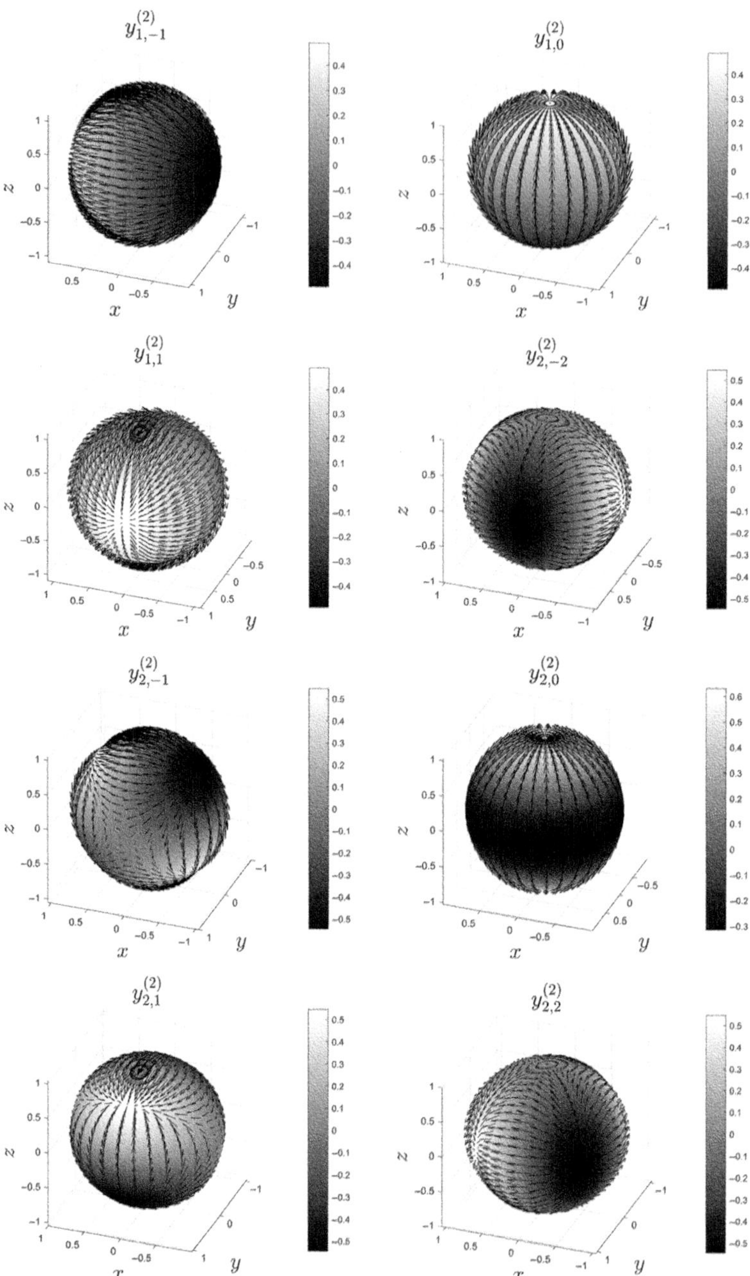

Figure 4.10 Vectorial Morse–Feshbach spherical harmonics $y^{(i)}_{n,j}$ of type $i = 2$, degrees $n \in \{1, 2\}$, and orders $j \in \{-n, \dots, n\}$ (based on the scalar fully normalized spherical harmonics, which are represented by the shading): note that the functions of type $i = 2$ are tangential and surface-curl-free fields to the sphere.

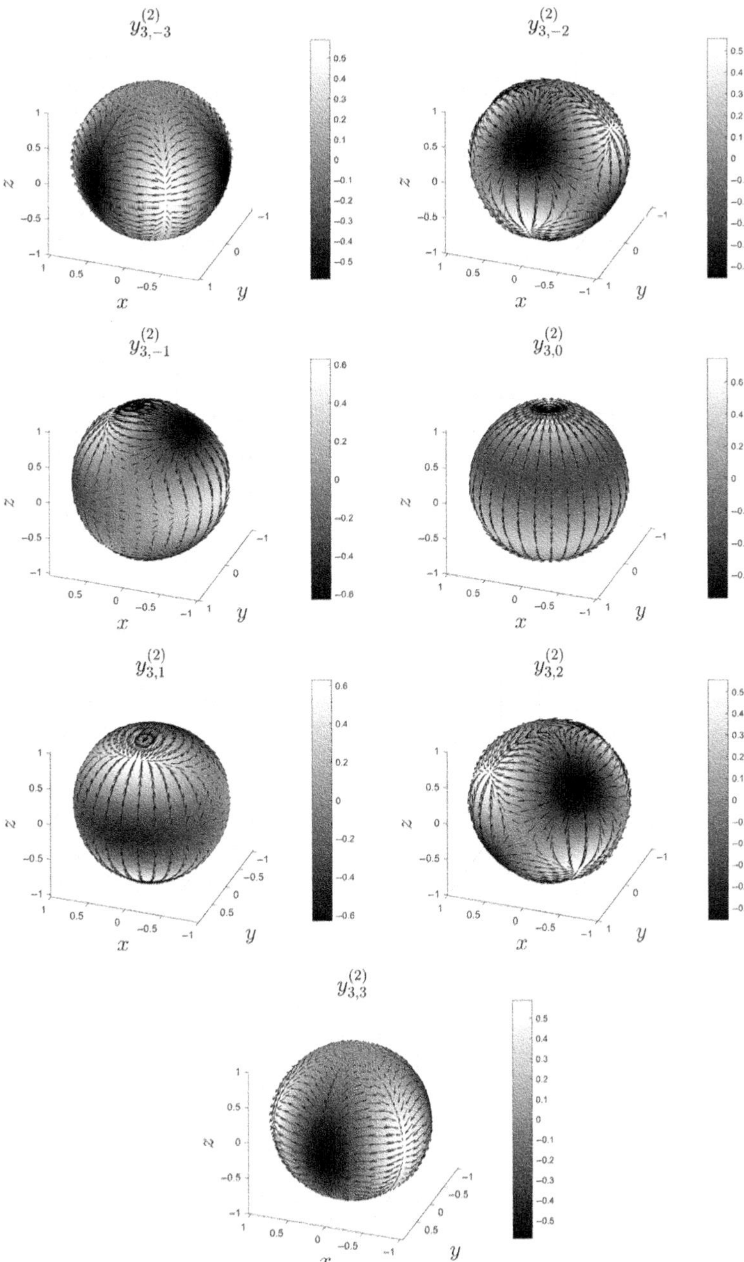

Figure 4.11 Vectorial Morse–Feshbach spherical harmonics $y^{(i)}_{n,j}$ of type $i = 2$, degree $n = 3$, and orders $j \in \{-n, \ldots, n\}$ (based on the scalar fully normalized spherical harmonics, which are represented by the shading): note that the functions of type $i = 2$ are tangential and surface-curl-free fields to the sphere.

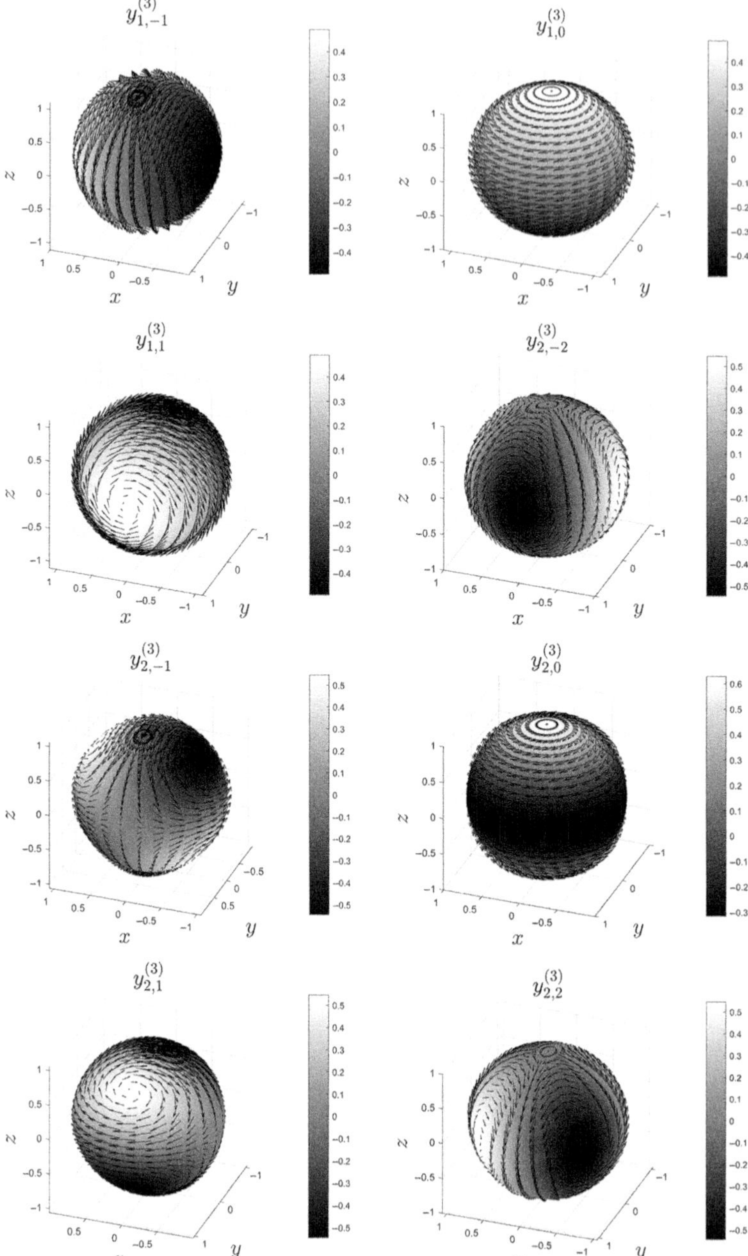

Figure 4.12 Vectorial Morse–Feshbach spherical harmonics $y^{(i)}_{n,j}$ of type $i = 3$, degrees $n \in \{1,2\}$, and orders $j \in \{-n,\dots,n\}$ (based on the scalar fully normalized spherical harmonics, which are represented by the shading): note that the functions of type $i = 3$ are tangential and surface-divergence-free fields to the sphere.

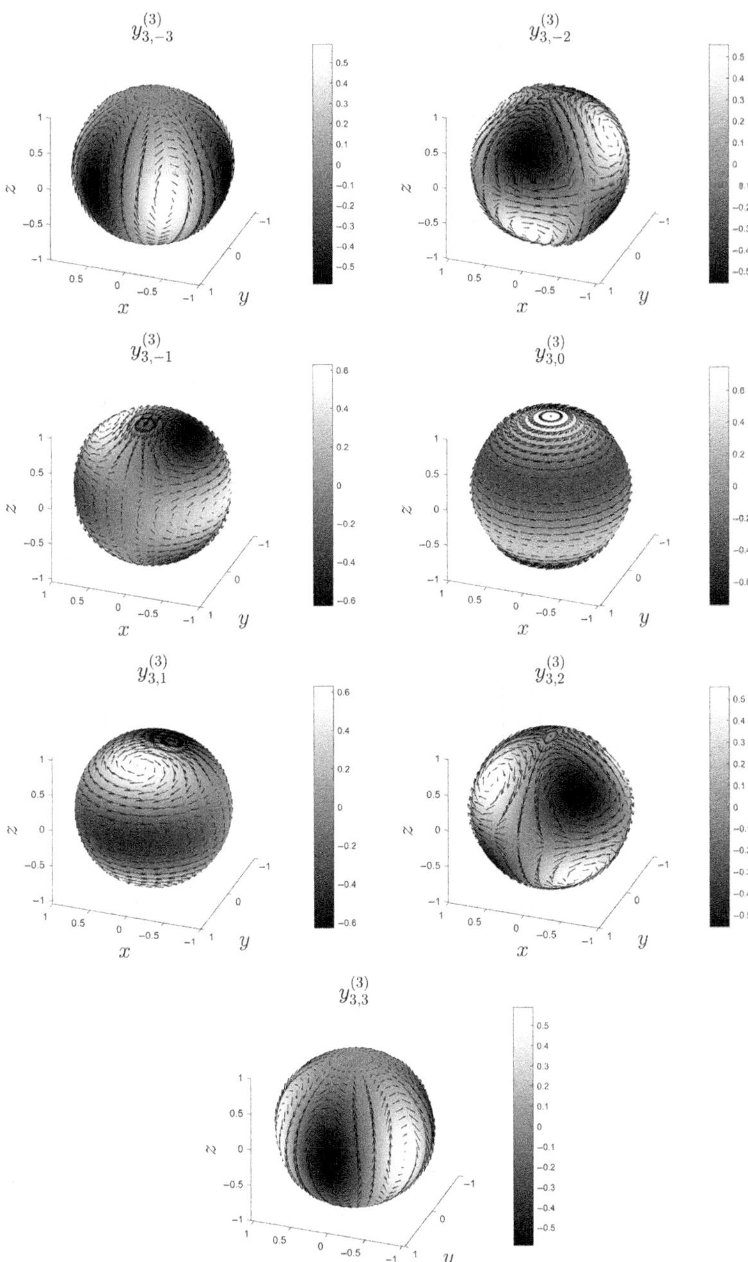

Figure 4.13 Vectorial Morse–Feshbach spherical harmonics $y^{(i)}_{n,j}$ of type $i = 3$, degree $n = 3$, and orders $j \in \{-n, \ldots, n\}$ (based on the scalar fully normalized spherical harmonics, which are represented by the shading): note that the functions of type $i = 3$ are tangential and surface-divergence-free fields to the sphere.

where

$$\mu_n^{(i)} := \begin{cases} 1, & \text{if } i = 1, \\ n(n+1), & \text{if } i = 2 \text{ or } i = 3. \end{cases}$$

Hence, in the sense of Definition 4.1.2, we get, due to the scalar addition theorem, the identity

$$\sum_{j=-n}^{n} y_{n,j}^{(i_1)}(\xi) \otimes y_{n,j}^{(i_2)}(\eta) = \left(\mu_n^{(i_1)} \mu_n^{(i_2)}\right)^{-1/2} o_\xi^{(i_1)} \otimes o_\eta^{(i_2)} P_n(\xi \cdot \eta) \frac{2n+1}{4\pi}.$$

Theorem 4.3.4 (Addition Theorem for Vector Spherical Harmonics I) *The Morse–Feshbach vector spherical harmonics satisfy*

$$\sum_{j=-n}^{n} y_{n,j}^{(i_1)}(\xi) \otimes y_{n,j}^{(i_2)}(\eta) = \frac{2n+1}{4\pi} \mathbf{p}_n^{(i_1,i_2)}(\xi,\eta), \quad \xi,\eta \in \Omega, \tag{4.25}$$

for all $i_1, i_2 \in \{1,2,3\}$ *and all* $n \in \mathbb{N}_0$ *with* $n \geq \max\left(0_{i_1}, 0_{i_2}\right)$, *where the* (i_1,i_2)*-**Legendre-tensor field** of degree n is defined by*

$$\mathbf{p}_n^{(i_1,i_2)}(\xi,\eta) := \left(\mu_n^{(i_1)} \mu_n^{(i_2)}\right)^{-1/2} o_\xi^{(i_1)} \otimes o_\eta^{(i_2)} P_n(\xi \cdot \eta), \quad \xi,\eta \in \Omega.$$

Furthermore,

$$\sum_{j=-n}^{n} \left| y_{n,j}^{(i)}(\xi) \right|^2 = \frac{2n+1}{4\pi}, \quad \xi \in \Omega, \tag{4.26}$$

for all $i \in \{1,2,3\}$ *and all* $n \in \mathbb{N}_0$ *with* $n \geq 0_i$.

Proof We have already proved (4.25). For (4.26), we need the case $i_1 = i_2$ only and set $i := i_1 = i_2$. Moreover, $|y_{n,j}^{(i)}(\xi)|^2 = \mathrm{tr}(y_{n,j}^{(i)}(\xi) \otimes y_{n,j}^{(i)}(\xi))$, where tr stands for the trace of a second-rank tensor. For $i = 1$, the calculations are simple: $|y_{n,j}^{(1)}(\xi)|^2 = \mathrm{tr}(\xi \otimes \xi) Y_{n,j}(\xi) Y_{n,j}(\xi)$ such that

$$\sum_{j=-n}^{n} \left| y_{n,j}^{(1)}(\xi) \right|^2 = |\xi|^2 \frac{2n+1}{4\pi} P_n(\xi \cdot \xi) = \frac{2n+1}{4\pi}.$$

We used here Theorems 4.2.9 and 4.2.11. For $i = 2$, we apply Corollary 4.1.5 and Theorem 4.1.3 and get

$$\begin{aligned} \nabla_\xi^* \otimes \nabla_\eta^* P_n(\xi \cdot \eta) &= \nabla_\xi^* \otimes \left\{ P_n'(\xi \cdot \eta)[\xi - (\xi \cdot \eta)\eta] \right\} \\ &= P_n''(\xi \cdot \eta)[\eta - (\xi \cdot \eta)\xi] \otimes [\xi - (\xi \cdot \eta)\eta] \\ &\quad + P_n'(\xi \cdot \eta) \{\mathbf{i} - \xi \otimes \xi - [\eta - (\xi \cdot \eta)\xi] \otimes \eta\}. \end{aligned}$$

Hence, Theorem 4.2.11 yields

$$\sum_{j=-n}^{n} \left| y_{n,j}^{(2)}(\xi) \right|^2 = \frac{2n+1}{4\pi} \frac{1}{n(n+1)} P_n'(1)(3 - |\xi|^2) = \frac{2n+1}{4\pi}.$$

With the same corollary and theorems, we can handle the case $i = 3$:

$$\mathrm{L}_\xi^* \otimes \mathrm{L}_\eta^* P_n(\xi \cdot \eta) = \mathrm{L}_\xi^* \otimes \left[P_n'(\xi \cdot \eta)\, \eta \times \xi \right]$$
$$= P_n''(\xi \cdot \eta)(\xi \times \eta) \otimes (\eta \times \xi) + P_n'(\xi \cdot \eta)\left[-\eta \otimes \xi + (\xi \cdot \eta)\mathbf{i} \right]$$

and consequently

$$\sum_{j=-n}^{n} \left| y_{n,j}^{(3)}(\xi) \right|^2 = \frac{2n+1}{4\pi} \frac{1}{n(n+1)} P_n'(\xi \cdot \xi)\left(-|\xi|^2 + 3|\xi|^2 \right) = \frac{2n+1}{4\pi}. \qquad \square$$

Within the proof of Theorem 4.3.4, we derived formulae for the Legendre-tensor fields $\mathbf{p}_n^{(i,i)}$ for $i = 1, 2, 3$. For a complete list of all nine formulae (including mixed types i_1 and i_2); see, for example Freeden et al. (1998, Theorem 12.6.3).

The scalar addition theorem allowed us to derive further interesting propositions such as an estimate for the maximum of spherical harmonics. This can also be done in the vectorial case.

Theorem 4.3.5 *Let* $i \in \{1, 2, 3\}$ *and* $n \in \mathbb{N}_0$ *with* $n \geq 0_i$. *If we have a function* $f \in \operatorname{span}\{y_{n,j}^{(i)}\}_{j=-n,\ldots,n}$, *then*

$$\|f\|_{\mathrm{c}(\Omega)} \leq \sqrt{\frac{2n+1}{4\pi}} \|f\|_{\mathrm{l}^2(\Omega)}.$$

In particular,

$$\left\| y_{n,j}^{(i)} \right\|_{\mathrm{c}(\Omega)} \leq \sqrt{\frac{2n+1}{4\pi}} \qquad \text{for all } j = -n, \ldots, n.$$

Proof Due to the orthonormality of the $y_{n,j}^{(i)}$ (see Theorem 4.3.2), we get, for each $\xi \in \Omega$, that

$$|f(\xi)| = \left| \sum_{j=-n}^{n} \left\langle f, y_{n,j}^{(i)} \right\rangle_{\mathrm{l}^2(\Omega)} y_{n,j}^{(i)}(\xi) \right|$$
$$\leq \left(\sum_{j=-n}^{n} \left\langle f, y_{n,j}^{(i)} \right\rangle_{\mathrm{l}^2(\Omega)}^2 \right)^{1/2} \left(\sum_{j=-n}^{n} \left| y_{n,j}^{(i)}(\xi) \right|^2 \right)^{1/2} = \|f\|_{\mathrm{l}^2(\Omega)} \sqrt{\frac{2n+1}{4\pi}},$$

where we used the Cauchy–Schwarz inequality, the Parseval identity, and the vectorial addition theorem. $\square$

An alternative system of vector spherical harmonics is popular in geomagnetics and is due to Edmonds (see Edmonds, 1957).

Definition 4.3.6 The **Edmonds vector spherical harmonics** $\tilde{y}_{n,j}^{(i)}$ are defined by

$$\tilde{y}_{n,j}^{(1)} := \sqrt{\frac{n+1}{2n+1}}\, y_{n,j}^{(1)} - \sqrt{\frac{n}{2n+1}}\, y_{n,j}^{(2)},$$

$$\tilde{y}^{(2)}_{n,j} := \sqrt{\frac{n}{2n+1}}\, y^{(1)}_{n,j} + \sqrt{\frac{n+1}{2n+1}}\, y^{(2)}_{n,j}, \qquad \tilde{y}^{(3)}_{n,j} := y^{(3)}_{n,j}$$

for $n \in \mathbb{N}$ and $j \in \{-n, \ldots, n\}$. Moreover, $\tilde{y}^{(1)}_{0,0} := y^{(1)}_{0,0}$. In general, we can interpret that formulae such as $n\, y^{(i)}_{n,j}$ are also defined for $n = 0$ and $i \neq 1$ in the sense that such terms vanish.

It is clear that $\langle \tilde{y}^{(i)}_{n,j}, \tilde{y}^{(3)}_{m,k}\rangle_{\mathrm{l}^2(\Omega)} = 0$ for $i \in \{1,2\}$ due to our previous considerations. In the same manner, it is clear that $\left\langle \tilde{y}^{(i_1)}_{n,j}, \tilde{y}^{(i_2)}_{m,k}\right\rangle_{\mathrm{l}^2(\Omega)} = 0$, if $(n,j) \neq (m,k)$. The remaining combinations can be handled as follows:

$$\left\langle \tilde{y}^{(1)}_{n,j}, \tilde{y}^{(2)}_{n,j}\right\rangle_{\mathrm{l}^2(\Omega)} = \frac{\sqrt{n(n+1)}}{2n+1} - \frac{\sqrt{n(n+1)}}{2n+1} = 0,$$

$$\left\langle \tilde{y}^{(1)}_{n,j}, \tilde{y}^{(1)}_{n,j}\right\rangle_{\mathrm{l}^2(\Omega)} = \frac{n+1}{2n+1} + \frac{n}{2n+1} = 1,$$

$$\left\langle \tilde{y}^{(2)}_{n,j}, \tilde{y}^{(2)}_{n,j}\right\rangle_{\mathrm{l}^2(\Omega)} = \frac{n}{2n+1} + \frac{n+1}{2n+1} = 1,$$

$$\left\langle \tilde{y}^{(3)}_{n,j}, \tilde{y}^{(3)}_{n,j}\right\rangle_{\mathrm{l}^2(\Omega)} = \left\| y^{(3)}_{n,j} \right\|^2_{\mathrm{l}^2(\Omega)} = 1.$$

Theorem 4.3.7 *The functions known as Edmonds vector spherical harmonics* $\{\tilde{y}^{(i)}_{n,j}\}_{i\in\{1,2,3\},\, n\geq 0_i,\, j\in\{-n,\ldots,n\}}$ *constitute a complete orthonormal system in* $\mathrm{l}^2(\Omega)$.

Proof We have just proved the orthonormality. For the completeness, we use that $\operatorname{span}\{y^{(i)}_{n,j}\}_{i\in\{1,2,3\},\, j\in\{-n,\ldots,n\}} = \operatorname{span}\{\tilde{y}^{(i)}_{n,j}\}_{i\in\{1,2,3\},\, j\in\{-n,\ldots,n\}}$ for all $n \geq 1$ and that $\tilde{y}^{(1)}_{0,0} = y^{(1)}_{0,0}$. □

Note that (4.22) shows that gradients of scalar functions, as we have to deal with, for example, when investigating the gravitational force, can be represented in types $i = 1$ and $i = 2$ of vector spherical harmonics – no matter, if we use the Edmonds version or the Morse–Feshbach version. As a consequence, in both cases, type $i = 3$ corresponds to a surface-divergence-free field, that is, $\nabla^* \cdot y^{(3)}_{n,j} = 0 = \nabla^* \cdot \tilde{y}^{(3)}_{n,j}$ due to Theorem 4.1.6.

In analogy to the case of Morse–Feshbach vector spherical harmonics, one can also define operators $\tilde{o}^{(i)}_n$, $i = 1,2,3$, but now with an index $n \geq 0_i$ such that

$$\left(\tilde{o}^{(1)}_n\right)_\xi F(\xi) := (n+1) o^{(1)}_\xi F(\xi) - o^{(2)}_\xi F(\xi),$$

$$\left(\tilde{o}^{(2)}_n\right)_\xi F(\xi) := n\, o^{(1)}_\xi F(\xi) + o^{(2)}_\xi F(\xi), \qquad \left(\tilde{o}^{(3)}_n\right)_\xi F(\xi) := o^{(3)}_\xi F(\xi),$$

$\xi \in \Omega$, where $F \in \mathrm{C}^{(1)}(\Omega)$. Hence, $\tilde{y}^{(i)}_{n,j}(\xi) = \left(\tilde{\mu}^{(i)}_n\right)^{-1/2} \left(\tilde{o}^{(i)}_n\right)_\xi Y_{n,j}(\xi)$, $\xi \in \Omega$, for $i = 1, 2, 3$, $n \geq 0_i$, $j = -n, \ldots, n$, and

$$\tilde{\mu}^{(i)}_n = \begin{cases} (n+1)(2n+1), & \text{if } i = 1, \\ n(2n+1), & \text{if } i = 2, \\ n(n+1), & \text{if } i = 3. \end{cases}$$

With this nomenclature, we are also able to derive an addition theorem for this type of vector spherical harmonics.

Theorem 4.3.8 (Addition Theorem for Vector Spherical Harmonics II) *The Edmonds vector spherical harmonics satisfy*

$$\sum_{j=-n}^{n} \tilde{y}_{n,j}^{(i_1)}(\xi) \otimes \tilde{y}_{n,j}^{(i_2)}(\eta) = \frac{2n+1}{4\pi} \tilde{\boldsymbol{p}}_n^{(i_1,i_2)}(\xi,\eta), \quad \xi,\eta \in \Omega, \tag{4.27}$$

for all $i_1, i_2 \in \{1,2,3\}$ and all $n \in \mathbb{N}_0$ with $n \geq \max(0_{i_1}, 0_{i_2})$, where

$$\tilde{\boldsymbol{p}}_n^{(i_1,i_2)}(\xi,\eta) = \left(\tilde{\mu}_n^{(i_1)} \tilde{\mu}_n^{(i_2)}\right)^{-1/2} \left(\tilde{o}_n^{(i_1)}\right)_\xi \otimes \left(\tilde{o}_n^{(i_2)}\right)_\eta P_n(\xi \cdot \eta), \quad \xi,\eta \in \Omega.$$

Furthermore,

$$\sum_{j=-n}^{n} \left| \tilde{y}_{n,j}^{(i)}(\xi) \right|^2 = \frac{2n+1}{4\pi}, \quad \xi \in \Omega, \tag{4.28}$$

for all $i \in \{1,2,3\}$ and all $n \in \mathbb{N}_0$ with $n \geq 0_i$.

Proof Equation (4.27) is a direct consequence of the previous considerations. For (4.28), we have to look again at the trace $\left|\tilde{y}_{n,j}^{(i)}(\xi)\right|^2 = \operatorname{tr}\left(\tilde{y}_{n,j}^{(i)}(\xi) \otimes \tilde{y}_{n,j}^{(i)}(\xi)\right)$. The case $i = 3$ coincides with the definition by Morse–Feshbach, which requires, therefore, no further effort for us.

For $i = 1$, we get, due to Theorem 4.1.6, that

$$\begin{aligned}
\left|\tilde{y}_{n,j}^{(1)}(\xi)\right|^2 &= \frac{n+1}{2n+1} \operatorname{tr}(\xi \otimes \xi) \left(Y_{n,j}(\xi)\right)^2 + \frac{n}{2n+1} \operatorname{tr}\left(y_{n,j}^{(2)}(\xi) \otimes y_{n,j}^{(2)}(\xi)\right) \\
&\quad - \frac{\sqrt{n(n+1)}}{2n+1} \operatorname{tr}\left(\xi \otimes y_{n,j}^{(2)}(\xi) + y_{n,j}^{(2)}(\xi) \otimes \xi\right) Y_{n,j}(\xi) \\
&= \frac{n+1}{2n+1} \left(Y_{n,j}(\xi)\right)^2 + \frac{n}{2n+1} \left|y_{n,j}^{(2)}(\xi)\right|^2, \quad \xi \in \Omega,
\end{aligned}$$

and for $i = 2$, we have, analogously,

$$\left|\tilde{y}_{n,j}^{(2)}(\xi)\right|^2 = \frac{n}{2n+1} \left(Y_{n,j}(\xi)\right)^2 + \frac{n+1}{2n+1} \left|y_{n,j}^{(2)}(\xi)\right|^2, \quad \xi \in \Omega.$$

Hence, Theorems 4.2.9 and 4.3.4 yield, for $i = 1$ and $i = 2$,

$$\sum_{j=-n}^{n} \left|\tilde{y}_{n,j}^{(i)}(\xi)\right|^2 = \frac{n}{2n+1} \cdot \frac{2n+1}{4\pi} + \frac{n+1}{2n+1} \cdot \frac{2n+1}{4\pi} = \frac{2n+1}{4\pi} \qquad \square$$

We conclude this section by looking back at one of the motivations for considering vector spherical harmonics: in multiple applications, gradient fields play an important role. For example, the gravitational force is a gradient field, and modern spaceborne measurement technologies such as satellite-to-satellite-tracking (SST) as it was or is currently used in the CHAMP, GRACE, and GRACE-Follow-On missions indeed yield first-order derivatives of the gravitational potential; see, for example, Eicker et al. (2005), Flechtner et al. (2014), Freeden et al. (2002), Ilk et al. (2005), Jekeli (1999), Reigber et al. (2005), Rummel (2003), and Tapley et al. (2004a,b) for further details on the satellite missions and on SST.

From the decomposition of the gradient, as in (4.22), we see how we can write gradient fields in terms of vector spherical harmonics: if D is an open spherical shell centred at zero, $F \in \mathrm{C}^{(1)}(D)$, and the following interchanging of differentiation and limits is possible for F, then (with again $x = r\xi$, $\xi \in \Omega$)

$$
\begin{aligned}
&\nabla_x F(x) \\
&= \nabla_{r\xi}\left(\sum_{n=0}^{\infty}\sum_{j=-n}^{n} (F(r\cdot))^\wedge(n,j)Y_{n,j}(\xi)\right) \\
&= \sum_{n=0}^{\infty}\sum_{j=-n}^{n}\left(\xi\frac{\partial}{\partial r}+\frac{1}{r}\nabla_\xi^*\right)\left[(F(r\cdot))^\wedge(n,j)Y_{n,j}(\xi)\right] \\
&= \sum_{n=0}^{\infty}\sum_{j=-n}^{n}\left[\left(\frac{\partial F}{\partial r}(r\cdot)\right)^\wedge(n,j)y_{n,j}^{(1)}(\xi)+\frac{\sqrt{n(n+1)}}{r}(F(r\cdot))^\wedge(n,j)y_{n,j}^{(2)}(\xi)\right] \qquad (4.29)\\
&= \sum_{n=0}^{\infty}\sum_{j=-n}^{n}\left(\frac{\partial F}{\partial r}(r\cdot)\right)^\wedge(n,j)\left(\frac{n+1}{2n+1}+\frac{n}{2n+1}\right)y_{n,j}^{(1)}(\xi) \\
&\quad+\sum_{n=1}^{\infty}\sum_{j=-n}^{n}\left(\frac{\partial F}{\partial r}(r\cdot)\right)^\wedge(n,j)\left(-\frac{\sqrt{n(n+1)}}{2n+1}+\frac{\sqrt{n(n+1)}}{2n+1}\right)y_{n,j}^{(2)}(\xi) \\
&\quad+\sum_{n=0}^{\infty}\sum_{j=-n}^{n}\frac{(F(r\cdot))^\wedge(n,j)}{r}\left(-\frac{n(n+1)}{2n+1}+\frac{n(n+1)}{2n+1}\right)y_{n,j}^{(1)}(\xi) \\
&\quad+\sum_{n=1}^{\infty}\sum_{j=-n}^{n}\frac{(F(r\cdot))^\wedge(n,j)}{r}\left(\frac{n\sqrt{n(n+1)}}{2n+1}+\frac{(n+1)\sqrt{n(n+1)}}{2n+1}\right)y_{n,j}^{(2)}(\xi) \\
&= \sum_{n=0}^{\infty}\sum_{j=-n}^{n}\left[\sqrt{\frac{n+1}{2n+1}}\left(\frac{\partial F}{\partial r}(r\cdot)\right)^\wedge(n,j)-n\sqrt{\frac{n+1}{2n+1}}\frac{1}{r}(F(r\cdot))^\wedge(n,j)\right] \\
&\quad\times\left[\sqrt{\frac{n+1}{2n+1}}y_{n,j}^{(1)}(\xi)-\sqrt{\frac{n}{2n+1}}y_{n,j}^{(2)}(\xi)\right] \\
&\quad+\sum_{n=1}^{\infty}\sum_{j=-n}^{n}\left[\sqrt{\frac{n}{2n+1}}\left(\frac{\partial F}{\partial r}(r\cdot)\right)^\wedge(n,j)+\sqrt{\frac{n}{2n+1}}\frac{n+1}{r}(F(r\cdot))^\wedge(n,j)\right] \\
&\quad\times\left[\sqrt{\frac{n}{2n+1}}y_{n,j}^{(1)}(\xi)+\sqrt{\frac{n+1}{2n+1}}y_{n,j}^{(2)}(\xi)\right] \\
&= \sum_{n=0}^{\infty}\sum_{j=-n}^{n}\left[\sqrt{\frac{n+1}{2n+1}}\left(\frac{\partial F}{\partial r}(r\cdot)\right)^\wedge(n,j)-n\sqrt{\frac{n+1}{2n+1}}\frac{1}{r}(F(r\cdot))^\wedge(n,j)\right]\tilde{y}_{n,j}^{(1)}(\xi) \\
&\quad+\sum_{n=1}^{\infty}\sum_{j=-n}^{n}\left[\sqrt{\frac{n}{2n+1}}\left(\frac{\partial F}{\partial r}(r\cdot)\right)^\wedge(n,j)+\sqrt{\frac{n}{2n+1}}\frac{n+1}{r}(F(r\cdot))^\wedge(n,j)\right]\tilde{y}_{n,j}^{(2)}(\xi).
\end{aligned}
$$

(4.30)

Hence, we have representations of ∇F in terms of the Morse–Feshbach vector spherical harmonics, namely (4.29), and the Edmonds vector spherical harmonics, namely (4.30).

If we now have an inner harmonic $H_{n,j}^{\mathrm{int}}(r\xi) = r^n Y_{n,j}(\xi)$, then

$$\left(\frac{\partial H_{n,j}^{\mathrm{int}}}{\partial r}(r\cdot)\right)^{\wedge}(n,j) = n\, r^{n-1} \text{ and } \left(H_{n,j}^{\mathrm{int}}(r\cdot)\right)^{\wedge}(n,j) = r^n$$

such that $\nabla_x H_{n,j}^{\mathrm{int}}(x) = \sqrt{n(2n+1)}\, r^{n-1} \tilde{y}_{n,j}^{(2)}(\xi)$.

For an outer harmonic $H_{-n-1,j}^{\mathrm{ext}}(r\xi) = r^{-n-1} Y_{n,j}(\xi)$, we obtain

$$\left(\frac{\partial H_{-n-1,j}^{\mathrm{ext}}}{\partial r}(r\cdot)\right)^{\wedge}(n,j) = -(n+1) r^{-n-2}$$

and $\left(H_{-n-1,j}^{\mathrm{ext}}(r\cdot)\right)^{\wedge}(n,j) = r^{-n-1}$, which leads us to a representation of its gradient as $\nabla_x H_{-n-1,j}^{\mathrm{ext}}(x) = -\sqrt{(n+1)(2n+1)}\, r^{-n-2} \tilde{y}_{n,j}^{(1)}(\xi)$.

The latter results show why Edmonds vector spherical harmonics also have their justification. In particular, in cases where gradient fields of two different potentials – one originating from an inner source and one originating from an outer source – occur, the Edmonds vector spherical harmonics allow an orthogonal decomposition of these two fields. For instance, in geomagnetics this is a valuable tool, as discussed in Chapter 6.

Having said that SST is a standard technique for spaceborne gravity measurements, we should mention that the GOCE mission allowed the determination of second-order derivatives of the gravitational potential, see, for example, Abrikosov and Schwintzer (2004), Eicker et al. (2005), Freeden et al. (2002), Pail and Wermuth (2003), Pail et al. (2011), and Rummel et al. (2001). This makes the investigation of the Hessian of the potential necessary, which is also called the Marussi tensor in geodesy. A standard technique for approximating tensorial functions on the sphere relies on tensor spherical harmonics, which we will further explain in the following section.

4.4 Tensor Spherical Harmonics

Our considerations at the beginning of Section 4.3 can be done here analogously: it can be expected that (almost every) $Y_{n,j}$ needs to be the origin for (in the tensorial case) nine new basis functions. And, of course, if one has three orthonormal vectors b^1, b^2, b^3 in $\mathbb{R}^3$ such as $\varepsilon^1, \varepsilon^2, \varepsilon^3$ or $\varepsilon^r, \varepsilon^\varphi, \varepsilon^t$, then one could easily choose $\{b^i \otimes b^k\, Y_{n,j}\}_{i,k=1,2,3,n\in\mathbb{N}_0,j=-n,\ldots,n}$ as a complete orthonormal system in $\mathbf{l}^2(\Omega)$. Remember that we use the abbreviation $\mathbf{l}^2(\Omega) = \mathrm{L}^2(\Omega, \mathbb{R}^{3\times 3})$. However, also in the tensorial case, it turned out that some other choices of basis functions can be more helpful in specific applications. We will give here a short introduction into this topic, which is unfortunately always cursed by lengthy formulae. The considerations here are based on Freeden et al. (1994, 1998), Schreiner (1994), and Seibert (2018), which should also be consulted for further details. Note that there are more possible definitions of tensor spherical harmonics than those presented here. For instance, such systems are introduced in Mathews (1962), Sandberg (1978), and Zerilli (1970a,b). Note that, in these cases, as in our case, second-rank tensors (which can be associated to matrices) are considered. Higher-rank tensor spherical harmonics are constructed in James

(1976) and Winter (1982). Besides such a generalization to higher ranks of tensors, the dimension of the domain can also be generalized. In Sandberg (1978) and Tomita (1982), tensor spherical harmonics are defined on the unit sphere $\mathbb{S}^3$ in $\mathbb{R}^4$. General dimensions, that is, $\mathbb{S}^{n-1}$ in $\mathbb{R}^n$, are used in Higuchi (1987) and Rubin and Ordóñez (1984). These lists can certainly only serve as examples; see also the further references given in Martinec (2003).

We continue here with the construction of those (second-rank) tensor spherical harmonics (on $\mathbb{S}^2 = \Omega$) which were introduced in Freeden et al. (1994, 1998), and Schreiner (1994).

Definition 4.4.1 For functions $F \in \mathrm{C}(\Omega)$, $G \in \mathrm{C}^{(1)}(\Omega)$, and $H \in \mathrm{C}^{(2)}(\Omega)$, the operators $\mathbf{o}^{(i,k)}$ are defined as follows (with $\xi \in \Omega$ arbitrary):

$$
\begin{aligned}
\mathbf{o}_\xi^{(1,1)} F(\xi) &:= \xi \otimes \xi F(\xi),\\
\mathbf{o}_\xi^{(1,2)} G(\xi) &:= \xi \otimes \nabla_\xi^* G(\xi),\\
\mathbf{o}_\xi^{(1,3)} G(\xi) &:= \xi \otimes \mathrm{L}_\xi^* G(\xi),\\
\mathbf{o}_\xi^{(2,1)} G(\xi) &:= \left(\nabla_\xi^* G(\xi)\right) \otimes \xi,\\
\mathbf{o}_\xi^{(3,1)} G(\xi) &:= \left(\mathrm{L}_\xi^* G(\xi)\right) \otimes \xi,\\
\mathbf{o}_\xi^{(2,2)} F(\xi) &:= \mathbf{i}_{\mathrm{tan}}(\xi) F(\xi),\\
\mathbf{o}_\xi^{(2,3)} H(\xi) &:= \left(\nabla_\xi^* \otimes \nabla_\xi^* - \mathrm{L}_\xi^* \otimes \mathrm{L}_\xi^*\right) H(\xi) + 2\left(\nabla_\xi^* H(\xi)\right) \otimes \xi,\\
\mathbf{o}_\xi^{(3,2)} H(\xi) &:= \left(\nabla_\xi^* \otimes \mathrm{L}_\xi^* + \mathrm{L}_\xi^* \otimes \nabla_\xi^*\right) H(\xi) + 2\left(\mathrm{L}_\xi^* H(\xi)\right) \otimes \xi,\\
\mathbf{o}_\xi^{(3,3)} F(\xi) &:= \mathbf{j}_{\mathrm{tan}}(\xi) F(\xi).
\end{aligned}
$$

Note that some of the scalar spherical harmonics $Y_{n,j}$ are mapped to zero by some $\mathbf{o}^{(i,k)}$-operators. For a similar reason, the 'indexed zero' 0_i was introduced when vector spherical harmonics were defined. In analogy, we now define $0_{i,k}$.

Definition 4.4.2 The notation $0_{i,k}$ stands for

$$
0_{i,k} := \begin{cases} 0, & \text{if } (i,k) \in \{(1,1),(2,2),(3,3)\},\\ 1, & \text{if } (i,k) \in \{(1,2),(1,3),(2,1),(3,1)\},\\ 2, & \text{if } (i,k) \in \{(2,3),(3,2)\}. \end{cases}
$$

Definition 4.4.3 The functions $\mathbf{y}_{n,j}^{(i,k)} \colon \Omega \to \mathbb{R}^{3\times 3}$ are defined by

$$
\mathbf{y}_{n,j}^{(i,k)}(\xi) := \left(\mu_n^{(i,k)}\right)^{-1/2} \mathbf{o}_\xi^{(i,k)} Y_{n,j}(\xi), \quad \xi \in \Omega,
$$

where $n \geq 0_{i,k}$, $j \in \{-n,\ldots,n\}$, and $i,k \in \{1,2,3\}$. Moreover,

$$
\mu_n^{(i,k)} := \left\| \mathbf{o}^{(i,k)} Y_{n,j} \right\|_{\mathbf{l}^2(\Omega)}^2 .
$$

We call these functions here the **Freeden–Gervens–Schreiner tensor spherical harmonics**.

For some of these functions, the factors $\mu_n^{(i,k)}$ for their normalization are easy to compute, for instance,

$$\left|\mathbf{o}_\xi^{(1,1)} Y_{n,j}(\xi)\right|^2 = \left(\mathbf{o}_\xi^{(1,1)} Y_{n,j}(\xi)\right) : \left(\mathbf{o}_\xi^{(1,1)} Y_{n,j}(\xi)\right) = (\xi \otimes \xi) : (\xi \otimes \xi)(Y_{n,j}(\xi))^2$$

such that $\mu_n^{(1,1)} = \|Y_{n,j}\|^2_{\mathrm{L}^2(\Omega)} = 1$. In the same manner, we can use what we have learned about the vector spherical harmonics to deduce that

$$\mu_n^{(2,1)} = \mu_n^{(1,2)} = \int_\Omega |\xi|^2 \left|\nabla_\xi^* Y_{n,j}(\xi)\right|^2 \, \mathrm{d}\omega(\xi) = n(n+1)$$

$$\mu_n^{(3,1)} = \mu_n^{(1,3)} = \int_\Omega |\xi|^2 \left|\mathrm{L}_\xi^* Y_{n,j}(\xi)\right|^2 \, \mathrm{d}\omega(\xi) = n(n+1).$$

For some other pairs (i,k) such as $(2,3)$ and $(3,2)$, the calculations are a bit more lengthy. Similarly, the calculation of the additional pairwise inner products (which are $\binom{9}{2} = 36$ in total, if we use the symmetry) is also not difficult but it is time and pages consuming. Therefore, we only quote the result here from Freeden et al. (1998) and Schreiner (1994).

Theorem 4.4.4 *We have*

$$\mu_n^{(i,k)} = \begin{cases} 1, & \text{if } (i,k) = (1,1), \\ 2, & \text{if } (i,k) \in \{(2,2),(3,3)\}, \\ n(n+1), & \text{if } (i,k) \in \{(1,2),(1,3),(2,1),(3,1)\}, \\ 2n(n+1)(n^2+n-2), & \text{if } (i,k) \in \{(2,3),(3,2)\}. \end{cases}$$

Theorem 4.4.5 *The Freeden–Gervens–Schreiner tensor spherical harmonics are a complete orthonormal system in* $\left(\mathbf{l}^2(\Omega), \langle\cdot,\cdot\rangle_{\mathbf{l}^2(\Omega)}\right)$ *and they are closed (in the sense of the approximation theory) in* $\left(\mathbf{c}(\Omega), \|\cdot\|_{\mathbf{c}(\Omega)}\right)$.

Also an addition theorem can be formulated for all kinds of combinations $\sum_{j=-n}^n \mathbf{y}_{n,j}^{(i_1,k_1)} \otimes \mathbf{y}_{n,j}^{(i_2,k_2)}$, where the tensor product yields here a fourth-rank-tensor. In total, $9 \cdot 9 = 81$ combinations exist, where only 45 need to be calculated and the rest can be obtained by transposition. For details, see Freeden et al. (1998, section 14.6). We quote here one of the essential results.

Theorem 4.4.6 *We have*

$$\sum_{j=-n}^{n} \left(\mathbf{y}_{n,j}^{(i,k)}(\xi)\right) : \left(\mathbf{y}_{n,j}^{(i,k)}(\xi)\right) = \frac{2n+1}{4\pi}$$

for all $\xi \in \Omega, (i,k) \in \{1,2,3\}^2$, *and* $n \geq 0_{i,k}$. *Moreover,*

$$\left\|\mathbf{y}_{n,j}^{(i,k)}\right\|_{\mathbf{c}(\Omega)} \leq \sqrt{\frac{2n+1}{4\pi}}$$

for all $(i,k) \in \{1,2,3\}^2$, $n \geq 0_{i,k}$, *and* $j = -n,\ldots,n$. *More generally, for* $\mathbf{f} \in \operatorname{span}\{\mathbf{y}_{n,j}^{(i,k)}\}_{j=-n,\ldots,n}$, *we find*

$$\|\mathbf{f}\|_{\mathrm{c}(\Omega)} \leq \sqrt{\frac{2n+1}{4\pi}}\, \|\mathbf{f}\|_{\mathrm{l}^2(\Omega)}.$$

We avoid here to plot some tensor spherical harmonics since there is not a really useful way to plot a tensorial function on the sphere. One could plot the norm $|\mathbf{y}_{n,j}^{(i,k)}(\xi)|$ instead. However, the gain of information from such plots is limited. For instance, the type-$(1,1)$-functions satisfy

$$\left|y_{n,j}^{(1,1)}(\xi)\right| = \sqrt{(\xi \otimes \xi) : (\xi \otimes \xi)}\, \left|Y_{n,j}(\xi)\right| = \left|Y_{n,j}(\xi)\right|$$

such that merely the absolute value of the scalar spherical harmonics would be plotted.

Another difficulty, which has, however, been overcome in Freeden et al. (1994, 1998) and Schreiner (1994), is the characterization of tangential and normal fields in the case of tensor fields. In fact, the Freeden–Gervens–Schreiner tensor spherical harmonics have the advantage that they allow such a distinction.

Definition 4.4.7 Let $\mathbf{f}\colon \Omega \to \mathbb{R}^{3\times 3}$ be an arbitrary tensor field on the sphere. We call $\mathbf{f}$

(a) a **left-normal tensor field**, if $\xi \otimes \left(\xi^{\mathrm{T}}\mathbf{f}(\xi)\right)^{\mathrm{T}} = \mathbf{f}(\xi)$ for all $\xi \in \Omega$.
(b) a **right-normal tensor field**, if $(\mathbf{f}(\xi)\xi) \otimes \xi = \mathbf{f}(\xi)$ for all $\xi \in \Omega$.
(c) a **left-tangential tensor field**, if $\xi^{\mathrm{T}}\mathbf{f}(\xi) = 0$ for all $\xi \in \Omega$.
(d) a **right-tangential tensor field**, if $\mathbf{f}(\xi)\xi = 0$ for all $\xi \in \Omega$.
(e) a **normal tensor field**, if it is left-normal and right-normal.
(f) a **tangential tensor field**, if it is left-tangential and right-tangential.

Theorem 4.4.8 *Let* $F \in \mathrm{C}(\Omega)$, $G \in \mathrm{C}^{(1)}(\Omega)$, *and* $H \in \mathrm{C}^{(2)}(\Omega)$. *Then the following holds true:*

- $\mathbf{o}^{(1,1)}F$ *is a normal tensor field.*
- $\mathbf{o}^{(2,2)}F$, $\mathbf{o}^{(2,3)}H$, $\mathbf{o}^{(3,2)}H$, *and* $\mathbf{o}^{(3,3)}F$ *are tangential tensor fields.*
- $\mathbf{o}^{(1,2)}G$ *and* $\mathbf{o}^{(1,3)}G$ *are left-normal and right-tangential tensor fields.*
- $\mathbf{o}^{(2,1)}G$ *and* $\mathbf{o}^{(3,1)}G$ *are left-tangential and right-normal tensor fields.*

Proof Clearly, $\xi \otimes [\xi^{\mathrm{T}}(\xi \otimes \xi)]^{\mathrm{T}} = \xi \otimes (\xi^{\mathrm{T}}\xi\xi^{\mathrm{T}})^{\mathrm{T}} = \xi \otimes \xi$ and $[(\xi \otimes \xi)\xi] \otimes \xi = (\xi\xi^{\mathrm{T}}\xi) \otimes \xi = \xi \otimes \xi$ such that $\mathbf{o}^{(1,1)}F$ is a normal tensor field. Analogously, the left-normal or right-normal properties of $\mathbf{o}^{(1,2)}G$, $\mathbf{o}^{(1,3)}G$, $\mathbf{o}^{(2,1)}G$, and $\mathbf{o}^{(3,1)}G$, respectively, can be shown. The left-tangential or right-tangential properties are a consequence of the tangential property of ∇^* and L^* (see Theorem 4.1.6). For the tangential tensor fields, we recall the following (see Definition 4.1.4):

$$\mathbf{i}_{\tan}(\xi) = \mathbf{i} - \xi \otimes \xi \quad \text{and} \quad \mathbf{j}_{\tan}(\xi) = \sum_{j=1}^{3} (\xi \times \varepsilon^j) \otimes \varepsilon^j.$$

Since $\xi^{\mathrm{T}}\mathbf{i}_{\tan}(\xi) = \xi^{\mathrm{T}} - \xi^{\mathrm{T}}\xi\,\xi^{\mathrm{T}} = 0$ and $\mathbf{i}_{\tan}(\xi)\xi = \xi - \xi\,\xi^{\mathrm{T}}\xi = 0$, we have that $\mathbf{o}^{(2,2)}F$ is tangential. Moreover,

$$\xi^{\mathrm{T}}\mathbf{j}_{\tan}(\xi) = \sum_{j=1}^{3} \xi^{\mathrm{T}}(\xi \times \varepsilon^j)(\varepsilon^j)^{\mathrm{T}} = 0,$$

$$\mathbf{j}_{\tan}(\xi)\,\xi = \sum_{j=1}^{3}(\xi \times \varepsilon^j)(\varepsilon^j)^{\mathrm{T}}\,\xi = \xi \times \left(\sum_{j=1}^{3} \varepsilon^j \xi_j\right) = \xi \times \xi = 0$$

such that $\mathbf{o}^{(3,3)}F$ is also tangential. For the remaining two tangential fields, some more work has to be done. We recall that $\nabla^* H$ and $\mathrm{L}^* H$ are composed of ε^φ and ε^t (see (4.3) and (4.4)). With Corollary 3.4.4, we obtain

$$\begin{aligned}
\nabla^* \otimes \varepsilon^\varphi &= \frac{1}{\sqrt{1-t^2}}\,\varepsilon^\varphi \otimes \left(\frac{\partial}{\partial\varphi}\,\varepsilon^\varphi\right) + \sqrt{1-t^2}\,\varepsilon^t \otimes \left(\frac{\partial}{\partial t}\,\varepsilon^\varphi\right) \\
&= \varepsilon^\varphi \otimes \left(-\varepsilon^r + \frac{t}{\sqrt{1-t^2}}\,\varepsilon^t\right),
\end{aligned} \tag{4.31}$$

$$\begin{aligned}
\nabla^* \otimes \varepsilon^t &= \frac{1}{\sqrt{1-t^2}}\,\varepsilon^\varphi \otimes (-t\varepsilon^\varphi) + \sqrt{1-t^2}\,\varepsilon^t \otimes \left(-\frac{1}{\sqrt{1-t^2}}\,\varepsilon^r\right) \\
&= -\frac{t}{\sqrt{1-t^2}}\,\varepsilon^\varphi \otimes \varepsilon^\varphi - \varepsilon^t \otimes \varepsilon^r,
\end{aligned} \tag{4.32}$$

$$\begin{aligned}
\mathrm{L}^* \otimes \varepsilon^\varphi &= \frac{1}{\sqrt{1-t^2}}\,\varepsilon^t \otimes \left(\frac{\partial}{\partial\varphi}\,\varepsilon^\varphi\right) - \sqrt{1-t^2}\,\varepsilon^\varphi \otimes \left(\frac{\partial}{\partial t}\,\varepsilon^\varphi\right) \\
&= \varepsilon^t \otimes \left(-\varepsilon^r + \frac{t}{\sqrt{1-t^2}}\,\varepsilon^t\right),
\end{aligned} \tag{4.33}$$

$$\begin{aligned}
\mathrm{L}^* \otimes \varepsilon^t &= \frac{1}{\sqrt{1-t^2}}\,\varepsilon^t \otimes (-t\varepsilon^\varphi) - \sqrt{1-t^2}\,\varepsilon^\varphi \otimes \left(-\frac{1}{\sqrt{1-t^2}}\,\varepsilon^r\right) \\
&= -\frac{t}{\sqrt{1-t^2}}\,\varepsilon^t \otimes \varepsilon^\varphi + \varepsilon^\varphi \otimes \varepsilon^r.
\end{aligned} \tag{4.34}$$

Thus, $\nabla^* \otimes \nabla^* H$, $\nabla^* \otimes \mathrm{L}^* H$, $\mathrm{L}^* \otimes \nabla^* H$, and $\mathrm{L}^* \otimes \mathrm{L}^* H$ are left-tangential tensor fields. Moreover, with Theorem 4.1.6, we get $\xi^{\mathrm{T}}\big(\nabla^*_\xi H(\xi) \otimes \xi\big) = 0$ and $\xi^{\mathrm{T}}\big(\mathrm{L}^*_\xi H(\xi) \otimes \xi\big) = 0$ such that $\mathbf{o}^{(2,3)}H$ and $\mathbf{o}^{(3,2)}H$ are left-tangential. It remains to show that they are also right-tangential. For this purpose, we calculate

$$\begin{aligned}
\left[\nabla^*_\xi \otimes \nabla^*_\xi H(\xi)\right]\xi &= \left[\nabla^*_\xi \otimes \left(\varepsilon^\varphi \frac{1}{\sqrt{1-t^2}}\frac{\partial H}{\partial\varphi} + \varepsilon^t\sqrt{1-t^2}\,\frac{\partial H}{\partial t}\right)\right]\xi \\
&= -\varepsilon^\varphi \frac{1}{\sqrt{1-t^2}}\frac{\partial H}{\partial\varphi} + \left[\nabla^*_\xi\left(\frac{1}{\sqrt{1-t^2}}\frac{\partial H}{\partial\varphi}\right)\right]\underbrace{(\varepsilon^\varphi)^{\mathrm{T}}\xi}_{=0} \\
&\quad - \varepsilon^t\sqrt{1-t^2}\,\frac{\partial H}{\partial t} + \left[\nabla^*_\xi\left(\sqrt{1-t^2}\,\frac{\partial H}{\partial t}\right)\right]\underbrace{(\varepsilon^t)^{\mathrm{T}}\xi}_{=0} \\
&= -\nabla^*_\xi H(\xi),
\end{aligned}$$

$$\left[\mathrm{L}_\xi^* \otimes \nabla_\xi^* H(\xi)\right]\xi = -\varepsilon^t \frac{1}{\sqrt{1-t^2}} \frac{\partial H}{\partial \varphi} + \varepsilon^\varphi \sqrt{1-t^2} \frac{\partial H}{\partial t} = -\mathrm{L}_\xi^* H(\xi),$$

$$\left[\nabla_\xi^* \otimes \mathrm{L}_\xi^* H(\xi)\right]\xi = \left[\nabla_\xi^* \otimes \left(\varepsilon^t \frac{1}{\sqrt{1-t^2}} \frac{\partial H}{\partial \varphi} - \varepsilon^\varphi \sqrt{1-t^2} \frac{\partial H}{\partial t}\right)\right]\xi$$
$$= -\varepsilon^t \frac{1}{\sqrt{1-t^2}} \frac{\partial H}{\partial \varphi} + \varepsilon^\varphi \sqrt{1-t^2} \frac{\partial H}{\partial t} = -\mathrm{L}_\xi^* H(\xi),$$

$$\left[\mathrm{L}_\xi^* \otimes \mathrm{L}_\xi^* H(\xi)\right]\xi = \varepsilon^\varphi \frac{1}{\sqrt{1-t^2}} \frac{\partial H}{\partial \varphi} + \varepsilon^t \sqrt{1-t^2} \frac{\partial H}{\partial t} = \nabla_\xi^* H(\xi).$$

Consequently, we get

$$\left[\mathbf{o}_\xi^{(2,3)} H(\xi)\right]\xi = -\nabla_\xi^* H(\xi) - \nabla_\xi^* H(\xi) + 2\nabla_\xi^* H(\xi) = 0,$$
$$\left[\mathbf{o}_\xi^{(3,2)} H(\xi)\right]\xi = -\mathrm{L}_\xi^* H(\xi) - \mathrm{L}_\xi^* H(\xi) + 2\mathrm{L}_\xi^* H(\xi) = 0$$

such that $\mathbf{o}^{(2,3)} H$ and $\mathbf{o}^{(3,2)} H$ are also right-tangential. □

Note that the previous proof also shows that the terms $2\left(\nabla_\xi^* H(\xi)\right) \otimes \xi$ and $2\left(\mathrm{L}_\xi^* H(\xi)\right) \otimes \xi$ are included in the definition of $\mathbf{o}^{(2,3)} H$ and $\mathbf{o}^{(3,2)} H$, respectively, in order to obtain tangential fields.

From the preceding calculations, we also get that

$$\nabla_\xi^* \otimes \nabla_\xi^* H(\xi) = \left(\nabla^* \otimes \varepsilon^\varphi\right) \frac{1}{\sqrt{1-t^2}} \frac{\partial H}{\partial \varphi} + \nabla_\xi^* \left(\frac{1}{\sqrt{1-t^2}} \frac{\partial H}{\partial \varphi}\right) \otimes \varepsilon^\varphi$$
$$+ \left(\nabla^* \otimes \varepsilon^t\right) \sqrt{1-t^2} \frac{\partial H}{\partial t} + \nabla_\xi^* \left(\sqrt{1-t^2} \frac{\partial H}{\partial t}\right) \otimes \varepsilon^t,$$

$$\mathrm{L}_\xi^* \otimes \mathrm{L}_\xi^* H(\xi) = \left(\mathrm{L}^* \otimes \varepsilon^t\right) \frac{1}{\sqrt{1-t^2}} \frac{\partial H}{\partial \varphi} + \mathrm{L}_\xi^* \left(\frac{1}{\sqrt{1-t^2}} \frac{\partial H}{\partial \varphi}\right) \otimes \varepsilon^t$$
$$- \left(\mathrm{L}^* \otimes \varepsilon^\varphi\right) \sqrt{1-t^2} \frac{\partial H}{\partial t} - \mathrm{L}_\xi^* \left(\sqrt{1-t^2} \frac{\partial H}{\partial t}\right) \otimes \varepsilon^\varphi,$$

where

$$\nabla_\xi^* \left(\frac{1}{\sqrt{1-t^2}} \frac{\partial H}{\partial \varphi}\right) = \varepsilon^\varphi \frac{1}{1-t^2} \frac{\partial^2 H}{\partial \varphi^2} + \varepsilon^t \left(\frac{t}{1-t^2} \frac{\partial H}{\partial \varphi} + \frac{\partial^2 H}{\partial t \partial \varphi}\right),$$

$$\mathrm{L}_\xi^* \left(\sqrt{1-t^2} \frac{\partial H}{\partial t}\right) = \varepsilon^t \frac{\partial^2 H}{\partial \varphi \partial t} - \varepsilon^\varphi \left(-t \frac{\partial H}{\partial t} + (1-t^2) \frac{\partial^2 H}{\partial t^2}\right),$$

$$\nabla_\xi^* \left(\sqrt{1-t^2} \frac{\partial H}{\partial t}\right) = \varepsilon^\varphi \frac{\partial^2 H}{\partial \varphi \partial t} + \varepsilon^t \left(-t \frac{\partial H}{\partial t} + (1-t^2) \frac{\partial^2 H}{\partial t^2}\right),$$

$$\mathrm{L}_\xi^* \left(\frac{1}{\sqrt{1-t^2}} \frac{\partial H}{\partial \varphi}\right) = \varepsilon^t \frac{1}{1-t^2} \frac{\partial^2 H}{\partial \varphi^2} - \varepsilon^\varphi \left(\frac{t}{1-t^2} \frac{\partial H}{\partial \varphi} + \frac{\partial^2 H}{\partial t \partial \varphi}\right).$$

With (4.31), (4.32), (4.33), and (4.34), we obtain now, for $H \in \mathrm{C}^{(2)}(\Omega)$, that

$$
\begin{aligned}
&\nabla^*_\xi \otimes \nabla^*_\xi H(\xi) + \mathrm{L}^*_\xi \otimes \mathrm{L}^*_\xi H(\xi) \\
&= \varepsilon^\varphi \otimes \varepsilon^r \left(-\frac{1}{\sqrt{1-t^2}} \frac{\partial H}{\partial \varphi} + \frac{1}{\sqrt{1-t^2}} \frac{\partial H}{\partial \varphi} \right) \\
&\quad + \varepsilon^\varphi \otimes \varepsilon^t \left(\frac{t}{1-t^2} \frac{\partial H}{\partial \varphi} + \frac{\partial^2 H}{\partial \varphi \partial t} - \frac{t}{1-t^2} \frac{\partial H}{\partial \varphi} - \frac{\partial^2 H}{\partial t \partial \varphi} \right) \\
&\quad + \varepsilon^\varphi \otimes \varepsilon^\varphi \left(\frac{1}{1-t^2} \frac{\partial^2 H}{\partial \varphi^2} - t \frac{\partial H}{\partial t} - t \frac{\partial H}{\partial t} + (1-t^2) \frac{\partial^2 H}{\partial t^2} \right) \\
&\quad + \varepsilon^t \otimes \varepsilon^\varphi \left(\frac{t}{1-t^2} \frac{\partial H}{\partial \varphi} + \frac{\partial^2 H}{\partial t \partial \varphi} - \frac{t}{1-t^2} \frac{\partial H}{\partial \varphi} - \frac{\partial^2 H}{\partial \varphi \partial t} \right) \\
&\quad + \varepsilon^t \otimes \varepsilon^r \left(-\sqrt{1-t^2} \frac{\partial H}{\partial t} + \sqrt{1-t^2} \frac{\partial H}{\partial t} \right) \\
&\quad + \varepsilon^t \otimes \varepsilon^t \left(-t \frac{\partial H}{\partial t} + (1-t^2) \frac{\partial^2 H}{\partial t^2} + \frac{1}{1-t^2} \frac{\partial^2 H}{\partial \varphi^2} - t \frac{\partial H}{\partial t} \right) \\
&= \left(\varepsilon^\varphi \otimes \varepsilon^\varphi + \varepsilon^t \otimes \varepsilon^t \right) \Delta^*_\xi H(\xi).
\end{aligned}
$$

Remember the definition of Δ^* in Theorem 3.4.3. This leads us immediately to a well-known identity. There is a similar second identity, whose analogous proof is omitted here.

Theorem 4.4.9 *For all $H \in \mathrm{C}^{(2)}(\Omega)$, we have*

$$
\begin{aligned}
\nabla^* \otimes \nabla^* H + \mathrm{L}^* \otimes \mathrm{L}^* H &= \mathbf{i}_{\tan} \Delta^* H, \\
\mathrm{L}^* \otimes \nabla^* H - \nabla^* \otimes \mathrm{L}^* H &= \mathbf{j}_{\tan} \Delta^* H.
\end{aligned}
$$

Also similar calculations lead to the following results.

Theorem 4.4.10 *Let $F \in \mathrm{C}(\Omega)$, $G \in \mathrm{C}^{(1)}(\Omega)$, and $H \in \mathrm{C}^{(2)}(\Omega)$.*

(a) Then the following identity holds true:

$$
\begin{aligned}
&\left(\nabla^*_\xi \otimes \nabla^*_\xi H(\xi) - \mathrm{L}^*_\xi \otimes \mathrm{L}^*_\xi H(\xi) \right) - \left(\nabla^*_\xi \otimes \nabla^*_\xi H(\xi) - \mathrm{L}^*_\xi \otimes \mathrm{L}^*_\xi H(\xi) \right)^{\mathrm{T}} \\
&\quad = 2\left[\xi \otimes \nabla^*_\xi H(\xi) - \left(\nabla^*_\xi H(\xi) \right) \otimes \xi \right].
\end{aligned}
$$

(b) $\mathbf{o}^{(1,1)} F$, $\mathbf{o}^{(2,2)} F$, $\mathbf{o}^{(2,3)} H$, and $\mathbf{o}^{(3,2)} H$ are symmetric.
(c) $\left(\mathbf{o}^{(1,2)} G \right)^{\mathrm{T}} = \mathbf{o}^{(2,1)} G$ and $\left(\mathbf{o}^{(1,3)} G \right)^{\mathrm{T}} = \mathbf{o}^{(3,1)} G$.
(d) $\left(\mathbf{o}^{(3,3)} H \right)^{\mathrm{T}} = -\mathbf{o}^{(3,3)} H$.

We saw how gradient fields can be represented in terms of vector spherical harmonics in Section 4.3. We show here now the well-known expansion of a Hessian in terms of tensor spherical harmonics. Let $F\colon D \to \mathbb{R}$ be twice continuously differentiable on the open set $D \subset \mathbb{R}^3$. We have, with $x = r\xi$, $r = |x|$, $\xi \in \Omega$, that (remember Theorem 4.1.3)

$$\nabla_x F(x) = \left(\xi \frac{\partial}{\partial r} + \frac{1}{r} \nabla_\xi^*\right) F(r\xi)$$

such that the Hessian of F satisfies

$$\begin{aligned}
(\nabla_x \otimes \nabla_x) F(x) &= \left(\xi \frac{\partial}{\partial r} + \frac{1}{r} \nabla_\xi^*\right) \otimes \left(\xi \frac{\partial}{\partial r} + \frac{1}{r} \nabla_\xi^*\right) F(r\xi) \\
&= \xi \otimes \xi \frac{\partial^2}{\partial r^2} F(r\xi) - \frac{1}{r^2} \xi \otimes \nabla_\xi^* F(r\xi) + \frac{1}{r} \xi \otimes \nabla_\xi^* \left(\frac{\partial}{\partial r} F(r\xi)\right) \\
&\quad + \mathbf{i}_{\tan}(\xi) \frac{1}{r} \frac{\partial}{\partial r} F(r\xi) + \frac{1}{r} \left[\nabla_\xi^* \left(\frac{\partial}{\partial r} F(r\xi)\right)\right] \otimes \xi + \frac{1}{r^2} \nabla_\xi^* \otimes \nabla_\xi^* F(r\xi) \\
&= \xi \otimes \xi \frac{\partial^2}{\partial r^2} F(r\xi) + \xi \otimes \nabla_\xi^* \left(\frac{1}{r} \frac{\partial}{\partial r} - \frac{1}{r^2}\right) F(r\xi) \\
&\quad + \left[\nabla_\xi^* \left(\frac{1}{r} \frac{\partial}{\partial r} F(r\xi)\right)\right] \otimes \xi + \mathbf{i}_{\tan}(\xi) \frac{1}{r} \frac{\partial}{\partial r} F(r\xi) \\
&\quad + \frac{1}{2} (2\nabla_\xi^* \otimes \nabla_\xi^* - \mathrm{L}_\xi^* \otimes \mathrm{L}_\xi^* + \mathrm{L}_\xi^* \otimes \mathrm{L}_\xi^*) \left(\frac{1}{r^2} F(r\xi)\right) \\
&\quad + \frac{1}{2r^2} (2\nabla_\xi^* F(r\xi) - 2\nabla_\xi^* F(r\xi)) \otimes \xi.
\end{aligned}$$

Moreover, note that

$$\frac{\partial}{\partial r} \left(\frac{1}{r} F(r\xi)\right) = -\frac{1}{r^2} F(r\xi) + \frac{1}{r} \frac{\partial}{\partial r} F(r\xi).$$

If we now assume that $F(x) = \sum_{n=0}^{\infty} \sum_{j=-n}^{n} F(r\cdot)^\wedge(n,j) Y_{n,j}(\xi)$ and the series converges 'sufficiently nicely' such that all differentiation operators which are involved may be interchanged with the limit of the series and the inner product, then we get (using Theorem 4.4.9) the identity

$$\begin{aligned}
(\nabla_x \otimes \nabla_x) F(x) &= \sum_{n=0}^{\infty} \sum_{j=-n}^{n} \left(\frac{\partial^2}{\partial r^2} F(r\cdot)\right)^\wedge (n,j) \mathbf{y}_{n,j}^{(1,1)}(\xi) \\
&\quad + \sum_{n=1}^{\infty} \sum_{j=-n}^{n} \sqrt{n(n+1)} \left[\frac{\partial}{\partial r} \left(\frac{1}{r} F(r\cdot)\right)\right]^\wedge (n,j) \left(\mathbf{y}_{n,j}^{(1,2)}(\xi) + \mathbf{y}_{n,j}^{(2,1)}(\xi)\right) \\
&\quad + \sum_{n=0}^{\infty} \sum_{j=-n}^{n} \sqrt{2} \left[\left(\frac{1}{r} \frac{\partial}{\partial r} - \frac{n(n+1)}{2r^2}\right) F(r\cdot)\right]^\wedge (n,j) \mathbf{y}_{n,j}^{(2,2)}(\xi) \\
&\quad + \sum_{n=2}^{\infty} \sum_{j=-n}^{n} \sqrt{2n(n+1)(n^2+n-2)} \frac{1}{2r^2} F(r\cdot)^\wedge(n,j) \mathbf{y}_{n,j}^{(2,3)}(\xi).
\end{aligned}$$

(4.35)

Note that the Hessian of a twice continuously differentiable function is always symmetric due to Schwarz's theorem. In this respect, it is worth mentioning that some of the tensor spherical harmonics used in (4.35) are not symmetric, namely $\mathbf{y}_{n,j}^{(1,2)}$ and $\mathbf{y}_{n,j}^{(2,1)}$. Since they appear anyway in a dependent way, this matter can easily be repaired. We define the following (the variation in sign of $\mathbf{y}_{n,j}^{(2,2)}$ has historical reasons):

$$\mathbf{z}_{n,j}^{(1)} := \mathbf{y}_{n,j}^{(1,1)}, \qquad \mathbf{z}_{n,j}^{(2)} := \frac{1}{\sqrt{2}}\left(\mathbf{y}_{n,j}^{(1,2)} + \mathbf{y}_{n,j}^{(2,1)}\right),$$

$$\mathbf{z}_{n,j}^{(3)} := \mathbf{y}_{n,j}^{(2,3)}, \qquad \mathbf{z}_{n,j}^{(4)} := -\mathbf{y}_{n,j}^{(2,2)}.$$

Hence, (4.35) can easily be rewritten in this system. It can also be supplemented to a basis of all symmetric tensor fields, which obviously requires six types of functions:

$$\mathbf{z}_{n,j}^{(5)} := \frac{1}{\sqrt{2}}\left(\mathbf{y}_{n,j}^{(1,3)} + \mathbf{y}_{n,j}^{(3,1)}\right), \qquad \mathbf{z}_{n,j}^{(6)} := \mathbf{y}_{n,j}^{(3,2)}.$$

We also set

$$\tilde{0}_i := \begin{cases} 0, & \text{if } i = 1 \text{ or } i = 4, \\ 1, & \text{if } i = 2 \text{ or } i = 5, \\ 2, & \text{if } i = 3 \text{ or } i = 6. \end{cases}$$

This system was introduced in Zerilli (1970b) based on preliminary work by Regge and Wheeler (1957). It was used, for example, for the representation of a Hessian in Martinec (2003), where a similar notation was used. The following property of this system is clear due to our previous considerations.

Theorem 4.4.11 *The system* $\left\{\mathbf{z}_{n,j}^{(i)}\right\}_{i=1,\ldots,6,\, n\geq\tilde{0}_i,\, j=-n,\ldots,n}$ *is a complete orthonormal basis for all symmetric tensor fields in* $\mathbf{l}^2(\Omega)$.

4.5 On the Multitude of Trial Functions

Spherical harmonics are not the only kind of basis functions on the sphere. They have become a commonly used system not only in Earth sciences due to the huge amount of knowledge which has been gained on their theoretical and numerical properties. From the mathematical point of view, they are orthogonal polynomials with all their pros and cons. On the 'pro'-side, we have the features associated to the fact that these functions constitute an orthonormal basis of $\mathrm{L}^2(\Omega)$. Hence, we may use the expansion in a Fourier series and the Parseval identity. Moreover, we can decompose signals into orthogonal components associated, for example, to different degree bands. A disadvantage, however, is given (in the case of some applications) by the 'global' structure of these functions, which is also visible in the figures of the scalar and vector spherical harmonics in this chapter. This means that spherical harmonics 'live everywhere'. A decomposition of a signal into an expansion in spherical harmonics loses any spatial references. The expansion coefficients are spherical averages which reflect a pure degree dependence, which can here also be interpreted as a frequency dependence. Indeed, there are several versions of an uncertainty principle on the sphere, which tell us that perfect space and frequency localization are mutually exclusive. For further details (including also the more general case of a d-dimensional sphere), see Dai and Xu (2014); Dang et al. (2017); Freeden et al. (1998, 2018); Iglewska-Nowak (2016); Narcowich and Ward (1996). In this respect, spherical harmonics represent one extremal case in the multitude of spherical trial functions which have been developed so far: they enable a perfect frequency localization, but they completely lack a space localization.

The different systems of trial functions have already been categorized according to their different properties and ways of their construction. We will introduce these terminologies here. We first assume that we seek to approximate functions in a Banach space $\mathcal{H}(\Omega) \subset \mathrm{L}^2(\Omega)$ with $\|F\|_{\mathrm{L}^2(\Omega)} \leq c\|F\|_{\mathcal{H}(\Omega)}$ for all $F \in \mathcal{H}(\Omega)$ and a fixed constant $c \in \mathbb{R}^+$. We consider a function system $\{F_k\}_{k\in\kappa}$ in $\mathcal{H}(\Omega)$ which is supposed to be a basis. Either $\mathcal{H}(\Omega)$ is finite dimensional, that is, there is $N \in \mathbb{N}$ such that $\kappa = \{1, \ldots, N\}$, or $\kappa = \mathbb{N}$ (note that the functions $Y_{n,j}$ can be rearranged such that they have a single index of positive integers). In the former case, this means that $N = \dim \mathcal{H}(\Omega)$ and the functions F_k are linearly independent. In the latter case, we have the linear independence of every finite subsystem of $\{F_k\}_{k\in\kappa}$ and the property $\overline{\operatorname{span}\{F_k\}_{k\in\kappa}}^{\|\cdot\|_{\mathcal{H}(\Omega)}} = \mathcal{H}(\Omega)$. An example of a finite-dimensional space is $\mathcal{H}(\Omega) = \mathrm{Harm}_{0\ldots d}(\Omega)$ for an arbitrary $d \in \mathbb{N}_0$ with $\|\cdot\|_{\mathcal{H}(\Omega)} = \|\cdot\|_{\mathrm{L}^2(\Omega)}$. A simple infinite-dimensional case is $\mathrm{L}^2(\Omega)$ itself. We will get to know further examples later on.

We start with a distinction of some trial functions which is related to the uncertainty principle.

Definition 4.5.1 The basis $\{F_k\}_{k\in\kappa}$ is

(a) **bandlimited**, if, for each $k \in \kappa$, there exists $d_k \in \mathbb{N}_0$ such that the following inclusion holds true: $F_k \in \mathrm{Harm}_{0\ldots d_k}(\Omega)$.

(b) **spacelimited** or **locally supported**, if, for each $k \in \kappa$, there exists a proper measurable subset $S_k \subset \Omega$ in the sense that $\lambda(S_k) < 4\pi$ (λ is the Lebesgue measure) and $F_k(\xi) = 0$ for (almost) all $\xi \in \Omega \setminus S_k$.

Obviously, every basis of $\mathrm{Harm}_{0\ldots d}(\Omega)$ is bandlimited. Moreover, since each F_k is an element of $\mathrm{L}^2(\Omega)$, we may expand them into the Fourier series $F_k = \sum_{n=0}^{\infty}\sum_{j=-n}^{n} F_k^\wedge(n,j) Y_{n,j}$, which converges (at least) in the sense of $\|\cdot\|_{\mathrm{L}^2(\Omega)}$. Hence, the basis is bandlimited if and only if for each $k \in \kappa$ there exists d_k such that $F_k^\wedge(n,j) = 0$ for all $n > d_k$ and all $j \in \{-n, \ldots, n\}$. Clearly, such functions are polynomials.

Without knowledge of the uncertainty principle, we can easily see that 'bandlimited' and 'spacelimited' are properties of bases which are mutually exclusive, because polynomials only have a finite number of roots (except for the zero polynomial, which is, however, always linearly dependent).

The spherical harmonics basis has the outstanding property that $F_k^\wedge(n,j)$ vanishes for all pairs (n, j) except for one single pair. The opposite extremal situation is a function which is limited to a very small subdomain of Ω. There is, in fact, no optimal choice since this would lead to a function which is only non-vanishing at a single point. Such a function F_k would, however, be identified with the zero function due to $\|F_k\|_{\mathrm{L}^2(\Omega)} = 0$. This matter is connected to the non-existent Dirac delta function.

Nevertheless, in the bandlimited case, Slepian functions provide us with a well-known concept of optimal space localization.

Definition 4.5.2 Let $R \subset \Omega$ be a measurable subdomain and $d \in \mathbb{N}_0$ be a fixed integer. If $F \in \mathrm{Harm}_{0\ldots d}(\Omega)$ maximizes $\lambda_R(F) := \|F\|^2_{\mathrm{L}^2(R)} \|F\|^{-2}_{\mathrm{L}^2(\Omega)}$ among all functions in $\mathrm{Harm}_{0\ldots d}(\Omega)$, then F is called an **optimally space-localizing Slepian function** to the subdomain R and the maximal degree d. Instead of 'space-localizing', also 'space-concentrated' is common.

Clearly, it only makes sense that neither R nor $\Omega \setminus R$ should have measure zero.

We will see later that the maximization of $\lambda_R(F)$ leads us to an eigenvalue problem which enables us to construct a basis of $\mathrm{Harm}_{0\ldots d}(\Omega)$ which consists of functions with descending localization in R (i.e., decreasing $\lambda_R(F)$).

Another criterion for trial functions are their different kinds of isotropic or anisotropic variation on the sphere, that is, the distinction of rotationally symmetric functions from other functions (remember Definition 4.1.8 and the fact that $|\xi - \eta| = \sqrt{2(1 - \xi \cdot \eta)}$ for all $\xi, \eta \in \Omega$).

Definition 4.5.3 Functions from a basis system $\{F_k\}_{k \in \kappa}$ are called (spherical) **radial basis functions** if each F_k is a zonal function. In other words, for every $k \in \kappa$, there exists a point $\eta^{(k)} \in \Omega$ and a function $G_k \in \mathrm{L}^2[-1,1]$ such that $F_k(\xi) = G_k(\xi \cdot \eta^{(k)})$ for (almost) all $\xi \in \Omega$.

Often we have one generating function $G \in \mathrm{L}^2[-1,1]$ such that $F_k(\xi) = G(\xi \cdot \eta^{(k)})$ for (almost) all $\xi \in \Omega$. This is, however, not necessarily the case. Furthermore, it is clear that, in such a case, the points $\eta^{(k)}$ have to be pairwise distinct due to the required basis property of the F_k.

Since $G_k \in \mathrm{L}^2[-1,1]$, which is a reasonable assumption because of Theorem 4.1.9, we can expand F_k in terms of Legendre polynomials (remember Theorem 4.2.10) by

$$\begin{aligned} F_k(\xi) &= \sum_{n=0}^{\infty} \frac{2n+1}{2} \langle G_k, P_n \rangle_{\mathrm{L}^2[-1,1]} P_n\left(\xi \cdot \eta^{(k)}\right) \\ &= \sum_{n=0}^{\infty} \frac{2n+1}{4\pi} \left\langle G_k\left(\eta^{(k)} \cdot\right), P_n\left(\eta^{(k)} \cdot\right)\right\rangle_{\mathrm{L}^2(\Omega)} P_n\left(\xi \cdot \eta^{(k)}\right) \\ &= \sum_{n=0}^{\infty} \sum_{j=-n}^{n} \left\langle F_k, P_n\left(\eta^{(k)} \cdot\right)\right\rangle_{\mathrm{L}^2(\Omega)} Y_{n,j}\left(\eta^{(k)}\right) Y_{n,j}(\xi), \end{aligned}$$

where the convergence of the series holds true in the sense of $\mathrm{L}^2(\Omega)$, that is, not necessarily pointwise. This result shows us that the expansion coefficients of spherical radial basis functions must have the form

$$F_k^{\wedge}(n,j) = \left\langle F_k, P_n\left(\eta^{(k)} \cdot\right)\right\rangle_{\mathrm{L}^2(\Omega)} Y_{n,j}\left(\eta^{(k)}\right).$$

We will discuss the different kinds of basis functions briefly in the next sections.

4.6 Slepian Functions on the Sphere

The idea of maximizing the fraction $\|F\|_{\mathrm{L}^2(R)} / \|F\|_{\mathrm{L}^2(D)}$ to obtain functions F which are highly concentrated in the subdomain R of D was first developed for the case $D = \mathbb{R}$ in Landau and Pollak (1961); Slepian (1964); and Slepian and Pollak (1961). Later, also the spherical case, that is, the case where $D = \Omega$, was considered in Albertella et al. (1999), Dahlen and Simons (2008), Simons (2010), Simons and Dahlen (2006), Simons et al. (2006), Wieczorek and Simons (2005), and Wieczorek and Simons (2007). While scalar functions were considered first, vectorial Slepian functions on the sphere were later constructed in Jahn and Bokor (2012, 2013), Plattner et al. (2012), and Plattner and Simons (2014). Moreover, also tensorial Slepian functions have been investigated in Eshagh (2009), Michel et al. (2021), and Seibert (2018). The approach in Michel et al. (2021) and Seibert (2018) uses a generalized system of spherical harmonics, the so-called spin-weighted spherical harmonics, which have certain applications in quantum mechanics. The use of this system essentially enables the numerical implementation of the concept of Slepian functions, at least for the case that R is a spherical cap. It also provides a unified concept which includes the scalar, vector, and tensor case (with applicability in geosciences and cosmology) and also shows, therefore, a way to construct Slepian functions which are tensors of higher order.

Eventually, we want to mention that also Slepian functions on the ball, that is, for the case $D = B_R(0)$, were recently developed. For further details, the reader is referred to Khalid et al. (2016), Leweke et al. (2018b), and Maniar and Mitra (2005). Furthermore, a theory of Slepian functions in a general Hilbert space setting, which includes the choice of arbitrary (measurable) domains and the possibility to solve inverse problems instead of pure approximation problems, is derived in Michel and Simons (2017) based on the article by Plattner and Simons (2017), where the downward continuation of gravity or magnetic field data from regionally given gradients is solved by means of spherical Slepian functions.

Within this section, we show, as an example, the scalar spherical case. For the other cases, the reader is referred to the preceding listed references.

We consider the following scenario: we have a measurable subdomain $R \subset \Omega$ of the unit sphere and a chosen bandlimit $L \in \mathbb{N}$. For all $F \in \mathrm{Harm}_{0\ldots L}(\Omega)$, we calculate the **energy ratio** $\lambda_R(F) = \|F\|^2_{\mathrm{L}^2(R)}\|F\|^{-2}_{\mathrm{L}^2(\Omega)}$. Obviously, we have $0 \leq \lambda_R(F) \leq 1$.

With the expansion of F in a spherical harmonics basis, where we use the abbreviations $F_{n,j} := \langle F, Y_{n,j}\rangle_{\mathrm{L}^2(\Omega)}$ and $f := \left(F_{0,0}, F_{1,-1,\ldots}, F_{L,L}\right)^{\mathrm{T}}$, we get, due to the Parseval identity in $\mathrm{L}^2(\Omega)$, that

$$\lambda_R(F) = \frac{\sum_{n=0}^{L}\sum_{j=-n}^{n}\sum_{m=0}^{L}\sum_{k=-m}^{m} F_{n,j}F_{m,k}\langle Y_{n,j}, Y_{m,k}\rangle_{\mathrm{L}^2(R)}}{\sum_{n=0}^{L}\sum_{j=-n}^{n} F_{n,j}^2} = \frac{f^{\mathrm{T}}Mf}{|f|^2}$$

with the matrix $M \in \mathbb{R}^{(L+1)^2\times(L+1)^2}$ given by

$$M = \left(\langle Y_{n,j}, Y_{m,k}\rangle_{\mathrm{L}^2(R)}\right)_{\substack{n=0,\ldots,L;\, j=-n,\ldots,n\\ m=0,\ldots,L;\, k=-m,\ldots,m}}. \tag{4.36}$$

As a result, we have to determine the eigenvectors of the matrix M, which is symmetric because of the symmetry of the inner product.

Theorem 4.6.1 *The aforementioned matrix M is symmetric and positive definite. We can choose an orthonormal basis of $\mathbb{R}^{(L+1)^2}$ which consists of eigenvectors $g^{(1)}, \ldots, g^{(L+1)^2}$ of M. All eigenvalues of M are real and positive. Moreover, if $g^{(k)}$ is such an eigenvector of M to the eigenvalue λ_k, and if we define the functions $G_k(\xi) := \sum_{n=0}^{L} \sum_{j=-n}^{n} g_{n,j}^{(k)} Y_{n,j}(\xi)$, $\xi \in \Omega$, where $g^{(k)} = \left(g_{0,0}^{(k)}, g_{1,-1}^{(k)}, \ldots, g_{L,L}^{(k)}\right)^{\mathrm{T}}$, then $\lambda_R(G_k) = \lambda_k$. Furthermore, the set $\left\{G_1, \ldots, G_{(L+1)^2}\right\}$ is an orthonormal basis of $\left(\mathrm{Harm}_{0\ldots L}(\Omega), \langle\cdot,\cdot\rangle_{\mathrm{L}^2(\Omega)}\right)$.*

Proof It is a basic result of linear algebra that M is, as a Gramian matrix, symmetric and positive definite and hence has an orthonormal system of $(L+1)^2$ eigenvectors with real, non-negative eigenvalues. Thus, we have

$$\lambda_R(G_k) = \frac{\left(g^{(k)}\right)^{\mathrm{T}} M g^{(k)}}{\left|g^{(k)}\right|^2} = \left(g^{(k)}\right)^{\mathrm{T}} \left(\lambda_k g^{(k)}\right) = \lambda_k \left(g^{(k)}\right)^{\mathrm{T}} g^{(k)} = \lambda_k.$$

Eventually, the Parseval identity yields $\langle G_k, G_l\rangle_{\mathrm{L}^2(\Omega)} = \left(g^{(k)}\right)^{\mathrm{T}} g^{(l)} = \delta_{kl}$. □

Definition 4.6.2 By $g^{(1)}, \ldots, g^{(N)}$ with $N = (L+1)^2$, we denote an orthonormal system of eigenvectors of the matrix M in (4.36). The associated eigenvalues are denoted by $\lambda_1, \ldots, \lambda_N$. Eigenvectors and eigenvalues are assumed to be sorted such that $\lambda_1 \geq \cdots \geq \lambda_N > 0$. Furthermore, we define the **Slepian basis** $\{G_1, \ldots, G_N\}$ by $G_k(\xi) = \sum_{n=0}^{L} \sum_{j=-n}^{n} g_{n,j}^{(k)} Y_{n,j}(\xi)$, $\xi \in \Omega$, $k = 1, \ldots, N$. Here, we assume an appropriate rearrangement of the double index of $g^{(k)}$ to a single index in $\{1, \ldots, N\}$, for example, like in Theorem 4.6.1.

Clearly, G_1 is an optimally space-localizing Slepian function (see Definition 4.5.2). Slepian functions are particularly important if we only have data of an unknown signal F in the subdomain R or if we are only interested in analyzing F in R. The former occurs often in Earth sciences due to the non-uniform distribution of observation stations. A typical example of the latter is the investigation of the increasing or decreasing mass of an ice shield, for instance, of Greenland or Antarctica.

For both cases, it is also helpful to know the following theorem.

Theorem 4.6.3 *The Slepian basis $G_1, \ldots, G_N$ is also orthogonal in $\mathrm{L}^2(R)$. Moreover, if $F \in \mathrm{Harm}_{0\ldots L}(\Omega)$ is an arbitrary function, then its restriction to R can be represented by*

$$F(\xi) = \sum_{k=1}^{N} \frac{1}{\lambda_k} \int_R F(\eta) G_k(\eta)\, \mathrm{d}\omega(\eta)\, G_k(\xi), \quad \xi \in R.$$

Proof Due to the eigenvalue problem and the Parseval identity, we have, for the restrictions of the Slepian basis to R, that

$$\begin{aligned}\langle G_k, G_l\rangle_{\mathrm{L}^2(R)} &= \sum_{n,m=0}^{L} \sum_{i=-n}^{n} \sum_{j=-m}^{m} g_{n,i}^{(k)} g_{m,j}^{(l)} \left\langle Y_{n,i}, Y_{m,j}\right\rangle_{\mathrm{L}^2(R)} \\ &= \left(g^{(k)}\right)^{\mathrm{T}} M g^{(l)} = \lambda_l \left(g^{(k)}\right)^{\mathrm{T}} g^{(l)} = \lambda_l \delta_{kl}. \end{aligned} \tag{4.37}$$

Hence, if $F \in \mathrm{Harm}_{0\ldots L}(\Omega)$, then its restriction $F|_R$ to R can be represented in the $\mathrm{L}^2(R)$-orthogonal system $G_1|_R, \ldots, G_N|_R$ by

$$F(\xi) = \sum_{k=1}^{N} \left\langle F, \frac{G_k}{\|G_k\|_{\mathrm{L}^2(R)}} \right\rangle_{\mathrm{L}^2(R)} \frac{G_k(\xi)}{\|G_k\|_{\mathrm{L}^2(R)}}, \qquad \xi \in R.$$

The $\mathrm{L}^2(R)$-norm of each G_k can be found in (4.37), which leads us to the desired result. □

Hence, the Slepian basis is not only an $\mathrm{L}^2(\Omega)$-orthonormal basis of the bandlimited space $\mathrm{Harm}_{0\ldots L}(\Omega)$, it is also an $\mathrm{L}^2(R)$-orthogonal basis for the restrictions of all such bandlimited functions to R.

We have seen so far that the Slepian functions have coefficient vectors which are eigenvectors of the matrix M in (4.36). In fact, the Slepian functions are also eigenfunctions of a Fredholm integral operator of the first kind.

Theorem 4.6.4 *The Fredholm integral operator* $\mathcal{T}_K : \mathcal{X} \to \mathcal{X}$ *with* $\mathcal{X} = \mathrm{Harm}_{0\ldots L}(\Omega)$, $\langle\cdot,\cdot\rangle_{\mathcal{X}} = \langle\cdot,\cdot\rangle_{\mathrm{L}^2(\Omega)}$ *and* $(\mathcal{T}_K F)(\xi) := \int_R K(\xi \cdot \eta) F(\eta)\,\mathrm{d}\omega(\eta)$ *for all* $\xi \in \Omega$, *where* $K(t) = \sum_{n=0}^{L} \frac{2n+1}{4\pi} P_n(t)$, $t \in [-1,1]$, *has an orthonormal system of eigenfunctions which constitute a basis of* $\mathcal{X}$. *Moreover, all eigenvalues are real and positive. Every such eigenfunction basis is a Slepian basis and vice versa.*

Proof Every F in $\mathcal{X}$ can be represented by $F(\xi) = \sum_{n=0}^{L} \sum_{j=-n}^{n} F_{n,j} Y_{n,j}(\xi)$, $\xi \in \Omega$, such that we get, by using the addition theorem (Theorem 4.2.9),

$$\begin{aligned}(\mathcal{T}_K F)(\xi) &= \int_R \left(\sum_{m=0}^{L} \sum_{k=-m}^{m} Y_{m,k}(\xi) Y_{m,k}(\eta) \right) \left(\sum_{n=0}^{L} \sum_{j=-n}^{n} F_{n,j} Y_{n,j}(\eta) \right) \mathrm{d}\omega(\eta) \\ &= \sum_{m,n=0}^{L} \sum_{k=-m}^{m} \sum_{j=-n}^{n} Y_{m,k}(\xi) F_{n,j} \int_R Y_{m,k}(\eta) Y_{n,j}(\eta)\,\mathrm{d}\omega(\eta).\end{aligned}$$

Hence, we have $\mathcal{T}_K F = \lambda F$ if and only if we have

$$\sum_{n=0}^{L} \sum_{j=-n}^{n} \int_R Y_{m,k}(\eta) Y_{n,j}(\eta)\,\mathrm{d}\omega(\eta)\, F_{n,j} = \lambda F_{m,k}$$

for all $m = 0, \ldots, L$ and all $k = -m, \ldots, m$. The latter is exactly the eigenvalue problem which leads to the Slepian basis. □

Remark 4.6.5 In the literature, the following alternative has also been discussed: instead of using a bandlimit, we seek a (to R) spacelimited function $F \in \mathrm{L}^2(\Omega)$, that is, $F(\xi) = 0$ for almost every $\xi \in \Omega \setminus R$, which is **optimally band-localizing** (or: optimally band-concentrated), that is, the ratio

$$\tilde{\lambda}_L(F) := \frac{\sum_{n=0}^{L} \sum_{j=-n}^{n} \left\langle F, Y_{n,j} \right\rangle_{\mathrm{L}^2(\Omega)}^2}{\sum_{n=0}^{\infty} \sum_{j=-n}^{n} \left\langle F, Y_{n,j} \right\rangle_{\mathrm{L}^2(\Omega)}^2} \tag{4.38}$$

has to be maximized. It is well known that this optimization problem is actually not new but easily solvable with the previous results. To see this, we observe that

$$\begin{aligned}
&\sum_{n=0}^{L} \sum_{j=-n}^{n} \left\langle F, Y_{n,j} \right\rangle_{\mathrm{L}^2(\Omega)}^2 \\
&\quad = \sum_{n=0}^{L} \sum_{j=-n}^{n} \int_R F(\xi) Y_{n,j}(\xi) \, \mathrm{d}\omega(\xi) \int_R F(\eta) Y_{n,j}(\eta) \, \mathrm{d}\omega(\eta) \\
&\quad = \int_R F(\xi) \int_R \left(\sum_{n=0}^{L} \sum_{j=-n}^{n} Y_{n,j}(\xi) Y_{n,j}(\eta) \right) F(\eta) \, \mathrm{d}\omega(\eta) \, \mathrm{d}\omega(\xi).
\end{aligned} \tag{4.39}$$

Moreover, the denominator in (4.38) is the same as $\|F\|_{\mathrm{L}^2(\Omega)}^2$. If we assume that this norm equals 1, we are left with the maximization of (4.39). Consulting Theorem 4.6.4, we can use the operator $\mathcal{T}_K$ (stated now for $\mathcal{X} = \mathrm{L}^2(\Omega)$) to say that $\langle F, \mathcal{T} F \rangle_{\mathrm{L}^2(R)}$ has to be maximized. This is again an eigenvalue problem, which leads us to $G_1|_R, \ldots, G_N|_R$ as (some) eigenfunctions.

Numerical experiments reported in previous publications on Slepian functions show that the majority of the eigenvalues λ_k can be split up into values slightly below 1 and the rest which are slightly above 0. On the one hand, this shows that a clear distinction of the obtained Slepian functions into almost only well spatially concentrated or badly spatially concentrated functions is possible, which is appropriate for applications with regional aspects. On the other hand, this distribution of the eigenvalues causes a large condition number of the matrix M and makes it, therefore, numerically challenging to solve the corresponding eigenvalue problem. However, for some particular regions R such as spherical caps, it has been shown that the eigenvectors of M are also eigenvectors of an alternative eigenvalue problem with a more 'harmless' distribution of the eigenvalues. For the case of scalar functions on the sphere, this was elaborated in Grünbaum et al. (1982), Simons (2010), and Simons et al. (2006). Moreover, there exists a unified concept in Michel et al. (2021) and Seibert (2018) which includes this scalar case, the also known vectorial case, and a new result for the tensorial case. We restrict also here our considerations to the scalar case, but follow the approach of Michel et al. (2021) and Seibert (2018).

Before we proceed, we need a lemma about Legendre functions; see also Robin (1957, pp. 100–101).

Lemma 4.6.6 *The Legendre functions satisfy*

$$(t^2 - 1) P_{n,j}'(t) = n t P_{n,j}(t) - (n + j) P_{n-1,j}(t), \tag{4.40}$$

$t \in [-1,1], n \in \mathbb{N}, j \in \{0, \ldots, n\}$, where $P_{n-1,n}(t) := 0$ for all t. Hence, for the fully normalized spherical harmonics, the identity

$$(t^2-1)\frac{\partial}{\partial t} Y_{n,j}(\xi(\varphi,t)) \\ = ntY_{n,j}(\xi(\varphi,t)) - \sqrt{\frac{(n^2-j^2)(2n+1)}{2n-1}}\, Y_{n-1,j}(\xi(\varphi,t)) \tag{4.41}$$

is valid, where $\xi(\varphi,t)$ is the representation of $\xi \in \Omega$ in the polar coordinates (φ,t) and we have $t \in [-1,1]$, $n \in \mathbb{N}$, $j \in \{-n, \ldots, n\}$.

Proof Remember the following (see Corollary 3.4.16):

$$P_{n,j}(t) = (1-t^2)^{j/2} \frac{\mathrm{d}^j}{\mathrm{d}t^j} P_n(t), \quad t \in [-1,1]. \tag{4.42}$$

By differentiating this identity, we obtain

$$P'_{n,j}(t) = \frac{j}{2}(1-t^2)^{(j-2)/2} \cdot (-2t) \cdot \frac{\mathrm{d}^j}{\mathrm{d}t^j} P_n(t) + (1-t^2)^{j/2} \frac{\mathrm{d}^{j+1}}{\mathrm{d}t^{j+1}} P_n(t)$$

such that

$$(t^2-1)P'_{n,j}(t) = jtP_{n,j}(t) - \sqrt{1-t^2}\, P_{n,j+1}(t). \tag{4.43}$$

Note that this is also valid in the case $j = n$.

We now have a look at Corollary 3.4.21. Equation (3.117) tells us that

$$(2n+1)(t^2-1)P'_n(t) = n(n+1)\left(P_{n+1}(t) - P_{n-1}(t)\right)$$

for all $t \in [-1,1]$ and $n \in \mathbb{N}$. Then we differentiate this equation j times and take into account the Leibniz rule (Theorem 2.3.6). We get

$$(2n+1)\left[(t^2-1)\frac{\mathrm{d}^{j+1}}{\mathrm{d}t^{j+1}} + 2jt\frac{\mathrm{d}^j}{\mathrm{d}t^j} + \frac{j(j-1)}{2} \cdot 2 \cdot \frac{\mathrm{d}^{j-1}}{\mathrm{d}t^{j-1}}\right] P_n(t) \\ = n(n+1)\left(\frac{\mathrm{d}^j}{\mathrm{d}t^j} P_{n+1}(t) - \frac{\mathrm{d}^j}{\mathrm{d}t^j} P_{n-1}(t)\right). \tag{4.44}$$

Note that the coefficient of the $(j-1)$th derivative vanishes for $j = 0$ such that (4.44) is also valid in this case. We now apply another equation from Corollary 3.4.21, namely (3.113), which implies that

$$(2n+1)\frac{\mathrm{d}^{j-1}}{\mathrm{d}t^{j-1}} P_n(t) = \frac{\mathrm{d}^j}{\mathrm{d}t^j} P_{n+1}(t) - \frac{\mathrm{d}^j}{\mathrm{d}t^j} P_{n-1}(t),$$

if $j \geq 1$. Hence, we rewrite (4.44) as

$$(2n+1)\left[(t^2-1)\frac{\mathrm{d}^{j+1}}{\mathrm{d}t^{j+1}}P_n(t)+2jt\frac{\mathrm{d}^j}{\mathrm{d}t^j}P_n(t)\right] + j(j-1)\left(\frac{\mathrm{d}^j}{\mathrm{d}t^j}P_{n+1}(t)-\frac{\mathrm{d}^j}{\mathrm{d}t^j}P_{n-1}(t)\right) = n(n+1)\left(\frac{\mathrm{d}^j}{\mathrm{d}t^j}P_{n+1}(t)-\frac{\mathrm{d}^j}{\mathrm{d}t^j}P_{n-1}(t)\right),$$

which we simplify by first multiplying with $(1-t^2)^{j/2}$ and then applying (4.42) to

$$(2n+1)\left(-\sqrt{1-t^2}\,P_{n,j+1}(t)+2jt\,P_{n,j}(t)\right) = [n(n+1)-j(j-1)]\left(P_{n+1,j}(t)-P_{n-1,j}(t)\right). \tag{4.45}$$

As the next steps, we observe that $n(n+1)-j(j-1)=n^2+n-j^2+j=(n+j)(n-j+1)$ and insert (4.43) into (4.45). This leads us to

$$(2n+1)\left[(t^2-1)P'_{n,j}(t)+jt\,P_{n,j}(t)\right] = (n+j)(n-j+1)\left(P_{n+1,j}(t)-P_{n-1,j}(t)\right). \tag{4.46}$$

Consulting Theorem 4.2.22, we see that

$$(n+1-j)P_{n+1,j}(t)-(2n+1)t\,P_{n,j}(t)=-(n+j)P_{n-1,j}(t).$$

By subtracting $(n+1-j)P_{n-1,j}(t)$, we obtain

$$(n+1-j)(P_{n+1,j}(t)-P_{n-1,j}(t))=(2n+1)(t\,P_{n,j}(t)-P_{n-1,j}(t)).$$

Let us now insert this identity into (4.46). Then we see that

$$(2n+1)\left[(t^2-1)P'_{n,j}(t)+jt\,P_{n,j}(t)\right] = (2n+1)(n+j)\left(t\,P_{n,j}(t)-P_{n-1,j}(t)\right).$$

Hence, $(t^2-1)P'_{n,j}(t)=nt\,P_{n,j}(t)-(n+j)P_{n-1,j}(t)$. This is exactly (4.40), which is proved now.

For proving (4.41), we remember the definition of the fully normalized spherical harmonics (Definition 3.4.24). It is clear that we only have to take care of the normalizing factor. Comparing these factors for degrees n and $n-1$, we see that

$$\sqrt{\frac{(2n+1)(n-|j|)!\,(2-\delta_{j0})}{4\pi(n+|j|)!}} = \sqrt{\frac{(2n-1)(n-1-|j|)!\,(2-\delta_{j0})}{4\pi(n-1+|j|)!}}\sqrt{\frac{2n+1}{2n-1}\frac{n-|j|}{n+|j|}}.$$

Hence, we get the following (where ξ stands for $\xi(\varphi,t)$ here):

$$(t^2-1)\frac{\partial}{\partial t}Y_{n,j}(\xi) = ntY_{n,j}(\xi) - (n+|j|)\sqrt{\frac{2n+1}{2n-1}\frac{n-|j|}{n+|j|}}\,Y_{n-1,j}(\xi)$$
$$= ntY_{n,j}(\xi) - \sqrt{\frac{2n+1}{2n-1}(n^2-j^2)}\,Y_{n-1,j}(\xi).$$

Hence, also the proof of (4.41) is complete. Note that the coefficient of $Y_{n-1,j}$ vanishes for $|j|=n$. □

Now we can draw our attention towards Slepian functions on a spherical cap.

Definition 4.6.7 A **spherical cap** corresponding to a centre $\eta \in \Omega$ and a parameter $b \in [-1,1[$ is given by $C_{\eta,b} := \{\xi \in \Omega \mid \xi \cdot \eta \geq b\}$.

Due to the rotational symmetry of the sphere, it suffices to consider the case where $\eta = \varepsilon^3$. Then the cap $C_{\varepsilon^3,b}$ is represented in polar coordinates by $\varphi \in [0,2\pi[$ and $\eta \cdot \xi = t \in [b,1]$.

We come now to the main result regarding the alternative eigenvalue problem.

Theorem 4.6.8 *Let the kernel* $K\colon [-1,1] \to \mathbb{R}$ *be given as before by* $K(t) = \sum_{n=0}^{L} \frac{2n+1}{4\pi} P_n(t)$, $t \in [-1,1]$, *the region* R *be chosen as the spherical cap* $R = C_{\varepsilon^3,b}$ *with* $b \in]-1,1[$, *and the operator* $\mathcal{J}\colon \mathrm{C}^{(2)}(\Omega) \to \mathrm{C}(\Omega)$ *be defined by*

$$\mathcal{J}_\xi F(\xi) := (b-t)\Delta_\xi^* F(\xi) + (t^2-1)\frac{\partial}{\partial t}F(\xi(\varphi,t)) - L(L+2)tF(\xi),$$

where $\xi(\varphi,t)$ *is the representation of* $\xi \in \Omega$ *in polar coordinates. For the consequent part, let either* $D = \Omega$ *or* $D = R$ *be chosen. Then* $\mathcal{J}$ *is self-adjoint with respect to* $\langle\cdot,\cdot\rangle_{\mathrm{L}^2(D)}$. *Moreover,*

$$\mathcal{J}_\xi \int_D K(\xi\cdot\eta)F(\eta)\,\mathrm{d}\omega(\eta) = \int_D \mathcal{J}_\xi K(\xi\cdot\eta)F(\eta)\,\mathrm{d}\omega(\eta) \tag{4.47}$$

for all $\xi \in \Omega$ *and all* $F \in \mathrm{L}^2(D)$. *As a result, the operators* $\mathcal{J}$ *and* $\mathcal{T}_K$ *with* $(\mathcal{T}_K F)(\xi) := \langle K(\xi\cdot), F\rangle_{\mathrm{L}^2(D)}$ *satisfy* $\mathcal{T}_K\mathcal{J} = \mathcal{J}\mathcal{T}_K$ *on* $\mathrm{C}^{(2)}(\Omega)$. *Because of these properties, the operator* $\mathcal{J}$ *is called a* ***commuting differential operator***.

Proof We prove the properties step by step.

(1) The operator is self-adjoint:

Let $D := C_{\varepsilon^3,\tau}$ be an arbitrary spherical cap where $\tau = b$ represents our considered particular cap and $\tau = -1$ stands for the whole sphere. We now separate the integral

$$\int_D F(\xi)\mathcal{J}_\xi G(\xi)\,\mathrm{d}\omega(\xi) \tag{4.48}$$
$$= \int_0^{2\pi}\int_\tau^1 F(\xi)\left[(b-t)\Delta_\xi^* G(\xi) + (t^2-1)\frac{\partial G(\xi)}{\partial t} - L(L+2)tG(\xi)\right]\mathrm{d}t\,\mathrm{d}\varphi$$

into several parts. For the first part, we apply Green's second surface identity (Theorem 4.1.11) and get

$$\begin{aligned}
&\int_D F(\xi)(b-t)\Delta_\xi^* G(\xi)\,\mathrm{d}\omega(\xi) \\
&\quad = \int_D \Delta_\xi^* [F(\xi)(b-t)]\, G(\xi)\,\mathrm{d}\omega(\xi) \\
&\qquad - \int_{\partial D} \frac{\partial}{\partial \nu}[F(\xi)(b-t)]\, G(\xi) - F(\xi)(b-t)\frac{\partial G}{\partial \nu}(\xi)\,\mathrm{dl}(\xi)\,. \qquad (4.49)
\end{aligned}$$

In the case $D = \Omega$, the preceding boundary integral vanishes. For the case of a (proper) spherical cap, the outer unit normal is $-\varepsilon^t$ and the curve ∂D is parameterized by $\varphi \in [0, 2\pi]$, $t = \tau$. Moreover, the radius of this circle is $\sqrt{1-\tau^2}$. Hence, the parameterization satisfies

$$\xi = \begin{pmatrix} \sqrt{1-\tau^2}\cos\varphi \\ \sqrt{1-\tau^2}\sin\varphi \\ \tau \end{pmatrix}, \qquad \left|\frac{\partial}{\partial \varphi}\xi\right| = \sqrt{1-\tau^2},$$

and we have $\frac{\partial}{\partial \nu} = -\varepsilon^t \cdot \nabla^* = -\sqrt{1-t^2}\,\frac{\partial}{\partial t}$. With this constellation and the use of Theorem 3.4.1, the boundary integral can be calculated as

$$\begin{aligned}
&\int_{\partial D} \frac{\partial}{\partial \nu}[F(\xi)(b-t)]\,G(\xi) - F(\xi)(b-t)\frac{\partial G}{\partial \nu}(\xi)\,\mathrm{dl}(\xi) \\
&\quad = \int_0^{2\pi} -(1-t^2)\frac{\partial}{\partial t}[F(\xi)(b-t)]\,G(\xi)\,\mathrm{d}\varphi\Big|_{t=\tau} \\
&\qquad + \int_0^{2\pi} F(\xi)(b-t)(1-t^2)\,\frac{\partial G}{\partial t}(\xi)\,\mathrm{d}\varphi\Big|_{t=\tau} \\
&\quad = \int_0^{2\pi} (1-t^2)\left[F(\xi)(b-t)\frac{\partial G}{\partial t}(\xi) - \frac{\partial F}{\partial t}(\xi)(b-t)G(\xi)\right]\mathrm{d}\varphi\Big|_{t=\tau} \\
&\qquad + \int_0^{2\pi} (1-t^2)F(\xi)G(\xi)\,\mathrm{d}\varphi\Big|_{t=\tau}\,.
\end{aligned}$$

For the spherical cap, we have $\tau = b$ such that (note the minus!)

$$\begin{aligned}
&-\int_{\partial D} \frac{\partial}{\partial \nu}[F(\xi)(b-t)]\,G(\xi) - F(\xi)(b-t)\,\frac{\partial G}{\partial \nu}(\xi)\,\mathrm{dl}(\xi) \\
&\quad = \left[(t^2-1)\int_0^{2\pi} F(\xi)G(\xi)\,\mathrm{d}\varphi\right]_{t=\tau}\,. \qquad (4.50)
\end{aligned}$$

The right-hand side vanishes in the case of the sphere, where $\tau = -1$, such that we can also use (4.50) for the cap and the whole sphere.

We continue now with an auxiliary calculation:

$$
\begin{aligned}
&\Delta^*_\xi\left[F(\xi)(b-t)\right]\\
&= (-2t)\frac{\partial}{\partial t}\left[F(\xi)(b-t)\right] + (1-t^2)\frac{\partial^2}{\partial t^2}\left[F(\xi)(b-t)\right] + \frac{b-t}{1-t^2}\frac{\partial^2}{\partial \varphi^2}F(\xi)\\
&= (-2t)\left[\frac{\partial}{\partial t}F(\xi)(b-t) - F(\xi)\right]\\
&\quad + (1-t^2)\left[\frac{\partial^2}{\partial t^2}F(\xi)(b-t) - 2\frac{\partial}{\partial t}F(\xi)\right] + \frac{b-t}{1-t^2}\frac{\partial^2}{\partial \varphi^2}F(\xi)\\
&= (b-t)\frac{\partial}{\partial t}\left[(1-t^2)\frac{\partial}{\partial t}F(\xi)\right] + 2tF(\xi) + 2(t^2-1)\frac{\partial}{\partial t}F(\xi)\\
&\quad + \frac{b-t}{1-t^2}\frac{\partial^2}{\partial \varphi^2}F(\xi)\\
&= (b-t)\Delta^*_\xi F(\xi) + 2tF(\xi) + 2(t^2-1)\frac{\partial}{\partial t}F(\xi)\\
&= (b-t)\Delta^*_\xi F(\xi) + \frac{\partial}{\partial t}\left[(t^2-1)F(\xi)\right] + (t^2-1)\frac{\partial}{\partial t}F(\xi).
\end{aligned}
\tag{4.51}
$$

For the second term on the right-hand side of (4.48), integration by parts leads us to

$$
\begin{aligned}
&\int_\tau^1 F(\xi)(t^2-1)\frac{\partial}{\partial t}G(\xi)\,\mathrm{d}t \\
&\qquad = F(\xi)(t^2-1)G(\xi)\Big|_{t=\tau}^{t=1} - \int_\tau^1 \frac{\partial}{\partial t}\left[F(\xi)(t^2-1)\right]G(\xi)\,\mathrm{d}t.
\end{aligned}
\tag{4.52}
$$

We now use (4.49), (4.51), (4.50), and (4.52) in (4.48). Hence,

$$
\begin{aligned}
&\int_D F(\xi)\mathcal{J}_\xi G(\xi)\,\mathrm{d}\omega(\xi)\\
&\quad = \int_D G(\xi)(b-t)\Delta^*_\xi F(\xi)\,\mathrm{d}\omega(\xi) + \int_D G(\xi)\frac{\partial}{\partial t}\left[(t^2-1)F(\xi)\right]\mathrm{d}\omega(\xi)\\
&\quad\quad + \int_D G(\xi)(t^2-1)\frac{\partial}{\partial t}F(\xi) - \frac{\partial}{\partial t}\left[F(\xi)(t^2-1)\right]G(\xi)\,\mathrm{d}\omega(\xi)\\
&\quad\quad - \int_D L(L+2)t\,F(\xi)G(\xi)\,\mathrm{d}\omega(\xi)\\
&\quad = \int_D G(\xi)\mathcal{J}_\xi F(\xi)\,\mathrm{d}\omega(\xi).
\end{aligned}
$$

Thus, $\mathcal{J}$ is self-adjoint in the $\mathrm{L}^2(R)$- and in the $\mathrm{L}^2(\Omega)$-sense.

(2) Interchanging the operator and the integral:

We want to prove (4.47). This can be done by simply inserting the explicit formula of K and the addition theorem for spherical harmonics (Theorem 4.2.9):

$$\begin{aligned}\mathcal{J}_\xi \int_D K(\xi\cdot\eta)F(\eta)\,\mathrm{d}\omega(\eta) &= \mathcal{J}_\xi \int_D \sum_{n=0}^{L}\sum_{j=-n}^{n} Y_{n,j}(\xi)Y_{n,j}(\eta)F(\eta)\,\mathrm{d}\omega(\eta)\\ &= \sum_{n=0}^{L}\sum_{j=-n}^{n} \mathcal{J}_\xi Y_{n,j}(\xi)\int_D Y_{n,j}(\eta)F(\eta)\,\mathrm{d}\omega(\eta)\\ &= \sum_{n=0}^{L}\sum_{j=-n}^{n} \int_D \mathcal{J}_\xi Y_{n,j}(\xi)Y_{n,j}(\eta)F(\eta)\,\mathrm{d}\omega(\eta)\\ &= \int_D \sum_{n=0}^{L}\sum_{j=-n}^{n} \mathcal{J}_\xi Y_{n,j}(\xi)Y_{n,j}(\eta)F(\eta)\,\mathrm{d}\omega(\eta)\\ &= \int_D \mathcal{J}_\xi K(\xi\cdot\eta)F(\eta)\,\mathrm{d}\omega(\eta).\end{aligned}$$

(3) A symmetry property:

Before we can proceed to the commutation of the operators, we first have to prove that $\mathcal{J}_\xi K(\xi\cdot\eta) = \mathcal{J}_\eta K(\xi\cdot\eta)$. We remember the addition theorem for spherical harmonics (Theorem 4.2.9) and choose the fully normalized spherical harmonics as a particular basis in $K(\xi\cdot\eta) = \sum_{n=0}^{L}\sum_{j=-n}^{n} Y_{n,j}(\xi)Y_{n,j}(\eta)$, $\xi,\eta\in\Omega$. The application of $\mathcal{J}$ yields $\mathcal{J}_\xi K(\xi\cdot\eta) = \sum_{n=0}^{L}\sum_{j=-n}^{n}\left(\mathcal{J}Y_{n,j}\right)(\xi)Y_{n,j}(\eta)$. In view of the definition of $\mathcal{J}$, it is obviously useful to apply the eigenfunction property from Theorem 4.2.19 and the recurrence formula (4.41) from Lemma 4.6.6. We get (with polar coordinates (φ_ξ,t_ξ) and (φ_η,t_η) of ξ and η, respectively) that

$$\begin{aligned}\mathcal{J}_\xi Y_{n,j}(\xi) &= (b-t_\xi)\Delta_\xi^* Y_{n,j}(\xi) + \left(t_\xi^2-1\right)\frac{\partial}{\partial t_\xi}Y_{n,j}(\xi) - L(L+2)t_\xi Y_{n,j}(\xi)\\ &= -(b-t_\xi)n(n+1)Y_{n,j}(\xi) - L(L+2)t_\xi Y_{n,j}(\xi)\\ &\quad + nt_\xi Y_{n,j}(\xi) - (2n+1)c_{n,j}Y_{n-1,j}(\xi),\\ \mathcal{J}_\eta Y_{n,j}(\eta) &= -(b-t_\eta)n(n+1)Y_{n,j}(\eta) - L(L+2)t_\eta Y_{n,j}(\eta)\\ &\quad + nt_\eta Y_{n,j}(\eta) - (2n+1)c_{n,j}Y_{n-1,j}(\eta),\end{aligned}$$

where we abbreviated $c_{n,j} := \sqrt{(n^2-j^2)(2n+1)^{-1}(2n-1)^{-1}}$. Hence,

$$
\begin{aligned}
&\mathcal{J}_\xi K(\xi\cdot\eta) - \mathcal{J}_\eta K(\xi\cdot\eta)\\
&\quad= \sum_{n=0}^{L}\sum_{j=-n}^{n}\left[(\mathcal{J}Y_{n,j})(\xi)Y_{n,j}(\eta) - Y_{n,j}(\xi)(\mathcal{J}Y_{n,j})(\eta)\right]\\
&\quad= (t_\xi - t_\eta)\sum_{n=0}^{L}\sum_{j=-n}^{n}\left[n(n+1) - L(L+2) + n\right]Y_{n,j}(\xi)Y_{n,j}(\eta)\\
&\qquad -\sum_{n=0}^{L}\sum_{j=-n}^{n}(2n+1)c_{n,j}\left(Y_{n-1,j}(\xi)Y_{n,j}(\eta) - Y_{n,j}(\xi)Y_{n-1,j}(\eta)\right).
\end{aligned}
$$

The last term has the form of the right-hand side of the Christoffel–Darboux formula (see Theorem 4.2.24). Together with minor simplifications of the other summation term, we get

$$
\begin{aligned}
&\mathcal{J}_\xi K(\xi\cdot\eta) - \mathcal{J}_\eta K(\xi\cdot\eta)\\
&\quad= (t_\xi - t_\eta)\sum_{n=0}^{L-1}\sum_{j=-n}^{n}\left[n(n+2) - L(L+2)\right]Y_{n,j}(\xi)Y_{n,j}(\eta)\\
&\qquad + (t_\xi - t_\eta)\sum_{n=0}^{L}\sum_{j=-n}^{n}(2n+1)\sum_{k=|j|}^{n-1}Y_{k,j}(\xi)Y_{k,j}(\eta).
\end{aligned}
$$

With lengthy but easy calculations (see, e.g., Seibert, 2018, corollary 8.2.6), one can deduce for the triple summation $\sum_{n=0}^{L}\sum_{j=-n}^{n}\sum_{k=|j|}^{n-1}$ that it can be replaced by the summation $\sum_{k=0}^{L-1}\sum_{j=-k}^{k}\sum_{n=k+1}^{L}$. Hence,

$$
\begin{aligned}
&\mathcal{J}_\xi K(\xi\cdot\eta) - \mathcal{J}_\eta K(\xi\cdot\eta)\\
&\quad= (t_\xi - t_\eta)\sum_{n=0}^{L-1}\sum_{j=-n}^{n}\left[n(n+2) - L(L+2)\right]Y_{n,j}(\xi)Y_{n,j}(\eta)\\
&\qquad + (t_\xi - t_\eta)\sum_{k=0}^{L-1}\sum_{j=-k}^{k}\sum_{n=k+1}^{L}(2n+1)Y_{k,j}(\xi)Y_{k,j}(\eta).
\end{aligned}
$$

Note that, in the last term, the summation with respect to n can be separated from the other sums. Moreover, we can apply the addition theorem for spherical harmonics (Theorem 4.2.9). We get

$$
\begin{aligned}
&\mathcal{J}_\xi K(\xi\cdot\eta) - \mathcal{J}_\eta K(\xi\cdot\eta)\\
&\quad= (t_\xi - t_\eta)\sum_{n=0}^{L-1}\left[n(n+2) - L(L+2)\right]\frac{2n+1}{4\pi}P_n(\xi\cdot\eta)\\
&\qquad + (t_\xi - t_\eta)\sum_{k=0}^{L-1}\left[\sum_{n=k+1}^{L}(2n+1)\right]\frac{2k+1}{4\pi}P_k(\xi\cdot\eta).
\end{aligned}
$$

This identity is now independent of the chosen orthonormal system in $\mathrm{L}^2(\Omega)$, since it only uses Legendre polynomials. The fully normalized spherical harmonics only served as a tool in the intermediate derivations. Now we can finish this part of the proof (note Corollary 4.2.4):

$$\begin{aligned}
&\mathcal{J}_\xi K(\xi\cdot\eta) - \mathcal{J}_\eta K(\xi\cdot\eta)\\
&= (t_\xi - t_\eta)\sum_{n=0}^{L-1}\left(n^2+2n-L^2-2L\right)\frac{2n+1}{4\pi}P_n(\xi\cdot\eta)\\
&\quad + (t_\xi - t_\eta)\sum_{k=0}^{L-1}\underbrace{\left[(L+1)^2-(k+1)^2\right]}_{=L^2+2L+1-k^2-2k-1}\frac{2k+1}{4\pi}P_k(\xi\cdot\eta)\\
&= 0.
\end{aligned}$$

Hence, $\mathcal{J}_\xi K(\xi\cdot\eta) = \mathcal{J}_\eta K(\xi\cdot\eta)$ for all $\xi,\eta\in\Omega$.

(4) The commuting operators:

Let $F\in \mathrm{C}^{(2)}(\Omega)$. Then our previous results in this proof lead us to

$$\begin{aligned}
(\mathcal{T}_K\mathcal{J}F)(\xi) &= \int_D K(\xi\cdot\eta)(\mathcal{J}F)(\eta)\,\mathrm{d}\omega(\eta) = \int_D \mathcal{J}_\eta K(\xi\cdot\eta)F(\eta)\,\mathrm{d}\omega(\eta)\\
&= \int_D \mathcal{J}_\xi K(\xi\cdot\eta)F(\eta)\,\mathrm{d}\omega(\eta) = \mathcal{J}_\xi\int_D K(\xi\cdot\eta)F(\eta)\,\mathrm{d}\omega(\eta)\\
&= (\mathcal{J}\mathcal{T}_K F)(\xi) \qquad \text{for all } \xi\in\Omega.
\end{aligned}$$

□

As we have already mentioned, the purpose of the operator $\mathcal{J}$ is to have an alternative eigenvalue problem in comparison to the typical eigenvalue problem of Slepian functions. For further investigating this, we choose again a bandlimit L and look for eigenfunctions $F\in \mathrm{Harm}_{0\ldots L}(\Omega)$ of $\mathcal{J}$. If χ is the eigenvalue to the eigenfunction F, then $\mathcal{J}F = \chi F$ is equivalent to

$$\sum_{n=0}^{L}\sum_{j=-n}^{n} F^\wedge(n,j)\mathcal{J}_\xi Y_{n,j}(\xi) = \chi\sum_{n=0}^{L}\sum_{j=-n}^{n} F^\wedge(n,j)Y_{n,j}(\xi), \quad \xi\in\Omega. \tag{4.53}$$

Let us now calculate the Fourier coefficients of both sides. Then (4.53) is equivalent to

$$\int_\Omega \sum_{n=0}^{L}\sum_{j=-n}^{n} F^\wedge(n,j)\left(\mathcal{J}_\xi Y_{n,j}(\xi)\right)Y_{m,k}(\xi)\,\mathrm{d}\omega(\xi) = \chi F^\wedge(m,k)$$

for all $m=0,\ldots,L$ and $j=-m,\ldots,m$. Note that the right-hand side vanishes for $m>L$ such that this must also be the case for the left-hand side. These considerations lead us to the following result.

Theorem 4.6.9 *A function $F \in \mathrm{Harm}_{0\ldots L}(\Omega)$ is an eigenfunction of the operator $\mathcal{J}$ (see Theorem 4.6.8) if and only if its vector $f \in \mathbb{R}^{(L+1)^2}$ of Fourier coefficients is an eigenvector of the matrix*

$$J := \left(\int_\Omega Y_{m,k}(\xi) \mathcal{J}_\xi Y_{n,j}(\xi)\, \mathrm{d}\omega(\xi) \right)_{\substack{(m,k)\in I_L \\ (n,j)\in I_L}} . \tag{4.54}$$

where $I_L := \{(p,q) \mid p = 0, \ldots, L$ and $q = -p, \ldots, p\}$.

Theorem 4.6.10 *The matrices M from* (4.36) *and J from* (4.54) *commute, that is, $MJ = JM$.*

Proof We look at arbitrary components of the matrix products:

$$\begin{aligned} (MJ)_{\substack{(m,k)\\(n,j)}} &= \sum_{p=0}^{L} \sum_{q=-p}^{p} \langle Y_{m,k}, Y_{p,q} \rangle_{\mathrm{L}^2(R)} \langle Y_{p,q}, \mathcal{J} Y_{n,j} \rangle_{\mathrm{L}^2(\Omega)} \\ &= \int_R Y_{m,k}(\xi) \int_\Omega \sum_{p=0}^{L} \sum_{q=-p}^{p} Y_{p,q}(\xi) Y_{p,q}(\eta) \mathcal{J}_\eta Y_{n,j}(\eta)\, \mathrm{d}\omega(\eta)\, \mathrm{d}\omega(\xi). \end{aligned}$$

Hence, we can now use Theorem 4.6.8 and the $\mathrm{L}^2(\Omega)$-orthonormality of the $Y_{n,j}$ to get

$$\begin{aligned} (MJ)_{\substack{(m,k)\\(n,j)}} &= \int_R Y_{m,k}(\xi) \int_\Omega K(\xi \cdot \eta) \mathcal{J}_\eta Y_{n,j}(\eta)\, \mathrm{d}\omega(\eta)\, \mathrm{d}\omega(\xi) \\ &= \int_R Y_{m,k}(\xi) \left(\mathcal{T}_K \mathcal{J} Y_{n,j} \right)(\xi)\, \mathrm{d}\omega(\xi) = \int_R Y_{m,k}(\xi) \left(\mathcal{J} \mathcal{T}_K Y_{n,j} \right)(\xi)\, \mathrm{d}\omega(\xi) \\ &= \int_R Y_{m,k}(\xi) \mathcal{J}_\xi \int_\Omega K(\xi \cdot \eta) Y_{n,j}(\eta)\, \mathrm{d}\omega(\eta)\, \mathrm{d}\omega(\xi) \\ &= \int_R Y_{m,k}(\xi) \left(\mathcal{J} Y_{n,j} \right)(\xi)\, \mathrm{d}\omega(\xi), \end{aligned}$$

where here $\mathcal{T}_K$ acts on $\mathrm{L}^2(\Omega)$. For the interchanged factors, we obtain with Theorem 4.6.8 and Fubini's theorem that

$$\begin{aligned} (JM)_{\substack{(m,k)\\(n,j)}} &= \sum_{p=0}^{L} \sum_{q=-p}^{p} \langle Y_{m,k}, \mathcal{J} Y_{p,q} \rangle_{\mathrm{L}^2(\Omega)} \langle Y_{p,q}, Y_{n,j} \rangle_{\mathrm{L}^2(R)} \\ &= \int_\Omega Y_{m,k}(\xi) \int_R \sum_{p=0}^{L} \sum_{q=-p}^{p} \left(\mathcal{J}_\xi Y_{p,q}(\xi) \right) Y_{p,q}(\eta) Y_{n,j}(\eta)\, \mathrm{d}\omega(\eta)\, \mathrm{d}\omega(\xi) \\ &= \int_\Omega Y_{m,k}(\xi) \int_R \mathcal{J}_\xi K(\xi \cdot \eta) Y_{n,j}(\eta)\, \mathrm{d}\omega(\eta)\, \mathrm{d}\omega(\xi) \\ &= \langle Y_{m,k}, \mathcal{J} \mathcal{T}_K Y_{n,j} \rangle_{\mathrm{L}^2(\Omega)} = \langle Y_{m,k}, \mathcal{T}_K \mathcal{J} Y_{n,j} \rangle_{\mathrm{L}^2(\Omega)} \\ &= \int_\Omega Y_{m,k}(\eta) \int_R K(\eta \cdot \xi)(\mathcal{J} Y_{n,j})(\xi)\, \mathrm{d}\omega(\xi)\, \mathrm{d}\omega(\eta) \end{aligned}$$

$$= \int_R (\mathcal{J} Y_{n,j})(\xi) \int_\Omega Y_{m,k}(\eta) K(\eta \cdot \xi) \, d\omega(\eta) \, d\omega(\xi)$$

$$= \int_R (\mathcal{J} Y_{n,j})(\xi) Y_{m,k}(\xi) \, d\omega(\xi),$$

where here $\mathcal{T}_K$ acts on $L^2(R)$ and we used again the orthonormality of the $Y_{n,j}$. Hence, $MJ = JM$. □

For the entries of the matrix J, an explicit formula is known. The derivation is rather technical and is omitted here. For a proof, see, for example, Seibert (2018, Lemma 8.3.2).

Lemma 4.6.11 *The matrix J from* (4.54) *has, in the case of fully normalized spherical harmonics, the following entries:*

$$J_{(n,j),(n,j)} = -n(n+1)b,$$

$$J_{(n,j),(n+1,j)} = [n(n+2) - L(L+2)] \sqrt{\frac{(n+1-j)(n+1+j)}{(2n+1)(2n+3)}},$$

$$J_{(n+1,j),(n,j)} = J_{(n,j),(n+1,j)},$$

$$J_{(m,k),(n,j)} = 0 \quad \text{else},$$

where L is the chosen bandlimit and b is the parameter which defines the size of the spherical cap.

Lemma 4.6.11 tells us that J is a block-tridiagonal matrix, if we arrange the components appropriately. This can be done, for example, by choosing indices p and q of $J_{p,q}$ by

- $p := 0$
- for $k := -L, \ldots, L$
  - for $m := |k|, \ldots, L$
    - increase p by 1
    - $q := 0$
    - for $j := -L, \ldots, L$
      - for $n := |j|, \ldots, L$
        - increase q by 1
        - calculate $J_{p,q}$ to the indices $(m,k),(n,j)$
      - end
    - end
  - end
- end

Note that M also has a block structure in the case of fully normalized spherical harmonics, because we integrate over the full longitude $\varphi \in [0, 2\pi]$ if R is a cap around ε^3, and thus $Y_{n,j}$ is $L^2(R)$-orthogonal to $Y_{m,k}$ if $j \neq k$. However, M is not tridiagonal.

Within each block of the matrix J, we see that $J_{(n,j),(n+1,j)}$ cannot vanish because otherwise $n+1 = |j|$, but this constellation of degree and order does not occur. For such

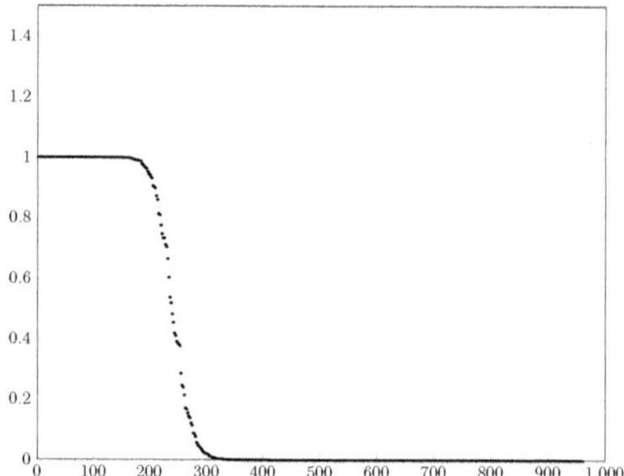

Figure 4.14 This image shows the distribution of the eigenvalues of the matrix M for the Slepian eigenvalue problem with bandlimit $L = 30$ and a spherical cap with an angle of 60°. The eigenvalues are sorted in descending order.

a tridiagonal matrix, it is known from linear algebra (see, e.g., Fox and Johnson, 1966) that the eigenvalues are non-degenerate. This means that each block $(J_{(n,j),(m,j)})_{n,m=|j|,\ldots,L}$ has $d := L - |j| + 1$ pairwise distinct eigenvalues $\mu_1, \ldots, \mu_d$ and corresponding eigenvectors $v^{(1)}, \ldots, v^{(d)} \in \mathbb{R}^d$. Hence, the blocks of M and J which are associated to order j and which we denote here by $M^{(j)}$ and $J^{(j)}$ satisfy $M^{(j)}J^{(j)}v^{(k)} = \mu_k M^{(j)}v^{(k)}$ and $M^{(j)}J^{(j)}v^{(k)} = J^{(j)}M^{(j)}v^{(k)}$ such that $J^{(j)}M^{(j)}v^{(k)} = \mu_k M^{(j)}v^{(k)}$. Moreover, since $v^{(k)} \neq 0$ (as an eigenvector) and $M^{(j)}$ is a regular matrix (see Theorem 4.6.1), $M^{(j)}v^{(k)} \neq 0$ also must hold true. The one-dimensional eigenspaces (i.e., the non-degenerate eigenvalues) of $J^{(j)}$ now imply that the eigenvector $M^{(j)}v^{(k)}$ of $J^{(j)}$ to the eigenvalue μ_k must be collinear with the eigenvector $v^{(k)}$ of $J^{(j)}$ to the same eigenvalue. In other words, there exists $\lambda_k^{(j)} \in \mathbb{R}$ such that $M^{(j)}v^{(k)} = \lambda_k^{(j)} v^{(k)}$ with $v^{(k)} \neq 0$.

Theorem 4.6.12 *Let $M^{(j)}$ and $J^{(j)}$ be the blocks of the matrices M and J corresponding to the order j of the fully normalized spherical harmonics. Then every orthonormal basis of eigenvectors $v^{(1)}, \ldots, v^{(L-|j|+1)}$ of $J^{(j)}$ is also an orthonormal basis of eigenvectors of $M^{(j)}$.*

For the purpose of illustration, we consider the case of the bandlimit $L = 30$ and a spherical cap of 60°, that is, the cap of all $\xi \in \Omega$ with $\xi \cdot \varepsilon^3 \geq \frac{1}{2}$. Since $\dim \mathrm{Harm}_{0\ldots30}(\Omega) = 31^2 = 961$, we obtain 961 orthonormal basis functions out of the Slepian eigenvalue problem. Figure 4.14 shows the distribution of the eigenvalues of the matrix M. Some eigenfunctions associated to particularly large and small eigenvalues are plotted in Figures 4.15 and 4.16. Note that orders j and $-j$ have eigenfunctions which are rotated copies of each other. The author gratefully acknowledges the support by Katrin Seibert in the production of these images.

The distribution of the eigenvalues shows an almost sharp transition from values close to 1 (very good localization in the cap) to values slightly above 0 (very bad localization in the cap). Indeed, such a behaviour has often been reported for Slepian functions in the literature.

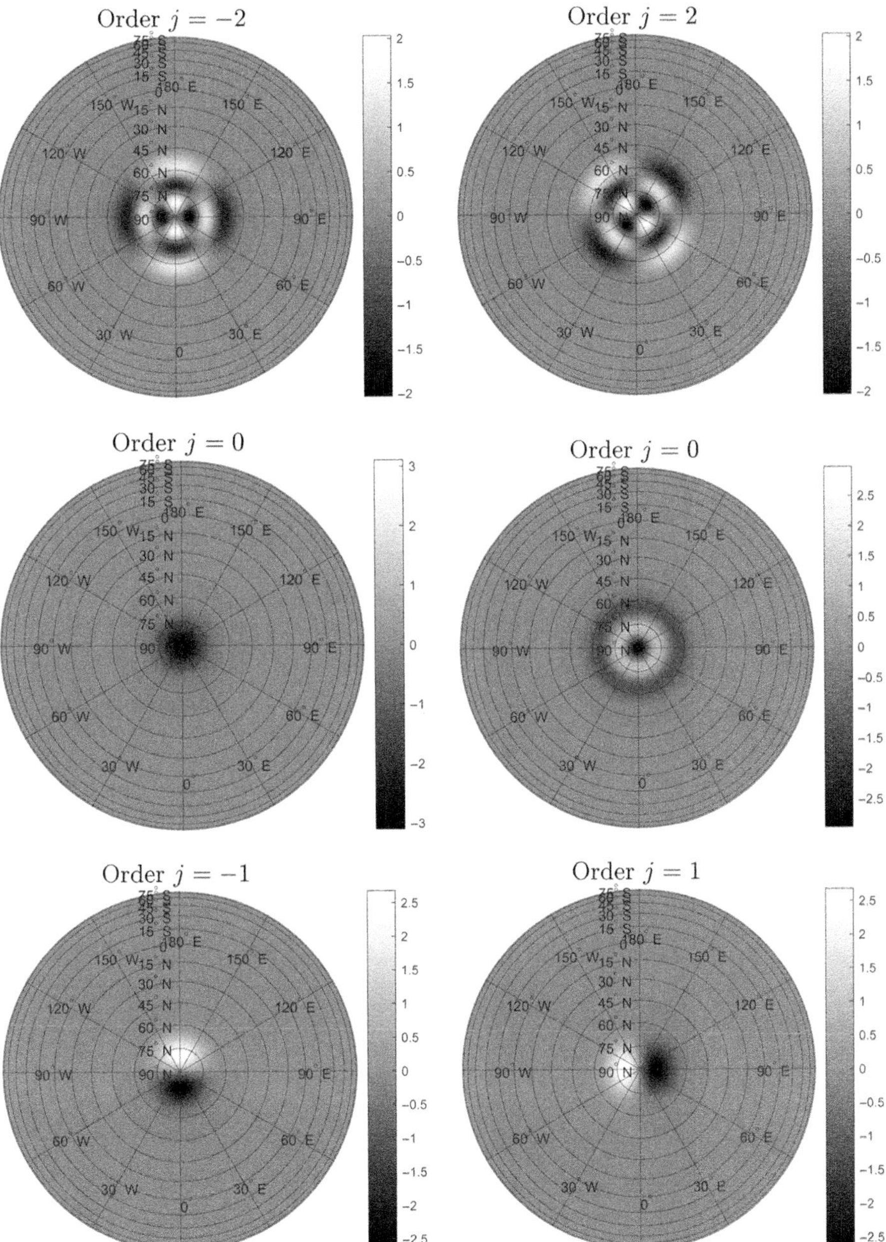

Figure 4.15 Six Slepian basis functions (i.e., eigenfunctions) with eigenvalues close to 1 are shown for the bandlimit 30 and the spherical cap with an angle of 60°. Due to the particular block structure of the matrix, each Slepian function is associated to a single spherical harmonic order j and has a rotated analogue for order $-j$ (as can be seen here for $j = \pm 1$ and $j = \pm 2$). For $j = 0$, two different Slepian functions are shown.

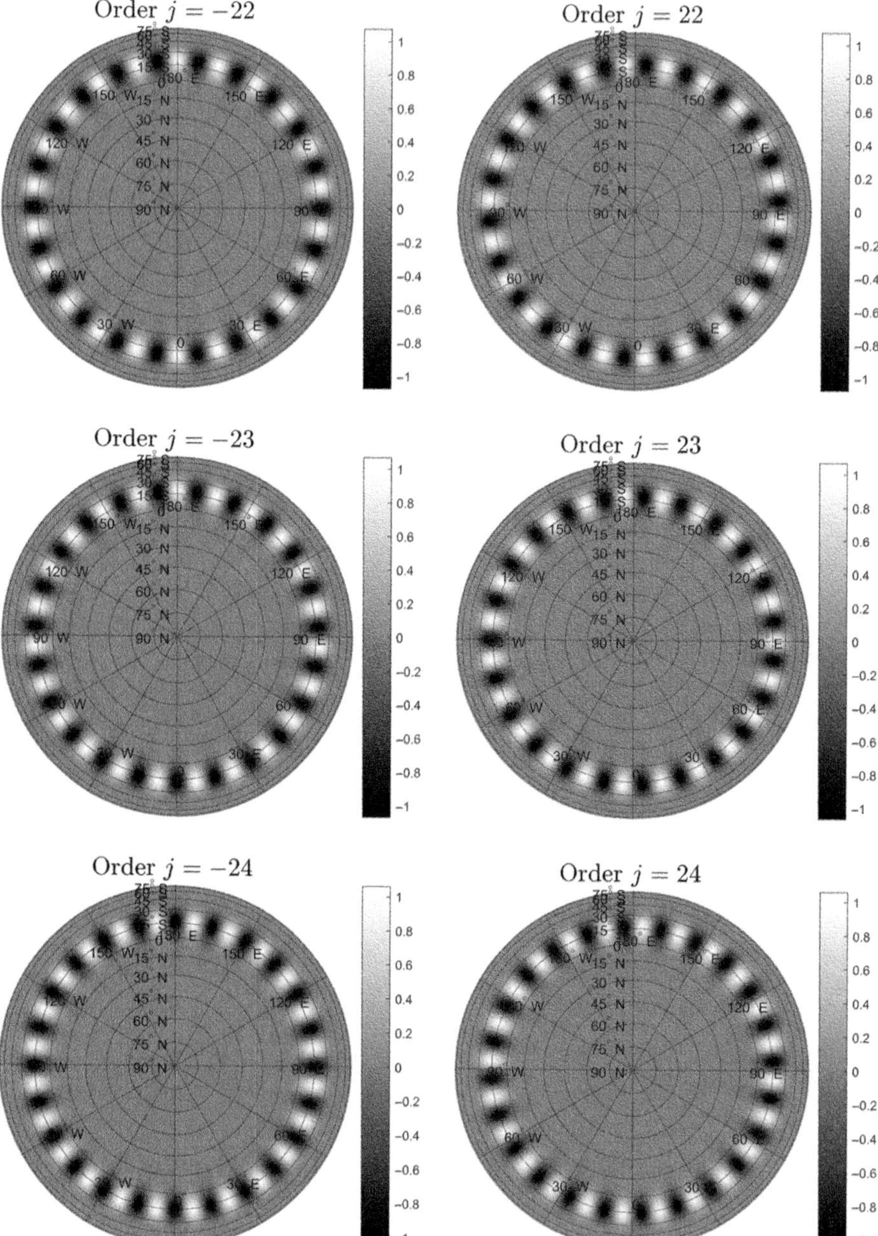

Figure 4.16 For the same scenario as in Figure 4.15, eigenfunctions corresponding to very low eigenvalues (close to 0) are shown. Clearly, these functions have only large values away from the chosen spherical cap.

It is certainly useful for our purposes, since we can rather easily separate a signal into its components associated to the subdomain R, which is here the cap, and the remainder of the signal. To define a kind of a cut in the Slepian basis in this respect, the **Shannon number** was introduced in the literature. The idea is as follows: in the idealized case where we only have eigenvalues 1 and 0, the sum of the eigenvalues would equal the number of well-localized Slepian functions. Since the trace of a matrix is invariant to basis transformations, this trace equals the sum of all eigenvalues in the case of a symmetric matrix. Hence, the trace of the matrix M is defined as the Shannon number and is often used as an approximate measure for the number of eigenfunctions concentrated to R. However, one should be aware that this is only a rough estimate and, in practice, one should try also slightly increased or decreased numbers of eigenfunctions.

In the case of a spherical cap, the Shannon number can be easily calculated. We have, due to the addition theorem for spherical harmonics, that

$$\operatorname{tr} M = \sum_{n=0}^{L} \sum_{j=-n}^{n} \int_R \left(Y_{n,j}(\xi)\right)^2 \mathrm{d}\omega(\xi) = \sum_{n=0}^{L} \frac{2n+1}{4\pi} \int_R \left(P_n(\xi \cdot \xi)\right)^2 \mathrm{d}\omega(\xi).$$

Due to Theorem 4.2.13, we consequently, get

$$\operatorname{tr} M = \sum_{n=0}^{L} \frac{2n+1}{4\pi} \lambda(R) = (L+1)^2 \frac{\lambda(R)}{4\pi},$$

where $\lambda(R)$ is the surface area of R, which is here $\lambda(R) = 2\pi(1-b)$ if the cap consists of all $\xi \in \Omega$ with $\xi \cdot \eta \geq b$ for a fixed point $\eta \in \Omega$ and a fixed parameter $b \in]-1,1[$. In our numerical example, we have $b = \frac{1}{2}$ and, hence, $\operatorname{tr} M = (L+1)^2/4 = 240.25$. Indeed, Figure 4.14 shows that the jump from large to small eigenvalues occurs in the neighbourhood of the Shannon number.

4.7 Radial Basis Functions and Sobolev Spaces

Recapitulating parts of Section 4.5, we know that a radial basis function (RBF) is a basis element $F_k \colon \Omega \to \mathbb{R}$ for which a point $\eta^{(k)} \in \Omega$ and a function $G_k \colon [-1,1] \to \mathbb{R}$ exist such that $F_k(\xi) = G_k(\eta^{(k)} \cdot \xi)$ for all $\xi \in \Omega$. In the case of L^2-spaces, we have to replace 'for all' here by 'for almost every'. Moreover, we learned that $\langle F_k, Y_{n,j}\rangle_{\mathrm{L}^2(\Omega)} = \left\langle F_k, P_n\left(\eta^{(k)}\cdot\right)\right\rangle_{\mathrm{L}^2(\Omega)} Y_{n,j}\left(\eta^{(k)}\right)$ for all $n \in \mathbb{N}_0$ and all $j = -n, \ldots, n$, where P_n is the Legendre polynomial of degree n.

When we want to study the properties of RBFs, it is useful to have spherical Sobolev spaces as a tool in the way how they were introduced in Freeden (1981a,b), and Freeden et al. (1998). Note that there are slight variations in the definitions in the literature. The only difference is, however, in the use of A_n^2 or A_n^{-2} in (4.56). Therefore, the reader is advised to check the precise definition of the spherical Sobolev space when different literature is used.

Definition 4.7.1 Let $(A_n)_{n\in\mathbb{N}_0}$ be a real sequence. By $\mathcal{E}((A_n);\Omega)$, we denote the subset of $\mathrm{C}^{(\infty)}(\Omega)$ which consists of all $F \in \mathrm{C}^{(\infty)}(\Omega)$ for which

$$\langle F, Y_{n,j}\rangle_{\mathrm{L}^2(\Omega)} = 0 \qquad \text{whenever } A_n = 0 \tag{4.55}$$

and

$$\sum_{\substack{n=0\\ A_n\neq 0}}^{\infty} \sum_{j=-n}^{n} A_n^{-2} \langle F, Y_{n,j}\rangle_{\mathrm{L}^2(\Omega)}^2 < +\infty. \tag{4.56}$$

This linear space is turned into an inner product space by defining

$$\langle F, G\rangle_{\mathcal{H}} := \langle F, G\rangle_{\mathcal{H}((A_n);\Omega)} := \sum_{\substack{n=0\\ A_n\neq 0}}^{\infty} \sum_{j=-n}^{n} A_n^{-2} \langle F, Y_{n,j}\rangle_{\mathrm{L}^2(\Omega)} \langle G, Y_{n,j}\rangle_{\mathrm{L}^2(\Omega)}$$

for all $F, G \in \mathcal{E}((A_n);\Omega)$. Eventually, the completion of $\mathcal{E}((A_n);\Omega)$ with respect to $\langle\cdot,\cdot\rangle_{\mathcal{H}}$ is denoted by $\mathcal{H}((A_n);\Omega)$ and called a **Sobolev space on the sphere**. If the chosen sequence is clear from the context, then we will simply write $\mathcal{H}$ instead of $\mathcal{H}((A_n);\Omega)$. The induced norm is denoted by $\|\cdot\|_{\mathcal{H}} := \|\cdot\|_{\mathcal{H}((A_n);\Omega)}$.

The sequence $(A_n)_{n\in\mathbb{N}_0}$ controls the smoothness of the functions in the space $\mathcal{H}((A_n);\Omega)$. Commonly, $(A_n)_{n\in\mathbb{N}_0}$ is chosen to converge to 0 (or to absolutely diverge to ∞, if A_n^2 is used instead of A_n^{-2}). The faster this convergence, the stronger is the condition (4.56). The Fourier coefficients $\langle F, Y_{n,j}\rangle_{\mathrm{L}^2(\Omega)}$ have to decay rapidly as $n \to \infty$ in order to yield a convergent series. Hence, the high-degree parts (i.e., the parts which contribute to highly frequent oscillations) are suppressed by a rapidly divergent sequence. More precisely, we have the following propositions, which obviously hold true.

Theorem 4.7.2 *For the spherical Sobolev spaces, we have the following:*

(a) If $(A_n)_{n\in\mathbb{N}_0}$, $(B_n)_{n\in\mathbb{N}_0}$ are real sequences with $|A_n| \leq |B_n|$ for all $n \in \mathbb{N}_0$, then $\mathcal{H}((A_n);\Omega) \subset \mathcal{H}((B_n);\Omega)$.
(b) For every constant sequence $A_n \equiv c \in \mathbb{R}\setminus\{0\}$, we have $\mathcal{H}((c);\Omega) = \mathrm{L}^2(\Omega)$.
(c) If $(A_n)_{n\in\mathbb{N}_0}$ is a real sequence with $\sup_{n\in\mathbb{N}_0}|A_n| < +\infty$, then $\mathcal{H}((A_n);\Omega) \subset \mathrm{L}^2(\Omega)$.
(d) Every space $(\mathcal{H}, \langle\cdot,\cdot\rangle_{\mathcal{H}})$ is a Hilbert space.

Remark 4.7.3 The Sobolev spaces can analogously be defined on spheres $S_R(0)$ by replacing the unit-sphere-based $Y_{n,j}$ by the orthonormal basis functions $S_R(0) \ni x \mapsto R^{-1}Y_{n,j}(x/R)$, $n \in \mathbb{N}_0$, $j = -n, \ldots, n$, of $\mathrm{L}^2(S_R(0))$.

As we mentioned earlier, the sequence $(A_n)_{n\in\mathbb{N}_0}$ is usually chosen to converge to 0. More precisely, an even stronger condition is required.

Definition 4.7.4 $(A_n)_{n\in\mathbb{N}_0} \subset \mathbb{R}$ is called **(spherically) summable** if

$$\sum_{n=0}^{\infty} \frac{2n+1}{4\pi} A_n^2 < +\infty. \tag{4.57}$$

Let now $F \in \mathcal{H}((A_n);\Omega)$ in the case of a summable sequence $(A_n)_{n\in\mathbb{N}_0}$. If $\xi \in \Omega$ and $N \in \mathbb{N}_0$, then the Cauchy–Schwarz inequality yields in combination with the addition theorem for spherical harmonics that

$$\begin{aligned}
&\left|\sum_{n=N}^{\infty}\sum_{j=-n}^{n}\langle F, Y_{n,j}\rangle_{\mathrm{L}^2(\Omega)} Y_{n,j}(\xi)\right| \\
&= \left|\sum_{\substack{n=N\\ A_n\neq 0}}^{\infty}\sum_{j=-n}^{n} A_n^{-1}\langle F, Y_{n,j}\rangle_{\mathrm{L}^2(\Omega)} A_n Y_{n,j}(\xi)\right| \\
&\leq \left(\sum_{\substack{n=N\\ A_n\neq 0}}^{\infty}\sum_{j=-n}^{n} A_n^{-2}\langle F, Y_{n,j}\rangle_{\mathrm{L}^2(\Omega)}^2\right)^{1/2}\left(\sum_{n=N}^{\infty} A_n^2\frac{2n+1}{4\pi}P_n(\underbrace{\xi\cdot\xi}_{=1})\right)^{1/2},
\end{aligned}$$

where we also used that $P_n(1) = 1$ (see Theorem 4.2.11).

For $N \to \infty$, we can use (4.56) and (4.57) to see that the right-hand side converges uniformly to zero. For $N = 0$, we obtain an estimate for $\|F\|_{\mathrm{C}(\Omega)}$.

Theorem 4.7.5 (Sobolev Lemma) *Let $(A_n)_{n\in\mathbb{N}_0}$ be summable. Then every function $F \in \mathcal{H}((A_n);\Omega)$ is continuous, has a uniformly convergent Fourier series $F(\xi) = \sum_{n=0}^{\infty}\sum_{j=-n}^{n}\langle F, Y_{n,j}\rangle_{\mathrm{L}^2(\Omega)} Y_{n,j}(\xi)$, $\xi \in \Omega$, and*

$$\|F\|_{\mathrm{C}(\Omega)} \leq \|F\|_{\mathcal{H}}\left(\sum_{n=0}^{\infty}\frac{2n+1}{4\pi}A_n^2\right)^{1/2}. \tag{4.58}$$

Corollary 4.7.6 *Let, for each $k \in \mathbb{N}_0$, a summable sequence $\left(A_n^{(k)}\right)_{n\in\mathbb{N}_0}$ be given. If each of the conditions*

(a) $\left|A_n^{(k)}\right| \leq \left|A_n^{(k+1)}\right|$ for all $n, k \in \mathbb{N}_0$,
(b) $\lim_{k\to\infty}\left|A_n^{(k)}\right| =: A_n^{(\infty)} \in \mathbb{R}_0^+$ exists for all $n \in \mathbb{N}_0$

*holds true, then the spaces $\mathcal{H}^{(k)} := \mathcal{H}\left(\left(A_n^{(k)}\right)_{n\in\mathbb{N}_0};\Omega\right)$, $k \in \mathbb{N}_0$, constitute a **multiresolution analysis** (MRA) on the sphere in the sense that*

(i) $\mathcal{H}^{(k)} \subset \mathcal{H}^{(k+1)} \subset \mathcal{H}^{(\infty)}$ for all $k \in \mathbb{N}_0$,
(ii) $\overline{\bigcup_{k\in\mathbb{N}_0}\mathcal{H}^{(k)}}^{\|\cdot\|_{\mathcal{H}^{(\infty)}}} = \mathcal{H}^{(\infty)}$,

where $\mathcal{H}^{(\infty)} := \mathcal{H}\left(\left(A_n^{(\infty)}\right)_{n\in\mathbb{N}_0};\Omega\right)$.

Proof Property (i) is a consequence of condition a in combination with part a of Theorem 4.7.2. For proving part (ii) of the current theorem, we choose an arbitrary $F \in \mathcal{H}^{(\infty)}$. We construct a corresponding sequence of functions $F_k\colon \Omega \to \mathbb{R}$ with the series $F_k = \sum_{n=0}^{\infty}\sum_{j=-n}^{n}\langle F_k, Y_{n,j}\rangle_{\mathrm{L}^2(\Omega)} Y_{n,j}$, which is considered as a $\|\cdot\|_{\mathcal{H}^{(k)}}$-limit, and the coefficients

$$\langle F_k, Y_{n,j}\rangle_{L^2(\Omega)} := \begin{cases} 0, & \text{if } A_n^{(\infty)} = 0, \\ \left|A_n^{(k)}\right| \left(A_n^{(\infty)}\right)^{-1} \langle F, Y_{n,j}\rangle_{L^2(\Omega)}, & \text{if } A_n^{(\infty)} \neq 0. \end{cases}$$

We have indeed $F_k \in \mathcal{H}^{(k)}$, since (4.55) is satisfied by construction of F_k and (4.56) is a consequence of the estimate (note that $A_n^{(\infty)} \neq 0$, if $A_n^{(k)} \neq 0$)

$$\begin{aligned} \sum_{\substack{n=0\\ A_n^{(k)}\neq 0}}^{\infty} \sum_{j=-n}^{n} \left(A_n^{(k)}\right)^{-2} \langle F_k, Y_{n,j}\rangle^2_{L^2(\Omega)} &= \sum_{\substack{n=0\\ A_n^{(k)}\neq 0}}^{\infty} \sum_{j=-n}^{n} \left(A_n^{(\infty)}\right)^{-2} \langle F, Y_{n,j}\rangle^2_{L^2(\Omega)} \\ &\leq \sum_{\substack{n=0\\ A_n^{(\infty)}\neq 0}}^{\infty} \sum_{j=-n}^{n} \left(A_n^{(\infty)}\right)^{-2} \langle F, Y_{n,j}\rangle^2_{L^2(\Omega)} \end{aligned}$$

and the fact that $F \in \mathcal{H}^{(\infty)}$. Hence, the sequence (F_k) is contained in $\bigcup_{k\in\mathbb{N}_0} \mathcal{H}^{(k)}$. Furthermore,

$$\begin{aligned} \|F - F_k\|^2_{\mathcal{H}^{(\infty)}} &= \sum_{\substack{n=0\\ A_n^{(\infty)}\neq 0}}^{\infty} \sum_{j=-n}^{n} \left(A_n^{(\infty)}\right)^{-2} \left(\langle F, Y_{n,j}\rangle_{L^2(\Omega)} - \langle F_k, Y_{n,j}\rangle_{L^2(\Omega)}\right)^2 \\ &= \sum_{\substack{n=0\\ A_n^{(\infty)}\neq 0}}^{\infty} \sum_{j=-n}^{n} \left(A_n^{(\infty)}\right)^{-2} \left[1 - \left|A_n^{(k)}\right| \left(A_n^{(\infty)}\right)^{-1}\right]^2 \langle F, Y_{n,j}\rangle^2_{L^2(\Omega)}. \end{aligned}$$

Since $0 \leq \left[1 - \left|A_n^{(k)}\right| \left(A_n^{(\infty)}\right)^{-1}\right]^2 \leq 1$ for all $k, n \in \mathbb{N}_0$ and $F \in \mathcal{H}^{(\infty)}$, the preceding series is uniformly convergent with respect to $k \in \mathbb{N}_0$. Thus, we are allowed to conclude that

$$\lim_{k\to\infty} \|F - F_k\|^2_{\mathcal{H}^{(\infty)}} = \sum_{\substack{n=0\\ A_n^{(\infty)}\neq 0}}^{\infty} \sum_{j=-n}^{n} \left(A_n^{(\infty)}\right)^{-2} \lim_{k\to\infty} \left(1 - \frac{\left|A_n^{(k)}\right|}{A_n^{(\infty)}}\right)^2 \langle F, Y_{n,j}\rangle^2_{L^2(\Omega)} = 0.$$

This implies that $F \in \overline{\bigcup_{k\in\mathbb{N}_0} \mathcal{H}^{(k)}}^{\|\cdot\|_{\mathcal{H}^{(\infty)}}}$. □

Such spaces are strongly related to the scale spaces of the spherical wavelet theory due to Freeden; see, for example, Freeden et al. (1998).

Theorem 4.7.7 *If $(A_n)_{n\in\mathbb{N}_0}$ is a summable real sequence with $A_n \neq 0$ for all $n \in \mathbb{N}_0$, then $\mathcal{H} = \mathcal{H}((A_n); \Omega)$ is dense in $(\mathrm{C}(\Omega), \|\cdot\|_{\mathrm{C}(\Omega)})$ and in $(\mathrm{L}^2(\Omega), \|\cdot\|_{\mathrm{L}^2(\Omega)})$.*

Proof From the Sobolev lemma (Theorem 4.7.5) and Theorem 2.5.20, we have that $\mathcal{H} \subset \mathrm{C}(\Omega) \subset \mathrm{L}^2(\Omega)$. Moreover, the construction of $\mathcal{H}$ (see Definition 4.7.1) immediately implies that $\operatorname{span}\{Y_{n,j}\}_{n\in\mathbb{N}_0;\, j=-n,\dots,n} \subset \mathcal{H}$, as a span only contains *finite* linear combinations. Since the functions $Y_{n,j}$ are closed (in the sense of the approximation theory) in the spaces $(\mathrm{C}(\Omega), \|\cdot\|_{\mathrm{C}(\Omega)})$ and $(\mathrm{L}^2(\Omega), \|\cdot\|_{\mathrm{L}^2(\Omega)})$ (see Theorems 4.2.15 and 4.2.17), we obtain the desired result. □

Note that we need $A_n \neq 0$ for all $n \in \mathbb{N}_0$ in the proof of Theorem 4.7.7, since otherwise not all $Y_{n,j}$ are elements of $\mathcal{H}$ due to (4.55).

The Sobolev spaces on the sphere which are used here have also proved to be useful because of their connection to reproducing kernels. These kernels are common as a building block for a spherical spline method due to Freeden and Wahba (see Freeden, 1981a,b, Freeden et al., 1998 and Wahba, 1981). Before we continue, a very short introduction into reproducing kernels may be helpful. For further details, see also Davis (1975) and Michel (2013).

Definition 4.7.8 Let $(\mathcal{X}, \langle\cdot,\cdot\rangle_{\mathcal{X}})$ be a Hilbert space and a $\mathbb{K}$-vector space ($\mathbb{K} \in \{\mathbb{R}, \mathbb{C}\}$) whose elements are functions with a joint domain $D \subset \mathbb{R}^n$. If there exists a kernel function $K_{\mathcal{X}} : D \times D \to \mathbb{K}$ such that

(a) $K_{\mathcal{X}}(x,\cdot) \in \mathcal{X}$ for all $x \in D$,
(b) $\langle F, K_{\mathcal{X}}(x,\cdot)\rangle_{\mathcal{X}} = F(x)$ for all $x \in D$ and all $F \in \mathcal{X}$,

then $K_{\mathcal{X}}$ is called a **reproducing kernel** of $\mathcal{X}$, and $\mathcal{X}$ is called a **reproducing kernel Hilbert space**.

Theorem 4.7.9 *Let $(\mathcal{X}, \langle\cdot,\cdot\rangle_{\mathcal{X}})$ be a reproducing kernel Hilbert space with functions $F: D \to \mathbb{K}$, where $D \subset \mathbb{R}^n$. Then its reproducing kernel $K_{\mathcal{X}}$ is unique and conjugately symmetric: $K_{\mathcal{X}}(x,y) = \overline{K_{\mathcal{X}}(y,x)}$ for all $x, y \in D$.*

Proof Suppose that $K_{\mathcal{X}}$ and $L_{\mathcal{X}}$ are both reproducing kernels of $\mathcal{X}$. Then the properties a and b from Definition 4.7.8 and (IP2) from Definition 2.5.1 imply that, for all $x, y \in D$, we have

$$\begin{aligned} K_{\mathcal{X}}(x,y) &= \langle K_{\mathcal{X}}(x,\cdot), K_{\mathcal{X}}(y,\cdot)\rangle_{\mathcal{X}} = \overline{\langle K_{\mathcal{X}}(y,\cdot), K_{\mathcal{X}}(x,\cdot)\rangle_{\mathcal{X}}} = \overline{K_{\mathcal{X}}(y,x)},\\ L_{\mathcal{X}}(x,y) &= \langle L_{\mathcal{X}}(x,\cdot), K_{\mathcal{X}}(y,\cdot)\rangle_{\mathcal{X}} = \overline{\langle K_{\mathcal{X}}(y,\cdot), L_{\mathcal{X}}(x,\cdot)\rangle_{\mathcal{X}}} = \overline{K_{\mathcal{X}}(y,x)}\\ &= K_{\mathcal{X}}(x,y). \end{aligned}$$

□

Let us now direct our attention to the spherical Sobolev spaces.

Theorem 4.7.10 *Let $(A_n)_{n\in\mathbb{N}_0}$ be a summable real sequence. Then $\mathcal{H} = \mathcal{H}((A_n);\Omega)$ is a reproducing kernel Hilbert space with a unique reproducing kernel. The latter is represented, for all $\xi, \eta \in \Omega$, by*

$$K_{\mathcal{H}}(\xi,\eta) = \sum_{n=0}^{\infty}\sum_{j=-n}^{n} A_n^2 Y_{n,j}(\xi) Y_{n,j}(\eta) = \sum_{n=0}^{\infty} \frac{2n+1}{4\pi} A_n^2 P_n(\xi\cdot\eta). \tag{4.59}$$

Proof The alternative representation in (4.59) is a direct consequence of the addition theorem (Theorem 4.2.9), and the uniqueness has already been proved in Theorem 4.7.9. It remains to verify properties a and b from Definition 4.7.8. Clearly, we have $\langle K_{\mathcal{H}}(\xi,\cdot), Y_{n,j}\rangle_{\mathrm{L}^2(\Omega)} = A_n^2 Y_{n,j}(\xi)$. Consequently,

$$\sum_{\substack{n=0\\A_n\neq 0}}^{\infty}\sum_{j=-n}^{n} A_n^{-2}\langle K_{\mathcal{H}}(\xi,\cdot), Y_{n,j}\rangle_{\mathrm{L}^2(\Omega)}^2 = \sum_{\substack{n=0\\A_n\neq 0}}^{\infty}\sum_{j=-n}^{n} A_n^2 (Y_{n,j}(\xi))^2 = \sum_{\substack{n=0\\A_n\neq 0}}^{\infty} \frac{2n+1}{4\pi} A_n^2 < +\infty,$$

where we used the addition theorem again, the fact that $P_n(1) = 1$ (see Theorem 4.2.11), and the summability condition (4.57). Hence, $K_{\mathcal{H}}(\xi, \cdot) \in \mathcal{H}$ (see Definition 4.7.1). Eventually, let $F \in \mathcal{H}$ be an arbitrary function, which is continuous due to the Sobolev lemma (Theorem 4.7.5). Then

$$\begin{aligned}\langle F, K_{\mathcal{H}}(\xi, \cdot)\rangle_{\mathcal{H}} &= \sum_{\substack{n=0\\ A_n \neq 0}}^{\infty} \sum_{j=-n}^{n} A_n^{-2} \langle F, Y_{n,j}\rangle_{\mathrm{L}^2(\Omega)} A_n^2 Y_{n,j}(\xi) \\ &= \sum_{n=0}^{\infty} \sum_{j=-n}^{n} \langle F, Y_{n,j}\rangle_{\mathrm{L}^2(\Omega)} Y_{n,j}(\xi) = F(\xi)\end{aligned}$$

for all $\xi \in \Omega$, where we used the construction of $\mathcal{H}$ (i.e., we had required $\langle F, Y_{n,j}\rangle_{\mathrm{L}^2(\Omega)} = 0$ if $A_n = 0$; see Definition 4.7.1) and the uniform convergence of the Fourier series of F (see the Sobolev lemma, Theorem 4.7.5). □

Since $K_{\mathcal{H}}$ is a zonal function, we will also write $K_{\mathcal{H}}(\xi \cdot \eta)$ instead of $K_{\mathcal{H}}(\xi, \eta)$.

Depending on the properties we require from the considered function on the sphere, we can choose the appropriate Sobolev space. One way of installing scales among Sobolev spaces on the sphere is given in the following way.

Definition 4.7.11 For each $s \in \mathbb{R}$, we define the spherical Sobolev space $\mathcal{H}_s(\Omega) := \mathcal{H}\left(\left(\left(n + \frac{1}{2}\right)^{-s}\right); \Omega\right)$.

Several properties can be proved for these spaces, for example concerning the applicability of certain (pseudo-)differential operators. For further details, see, for example, Freeden et al. (1998). We will only summarize a small selection here.

Theorem 4.7.12 *The spaces $\mathcal{H}_s(\Omega)$, $s \in \Omega$, satisfy*

(a) $\mathcal{H}_0(\Omega) = \mathrm{L}^2(\Omega)$,
(b) $\mathcal{H}_s(\Omega) \subset \mathcal{H}_t(\Omega)$, *if* $s \geq t$,
(c) $\mathcal{H}_s(\Omega) \subset \mathrm{C}(\Omega)$, *if* $s > 1$,
(d) $\mathcal{H}_s(\Omega) \subset \mathrm{C}^{(1)}(\Omega)$, *if* $s > 2$.

Proof Part a is trivial (see part b of Theorem 4.7.2). Part b is a consequence of part a of Theorem 4.7.2. Part c is valid due to the Sobolev lemma (Theorem 4.7.5), since the sequence $\left((n+1/2)^{-s}\right)_{n \in \mathbb{N}_0}$ is summable if and only if $s > 1$. For the proof of part d, we refer to, for example, Michel (2013, Theorem 6.16). □

Let us now prove how radial basis functions can be obtained on the sphere.

Theorem 4.7.13 *Let $(A_n)_{n \in \mathbb{N}_0}$ be a given summable real sequence and let $\left\{\eta^{(k)}\right\}_{k \in \mathbb{N}_0} \subset \Omega$ be a* dense *set of pairwise distinct points on the unit sphere. Corresponding to this point set, we define the sequence of zonal functions $F_k(\xi) := K_{\mathcal{H}}(\eta^{(k)} \cdot \xi)$, $\xi \in \Omega$, $k \in \mathbb{N}_0$, where $K_{\mathcal{H}}$ is the reproducing kernel of $\mathcal{H} = \mathcal{H}((A_n); \Omega)$. Then the following holds true:*

(a) $\mathrm{span}\{F_k\}_{k \in \mathbb{N}_0}$ *is dense in $\mathcal{H}$ (i.e., the F_k are radial basis functions).*
(b) $\mathrm{span}\{F_k\}_{k \in \mathbb{N}_0}$ *is dense in $(\mathrm{C}(\Omega), \|\cdot\|_{\mathrm{C}(\Omega)})$ and in $(\mathrm{L}^2(\Omega), \langle\cdot,\cdot\rangle_{\mathrm{L}^2(\Omega)})$ if and only if $A_n \neq 0$ for all $n \in \mathbb{N}_0$.*

Proof We prove each part separately.

(1) Proof of part a:

We define $h := \operatorname{span}\{F_k\}_{k\in\mathbb{N}_0}$. Since $\overline{h}^{\|\cdot\|_{\mathcal{H}}}$ is a closed linear subspace of $\mathcal{H}$, we can decompose $\mathcal{H}$ into $\mathcal{H} = \overline{h}^{\|\cdot\|_{\mathcal{H}}} \oplus \left(\overline{h}^{\|\cdot\|_{\mathcal{H}}}\right)^{\perp}$ due to Theorem 2.5.71. If now $F \in \left(\overline{h}^{\|\cdot\|_{\mathcal{H}}}\right)^{\perp}$, then $\left\langle K_{\mathcal{H}}\left(\eta^{(k)}\cdot\right), F\right\rangle_{\mathcal{H}} = 0$ since $K_{\mathcal{H}}\left(\eta^{(k)}\cdot\right) \in h$. However, the properties of a reproducing kernel (see Definition 4.7.8) yield that $F(\eta^{(k)}) = 0$ for all $k \in \mathbb{N}_0$. Since $(A_n)_{n\in\mathbb{N}_0}$ is summable, F must be continuous due to the Sobolev lemma (Theorem 4.7.5). If $\left\{\eta^{(k)}\right\}_{k\in\mathbb{N}_0}$ is dense in Ω, then we must get $F \equiv 0$.

(2) Proof of part b:

Our previous conclusions are still valid. Let now $A_n \neq 0$ for all $n \in \mathbb{N}_0$. If $F \in \mathrm{C}(\Omega)$ or $F \in \mathrm{L}^2(\Omega)$, then Theorem 4.7.7 tells us that there exists a function $G \in \mathcal{H}$ in an ε-neighbourhood of F, where the corresponding norm of $\mathrm{C}(\Omega)$ or $\mathrm{L}^2(\Omega)$ is used. Due to the previous considerations, there is $H \in h$ in a neighbourhood of G, measured in the norm of $\mathcal{H}$. Due to part b of Theorem 2.5.20 and (4.58), we find also a function $\tilde{H} \in h$ in an ε-neighbourhood of G, measured in the norm of $\mathrm{C}(\Omega)$ or $\mathrm{L}^2(\Omega)$, respectively. Hence, $\tilde{H}$ is in a 2ε-neighbourhood of F in terms of the $\mathrm{C}(\Omega)$-norm or the $\mathrm{L}^2(\Omega)$-norm, respectively.

Vice versa, let us assume that there exists $\nu \in \mathbb{N}_0$ with $A_\nu = 0$. Then we have, for example, $\left\langle Y_{\nu,0}, K_{\mathcal{H}}\left(\eta^{(k)}\cdot\right)\right\rangle_{\mathrm{L}^2(\Omega)} = 0$ for all $k \in \mathbb{N}_0$ due to (4.59). Hence, $Y_{\nu,0}$ is orthogonal to the entire basis of $\overline{h}^{\|\cdot\|_{\mathrm{L}^2(\Omega)}}$, but $Y_{\nu,0} \neq 0$. Thus, h is not dense in $\mathrm{L}^2(\Omega)$. Since $\overline{h}^{\|\cdot\|_{\mathrm{C}(\Omega)}} \subset \overline{h}^{\|\cdot\|_{\mathrm{L}^2(\Omega)}}$, the function is also not in $\overline{h}^{\|\cdot\|_{\mathrm{C}(\Omega)}}$, although $Y_{\nu,0} \in \mathrm{C}(\Omega) \subset \mathrm{L}^2(\Omega)$.

Note that the latter argumentation is not possible in $\mathcal{H}$, because $Y_{\nu,0} \notin \mathcal{H}$ if $A_\nu = 0$, because of (4.55). □

It is important that $\left\{\eta^{(k)}\right\}_{k\in\mathbb{N}_0}$ is dense in Ω. Otherwise, there would exist a ball $B_\varepsilon(\eta)$ with $\eta \in \Omega$ and $\varepsilon > 0$ such that $\eta^{(k)} \notin B_\varepsilon(\eta)$ for all k. Then, in the proof of part a, we would not be able to make conclusions about F on $B_\varepsilon(\eta)$.

Example 4.7.14 Let us consider some examples of spherical radial basis functions.

(a) We have already encountered the **Abel–Poisson kernel**. In Corollary 3.4.20, we found out that the kernel can be expanded in Legendre polynomials by

$$\frac{1}{4\pi}\,\frac{R^2 - r^2}{(R^2 + r^2 - 2rRx)^{3/2}} = \sum_{n=0}^{\infty} \frac{2n+1}{4\pi R}\left(\frac{r}{R}\right)^n P_n(x),$$

$x \in [-1, 1]$, where $0 \leq r < R$. We can easily turn this into a zonal function by setting

$$K_h(\xi\cdot\eta) := \frac{1}{4\pi}\,\frac{1 - h^2}{(1 + h^2 - 2h\,\xi\cdot\eta)^{3/2}} = \sum_{n=0}^{\infty} \frac{2n+1}{4\pi}\,h^n P_n(\xi\cdot\eta),$$

$\xi, \eta \in \Omega$, $h \in [0,1[$. We get non-bandlimited radial basis functions by choosing a fixed $h \in [0,1[$ and a dense but countable subset $\{\eta^{(k)}\}_{k \in \mathbb{N}_0} \subset \Omega$. Then $F_k(\xi) := K_h(\eta^{(k)} \cdot \xi)$, $\xi \in \Omega$, satisfies the required properties from Theorem 4.7.13. In view of Theorem 4.7.10, the corresponding Sobolev space is associated to the sequence $A_n := h^{n/2}$, $n \in \mathbb{N}_0$, which is, due to its exponential character, obviously summable; see (4.57). Though these radial basis functions vanish nowhere, they show a spatial concentration around their maximum at $\xi = \eta^{(k)}$, as shown in Figure 4.17. The location of this maximum is obvious, because $|P_n(t)| \leq P_n(1)$ for all $t \in [-1,1]$ and all $n \in \mathbb{N}_0$; see Theorem 4.2.13. This concentration gets stronger as h increases.

Due to the closed representation, the Abel–Poisson kernel has a rare outstanding status among the non-bandlimited radial basis functions and is, therefore, frequently used in numerical computations.

(b) Note that bandlimited kernels are not covered by part b of Theorem 4.7.13. A class of such kernels can be constructed by setting the sequence $A_n := (\varphi_J(n))^{1/2}$, where $J \in \mathbb{N}_0$ is a fixed scale and φ_J is a generator of a (spherical) bandlimited scaling function; see, for example, Freeden et al. (1998, pp. 279–296) or Michel (2013, pp. 195–199). We pick out two examples: the **Shannon scaling function** corresponds to

$$\varphi_J(n) := \begin{cases} 1, & 0 \leq n < 2^J, \\ 0, & n \geq 2^J \end{cases}$$

and, consequently,

$$K_J(\xi \cdot \eta) := K_{\mathcal{H}}(\xi \cdot \eta) = \sum_{n=0}^{2^J-1} \frac{2n+1}{4\pi} P_n(\xi \cdot \eta); \quad \xi, \eta \in \Omega;$$

and the **cubic polynomial (cp) scaling function** is characterized by

$$\varphi_J(n) := \begin{cases} \left(1 - 2^{-J} n\right)^2 \left(1 + 2^{1-J} n\right), & 0 \leq n < 2^J, \\ 0, & n \geq 2^J \end{cases}$$

such that

$$K_J(\xi \cdot \eta) := K_{\mathcal{H}}(\xi \cdot \eta) = \sum_{n=0}^{2^J-1} \left(1 - 2^{-J} n\right)^2 \left(1 + 2^{1-J} n\right) \frac{2n+1}{4\pi} P_n(\xi \cdot \eta);$$

$\xi, \eta \in \Omega$. The associated radial basis functions are illustrated in Figures 4.18 and 4.19. The Shannon basis functions show some typical oscillatory behaviour which is not the case for the cp basis functions. For this reason, the latter can be expected to cause a lower risk of numerical artefacts.

Since bandlimited radial basis functions (i.e., $A_n \neq 0$ for only a finite number of n) only span a finite-dimensional space, only a finite number of points suffices to obtain a

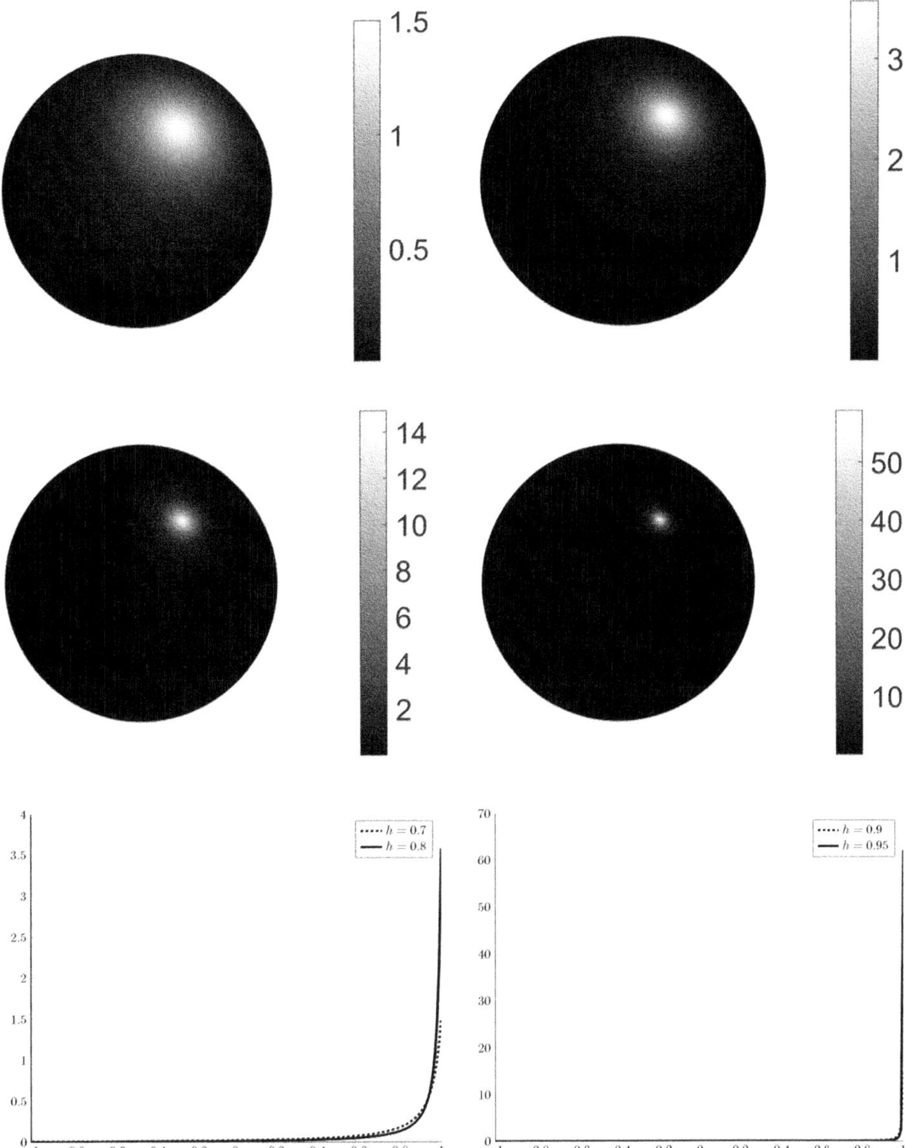

Figure 4.17 The Abel–Poisson kernel $\Omega \ni \xi \mapsto K_h(\eta \cdot \xi)$ with a fixed point η (location of the maximum) can be used to construct non-bandlimited radial basis functions by choosing a dense but countable subset $\left\{\eta^{(k)}\right\}_{k \in \mathbb{N}_0} \subset \Omega$ for the centres η. In this figure, the kernels are shown for one choice of η and different parameters $h = 0.7$ (top-left), $h = 0.8$ (top-right), $h = 0.9$ (middle-left), and $h = 0.95$ (middle-right). The bottom row shows the functions as zonal functions $[-1, 1] \ni t \mapsto K_h(t)$. The parameter h controls the localization in a sense that the 'hats' become higher and narrower with increasing h.

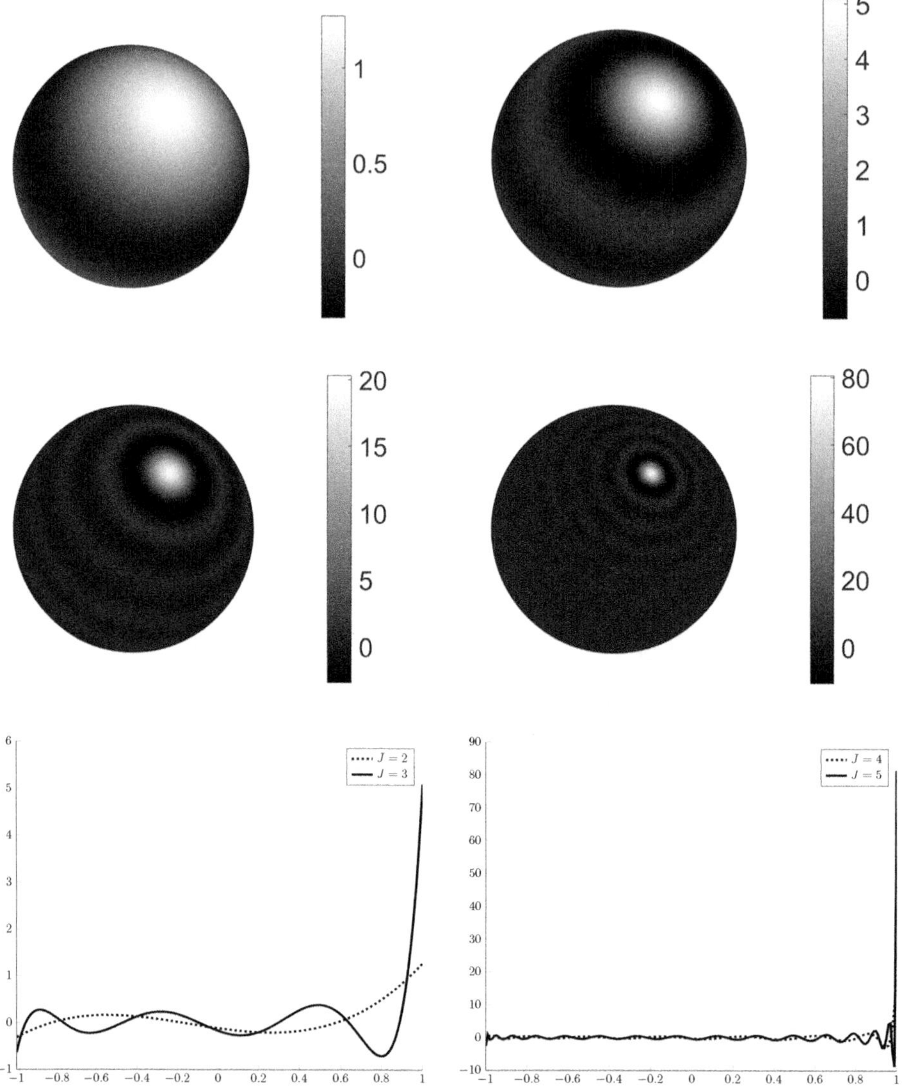

Figure 4.18 The Shannon scaling function $\Omega \ni \xi \mapsto K_J(\eta \cdot \xi)$ with a fixed point η (location of the maximum) can be used to construct bandlimited radial basis functions by choosing an appropriate subset $\left\{\eta^{(k)}\right\}_k \subset \Omega$ for the centres η. In this figure, the kernels are shown for one choice of η and different parameters $J = 2$ (top-left), $J = 3$ (top-right), $J = 4$ (middle-left), and $J = 5$ (middle-right). The bottom row shows the functions as zonal functions $[-1, 1] \ni t \mapsto K_J(t)$. In analogy to the parameter h for the Abel–Poisson kernel (see Figure 4.17), the so-called scale J controls the localization of the Shannon kernel. Shannon scaling functions have typical oscillatory patterns.

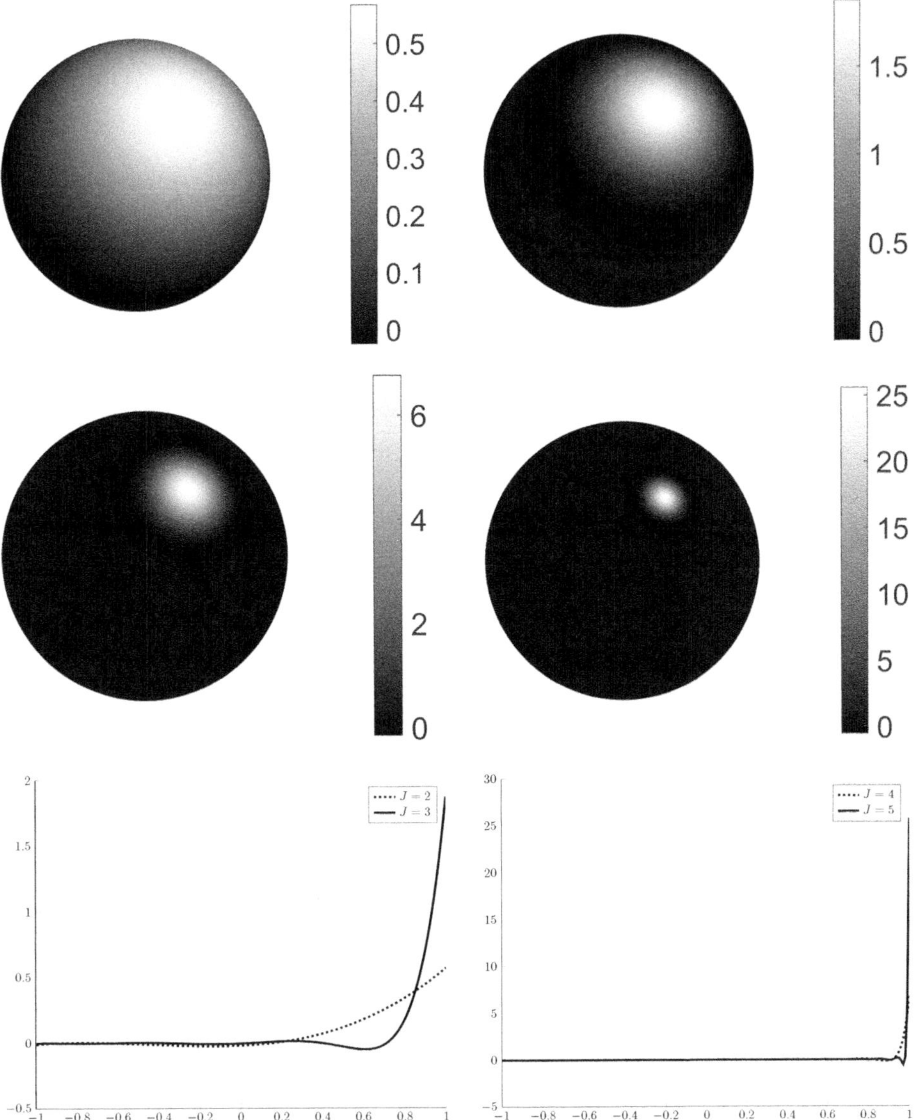

Figure 4.19 In analogy to the Shannon scaling function, shown in Figure 4.18, the cp scaling function is plotted here for different scales J. Again, the scale controls the localization of the trial function. However, the radial basis functions based on the cp scaling function are essentially smoother (i.e., less oscillatory) than the Shannon scaling function.

(theoretical) maximal coverage of the corresponding linear space $\mathcal{H}$. If $F \in \mathcal{H}$ and we want to find coefficients $a_1, \ldots, a_N$ such that

$$F(\xi) = \sum_{n=1}^{N} a_n K_{\mathcal{H}} \left(\xi \cdot \eta^{(n)} \right) \qquad \text{for all } \xi \in \Omega,$$

then we get the system of linear equations

$$F\left(\eta^{(k)}\right) = \sum_{n=1}^{N} a_n K_{\mathcal{H}}\left(\eta^{(k)} \cdot \eta^{(n)}\right), \quad k = 1, \ldots, N,$$

with the quadratic matrix $M := (K_{\mathcal{H}}(\eta^{(k)} \cdot \eta^{(n)}))_{k,n=1,\ldots,N}$. From the construction of $\mathcal{H}$, we have $\mathcal{H} = \operatorname{span}\{Y_{n,j} \mid A_n \neq 0, n \in \mathbb{N}_0, j = -n, \ldots, n\}$. Hence, if $N = \dim \mathcal{H}$ and the Gramian matrix

$$M = \left(\left\langle K_{\mathcal{H}}\left(\eta^{(k)} \cdot\right), K_{\mathcal{H}}\left(\eta^{(n)} \cdot\right)\right\rangle_{\mathcal{H}}\right)_{k,n=1,\ldots,N}$$

is regular, then the functions $\left\{K_{\mathcal{H}}\left(\eta^{(k)} \cdot\right)\right\}_{k=1,\ldots,N}$ constitute a basis of $\mathcal{H}$.

Theorem 4.7.15 *Let $(A_n)_{n\in\mathbb{N}_0}$ be a given real sequence where $A_n \neq 0$ for only a finite number of indices n. Furthermore, let $N := \sum_{n=0,\ A_n\neq 0}^{\infty}(2n+1)$. Then $N = \dim \mathcal{H}\ ((A_n); \Omega)$. If $\left\{\eta^{(k)}\right\}_{k=1,\ldots,N} \subset \Omega$ is a finite set of pairwise distinct points such that the matrix $M = \left(K_{\mathcal{H}}\left(\eta^{(k)} \cdot \eta^{(n)}\right)\right)_{k,n=1,\ldots,N}$ is regular, then the radial basis functions $F_k(\xi) := K_{\mathcal{H}}\left(\eta^{(k)} \cdot \xi\right)$, $\xi \in \Omega$, $k = 1, \ldots, N$, span the Sobolev space $\mathcal{H} = \mathcal{H}\ ((A_n); \Omega)$.*

In Section 4.5, we also mentioned the case of spacelimited basis functions (see Definition 4.5.1). Therefore, we discuss also an example of such trial functions. This example and the properties we recapitulate can be found in Freeden et al. (1998, section 5.8.2), which is, together with the references therein, recommended for further details. Moveover, further discussions can also be found in Freeden and Hesse (2002).

Definition 4.7.16 For each $h \in]0,1[$ and each $k \in \mathbb{N}_0$, we define the function $B_h^{(k)}\colon [-1,1] \to \mathbb{R}$ by

$$B_h^{(k)}(t) := \begin{cases} \frac{(t-h)^k}{(1-h)^k}, & \text{if } t \geq h, \\ 0, & \text{if } t < h, \end{cases} \qquad t \in [-1,1].$$

The corresponding zonal functions $\Omega \ni \xi \mapsto B_h^{(k)}(\eta \cdot \xi)$ for fixed $\eta \in \Omega$ are called **smoothed Haar scaling functions on the sphere** or **finite elements on the sphere**.

The functions are shown in Figures 4.20–4.23. For a better illustration of the different orders of smoothness, the 1D-functions $B_h^{(k)}$ are also shown in Figure 4.24.

Is Theorem 4.7.13 applicable to the functions $B_h^{(k)}$? We have to interpret them as reproducing kernels in order to clarify this. By using Theorem 4.2.10, we can apply an orthonormal basis of $\mathrm{L}^2[-1,1]$ which is based on the Legendre polynomials, since each $B_h^{(k)}$ is obviously an element of $\mathrm{L}^2[-1,1]$. Moreover, $B_h^{(k)} \in \mathrm{C}^{(k-1)}[-1,1]$ for all $k \geq 1$. We have

$$B_h^{(k)} = \sum_{n=0}^{\infty} \left\langle B_h^{(k)}, P_n \right\rangle_{\mathrm{L}^2[-1,1]} \frac{2n+1}{2} P_n = \sum_{n=0}^{\infty} \left(B_h^{(k)}\right)^{\wedge}(n) \frac{2n+1}{4\pi} P_n \tag{4.60}$$

in the sense of $\mathrm{L}^2[-1,1]$, where we used a common abbreviation.

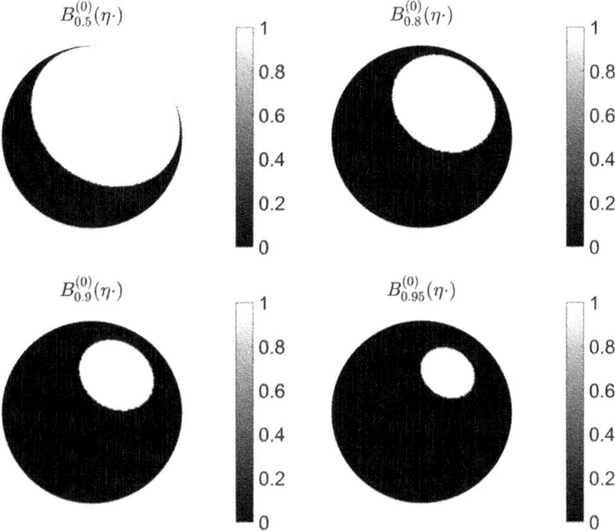

Figure 4.20 The spacelimited radial basis functions $\Omega \ni \xi \mapsto B_h^{(k)}(\eta \cdot \xi)$ with a fixed point η have $\operatorname{supp} B_h^{(k)}(\eta\cdot) = \{\xi \in \Omega \,|\, \eta \cdot \xi \geq h\}$ as their support. In this figure, the kernels are shown for different parameters h and the fixed parameter $k = 0$. The parameter h controls the support, that is, the extent of the spacelimitation, whereas the parameter k controls the smoothness of the function. In general, $B_h^{(k)}(\eta\cdot) \in \mathrm{C}^{(k-1)}(\Omega)$ for $k \geq 1$. In the case $k = 0$, the functions are indeed discontinuous at the boundary of the support.

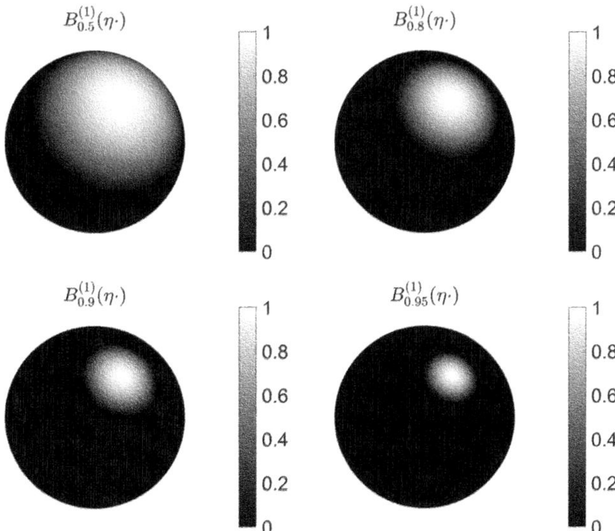

Figure 4.21 In analogy to Figure 4.20, the spacelimited radial basis functions $\Omega \ni \xi \mapsto B_h^{(k)}(\eta \cdot \xi)$ with a fixed point η are shown for different parameters h and the fixed parameter $k = 1$. Note that these functions are continuous on the whole sphere but not differentiable at the boundary of their support.

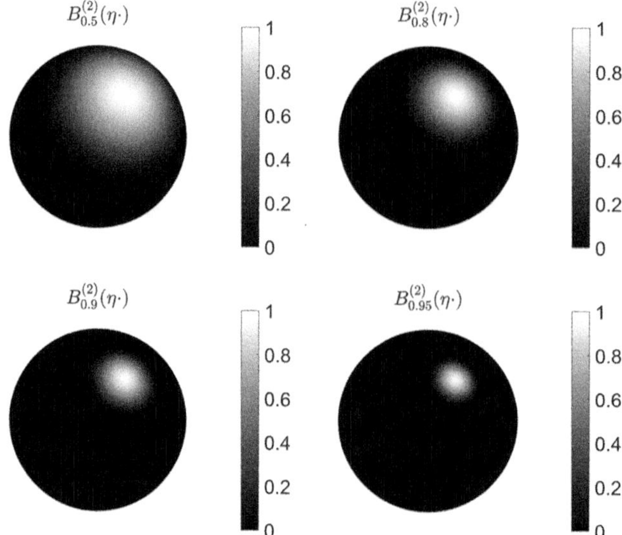

Figure 4.22 In analogy to Figures 4.20 and 4.21, the spacelimited radial basis functions $\Omega \ni \xi \mapsto B_h^{(k)}(\eta \cdot \xi)$ with a fixed point η are shown for different parameters h and the fixed parameter $k = 2$. Note that these functions are continuously differentiable on the whole sphere but not twice differentiable at the boundary of their support.

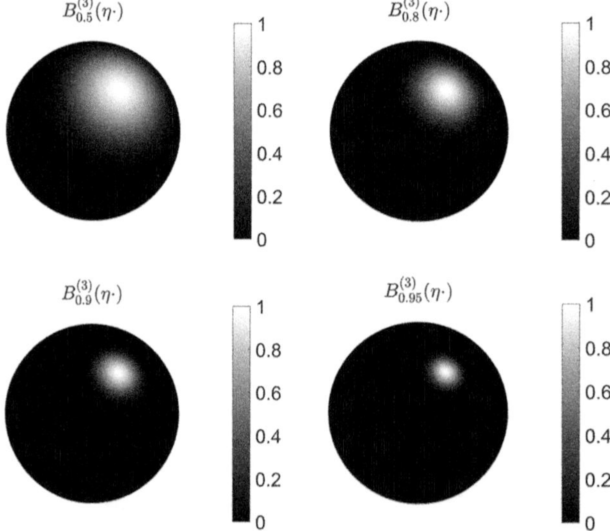

Figure 4.23 In analogy to Figures 4.20–4.22, the spacelimited radial basis functions $\Omega \ni \xi \mapsto B_h^{(k)}(\eta \cdot \xi)$ with a fixed point η are shown for different parameters h and the fixed parameter $k = 3$. Note that these functions are twice continuously differentiable on the whole sphere but not three times differentiable at the boundary of their support.

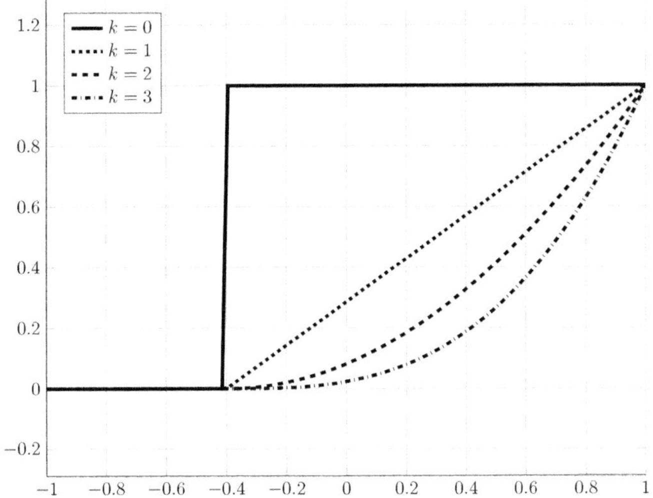

Figure 4.24 The 1D-functions $B_h^{(k)}$ are shown for a fixed parameter $h = -0.4$ and varying parameters $k = 0, \ldots, 3$. Note that $B_h^{(k)} \in \mathrm{C}^{(k-1)}[-1,1]$ for $k \geq 1$.

Definition 4.7.17 For each $G \in \mathrm{L}^2[-1,1]$, the **Legendre coefficients** are defined by $G^\wedge(n) := 2\pi \int_{-1}^{1} G(t)P_n(t)\,\mathrm{d}t$, $n \in \mathbb{N}_0$, where P_n is the Legendre polynomial of degree n.

Due to the obvious link to the representation of a reproducing kernel on the sphere (see Theorem 4.7.10), we have to further investigate these Legendre coefficients.

There is a known recurrence formula for the Legendre coefficients of $B_h^{(k)}$, whose proof we will recapitulate here.

Theorem 4.7.18 *For each $h \in]0,1[$ and $k \in \mathbb{N}_0$ as well as all $n \in \mathbb{N}_0$, the Legendre coefficients of $B_h^{(k)}$ satisfy*

$$\left(B_h^{(k)}\right)^\wedge(0) = 2\pi\,\frac{1-h}{k+1}, \tag{4.61}$$

$$\left(B_h^{(k)}\right)^\wedge(1) = 2\pi\,\frac{(1-h)(k+1+h)}{(k+1)(k+2)}, \tag{4.62}$$

$$(n+k+3)\left(B_h^{(k)}\right)^\wedge(n+2) = (2n+3)h\left(B_h^{(k)}\right)^\wedge(n+1) + (k-n)\left(B_h^{(k)}\right)^\wedge(n). \tag{4.63}$$

Proof First of all, we notice that the local support of $B_h^{(k)}$ causes that

$$\left(B_h^{(k)}\right)^\wedge(n) = \frac{2\pi}{(1-h)^k}\int_h^1 (t-h)^k P_n(t)\,\mathrm{d}t. \tag{4.64}$$

For $n = 0$ and $n = 1$, this is easy to calculate:

$$\left(B_h^{(k)}\right)^\wedge(0) = \frac{2\pi}{(1-h)^k}\int_h^1 (t-h)^k\,\mathrm{d}t = \frac{2\pi}{(1-h)^k}\,\frac{(1-h)^{k+1}}{k+1} = 2\pi\,\frac{1-h}{k+1},$$

$$\begin{aligned}
\left(B_h^{(k)}\right)^\wedge(1) &= \frac{2\pi}{(1-h)^k}\int_h^1 (t-h)^k t\,\mathrm{d}t \\
&= \frac{2\pi}{(1-h)^k}\left[\frac{(t-h)^{k+1}}{k+1}\,t\Big|_{t=h}^{t=1} - \int_h^1 \frac{(t-h)^{k+1}}{k+1}\,\mathrm{d}t\right] \\
&= 2\pi\,\frac{1-h}{k+1} - 2\pi\,\frac{1-h}{k+1}\,\frac{1-h}{k+2} \\
&= 2\pi\,\frac{1-h}{(k+1)(k+2)}\,(k+1+h).
\end{aligned}$$

Hence, (4.61) and (4.62) are valid.

For the recurrence relation of the coefficients, we use one of the known recurrence formulae for the Legendre polynomials, namely (3.110).

$$\begin{aligned}
&(n+2)\int_h^1 (t-h)^k P_{n+2}(t)\,\mathrm{d}t \\
&\quad= (2n+3)\int_h^1 (t-h)^k t\,P_{n+1}(t)\,\mathrm{d}t - (n+1)\int_h^1 (t-h)^k P_n(t)\,\mathrm{d}t \\
&\quad= (2n+3)\int_h^1 (t-h)^{k+1} P_{n+1}(t)\,\mathrm{d}t + (2n+3)h\int_h^1 (t-h)^k P_{n+1}(t)\,\mathrm{d}t \\
&\qquad - (n+1)\int_h^1 (t-h)^k P_n(t)\,\mathrm{d}t.
\end{aligned} \tag{4.65}$$

We apply now (3.113) and integration by parts to the first integral on the right-hand side of (4.65) and obtain

$$\begin{aligned}
&(2n+3)\int_h^1 (t-h)^{k+1} P_{n+1}(t)\,\mathrm{d}t \\
&= \int_h^1 (t-h)^{k+1}\left(P'_{n+2}(t) - P'_n(t)\right)\,\mathrm{d}t \\
&= (1-h)^{k+1}\underbrace{\left(P_{n+2}(1) - P_n(1)\right)}_{=0} - (k+1)\int_h^1 (t-h)^k\left(P_{n+2}(t) - P_n(t)\right)\,\mathrm{d}t.
\end{aligned} \tag{4.66}$$

When we insert this into (4.65), we obtain using (4.64) the result

$$\begin{aligned}
(n+2)\left(B_h^{(k)}\right)^\wedge(n+2) &= -(k+1)\left[\left(B_h^{(k)}\right)^\wedge(n+2) - \left(B_h^{(k)}\right)^\wedge(n)\right] \\
&\quad + (2n+3)h\left(B_h^{(k)}\right)^\wedge(n+1) - (n+1)\left(B_h^{(k)}\right)^\wedge(n).
\end{aligned}$$

We can easily summarize this to get (4.63). □

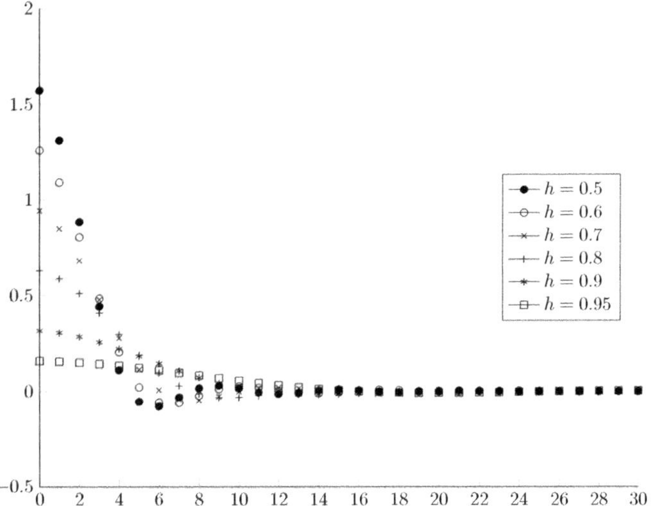

Figure 4.25 Legendre coefficients of the locally supported trial functions $B_h^{(1)}$ for different values of h: note that some of the coefficients are negative.

The recurrence relation allows the numerical calculation of the Legendre coefficients of the spherical finite elements. Some coefficients are shown in Figure 4.25. In this figure, we already see that the $B_h^{(k)}$ cannot be reproducing kernels. $\left(B_h^{(k)}\right)^\wedge(n) = A_n^2$ needs to hold, but some of the $\left(B_h^{(k)}\right)^\wedge(n)$ are negative.

This known drawback of the spherical Haar functions was tackled in the literature by using the concept of the convolution of spherical radial basis functions. We will only use this as a tool here such that we merely summarize what is important for the further considerations but omit the proofs. For details, see, for example, Freeden et al. (1998), Freeden and Schreiner (2009), and Michel (2013).

Theorem 4.7.19 *Let $G, H \in \mathrm{L}^2[-1,1]$. Then the function $K\colon \Omega^2 \to \mathbb{R}$, which is defined by*

$$K(\xi,\eta) := \int_\Omega G(\xi \cdot \zeta)\, H(\zeta \cdot \eta)\, \mathrm{d}\omega(\zeta), \quad \xi,\eta \in \Omega,$$

is a zonal function, that is, there is $\widetilde{K} \in \mathrm{L}^2[-1,1]$ such that $\widetilde{K}(\xi\cdot\eta) = K(\xi,\eta)$ almost everywhere. This function is denoted by $G * H := \widetilde{K}$ and is called a ***convolution of zonal functions***. *It has the property that $(G * H)^\wedge(n) = G^\wedge(n)H^\wedge(n)$ for all $n \in \mathbb{N}_0$.*

Definition 4.7.20 Let $G \in \mathrm{L}^2[-1,1]$. Then the **iterated convolution** $G^{(k)}$, $k \in \mathbb{N}$, is defined by $G^{(1)} := G$ and $G^{(k+1)} := G^{(k)} * G$ for all $k \in \mathbb{N}$.

We will now use the iterated convolution $\left(B_h^{(k)}\right)^{(2)}$, as it has only non-negative Legendre coefficients. This function can be calculated by using (a truncation of) the expansion $\left(B_h^{(k)}\right)^{(2)} = \sum_{n=0}^\infty \left[\left(B_h^{(k)}\right)^\wedge(n)\right]^2 [(2n+1)/(4\pi)] P_n$. Nevertheless, the question whether these functions are reproducing kernels cannot be answered with the facts that we have recapitulated so far. For continuing, we need a known property of Legendre polynomials which we will only quote here (see Freeden et al., 1998, (3.2.44)).

Lemma 4.7.21 *Each Legendre polynomial satisfies the estimate*

$$|P_n(t)| \le \sqrt{\frac{2}{\pi(n+1/2)}}\,(1-t^2)^{-1/4}, \qquad \textit{if } -1 < t < 1.$$

Now we come to the result which we need for deciding about the occurrence of a reproducing kernel. Note that we show here a stronger result in comparison to Freeden et al. (1998, p. 126).

Theorem 4.7.22 *The Legendre coefficients of the spherical finite elements, for fixed $k \in \mathbb{N}_0$ and $h \in]0,1[$, have the asymptotic behaviour*

$$\left|\left(B_h^{(k)}\right)^\wedge(n)\right| = \mathcal{O}\left(n^{-k-5/2}\right) \quad \textit{as } n \to \infty.$$

Proof In the case $k = 0$, we have, due to (3.113), the identity

$$\left(B_h^{(0)}\right)^\wedge(n) = 2\pi \int_h^1 P_n(t)\,\mathrm{d}t = \frac{2\pi}{2n+1}\left(P_{n-1}(h) - P_{n+1}(h)\right)$$

for all $h \in]0,1[$ and all $n \in \mathbb{N}$. Then Lemma 4.7.21 enables us to conclude that

$$\begin{aligned}
\left|\left(B_h^{(0)}\right)^\wedge(n)\right| &\le \frac{\sqrt{8\pi}}{2n+1}(1-h^2)^{-1/4}\left(\frac{1}{\sqrt{n-1/2}} - \frac{1}{\sqrt{n+3/2}}\right)\\
&= \frac{\sqrt{8\pi}}{2n+1}(1-h^2)^{-1/4}\frac{\sqrt{n+3/2}-\sqrt{n-1/2}}{\sqrt{n^2+n-3/4}}\\
&= \frac{\sqrt{8\pi}}{2n+1}(1-h^2)^{-1/4}\frac{2}{\sqrt{n^2+n-3/4}\left(\sqrt{n+3/2}+\sqrt{n-1/2}\right)}\\
&= \mathcal{O}\left(n^{-5/2}\right)
\end{aligned}$$

as $n \to \infty$. Now we have a look at (4.66), according to which we have

$$(1-h)(2n+3)\left(B_h^{(k+1)}\right)^\wedge(n+1) = (k+1)\left[\left(B_h^{(k)}\right)^\wedge(n) - \left(B_h^{(k)}\right)^\wedge(n+2)\right].$$

Hence, an easy induction yields that

$$\left|\left(B_h^{(k)}\right)^\wedge(n)\right| = \mathcal{O}\left(n^{-k}n^{-5/2}\right) \quad \text{as } n \to \infty. \qquad \square$$

Now we combine Theorems 4.7.10 and 4.7.22, Definition 4.7.4, and (4.60), and our previous question is answered.

Theorem 4.7.23 *Let $h \in]0,1[$ and $k \in \mathbb{N}_0$ be arbitrary. Then $(B_h^{(k)})^{(2)}$ is the reproducing kernel of the Sobolev space $\mathcal{H}\ ((A_n)\,;\Omega)$ associated to*

$$A_n := \begin{cases} \left|\left(B_h^{(k)}\right)^{\wedge}(n)\right|, & \text{if } \left(B_h^{(k)}\right)^{\wedge}(n) \neq 0, \\ 0 & \text{else}. \end{cases}$$

Hence, we can use the iterated spherical finite elements $\left(B_h^{(k)}\right)^{(2)}$ to construct spacelimited radial basis functions based on Theorems 4.7.13 and 4.7.15, respectively.

In this context, the following theorem, which was proved in Freeden et al. (1998, corollary 5.8.6), is worth knowing.

Theorem 4.7.24 *Let $k \in \mathbb{N}_0$ be chosen arbitrarily.*

(a) There exist parameters $h \in]0,1[$ for which $\left(B_h^{(k)}\right)^{\wedge}(n) \neq 0$ for all $n \in \mathbb{N}_0$.
(b) For each $m \in \mathbb{N}_0$, there exists $h_0 \in]0,1[$ such that $\left(B_h^{(k)}\right)^{\wedge}(n) \neq 0$ for all $h \in]h_0,1[$ and all $n \leq m$.

Since not only radial basis functions but also Sobolev spaces on the sphere comprise the topics of this section, we also discuss here a certain aspect of differential operators and equations, respectively, on the sphere. However, we skip the general abstract theory and consider the example of the Beltrami operator instead, which we will need later on.

In view of Theorem 4.2.19, we can define the Beltrami operator in an alternative way.

Definition 4.7.25 The **Beltrami operator** can also be defined as the operator $\Delta^*\colon \mathcal{H}\ \left((A_n)_{n\in\mathbb{N}_0}\,;\Omega\right) \to \mathrm{L}^2(\Omega)$ with

$$\Delta^* V := \sum_{n=1}^{\infty}\sum_{j=-n}^{n} [-n(n+1)]\,\langle V, Y_{n,j}\rangle_{\mathrm{L}^2(\Omega)} Y_{n,j},$$

$A_0 := 0$, and $A_n := [n(n+1)]^{-1}$ for $n \in \mathbb{N}$. We write $\mathcal{H}_{\Delta^*}$ for this Sobolev space.

Indeed, $\Delta^* V \in \mathrm{L}^2(\Omega)$, because

$$\|\Delta^* V\|_{\mathrm{L}^2(\Omega)}^2 = \sum_{n=1}^{\infty}\sum_{j=-n}^{n} [n(n+1)]^2 \langle V, Y_{n,j}\rangle_{\mathrm{L}^2(\Omega)}^2 = \|V\|_{\mathcal{H}\ ((A_n)\,;\Omega)}^2$$

for all $V \in \mathcal{H}\ ((A_n)\,;\Omega)$. This also shows that the linear operator is bounded.

Theorem 4.7.26 *The Beltrami operator from Definition 4.7.25 is linear, continuous, and isometric.*

In the following, we investigate the solutions of the partial differential equation

$$\Delta_\xi^* U(\xi) = F(\xi), \quad \xi \in \Omega, \tag{4.67}$$

where $F \in \mathrm{L}^2(\Omega)$ is given and $U \in \mathcal{H}_{\Delta^*}$ is unknown. The considerations here are based on Backus (1986). Let us first state a theorem on the solutions of (4.67) and then prove it.

Theorem 4.7.27 *Let $F \in \mathrm{L}^2(\Omega)$ be an arbitrary function. Then* (4.67) *is solvable if and only if*

$$\int_\Omega F(\xi)\, \mathrm{d}\omega(\xi) = 0. \tag{4.68}$$

Moreover, if $U \in \mathrm{C}^{(2)}(\Omega)$ solves (4.67)*, then all solutions of the equation must have the form $U + c$, where $c \in \mathbb{R}$ is an arbitrary constant. The solution is unique if one additionally requires that*

$$\int_\Omega U(\xi)\, \mathrm{d}\omega(\xi) = 0. \tag{4.69}$$

Note that, for $U \in \mathcal{H}_{\Delta^*}$, condition (4.69) is satisfied anyway.

Proof The proof is a bit lengthy and therefore subdivided.

(1) Solution space:

It is clear that $U + c$ solves (4.67) if U is a solution. Let us now assume that U_1 and U_2 both solve the equation. Then $\Delta^*(U_1 - U_2) = 0$. Hence, Theorem 4.2.19 tells us that $U_1 - U_2 \in \mathrm{Harm}_0(\Omega)$, that is, $U_1 - U_2$ is constant on Ω.

(2) Uniqueness:

This fact is clear, since $\int_\Omega (U(\xi) + c)\, \mathrm{d}\omega(\xi) = \int_\Omega U(\xi)\, \mathrm{d}\omega(\xi) + 4\pi c = 0$ can only be achieved by one constant c.

(3) Necessary condition for the existence:

With Green's first surface identity, we immediately obtain that

$$\int_\Omega \underbrace{(\nabla_\xi^* 1)}_{=0} \cdot \left(\nabla_\xi^* U(\xi)\right) \mathrm{d}\omega(\xi) + \int_\Omega 1 \cdot \Delta_\xi^*\, U(\xi)\, \mathrm{d}\omega(\xi) = 0$$

such that (4.68) is necessary for the existence of a solution. Besides, if $U \in \mathcal{H}_{\Delta^*}$ and $\Delta^* U = F \in \mathrm{L}^2(\Omega)$, then Definition 4.7.25 immediately yields (4.68).

(4) Existence of the solution:

Since $F \in \mathrm{L}^2(\Omega)$, Theorem 4.2.18 allows us to expand F in spherical harmonics as $F = \sum_{n=0}^\infty \sum_{j=-n}^n \langle F, Y_{n,j} \rangle_{\mathrm{L}^2(\Omega)} Y_{n,j}$ in the sense of $\mathrm{L}^2(\Omega)$. Since $Y_{0,0}$ is the constant basis function, the condition in (4.68) implies that $\langle F, Y_{0,0} \rangle_{\mathrm{L}^2(\Omega)} = 0$. Let us now check the following candidate for a solution:

$$U(\xi) := \sum_{n=1}^\infty \sum_{j=-n}^n \frac{-1}{n(n+1)} \langle F, Y_{n,j} \rangle_{\mathrm{L}^2(\Omega)} Y_{n,j}(\xi), \quad \xi \in \Omega.$$

Then, obviously, $\langle U, Y_{0,0} \rangle_{\mathrm{L}^2(\Omega)} = 0$, that is, (4.69) holds true. It is clear that $U \in \mathcal{H}_{\Delta^*}$, because

$$\sum_{n=1}^\infty \sum_{j=-n}^n [n(n+1)]^2 \langle U, Y_{n,j} \rangle_{\mathrm{L}^2(\Omega)}^2 = \|F\|_{\mathrm{L}^2(\Omega)}^2 < +\infty.$$

Furthermore, we define

$$F_N(\xi) := \sum_{n=1}^{N} \sum_{j=-n}^{n} \langle F, Y_{n,j} \rangle_{\mathrm{L}^2(\Omega)} Y_{n,j}(\xi),$$

$$U_N(\xi) := \sum_{n=1}^{N} \sum_{j=-n}^{n} \frac{-1}{n(n+1)} \langle F, Y_{n,j} \rangle_{\mathrm{L}^2(\Omega)} Y_{n,j}(\xi),$$

$\xi \in \Omega$, where clearly $U_N \in \mathcal{H}_{\Delta^*}$ and $F_N \in \mathrm{L}^2(\Omega)$ for all $N \in \mathbb{N}$. Moreover, we have $\|F - F_N\|_{\mathrm{L}^2(\Omega)} \to 0$ as $N \to \infty$ (for trivial reasons) and $\|U - U_N\|_{\mathcal{H}_{\Delta^*}} = \|F - F_N\|_{\mathrm{L}^2(\Omega)} \to 0$. Since $\Delta^* U_N = F_N$ due to Theorem 4.2.19 (or Definition 4.7.25) and the operator $\Delta^* \colon \mathcal{H}_{\Delta^*} \to \mathrm{L}^2(\Omega)$ is continuous (see Theorem 4.7.26), we are now able to conclude that $\Delta^* U = {}^d\lim_{n\to\infty} \Delta^* U_N = F$ in the sense of $\mathrm{L}^2(\Omega)$, that is, d is the metric induced by $\|\cdot\|_{\mathrm{L}^2(\Omega)}$. □

Consequently, we can define an inverse operator.

Definition 4.7.28 The operator

$$(\Delta^*)^{-1} \colon \left\{ F \in \mathrm{L}^2(\Omega) \,|\, \langle F, Y_{0,0} \rangle_{\mathrm{L}^2(\Omega)} = 0 \right\} \to \mathcal{H}_{\Delta^*}$$

is defined by

$$(\Delta^*)^{-1} F := \sum_{n=1}^{\infty} \sum_{j=-n}^{n} \frac{-1}{n(n+1)} \langle F, Y_{n,j} \rangle_{\mathrm{L}^2(\Omega)} Y_{n,j}$$

for all $F \in \mathrm{L}^2(\Omega)$ with $\int_\Omega F(\xi)\, \mathrm{d}\omega(\xi) = 0$.

Theorem 4.7.29 *Let the kernel $K \colon [-1,1] \to \mathbb{R}$ be defined by*

$$K(\xi \cdot \eta) = \sum_{n=1}^{\infty} \frac{-2n-1}{n(n+1)4\pi} P_n(\xi \cdot \eta) = \sum_{n=1}^{\infty} \sum_{j=-n}^{n} \frac{-1}{n(n+1)} Y_{n,j}(\xi)\, Y_{n,j}(\eta),$$

$\xi, \eta \in \Omega$. Then $(\Delta^*)^{-1}$ can be represented as the Fredholm integral operator

$$\left[(\Delta^*)^{-1} F\right](\xi) = \int_\Omega K(\xi \cdot \eta) F(\eta)\, \mathrm{d}\omega(\eta).$$

Proof Clearly, $K(\xi \cdot) \in \mathrm{L}^2(\Omega)$ for all $\xi \in \Omega$, because

$$\|K(\xi \cdot)\|^2_{\mathrm{L}^2(\Omega)} = \sum_{n=1}^{\infty} \sum_{j=-n}^{n} \frac{1}{n^2(n+1)^2} \left(Y_{n,j}(\xi)\right)^2 = \sum_{n=1}^{\infty} \frac{2n+1}{n^2(n+1)^2 4\pi} < +\infty.$$

Hence, the Parseval identity yields, for all $\xi \in \Omega$,

$$\begin{aligned} \int_\Omega K(\xi \cdot \eta) F(\eta)\, \mathrm{d}\omega(\eta) &= \langle K(\xi \cdot), F \rangle_{\mathrm{L}^2(\Omega)} \\ &= \sum_{n=1}^{\infty} \sum_{j=-n}^{n} \frac{-1}{n(n+1)} Y_{n,j}(\xi)\, \langle F, Y_{n,j} \rangle_{\mathrm{L}^2(\Omega)} \\ &= \left[(\Delta^*)^{-1} F\right](\xi). \end{aligned}$$

□

In terms of differentiability, that is, the discussion of $\Delta^* U = F$ for the Beltrami operator on $\mathrm{C}^{(2)}(\Omega)$, there is a theorem which is the spherical analogue to the theorem for the Poisson equation in Euclidean space (see Theorem 3.1.12). For the proof of the following theorem, the reader is referred to (Aubin, 1982, theorem 4.7).

Theorem 4.7.30 *Let $F \in \mathrm{C}^{(k,\alpha)}(\Omega)$ with $k \in \mathbb{N}_0$ and $0 < \alpha < 1$. If $\int_\Omega F(\xi)\,\mathrm{d}\omega(\xi) = 0$, then there exists a function $U \in \mathrm{C}^{(k+2,\alpha)}(\Omega)$ such that $\Delta^* U = F$. If $F \in \mathrm{C}^{(\infty)}(\Omega)$, then $U \in \mathrm{C}^{(\infty)}(\Omega)$.*

4.8 Basis Functions on the 3D Ball – Briefly

Basis functions on the three-dimensional ball are also of importance in Earth sciences. Such systems are needed whenever the Earth's interior comes into play. This can be the case for tomographic inverse problems, where a connection to medical imaging is obvious, or for the modelling of processes in the mantle or the core. In this book, the most important topics are only summarized, but basically without proofs for the sake of brevity. The reader shall be provided with the tools which he or she might need, whereas the theory has already been summarized and elaborated in Michel (2013, part III). For more recent results, the reader is also referred to Leweke et al. (2018a) and Michel and Orzlowski (2016). For a generalization to d-dimensional balls, see Ishtiaq (2018) and Ishtiaq et al. (2019). Since the Earth's interior approximately consists of layers with boundaries which are concentric spheres, it is canonical to seek basis functions G on the ball $\mathcal{B} := \{x \in \mathbb{R}^3 \mid |x| \le \beta\} = \overline{B_\beta(0)}$ which have the form $G(x) = F(|x|)Y(x/|x|)$. The considerations here are consequently restricted to such basis functions. For orthogonal basis functions on the ball where Cartesian coordinates are used, see Dunkl and Xu (2001).

The ansatz $G(x) = F(|x|)Y(x/|x|)$ leads to the following orthonormality condition for two functions $G, \widetilde{G} \in \mathrm{L}^2(\mathcal{B})$:

$$\int_{\mathcal{B}} G(x)\widetilde{G}(x)\,\mathrm{d}x = \begin{cases} 0, & \text{if } G \neq \widetilde{G}, \\ 1, & \text{if } G = \widetilde{G}, \end{cases}$$

where

$$\int_{\mathcal{B}} G(x)\widetilde{G}(x)\,\mathrm{d}x = \int_0^\beta r^2 F(r)\widetilde{F}(r)\,\mathrm{d}r \int_\Omega Y(\xi)\widetilde{Y}(\xi)\,\mathrm{d}\omega(\xi).$$

Obviously, the system $\{Y_{n,j}\}_{n\in\mathbb{N}_0;\, j=-n,\dots,n}$ of spherical harmonics is a good choice for the angular dependence of G. For the radially dependent part, we need functions which are orthogonal with respect to a weighted L^2-inner product. Depending on the application, different choices for F have been investigated in the literature. In each case, Jacobi polynomials are used.

Theorem 4.8.1 *Let $\alpha, \beta > -1$ be fixed real numbers. Then there exists one and only one sequence of polynomials $\left\{P_m^{(\alpha,\beta)}\right\}_{m\in\mathbb{N}_0}$ with the following properties:*

(a) Each $P_m^{(\alpha,\beta)}$ has the degree m.
(b) If $m_1 \neq m_2$, then

$$\int_{-1}^{1} (1-x)^{\alpha}(1+x)^{\beta}\, P_{m_1}^{(\alpha,\beta)}(x)\, P_{m_2}^{(\alpha,\beta)}(x)\,\mathrm{d}x = 0.$$

(c) *For every $m \in \mathbb{N}_0$, we have $P_m^{(\alpha,\beta)}(1) = \binom{m+\alpha}{m}$, where the binomial coefficient is defined as in Definition 2.3.7.*

*These unique functions are called the **Jacobi polynomials**.*

Note that the Legendre polynomials represent a particular case of the Jacobi polynomials since $P_n = P_n^{(0,0)}$. For further details on such 1D-orthogonal polynomials, see, for example, Chihara (1978), Michel (2013), and Szegö (1975). For normalizing the basis on the ball, we also need the weighted L^2-norm of the Jacobi polynomials, whose proof is omitted here.

Theorem 4.8.2 *For each $\alpha, \beta > -1$ and all $m \in \mathbb{N}_0$, we have*

$$\int_{-1}^{1} (1-x)^{\alpha}(1+x)^{\beta}\left[P_m^{(\alpha,\beta)}(x)\right]^2 \mathrm{d}x$$
$$= \frac{2^{\alpha+\beta+1}}{2m+\alpha+\beta+1}\;\frac{\Gamma(m+\alpha+1)\Gamma(m+\beta+1)}{m!\;\Gamma(m+\alpha+\beta+1)},$$

where Γ is the gamma function.

Note that this coincides with the particular result for the Legendre polynomials if $\alpha = \beta = 0$; see Theorem 3.4.22.

Theorem 4.8.3 (Orthonormal Basis on the Ball) *Let $\mathcal{B} := \overline{B_\beta(0)}$. Then each of the following systems is an orthonormal basis of $\mathrm{L}^2(\mathcal{B})$.*

(a) *Let $(\ell_n)_{n\in\mathbb{N}_0}$ be a fixed sequence of non-negative real numbers. Then we set*

$$G_{m,n,j}^{\mathrm{I}}(x) := \sqrt{\frac{4m+2\ell_n+3}{\beta^3}}\; P_m^{(0,\ell_n+1/2)}\left(2\frac{|x|^2}{\beta^2}-1\right)\frac{|x|^{\ell_n}}{\beta^{\ell_n}}\, Y_{n,j}\left(\frac{x}{|x|}\right),$$

$x \in \mathcal{B}\setminus\{0\}$, $m,n \in \mathbb{N}_0$, $j \in \{-n,\dots,n\}$.

(b) *We define*

$$G_{m,n,j}^{\mathrm{II}}(x) := \sqrt{\frac{2m+3}{\beta^3}}\; P_m^{(0,2)}\left(2\frac{|x|}{\beta}-1\right) Y_{n,j}\left(\frac{x}{|x|}\right),$$

$x \in \mathcal{B}\setminus\{0\}$, $m,n \in \mathbb{N}_0$, $j \in \{-n,\dots,n\}$.

If propositions are formulated which are valid for both types of basis systems, then we will simply write $G_{m,n,j}$ to represent the arbitrary choice.

Remark 4.8.4 The different basis systems have different properties and different applications:

(a) If we set $\ell_n := n$ for all $n \in \mathbb{N}_0$ in type I, then the functions play an important role in the context of the inverse gravimetric problem; see also Section 5.2. Indeed, this system was developed in Ballani et al. (1993) and Dufour (1977) for this particular application. A closer look at the basis functions reveals that $\left\{G^{\mathrm{I}}_{0,n,j}\right\}_{n\in\mathbb{N}_0;\,j=-n,\ldots,n}$ is a basis of all harmonic functions (the basis of inner harmonics) on $\mathcal{B}$; see also the discussions after Theorem 3.4.25. The elements of the $\mathrm{L}^2(\mathcal{B})$-orthogonal complement, which is spanned by $\left\{G^{\mathrm{I}}_{m,n,j}\right\}_{m\in\mathbb{N},n\in\mathbb{N}_0;\,j=-n,\ldots,n}$, are called **anharmonic functions**.

These functions $G^{\mathrm{I}}_{m,n,j}$ with $\ell_n := n$ for all n also have another nice property: since we can write them as the product of a polynomial in $|x|^2 = x_1^2 + x_2^2 + x_3^2$ and an inner harmonic, all $G^{\mathrm{I}}_{m,n,j}$ of this kind are algebraic polynomials in x_1, x_2, x_3 and can, particularly, be extended to a $\mathrm{C}^{(\infty)}(\mathcal{B})$-function by setting $G^{\mathrm{I}}_{m,n,j}(0) := 0$ if $n > 0$, and $G^{\mathrm{I}}_{m,0,0}(0) := \sqrt{(4m+3)/(4\pi\beta^3)}\, P_m^{(0,1/2)}(-1)$, where the well-known theory of Jacobi polynomials tells us that $P_m^{(0,1/2)}(-1) = (-1)^m \binom{m+1/2}{m}$.

(b) The case $\ell_n := n - 1$ for all $n \in \mathbb{N}$, where the functions $G^{\mathrm{I}}_{m,n,j}$ are only considered for $n > 0$, plays an important role in medical imaging; see Leweke (2018) and Leweke et al. (2020) for further details.

(c) The functions of type II go back to Tscherning (1996). Their advantage is the complete decoupling of the radial and the angular part, whereas type I requires (if $\ell_n \neq \ell_\nu$ for $n \neq \nu$) calculating a set of Jacobi polynomials for each angular degree n. On the other hand, type II is not extendable to $x = 0$, if $n > 0$, because then $\lim_{x\to 0} G^{\mathrm{II}}_{m,n,j}(x)$ does not exist.

With Theorem 2.5.27, we also get the following property.

Theorem 4.8.5 *The functions* $\left\{G^{\mathrm{I}}_{m,n,j}\right\}_{m,n\in\mathbb{N}_0;\,j=-n,\ldots,n}$ *are closed (in the sense of the approximation theory) in* $\left(\mathrm{C}(\mathcal{B}), \langle\cdot,\cdot\rangle_{\mathrm{L}^2(\mathcal{B})}\right)$ *if* $\ell_n = n$ *for all* $n \in \mathbb{N}_0$.

Moreover, the Stone–Weierstraß approximation theorem (Theorem 2.5.23) says that the polynomials on $\mathcal{B}$ are dense in $\left(\mathrm{C}(\mathcal{B}), \|\cdot\|_{\mathrm{C}(\mathcal{B})}\right)$. Since the functions $\left\{G^{\mathrm{I}}_{m,n,j}\right\}_{m,n\in\mathbb{N}_0;\,j=-n,\ldots,n}$ with $\ell_n = n$ for all $n \in \mathbb{N}_0$ constitute a basis for all polynomials on $\mathcal{B}$ (see Michel, 1999, lemma 2.3.7 and theorem 2.3.11), we are able to add another conclusion.

Theorem 4.8.6 *Let* $\ell_n := n$ *be chosen for all* $n \in \mathbb{N}_0$. *Then the previously defined functions* $\left\{G^{\mathrm{I}}_{m,n,j}\right\}_{m,n\in\mathbb{N}_0;\,j=-n,\ldots,n}$ *are closed (in the sense of the approximation theory) in* $\left(\mathrm{C}(\mathcal{B}), \|\cdot\|_{\mathrm{C}(\mathcal{B})}\right)$.

Theorem 4.8.7 *Let the operator* ${}^*\Delta^X$ *acting on sufficiently differentiable functions on* $\mathcal{B} \setminus \{0\}$ *be defined as the concatenation* ${}^*\Delta^X_x := D^X_{|x|} \circ \Delta^*_{x/|x|}$, *where* Δ^* *is the usual Beltrami operator and*

(a) in the case of type $X = \mathrm{I}$, *the operator* D *is defined by*

$$D^{\mathrm{I}}_r := (\beta^2 - r^2)\frac{\mathrm{d}^2}{\mathrm{d}r^2} + 2\left(1 - \frac{2r^2}{\beta^2}\right)\frac{\beta^2}{r}\frac{\mathrm{d}}{\mathrm{d}r} - \ell_n(\ell_n + 1)\frac{\beta^2}{r^2},$$

(b) in the case of type $X = \mathrm{II}$, we set

$$D_r^{\mathrm{II}} := r(\beta - r)\frac{\mathrm{d}^2}{\mathrm{d}r^2} + (3\beta - 4r)\frac{\mathrm{d}}{\mathrm{d}r}.$$

Then the functions $G_{m,n,j}^X$ are eigenfunctions of $^*\Delta^X$ in the sense that

$$^*\Delta_x^X G_{m,n,j}^X(x) = \left(^*\Delta^X\right)^\wedge(m,n)\, G_{m,n,j}^X(x) \quad \textit{for all } x \in \mathcal{B} \setminus \{0\},$$

where

$$\left(^*\Delta^X\right)^\wedge(m,n) = \begin{cases} \left[\ell_n(\ell_n+3) + 4m\left(m + \ell_n + \frac{3}{2}\right)\right] n(n+1), & \textit{if } X = \mathrm{I}, \\ m(m+3)n(n+1), & \textit{if } X = \mathrm{II}. \end{cases}$$

In analogy to the spherical case, Sobolev spaces and their reproducing kernels can be constructed. Also this topic will only be briefly summarized here. For further details and proofs, see Akram et al. (2011), Amirbekyan (2007), Amirbekyan and Michel (2008), Ishtiaq (2018), Michel (2013), and Michel and Orzlowski (2016). For corresponding structures built out of vectorial functions on the ball, see Leweke (2018).

Definition 4.8.8 Let $X \in \{\mathrm{I}, \mathrm{II}\}$. Moreover, let $(A_{m,n})_{m,n\in\mathbb{N}_0}$ be a countable set of real numbers. In the set $\mathcal{E}((A_{m,n}); X; \mathcal{B})$, we collect all $F \in \mathrm{C}^{(\infty)}(\mathcal{B})$ for which

$$\left\langle F, G_{m,n,j}^X \right\rangle_{\mathrm{L}^2(\mathcal{B})} = 0 \quad \text{whenever } A_{m,n} = 0$$

and

$$\sum_{\substack{m,n=0\\ A_{m,n}\neq 0}}^{\infty} \sum_{j=-n}^{n} A_{m,n}^{-2} \left\langle F, G_{m,n,j}^X \right\rangle_{\mathrm{L}^2(\mathcal{B})}^2 < +\infty.$$

We equip this space with the inner product

$$\langle F_1, F_2 \rangle_{\mathcal{H}} := \sum_{\substack{m,n=0\\ A_{m,n}\neq 0}}^{\infty} \sum_{j=-n}^{n} A_{m,n}^{-2} \left\langle F_1, G_{m,n,j}^X \right\rangle_{\mathrm{L}^2(\mathcal{B})} \left\langle F_2, G_{m,n,j}^X \right\rangle_{\mathrm{L}^2(\mathcal{B})}.$$

The completion of $\mathcal{E}((A_{m,n}); X; \mathcal{B})$ with respect to this inner product is denoted by $\mathcal{H}((A_{m,n}); X; \mathcal{B})$ and is called a **Sobolev space on the ball**. If no confusion is likely to arise, we will simply write $\mathcal{H}$ for this Sobolev space.

Remark 4.8.9 Two short comments are worth adding.

(a) Obviously, $\mathcal{H}$ is a Hilbert space.
(b) Since we use an L^2-inner product, it is negligible that the functions $G_{m,n,j}^X$ are, in general, not defined in $x = 0$.

Definition 4.8.10 We say that $(A_{m,n})_{m,n\in\mathbb{N}_0}$ is **I-summable** if

$$\sum_{m,n=0}^{\infty} A_{m,n}^2 (2n+1)(4m + 2\ell_n + 3)\frac{(\ell_n + m + 1/2)^{2m}}{(m!)^2} < +\infty,$$

and **II-summable** if

$$\sum_{m,n=0}^{\infty} A_{m,n}^2 (2n+1)(2m+3)^5 < +\infty.$$

Note that in some earlier publications, such as Michel (2013), a minor error occurred in the summability condition for the ball. The preceding conditions are the corrected versions and have already been mentioned in Ishtiaq (2018) and Ishtiaq and Michel (2017).

Theorem 4.8.11 (Sobolev Lemma) *If $(A_{m,n})_{m,n\in\mathbb{N}_0}$ is X-summable, then every $F \in \mathcal{H}((A_{m,n}); X; \mathcal{B})$ has a uniformly convergent series*

$$F(x) = \sum_{m,n=0}^{\infty} \sum_{j=-n}^{n} \left\langle F, G_{m,n,j}^X \right\rangle_{\mathrm{L}^2(\mathcal{B})} G_{m,n,j}^X(x), \quad x \in \mathcal{B} \setminus \{0\},$$

and is continuous on $\mathcal{B} \setminus \{0\}$. If $X = \mathrm{I}$ and $\ell_n = n$ for all $n \in \mathbb{N}_0$, then the uniform convergence and the continuity is even given on $\mathcal{B}$.

Theorem 4.8.12 *If $(A_{m,n})_{m,n\in\mathbb{N}_0}$ is X-summable, then $\mathcal{H}((A_{m,n}); X; \mathcal{B})$ is a reproducing kernel Hilbert space. Its unique reproducing kernel obeys the representation*

$$K_{\mathcal{H}}(x,y) = \sum_{m,n=0}^{\infty} \sum_{j=-n}^{n} A_{m,n}^2 G_{m,n,j}^X(x) G_{m,n,j}^X(y); \quad x, y \in \mathcal{B} \setminus \{0\}. \tag{4.70}$$

If $X = \mathrm{I}$ and $\ell_n = n$ for all $n \in \mathbb{N}_0$, then (4.70) is valid on $\mathcal{B} \times \mathcal{B}$.

Moreover, we can also conclude that analogous propositions to Theorems 4.7.2, 4.7.7, 4.7.13, and 4.7.15 as well as Corollary 4.7.6 hold true, particularly for $X = \mathrm{I}$ and $\ell_n = n$ for all $n \in \mathbb{N}_0$. In the other cases, limitations occur; note the restrictions on the embedding of the basis functions on the ball into spaces of continuous functions.

Theorem 4.8.13 *The Sobolev spaces on the ball have the following properties (always $X \in \{\mathrm{I}, \mathrm{II}\}$):*

(a) If $(A_{m,n})_{m,n\in\mathbb{N}_0}$, $(B_{m,n})_{m,n\in\mathbb{N}_0}$ are real sequences with $|A_{m,n}| \le |B_{m,n}|$ for all $m,n \in \mathbb{N}_0$, then $\mathcal{H}((A_{m,n}); X; \mathcal{B}) \subset \mathcal{H}((B_{m,n}); X; \mathcal{B})$.

(b) For every constant sequence $A_{m,n} \equiv c \in \mathbb{R} \setminus \{0\}$, we have $\mathcal{H}((c); X; \mathcal{B}) = \mathrm{L}^2(\mathcal{B})$.

(c) If $(A_{m,n})_{m,n\in\mathbb{N}_0}$ is a real sequence with $\sup_{m,n\in\mathbb{N}_0} |A_{m,n}| < +\infty$, then $\mathcal{H}((A_{m,n}); X; \mathcal{B}) \subset \mathrm{L}^2(\mathcal{B})$.

(d) Let, for each $k \in \mathbb{N}_0$, a summable sequence $\left(A_{m,n}^{(k)}\right)_{m,n\in\mathbb{N}_0}$ be given. If each of the conditions

(i) $\left|A_{m,n}^{(k)}\right| \le \left|A_{m,n}^{(k+1)}\right|$ for all $m,n,k \in \mathbb{N}_0$,

(ii) $\lim_{k\to\infty} \left|A_{m,n}^{(k)}\right| =: A_{m,n}^{(\infty)} \in \mathbb{R}_0^+$ exists for all $m,n \in \mathbb{N}_0$

holds true, then the spaces $\mathcal{H}^{(k)} := \mathcal{H}\left(\left(A_{m,n}^{(k)}\right); X; \mathcal{B}\right)$, $k \in \mathbb{N}_0$, constitute a multi-resolution analysis on the ball in the sense that

- $\mathcal{H}^{(k)} \subset \mathcal{H}^{(k+1)} \subset \mathcal{H}^{(\infty)} := \mathcal{H}\left(\left(A_{m,n}^{(\infty)}\right); X; \mathcal{B}\right)$ *for all* $k \in \mathbb{N}_0$,
- $\overline{\bigcup_{k\in\mathbb{N}_0} \mathcal{H}^{(k)}}^{\|\cdot\|_{\mathcal{H}^{(\infty)}}} = \mathcal{H}^{(\infty)}$.

(e) If $(A_{m,n})_{m,n\in\mathbb{N}_0}$ *is summable and* $A_{m,n} \neq 0$ *for all* $m,n \in \mathbb{N}_0$, *then* $\mathcal{H}((A_{m,n}); X; \mathcal{B})$ *is dense in* $\left(\mathrm{L}^2(\mathcal{B}), \|\cdot\|_{\mathrm{L}^2(\mathcal{B})}\right)$. *If additionally* $\ell_n = n$ *for all* $n \in \mathbb{N}_0$, *then* $\mathcal{H}((A_{m,n}); \mathrm{I}; \mathcal{B})$ *is dense in* $(\mathrm{C}(\mathcal{B}), \|\cdot\|_{\mathrm{C}(\mathcal{B})})$.

(f) If $(A_{m,n})_{m,n\in\mathbb{N}_0}$ *is summable and* $\left\{x^{(k)}\right\}_{k\in\mathbb{N}_0} \subset \mathcal{B}$ *is a set of pairwise distinct points in the ball* $\mathcal{B}$ *(where, except for* $X = \mathrm{I}$ *and* $\ell_n = n$ *for all* $n \in \mathbb{N}_0$, *the point* $x = 0$ *must be contained in this point set), then the functions* $F_k(y) := K_{\mathcal{H}}\left(x^{(k)}, y\right)$, $y \in \mathcal{B}$, $k \in \mathbb{N}_0$, *constructed out of the reproducing kernel* $K_{\mathcal{H}}$ *of* $\mathcal{H} = \mathcal{H}((A_{m,n}); X; \mathcal{B})$, *satisfy the following properties:*

  (i) If $\left\{x^{(k)}\right\}_{k\in\mathbb{N}_0}$ *is dense in* $\mathcal{B}$, *then* $\mathrm{span}\{F_k\}_{k\in\mathbb{N}_0}$ *is dense in* $\mathcal{H}$,

  (ii) $\mathrm{span}\{F_k\}_{k\in\mathbb{N}_0}$ *is dense (in* $(\mathrm{C}(\mathcal{B}), \|\cdot\|_{\mathrm{C}(\mathcal{B})})$ *if* $X = \mathrm{I}$ *and* $\ell = n$ *for all* $n \in \mathbb{N}_0$ *and) in* $(\mathrm{L}^2(\mathcal{B}), \langle\cdot,\cdot\rangle_{\mathrm{L}^2(\mathcal{B})})$, *if* $\{x^{(k)}\}_{k\in\mathbb{N}_0}$ *is dense in* $\mathcal{B}$, *and* $A_{m,n} \neq 0$ *for all* $m,n \in \mathbb{N}_0$.

(g) Let $(A_{m,n})$ *be a summable sequence with only a finite number of non-vanishing components. We set* $d := \sum_{m,n=0;\, A_{m,n}\neq 0}^{\infty}(2n+1)$. *Then* $\mathcal{H} = \mathcal{H}((A_{m,n}); X; \mathcal{B})$ *has dimension* d. *If* $\{x^{(k)}\}_{k=1,\ldots,d} \subset \mathcal{B}$ *is a finite set of pairwise distinct points such that the matrix* $M = \left(K_{\mathcal{H}}\left(x^{(k)}, x^{(\ell)}\right)\right)_{k,\ell=1,\ldots,d}$ *is regular, then the functions* $F_k(y) := K_{\mathcal{H}}\left(x^{(k)}, y\right)$, $y \in \mathcal{B}$, $k = 1, \ldots, d$, *represent a basis of* $\mathcal{H}$.

Proof The proofs are basically completely analogous to the spherical case. We only need to consider that, except for $X = \mathrm{I}$ and $\ell_n = n$ for all $n \in \mathbb{N}_0$, the functions in $\mathcal{H}$ are not necessarily continuous on $\mathcal{B}$. Therefore, the conclusion

$$\left\langle F, K_{\mathcal{H}}\left(x^{(k)}, \cdot\right)\right\rangle_{\mathcal{H}} = 0 \quad \text{for all } k \quad \Rightarrow \quad F \equiv 0$$

requires that the conclusion $F(0) = \langle F, K_{\mathcal{H}}(0,\cdot)\rangle_{\mathcal{H}} = 0$ can be made directly. Moreover, $\mathcal{B} \setminus \{0\}$ is not compact and, thus, there is no norm '$\|\cdot\|_{\mathrm{C}(\mathcal{B}\setminus\{0\})}$'. □

It should eventually be remarked that there also exist spacelimited trial functions on the ball which were developed in Akram (2008) and Akram and Michel (2010).

4.9 A Best Basis Algorithm

When we try to model a geoscientifically relevant function, we are confronted with the *embarras de richesses* of available trial functions. We have seen here only a small selection of basis functions on the sphere and their counterparts on the ball. All have their pros and cons. For example, spherical harmonics are probably the most established basis functions and are accompanied by a vast amount of efficient numerical tools. Their disadvantages can become visible, when high-resolution models are desired and, for example, the data coverage is not of a spatially uniform quality or quantity. This gap can be filled by adding radial basis functions into the toolbox. However, then we lose the orthonormality of the trial functions and, at least in theory, even their linear independence. We now have an overcomplete set of trial functions. Such a general set of trial functions (the 'toolbox', as we called it previously) from which we pick those trial functions which turn out to be useful is called a **dictionary** in signal analysis.

A 'literal' dictionary represents a collection of the words which are needed to explain semantic contents in a particular language (see also the explanations in the introduction of Mallat and Zhang, 1993, for the connections between a 'literal' dictionary and a mathematical dictionary). Usually, not all words are needed for a particular purpose; for example, this book does not contain all words which are listed in the *Cambridge Advanced Learner's Dictionary* (Walter et al., 2005). Other books will most likely use a different subset of this dictionary. Moreover, there are several words with similar meanings but sometimes slightly different nuances in their usage and interpretation. Hence, the same particular issue can be explained in different ways. Which one of the different ways is most desired depends on the situation and on the specific intentions of the speaker. In this sense, a dictionary is redundant. We could say, 'this object is a geometrical two-dimensional structure with four sides of equal length and perpendicular vertices'. Or we might say, 'this object is a square'. Both sentences describe exactly the same fact. In this respect, we could remove the word 'square' from all English dictionaries, because it is unnecessary and we are happy with the first sentence. Fortunately, this has not been done, because conciseness (in signal analysis, we would say 'sparsity') of a description has advantages by itself.

In an analogous manner, a mathematical dictionary collects all functions which might be useful building blocks for constructing an approximation to an unknown function. Not all of them will be needed for a particular signal under consideration, but other signals might require different functions, which are also available in the dictionary. Moreover, there is not a unique way of representing the unknown function by elements of an overcomplete dictionary. However, among those possibilities there is often a 'best' choice, which depends on the particular objectives of the researcher. The advantage of the overcompleteness is this possibility of choosing among different options of representing the solution of the problem under consideration. Moreover, an infinite basis always needs to be reduced to a finite subsystem in numerical calculations. For instance, a radial basis function in general or a point mass can be expanded into spherical harmonics, but we need all (infinitely many) spherical harmonics for this purpose. Thus, combining a bandlimited system of spherical harmonics with a system of radial basis functions allows us to cover structures in the solution which could not be covered otherwise in numerics.

There are several algorithms which were designed for approximating signals based on a dictionary. We will present here the Regularized Functional Matching Pursuit (RFMP), which was developed with a particular focus on its applicability to direct and inverse problems in Earth sciences. It is based on the Matching Pursuit (MP) algorithm due to Mallat and Zhang (1993) and Vincent and Bengio (2002). The RFMP was constructed, theoretically investigated, and applied to various problems in Fischer (2011), Fischer and Michel (2012), Fischer and Michel (2013a,b), Gutting et al. (2017), Leweke (2018), Michel (2015), Michel and Orzlowski (2017), Michel and Schneider (2020), and Michel and Telschow (2014).

The MP has several variants, in particular, the Orthogonal Matching Pursuit (OMP) (see Pati et al., 1993, and Vincent and Bengio, 2002) and the Weak Matching Pursuit (WMP) (see Temlyakov, 2000). Also for these algorithms, corresponding extensions exist: the Regularized Orthogonal Functional Matching Pursuit (ROFMP) (see Michel and Telschow, 2016, and Telschow, 2014) and the Regularized Weak Functional Matching Pursuit (RWFMP) (see Kontak, 2018, and Kontak and Michel, 2019). The novelties of RFMP, ROFMP, and RWFMP, in comparison to MP, OMP, and WMP, are the following:

- The data need not originate from the same space where the unknown function is contained; in particular, inverse problems may be considered.
- For the handling of ill-posed inverse problems (see also Chapter 5), a regularization is included into the algorithm.
- In the details of the implementation and in the numerical experiments, a focus was set on problems where the domain of the unknown function is a sphere or a ball. As a consequence, most of the applications can be found in Earth sciences, but it was also shown in Leweke (2018) that the RFMP and the ROFMP yield very good results in medical imaging as well.

Note that these methods do not use particular sparsity-promoting penalty terms as some other signal analysis methods do (see, e.g., Candès et al., 2006; Donoho, 2006; Grasmair and Naumova, 2016; Peter et al., 2015; and Zhang and Xin, 2018) where sparsity is the primary objective. However, numerical experiments showed that, due to their iterative character, the RFMP and its variants are able to construct equally (or more) accurate approximations while choosing essentially fewer trial functions out of the dictionary, in comparison to previously established approximation methods.

In this section, we only discuss the unregularized versions FMP, OFMP, and WFMP. In Chapter 5, we will discuss the basics on inverse problems and their regularization. This will then enable us to explain the RFMP, the ROFMP, and the RWFMP in an appropriate context.

The (inverse) problem which we want to solve by a variant of the MP is as follows: there are two real Hilbert spaces $(\mathcal{X}, \langle\cdot,\cdot\rangle_{\mathcal{X}})$ and $(\mathcal{Y}, \langle\cdot,\cdot\rangle_{\mathcal{Y}})$ and a linear and continuous operator $\mathcal{T}\colon \mathcal{X} \to \mathcal{Y}$. While the right-hand side $y \in \mathcal{Y}$ is given, we seek $F \in \mathcal{X}$ such that $\mathcal{T}F = y$. This setting allows finite data in the sense that $\mathcal{Y} = \mathbb{R}^l$. It also includes the setting for the pure approximation of a function $F\colon D \to \mathbb{R}$, as it is used for the MP, the OMP, and the WMP, because we can use the identity operator $\mathcal{T} = \mathcal{I}$ with $\mathcal{X} = \mathcal{Y}$ or a sampling operator $\mathcal{T}\colon \mathcal{X} \to \mathbb{R}^l$ with $\mathcal{T}F := (F(x_n))_{n=1,\ldots,l} \in \mathbb{R}^l$ and $\{x_n\}_{n=1,\ldots,l} \subset D$. For further details on inverse problems, see also Chapter 5.

We choose a **dictionary** $\mathcal{D} \subset \mathcal{X} \setminus \ker\mathcal{T}$ and do not require any further conditions regarding $\mathcal{D}$ at the moment. For reasons of brevity of the occurring formulae, we have the freedom to require additionally that each $d \in \mathcal{D}$ is normalized in the sense that either $\|d\|_{\mathcal{X}} = 1$ for all $d \in \mathcal{D}$ (we call this an **$\mathcal{X}$-normalized dictionary** here) or $\|\mathcal{T}d\|_{\mathcal{Y}} = 1$ for all $d \in \mathcal{D}$ (for which we use the nomenclature of an **image-normalized dictionary** or, briefly, **i.n.d.**).

Certainly, the choice of $\mathcal{D}$ can influence the obtained result. In a previous and an ongoing research project, it has, therefore, been investigated how the dictionary can be learned automatically for the given problem (for further details and results, see Michel and Schneider, 2020, and Schneider, 2020).

The basic idea of all kinds of matching pursuits is to iteratively construct a sequence of approximations $(F_n)_{n\in\mathbb{N}}$, where n denotes the number of summands of F_n. More precisely, $F_n = \sum_{k=1}^{n} \alpha_k d_k$, where $\alpha_k \in \mathbb{R}$ and $d_k \in \mathcal{D}$ for all k.

The initial step of the algorithm is to choose a first approximation F_0. However, we need not implement this into the algorithm itself, because the linearity of $\mathcal{T}$ allows us to replace the right-hand side by $y - \mathcal{T}F_0$. Therefore, we set $F_0 := 0$. In this sense, there is no difference between finding $F_{n+1} = F_n + \alpha_{n+1}d_{n+1}$ for a given F_n and finding the first approximation $F_1 = \alpha_1 d_1$.

4.9.1 Functional Matching Pursuit (FMP)

We start with the construction of the FMP. The objective in every iteration step is here to minimize the (squared) data misfit

$$\mathcal{M} : (\alpha, d) \mapsto \|y - \mathcal{T}(F_n + \alpha d)\|_{\mathcal{Y}}^2$$

among all $(\alpha, d) \in \mathbb{R} \times \mathcal{D}$. Clearly,

$$\begin{aligned} \mathcal{M}(\alpha,d) &= \langle y - \mathcal{T}(F_n + \alpha d), y - \mathcal{T}(F_n + \alpha d)\rangle_{\mathcal{Y}} \\ &= \|y - \mathcal{T}F_n\|_{\mathcal{Y}}^2 - 2\alpha\langle y - \mathcal{T}F_n, \mathcal{T}d\rangle_{\mathcal{Y}} + \alpha^2\|\mathcal{T}d\|_{\mathcal{Y}}^2. \end{aligned} \tag{4.71}$$

A necessary condition for a minimizer is given by

$$\frac{\partial}{\partial \alpha}\mathcal{M}(\alpha,d) = 0 \quad \Leftrightarrow \quad -\langle y - \mathcal{T}F_n, \mathcal{T}d\rangle_{\mathcal{Y}} + \alpha\|\mathcal{T}d\|_{\mathcal{Y}}^2 = 0.$$

Since $\ker\mathcal{T} \cap \mathcal{D} = \emptyset$, we get

$$\alpha = \frac{\langle y - \mathcal{T}F_n, \mathcal{T}d\rangle_{\mathcal{Y}}}{\|\mathcal{T}d\|_{\mathcal{Y}}^2}. \tag{4.72}$$

If we insert this into the data misfit (4.71), then we get

$$\begin{aligned} \mathcal{M}(\alpha,d) &= \|y - \mathcal{T}F_n\|_{\mathcal{Y}}^2 - 2\frac{\langle y - \mathcal{T}F_n, \mathcal{T}d\rangle_{\mathcal{Y}}^2}{\|\mathcal{T}d\|_{\mathcal{Y}}^2} + \frac{\langle y - \mathcal{T}F_n, \mathcal{T}d\rangle_{\mathcal{Y}}^2}{\|\mathcal{T}d\|_{\mathcal{Y}}^2} \\ &= \|y - \mathcal{T}F_n\|_{\mathcal{Y}}^2 - \frac{\langle y - \mathcal{T}F_n, \mathcal{T}d\rangle_{\mathcal{Y}}^2}{\|\mathcal{T}d\|_{\mathcal{Y}}^2}. \end{aligned} \tag{4.73}$$

Hence, (α,d) minimizes $\mathcal{M}(\alpha,d)$ only if α satisfies (4.72) and d maximizes the term $\langle y - \mathcal{T}F_n, \mathcal{T}d\rangle_{\mathcal{Y}}^2 \|\mathcal{T}d\|_{\mathcal{Y}}^{-2}$. This leads us to the following algorithm.

Algorithm 4.9.1 (Functional Matching Pursuit (FMP)) *Let an operator* $\mathcal{T} \in \mathcal{L}(\mathcal{X},\mathcal{Y}) \setminus \{0\}$ *and a right-hand side* $y \in \mathcal{Y}$ *be given.*

(1) Set $n := 0$ *and* $F_0 := 0$ *and choose a stopping criterion. Select a dictionary* $\mathcal{D} \subset \mathcal{X} \setminus \ker\mathcal{T}$.

(2) *Find* $d_{n+1} \in \mathcal{D}$ *such that*

$$\frac{\left|\langle y - \mathcal{T}F_n, \mathcal{T}d_{n+1}\rangle_{\mathcal{Y}}\right|}{\|\mathcal{T}d_{n+1}\|_{\mathcal{Y}}} \geq \frac{\left|\langle y - \mathcal{T}F_n, \mathcal{T}d\rangle_{\mathcal{Y}}\right|}{\|\mathcal{T}d\|_{\mathcal{Y}}} \quad \text{for all } d \in \mathcal{D} \tag{4.74}$$

and set

$$\alpha_{n+1} := \frac{\langle y - \mathcal{T}F_n, \mathcal{T}d_{n+1}\rangle_{\mathcal{Y}}}{\|\mathcal{T}d_{n+1}\|_{\mathcal{Y}}^2} \tag{4.75}$$

as well as $F_{n+1} := F_n + \alpha_{n+1}d_{n+1}$.

(3) *If the stopping criterion is fulfilled, then* F_{n+1} *is the output as an approximate solution of* $\mathcal{T}F = y$. *Otherwise, increase n by* 1 *and go to step* 2.

In the case of an i.n.d., (4.74) and (4.75) are simplified to

$$\left|\langle y - \mathcal{T}F_n, \mathcal{T}d_{n+1}\rangle_{\mathcal{Y}}\right| \geq \left|\langle y - \mathcal{T}F_n, \mathcal{T}d\rangle_{\mathcal{Y}}\right| \quad \text{for all } d \in \mathcal{D} \tag{4.76}$$

and

$$\alpha_{n+1} := \langle y - \mathcal{T}F_n, \mathcal{T}d_{n+1}\rangle_{\mathcal{Y}}. \tag{4.77}$$

This shows that α_{n+1} is exactly the minimum value found in (4.76) for an i.n.d.

The following are possible stopping criteria:

(a) A fixed number N of iterations is given, and the algorithms stops if $n + 1 > N$.
(b) A threshold ε for the residual is given such that the algorithm stops as soon as $\|y - \mathcal{T}F_n\|_{\mathcal{Y}} < \varepsilon$.
(c) A threshold δ for the expansion coefficients is given, and the algorithm stops if $|\alpha_{n+1}| < \delta$.

As long as nothing different is stated, we assume for the theoretical considerations that the algorithm is not stopped, that is, we get a sequence $(F_n)_{n\in\mathbb{N}}$. The stopping criteria are for numerical purposes where we can (or better: need to) accept sufficiently close approximations to the exact result by applying only a finite number of iterations.

Example 4.9.2 We demonstrate the FMP in the following example: we take values of the Earth Gravitational Model 2008 (EGM2008, Pavlis et al., 2012), where we use the degrees 3 to 2,190 and orders 0 to 2,159 (we start at degree 3 to skip the dominant effect of the Earth's ellipticity) at an equidistributed spherical grid $(\xi^{(i)})_i$ with 12,684 points (Reuter grid with parameter 100; see the original reference and the textbook, respectively: Reuter, 1982; Michel, 2013) as the components of the given data vector $y \in \mathbb{R}^{12,684} =: \mathcal{Y}$. The operator $\mathcal{T}$ is simply the sampling operator on $\mathcal{H}_2(\Omega) =: \mathcal{X}$, that is, $\mathcal{T}F := (F(\xi^{(i)}))_{i=1,\ldots,12,684}$ yields the values of the gravitational potential F at the Earth's surface. The problem of finding F is, therefore, the problem of interpolating the gravitational potential from grid-based terrestrial data.

The given data are contaminated by 5% Gaussian noise, that is, we take values ε_i from an array of standard normally distributed pseudo-random numbers and replace y_i by $y_i(1+0.05\varepsilon_i)$. We use the following dictionary:

$$\begin{aligned}\mathcal{D}_{\text{test}} := &\left\{Y_{n,j}\,\middle|\, n=0,\ldots,25;\ j=-n,\ldots,n\right\}\cup\\ &\left\{P_h\left(\eta^{(k)},\cdot\right)\middle|\, h\in H;\ k=1,\ldots,4{,}551\right\}\cup\\ &\left\{W_h\left(\eta^{(k)},\cdot\right)\middle|\, h\in H;\ k=1,\ldots,4{,}551\right\}\cup\\ &\{S(\theta,\alpha,\beta,\gamma;\cdot)\,|\,\theta=\pi/4,\pi/2;\ \alpha,\gamma=0,\pi/2,\pi,3\pi/2;\ \beta=0,\pi/2,\pi\}\end{aligned}$$

with

$$H := \{0.75,0.8,0.85,0.89,0.91,0.93,0.94,0.95,0.96,0.97\},$$

where the notation has to be understood as follows: as usual, $Y_{n,j}$ stands for a fully normalized spherical harmonic of degree n and order j (see Definition 3.4.24). Furthermore, we use the Abel–Poisson kernels (see Definition 3.3.39 and Theorem 4.2.14)

$$P_h(\eta,\zeta) := \frac{1-h^2}{4\pi\left(1+h^2-2h\eta\cdot\zeta\right)^{3/2}}$$

with different parameters $h\in H$, which control the 'hat-width' of these radial basis functions, and different centres $\eta^{(k)}$, which are chosen from a Reuter grid with parameter 60. The Abel–Poisson wavelets are defined as differences of Abel–Poisson kernels by $W_h := P_h - P_{h^2}$. Eventually, the dictionary also contains some Slepian functions (see Section 4.6) with bandlimit 5. The parameters of the Slepian function $S(\theta,\alpha,\beta,\gamma;\cdot)$ represent the 'angular size' $\theta=\arccos b$ of the spherical cap to which the Slepian function is concentrated (see Definition 4.6.7) and Euler angles (α,β,γ), that is, each Slepian function is calculated in a preprocessing for the spherical cap centred around the North pole ε^3 and is then rotated to a different location on the sphere by use of the three Euler angles (for further details on rotations represented in Euler angles, see, for example, Dahlen and Tromp, 1998, appendix C.8).

The FMP stops if at least one of the following criteria is satisfied: the number of iterations reaches 200,000 or $\|R^n\|_{\mathcal{Y}}/\|y\|_{\mathcal{Y}}\le 0.05$ (or ≥ 2 as a 'negative' criterion for failures due to inappropriate regularization parameters). The obtained approximating function F_n is evaluated at an equiangular latitude-longitude grid of $181\times 361 = 65{,}341$ points $(\zeta^{(i)})_i$ on the unit sphere Ω. This grid is used for plotting the result and for calculating the relative approximation error

$$\sqrt{\frac{\sum_{i=1}^{65,341}\left(F_n\left(\zeta^{(i)}\right)-F_{\text{exact}}\left(\zeta^{(i)}\right)\right)^2}{65{,}341\sum_{i=1}^{65,341}\left(F_{\text{exact}}\left(\zeta^{(i)}\right)\right)^2}},$$

where F_{exact} is the exact solution given by EGM2008. Figures 4.26 and 4.27 show the obtained result. Note that all plots of the solution and its (complete and partial) approximations were produced with the same colour scale for the sake of comparability. Only for the approximation error, the colour scale was adapted to the minimum and the maximum of the specific plot.

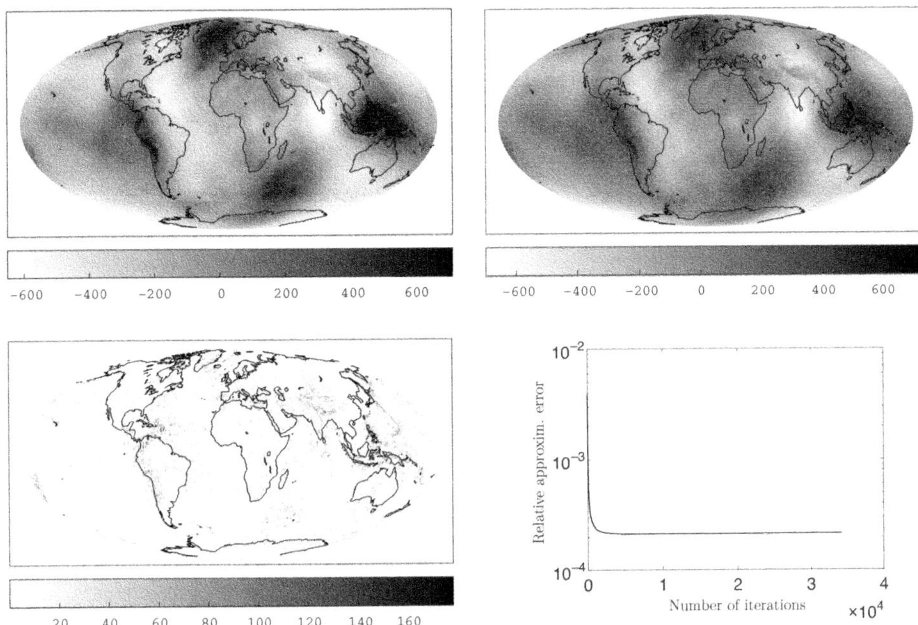

Figure 4.26 The FMP is used to interpolate grid-based data from EGM2008. The exact solution (given by EGM2008) is shown at the top on the left-hand side. The obtained approximation by the FMP (top-right) is very close to the exact solution. The absolute difference (bottom-left) shows that only small deviations, particularly in areas with fine local details, occur. These deviations become visible on the finer plot grid but are not captured by the coarser data grid, in particular in the presence of noise, as was the case for this demonstrating example. All values of these plots are in m^2s^{-2}. Moreover, the curve on the right-hand side at the bottom shows the rapid decay of the relative approximation error during the iterations.

The algorithm stopped after 34, 130 iterations, because the residual had reached the given threshold (which was chosen in accordance with the noise level). The relative approximation error of the result is 0.000215. The FMP is able to construct a very close approximation to the exact, noise-free solution. Moreover, due to the use of the dictionary, it provides us with the possibility to decorrelate the signal into, for example, global and local structures. However, a very critical and careful look at the images shows that some small peak-like perturbations occur due to a few (apparently inappropriately) chosen radial basis functions. Therefore, it is reasonable to add a smoothing effect to the algorithm. This can be achieved by including a regularization, which is also here useful, though the problem is theoretically not an unstable inverse problem. For further details, see Section 5.4.

All numerical examples for matching pursuits in this book were calculated by Naomi Schneider (Geomathematics Group Siegen), whose support for this book is gratefully acknowledged by the author.

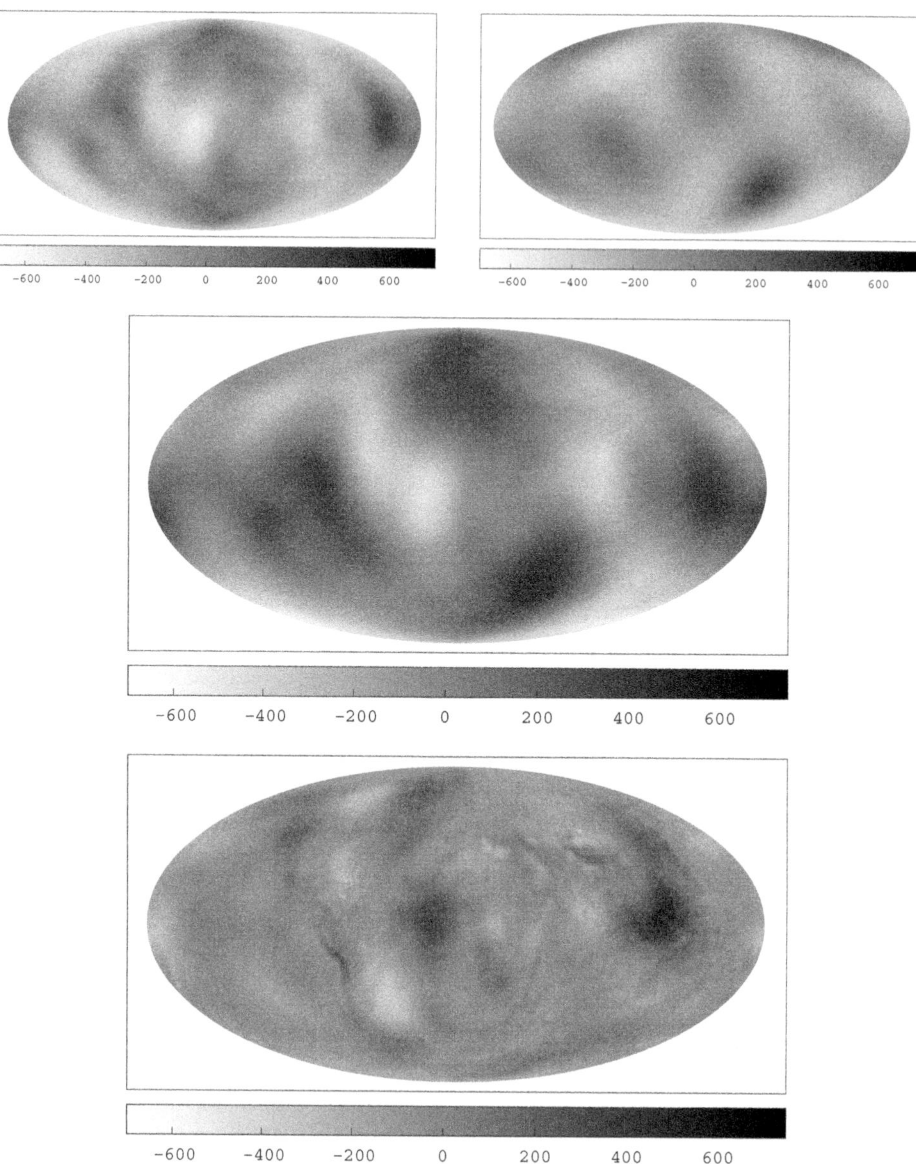

Figure 4.27 The dictionary elements which are chosen by the FMP in Example 4.9.2 are separated: we have a look at the spherical harmonics alone (top-left), the Slepian functions alone (top-right) and their sum (middle). These basis functions provide a coarse approximation of the large global structures of the solution, while the sum of all chosen radial basis functions (bottom), that is, the Abel–Poisson kernels and wavelets, reveals more localized structures in the gravitational field such as the signals due to the Andes or the Himalayas. For a better visibility of these anomalies, the coast lines were omitted in the plots.

Let us now investigate some theoretical properties of the FMP. From (4.73), we directly get the following result.

Lemma 4.9.3 *Let the sequence $(F_n)_n$ be produced by the FMP (Algorithm 4.9.1). Then the sequence of (norms of the) residuals $(\|y - \mathcal{T} F_n\|_{\mathcal{Y}})_n$ is monotonically decreasing.*

Since the residual is bounded from below by zero, we get another immediate conclusion.

Lemma 4.9.4 *Let the sequence $(F_n)_n$ be produced by the FMP (Algorithm 4.9.1). Then the sequence of (norms of the) residuals $(\|y - \mathcal{T} F_n\|_{\mathcal{Y}})_n$ is convergent.*

The following lemma is due to Jones (1987, lemma 2).

Lemma 4.9.5 *Let (a_k) be a square-summable sequence of non-negative real numbers. Then $\liminf_{n\to\infty} \left(a_n \sum_{k=1}^n a_k\right) = 0$.*

Proof We choose an arbitrary $\varepsilon > 0$ and observe that, due to the square-summability of (a_k), there exists an integer K_1 such that

$$\sum_{k=K_1}^{\infty} a_k^2 < \frac{\varepsilon}{2}. \tag{4.78}$$

Let $m > K_1$ be an arbitrary but fixed integer. Since $a_k \to 0$ is necessary for the square-summability, there exists $K_2 = K_2(m) > m$ such that

$$a_{K_2} \sum_{k=1}^{m} a_k < \frac{\varepsilon}{2}. \tag{4.79}$$

Eventually, we choose an integer $K_3 = K_3(m) \in \{m+1, \ldots, K_2\}$ such that $a_{K_3} = \min\{a_k \mid k = m+1, \ldots, K_2\}$. If we now combine this with the non-negativity of the a_k and Equations (4.78) and (4.79), then we get

$$\begin{aligned}
\inf_{n \geq m} \left(a_n \sum_{k=1}^{n} a_k\right) &\leq a_{K_3} \sum_{k=1}^{K_3} a_k = a_{K_3} \sum_{k=1}^{m} a_k + \sum_{k=m+1}^{K_3} a_{K_3} a_k \\
&\leq a_{K_2} \sum_{k=1}^{m} a_k + \sum_{k=m+1}^{K_3} a_k^2 \leq a_{K_2} \sum_{k=1}^{m} a_k + \sum_{k-K_1}^{\infty} a_k^2 < \frac{\varepsilon}{2} + \frac{\varepsilon}{2}.
\end{aligned}$$

Hence, since $m > K_1$ was arbitrary, $a_n \sum_{k=1}^n a_k$ is in $[0, \varepsilon]$ for an infinite number of integers n. □

Definition 4.9.6 We define the abbreviation $R^n := y - \mathcal{T} F_n$ for the residuals in all considered matching pursuits.

Lemma 4.9.7 *Let the sequence $(F_n)_n$ be produced by the FMP (Algorithm 4.9.1). Then the sequence of residuals $(R^n)_n = (y - \mathcal{T} F_n)_n$ is convergent in the Hilbert space $\mathcal{Y}$. Moreover, the sequence $(\alpha_n)_n$ is square-summable.*

Proof Considering an i.n.d. does not influence the determination of d_{n+1} (see (4.74)) and it also does not influence the term $\alpha_{n+1}d_{n+1}$ (see (4.74) and (4.75)) in $F_{n+1} := F_n + \alpha_{n+1}d_{n+1}$. Therefore, we assume, without loss of generality, that the dictionary is image-normalized.

We prove the proposition of the theorem by contradiction: We assume that $(R^n)_n$ is not a Cauchy sequence. For this purpose, let us assume that there exists an $\varepsilon > 0$ such that, for all $N_0 \in \mathbb{N}_0$, there exist integers $m = m(N_0), n = n(N_0) \geq N_0$ with

$$\left\| R^n - R^m \right\|_{\mathcal{Y}} > \varepsilon. \tag{4.80}$$

Moreover, let $c > 0$ be an arbitrary constant. Since $(\|R^k\|_{\mathcal{Y}})_k$ is convergent due to Lemma 4.9.4, we are able to find an index $N_1 \in \mathbb{N}_0$ such that $\|R^{N_1}\|_{\mathcal{Y}}^2 < r^2 + c$, where

$$r := \lim_{k\to\infty} \left\| R^k \right\|_{\mathcal{Y}} = \inf_k \left\| R^k \right\|_{\mathcal{Y}}. \tag{4.81}$$

Due to the monotonicity (see Lemma 4.9.3) of $(\|R^k\|_{\mathcal{Y}})_k$, we automatically get that

$$\left\| R^{m(N_1)} \right\|_{\mathcal{Y}}^2 < r^2 + c \qquad \text{and} \qquad \left\| R^{n(N_1)} \right\|_{\mathcal{Y}}^2 < r^2 + c. \tag{4.82}$$

From (4.73), we deduce by means of a telescoping sum that

$$\begin{aligned} \|R^0\|_{\mathcal{Y}}^2 &= \sum_{l=0}^{k-1} \left(\|R^l\|_{\mathcal{Y}}^2 - \|R^{l+1}\|_{\mathcal{Y}}^2 \right) + \|R^k\|_{\mathcal{Y}}^2 \\ &= \sum_{l=0}^{k-1} \langle R^l, \mathcal{T} d_{l+1} \rangle_{\mathcal{Y}}^2 + \|R^k\|_{\mathcal{Y}}^2 = \sum_{l=0}^{k-1} \alpha_{l+1}^2 + \|R^k\|_{\mathcal{Y}}^2 \end{aligned}$$

for all $k \in \mathbb{N}_0$, where $R^0 = y$. Clearly, the limit $k \to \infty$ must yield a convergent series $\sum_{l=0}^{\infty} \alpha_{l+1}^2$ because of Lemma 4.9.4.

The square-summability of the α_l in combination with Lemma 4.9.5 leads us to the existence of an integer $p > \max(m(N_1), n(N_1))$ for which

$$\left|\alpha_{p+1}\right| \sum_{k=1}^{p+1} |\alpha_k| < c. \tag{4.83}$$

Furthermore, we consult (4.80) and use the triangle inequality

$$\varepsilon < \left\| R^{n(N_1)} - R^{m(N_1)} \right\|_{\mathcal{Y}} \leq \left\| R^{n(N_1)} - R^p \right\|_{\mathcal{Y}} + \left\| R^p - R^{m(N_1)} \right\|_{\mathcal{Y}}$$

and see that, for $q = n(N_1)$ or $q = m(N_1)$, we have

$$\left\| R^q - R^p \right\|_{\mathcal{Y}} > \varepsilon/2. \tag{4.84}$$

Remember that $q < p$.

In the next step, we use the underlying inner product of $\mathcal{Y}$, the representation $R^l = y - \mathcal{T} F_l = y - \sum_{k=1}^{l} \alpha_k \mathcal{T} d_k$, the inequality $-x \leq |x|$ for all $x \in \mathbb{R}$, the triangle inequality, the bilinearity of the inner product, the algorithmic criterion for d_{k+1} (see (4.76)); Equations (4.81)–(4.83), and the formula $\alpha_{k+1} = \langle R^k, \mathcal{T} d_{k+1}\rangle_{\mathcal{Y}}$ (see (4.77)).

$$
\begin{aligned}
\| R^q - R^p \|_{\mathcal{Y}}^2 &= \| R^q \|_{\mathcal{Y}}^2 + \| R^p \|_{\mathcal{Y}}^2 - 2\langle R^p, R^q\rangle_{\mathcal{Y}} \\
&= \| R^q \|_{\mathcal{Y}}^2 + \| R^p \|_{\mathcal{Y}}^2 - 2\left\langle R^p, R^p + \sum_{k=q+1}^{p} \alpha_k \mathcal{T} d_k \right\rangle_{\mathcal{Y}} \\
&\leq \| R^q \|_{\mathcal{Y}}^2 - \| R^p \|_{\mathcal{Y}}^2 + 2 \sum_{k=q+1}^{p} |\alpha_k| \left|\langle R^p, \mathcal{T} d_k\rangle_{\mathcal{Y}}\right| \\
&\leq \| R^q \|_{\mathcal{Y}}^2 - \| R^p \|_{\mathcal{Y}}^2 + 2 \sum_{k=q+1}^{p} |\alpha_k| \left|\langle R^p, \mathcal{T} d_{p+1}\rangle_{\mathcal{Y}}\right| \qquad (4.85) \\
&\leq r^2 + c - r^2 + 2 \left|\langle R^p, \mathcal{T} d_{p+1}\rangle_{\mathcal{Y}}\right| \sum_{k=q+1}^{p} |\alpha_k| \leq 3c.
\end{aligned}
$$

Since we may choose $c < \varepsilon^2/12$, we get a contradiction to (4.84). Hence, $(R^k)_{k\in\mathbb{N}}$ is a Cauchy sequence and, since $\mathcal{Y}$ is complete, it is also convergent. □

In the following theorem, we will see that the limit of the residuals is zero under certain reasonable constraints.

Theorem 4.9.8 *Let the dictionary used in the FMP (Algorithm 4.9.1) satisfy* $\overline{\operatorname{span}\{\mathcal{T} d \mid d \in \mathcal{D}\}}^{\|\cdot\|_{\mathcal{Y}}} = \overline{\mathcal{T}(\mathcal{X})}^{\|\cdot\|_{\mathcal{Y}}}$ *and let* (F_n) *be the sequence produced by the FMP for a given right-hand side* $y \in \overline{\mathcal{T}(\mathcal{X})}$. *Then we have* $\lim_{n\to\infty} \|y - \mathcal{T} F_n\|_{\mathcal{Y}} = 0$, *that is,* $(\mathcal{T} F_n)_n$ *strongly converges to* y.

Proof As in the proof of Lemma 4.9.7, we assume, without loss of generality, that $\mathcal{D}$ is an i.n.d.

(1) Square-summability of the α_n:
From Lemma 4.9.7, we have the square-summability of $(\alpha_n)_n$. This also implies that $\alpha_n \to 0$ as $n \to \infty$. Moreover, with (4.76) and (4.77), we are also able to conclude that

$$\lim_{n\to\infty} \langle R^n, \mathcal{T} d\rangle_{\mathcal{Y}} = 0 \quad \text{for all } d \in \mathcal{D}. \qquad (4.86)$$

(2) Weak convergence of the residuals to zero:
Since $(\|R^n\|_{\mathcal{Y}})_n$ is convergent due to Lemma 4.9.4, we have $\sigma := \sup_n \|R^n\|_{\mathcal{Y}} < +\infty$. Furthermore, (4.86) implies the following (remember that $\mathcal{T}$ is linear and an inner product is bilinear):

$$\lim_{n\to\infty} \langle R^n, \mathcal{T} d\rangle_{\mathcal{Y}} = 0 \quad \text{for all } d \in \operatorname{span} \mathcal{D}. \qquad (4.87)$$

We can also conclude that, for arbitrary $z \in \overline{\mathcal{T}(\mathcal{X})}$, there is (see the condition of the theorem) a sequence $(\tilde{d}_m)_m \subset \operatorname{span} \mathcal{D}$ such that $\mathcal{T}\tilde{d}_m \to z$ and $|\langle R^n, \mathcal{T}\tilde{d}_m - z\rangle_{\mathcal{Y}}| \leq \sigma \|\mathcal{T}\tilde{d}_m - z\|_{\mathcal{Y}} \to 0$ as $m \to \infty$, where the latter convergence is uniform with respect to n. Thus, we may apply the Moore–Osgood double limit theorem and obtain

$$\lim_{n\to\infty} \left\langle R^n, z\right\rangle_{\mathcal{Y}} = \lim_{n\to\infty} \lim_{m\to\infty} \left\langle R^n, \mathcal{T}\tilde{d}_m\right\rangle_{\mathcal{Y}} = \lim_{m\to\infty} \lim_{n\to\infty} \left\langle R^n, \mathcal{T}\tilde{d}_m\right\rangle_{\mathcal{Y}} = 0$$

due to (4.87).

Let now $z \in \mathcal{Y}$ be arbitrary. Then Theorem 2.5.71 yields a decomposition $z = z_{\|} + z_{\perp}$ with $z_{\|} \in \overline{\mathcal{T}(\mathcal{X})}$ and $z_{\perp} \in \overline{\mathcal{T}(\mathcal{X})}^{\perp}$. Since $R^n \in \overline{\mathcal{T}(\mathcal{X})}$ such that $\langle R^n, z_{\perp}\rangle_{\mathcal{Y}} = 0$, we arrive at

$$\lim_{n\to\infty} \left\langle R^n, z\right\rangle_{\mathcal{Y}} = \lim_{n\to\infty} \left(\left\langle R^n, z_{\|}\right\rangle_{\mathcal{Y}} + \left\langle R^n, z_{\perp}\right\rangle_{\mathcal{Y}}\right) = \lim_{n\to\infty} \left\langle R^n, z_{\|}\right\rangle_{\mathcal{Y}} = 0.$$

Consequently, $(R^n)_n$ weakly converges to zero in $\mathcal{Y}$.

From Lemma 4.9.7, we know that $(R^n)_n$ is (strongly) convergent in $\mathcal{Y}$. In combination with the weak convergence to zero (see Theorem 2.5.57), we get $\|R^n\|_{\mathcal{Y}} \to 0$. □

Corollary 4.9.9 *Let the conditions of Theorem 4.9.8 be satisfied, where $y \in \mathcal{T}(\mathcal{X})$. If the generated sequence $(F_n)_n$ converges in $\mathcal{X}$, then the limit $F_\infty := \sum_{k=1}^{\infty} \alpha_k d_k$ is an exact solution of $\mathcal{T}F = y$.*

Proof This corollary is an immediate consequence of Theorem 4.9.8 and the continuity of $\mathcal{T}$. □

The conditions of Theorem 4.9.8 are reasonable: if the dictionary is not able to cover the image space of $\mathcal{T}$, that is, if $\overline{\operatorname{span}\{\mathcal{T}d \mid d \in \mathcal{D}\}}^{\|\cdot\|_{\mathcal{Y}}} \subsetneq \overline{\mathcal{T}(\mathcal{X})}^{\|\cdot\|_{\mathcal{Y}}}$, then we cannot expect the sequence $(\mathcal{T}F_n)_n$ to approximate all kinds of data vectors y sufficiently. For instance, if $\mathcal{T}(\mathcal{X}) = \mathcal{Y} = \mathbb{R}^2$ and $\mathcal{D}$ is only chosen such that $\mathcal{T}d$ always has the form $(r,0)^{\mathrm{T}}$ for some $r \in \mathbb{R}$, then data vectors such as $(0,1)^{\mathrm{T}}$ cannot be approximated. A simple example is given by the identity operator $\mathcal{T} = \mathcal{I}$ on $\mathbb{R}^2$ and the dictionary $\mathcal{D} = \{(1,0)^{\mathrm{T}}\}$.

For similar reasons, right-hand sides $y \in \mathcal{Y} \setminus \overline{\mathcal{T}(\mathcal{X})}$ leave components $0 \neq \mathcal{P}_{\overline{\mathcal{T}(\mathcal{X})}^{\perp}} y =: y_{\perp} \in \overline{\mathcal{T}(\mathcal{X})}^{\perp}$, where $\langle y - \mathcal{T}F_n, \mathcal{T}d\rangle_{\mathcal{Y}} = \langle (y - y_{\perp}) - \mathcal{T}F_n, \mathcal{T}d\rangle_{\mathcal{Y}}$ for all $d \in \mathcal{D}$ shows that $y_{\perp}$ does not influence the choice of d_{n+1} (and, analogously, of α_{n+1}). Therefore, we can only expect that right-hand sides $y \in \overline{\mathcal{T}(\mathcal{X})}$ guarantee that the residual tends to zero.

Note that Corollary 4.9.9 requires the convergence of the sequence (F_n) produced by the FMP. We can, indeed, prove this converges under a particular condition on the dictionary.

Theorem 4.9.10 *Let $\mathcal{D}$ be a dictionary which satisfies the* ***semi-frame condition****: there exist a constant $c > 0$ and an integer $M \in \mathbb{N}$ such that, for all expansions $H = \sum_{k=1}^{\infty} \beta_k \tilde{d}_k$ with $\beta_k \in \mathbb{R}$ and $\tilde{d}_k \in \mathcal{D}$ where each dictionary element may occur multiple times but not more than M times (i.e., $|\{j \in \mathbb{N} \mid \tilde{d}_j = \tilde{d}_k\}| \leq M$ for all $k \in \mathbb{N}$), the inequality*

$$c\|H\|_{\mathcal{X}}^2 \leq \sum_{k=1}^{\infty} \beta_k^2 \tag{4.88}$$

is satisfied. If this dictionary is used in the FMP (Algorithm 4.9.1) and no dictionary element is chosen more than M times by the FMP, then the generated sequence $(F_n)_n$ *is convergent. Moreover, if*

$$\overline{\mathrm{span}\,\{\mathcal{T}d \mid d \in \mathcal{D}\}}^{\|\cdot\|_{\mathcal{Y}}} = \overline{\mathcal{T}(\mathcal{X})}^{\|\cdot\|_{\mathcal{Y}}}$$

and $y \in \mathcal{T}(\mathcal{X})$, *then the limit* $F_\infty = \sum_{k=1}^{\infty} \alpha_k d_k$ *satisfies* $\mathcal{T} F_\infty = y$.

Proof From Lemma 4.9.7, we know that $\sum_{k=1}^{\infty} \alpha_k^2 < +\infty$. Hence, the semi-frame condition (4.88) yields

$$\left\| \sum_{k=N+1}^{\infty} \alpha_k d_k \right\|_{\mathcal{X}}^2 \leq \frac{1}{c} \sum_{k=N+1}^{\infty} \alpha_k^2 \to 0 \quad \text{as } N \to \infty$$

such that $\sum_{k=0}^{N} \alpha_k d_k \to \sum_{k=0}^{\infty} \alpha_k d_k$ as $N \to \infty$. The consultation of Corollary 4.9.9 completes the proof. □

The semi-frame condition (4.88) is a bit inconvenient, because, so far, it could not be verified for all kinds of dictionaries which have been (successfully) used in numerical experiments. However, we will see later on that the (practically more relevant) regularized version RFMP does not require this condition any more.

It should also be noted that the terminus 'semi-frame condition' is not completely adequate, because the condition is rather one half of the condition for a Riesz basis than for a frame, though the differences are minor. See, for example, also Chui (1992, pp. 69–71). We will give here an abstract definition of a Riesz basis.

Definition 4.9.11 A system of elements $\{b_j\}_{j\in\mathbb{N}} \subset \mathcal{X}$ is called a **Riesz basis** of $\mathcal{X}$, if span $\{b_j\}_{j\in\mathbb{N}}$ is dense in $\mathcal{X}$ and if there exist constants $0 < A \leq B < +\infty$ such that

$$A\|c\|_2^2 \leq \left\| \sum_{j=1}^{\infty} c_j b_j \right\|_{\mathcal{X}}^2 \leq B\|c\|_2^2 \tag{4.89}$$

for all sequences $c \in \ell^2$. The constants A and B are called the **Riesz bounds** of the Riesz basis.

The following particular case of a Riesz basis is obvious due to the Parseval identity (see Theorem 2.5.25).

Theorem 4.9.12 *Every orthonormal basis is a Riesz basis.*

Despite the critical aspects of the semi-frame condition, let us at least show here that every Riesz basis satisfies the semi-frame condition.

Theorem 4.9.13 *Every Riesz basis satisfies the semi-frame condition.*

Proof Let the dictionary $\mathcal{D}$ be a Riesz basis and let $H = \sum_{k=1}^{\infty} \beta_k d_k$ be an expansion as it is considered for the semi-frame condition. Corresponding to H, let $(\tilde{d}_j)_{j\in\mathbb{N}} = \tilde{\mathcal{D}}$ be the sequence of all dictionary elements which occur in H, but now none of them is counted multiple times, that is, $\tilde{\mathcal{D}}$ is necessarily a subsystem of the Riesz basis $\mathcal{D}$ with pairwise disjoint elements. Then we get with the Riesz bounds A and B (see (4.89)) the inequality

$$\left\| \sum_{k=1}^{N} \beta_k d_k \right\|_{\mathcal{X}}^2 = \left\| \sum_{j=1}^{\infty} \sum_{\substack{1\le k\le N \\ d_k=\tilde{d}_j}} \beta_k \tilde{d}_j \right\|_{\mathcal{X}}^2 \le B \sum_{j=1}^{\infty} \left(\sum_{\substack{1\le k\le N \\ d_k=\tilde{d}_j}} \beta_k \right)^2$$
$$\le B \sum_{j=1}^{\infty} \left(\sum_{\substack{k\in\mathbb{N} \\ d_k=\tilde{d}_j}} |\beta_k| \right)^2 .$$

We observe that the term within the brackets in the previous line has at most M summands for each j, and we could add vanishing summands to get exactly M summands per j. With the corollary of Jensen's inequality (see Corollary 2.5.69), we eventually get

$$\left\| \sum_{k=1}^{N} \beta_k d_k \right\|_{\mathcal{X}}^2 \le B \sum_{j=1}^{\infty} M \sum_{\substack{k\in\mathbb{N} \\ d_k=\tilde{d}_j}} \beta_k^2 = BM \sum_{k=1}^{\infty} \beta_k^2,$$

which is the semi-frame condition (4.88) with $c = (BM)^{-1}$. □

We can also prove a convergence rate for the residual in the case of the FMP.

Theorem 4.9.14 *Let the sequence $(R^n)_n$ be produced by the FMP. Then*

$$\left\| R^n \right\|_{\mathcal{Y}} \le \|y\|_{\mathcal{Y}} \left[1 - I(\tau)^2\right]^{n/2},$$

*where $I(\tau) := \inf_{z\in\mathcal{Y}\setminus\{0\}} \tau(z)$ and the **correlation ratio** $\tau(z)$ is given by*

$$\tau(z) := \sup_{d\in\mathcal{D}} \frac{|\langle z, \mathcal{T}d\rangle_{\mathcal{Y}}|}{\|z\|_{\mathcal{Y}} \|\mathcal{T}d\|_{\mathcal{Y}}}, \quad z \in \mathcal{Y} \setminus \{0\}. \tag{4.90}$$

Proof Obviously, $0 \le \tau(z) \le 1$ for all $z \in \mathcal{Y} \setminus \{0\}$ due to the Cauchy–Schwarz inequality such that also $0 \le I(\tau) \le 1$. Due to the algorithmic criterion for d_{n+1} (see (4.74)), we get

$$\frac{\left|\langle R^n, \mathcal{T}d_{n+1}\rangle_{\mathcal{Y}}\right|}{\|\mathcal{T}d_{n+1}\|_{\mathcal{Y}}} \ge \sup_{d\in\mathcal{D}} \frac{\left|\langle R^n, \mathcal{T}d\rangle_{\mathcal{Y}}\right|}{\|\mathcal{T}d\|_{\mathcal{Y}}} = \tau\left(R^n\right) \left\| R^n \right\|_{\mathcal{Y}} .$$

We insert this now into (4.73) and obtain

$$\left\| R^{n+1} \right\|_{\mathcal{Y}}^2 = \left\| R^n \right\|_{\mathcal{Y}}^2 - \frac{\langle R^n, \mathcal{T}d_{n+1}\rangle_{\mathcal{Y}}^2}{\|\mathcal{T}d_{n+1}\|_{\mathcal{Y}}^2} \le \left\| R^n \right\|_{\mathcal{Y}}^2 - \tau\left(R^n\right)^2 \left\| R^n \right\|_{\mathcal{Y}}^2 .$$

A simple induction leads us eventually to

$$\left\| R^n \right\|_{\mathcal{Y}} \le \left\| R^{n-1} \right\|_{\mathcal{Y}} \left[1 - \tau\left(R^{n-1}\right)^2\right]^{1/2} \le \cdots \le \left\| R^0 \right\|_{\mathcal{Y}} \prod_{k=0}^{n-1} \left[1 - \tau\left(R^k\right)^2\right]^{1/2}$$

such that $\left\| R^n \right\|_{\mathcal{Y}} \le \left\| R^0 \right\|_{\mathcal{Y}} [1 - I(\tau)^2]^{n/2}$, which is the desired result (remember that $R^0 = y$). □

Clearly, the estimates made in this proof are rather rough. Nevertheless, the proved convergence is exponential. However, one should take into account that, on the one hand, $I(\tau) > 0$ for finite-dimensional data spaces $\mathcal{Y}$, as we will show in the next theorem, but, on the other hand, a value of $I(\tau)$ close to zero can nevertheless thwart the exponential decay.

Theorem 4.9.15 *If* $\dim \mathcal{Y} < +\infty$ *and if the dictionary* $\mathcal{D}$ *satisfies*

$$\overline{\operatorname{span}\{\mathcal{T} d \mid d \in \mathcal{D}\}}^{\|\cdot\|_{\mathcal{Y}}} = \mathcal{Y}, \tag{4.91}$$

then $I(\tau) > 0$.

Proof Let us assume the contrary. Then we find sequences $(z_n)_n \subset \mathcal{Y}\setminus\{0\}$ and $(\tau_n)_n \subset \mathbb{R}^+$ such that $\tau(z_n) \leq \tau_n$ for all $n \in \mathbb{N}$ and $\tau_n \to 0$ as $n \to \infty$. If we have a closer look at (4.90), then we see that $\tau(z)$ only depends on the normalized vector $z/\|z\|_{\mathcal{Y}}$. Hence, we may assume, without loss of generality, that $\|z_n\|_{\mathcal{Y}} = 1$ for all $n \in \mathbb{N}$.

We have $\dim \mathcal{Y} < +\infty$, which holds true if and only if the unit sphere $\mathcal{U}$ of $\mathcal{Y}$ is compact (see Theorem 2.5.49). Hence, (z_n) has a convergent subsequence $(z_{n_j})_j$. If we denote the limit by ζ, then $\|\zeta\|_{\mathcal{Y}} = 1$ and

$$\tau(\zeta) = \sup_{d\in\mathcal{D}} \frac{|\langle \zeta, \mathcal{T} d\rangle_{\mathcal{Y}}|}{\|\mathcal{T} d\|_{\mathcal{Y}}} = \lim_{j\to\infty} \sup_{d\in\mathcal{D}} \frac{\left|\langle z_{n_j}, \mathcal{T} d\rangle_{\mathcal{Y}}\right|}{\|\mathcal{T} d\|_{\mathcal{Y}}} \leq \lim_{j\to\infty} \tau_{n_j} = 0,$$

where the interchanging of the limit and the supremum is easy to verify. However, this means that $\langle \zeta, \mathcal{T} d\rangle_{\mathcal{Y}} = 0$ for all $d \in \operatorname{span} \mathcal{D}$, because $\mathcal{T}$ is linear. Due to (4.91), this is only possible, if $\zeta = 0$, which contradicts the previously derived identity $\|\zeta\|_{\mathcal{Y}} = 1$. □

From the previous proof, we can see that the finite dimension of $\mathcal{Y}$ plays an important role in the deduction of $I(\tau) > 0$. This is not essentially critical, because we usually have finite-dimensional data spaces in practice, for example, if the right-hand side of $\mathcal{T} F = y$ consists of samples of a function. For reasons of completeness of the theoretical discussions, we also show here a counterexample where $\dim \mathcal{Y} = +\infty$ and $I(\tau) = 0$.

Example 4.9.16 Let $\mathcal{Y}$ be an arbitrary Hilbert space with $\dim \mathcal{Y} = +\infty$, and let $\mathcal{D}$ consist of an orthonormal basis of $\mathcal{Y}$ (see Theorem 2.5.28 for the existence of such a basis). For instance, we could have $\mathcal{Y} = \mathrm{L}^2(\Omega)$ and $\mathcal{D} = \{Y_{n,j}\}_{n\in\mathbb{N}_0;\ j=-n,\ldots,n}$. Moreover, let $\mathcal{T} = \mathcal{I}$ be the identity. Hence, (4.91) is satisfied. We choose now a countable but infinite system $(u_n)_{n\in\mathbb{N}_0}$ out of $\mathcal{D}$ and a real sequence $(h_k)_{k\in\mathbb{N}_0} \subset]0,1[$ with $h_k \to 1$ as $k \to \infty$. Then we define the vectors $z_k := \sqrt{1-h_k^2}\, \sum_{n=0}^{\infty} h_k^n u_n$, $k \in \mathbb{N}_0$. Consequently, we obtain

$$\|z_k\|_{\mathcal{Y}}^2 = \left(1-h_k^2\right) \sum_{n=0}^{\infty} h_k^{2n} = \left(1-h_k^2\right) \frac{1}{1-h_k^2} = 1,$$

$$\tau\left(z_k\right) = \sup_{d\in\mathcal{D}} |\langle z_k, d\rangle_{\mathcal{Y}}| = \sup_{n\in\mathbb{N}_0} \left|\langle z_k, u_n\rangle_{\mathcal{Y}}\right| = \sqrt{1-h_k^2}\, h_k^0 = \sqrt{1-h_k^2}$$

for all $k \in \mathbb{N}_0$. Since $\tau(z_k) \to 0$ as $k \to \infty$, we get $I(\tau) = 0$.

The bottleneck of the FMP is to find d_{n+1} in (4.74) or (4.76). We have to test every $d \in \mathcal{D}$ in order to find a maximizer. This procedure can be accelerated by saving the inner products $\langle d, \tilde{d} \rangle_{\mathcal{X}}$ and $\langle \mathcal{T}d, \mathcal{T}\tilde{d} \rangle_{\mathcal{Y}}$ for all $d, \tilde{d} \in \mathcal{D}$ as entries of symmetric matrices in the preprocessing. Moreover, we save $\langle y - \mathcal{T}F_0, \mathcal{T}d \rangle_{\mathcal{Y}}$ for all $d \in \mathcal{D}$ in the preprocessing. Then, in every step, we can utilize the identity

$$\langle y - \mathcal{T}F_{n+1}, \mathcal{T}d \rangle_{\mathcal{Y}} = \langle y - \mathcal{T}F_n, \mathcal{T}d \rangle_{\mathcal{Y}} - \alpha_{n+1} \langle \mathcal{T}d_{n+1}, \mathcal{T}d \rangle_{\mathcal{Y}}$$

and 'update' the previously stored values (in the right-hand side). It needs to be taken into account that the available RAM then limits the size of the dictionary if this preprocessing is desired.

Some dictionary elements have continuous parameters, for example, the centre ξ of a radial basis function and, possibly, its associated localization parameter h (see, e.g., the Abel–Poisson kernel). In this case, we can also interpret the maximization of $|\langle y - \mathcal{T}F_n, \mathcal{T}d \rangle_{\mathcal{Y}}|$ as a continuous non-linear optimization problem with respect to the unknowns ξ and h. This approach is (a variant of) a **dictionary learning** and was recently elaborated for the FMP and several of its variants. For details, the reader is referred to Michel and Schneider (2020) and Schneider (2020). It can be seen that essentially fewer iterations, resulting in an essentially smaller 'best basis', suffice to achieve the same approximation accuracy in comparison to the use of a dictionary with discrete centres ξ and localization parameters h.

4.9.2 Weak Functional Matching Pursuit (WFMP)

Another alternative for accelerating the algorithm is to skip the requirement of hitting the exact maximum. This is implemented in the following algorithm, which is due to Kontak (2018) and Kontak and Michel (2019).

Algorithm 4.9.17 (Weak Functional Matching Pursuit (WFMP)) *Let an operator $\mathcal{T} \in \mathcal{L}(\mathcal{X}, \mathcal{Y}) \setminus \{0\}$ and a right-hand side $y \in \mathcal{Y}$ be given.*

(1) Set $n := 0$ and $F_0 := 0$ and choose a stopping criterion. Select a dictionary $\mathcal{D} \subset \mathcal{X} \setminus \ker \mathcal{T}$. Choose a real number $\varrho \in]0, 1]$.

(2) Find $d_{n+1} \in \mathcal{D}$ such that

$$\frac{|\langle y - \mathcal{T}F_n, \mathcal{T}d_{n+1} \rangle_{\mathcal{Y}}|}{\|\mathcal{T}d_{n+1}\|_{\mathcal{Y}}} \geq \varrho \frac{|\langle y - \mathcal{T}F_n, \mathcal{T}d \rangle_{\mathcal{Y}}|}{\|\mathcal{T}d\|_{\mathcal{Y}}} \quad \text{for all } d \in \mathcal{D} \tag{4.92}$$

and set

$$\alpha_{n+1} := \frac{\langle y - \mathcal{T}F_n, \mathcal{T}d_{n+1} \rangle_{\mathcal{Y}}}{\|\mathcal{T}d_{n+1}\|_{\mathcal{Y}}^2} \tag{4.93}$$

as well as $F_{n+1} := F_n + \alpha_{n+1} d_{n+1}$.

(3) If the stopping criterion is fulfilled, then F_{n+1} is the output as an approximate solution of $\mathcal{T}F = y$. Otherwise, increase n by 1 and go to step 2.

Correspondingly, for an i.n.d., (4.92) and (4.93) become

$$\left|\langle y - \mathcal{T} F_n, \mathcal{T} d_{n+1}\rangle_{\mathcal{Y}}\right| \geq \varrho \left|\langle y - \mathcal{T} F_n, \mathcal{T} d\rangle_{\mathcal{Y}}\right| \quad \text{for all } d \in \mathcal{D} \tag{4.94}$$

and

$$\alpha_{n+1} := \langle y - \mathcal{T} F_n, \mathcal{T} d_{n+1}\rangle_{\mathcal{Y}}.$$

The only difference, therefore, is that it suffices to reach ϱ-times the supremum. We will call ϱ the **weakness parameter** here. For $\varrho = 1$, the algorithm is the FMP as we know it. One drawback is that we need to know the supremum of the right-hand side in (4.92) and (4.94), respectively. This can be overcome by using the Cauchy–Schwarz inequality in

$$\frac{\left|\langle y - \mathcal{T} F_n, \mathcal{T} d\rangle_{\mathcal{Y}}\right|}{\|\mathcal{T} d\|_{\mathcal{Y}}} \leq \|y - \mathcal{T} F_n\|_{\mathcal{Y}}.$$

We, therefore, look for a $d_{n+1} \in \mathcal{D}$ such that

$$\frac{\left|\langle y - \mathcal{T} F_n, \mathcal{T} d_{n+1}\rangle_{\mathcal{Y}}\right|}{\|\mathcal{T} d_{n+1}\|_{\mathcal{Y}}} \geq \tilde{\varrho}\, \|y - \mathcal{T} F_n\|_{\mathcal{Y}},$$

where $\tilde{\varrho}$ needs to be smaller than ϱ due to the error caused by the Cauchy–Schwarz inequality. Numerical experiments in Kontak (2018), Kontak and Michel (2019), and Rennhack (2018) support the hope that the computation time can be essentially reduced while the approximation quality is not significantly influenced. However, for a scientific judgement, further tests have to be made in the future.

Let us now have a look at the theoretical properties of the WFMP. For this purpose, we follow step by step the derivation of the theory of the FMP. Equation (4.73) is caused by the choice of α, which is the same for the FMP and the WFMP. Hence, Lemmata 4.9.3 and 4.9.4 remain valid.

Lemma 4.9.18 *Let the sequence $(F_n)_n$ be produced by the WFMP (Algorithm 4.9.17). Then the sequence of (norms of the) residuals $(\|y - \mathcal{T} F_n\|_{\mathcal{Y}})_n$ is monotonically decreasing.*

Lemma 4.9.19 *Let the sequence $(F_n)_n$ be produced by the WFMP (Algorithm 4.9.17). Then the sequence of (norms of the) residuals $(\|y - \mathcal{T} F_n\|_{\mathcal{Y}})_n$ is convergent.*

The proof of Theorem 4.9.8 was lengthy and, therefore, requires a closer look. It turns out that it is also still valid.

Theorem 4.9.20 *Let the dictionary used in the WFMP (Algorithm 4.9.17) satisfy $\overline{\operatorname{span}\{\mathcal{T} d \mid d \in \mathcal{D}\}}^{\|\cdot\|_{\mathcal{Y}}} = \overline{\mathcal{T}(\mathcal{X})}^{\|\cdot\|_{\mathcal{Y}}}$ and let (F_n) be the sequence produced by the WFMP for a given right-hand side $y \in \overline{\mathcal{T}(\mathcal{X})}$. Then we have $\lim_{n\to\infty} \|y - \mathcal{T} F_n\|_{\mathcal{Y}} = 0$, that is, $(\mathcal{T} F_n)_n$ strongly converges to y.*

Proof We follow the proof of Theorem 4.9.8 and its preceding Lemma 4.9.7 and assume again that the dictionary is image normalized. Since (4.73) is still valid, we obtain again the square-summability of $(\alpha_n)_n$ and the implication that $\alpha_n \to 0$ as $n \to \infty$ and, thus,

$$0 = \lim_{n\to\infty} |\alpha_{n+1}| = \lim_{n\to\infty} \left|\langle R^n, \mathcal{T} d_{n+1}\rangle_{\mathcal{Y}}\right| \geq \varrho \lim_{n\to\infty} \left|\langle R^n, \mathcal{T} d\rangle_{\mathcal{Y}}\right|,$$

that is, $\lim_{n\to\infty} \langle R^n, \mathcal{T} d\rangle_{\mathcal{Y}} = 0$, for all $d \in \mathcal{D}$, which is the analogue of (4.86). The remaining derivations are almost completely analogous. We only have to take into account that, in (4.85), we get

$$\begin{aligned}\|R^q - R^p\|_{\mathcal{Y}}^2 &\leq \|R^q\|_{\mathcal{Y}}^2 - \|R^p\|_{\mathcal{Y}}^2 + 2 \sum_{k=q+1}^{p} |\alpha_k| \left|\langle R^p, \mathcal{T} d_k\rangle_{\mathcal{Y}}\right| \\ &\leq \|R^q\|_{\mathcal{Y}}^2 - \|R^p\|_{\mathcal{Y}}^2 + \frac{2}{\varrho} \sum_{k=q+1}^{p} |\alpha_k| \left|\langle R^p, \mathcal{T} d_{p+1}\rangle_{\mathcal{Y}}\right| \\ &\leq r^2 + c - r^2 + \frac{2}{\varrho} \left|\langle R^p, \mathcal{T} d_{p+1}\rangle_{\mathcal{Y}}\right| \sum_{k=q+1}^{p} |\alpha_k| \leq \left(1 + \frac{2}{\varrho}\right) c.\end{aligned}$$

We may still choose c sufficiently small; for example, in this case we could take an arbitrary value with $0 < c < (4 + 8/\varrho)^{-1}\varepsilon^2$ to cause the desired contradiction. □

We consequently also get the following propositions.

Corollary 4.9.21 *The sequence of coefficients $(\alpha_k)_k$ which is produced by the WFMP (Algorithm 4.9.17) is square-summable: $\sum_{k=1}^{\infty} \alpha_k^2 < +\infty$.*

Corollary 4.9.22 *Let the conditions of Theorem 4.9.20 be satisfied, where $y \in \mathcal{T}(\mathcal{X})$. If the generated sequence $(F_n)_n$ converges in $\mathcal{X}$, then the limit $F_\infty := \sum_{k=1}^{\infty} \alpha_k d_k$ is an exact solution of $\mathcal{T} F = y$.*

Concerning the convergence of the WFMP, we have to use the semi-frame condition again.

Theorem 4.9.23 *Let the dictionary $\mathcal{D}$ satisfy the semi-frame condition (see Theorem 4.9.10) with the associated integer $M \in \mathbb{N}$, where the WFMP chooses $(d_k)_k$ out of $\mathcal{D}$ but selects each element of $\mathcal{D}$ not more than M times. Then the sequence $(F_n)_n$ produced by the WFMP converges. If*

$$\overline{\operatorname{span}\{\mathcal{T} d \mid d \in \mathcal{D}\}}^{\|\cdot\|_{\mathcal{Y}}} = \overline{\mathcal{T}(\mathcal{X})}^{\|\cdot\|_{\mathcal{Y}}} \tag{4.95}$$

and $y \in \mathcal{T}(\mathcal{X})$, then the limit $F_\infty = \sum_{k=1}^{\infty} \alpha_k d_k$ solves the inverse problem $\mathcal{T} F_\infty = y$.

Proof Due to Corollary 4.9.21 and the semi-frame condition, we can also here conclude that

$$\left\| \sum_{k=N+1}^{\infty} \alpha_k d_k \right\|_{\mathcal{X}}^2 \leq \frac{1}{c} \sum_{k=N+1}^{\infty} \alpha_k^2 \to 0 \quad \text{as } N \to \infty,$$

which leads us to the existence of the limit F_∞. The fact that F_∞ solves the inverse problem under the condition (4.95) is implied by Corollary 4.9.22. □

Let us now focus on the convergence rate of the algorithm.

Theorem 4.9.24 *Let the sequence $(R^n)_n$ be produced by the WFMP. Then*

$$\|R^n\|_{\mathcal{Y}} \le \|y\|_{\mathcal{Y}} \left[1 - \varrho^2 I(\tau)^2\right]^{n/2},$$

where ϱ is the weakness parameter of the WFMP and $I(\tau)$ is the infimum of the correlation ratio (see Theorem 4.9.14).

Proof Because of (4.92), we now get

$$\frac{\left|\langle R^n, \mathcal{T} d_{n+1}\rangle_{\mathcal{Y}}\right|}{\|\mathcal{T} d_{n+1}\|_{\mathcal{Y}}} \ge \varrho \sup_{d\in\mathcal{D}} \frac{\left|\langle R^n, \mathcal{T} d\rangle_{\mathcal{Y}}\right|}{\|\mathcal{T} d\|_{\mathcal{Y}}} = \varrho\, \tau\left(R^n\right) \left\|R^n\right\|_{\mathcal{Y}}.$$

If we use this inequality in (4.73), then we arrive at

$$\left\|R^{n+1}\right\|_{\mathcal{Y}}^2 = \left\|R^n\right\|_{\mathcal{Y}}^2 - \frac{\langle R^n, \mathcal{T} d_{n+1}\rangle_{\mathcal{Y}}^2}{\|\mathcal{T} d_{n+1}\|_{\mathcal{Y}}^2} \le \left\|R^n\right\|_{\mathcal{Y}}^2 - \varrho^2\, \tau\left(R^n\right)^2 \left\|R^n\right\|_{\mathcal{Y}}^2.$$

As in the proof of Theorem 4.9.14, we can refer to a simple induction which leads us to $\|R^n\|_{\mathcal{Y}} \le \|R^0\|_{\mathcal{Y}}[1 - \varrho^2 I(\tau)^2]^{n/2}$, while $R^0 = y$. □

4.9.3 Orthogonal Functional Matching Pursuit (OFMP)

The FMP often chooses dictionary elements several times. This means that coefficients α_k of previously selected dictionary elements d_k are corrected by the algorithm at a later stage. The reason is the (usual) non-orthogonality of the images $\{\mathcal{F} d \mid d \in \mathcal{D}\}$. If $d_1, \ldots, d_n$ have already been chosen, then the next choice d_{n+1} is (by construction of the algorithm) best in approximating the residual by $\mathcal{F} d_{n+1}$. However, $\mathcal{F} d_{n+1}$ often has non-vanishing projections onto $\mathcal{F} d_1, \ldots, \mathcal{F} d_n$. Hence, adding $\alpha_{n+1} d_{n+1}$ to F_n improves the approximation but there are undesired contributions by $d_1, \ldots, d_n$. As a consequence, the algorithm later on identifies a notable error in these 'directions' and chooses them again for reducing the residual. This effect is illustrated in a simple example in Figures 4.28 and 4.29. The author gratefully acknowledges a discussion with Ingrid Daubechies and Doreen Fischer in this context, which eventually led to the introduction of the OFMP and the ROFMP (see Section 5.4.3) in a joint work with Roger Telschow (see Michel and Telschow, 2016; Telschow, 2014). OFMP and ROFMP are based on the OMP in Pati et al. (1993) and Vincent and Bengio (2002).

Let us further examine the situation mathematically. After step n, we have the residual $R^n = y - \mathcal{T} F_n$ and the approximation $F_n = \sum_{k=1}^n \alpha_k d_k$. Let us define $\mathcal{V}_n :=$ span $\{\mathcal{T} d_1, \ldots, \mathcal{T} d_n\}$ and its orthogonal complement $\mathcal{W}_n := \mathcal{V}_n^{\perp}$ with respect to $(\mathcal{Y}, \langle\cdot,\cdot\rangle_{\mathcal{Y}})$; see also Theorem 2.5.71.

If we assume that $\mathcal{T} F_n$ is the best approximation of y in $\mathcal{V}_n$ (see also Corollary 2.5.64), then $\mathcal{T} F_n = \mathcal{P}_{\mathcal{V}_n} y$, where $\mathcal{P}_{\mathcal{V}_n}$ is the orthogonal projection onto $\mathcal{V}_n$. This is clear, because

$$\begin{aligned}\|y - \mathcal{T} F_n\|_{\mathcal{Y}}^2 &= \left\|\mathcal{P}_{\mathcal{V}_n}(y - \mathcal{T} F_n)\right\|_{\mathcal{Y}}^2 + \left\|\mathcal{P}_{\mathcal{W}_n}(y - \mathcal{T} F_n)\right\|_{\mathcal{Y}}^2 \\ &= \left\|\mathcal{P}_{\mathcal{V}_n} y - \mathcal{T} F_n\right\|_{\mathcal{Y}}^2 + \left\|\mathcal{P}_{\mathcal{W}_n} y\right\|_{\mathcal{Y}}^2.\end{aligned}$$

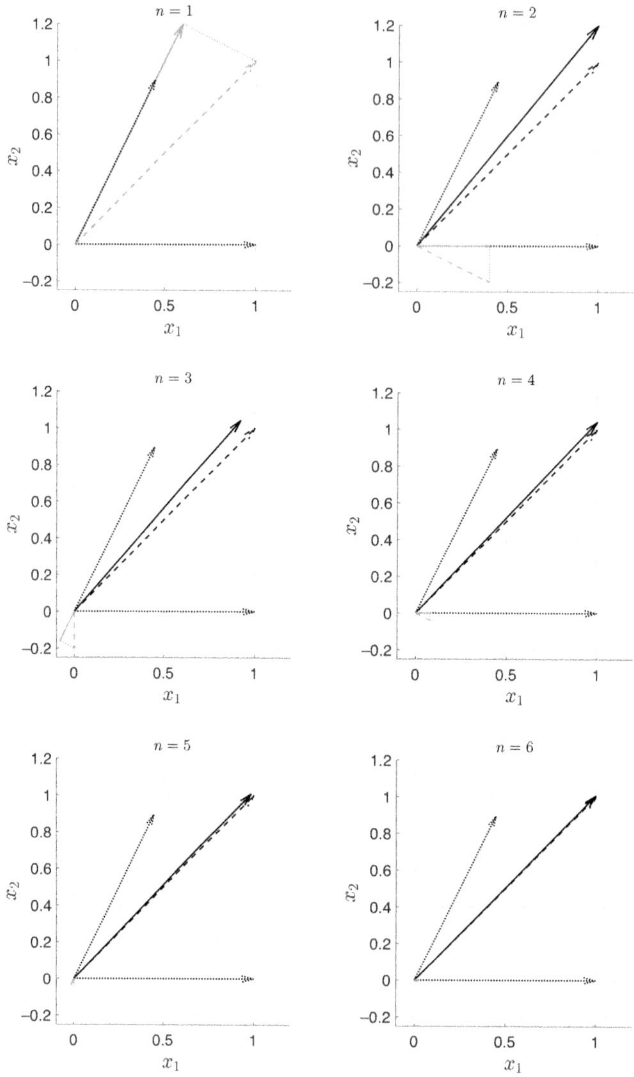

Figure 4.28 We show a simple example of the FMP where $\mathcal{Y} = \mathbb{R}^2$ and $\mathcal{T}\mathcal{D} := \{\mathcal{T}d \mid d \in \mathcal{D}\} = \{(1,0)^{\mathrm{T}}, 5^{-1/2}(1,2)^{\mathrm{T}}\}$ (dotted). The data vector $y = (1,1)^{\mathrm{T}}$ (dashed) is iteratively approximated. We have a look at the first six iteration steps (see the headings at the top of the graphs). In every step, the residual R^{n-1} (dashed grey; note that $R^0 = y$) is projected onto each $\mathcal{T}d$ and the largest projection (solid grey) is chosen as $\alpha_n d_n$. The new approximation is $F_n := F_{n-1} + \alpha_n d_n$ (we show $\mathcal{T}F_n$ in solid black for $n > 1$), and the residual is $R^n = y - \mathcal{T}F_n$ (dotted grey). We see that $R^n \perp \mathcal{T}d_n$ and, since no $\mathcal{T}d$ is orthogonal to $\mathcal{T}d_n$, there is no $\mathcal{T}d$ which is collinear to R^n. So, $\mathcal{T}d_{n+1}$ cannot cover R^n and $0 \neq R^{n+1} \perp \mathcal{T}d_{n+1}$, and so on. As a result, the algorithm only asymptotically approaches the data vector y. The norm of the residual after six iterations is 0.8%. See also Example 4.9.30.

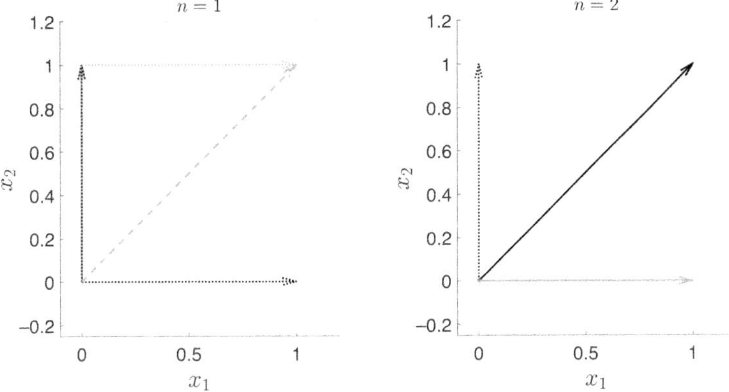

Figure 4.29 In order to illustrate that the non-orthogonality of $\mathcal{T}\mathcal{D}$ is the reason for the slow convergence in Figure 4.28, we modify the dictionary such that $\mathcal{T}\mathcal{D} = \left\{(1,0)^{\mathrm{T}}, (0,1)^{\mathrm{T}}\right\}$. Now, the FMP only needs two iterations in order to obtain an exact result: $\mathcal{T}F_2 = y$. Since non-orthogonal dictionaries cannot be avoided in practice (due to the desired over-completeness of the dictionary, this is partially also intended), this drawback can only be overcome by changing the algorithm, which leads to the OFMP.

The last summand cannot be changed, and the penultimate one is minimal, if and only if $\mathcal{P}_{\mathcal{V}_n} y = \mathcal{T}F_n$. This fact that the 'plumbline' shows the way to the best approximation is commonly known, also in Hilbert spaces.

In the search for d_{n+1}, we decompose $\mathcal{T}d_{n+1}$ into $\mathcal{P}_{\mathcal{V}_n}\mathcal{T}d_{n+1}$ and the orthogonal remainder $\mathcal{P}_{\mathcal{W}_n}\mathcal{T}d_{n+1}$. The component $\mathcal{P}_{\mathcal{V}_n}\mathcal{T}d_{n+1}$ is not desired, because the definition $F_{n+1} := F_n + \alpha_{n+1}d_{n+1}$ would cause a residual

$$R^{n+1} = y - \big(\underbrace{\mathcal{T}F_n + \alpha_{n+1}\mathcal{P}_{\mathcal{V}_n}\mathcal{T}d_{n+1}}_{\in\mathcal{V}_n} + \underbrace{\alpha_{n+1}\mathcal{P}_{\mathcal{W}_n}\mathcal{T}d_{n+1}}_{\in\mathcal{W}_n}\big),$$

that is, we would deteriorate a previously best-approximated component in $\mathcal{V}_n$. Therefore, we should only look for updates of the kind

$$R^{n+1} := R^n - \alpha_{n+1}\mathcal{P}_{\mathcal{W}_n}\mathcal{T}d_{n+1}.$$

We introduce now coefficients $\beta_1^{(n)}, \ldots, \beta_n^{(n)}$, $n \in \mathbb{N}$, with

$$\mathcal{P}_{\mathcal{V}_n}\mathcal{T}d = \sum_{j=1}^{n} \beta_j^{(n)}(d)\,\mathcal{T}d_j. \tag{4.96}$$

These coefficients need not be unique ($\mathcal{T}d_1, \ldots, \mathcal{T}d_n$ could be linearly dependent), but this is not a problem here. We will discuss further later how the $\beta_j^{(n)}$ can be calculated.

In the FMP, d_{n+1} is chosen to maximize $|\langle R^n, \mathcal{T}d_{n+1}\rangle_{\mathcal{Y}}|\,\|\mathcal{T}d_{n+1}\|_{\mathcal{Y}}^{-1}$, which is equivalent to minimizing the residual $\|R^n - \alpha_{n+1}\mathcal{T}d_{n+1}\|_{\mathcal{Y}}$; see (4.73) and (4.74). In our case, where we intend to minimize $\|R^n - \alpha_{n+1}\mathcal{P}_{\mathcal{W}_n}\mathcal{T}d_{n+1}\|_{\mathcal{Y}}$, the derivations lead analogously to the maximization of $|\langle R^n, \mathcal{P}_{\mathcal{W}_n}\mathcal{T}d_{n+1}\rangle_{\mathcal{Y}}| \, \|\mathcal{P}_{\mathcal{W}_n}\mathcal{T}d_{n+1}\|_{\mathcal{Y}}^{-1}$ and a correspondingly analogous formula for α_{n+1}. The new approximation is then chosen such that

$$\mathcal{T}F_{n+1} = \mathcal{T}F_n + \alpha_{n+1}\mathcal{P}_{\mathcal{W}_n}\mathcal{T}d_{n+1} = \mathcal{T}F_n + \alpha_{n+1}\mathcal{T}d_{n+1} - \alpha_{n+1}\mathcal{P}_{\mathcal{V}_n}\mathcal{T}d_{n+1}.$$

This can be achieved by

$$\begin{aligned} F_{n+1} &:= F_n + \alpha_{n+1}d_{n+1} - \alpha_{n+1}\sum_{j=1}^{n}\beta_j^{(n)}(d_{n+1})\,d_j \\ &= \sum_{j=1}^{n}\left(\alpha_j - \alpha_{n+1}\beta_j^{(n)}(d_{n+1})\right)d_j + \alpha_{n+1}d_{n+1}. \end{aligned}$$

Algorithm 4.9.25 (Orthogonal Functional Matching Pursuit (OFMP))
Let an operator $\mathcal{T} \in \mathcal{L}(\mathcal{X},\mathcal{Y}) \setminus \{0\}$ and a right-hand side $y \in \mathcal{Y}$ be given.

(1) Set $n := 0$ and $F_0 := 0$ and choose a stopping criterion. Select a dictionary $\mathcal{D} \subset \mathcal{X} \setminus \ker \mathcal{T}$. Moreover, set $\mathcal{V}_0 := \{0\}$ and $\mathcal{W}_0 := \mathcal{Y}$.

(2) Find $d_{n+1} \in \mathcal{D}$ such that

$$\frac{\left|\langle y - \mathcal{T}F_n, \mathcal{P}_{\mathcal{W}_n}\mathcal{T}d_{n+1}\rangle_{\mathcal{Y}}\right|}{\|\mathcal{P}_{\mathcal{W}_n}\mathcal{T}d_{n+1}\|_{\mathcal{Y}}} \geq \frac{\left|\langle y - \mathcal{T}F_n, \mathcal{P}_{\mathcal{W}_n}\mathcal{T}d\rangle_{\mathcal{Y}}\right|}{\|\mathcal{P}_{\mathcal{W}_n}\mathcal{T}d\|_{\mathcal{Y}}} \quad \text{for all } d \in \mathcal{D} \tag{4.97}$$

(where only $d \in \mathcal{D}$ with $\mathcal{P}_{\mathcal{W}_n}\mathcal{T}d \neq 0$ are considered) and set

$$\alpha_{n+1}^{(n+1)} := \frac{\langle y - \mathcal{T}F_n, \mathcal{P}_{\mathcal{W}_n}\mathcal{T}d_{n+1}\rangle_{\mathcal{Y}}}{\|\mathcal{P}_{\mathcal{W}_n}\mathcal{T}d_{n+1}\|_{\mathcal{Y}}^2}, \tag{4.98}$$

where $\mathcal{P}_{\mathcal{W}_n}$ is the orthogonal projection onto $\mathcal{W}_n$. With the coefficients $\beta_j^{(n)}$ from (4.96), we also set

$$\alpha_j^{(n+1)} := \alpha_j^{(n)} - \alpha_{n+1}^{(n+1)}\beta_j^{(n)}(d_{n+1}), \quad j = 1, \ldots, n, \tag{4.99}$$

$$F_{n+1} := \sum_{j=1}^{n+1}\alpha_j^{(n+1)}d_j, \tag{4.100}$$

$$\begin{aligned} R^{n+1} &:= R^n - \alpha_{n+1}^{(n+1)}\mathcal{P}_{\mathcal{W}_n}\mathcal{T}d_{n+1}, \\ \mathcal{V}_{n+1} &:= \mathcal{V}_n + \{r\mathcal{T}d_{n+1} \,|\, r \in \mathbb{R}\}, \\ \mathcal{W}_{n+1} &:= \mathcal{V}_{n+1}^{\perp} \quad (\text{orthogonal complement in } \mathcal{Y}). \end{aligned}$$

(3) If the stopping criterion is fulfilled (stop at least, if $R^n = 0$), then F_{n+1} is the output as an approximate solution of $\mathcal{T}F = y$. Otherwise, increase n by 1 and go to step 2.

Let us have a look at the properties of this algorithm.

Theorem 4.9.26 *Let $(F_n)_n$ be a sequence produced by the OFMP. Then $R^n \in \mathcal{W}_n$ and $\mathcal{T} F_n = \mathcal{P}_{\mathcal{V}_n} y$ for all $n \in \mathbb{N}_0$.*

Proof For $n = 0$, we have $R^n = y$, $F_0 = 0$, $\mathcal{P}_{\mathcal{V}_0} y = 0$, and $\mathcal{W}_n = \mathcal{Y}$. Let now $R^n \in \mathcal{W}_n$ and $\mathcal{T} F_n = \mathcal{P}_{\mathcal{V}_n} y$ for a fixed $n \in \mathbb{N}_0$. Then our previous discussions reveal that $\mathcal{T} F_{n+1} = \mathcal{P}_{\mathcal{V}_{n+1}} y$ such that $R^{n+1} = y - \mathcal{T} F_{n+1} = \mathcal{P}_{\mathcal{W}_{n+1}} y$. □

Lemma 4.9.27 *Let a dictionary $\mathcal{D}$ be chosen for the OFMP such that $\overline{\operatorname{span}\{\mathcal{T} d \mid d \in \mathcal{D}\}}^{\|\cdot\|_{\mathcal{Y}}} = \mathcal{Y}$. If $R^n \neq 0$ in an iteration step $n \in \mathbb{N}$, then $\mathcal{P}_{\mathcal{W}_n} \mathcal{T} d_{n+1} \neq 0$.*

Proof Let us assume that $\mathcal{P}_{\mathcal{W}_n} \mathcal{T} d_{n+1} = 0$. Then the criterion for choosing d_{n+1} in the OFMP (see (4.97)) implies that $\langle R^n, \mathcal{P}_{\mathcal{W}_n} \mathcal{T} d \rangle_{\mathcal{Y}} = 0$ for all $d \in \mathcal{D}$. In combination with Theorem 4.9.26, this implies $\langle R^n, \mathcal{T} d \rangle_{\mathcal{Y}} = 0$ for all $d \in \mathcal{D}$. Due to the condition on the dictionary, this is only valid if $R^n = 0$. Hence, $\mathcal{P}_{\mathcal{W}_n} \mathcal{T} d_{n+1} \neq 0$, if $R^n \neq 0$. □

Lemma 4.9.28 *Let a dictionary $\mathcal{D}$ be chosen for the OFMP such that $\overline{\operatorname{span}\{\mathcal{T} d \mid d \in \mathcal{D}\}}^{\|\cdot\|_{\mathcal{Y}}} = \mathcal{Y}$. If $R^n \neq 0$ in an iteration step $n \in \mathbb{N}$, then*

$$\dim \mathcal{V}_{n+1} = \dim \mathcal{V}_n + 1.$$

Proof Since $\mathcal{P}_{\mathcal{W}_n} \mathcal{T} d_{n+1} \neq 0$ due to Lemma 4.9.27, we have $\mathcal{T} d_{n+1} \notin \mathcal{V}_n$. Hence, the dimension of $\mathcal{V}_{n+1} = \operatorname{span}\{\mathcal{T} d_i \mid i = 1, \ldots, n+1\}$ is increased by 1 in comparison to $\dim \mathcal{V}_n$. □

Theorem 4.9.29 *Let $\mathcal{Y}$ have a finite dimension. Moreover, let the dictionary $\mathcal{D}$ satisfy $\operatorname{span}\{\mathcal{T} d \mid d \in \mathcal{D}\} = \mathcal{Y}$. Then the OFMP reaches an exact solution F_n, that is, $\mathcal{T} F_n = y$, after at most $n = \dim \mathcal{Y}$ steps.*

Proof In every iteration step n, there are two possibilities: either $R^n = 0$ (i.e., we have reached our aim) or $R^n \neq 0$. If the latter case keeps appearing, then Lemma 4.9.28 tells us that, after $\ell := \dim \mathcal{Y}$ steps, we must have $\dim \mathcal{V}_\ell = \ell$ and, thus, $\mathcal{V}_\ell = \mathcal{Y}$. Hence, $\mathcal{W}_\ell = \{0\}$, and Theorem 4.9.26 leaves us no other possibility than $R^\ell = 0$. □

Example 4.9.30 Let us have a look again at the simple problem in Figure 4.28. We had $\mathcal{T}\mathcal{D} = \left\{(1,0)^{\mathrm{T}}, 5^{-1/2}(1,2)^{\mathrm{T}}\right\}$ and $y = (1,1)^{\mathrm{T}}$. Without loss of generality, we assume that $\mathcal{T} = \mathcal{I}$, the identity operator. In the first step, we start with $R^0 = y$, $F_0 = 0$, $\mathcal{V}_0 = \{0\}$, and $\mathcal{W}_0 = \mathcal{Y} = \mathbb{R}^2$. Clearly, $\mathcal{P}_{\mathcal{W}_0} \mathcal{T} d = \mathcal{T} d$ for all $d \in \mathcal{D}$. Hence, the FMP and the OFMP do not differ in the first step. For choosing d_1, we observe the following (see (4.74) and (4.97) for FMP and OFMP, respectively):

$$\frac{\left\langle R^0, (1,0)^{\mathrm{T}} \right\rangle_{\mathbb{R}^2}}{\left\|(1,0)^{\mathrm{T}}\right\|_{\mathbb{R}^2}} = 1 < \frac{3}{\sqrt{5}} = \frac{\left\langle R^0, 5^{-1/2}(1,2)^{\mathrm{T}} \right\rangle_{\mathbb{R}^2}}{\left\|5^{-1/2}(1,2)^{\mathrm{T}}\right\|_{\mathbb{R}^2}}.$$

Hence, the FMP and the OFMP yield $d_1 = 5^{-1/2}(1,2)^{\mathrm{T}}$, $(\alpha_1 =) \alpha_1^{(1)} = 5^{-1/2} 3$, $F_1 = (3/5, 6/5)^{\mathrm{T}}$, $R^1 = (2/5, -1/5)^{\mathrm{T}}$.

Let us now have a look at the second step, where FMP and OFMP differ. With the FMP, we would now compare

$$\frac{\langle R^1,(1,0)^{\mathrm{T}}\rangle_{\mathbb{R}^2}}{\|(1,0)^{\mathrm{T}}\|_{\mathbb{R}^2}} = \frac{2}{5} > 0 = \frac{\langle R^1,5^{-1/2}(1,2)^{\mathrm{T}}\rangle_{\mathbb{R}^2}}{\|5^{-1/2}(1,2)^{\mathrm{T}}\|_{\mathbb{R}^2}} \tag{4.101}$$

and choose $d_2 = (1,0)^{\mathrm{T}}$ and $\alpha = 2/5$. However, the next approximation $F_2 = (3/5, 6/5)^{\mathrm{T}} + 2/5(1,0)^{\mathrm{T}} = (1,6/5)^{\mathrm{T}}$ and its residual $R^2 = (0, -1/5)^{\mathrm{T}}$ show us that we are not finished yet, as we have already seen in Figure 4.28.

We pursue the second step of the OFMP, which is also illustrated in Figure 4.30. We have $\mathcal{V}_1 = \operatorname{span}(1,2)^{\mathrm{T}}$, $\mathcal{W}_1 = \operatorname{span}(-2,1)^{\mathrm{T}}$, and (see (2.11) for the formula of a one-dimensional projection):

$$\mathcal{P}_{\mathcal{W}_1}\begin{pmatrix}1\\0\end{pmatrix} = \frac{1}{5}\left\langle\begin{pmatrix}1\\0\end{pmatrix},\begin{pmatrix}-2\\1\end{pmatrix}\right\rangle_{\mathbb{R}^2}\begin{pmatrix}-2\\1\end{pmatrix} = \frac{1}{5}\begin{pmatrix}4\\-2\end{pmatrix}, \quad \mathcal{P}_{\mathcal{W}_1}\frac{1}{\sqrt{5}}\begin{pmatrix}1\\2\end{pmatrix} = \begin{pmatrix}0\\0\end{pmatrix},$$

$$\mathcal{P}_{\mathcal{V}_1}\begin{pmatrix}1\\0\end{pmatrix} = \frac{1}{5}\left\langle\begin{pmatrix}1\\0\end{pmatrix},\begin{pmatrix}1\\2\end{pmatrix}\right\rangle_{\mathbb{R}^2}\begin{pmatrix}1\\2\end{pmatrix} = \frac{1}{\sqrt{5}}\frac{1}{\sqrt{5}}\begin{pmatrix}1\\2\end{pmatrix}, \quad \beta_1^{(1)}\left((1,0)^{\mathrm{T}}\right) = \frac{1}{\sqrt{5}}.$$

Hence, in contrast to (4.101), the OFMP now compares

$$\frac{\langle R^1,\mathcal{P}_{\mathcal{W}_1}(1,0)^{\mathrm{T}}\rangle_{\mathbb{R}^2}}{\|\mathcal{P}_{\mathcal{W}_1}(1,0)^{\mathrm{T}}\|_{\mathbb{R}^2}} = \frac{2/5}{2/\sqrt{5}} = \frac{1}{\sqrt{5}} > 0 = \frac{\langle R^1,\mathcal{P}_{\mathcal{W}_1}5^{-1/2}(1,2)^{\mathrm{T}}\rangle_{\mathbb{R}^2}}{\|\mathcal{P}_{\mathcal{W}_1}5^{-1/2}(1,2)^{\mathrm{T}}\|_{\mathbb{R}^2}}.$$

Like the FMP, it also chooses $d_2 = (1,0)^{\mathrm{T}}$, but it continues with the following (see (4.98) and (4.99) for the formulae):

$$\alpha_2^{(2)} = \frac{\langle R^1,\mathcal{P}_{\mathcal{W}_1}(1,0)^{\mathrm{T}}\rangle_{\mathbb{R}^2}}{\|\mathcal{P}_{\mathcal{W}_1}(1,0)^{\mathrm{T}}\|_{\mathbb{R}^2}^2} = \frac{2/5}{4/5} = \frac{1}{2}, \qquad \alpha_1^{(2)} = \frac{3}{\sqrt{5}} - \frac{1}{2}\frac{1}{\sqrt{5}} = \frac{1}{2}\sqrt{5}.$$

Eventually,

$$F_2 = \frac{1}{2}\sqrt{5}\frac{1}{\sqrt{5}}\begin{pmatrix}1\\2\end{pmatrix} + \frac{1}{2}\begin{pmatrix}1\\0\end{pmatrix} = \begin{pmatrix}1\\1\end{pmatrix} = y.$$

Indeed, the OFMP yields an exact solution after two steps in this two-dimensional example.

In the following, we will derive explicit formulae for the $\beta_j^{(n)}$, which were developed in Schneider (2020).

In step $n = 1$, we have $\mathcal{V}_0 = \{0\}$ and $\mathcal{W}_0 = \mathcal{Y}$, and no projection is necessary. In the next step, we have $\mathcal{V}_1 = \{c_1\mathcal{T}d_1 \mid c_1 \in \mathbb{R}\}$. For arbitrary $d \in \mathcal{D}$, we then get (see also (2.11)) the projections

$$\mathcal{P}_{\mathcal{V}_1}\mathcal{T}d = \underbrace{\frac{\langle \mathcal{T}d,\mathcal{T}d_1\rangle_{\mathcal{Y}}}{\|\mathcal{T}d_1\|_{\mathcal{Y}}^2}}_{=\beta_1^{(1)}(d)}\mathcal{T}d_1, \qquad \mathcal{P}_{\mathcal{W}_1}\mathcal{T}d = \mathcal{T}d - \beta_1^{(1)}(d)\mathcal{T}d_1. \tag{4.102}$$

If we then have $\mathcal{V}_2 = \{c_1\mathcal{T}d_1 + c_2\mathcal{T}d_2 \mid c_1, c_2 \in \mathbb{R}\}$, then we need to take into account that $\mathcal{T}d_1$ and $\mathcal{T}d_2$ are in general not orthogonal. Hence, we need to project an arbitrary $\mathcal{T}d$

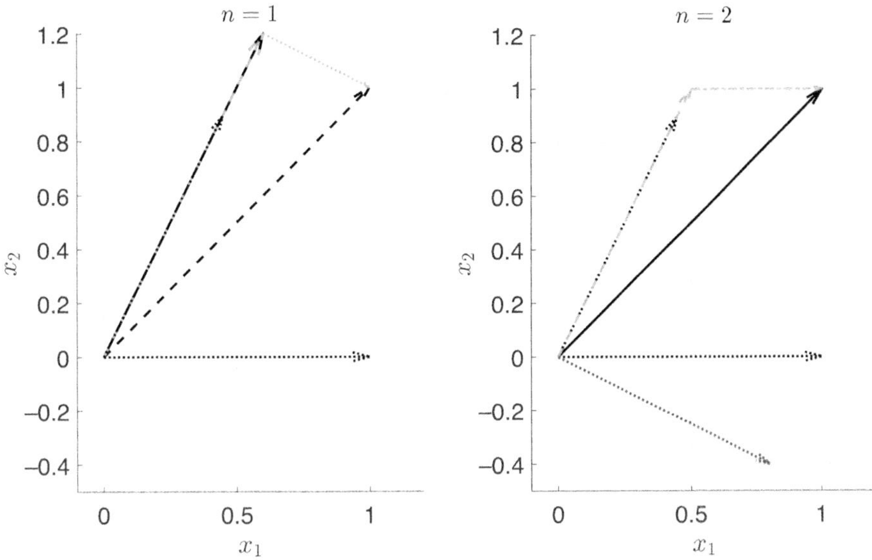

Figure 4.30 We solve the problem in Figure 4.28 with the OFMP; see also Example 4.9.30. Without loss of generality, let $\mathcal{T} = \mathcal{I}$. Again, the dictionary $\mathcal{D} = \mathcal{T}\mathcal{D}$ is dotted, the data vector y is dashed, $\mathcal{T} F_n$ is solid, and the new residual (after the OFMP step) is dotted grey. The iteration steps n are given in the titles of the images. Now we show $\alpha_1^{(n)} d_1$ in dashed grey and $\alpha_2^{(n)} d_2$ in dash-dotted grey. The first steps of FMP and OFMP are identical. However, in the second step, the OFMP considers projections of the residual onto $\mathcal{P}_{\mathcal{W}_n}\mathcal{T}d$ instead of $\mathcal{T}d$. We show $\mathcal{P}_{\mathcal{W}_1}\mathcal{T}d_2$ in dotted dark grey. As we can see, (4.99) causes a change of the approximation in the direction of d_1, and the obtained iterate F_2 calculated via (4.100) is an exact solution.

onto $\mathcal{T}d_1$ (i.e., onto $\mathcal{V}_1$) as in (4.102) and then determine the remainder in $\mathcal{V}_2$ which is orthogonal to $\mathcal{T}d_1$ (in other words, the projection of $\mathcal{T}d$ onto $\mathcal{P}_{\mathcal{W}_1}\mathcal{T}d_2$). Thus, we have

$$\mathcal{P}_{\mathcal{V}_2}\mathcal{T}d = \frac{\langle \mathcal{T}d, \mathcal{T}d_1\rangle_{\mathcal{Y}}}{\|\mathcal{T}d_1\|_{\mathcal{Y}}^2}\,\mathcal{T}d_1 + \frac{\left\langle \mathcal{T}d, \mathcal{P}_{\mathcal{W}_1}\mathcal{T}d_2\right\rangle_{\mathcal{Y}}}{\left\|\mathcal{P}_{\mathcal{W}_1}\mathcal{T}d_2\right\|_{\mathcal{Y}}^2}\,\mathcal{P}_{\mathcal{W}_1}\mathcal{T}d_2.$$

With the formula for $\mathcal{P}_{\mathcal{W}_1}\mathcal{T}d$ in (4.102), we obtain

$$\begin{aligned}\mathcal{P}_{\mathcal{V}_2}\mathcal{T}d &= \underbrace{\frac{\langle \mathcal{T}d, \mathcal{T}d_1\rangle_{\mathcal{Y}}}{\|\mathcal{T}d_1\|_{\mathcal{Y}}^2}}_{=\beta_1^{(1)}(d)}\,\mathcal{T}d_1 + \frac{\left\langle \mathcal{T}d, \mathcal{P}_{\mathcal{W}_1}\mathcal{T}d_2\right\rangle_{\mathcal{Y}}}{\left\|\mathcal{P}_{\mathcal{W}_1}\mathcal{T}d_2\right\|_{\mathcal{Y}}^2}\left(\mathcal{T}d_2 - \beta_1^{(1)}(d_2)\,\mathcal{T}d_1\right)\\ &= \underbrace{\left[\beta_1^{(1)}(d) - \frac{\left\langle \mathcal{T}d, \mathcal{P}_{\mathcal{W}_1}\mathcal{T}d_2\right\rangle_{\mathcal{Y}}}{\left\|\mathcal{P}_{\mathcal{W}_1}\mathcal{T}d_2\right\|_{\mathcal{Y}}^2}\,\beta_1^{(1)}(d_2)\right]}_{=\beta_1^{(2)}(d)}\mathcal{T}d_1 + \underbrace{\frac{\left\langle \mathcal{T}d, \mathcal{P}_{\mathcal{W}_1}\mathcal{T}d_2\right\rangle_{\mathcal{Y}}}{\left\|\mathcal{P}_{\mathcal{W}_1}\mathcal{T}d_2\right\|_{\mathcal{Y}}^2}}_{=\beta_2^{(2)}(d)}\,\mathcal{T}d_2.\end{aligned} \tag{4.103}$$

After considering the case in $\mathcal{V}_3$ and $\mathcal{W}_3$, which we omit here, we can find a conjecture for a formula of the $\beta_j^{(n)}$.

Theorem 4.9.31 *For all $n \in \mathbb{N}$ and all $j = 1, \ldots, n-1$ and every $d \in \mathcal{X}$, we have (provided that we do not divide by zero)*

$$\beta_n^{(n)}(d) = \frac{\langle \mathcal{T}d, \mathcal{P}_{\mathcal{W}_{n-1}} \mathcal{T}d_n \rangle_{\mathcal{Y}}}{\| \mathcal{P}_{\mathcal{W}_{n-1}} \mathcal{T}d_n \|_{\mathcal{Y}}^2}, \tag{4.104}$$

$$\beta_j^{(n)}(d) = \beta_j^{(n-1)}(d) - \beta_n^{(n)}(d)\, \beta_j^{(n-1)}(d_n) \tag{4.105}$$

such that

$$\mathcal{P}_{\mathcal{V}_n} \mathcal{T}d = \sum_{j=1}^{n} \beta_j^{(n)}(d)\, \mathcal{T}d_j, \tag{4.106}$$

$$\mathcal{P}_{\mathcal{W}_n} \mathcal{T}d = \mathcal{T}d - \mathcal{P}_{\mathcal{V}_n} \mathcal{T}d \tag{4.107}$$

for $\mathcal{V}_n = \operatorname{span}\{\mathcal{T}d_1, \ldots, \mathcal{T}d_n\}$, $\mathcal{W}_n = \mathcal{V}_n^{\perp}$, and an arbitrary *choice of elements $d_1, \ldots, d_n \in \mathcal{X}$.*

Proof The case $n = 1$ corresponds to (4.102). For $n = 2$, (4.104) and (4.105) are confirmed by (4.103). Let us now assume that (4.104) and (4.105) in combination with (4.106) and (4.107) are valid for an integer $n \in \mathbb{N}$ and all corresponding $j = 1, \ldots, n-1$. Then, with analogous considerations to those preceding this theorem, we observe that

$$\mathcal{P}_{\mathcal{V}_{n+1}} \mathcal{T}d = \mathcal{P}_{\mathcal{V}_n} \mathcal{T}d + \frac{\langle \mathcal{T}d, \mathcal{P}_{\mathcal{W}_n} \mathcal{T}d_{n+1} \rangle_{\mathcal{Y}}}{\| \mathcal{P}_{\mathcal{W}_n} \mathcal{T}d_{n+1} \|_{\mathcal{Y}}^2} \mathcal{P}_{\mathcal{W}_n} \mathcal{T}d_{n+1}.$$

With the assumption of the induction, we conclude that

$$\begin{aligned}
\mathcal{P}_{\mathcal{V}_{n+1}} \mathcal{T}d &= \sum_{j=1}^{n} \beta_j^{(n)}(d)\, \mathcal{T}d_j \\
&\quad + \frac{\langle \mathcal{T}d, \mathcal{P}_{\mathcal{W}_n} \mathcal{T}d_{n+1} \rangle_{\mathcal{Y}}}{\| \mathcal{P}_{\mathcal{W}_n} \mathcal{T}d_{n+1} \|_{\mathcal{Y}}^2} \left[\mathcal{T}d_{n+1} - \sum_{j=1}^{n} \beta_j^{(n)}(d_{n+1})\, \mathcal{T}d_j \right] \\
&= \sum_{j=1}^{n} \underbrace{\left[\beta_j^{(n)}(d) - \frac{\langle \mathcal{T}d, \mathcal{P}_{\mathcal{W}_n} \mathcal{T}d_{n+1} \rangle_{\mathcal{Y}}}{\| \mathcal{P}_{\mathcal{W}_n} \mathcal{T}d_{n+1} \|_{\mathcal{Y}}^2} \beta_j^{(n)}(d_{n+1}) \right]}_{=\beta_j^{(n+1)}(d)} \mathcal{T}d_j \\
&\quad + \underbrace{\frac{\langle \mathcal{T}d, \mathcal{P}_{\mathcal{W}_n} \mathcal{T}d_{n+1} \rangle_{\mathcal{Y}}}{\| \mathcal{P}_{\mathcal{W}_n} \mathcal{T}d_{n+1} \|_{\mathcal{Y}}^2}}_{=\beta_{n+1}^{(n+1)}(d)} \mathcal{T}d_{n+1}.
\end{aligned}$$

□

Schneider (2020) also showed that the recursion in (4.105) can be resolved. We also recapitulate this result here.

Theorem 4.9.32 *The coefficients* $\beta_j^{(n)}$ *in Theorem 4.9.31 satisfy*

$$\beta_j^{(n)}(d) = \beta_j^{(j)}(d) - \sum_{k=j+1}^{n} \beta_k^{(k)}(d)\, \beta_j^{(k-1)}(d_k)$$

for all $n \in \mathbb{N}$, $j = 1, \ldots, n$, *and* $d \in \mathcal{X}$.

Proof We keep $j \in \mathbb{N}$ fixed and prove the formula for all $n \in \{j, j+1, \ldots\}$. For $n = j$, the formula is trivial. Let now $n > j$. We iteratively apply (4.105):

$$\begin{aligned}
\beta_j^{(n)}(d) &= \beta_j^{(n-1)}(d) - \beta_n^{(n)}(d)\, \beta_j^{(n-1)}(d_n) \\
&= \beta_j^{(n-2)}(d) - \beta_{n-1}^{(n-1)}(d)\, \beta_j^{(n-2)}(d_{n-1}) - \beta_n^{(n)}(d)\, \beta_j^{(n-1)}(d_n) \\
&= \beta_j^{(n-2)}(d) - \sum_{k=n-1}^{n} \beta_k^{(k)}(d)\, \beta_j^{(k-1)}(d_k) \\
&= \beta_j^{(n-3)}(d) - \beta_{n-2}^{(n-2)}(d)\, \beta_j^{(n-3)}(d_{n-2}) - \sum_{k=n-1}^{n} \beta_k^{(k)}(d)\, \beta_j^{(k-1)}(d_k) \\
&= \beta_j^{(n-3)}(d) - \sum_{k=n-2}^{n} \beta_k^{(k)}(d)\, \beta_j^{(k-1)}(d_k) = \cdots \text{ (and so on)} \\
&= \beta_j^{(j)}(d) - \sum_{k=j+1}^{n} \beta_k^{(k)}(d)\, \beta_j^{(k-1)}(d_k).
\end{aligned}$$

□

For algorithmic details of an efficient implementation of the OFMP, see Telschow (2014). For the regularized versions of the algorithms FMP, OFMP, and WFMP, see Section 5.4.

Exercises

4.1 Prove that $\nabla^* \cdot \nabla^* = \Delta^* = \mathrm{L}^* \cdot \mathrm{L}^*$ on $\mathrm{C}^{(2)}(\Omega)$.

4.2 Derive explicit formulae for $Y_{0,0}$, $Y_{1,-1}$, $Y_{1,0}$, and $Y_{1,1}$ and their vectorial counterparts $y_{n,j}^{(i)}$. Express the results in polar coordinates φ, t and in Cartesian coordinates ξ_1, ξ_2, ξ_3. Simplify your results as far as possible.

4.3 Find the reproducing kernel of $(\mathrm{Harm}_n(\Omega), \langle \cdot, \cdot \rangle_{\mathrm{L}^2(\Omega)})$.

4.4 Show that the Legendre polynomials satisfy $|P_n(\xi \cdot \zeta) - P_n(\eta \cdot \zeta)| \le [n(n+1)/2]|\xi - \eta|$ for all $\xi, \eta, \zeta \in \Omega$.

4.5 It is known that functions in $\mathcal{H}_s(\Omega)$ with $s > 2$ are Lipschitz continuous. Show that, for a function $F \in \mathcal{H}_s(\Omega)$, the following finite term can be used as a Lipschitz constant:

$$C_F(s) := \left(\frac{1}{2} \sum_{n=1}^{\infty} \frac{2n+1}{4\pi} \frac{n(n+1)}{(n+1/2)^{2s}} \right)^{1/2} \|F\|_{\mathcal{H}_s(\Omega)}.$$

4.6 Prove part a of Theorem 4.4.10.

5
Inverse Problems

There are three purposes of this chapter: first, two examples of geomathematical inverse problems are discussed and their most important properties are summarized. These two problems will show us what the typical features of inverse problems are. We will also come across another example in the following chapters. Secondly, basics on the theory of inverse problems and their regularization will be explained. Thirdly, the matching-pursuit type algorithms for constructing an approximation in a best basis, which were discussed in Section 4.9, will be supplemented with their known regularized versions.

5.1 Example 1: Downward Continuation

We have already discussed the boundary-value problems for the Laplace equation in detail. One formula from this context is important for the following derivations: the unique solution of the EDP with given boundary values $F \in \mathrm{C}(\Omega)$ is given by the following (see the discussions after Theorem 3.4.25):

$$U(r\xi) = \sum_{n=0}^{\infty} \sum_{j=-n}^{n} \langle F, Y_{n,j} \rangle_{\mathrm{L}^2(\Omega)} r^{-n-1} Y_{n,j}(\xi), \quad r \geq 1, \xi \in \Omega. \tag{5.1}$$

This can easily be generalized to the case that the boundary is the sphere $\Omega_\varrho := \varrho\Omega = \{\varrho\xi \mid \xi \in \Omega\} = S_\varrho(0)$ of radius $\varrho > 0$. Then $F \in \mathrm{C}(\Omega_\varrho)$ is continued to a harmonic function by

$$U(r\xi) = \sum_{n=0}^{\infty} \sum_{j=-n}^{n} \left\langle F, Y_{n,j}^{\varrho} \right\rangle_{\mathrm{L}^2(\Omega_\varrho)} \left(\frac{r}{\varrho}\right)^{-n-1} Y_{n,j}^{\varrho}(\varrho\xi), \quad r \geq \varrho, \xi \in \Omega, \tag{5.2}$$

where $Y_{n,j}^{\varrho}(x) := \varrho^{-1} Y_{n,j}(x/|x|)$, $x \in \Omega_\varrho$, $n \in \mathbb{N}_0$, $j \in \{-n, \ldots, n\}$, is obviously an orthonormal basis of $(\mathrm{L}^2(\Omega_\varrho), \langle \cdot, \cdot \rangle_{\mathrm{L}^2(\Omega_\varrho)})$. It should be noted that the series in (5.1) and (5.2) are pointwise convergent outside the surface and convergent in the L^2-sense on the surface.

In practice, ϱ would be the radius of the approximately ball-shaped Earth (or any other planet) and U would be the gravitational potential. If we now evaluate the potential at an approximately spherical surface Ω_σ (containing, e.g., a satellite orbit) with radius $\sigma > \varrho$, then

$$\begin{aligned}
U(\sigma\xi) &= \sum_{n=0}^{\infty}\sum_{j=-n}^{n} \left\langle F, Y_{n,j}^{\varrho}\right\rangle_{\mathrm{L}^2(\Omega_\varrho)} \left(\frac{\sigma}{\varrho}\right)^{-n-1} Y_{n,j}^{\varrho}(\varrho\xi)\\
&= \sum_{n=0}^{\infty}\sum_{j=-n}^{n} \left\langle F, Y_{n,j}^{\varrho}\right\rangle_{\mathrm{L}^2(\Omega_\varrho)} \left(\frac{\sigma}{\varrho}\right)^{-n-1} \frac{1}{\varrho}\frac{\sigma}{\sigma} Y_{n,j}(\xi)\\
&= \sum_{n=0}^{\infty}\sum_{j=-n}^{n} \left\langle F, Y_{n,j}^{\varrho}\right\rangle_{\mathrm{L}^2(\Omega_\varrho)} \left(\frac{\varrho}{\sigma}\right)^{n} Y_{n,j}^{\sigma}(\sigma\xi), \quad \xi \in \Omega.
\end{aligned}$$

This way we can upward continue the gravitational potential U from the Earth's surface Ω_ϱ to a higher altitude σ. This relation can be represented by a linear and continuous operator.

Definition 5.1.1 The operator $\mathcal{T}^{\mathrm{up}}\colon \mathrm{L}^2(\Omega_\varrho) \to \mathrm{L}^2(\Omega_\sigma)$, where $\sigma > \varrho > 0$, with the singular-value decomposition

$$\mathcal{T}^{\mathrm{up}} F := \sum_{n=0}^{\infty}\sum_{j=-n}^{n} \left\langle F, Y_{n,j}^{\varrho}\right\rangle_{\mathrm{L}^2(\Omega_\varrho)} \left(\frac{\varrho}{\sigma}\right)^{n} Y_{n,j}^{\sigma}$$

is called the **upward continuation operator**.

Theorem 5.1.2 *The upward continuation operator is linear, continuous, and compact.*

Proof The linearity is obvious. Moreover, since the singular values $(\varrho\sigma^{-1})^n$ tend to zero as $n \to \infty$, the operator is also continuous and compact (see the considerations after Theorem 2.5.81). □

However, in practice, we have the opposite, that is the **inverse problem**. Gravitational data are given on a higher altitude due to airborne or spaceborne measurements. Thus, the question is: can we invert $\mathcal{T}^{\mathrm{up}}$? This is a typical example of an inverse problem. From Definition 3.3.2, we already have a classification of problems including inverse problems: an inverse problem $\mathcal{T}F = G$ is **well-posed in the sense of Hadamard** if each of the following criteria is satisfied:

(a) The problem is **solvable**, that is, $\mathcal{T}$ is surjective.
(b) There is **not more than one solution** to a given G, that is, $\mathcal{T}$ is injective.
(c) The **solution is stable**, that is, $\mathcal{T}^{-1}$ is continuous.

Otherwise, the problem is called **ill-posed**.

Let us investigate these requirements on well-posed problems in the case of the upward continuation operator.

Definition 5.1.3 The inverse problem of finding $F \in \mathrm{L}^2(\Omega_\varrho)$ such that $\mathcal{T}^{\mathrm{up}} F = G$ for a given $G \in \mathrm{L}^2(\Omega_\sigma)$ is called the **downward continuation problem**.

Question 1: is the problem $\mathcal{T}^{\mathrm{up}} F = G$ always solvable? Since $\mathcal{T}^{\mathrm{up}} F = G$ is equivalent to

$$\sum_{n=0}^{\infty}\sum_{j=-n}^{n} \left\langle F, Y_{n,j}^{\varrho}\right\rangle_{\mathrm{L}^2(\Omega_\varrho)} \left(\frac{\varrho}{\sigma}\right)^{n} Y_{n,j}^{\sigma} = \sum_{n=0}^{\infty}\sum_{j=-n}^{n} \left\langle G, Y_{n,j}^{\sigma}\right\rangle_{\mathrm{L}^2(\Omega_\sigma)} Y_{n,j}^{\sigma}$$

and both sides are expansions in the same orthonormal basis of $\mathrm{L}^2(\Omega_\sigma)$, we obtain that $\mathcal{T}^{\mathrm{up}} F = G$ if and only if

$$\left\langle F, Y_{n,j}^{\varrho}\right\rangle_{\mathrm{L}^2(\Omega_\varrho)} \left(\frac{\varrho}{\sigma}\right)^n = \left\langle G, Y_{n,j}^{\sigma}\right\rangle_{\mathrm{L}^2(\Omega_\sigma)} \quad \text{for all } n \in \mathbb{N}_0 \text{ and } j \in \{-n, \ldots, n\}.$$

For solvability, it is necessary that the function $F \in \mathrm{L}^2(\Omega_\varrho)$ with

$$\left\langle F, Y_{n,j}^{\varrho}\right\rangle_{\mathrm{L}^2(\Omega_\varrho)} = \left(\frac{\sigma}{\varrho}\right)^n \left\langle G, Y_{n,j}^{\sigma}\right\rangle_{\mathrm{L}^2(\Omega_\sigma)} \quad \text{for all } n \in \mathbb{N}_0 \text{ and } j \in \{-n, \ldots, n\} \tag{5.3}$$

exists. In other words, we need that

$$\sum_{n=0}^{\infty} \sum_{j=-n}^{n} \left(\frac{\sigma}{\varrho}\right)^{2n} \left\langle G, Y_{n,j}^{\sigma}\right\rangle_{\mathrm{L}^2(\Omega_\sigma)}^2 < +\infty \tag{5.4}$$

due to the Parseval identity. Hence, the answer is no: for solvability, it is necessary (and, as a matter of fact, also sufficient) that $G \in \mathcal{H}\left(((\varrho\sigma^{-1})^n); \Omega_\sigma\right)$, which is a proper subspace of $\mathrm{L}^2(\Omega_\sigma)$, as indicated in Theorem 4.7.2. For example, $G := \sum_{n=0}^{\infty} \sum_{j=-n}^{n} (\varrho\sigma^{-1})^{n/2} Y_{n,j}^{\sigma}$ is an element of $\mathrm{L}^2(\Omega_\sigma)$, since $0 < \varrho < \sigma$, but violates (5.4).

Question 2: if $G \in \mathcal{T}^{\mathrm{up}}(\mathrm{L}^2(\Omega_\varrho))$, is there only one F such that $\mathcal{T}^{\mathrm{up}} F = G$? Yes, F is given by (5.3). We do have a **downward continuation operator** $\mathcal{T}^{\mathrm{down}} := (\mathcal{T}^{\mathrm{up}})^{-1}$: $\mathcal{T}^{\mathrm{up}}(\mathrm{L}^2(\Omega_\varrho)) \to \mathrm{L}^2(\Omega_\varrho)$ with

$$\mathcal{T}^{\mathrm{down}} G = \sum_{n=0}^{\infty} \sum_{j=-n}^{n} \left(\frac{\sigma}{\varrho}\right)^n \left\langle G, Y_{n,j}^{\sigma}\right\rangle_{\mathrm{L}^2(\Omega_\sigma)} Y_{n,j}^{\varrho}.$$

Question 3: is $\mathcal{T}^{\mathrm{down}}$ continuous? Obviously not, because it has exponentially diverging singular values! For example, each $G_n := (n+1)^{-1} Y_{n,0}^{\sigma}$ is an element of $\mathcal{T}^{\mathrm{up}}(\mathrm{L}^2(\Omega_\varrho))$ due to its bandlimit. However, we can easily see that $\|G\|_{\mathrm{L}^2(\Omega_\sigma)} = (n+1)^{-1} \to 0$, but $\|\mathcal{T}^{\mathrm{down}} G_n\|_{\mathrm{L}^2(\Omega_\varrho)} = (\sigma\varrho^{-1})^n (n+1)^{-1} \to \infty$. For this reason, the downward continuation problem is considered to be an example of an exponentially ill-posed problem; see also Louis (1989, Definition 3.2.1).

Theorem 5.1.4 *The downward continuation problem stated in Definition 5.1.3 is ill-posed in the sense of Hadamard.*

The downward continuation problem is a well-known and intensively discussed inverse problem in geomathematics. When we summarize general properties of ill-posed problems, we will recognize also the properties that we found here for the downward continuation problem.

5.2 Example 2: Inverse Gravimetry

Rather early in this book, on page 51, we had our first look at Newton's gravitational potential in (3.3), namely

$$G \int_D \frac{F(x)}{|x-y|}\, \mathrm{d}x = V(y), \quad y \in \mathbb{R}^3, \tag{5.5}$$

where G is the gravitational constant. This equation is additionally the starting point for another inverse problem: can we determine the mass density distribution F out of data of the gravitational potential V?

Definition 5.2.1 The (linear) **inverse gravimetric problem** consists of inverting (5.5), that is, given V (and D), find F.

Note that there is also a non-linear inverse gravimetric problem, which is: given V and F, find D. We will not investigate this problem here, because, for reasons of brevity, the considerations in this chapter are limited to linear inverse problems. The non-linear inverse gravimetric problem is not only applicable to the determination of the topography of the Earth's surface but (more interestingly) also of boundaries inside the Earth such as the Mohorovičić discontinuity between the mantle and the crust. For further details on the non-linear inverse gravimetric problem, see, for example, Bagherbandi (2012), Hettlich and Rundell (1996), Isakov (1990, 2006), Kontak (2018), Kontak and Michel (2018), Novikoff (1938), and Weck (1972).

The linear inverse gravimetric problem has several applications. For example, gravitational data can be used in combination with other data to reveal structures inside the Earth or other celestial bodies; see, for example, Berkel (2009), Berkel and Michel (2010), Berkel et al. (2011), Blick et al. (2017), Bolton et al. (2017), Boulanger and Chouteau (2001), Fengler et al. (2006), Fischer (2011), Fischer and Michel (2012), Kaban et al. (2002), Last and Kubik (1983), Li and Oldenburg (1998), Michel (1999, 2002b, 2005), Michel and Wolf (2008), Tesauro et al. (2008), Yegorova and Starostenko (1999), Yegorova et al. (1995, 1997), and Zuber et al. (2013). Furthermore, temporal variations of the gravitational potential, which have been observed with the GRACE (Tapley et al., 2004b) and the GRACE follow-on (Flechtner et al., 2014) missions, provide us with possibilities to observe mass transports on the surface of the Earth. Such mass transports often have climatic reasons, either due to seasonal phenomena or due to climate change. Examples of such data analyses can be found in Baur and Sneeuw (2011), Chen et al. (2006), Fengler et al. (2007), Fersch et al. (2012), Fischer and Michel (2012, 2013b), Harig and Simons (2016), Jansen et al. (2009), Ramillien et al. (2006), Sasgen et al. (2012), and Velicogna and Wahr (2006, 2013). Moreover, non-climatic influences such as major seismic events can also reveal traces in the gravitational field, possibly even beforehand; see, for example, Panet et al. (2018).

The theory of the inverse gravimetric problem has already been intensively studied; see (in chronological order), in particular, Stokes (1867); Pizzetti (1909, 1910); Lauricella (1912); Novikoff (1938); Weck (1972); Buchheim (1975); Rubincam (1979); Marussi (1980); Moritz (1990); Michel (1999); Freeden and Michel (2004); the survey article Michel and Fokas (2008), including the references therein; and the generalized approach to a class of integral equations comprising the inverse gravimetric problem in Michel and Orzlowski (2016) and Leweke et al. (2018a). Note that Leweke et al. (2018a) also includes the case of reconstructing a surface mass distribution from gravitational data.

We will derive here again the most important properties of the inverse gravimetric problem. Before we start, we should further specify the problem: first of all, if we know V inside D presumably on a sufficiently dense grid, then the problem is rather easy. If the unknown function is Hölder continuous (which might not be realistic but could be a sufficiently close approximation), then V and F satisfy the Poisson equation (see Theorem 3.1.11) $\Delta V = -4\pi G F$ inside D; see also Blick et al. (2017), Moritz (1990), and Stokes (1867). Though numerical differentiation is not easy, this is still relatively simple in comparison to the case that V is only given on $\mathbb{R}^3 \setminus \overline{D}$, which we will consider now. Moreover, we will assume a ball-shaped Earth $D = B_R(0)$. Essentially, more properties can be proved for this particular case. Only some of them can be generalized to the case where D is the interior of a regular surface; see for instance, Freeden and Michel (2004, chapter 6). So, let us now further specify our inverse problem.

Definition 5.2.2 The **inverse gravimetric problem** which we will consider is as follows: given a function $V\colon \mathbb{R}^3 \setminus \overline{B_R(0)} \to \mathbb{R}$ which is harmonic and regular at infinity, find $F \in \mathrm{L}^2(B_R(0))$ such that

$$\left(\mathcal{T}^{\mathrm{gr}} F\right)(y) := G \int_{B_R(0)} \frac{F(x)}{|x-y|}\,\mathrm{d}x = V(y) \quad \text{for all } y \in \mathbb{R}^3 \setminus \overline{B_R(0)}. \tag{5.6}$$

We have already encountered the integral kernel $k(x,y) := |x-y|^{-1}$ in Corollary 3.4.18: if $|x| < |y|$, then

$$\frac{1}{|x-y|} = \sum_{n=0}^{\infty} \frac{|x|^n}{|y|^{n+1}} P_n\left(\frac{x}{|x|} \cdot \frac{y}{|y|}\right). \tag{5.7}$$

The condition $|x| < |y|$ for (5.7) is obvious, because otherwise the series is divergent, and is the reason why we would get into trouble if we did not assume a ball-shaped Earth. With $|x| < |y|$, we can say that, for fixed $y \in \mathbb{R}^3 \setminus \overline{B_R(0)}$, the series in (5.7) is uniformly convergent with respect to $x \in \overline{B_R(0)}$ and, thus, also convergent in the sense of $\mathrm{L}^2(B_R(0))$.

Furthermore, F needs to be represented in an appropriate basis. We slightly postpone this and only do half the work in this respect. We expect F to be expanded on (almost) every sphere $S_r(0)$ with $0 < r < R$ into a series of spherical harmonics:

$$F(r\xi) = \sum_{n=0}^{\infty} \sum_{j=-n}^{n} F_{n,j}(r) Y_{n,j}^r(r\xi), \quad r \in]0,R[,\ \xi \in \Omega, \tag{5.8}$$

where the (strong) convergence of the series is given in $\mathrm{L}^2(B_R(0))$ and $F_{n,j}(r) = \left\langle F(r\cdot), Y_{n,j}^r \right\rangle_{\mathrm{L}^2(S_r(0))}$. We insert now (5.7) and (5.8) into (5.6) and use the addition theorem for spherical harmonics (Theorem 4.2.9). Moreover, we observe that $(\mathcal{T}^{\mathrm{gr}}F)(y) = G\langle F, k(\cdot,y)\rangle_{\mathrm{L}^2(B_R(0))}$ and use the fact that the strong convergence of the series in (5.7) and (5.8) in $\mathrm{L}^2(B_R(0))$ implies their weak convergence (see Theorem 2.5.57). We arrive at

$$\begin{aligned} &\left(\mathcal{T}^{\mathrm{gr}} F\right)(y) \\ &= G \sum_{n=0}^{\infty} \sum_{j=-n}^{n} \sum_{n'=0}^{\infty} \int_0^R r^2 \int_\Omega F_{n,j}(r) \frac{1}{r} Y_{n,j}(\xi) \frac{r^{n'}}{|y|^{n'+1}} P_{n'}\left(\xi \cdot \frac{y}{|y|}\right) \mathrm{d}\omega(\xi)\,\mathrm{d}r \end{aligned}$$

$$= G \sum_{n=0}^{\infty} \sum_{j=-n}^{n} \sum_{n'=0}^{\infty} \sum_{j'=-n'}^{n'} \int_0^R r^{n'+1} F_{n,j}(r)\,\mathrm{d}r \int_\Omega Y_{n,j}(\xi)\, Y_{n',j'}(\xi)\,\mathrm{d}\omega(\xi)$$

$$\times \frac{1}{|y|^{n'+1}} Y_{n',j'}\left(\frac{y}{|y|}\right) \frac{4\pi}{2n'+1}$$

$$= G \sum_{n=0}^{\infty} \sum_{j=-n}^{n} \int_0^R r^{n+1} F_{n,j}(r)\,\mathrm{d}r \,\frac{1}{|y|^{n+1}}\, Y_{n,j}\left(\frac{y}{|y|}\right) \frac{4\pi}{2n+1}. \tag{5.9}$$

In the latter identity, we used the orthonormality of the $Y_{n,j}$-functions (see Theorem 4.2.18). Since V is harmonic in $\mathbb{R}^3 \setminus \overline{B_R(0)}$ and regular at infinity and since the EDP is uniquely solvable (see Corollary 3.3.7), it suffices to assume that V is given at $S_{R+\varepsilon}(0)$ for a sufficiently small $\varepsilon > 0$. Its expansion would be

$$V(y) = \sum_{n=0}^{\infty} \sum_{j=-n}^{n} V_{n,j} \left(\frac{R+\varepsilon}{|y|}\right)^{n+1} Y_{n,j}^{R+\varepsilon}(y), \quad |y| \geq R+\varepsilon, \tag{5.10}$$

where $V_{n,j} = \int_{S_{R+\varepsilon}(0)} V(y)\, Y_{n,j}^{R+\varepsilon}(y)\,\mathrm{d}\omega(y)$. A comparison of (5.9) and (5.10) shows us that

$$\frac{4\pi G}{2n+1} \int_0^R r^{n+1} F_{n,j}(r)\,\mathrm{d}r = V_{n,j}(R+\varepsilon)^n \tag{5.11}$$

must hold true for all $n \in \mathbb{N}_0$ and all $j \in \{-n, \ldots, n\}$. We could already conjecture here that the chances for getting a unique solution tend to zero. Apparently, there is only a weighted mean of each $F_{n,j}(\cdot)$ which contributes to $V_{n,j}$. We get a more precise answer if we use the basis functions $G_{m,n,j}^{\mathrm{I}}$ with $\ell_n := n$ for all $n \in \mathbb{N}_0$ (see Theorem 4.8.3) for expanding F in $\mathrm{L}^2(B_R(0))$. As a matter of fact, this basis was introduced by Ballani et al. (1993) and Dufour (1977) for exactly this purpose of resolving the non-uniqueness of the solution of the inverse gravimetric problem. The formula of the functions is

$$G_{m,n,j}^{\mathrm{I}}(x) = \sqrt{\frac{4m+2n+3}{R^3}}\, P_m^{(0,n+1/2)}\left(2\frac{|x|^2}{R^2} - 1\right) \left(\frac{|x|}{R}\right)^n Y_{n,j}\left(\frac{x}{|x|}\right).$$

If we now expand F by $F = \sum_{m=0}^{\infty} \sum_{n=0}^{\infty} \sum_{j=-n}^{n} \langle F, G_{m,n,j}^{\mathrm{I}} \rangle_{\mathrm{L}^2(B_R(0))} G_{m,n,j}^{\mathrm{I}}$ in the sense of $\mathrm{L}^2(B_R(0))$ and compare this ansatz with (5.8), then we observe that

$$\frac{1}{r} F_{n,j}(r) = \sum_{m=0}^{\infty} \langle F, G_{m,n,j}^{\mathrm{I}} \rangle_{\mathrm{L}^2(B_R(0))} \sqrt{\frac{4m+2n+3}{R^3}}\, P_m^{(0,n+1/2)}\left(2\frac{r^2}{R^2} - 1\right) \left(\frac{r}{R}\right)^n.$$

We insert this into (5.11) and obtain

$$\frac{4\pi G}{2n+1} \int_0^R \sum_{m=0}^{\infty} \langle F, G_{m,n,j}^{\mathrm{I}} \rangle_{\mathrm{L}^2(B_R(0))} \sqrt{\frac{4m+2n+3}{R^{3+2n}}}$$

$$\times P_m^{(0,n+1/2)}\left(2\frac{r^2}{R^2} - 1\right) r^{2n+2}\,\mathrm{d}r = V_{n,j}(R+\varepsilon)^n.$$

With the substitution $t := 2r^2R^{-2} - 1$, which implies that $\mathrm{d}t = 4rR^{-2}\,\mathrm{d}r$ and $r = R\sqrt{(t+1)/2}$, we get the equivalent formula

$$\frac{4\pi G}{2n+1}\int_{-1}^{1}\sum_{m=0}^{\infty}\left\langle F, G^{\mathrm{I}}_{m,n,j}\right\rangle_{\mathrm{L}^2(B_R(0))}\sqrt{\frac{4m+2n+3}{R^{3+2n}}} \times P_m^{(0,n+1/2)}(t)\,(t+1)^{n+1/2}\,\frac{R^{2n+1}}{2^{n+1/2}}\,\frac{R^2}{4}\,\mathrm{d}t = V_{n,j}(R+\varepsilon)^n\,. \tag{5.12}$$

Now it has become obvious why this particular basis was chosen for expanding F. The Jacobi polynomials $P_m^{(0,n+1/2)}$ are orthogonal with respect to the occurring inner product on $[-1,1]$ with the weight function $w(t) := (t+1)^{n+1/2}$; see Theorem 4.8.1. In combination with Theorem 4.8.2 on the norm of the Jacobi polynomials, this simplifies (5.12) essentially. F solves $\mathcal{T}^{\mathrm{gr}}F = V$, if and only if

$$\frac{4\pi G}{2n+1}\left\langle F, G^{\mathrm{I}}_{0,n,j}\right\rangle_{\mathrm{L}^2(B_R(0))}\sqrt{\frac{2n+3}{R^{3+2n}}}\,\frac{2^{n+3/2}}{n+3/2}\,\frac{R^{2n+3}}{2^{n+5/2}} = V_{n,j}(R+\varepsilon)^n$$

or, briefly,

$$\left\langle F, G^{\mathrm{I}}_{0,n,j}\right\rangle_{\mathrm{L}^2(B_R(0))} = V_{n,j}\,\frac{(R+\varepsilon)^n}{G R^{n+3/2}}\,\sqrt{2n+3}\,\frac{2n+1}{4\pi}$$

for all $n \in \mathbb{N}_0$ and all $j \in \{-n,\dots,n\}$. Clearly, the unknown expansion coefficients $\left\langle F, G^{\mathrm{I}}_{m,n,j}\right\rangle_{\mathrm{L}^2(B_R(0))}$ are free to choose for $m > 0$ and uniquely determined for $m = 0$. We have an ill-posed problem!

The rest of Hadamard's criteria are easy to handle. For solvability, we need that $F \in \mathrm{L}^2(B_R(0))$, that is,

$$\sum_{n=0}^{\infty}\sum_{j=-n}^{n}\left[V_{n,j}\left(\frac{R+\varepsilon}{R}\right)^n\sqrt{2n+3}\,(2n+1)\right]^2 < +\infty. \tag{5.13}$$

The factor $[(R+\varepsilon)/R]^n$ is a result of the downward continuation which is actually a part of the inverse gravimetric problem. This implies here that analogously (5.13) is not satisfied by all $V \in \mathrm{L}^2(S_{R+\varepsilon}(0))$ and apparently the inverse singular values exponentially diverge. An instability of polynomial order still remains in the limit $\varepsilon \to 0+$, since $\sqrt{2n+3}\,(2n+1)$ is also divergent.

Let us summarize the derived results.

Theorem 5.2.3 *The inverse gravimetric problem is ill-posed in the sense of Hadamard. More precisely, it violates all three criteria due to Hadamard. The operator* $\mathcal{T}^{\mathrm{gr}}$: $\mathrm{L}^2(B_R(0)) \to \mathrm{L}^2(S_{R+\varepsilon}(0))$ *has the singular-value decomposition*

$$\mathcal{T}^{\mathrm{gr}}F = \sum_{n=0}^{\infty}\sum_{j=-n}^{n}\frac{G R^{n+3/2}}{(R+\varepsilon)^n}\,\frac{4\pi}{\sqrt{2n+3}\,(2n+1)}\left\langle F, G^{\mathrm{I}}_{0,n,j}\right\rangle_{\mathrm{L}^2(B_R(0))} Y^{R+\varepsilon}_{n,j}$$

and its null space is characterized by

$$\ker\mathcal{T}^{\mathrm{gr}} = \overline{\operatorname{span}\left\{G^{\mathrm{I}}_{m,n,j}\,\middle|\, m \in \mathbb{N},\, n \in \mathbb{N}_0,\, j = -n,\dots,n\right\}}^{\|\cdot\|_{\mathrm{L}^2(B_R(0))}}.$$

There is, at last, one bit of good news. Due to the structure of the $G^{\mathrm{I}}_{m,n,j}$-functions, the non-uniqueness is limited to the radial dependence of F. This was also indicated by (5.11). Formulated in a positive manner, this means that we can recover the angular dependence of F completely. This is important, because it justifies the calculation of temporal surface mass density changes out of gravitational data (for explicit formulae, see, e.g., Leweke et al., 2018a). However, if one is interested in using gravitational data to infer anomalies inside the Earth, then this is impossible without additional types of data (e.g., magnetic or seismic data) or a priori information about the solution!

5.3 Basic Theory of Inverse Problems and Their Regularization

We will collect here the essential properties of inverse problems and the foundations of handling ill-posed problems by a so-called regularization. Inverse problems alone suffice as a topic for a book. Therefore, for a more detailed treatise, we refer the readers to standard books on inverse problems such as Engl et al. (1996), Hanke (2017), Hofmann (1986), Kirsch (1996), Louis (1989), Lu and Pereverzev (2013), and Rieder (2003), which serve as a basis for this section.

We assume that we have an inverse problem

$$\mathcal{T}F = G, \tag{5.14}$$

where $\mathcal{T}\colon \mathcal{X} \to \mathcal{Y}$ is a linear and continuous operator, $(\mathcal{X}, \langle\cdot,\cdot\rangle_{\mathcal{X}})$ and $(\mathcal{Y}, \langle\cdot,\cdot\rangle_{\mathcal{Y}})$ are Hilbert spaces, $G \in \mathcal{Y}$ is given, and $F \in \mathcal{X}$ is known. It should be mentioned here that one can also construct a concept of regularizations in Banach spaces; see, for example, Hofmann et al. (2007), Resmerita (2005), Schöpfer et al. (2006), and Schuster et al. (2012), but this topic is sacrificed here for reasons of brevity.

Often the interesting cases occur if $\mathcal{T}$ is also compact. The reason is that Theorem 2.5.50 tells us that, if $\dim \mathcal{X} = +\infty$, the operator $\mathcal{T}$ cannot have a continuous inverse. In other words, in the realistic case of infinite dimensions, the inverse problem (5.14) must be ill-posed in the sense of Hadamard, if $\mathcal{T}$ is compact. This also fits the examples which we developed in the previous sections.

Compact operators have another and this time useful property: they possess a singular-value decomposition (svd; see Theorem 2.5.81)

$$\mathcal{T}x = \sum_n \sigma_n \langle x, u_n\rangle_{\mathcal{X}}\, v_n, \tag{5.15}$$

where $\sigma_n > 0$ for all n and $(u_n)_n$ and $(v_n)_n$ are orthonormal systems in $\mathcal{X}$ and $\mathcal{Y}$, respectively. More precisely, it is obvious that $(u_n)_n$ is an orthonormal basis of $(\ker \mathcal{T})^{\perp}$ and $(v_n)_n$ is an orthonormal basis of $\overline{\mathcal{T}(\mathcal{X})}$.

The fact that an svd exists does not necessarily mean that we also know it. Fortunately, there are inverse problems in geomathematics with a known svd such as the examples in the previous sections. These examples can also be extended to the case that first- or second-order derivatives of the potential are given. We only have to combine the svds from

Definition 5.1.1 and Theorem 5.2.3, respectively, with the formulae for the gradient and the Hessian tensor in vector/tensor spherical harmonics in Sections 4.3 and 4.4.

Moreover, for finite dimensions, there exist numerical techniques for determining an svd of a linear operator $\mathcal{T}\colon \mathcal{X} \to \mathcal{Y}$ (which is automatically compact due to Theorem 2.5.47) with $\dim \mathcal{X} < +\infty$ (or, at least, $\dim(\ker \mathcal{T})^{\perp} < +\infty$). One of the techniques is based on the idea of the Slepian functions. The reason is that, in practice, the situation of an unknown svd also occurs if the data are only given on a subdomain. Our examples from the previous sections also rely on the fact that the data are given on an entire sphere and we can use spherical harmonics as a basis. If we have data only on a part of the sphere, then Slepian functions can be constructed as a basis, but not every orthonormal basis yields an svd. The construction, therefore, has to be done in a more sophisticated way. We explain here the method which was developed in Michel and Simons (2017), which is also applicable to other cases of unknown svds.

We need to know the operator $\mathcal{T}\colon \mathcal{X} \to \mathcal{Y}$ with its finite-dimensional svd $\mathcal{T}x = \sum_{n=1}^{N} \sigma_n \langle x, u_n\rangle_{\mathcal{X}}\, v_n$, and we have an isometric (i.e., $\langle \iota(F_1), \iota(F_2)\rangle_{\mathcal{Y}} = \langle F_1, F_2\rangle_{\mathcal{Z}}$ for all $F_1, F_2 \in \mathcal{Z}$) embedding $\iota\colon \mathcal{Z} \hookrightarrow \mathcal{Y}$, where $(\mathcal{Z}, \langle\cdot,\cdot\rangle_{\mathcal{Z}})$ is another Hilbert space, such that we may interpret $\mathcal{Z}$ as a subspace of $\mathcal{Y}$ (e.g., $\mathcal{Z} = \mathrm{L}^2(R)$ and $\mathcal{Y} = \mathrm{L}^2(\Omega)$, where R is a surface in $\mathbb{R}^3$ with $R \subset \Omega$). Moreover, we assume that there is a projection $\mathcal{P}: \mathcal{Y} \to \mathcal{Z}$, that is, $\mathcal{P}\mathcal{P}G = \mathcal{P}G$ for all $G \in \mathcal{Y}$, such that $\mathcal{P} \circ \iota = \mathrm{id}_{\mathcal{Z}}$ (in the mentioned example where $\mathcal{Z} = \mathrm{L}^2(R)$ and $\mathcal{Y} = \mathrm{L}^2(\Omega)$, $\mathcal{P}$ could be the restriction operator $F \mapsto F|_R$, $F \in \mathrm{L}^2(\Omega)$ and ι could be a continuation by zero, i.e., $(\iota G)(x) := G(x)$ for all $x \in R$ and $(\iota G)(x) := 0$ for all $x \in \Omega \setminus R$, if $G \in \mathrm{L}^2(R)$).

We seek an svd for the restricted operator $\mathcal{P}\mathcal{T}\colon \mathcal{X} \to \mathcal{Z}$ in order to solve the inverse problem $\mathcal{P}\mathcal{T}F = \tilde{G}$ where 'only' $\tilde{G} \in \mathcal{Z}$ is known. For this purpose, we calculate the Gramian matrix $A := (\sigma_m \langle \mathcal{P}v_n, \mathcal{P}v_m\rangle_{\mathcal{Z}}\, \sigma_n)_{m,n=1,\ldots,N}$, its eigenvalues $(\lambda_n)_{n=1,\ldots,N}$ and an associated orthonormal system of eigenvectors $(w^{(n)})_{n=1,\ldots,N} \subset \mathbb{C}^N$ such that $Aw^{(n)} = \lambda_n w^{(n)}$ for all $n = 1, \ldots, N$. Since A is self-adjoint and positive semi-definite, all eigenvalues are real and non-negative and an orthonormal basis of eigenvectors in $\mathbb{C}^N$ exists. We now use the components of the eigenvectors as expansion coefficients and set

$$g_n := \sum_{k=1}^{N} w_k^{(n)} u_k \in \mathcal{X}, \quad n = 1, \ldots, N.$$

Due to the Parseval identity, we have $\langle g_n, g_m\rangle_{\mathcal{X}} = \left\langle w^{(n)}, w^{(m)}\right\rangle_{\mathbb{C}^N} = \delta_{nm}$. Hence, the functions g_n, $n = 1, \ldots, N$, represent an alternative to u_n, $n = 1, \ldots, N$, as an orthonormal basis of $(\ker \mathcal{T})^{\perp}$. The advantage of this new basis is as follows: if we take into account that $\mathcal{T}g_n = \sum_{k=1}^{N} \sigma_k w_k^{(n)} v_k$ for all $n = 1, \ldots, N$, then

$$\begin{aligned}\langle \mathcal{P}\mathcal{T}g_n, \mathcal{P}\mathcal{T}g_m\rangle_{\mathcal{Z}} &= \sum_{k=1}^{N}\sum_{l=1}^{N} \sigma_k \sigma_l w_k^{(n)} \overline{w_l^{(m)}} \langle \mathcal{P}v_k, \mathcal{P}v_l\rangle_{\mathcal{Z}} = \left(w^{(n)}\right)^{\mathrm{T}} \overline{Aw^{(m)}} \\ &= \left(w^{(n)}\right)^{\mathrm{T}} \overline{\lambda_m w^{(m)}} = \lambda_m \left\langle w^{(n)}, w^{(m)}\right\rangle_{\mathbb{C}^N} = \lambda_n \delta_{nm}.\end{aligned}$$

Now we rearrange the functions g_n and the eigenvalues λ_n such that $\lambda_n = 0 \Leftrightarrow n > \tilde{N}$. If we set $h_n := \lambda_n^{-1/2} \mathcal{P}\mathcal{T} g_n$ and $\tau_n := \lambda_n^{1/2}$ for $n = 1, \ldots, \tilde{N}$, then the operator $\mathcal{P}\mathcal{T}\colon \mathcal{X} \to \mathcal{Z}$ satisfies $\mathcal{P}\mathcal{T}F = \mathcal{P}\mathcal{T} \sum_{n=1}^{\tilde{N}} \langle F, g_n \rangle_{\mathcal{X}} g_n$ for all $F \in \mathcal{X}$ and, consequently, has the svd $\mathcal{P}\mathcal{T}F = \sum_{n=1}^{\tilde{N}} \langle F, g_n \rangle_{\mathcal{X}} \tau_n h_n$ with the singular values $(\tau_n)_{n=1,\ldots,\tilde{N}}$ and the orthonormal bases $(g_n)_{n=1,\ldots,\tilde{N}}$ and $(h_n)_{n=1,\ldots,\tilde{N}}$ of $(\ker \mathcal{P}\mathcal{T})^{\perp}$ and $(\mathcal{P}\mathcal{T})(\mathcal{X})$, respectively. For further details and numerical aspects of this ansatz, the reader is referred to Michel and Simons (2017).

If we have an svd of the form in (5.15), that is,

$$\mathcal{T}x = \sum_n \sigma_n \langle x, u_n \rangle_{\mathcal{X}} v_n \tag{5.16}$$

with $\sigma_n > 0$ for all n, where we allow now finite as well as countably infinite summation ranges, then we can address the criteria due to Hadamard in analogy to the considerations which we made for the particular examples in the previous sections.

Theorem 5.3.1 *The inverse problem* (5.14) *with svd* (5.16) *is solvable if and only if* $G \in \overline{\mathcal{T}(\mathcal{X})}$ *and* G *satisfies the* ***Picard condition***

$$\sum_n \frac{\left|\langle G, v_n \rangle_{\mathcal{Y}}\right|^2}{\sigma_n^2} < +\infty. \tag{5.17}$$

The Picard condition (5.17) is certainly only relevant if the summation range is infinite.

Concerning the other two criteria by Hadamard, we can again analogously state the following:

- The problem has not more than one solution if and only if $(u_n)_n$ is complete in $\mathcal{X}$.
- The solution is stable if and only if $(\sigma_n^{-1})_n$ is bounded.

Compact operators on infinite-dimensional spaces must violate at least one of these three conditions. In the case of the downward continuation, the stability and the Picard condition were violated. For the inverse gravimetric problem, all three criteria failed.

Let us talk a bit more about compact operators. There is one thing which often confuses students at this point: let us assume that we have an injective, compact, and linear operator $\mathcal{T}\colon \mathcal{X} \to \mathcal{Y}$ (such as the upward continuation operator $\mathcal{T}^{\mathrm{up}}$) with $\dim \mathcal{X} = +\infty$ (e.g. $\mathcal{X} = \mathrm{L}^2(\Omega_\varrho)$). Then the simple modification $\tilde{\mathcal{T}}\colon \mathcal{X} \to \mathcal{T}(\mathcal{X})$, $F \mapsto \mathcal{T}F$, is bijective but still compact. It is invertible, but the inverse $\tilde{\mathcal{T}}^{-1}$ is definitely not continuous (because, again, compact operators on infinite-dimensional spaces never have a continuous inverse; see Theorem 2.5.50). However, is this not a contradiction to the inverse mapping theorem (Theorem 2.5.41), where linearity, (mere) continuity, and bijectivity suffice for a continuous inverse?

As a matter of fact, this is *not* a contradiction, because the inverse mapping theorem is only valid for mappings between Banach spaces, but $\mathcal{T}(\mathcal{X})$ is not complete. So, the inverse mapping theorem is not applicable to $\tilde{\mathcal{T}}$. Actually, $\tilde{\mathcal{T}}$ can serve as a counterexample which shows that the inverse mapping theorem fails if the image space is not complete.

Theorem 5.3.2 *Let $(\mathcal{X}, \langle\cdot,\cdot\rangle_{\mathcal{X}})$ and $(\mathcal{Y}, \langle\cdot,\cdot\rangle_{\mathcal{Y}})$ be Hilbert spaces and let $\mathcal{T} \in \mathcal{L}(\mathcal{X},\mathcal{Y})$. The bijective mapping $\mathcal{T}|_{(\ker \mathcal{T})^\perp} : (\ker \mathcal{T})^\perp \to \mathcal{T}(\mathcal{X})$ has a continuous inverse if and only if the image $\mathcal{T}(\mathcal{X})$ is closed in $\mathcal{Y}$.*

Proof We prove both directions of the conclusions.

(1) '$\Rightarrow$':

Let us assume that the inverse $\mathcal{S} := \left(\mathcal{T}|_{(\ker \mathcal{T})^\perp}\right)^{-1} : \mathcal{T}(\mathcal{X}) \to (\ker \mathcal{T})^\perp$ is continuous. Moreover, let $(x_n)_n$ be a Cauchy sequence in $\mathcal{T}(\mathcal{X})$. Since the continuity of $\mathcal{S}$ implies that $\|\mathcal{S}x_n - \mathcal{S}x_m\|_{\mathcal{X}} \le \|\mathcal{S}\|_{\mathcal{L}} \, \|x_n - x_m\|_{\mathcal{Y}}$, the sequence $(\mathcal{S}x_n)_n$ is a Cauchy sequence in $(\ker \mathcal{T})^\perp \subset \mathcal{X}$. Hence, there is a limit $y := \lim_{n\to\infty} \mathcal{S}x_n \in \mathcal{X}$. Consulting Theorem 2.5.71 and the considerations after Theorem 2.5.72, we observe that $(\ker \mathcal{T})^\perp$ is closed such that $y \in (\ker \mathcal{T})^\perp$. Let now $z := \mathcal{T}y$. Then the continuity of $\mathcal{T}$ yields

$$\begin{aligned} \|x_n - z\|_{\mathcal{Y}} &= \|\mathcal{T}\mathcal{S}x_n - \mathcal{T}\mathcal{S}z\|_{\mathcal{Y}} \le \|\mathcal{T}\|_{\mathcal{L}} \, \|\mathcal{S}x_n - \mathcal{S}z\|_{\mathcal{X}} \\ &= \|\mathcal{T}\|_{\mathcal{L}} \, \|\mathcal{S}x_n - y\|_{\mathcal{X}} \longrightarrow 0 \qquad \text{as } n \to \infty. \end{aligned}$$

Hence, (x_n) is convergent and $\mathcal{T}(\mathcal{X})$ is complete and thus closed.

(2) '$\Leftarrow$':

If $\mathcal{T}(\mathcal{X})$ is closed, then it is complete (see Theorem 2.5.16). With the same argumentation, $(\ker \mathcal{T})^\perp$ is, as a closed subset of $\mathcal{X}$, also complete. Hence, we may apply the inverse mapping theorem (Theorem 2.5.41) such that the inverse $\mathcal{S}$ is continuous. □

The connection between the stability of the solution of an inverse problem and the image of the operator is used for an alternative definition of an ill-posed problem which goes back to Nashed (1987).

Definition 5.3.3 Let $(\mathcal{X}, \langle\cdot,\cdot\rangle_{\mathcal{X}})$ and $(\mathcal{Y}, \langle\cdot,\cdot\rangle_{\mathcal{Y}})$ be Hilbert spaces and let $\mathcal{T} \in \mathcal{L}(\mathcal{X},\mathcal{Y})$. The inverse problem $\mathcal{T}F = G$ is called **well-posed in the sense of Nashed** if $\mathcal{T}(\mathcal{X})$ is closed in $\mathcal{Y}$. Otherwise, it is called **ill-posed in the sense of Nashed**.

Example 5.3.4 From our previous considerations, it is obvious that the **downward continuation problem** and the **inverse gravimetric problem** both are also ill-posed in the sense of Nashed.

We come back now to the particular cases where $(\mathcal{X}, \langle\cdot,\cdot\rangle_{\mathcal{X}})$ and $(\mathcal{Y}, \langle\cdot,\cdot\rangle_{\mathcal{Y}})$ are Hilbert spaces. And still we have an inverse problem $\mathcal{T}F = G$ with an operator $\mathcal{T} \in \mathcal{L}(\mathcal{X},\mathcal{Y})$, given $G \in \mathcal{Y}$, and unknown $F \in \mathcal{X}$.

One of the questions that arise if we have to solve an ill-posed inverse problem is: what can we do if $G \in \mathcal{Y} \setminus \mathcal{T}(\mathcal{X})$? For instance, we could try to solve the unsolvable equation $\mathcal{T}F = G$ 'as well as possible', that is, we seek $F \in \mathcal{X}$ such that the **residual** $\|\mathcal{T}F - G\|_{\mathcal{Y}}$ is minimized. This is, indeed, a common ansatz.

Theorem 5.3.5 *The following propositions are equivalent:*

(a) $F \in \mathcal{X}$ *solves* $\mathcal{T}F = \mathcal{P}_{\overline{\mathcal{T}(\mathcal{X})}}G$, *where* $\mathcal{P}_{\overline{\mathcal{T}(\mathcal{X})}}$ *is the orthogonal projector onto* $\overline{\mathcal{T}(\mathcal{X})}$*: see Theorem 2.5.71.*

(b) $F \in \mathcal{X}$ *minimizes the residual, that is,* $\|\mathcal{T}F - G\|_{\mathcal{Y}} \le \|\mathcal{T}H - G\|_{\mathcal{Y}}$ *for all* $H \in \mathcal{X}$.

(c) $F \in \mathcal{X}$ *solves the* ***normal equation*** $\mathcal{T}^*\mathcal{T}F = \mathcal{T}^*G$, *where* $\mathcal{T}^*$ *is the adjoint operator to* $\mathcal{T}$*; see Theorem 2.5.59.*

Proof We prove the theorem by deriving the equivalences a⇔b and a⇔c.

(1) a ⇔ b:

$\overline{\mathcal{T}(\mathcal{X})}$ is, as a closed subset of the complete space $\mathcal{Y}$, also complete (see Theorem 2.5.16). With Theorem 2.5.71, we get

$$\mathcal{T}F - G = \underbrace{\mathcal{P}_{\overline{\mathcal{T}(\mathcal{X})}}\mathcal{T}F}_{=\mathcal{T}F} + \underbrace{\mathcal{P}_{\overline{\mathcal{T}(\mathcal{X})}^{\perp}}\mathcal{T}F}_{=0} - \mathcal{P}_{\overline{\mathcal{T}(\mathcal{X})}}G - \mathcal{P}_{\overline{\mathcal{T}(\mathcal{X})}^{\perp}}G$$

such that

$$\|\mathcal{T}F - G\|_{\mathcal{Y}}^2 = \left\|\mathcal{T}F - \mathcal{P}_{\overline{\mathcal{T}(\mathcal{X})}}G\right\|_{\mathcal{Y}}^2 + \left\|\mathcal{P}_{\overline{\mathcal{T}(\mathcal{X})}^{\perp}}G\right\|_{\mathcal{Y}}^2 .$$

We cannot influence $\mathcal{P}_{\overline{\mathcal{T}(\mathcal{X})}^{\perp}}G$ by choosing F. Hence, the residual is minimal if and only if $\mathcal{T}F = \mathcal{P}_{\overline{\mathcal{T}(\mathcal{X})}}G$, provided that the latter equation is solvable. This makes clear that 'a ⇒ b' holds true. We should add one more argument for 'a ⇐ b'. If $\mathcal{T}F = \mathcal{P}_{\overline{\mathcal{T}(\mathcal{X})}}G$ is not solvable, then we set $z := \mathcal{P}_{\overline{\mathcal{T}(\mathcal{X})}}G$ and observe that there must exist a sequence $(z_n)_n \subset \mathcal{T}(\mathcal{X})$ with $\|z_n - z\|_{\mathcal{Y}} \to 0$. We can definitely solve $\mathcal{T}F_n = z_n$, which yields $\|\mathcal{T}F_n - \mathcal{P}_{\overline{\mathcal{T}(\mathcal{X})}}G\|_{\mathcal{Y}} \longrightarrow 0$ as $n \to \infty$, that is, $\inf_{F\in\mathcal{X}} \|\mathcal{T}F - \mathcal{P}_{\overline{\mathcal{T}(\mathcal{X})}}G\|_{\mathcal{Y}} = 0$, whereas there is no $F \in \mathcal{X}$ such that $\|\mathcal{T}F - \mathcal{P}_{\overline{\mathcal{T}(\mathcal{X})}}G\|_{\mathcal{Y}} = 0$. Hence, there is no minimizer of the residual. Therefore, we also have 'not a ⇒ not b'.

(2) a ⇔ c:

Let F solve the normal equation. We get

$$\begin{aligned}
\mathcal{T}^*\mathcal{T}F = \mathcal{T}^*G &\Leftrightarrow \langle \mathcal{T}^*\mathcal{T}F - \mathcal{T}^*G, H\rangle_{\mathcal{X}} = 0 \quad \text{for all } H \in \mathcal{X}\\
&\Leftrightarrow \langle \mathcal{T}F - G, \mathcal{T}H\rangle_{\mathcal{Y}} = 0 \quad \text{for all } H \in \mathcal{X}\\
&\Leftrightarrow \langle \mathcal{T}F - G, J\rangle_{\mathcal{Y}} = 0 \quad \text{for all } J \in \mathcal{T}(\mathcal{X})\\
&\Leftrightarrow \langle \mathcal{T}F - G, J\rangle_{\mathcal{Y}} = 0 \quad \text{for all } J \in \overline{\mathcal{T}(\mathcal{X})}.
\end{aligned}$$

In the latter equivalence, the '⇐' part is trivial. The '⇒' part can be achieved by choosing, for an arbitrary $J \in \overline{\mathcal{T}(\mathcal{X})}$, a sequence $(J_n)_n \subset \mathcal{T}(\mathcal{X})$ with $J_n \to J$. Since then also $J_n \rightharpoonup J$ (see Theorem 2.5.57), we get the desired conclusion.

We now decompose G into $G = \mathcal{P}_{\overline{\mathcal{T}(\mathcal{X})}}G + \mathcal{P}_{\overline{\mathcal{T}(\mathcal{X})}^{\perp}}G$. This enables us to continue our chain of equivalences by

$$\begin{aligned}
&\mathcal{T}^*\mathcal{T}F = \mathcal{T}^*G \\
&\quad\Leftrightarrow \langle \mathcal{T}F, J\rangle_{\mathcal{Y}} = \left\langle \mathcal{P}_{\overline{\mathcal{T}(\mathcal{X})}}G, J\right\rangle_{\mathcal{Y}} + \underbrace{\left\langle \mathcal{P}_{\overline{\mathcal{T}(\mathcal{X})}^{\perp}}G, J\right\rangle_{\mathcal{Y}}}_{=0} \quad \text{for all } J \in \overline{\mathcal{T}(\mathcal{X})} \\
&\quad\Leftrightarrow \left\langle \mathcal{T}F - \mathcal{P}_{\overline{\mathcal{T}(\mathcal{X})}}G, J\right\rangle_{\mathcal{Y}} = 0 \quad \text{for all } J \in \overline{\mathcal{T}(\mathcal{X})} \\
&\quad\Leftrightarrow \mathcal{T}F - \mathcal{P}_{\overline{\mathcal{T}(\mathcal{X})}}G \in \overline{\mathcal{T}(\mathcal{X})}^{\perp} \\
&\quad\Leftrightarrow \mathcal{T}F - \mathcal{P}_{\overline{\mathcal{T}(\mathcal{X})}}G = 0.
\end{aligned}$$

The latter holds true because $\overline{\mathcal{T}(\mathcal{X})} \cap \overline{\mathcal{T}(\mathcal{X})}^{\perp} = \{0\}$. □

Apparently, the normal equation is a key to handling overdetermined problems, that is, problems where $\mathcal{Y}$ is larger than the range $\mathcal{T}(\mathcal{X})$. The question is, therefore: can we solve the normal equation $\mathcal{T}^*\mathcal{T}F = \mathcal{T}^*G$ for more right-hand sides G than in the case $\mathcal{T}F = G$? Indeed, we can.

Theorem 5.3.6 *The following propositions hold true:*

(a) The normal equation $\mathcal{T}^*\mathcal{T}F = \mathcal{T}^*G$ is solvable if and only if the right-hand side fulfils $G \in \mathcal{T}(\mathcal{X}) \oplus \mathcal{T}(\mathcal{X})^{\perp}$.

(b) The solution space $\mathcal{S}(\mathcal{T}; G) := \{F \in \mathcal{X} \mid \mathcal{T}^*\mathcal{T}F = \mathcal{T}^*G\}$ is closed and convex.

Proof Remember that $\mathcal{U}^{\perp} = \overline{\mathcal{U}}^{\perp}$ for every linear subspace $\mathcal{U} \subset \mathcal{Y}$.

(1) Part a, conclusion '⇒':

If the normal equation is solved by $F \in \mathcal{X}$, then we use that $G = \mathcal{T}F + (G - \mathcal{T}F)$, while $\mathcal{T}F \in \mathcal{T}(\mathcal{X})$ and, for all $H \in \mathcal{X}$,

$$\langle G - \mathcal{T}F, \mathcal{T}H\rangle_{\mathcal{Y}} = \langle \mathcal{T}^*G - \mathcal{T}^*\mathcal{T}F, H\rangle_{\mathcal{X}} = 0,$$

that is, $G - \mathcal{T}F \in \mathcal{T}(\mathcal{X})^{\perp}$. Hence, $G \in \mathcal{T}(\mathcal{X}) \oplus \mathcal{T}(\mathcal{X})^{\perp}$.

(2) Part a, conclusion '⇐':

If $G \in \mathcal{T}(\mathcal{X}) \oplus \mathcal{T}(\mathcal{X})^{\perp}$, then there exists $F \in \mathcal{X}$ and $H \in \mathcal{T}(\mathcal{X})^{\perp}$ such that $G = \mathcal{T}F + H$. With the orthogonal projector (see Theorem 2.5.71) onto the closure of the image of $\mathcal{T}$, we obtain then $\mathcal{P}_{\overline{\mathcal{T}(\mathcal{X})}}G = \mathcal{T}F$. Consequently, F solves the normal equation due to Theorem 5.3.5.

(3) Part b, closed:

Let $(F_n) \subset \mathcal{S}(\mathcal{T}; G)$ be a convergent sequence with limit $F \in \mathcal{X}$. Due to the continuity of $\mathcal{T}$ and $\mathcal{T}^*$ (see Theorem 2.5.59), we directly get $\mathcal{T}^*\mathcal{T}F = \lim_{n\to\infty} \mathcal{T}^*\mathcal{T}F_n = \lim_{n\to\infty} \mathcal{T}^*G = \mathcal{T}^*G$. Hence, $F \in \mathcal{S}(\mathcal{T}; G)$.

(4) Part b, convex:

If $F_1, F_2 \in \mathcal{S}(\mathcal{T}; G)$ and $\lambda \in [0, 1]$ are arbitrary, then the linearity of $\mathcal{T}$ and $\mathcal{T}^*$ implies

$$\begin{aligned}\mathcal{T}^*\mathcal{T}\,[\lambda F_1 + (1-\lambda)F_2] &= \lambda\mathcal{T}^*\mathcal{T}F_1 + (1-\lambda)\mathcal{T}^*\mathcal{T}F_2\\ &= \lambda\mathcal{T}^*G + (1-\lambda)\mathcal{T}^*G = \mathcal{T}^*G\end{aligned}$$

such that $\lambda F_1 + (1-\lambda)F_2 \in \mathcal{S}(\mathcal{T};G)$ and $\mathcal{S}(\mathcal{T};G)$ is convex. □

Note that $\mathcal{T}(\mathcal{X}) = \overline{\mathcal{T}(\mathcal{X})}$ if and only if the inverse problem is well-posed in the sense of Nashed. Theorem 5.3.6 now tells us that the normal equation $\mathcal{T}^*\mathcal{T}F = \mathcal{T}^*G$ is solvable for all $G \in \mathcal{Y}$ if and only if the inverse problem $\mathcal{T}F = G$ is well-posed in the sense of Nashed. In combination with Theorem 5.3.5, we can add another conclusion.

Corollary 5.3.7 *The residual $\mathcal{X} \ni F \mapsto \|\mathcal{T}F - G\|_{\mathcal{Y}}$ has, for every fixed $G \in \mathcal{Y}$, a minimum if and only if the inverse problem $\mathcal{T}F = G$ is well-posed in the sense of Nashed.*

Moreover, part b of Theorem 5.3.6 in combination with Theorem 2.5.63 yields another immediate consequence.

Theorem 5.3.8 *If $G \in \mathcal{T}(\mathcal{X}) \oplus \mathcal{T}(\mathcal{X})^{\perp}$, then there exists one and only one $F^+ \in \mathcal{S}(\mathcal{T};G)$ such that $\|F^+\|_{\mathcal{X}} = \min\{\|F\|_{\mathcal{X}} : F \in \mathcal{S}(\mathcal{T};G)\}$. This F^+ is called the* ***minimum-norm solution*** *of the normal equation (but also of the inverse problem).*

This result motivates the following definition.

Definition 5.3.9 Let $\mathcal{D}(\mathcal{T}^+) := \mathcal{T}(\mathcal{X}) \oplus \mathcal{T}(\mathcal{X})^{\perp} \subset \mathcal{Y}$ be a given domain. The mapping $\mathcal{T}^+ \colon \mathcal{D}(\mathcal{T}^+) \to \mathcal{X}$ which maps every $G \in \mathcal{D}(\mathcal{T}^+)$ to the unique minimum-norm solution F^+ of $\mathcal{T}^*\mathcal{T}F = \mathcal{T}^*G$ is called the **generalized inverse** or the **Moore–Penrose inverse** of $\mathcal{T}$.

As we mentioned previously, one sometimes calls F^+ the minimum-norm solution of the inverse problem $\mathcal{T}F = G$. When using this nomenclature, one should also be aware that F^+ is not necessarily a solution of $\mathcal{T}F = G$, but it is a solution of $\mathcal{T}^*\mathcal{T}F = \mathcal{T}^*G$. Vice versa, every solution of $\mathcal{T}F = G$ solves the normal equation.

So much for the non-solvability of the inverse problem. Constraints from applications sometimes also suggest to look for an actual solution (of the inverse problem) with a minimal norm if the inverse problem has more than one solution. Indeed, the Moore–Penrose inverse can also be connected to the null space of $\mathcal{T}$.

Theorem 5.3.10 *For each $G \in \mathcal{D}(\mathcal{T}^+)$, the normal equation $\mathcal{T}^*\mathcal{T}F = \mathcal{T}^*G$ has one and only one solution F in $(\ker\mathcal{T})^{\perp}$. This solution is the minimum-norm solution $F^+ = \mathcal{T}^+G$.*

Proof We apply the principle of an orthogonal decomposition (see Theorem 2.5.71) to an arbitrary solution $F \in \mathcal{X}$ of $\mathcal{T}^*\mathcal{T}F = \mathcal{T}^*G$ and the closed subspace $\ker\mathcal{T} \subset \mathcal{X}$. This leads us to

$$\mathcal{T}^*G = \mathcal{T}^*\mathcal{T}\left(\mathcal{P}_{\ker\mathcal{T}}F + \mathcal{P}_{(\ker\mathcal{T})^{\perp}}F\right) = \mathcal{T}^*\mathcal{T}\mathcal{P}_{(\ker\mathcal{T})^{\perp}}F.$$

Since $\|F\|_{\mathcal{X}}^2 = \|\mathcal{P}_{\ker\mathcal{T}} F\|_{\mathcal{X}}^2 + \|\mathcal{P}_{(\ker\mathcal{T})^\perp} F\|_{\mathcal{X}}^2$, the minimum-norm solution F^+ of the normal equation is obviously in $(\ker\mathcal{T})^\perp$. Let us now look at two arbitrary solutions $\widetilde{F}_\perp$ and $\widehat{F}_\perp$ of

$$\mathcal{T}^*\mathcal{T} F_\perp = \mathcal{T}^* G, \quad F_\perp \in (\ker\mathcal{T})^\perp. \tag{5.18}$$

Then $\mathcal{T}^*\mathcal{T}(\widetilde{F}_\perp - \widehat{F}_\perp) = 0$, while $\ker\mathcal{T}^* = \mathcal{T}(\mathcal{X})^\perp$ due to Theorem 2.5.72. Hence, $\mathcal{T}(\widetilde{F}_\perp - \widehat{F}_\perp) \in \mathcal{T}(\mathcal{X}) \cap \mathcal{T}(\mathcal{X})^\perp = \{0\}$. However, this yields $\widetilde{F}_\perp - \widehat{F}_\perp \in \ker\mathcal{T} \cap (\ker\mathcal{T})^\perp = \{0\}$. Hence (5.18) is uniquely solvable and its solution must consequently be $F_\perp = F^+$. □

Example 5.3.11 In the case of the inverse gravimetric problem (see Section 5.2), $(\ker\mathcal{T}^{\mathrm{gr}})^\perp$ is the set of all harmonic functions on $B_R(0)$; see Theorem 5.2.3 and Remark 4.8.4. The unique minimum-norm solution of $\mathcal{T}^{\mathrm{gr}} F = V$ is, consequently, at the same time the unique harmonic solution of this inverse problem. Since no physical justification for a harmonic solution exists in this context, other uniqueness constraints have been discussed in the literature; see the survey article Michel and Fokas (2008).

Let us now look at some properties of the Moore–Penrose inverse.

Theorem 5.3.12 *Let $\mathcal{T}^+$ be the Moore–Penrose inverse to the operator $\mathcal{T} \in \mathcal{L}(\mathcal{X},\mathcal{Y})$. Then $\mathcal{T}^+$ has the following properties:*

(a) $\mathcal{D}(\mathcal{T}^+) = \mathcal{Y}$ if and only if $\mathcal{T}(\mathcal{X})$ is closed.
(b) $\mathcal{T}^+(\mathcal{D}(\mathcal{T}^+)) = (\ker\mathcal{T})^\perp$.
(c) $\mathcal{T}^+$ is a linear operator.
(d) $\mathcal{T}^+$ is continuous if and only if $\mathcal{T}(\mathcal{X})$ is closed.

Proof Part a is obvious due to Theorem 2.5.71. Let us take care of the rest.

(1) Part b:

From Theorem 5.3.10, we know that $F^+ = \mathcal{T}^+ G \in (\ker\mathcal{T})^\perp$ for all $G \in \mathcal{D}(\mathcal{T}^+)$. Hence, $\mathcal{T}^+(\mathcal{D}(\mathcal{T}^+)) \subset (\ker\mathcal{T})^\perp$. Vice versa, if $F \in (\ker\mathcal{T})^\perp$, then $G := \mathcal{T}F \in \mathcal{T}(\mathcal{X}) \subset \mathcal{D}(\mathcal{T}^+)$, and consequently we have $\mathcal{T}^*\mathcal{T}F = \mathcal{T}^*G$ with $F \in (\ker\mathcal{T})^\perp$. The latter implies, because of Theorem 5.3.10, that $F = F^+ = \mathcal{T}^+G \in \mathcal{T}^+(\mathcal{D}(\mathcal{T}^+))$. Hence, $\mathcal{T}^+(\mathcal{D}(\mathcal{T}^+)) = (\ker\mathcal{T})^\perp$.

(2) Part c:

Let $G_1, G_2 \in \mathcal{D}(\mathcal{T}^+)$ and $G_3 := G_1 + G_2$. We know that $F_k := \mathcal{T}^+ G_k$, $k \in \{1,2,3\}$, is the minimum-norm solution of $\mathcal{T}^*\mathcal{T} F_k = \mathcal{T}^* G_k$, that is,

$$\mathcal{T}^*\mathcal{T}\mathcal{T}^+ G_k = \mathcal{T}^* G_k. \tag{5.19}$$

Additionally, we observe that $\mathcal{T}\mathcal{T}^+ G_k \in \overline{\mathcal{T}(\mathcal{X})} = (\ker\mathcal{T}^*)^\perp$ due to Theorem 2.5.72 and $\mathcal{T}^*$ is injective on $(\ker\mathcal{T}^*)^\perp$. This shows us that (5.19) implies $\mathcal{T}\mathcal{T}^+ G_k = \left(\mathcal{T}^*|_{(\ker\mathcal{T}^*)^\perp}\right)^{-1}\mathcal{T}^* G_k$. With the orthogonal projectors (see Theorem 2.5.71), we can modify the latter equation into

$$\mathcal{T}\mathcal{T}^+ G_k = \left(\mathcal{T}^*|_{(\ker\mathcal{T}^*)^\perp}\right)^{-1}\mathcal{T}^*\mathcal{P}_{(\ker\mathcal{T}^*)^\perp} G_k = \mathcal{P}_{(\ker\mathcal{T}^*)^\perp} G_k = \mathcal{P}_{\overline{\mathcal{T}(\mathcal{X})}} G_k. \tag{5.20}$$

Since the orthogonal projectors are linear, we have

$$\mathcal{T}\mathcal{T}^+G_1+\mathcal{T}\mathcal{T}^+G_2=\mathcal{P}_{\overline{\mathcal{T}(\mathcal{X})}}G_1+\mathcal{P}_{\overline{\mathcal{T}(\mathcal{X})}}G_2=\mathcal{P}_{\overline{\mathcal{T}(\mathcal{X})}}(G_1+G_2)$$

and

$$\mathcal{T}\mathcal{T}^+(G_1+G_2)=\mathcal{T}\mathcal{T}^+G_3=\mathcal{P}_{\overline{\mathcal{T}(\mathcal{X})}}G_3=\mathcal{P}_{\overline{\mathcal{T}(\mathcal{X})}}(G_1+G_2)\,.$$

Since we have already proved part b, we may use that the image of $\mathcal{T}^+$ is $(\ker\mathcal{T})^\perp$ such that $\mathcal{T}$ is injective on the image of $\mathcal{T}^+$. Hence,

$$\mathcal{T}\mathcal{T}^+G_1+\mathcal{T}\mathcal{T}^+G_2=\mathcal{T}\mathcal{T}^+(G_1+G_2)$$

implies

$$\mathcal{T}^+G_1+\mathcal{T}^+G_2=\mathcal{T}^+(G_1+G_2)\,.$$

Eventually, let $\alpha\in\mathbb{K}$, where $\mathbb{K}$ is the field $\mathbb{R}$ or $\mathbb{C}$ corresponding to the Hilbert space $\mathcal{Y}$. In analogy to our preceding considerations, we get

$$\mathcal{T}\mathcal{T}^+(\alpha G)=\mathcal{P}_{\overline{\mathcal{T}(\mathcal{X})}}(\alpha G)=\alpha\mathcal{P}_{\overline{\mathcal{T}(\mathcal{X})}}G=\alpha\mathcal{T}\mathcal{T}^+G$$

and $\mathcal{T}^+(\alpha G)=\alpha\mathcal{T}^+G$ for all $G\in\mathcal{D}(\mathcal{T}^+)$.

(3) Part d, conclusion '⇒':

Let $\mathcal{T}^+$ be continuous. Clearly, $\mathcal{D}(\mathcal{T}^+)$ is dense in $\mathcal{Y}$. We define an operator $\mathcal{A}:\mathcal{Y}\to\mathcal{X}$ as follows: if $y\in\mathcal{Y}$ and $(\eta_k)_k\subset\mathcal{D}(\mathcal{T}^+)$ with $\eta_k\to y$, then we set $\mathcal{A}y:=\lim_{k\to\infty}\mathcal{T}^+\eta_k$. The limit exists, because $\mathcal{T}^+$ is continuous and $\mathcal{X}$ is complete. Moreover, $\mathcal{A}$ is well defined this way, because a different sequence $(\tilde{\eta}_k)_k\subset\mathcal{D}(\mathcal{T}^+)$ with the same limit $y=\lim\tilde{\eta}_k$ must satisfy

$$\mathcal{T}^+\eta_k-\mathcal{T}^+\tilde{\eta}_k=\mathcal{T}^+(\eta_k-\tilde{\eta}_k)\longrightarrow\mathcal{T}^+0=0,$$

since $\mathcal{T}^+$ is linear (see part c, which we have already proved) and continuous. Hence, $\lim\mathcal{T}^+\eta_k=\lim\mathcal{T}^+\tilde{\eta}_k$ and $\mathcal{A}$ is well defined.

Due to the construction of $\mathcal{A}$, the operator is linear and continuous. For proving that $\mathcal{T}(\mathcal{X})$ is closed, we choose an arbitrary $y\in\overline{\mathcal{T}(\mathcal{X})}$ and pick a sequence $(\eta_k)\subset\mathcal{D}(\mathcal{T}^+)$, which converges to y. With (5.20), we get $\mathcal{T}\mathcal{T}^+\eta_k=\mathcal{P}_{\overline{\mathcal{T}(\mathcal{X})}}\eta_k$ such that the continuity of the involved operators implies $\mathcal{T}\mathcal{A}y=\mathcal{P}_{\overline{\mathcal{T}(\mathcal{X})}}y=y$. Hence, $\overline{\mathcal{T}(\mathcal{X})}=\mathcal{T}(\mathcal{X})$.

(4) Part d, conclusion '⇐':

Let $\mathcal{T}(\mathcal{X})$ be closed. Clearly, $\hat{\mathcal{T}}:=\mathcal{T}|_{(\ker\mathcal{T})^\perp}:(\ker\mathcal{T})^\perp\to\mathcal{T}(\mathcal{X})$ is bijective, where $(\ker\mathcal{T})^\perp$ and $\mathcal{T}(\mathcal{X})$ are closed and thus complete (see Theorem 2.5.16) inner product spaces. Hence, $\hat{\mathcal{T}}^{-1}$ is continuous due to the inverse mapping theorem (Theorem 2.5.41). As a consequence, remembering that $\mathcal{T}^+y\in(\ker\mathcal{T})^\perp$ due to part b, we obtain

$$\|\mathcal{T}^+y\|_{\mathcal{X}}=\left\|\hat{\mathcal{T}}^{-1}\hat{\mathcal{T}}\mathcal{T}^+y\right\|_{\mathcal{X}}\le\left\|\hat{\mathcal{T}}^{-1}\right\|_{\mathcal{L}}\left\|\hat{\mathcal{T}}\mathcal{T}^+y\right\|_{\mathcal{Y}}=\left\|\hat{\mathcal{T}}^{-1}\right\|_{\mathcal{L}}\|\mathcal{T}\mathcal{T}^+y\|_{\mathcal{Y}}$$

for all $y \in \mathcal{D}(\mathcal{T}^+) = \mathcal{Y}$ (note that we assume that $\mathcal{T}(\mathcal{X})$ is closed). Hence, with (5.20), we get

$$\begin{aligned}\|y\|_{\mathcal{Y}} &= \sqrt{\left\|\mathcal{P}_{\mathcal{T}(\mathcal{X})}y\right\|_{\mathcal{Y}}^2 + \left\|\mathcal{P}_{\mathcal{T}(\mathcal{X})^\perp}y\right\|_{\mathcal{Y}}^2} \geq \left\|\mathcal{P}_{\mathcal{T}(\mathcal{X})}y\right\|_{\mathcal{Y}} = \left\|\mathcal{T}\mathcal{T}^+y\right\|_{\mathcal{Y}}\\ &\geq \left\|\hat{\mathcal{T}}^{-1}\right\|_{\mathcal{L}}^{-1} \left\|\mathcal{T}^+y\right\|_{\mathcal{X}}\end{aligned}$$

such that $\|\mathcal{T}^+y\|_{\mathcal{X}} \leq \|\hat{\mathcal{T}}^{-1}\|_{\mathcal{L}} \|y\|_{\mathcal{Y}}$. Thus, $\mathcal{T}^+$ is bounded and, since it is linear (see part c), it is also continuous. □

Corollary 5.3.13 *Let $\mathcal{T}^+$ be the Moore–Penrose inverse to the operator $\mathcal{T} \in \mathcal{L}(\mathcal{X},\mathcal{Y})$. Then $\mathcal{T}\mathcal{T}^+G = \mathcal{P}_{\overline{\mathcal{T}(\mathcal{X})}}G$ for all $G \in \mathcal{D}(\mathcal{T}^+)$.*

Theorem 5.3.14 *For each $\mathcal{T} \in \mathcal{L}(\mathcal{X},\mathcal{Y})$, there exists one and only one operator $\mathcal{S} : \mathcal{D}(\mathcal{T}^+) \to \mathcal{X}$ which fulfils the* ***Moore–Penrose axioms****:*

$$\mathcal{T}\mathcal{S}\mathcal{T} = \mathcal{T}, \qquad \mathcal{S}\mathcal{T}\mathcal{S} = \mathcal{S}, \qquad \mathcal{S}\mathcal{T} = \mathcal{P}_{\overline{\mathcal{T}^*(\mathcal{Y})}}, \qquad \mathcal{T}\mathcal{S} = \mathcal{P}_{\overline{\mathcal{T}(\mathcal{X})}}.$$

This operator is the Moore–Penrose inverse: $\mathcal{S} = \mathcal{T}^+$.

The proof is omitted for reasons of brevity.

As a short summary, we should observe here that problems which are well-posed in the sense of Nashed have a continuous Moore–Penrose inverse $\mathcal{T}^+$ which is defined on the whole space $\mathcal{Y}$. Vice versa, ill-posed problems in terms of Nashed's definition neither have a continuous Moore–Penrose inverse nor yield a minimum-norm solution $F^+ = \mathcal{T}^+G$ for every $G \in \mathcal{Y}$ but only for $G \in \mathcal{T}(\mathcal{X}) \oplus \mathcal{T}(\mathcal{X})^\perp$.

We will now have again a look at the particular but also typical case where $\mathcal{T}$ is a compact operator. Since $\mathcal{T}$ must have an svd in this case, some more useful propositions can be derived. Moreover, a well-known classification of inverse problems is based on the knowledge of an svd.

Definition 5.3.15 Let $\mathcal{T}$ be compact with $\dim(\ker \mathcal{T})^\perp = +\infty$ and let an svd of $\mathcal{T}$ be denoted by

$$\mathcal{T}x = \sum_{k=0}^{\infty} \sigma_k \langle x, u_k \rangle_{\mathcal{X}} v_k, \quad x \in \mathcal{X}. \tag{5.21}$$

(a) If $\sigma_k = \mathcal{O}(k^{-\alpha})$ as $k \to \infty$ for a fixed $\alpha > 0$, then the inverse problem $\mathcal{T}F = G$ is called **(polynomially) ill-posed of order α**.

(b) If there exist constants $k_0 \in \mathbb{N}_0$, $C > 0$, and $\beta > 0$ such that the singular values satisfy $|\log \sigma_k| \geq Ck^\beta$ for every $k \in \mathbb{N}_0$ with $k \geq k_0$, then the inverse problem $\mathcal{T}F = G$ is called **exponentially ill-posed**.

Example 5.3.16 We discussed examples of geomathematical inverse problems in Sections 5.1 and 5.2. For bringing their svds into the form of (5.21), we use a renumbering $k := n^2 + n + j$ for $n \in \mathbb{N}_0$, $j \in \{-n, \ldots, n\}$ such that $n = \lfloor \sqrt{k} \rfloor$ (see Definition 2.1.10 for the Gauß brackets).

In the case of the **downward continuation problem**, we have then the following (note that $(\sigma_k)_k$ are the singular values and $\sigma > \varrho$ is the radius of the satellite orbit):

$$|\log \sigma_k| = \left|\log \left(\frac{\varrho}{\sigma}\right)^{\lfloor \sqrt{k} \rfloor}\right| = \left\lfloor \sqrt{k} \right\rfloor \log \frac{\sigma}{\varrho} \geq \sqrt{k}\,\frac{1}{3}\,\log \frac{\sigma}{\varrho}.$$

Hence, the downward continuation problem is exponentially ill-posed, as we have already mentioned in Section 5.1.

For the **inverse gravimetric problem**, we can analogously derive that it is also exponentially ill-posed, because it includes the downward continuation problem. For surface-based data ($\varepsilon \to 0+$), the inverse gravimetric problem satisfies

$$\sigma_k = \frac{4\pi G R^{3/2}}{\sqrt{2\left\lfloor \sqrt{k}\right\rfloor + 3}\,\left(2\left\lfloor \sqrt{k}\right\rfloor + 1\right)} \leq \frac{4\pi G R^{3/2}}{\sqrt{\frac{2}{3}\sqrt{k} + 3}\,\left(\frac{2}{3}\sqrt{k} + 1\right)} \leq \frac{8\pi G R^{3/2}}{k^{3/4}}$$

and is, therefore, ill-posed of order 3/4.

Theorem 5.3.17 *Let $\mathcal{T}$ be compact and let an svd of $\mathcal{T}$ be denoted by*

$$\mathcal{T}x = \sum_n \sigma_n \langle x, u_n\rangle_{\mathcal{X}}\, v_n, \quad x \in \mathcal{X}. \tag{5.22}$$

Then the adjoint operator $\mathcal{T}^* \in \mathcal{K}(\mathcal{Y},\mathcal{X})$ and the Moore–Penrose inverse $\mathcal{T}^+ : \mathcal{D}(\mathcal{T}^+) \to \mathcal{X}$ of $\mathcal{T}$ have the following svds:

$$\mathcal{T}^* y = \sum_n \sigma_n \langle y, v_n\rangle_{\mathcal{Y}}\, u_n, \quad y \in \mathcal{Y}, \tag{5.23}$$

$$\mathcal{T}^+ y = \sum_n \frac{1}{\sigma_n} \langle y, v_n\rangle_{\mathcal{Y}}\, u_n, \quad y \in \mathcal{D}(\mathcal{T}^+). \tag{5.24}$$

Proof We use Theorem 2.5.72, according to which we have

$$\ker \mathcal{T} = \mathcal{T}^*(\mathcal{Y})^{\perp}, \qquad (\ker \mathcal{T})^{\perp} = \overline{\mathcal{T}^*(\mathcal{Y})}, \qquad \mathcal{T}(\mathcal{X})^{\perp} = \ker \mathcal{T}^*,$$

and the theorem on orthogonal decompositions (Theorem 2.5.71).

(1) SVD of $\mathcal{T}^*$:

The adjoint operator $\mathcal{T}^*$ must satisfy $\langle \mathcal{T}x, y\rangle_{\mathcal{Y}} = \langle x, \mathcal{T}^* y\rangle_{\mathcal{X}}$ for all $x \in \mathcal{X}$ and $y \in \mathcal{Y}$ (see Theorem 2.5.59). With the svd in (5.22) and the Parseval identity, we get

$$\langle \mathcal{T}x, y\rangle_{\mathcal{Y}} = \sum_n \sigma_n \langle x, u_n\rangle_{\mathcal{X}}\, \overline{\langle y, v_n\rangle_{\mathcal{Y}}}, \tag{5.25}$$

since the orthogonal projection of y onto $\mathcal{T}(\mathcal{X})^{\perp}$ is orthogonal to $\mathcal{T}x$. Moreover, with $x = \sum_n \langle x, u_n\rangle_{\mathcal{X}} u_n + \mathcal{P}_{\ker \mathcal{T}}\, x$ and $\ker \mathcal{T} = \mathcal{T}^*(\mathcal{Y})^{\perp}$, we obtain

$$\langle x, \mathcal{T}^* y\rangle_{\mathcal{X}} = \sum_n \langle x, u_n\rangle_{\mathcal{X}}\, \overline{\langle \mathcal{T}^* y, u_n\rangle_{\mathcal{X}}}. \tag{5.26}$$

Since (5.25) and (5.26) must be equal, in particular for $x = u_k$, we get $\sigma_k \langle y, v_k \rangle_{\mathcal{Y}} = \langle \mathcal{T}^* y, u_k \rangle_{\mathcal{X}}$ for all k and all $y \in \mathcal{Y}$. Eventually, we remember that $(u_n)_n$ is an orthonormal basis of the subspace $(\ker \mathcal{T})^\perp = \overline{\mathcal{T}^*(\mathcal{Y})}$, and consequently

$$\mathcal{T}^* y = \sum_k \langle \mathcal{T}^* y, u_k \rangle_{\mathcal{X}} u_k = \sum_k \sigma_k \langle y, v_k \rangle_{\mathcal{Y}} u_k \quad \text{for all } y \in \mathcal{Y}.$$

(2) SVD of $\mathcal{T}^+$:

If $y \in \mathcal{D}(\mathcal{T}^+)$, then there exist $x \in \mathcal{X}$ and $z \in \mathcal{T}(\mathcal{X})^\perp = \ker \mathcal{T}^*$ such that $y = \mathcal{T}x + z$. We set

$$f := \sum_n \frac{1}{\sigma_n} \langle y, v_n \rangle_{\mathcal{Y}} u_n. \tag{5.27}$$

In the case of an infinite number of summands, we need to clarify the convergence in (5.27). This is, however, a consequence of the Picard condition (see Theorem 5.3.1), since $(v_n)_n$ is an orthonormal basis of $\overline{\mathcal{T}(\mathcal{X})}$ and thus $\langle y, v_n \rangle_{\mathcal{Y}} = \langle \mathcal{T}x, v_n \rangle_{\mathcal{Y}}$ for all n.

With (5.22), (5.23), (5.27), and again (5.23), we obtain

$$\begin{aligned} \mathcal{T}^*\mathcal{T} f &= \mathcal{T}^* \Big(\sum_n \sigma_n \langle f, u_n \rangle_{\mathcal{X}} v_n \Big) = \sum_n \sigma_n^2 \langle f, u_n \rangle_{\mathcal{X}} u_n \\ &= \sum_n \sigma_n \langle y, v_n \rangle_{\mathcal{Y}} u_n = \mathcal{T}^* y. \end{aligned} \tag{5.28}$$

By construction, $f \in (\ker \mathcal{T})^\perp$. Moreover, (5.28) shows that f solves the normal equation. Hence, Theorem 5.3.10 implies that $f = \mathcal{T}^+ y$.

Thus, (5.23) and (5.24) represent the svds of $\mathcal{T}^*$ and $\mathcal{T}^+$. □

Example 5.3.18 Let us have a look again at two inverse problems which we have already discussed. For the **downward continuation problem** $\mathcal{T}^{\text{up}} F = V$ (see Section 5.1), we have

$$\begin{aligned} \left(\mathcal{T}^{\text{up}}\right)^* V &= \sum_{n=0}^{\infty} \sum_{j=-n}^{n} \left(\frac{\varrho}{\sigma}\right)^n \left\langle V, Y_{n,j}^{\sigma} \right\rangle_{\mathrm{L}^2(\Omega_\sigma)} Y_{n,j}^{\varrho}, && V \in \mathrm{L}^2(\Omega_\sigma), \\ \left(\mathcal{T}^{\text{up}}\right)^+ V &= \sum_{n=0}^{\infty} \sum_{j=-n}^{n} \left(\frac{\sigma}{\varrho}\right)^n \left\langle V, Y_{n,j}^{\sigma} \right\rangle_{\mathrm{L}^2(\Omega_\sigma)} Y_{n,j}^{\varrho} \\ &= \mathcal{T}^{\text{down}} V, && V \in \mathcal{T}^{\text{up}}\left(\mathrm{L}^2(\Omega_\varrho)\right), \end{aligned}$$

since here $\mathcal{T}^{\text{up}}(\mathrm{L}^2(\Omega_\varrho))$ is dense in $\mathrm{L}^2(\Omega_\sigma)$ and, thus, $\left[\mathcal{T}^{\text{up}}(\mathrm{L}^2(\Omega_\varrho))\right]^\perp = \{0\}$.

For the **inverse gravimetric problem** $\mathcal{T}^{\text{gr}} F = V$ (see Section 5.2), we obtain

$$\left(\mathcal{T}^{\text{gr}}\right)^* V = \sum_{n=0}^{\infty} \sum_{j=-n}^{n} \frac{G R^{n+3/2}}{(R+\varepsilon)^n} \frac{4\pi}{\sqrt{2n+3}\,(2n+1)} \left\langle V, Y_{n,j}^{R+\varepsilon} \right\rangle_{\mathrm{L}^2(\Omega_{R+\varepsilon})} G_{0,n,j}^{\mathrm{I}}$$

for $V \in \mathrm{L}^2(\Omega_{R+\varepsilon})$ and

$$\left(\mathcal{T}^{\mathrm{gr}}\right)^{+} V = \sum_{n=0}^{\infty} \sum_{j=-n}^{n} \frac{(R+\varepsilon)^n}{GR^{n+3/2}} \frac{\sqrt{2n+3}\,(2n+1)}{4\pi} \left\langle V, Y_{n,j}^{R+\varepsilon} \right\rangle_{\mathrm{L}^2(\Omega_{R+\varepsilon})} G_{0,n,j}^{\mathrm{I}}$$

for $V \in \mathcal{T}^{\mathrm{gr}}(\mathrm{L}^2(B_R(0))) \oplus \left[\mathcal{T}^{\mathrm{gr}}(\mathrm{L}^2(B_R(0)))\right]^{\perp}$. Note that

$$\left(\mathcal{T}^{\mathrm{gr}}\right)^{+} V = \left(\mathcal{T}^{\mathrm{gr}}\big|_{\mathrm{Harm}(B_R(0))}\right)^{-1} V \quad \text{for all } V \in \mathcal{T}^{\mathrm{gr}}\left(\mathrm{L}^2(B_R(0))\right),$$

where $\mathrm{Harm}(B_R(0))$ is the set of all harmonic functions on $B_R(0)$.

These examples and propositions show that the instability of the inverse problem is directly connected to the fact that the singular values of the forward operator $\mathcal{T}$ tend to 0 (again, in the case that $\mathcal{T}$ is compact). This is also obvious: while $\|v_n\|_{\mathcal{Y}} = 1$ for all n, we have $\|\mathcal{T}^+ v_n\|_{\mathcal{X}} = \sigma_n^{-1} \to \infty$. And, clearly, the faster $(\sigma_n^{-1})_n$ diverges, the more severe is the instability. This explains also why we distinguish different orders of ill-posedness (see Definition 5.3.15) depending on the speed of divergence of $(\sigma_n^{-1})_n$.

In Theorem 5.3.17, we saw that the svds of $\mathcal{T}^*$ and $\mathcal{T}^+$ can be obtained by applying simple algebraic operations to the singular values of $\mathcal{T}$ (and interchanging the orthonormal systems). Some regularization methods can also be associated to operators with svds where a particular function is applied to the singular values of $\mathcal{T}$. For this reason, the concept of **functional calculus** is useful for inverse problems.

Definition 5.3.19 Let the compact operator $\mathcal{T}$ have the svd

$$\mathcal{T}x = \sum_n \sigma_n \langle x, u_n \rangle_{\mathcal{X}} v_n, \quad x \in \mathcal{X}.$$

Furthermore, let $\varphi\colon \mathbb{R}_0^+ \to \mathbb{R}$ be a piecewise continuous function with at most finitely many discontinuities. At these discontinuities, the left-hand and the right-hand limit need to exist. Then we define the operator $\varphi(\mathcal{T}^*\mathcal{T})\colon \mathcal{X} \to \mathcal{X}$ by

$$\varphi\left(\mathcal{T}^*\mathcal{T}\right) x := \sum_n \varphi\left(\sigma_n^2\right) \langle x, u_n \rangle_{\mathcal{X}} u_n + \varphi(0) \mathcal{P}_{\ker \mathcal{T}}\, x, \tag{5.29}$$

where $\mathcal{P}_{\ker \mathcal{T}}$ is the orthogonal projector onto $\ker \mathcal{T}$ (see Theorem 2.5.71).

Theorem 5.3.20 *The operator $\varphi(\mathcal{T}^*\mathcal{T})$ is linear and continuous with*

$$\left\|\varphi\left(\mathcal{T}^*\mathcal{T}\right)\right\|_{\mathcal{L}} \leq \sup_{\sigma \in \left[0, \|\mathcal{T}\|_{\mathcal{L}}^2\right]} |\varphi(\sigma)|. \tag{5.30}$$

If $\dim(\ker \mathcal{T})^{\perp} = +\infty$, then the series in (5.29) converges strongly in $\mathcal{X}$.

Proof The linearity is obvious. Furthermore, since $(\sigma_n)_n$ is either a finite sequence or converges to 0 and is therefore bounded, the conditions on φ imply that $(\varphi(\sigma_n^2))_n$ is also bounded. Hence, the considerations after Theorem 2.5.81 together with Theorem 2.5.71

imply the continuity of the operator. The operator norm estimate (5.30) and the convergence of the series are a consequence of

$$\|\varphi(\mathcal{T}^*\mathcal{T})x\|_{\mathcal{X}}^2 = \sum_n \left[\varphi(\sigma_n^2)\right]^2 |\langle x,u_n\rangle_{\mathcal{X}}|^2 + [\varphi(0)]^2 \|\mathcal{P}_{\ker\mathcal{T}}\, x\|_{\mathcal{X}}^2$$

$$\leq \sup_{\sigma\in[0,\|\mathcal{T}\|_{\mathcal{L}}^2]} [\varphi(\sigma)]^2 \left(\sum_n |\langle x,u_n\rangle_{\mathcal{X}}|^2 + \|\mathcal{P}_{\ker\mathcal{T}}\, x\|_{\mathcal{X}}^2\right),$$

where the term in the latter brackets equals $\|x\|_{\mathcal{X}}^2$. □

Example 5.3.21 Let us consider some simple examples.

(a) If $\varphi \equiv 1$, then $\varphi(\mathcal{T}^*\mathcal{T})x = \sum_n \langle x,u_n\rangle_{\mathcal{X}} u_n + \mathcal{P}_{\ker\mathcal{T}}\, x = x$ such that $\varphi(\mathcal{T}^*\mathcal{T}) = \mathrm{Id}_{\mathcal{X}}$.
(b) If $\varphi(t) = t^k$ for all $t \in \mathbb{R}_0^+$ and a fixed $k \in \mathbb{N}$, then

$$\varphi(\mathcal{T}^*\mathcal{T})x = \sum_n \sigma_n^{2k}\langle x,u_n\rangle_{\mathcal{X}} u_n + 0 = (\mathcal{T}^*\mathcal{T})^k x.$$

In other words, the notation $\varphi(\mathcal{T}^*\mathcal{T})$ indeed makes sense, since polynomials $\varphi(t) = \sum_{j=0}^k a_j t^j$ yield $\varphi(\mathcal{T}^*\mathcal{T}) = \sum_{j=0}^k a_j(\mathcal{T}^*\mathcal{T})^j$, where the right-hand side may be understood in the conventional sense without functional calculus.
(c) If $\varphi(t) = \sqrt{t}$ for all $t \in \mathbb{R}_0^+$, then

$$\varphi(\mathcal{T}^*\mathcal{T})x = \sqrt{\mathcal{T}^*\mathcal{T}}\, x = \sum_n \sigma_n\langle x,u_n\rangle_{\mathcal{X}} u_n. \tag{5.31}$$

This operator is known from other contexts of functional analysis, because it has the property that $(\sqrt{\mathcal{T}^*\mathcal{T}})^2 = \mathcal{T}^*\mathcal{T}$. It is, therefore, also denoted by $|\mathcal{T}|$.

Definition 5.3.22 Let $\mathcal{T} \in \mathcal{K}(\mathcal{X},\mathcal{Y})$. Then we write $|\mathcal{T}| := \sqrt{\mathcal{T}^*\mathcal{T}}$, where the latter is understood in terms of functional calculus.

Note that $\mathcal{T}^{**} = \mathcal{T}$, and therefore we automatically also defined $|\mathcal{T}^*| = \sqrt{\mathcal{T}\mathcal{T}^*}$. With functional calculus, we may also formulate the following inequality.

Theorem 5.3.23 (Interpolation Inequality) *If $\mathcal{T}$ is a compact operator and some constants $\alpha,\beta \in \mathbb{R}^+$ are given, then*

$$\left\||\mathcal{T}|^\beta x\right\|_{\mathcal{X}} \leq \left\||\mathcal{T}|^{\alpha+\beta}x\right\|_{\mathcal{X}}^{\frac{\beta}{\alpha+\beta}} \|x\|_{\mathcal{X}}^{\frac{\alpha}{\alpha+\beta}} \quad \textit{for all } x \in \mathcal{X}.$$

The proof is omitted and left to the reader as an exercise.

Theorem 5.3.24 *Let $\mathcal{T}$ be compact. Then the images of $|\mathcal{T}|$ and $|\mathcal{T}^*|$ satisfy $|\mathcal{T}|(\mathcal{X}) = \mathcal{T}^*(\mathcal{Y})$ and $|\mathcal{T}^*|(\mathcal{Y}) = \mathcal{T}(\mathcal{X})$.*

Proof If suffices to prove one of the two identities, because $\mathcal{T} = \mathcal{T}^{**}$.

Let $\{(\sigma_n,u_n,v_n)\}_n$ be a singular system of $\mathcal{T}$. Due to Theorem 2.5.72, we have $\overline{\mathcal{T}^*(\mathcal{Y})} = (\ker\mathcal{T})^\perp$. Hence, if $x \in \mathcal{T}^*(\mathcal{Y})$, then $x \in (\ker\mathcal{T})^\perp$ and $\mathcal{T}x \in \mathcal{T}\mathcal{T}^*(\mathcal{Y})$. Vice versa, if

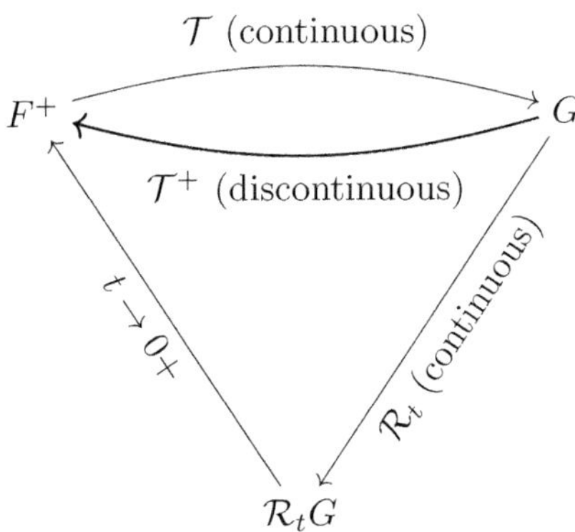

Figure 5.1 Illustration of the basic idea of a regularization $\{\mathcal{R}_t\}_t$.

$x \in (\ker \mathcal{T})^\perp$ and $\mathcal{T}x \in \mathcal{T}\mathcal{T}^*(\mathcal{Y})$, then there exists y such that $\mathcal{T}x = \mathcal{T}\mathcal{T}^*y$. Since $\mathcal{T}^*y \in \mathcal{T}^*(\mathcal{Y}) \subset (\ker \mathcal{T})^\perp$ and $\mathcal{T}$ is injective on $(\ker \mathcal{T})^\perp$, we get $x = \mathcal{T}^*y$, that is, $x \in \mathcal{T}^*(\mathcal{Y})$. By applying the Picard condition (Theorem 5.3.1) to the fact that $\mathcal{T}x \in \mathcal{T}\mathcal{T}^*(\mathcal{Y})$, we can formulate our preliminary deductions as follows (note that $\{(\sigma_n^2, v_n, v_n)\}_n$ is a singular system of $\mathcal{T}\mathcal{T}^*$):

$$x \in \mathcal{T}^*(\mathcal{Y}) \quad \Leftrightarrow \quad \sum_n \frac{\left|\langle \mathcal{T}x, v_n\rangle_{\mathcal{Y}}\right|^2}{\sigma_n^4} < +\infty \quad \text{and} \quad x \in (\ker \mathcal{T})^\perp.$$

From the expansion (5.31) of $|\mathcal{T}|x$, we deduce that $\overline{|\mathcal{T}|(\mathcal{X})} = (\ker \mathcal{T})^\perp$. Moreover, the svd of $\mathcal{T}$ yields

$$x \in \mathcal{T}^*(\mathcal{Y}) \quad \Leftrightarrow \quad \sum_n \frac{|\langle x, u_n\rangle_{\mathcal{X}}|^2}{\sigma_n^2} < +\infty \quad \text{and} \quad x \in \overline{|\mathcal{T}|(\mathcal{X})}.$$

The right-hand side is exactly the Picard condition (Theorem 5.3.1) for $x \in |\mathcal{T}|(\mathcal{X})$. □

Let us not forget that the main objective here is to handle ill-posed inverse problems. The keyword for methods which are able to solve ill-posed problems in a stable way is 'regularization'. The basic idea is that we have an operator $\mathcal{T}$ and a family $\{\mathcal{R}_t\}_t$ of operators. While it is not recommended to calculate the minimum-norm solution $F^+ = \mathcal{T}^+G$, if the problem is ill-posed in the sense of Nashed, the operators $\mathcal{R}_t$ are continuous. The price which we have to pay is that $\mathcal{R}_t G \neq \mathcal{T}^+G$ in general, but by choosing t appropriately we can get sufficiently close to $\mathcal{T}^+G$. This idea is illustrated in Figure 5.1.

The more precise definition of a regularization is commonly as follows.

Definition 5.3.25 Let $\mathcal{T}$ be, as usual, a linear and continuous operator between the Hilbert space $(\mathcal{X}, \langle\cdot,\cdot\rangle_{\mathcal{X}})$ and $(\mathcal{Y}, \langle\cdot,\cdot\rangle_{\mathcal{Y}})$. Moreover, let $\{\mathcal{R}_t\}_{t\in I}$ be a family of continuous operators $\mathcal{R}_t \colon \mathcal{Y} \to \mathcal{X}$ such that $\mathcal{R}_t 0 = 0$ for every $t \in I$, where $I \subset \mathbb{R}^+$ is a set which has (at least) the accumulation point 0. Furthermore, let there be a function $\gamma \colon \mathbb{R}^+ \times \mathcal{Y} \to I$ such that, for all $G \in \mathcal{T}(\mathcal{X})$,

$$\lim_{\varepsilon \to 0+} \sup_{G^\varepsilon \in \overline{B_\varepsilon(G)}} \gamma\left(\varepsilon, G^\varepsilon\right) = 0 \tag{5.32}$$

and

$$\lim_{\varepsilon \to 0+} \sup_{G^\varepsilon \in \overline{B_\varepsilon(G)}} \left\| \mathcal{T}^+ G - \mathcal{R}_{\gamma(\varepsilon, G^\varepsilon)} G^\varepsilon \right\|_{\mathcal{X}} = 0, \tag{5.33}$$

where $B_\varepsilon(G)$ is the ball in $\mathcal{Y}$ with centre G and radius ε. In this case, the pair $(\{\mathcal{R}_\iota\}_{\iota \in I}, \gamma)$ is called a **regularization** (or a **regularization method**) of (or for) $\mathcal{T}^+$, γ is called a **parameter choice**, and $\gamma(\varepsilon, G^\varepsilon)$ is called the **regularization parameter**. We set $\Gamma := \{\gamma(\varepsilon, G) \,|\, \varepsilon > 0,\, G \in \mathcal{T}(\mathcal{X})\}$. If each $\mathcal{R}_\iota$ is linear, then we refer to a **linear regularization**, otherwise a **non-linear regularization**. If the mapping γ is independent of the second argument, that is, if it only depends on ε, then γ is called an **a priori parameter choice**.

The parameter choice tells us how to choose the particular operator $\mathcal{R}_\iota$ among the family of operators depending on the noise level ε and the given (noisy) right-hand side G^ε. We expect that, if the noise level tends to 0, then the stable(!) approximations $\mathcal{R}_{\gamma(\varepsilon, G^\varepsilon)} G^\varepsilon$ tend to $\mathcal{T}^+ G$, the minimum-norm solution of $\mathcal{T} F = G$ with exact, unnoisy, right-hand side G. This is what (5.33) says.

In the following, we derive some fundamental properties of regularizations.

Lemma 5.3.26 *Condition* (5.33) *is equivalent to requiring that, for all* $F \in (\ker \mathcal{T})^\perp$, *we have* $\lim_{\varepsilon \to 0+} \sup_{G^\varepsilon \in \overline{B_\varepsilon(\mathcal{T} F)}} \left\| F - \mathcal{R}_{\gamma(\varepsilon, G^\varepsilon)} G^\varepsilon \right\|_{\mathcal{X}} = 0$.

Proof This equivalence is easy to show. Since G needs to be in $\mathcal{T}(\mathcal{X})$, we can replace it by $\mathcal{T} F$ with $F \in (\ker \mathcal{T})^\perp$. Due to Theorem 5.3.10, this F equals $F^+ = \mathcal{T}^+ G$. □

Lemma 5.3.27 *If* $(\{\mathcal{R}_\iota\}_{\iota \in I}, \gamma)$ *is a regularization of* $\mathcal{T}^+$, *then*

$$\lim_{\varepsilon \to 0+} \mathcal{R}_{\gamma(\varepsilon, G)} G = \mathcal{T}^+ G \quad \textit{for all } G \in \mathcal{T}(\mathcal{X}).$$

Proof This is necessary for (5.33), since $G \in \overline{B_\varepsilon(G)}$. □

Lemma 5.3.28 *If* $(\{\mathcal{R}_\iota\}_{\iota \in I}, \gamma)$ *is a regularization of* $\mathcal{T}^+$ *where* γ *continuously depends on* ε *and* I *is an interval of the form* $]0, c[$, *then*

$$\lim_{\lambda \to 0+} \mathcal{R}_\lambda G = \mathcal{T}^+ G \quad \textit{for all } G \in \mathcal{T}(\mathcal{X}).$$

Proof Due to the continuity of γ and (5.32), we get, for every λ sufficiently close to zero, an $\varepsilon > 0$ such that $\gamma(\varepsilon, G) = \lambda$. The rest is implied by Lemma 5.3.27. □

Note that the convergence in Lemma 5.3.28 is pointwise; it cannot be uniform, because then the continuity of $\mathcal{R}_\lambda$ would be inherited to $\mathcal{T}^+$, which need not be continuous.

Theorem 5.3.29 *If* $(\{\mathcal{R}_\iota\}_{\iota \in I}, \gamma)$ *is a* linear *regularization of* $\mathcal{T}^+$, *where* γ *is continuous in* ε *and* $I =]0, c[$, *and the inverse problem is ill-posed in the sense of Nashed, then it must hold true that* $\sup_{\iota \in I} \|\mathcal{R}_\iota\|_{\mathcal{L}} = +\infty$.

Proof We assume the contrary: let there be a constant $C > 0$ such that $\|\mathcal{R}_t\|_{\mathcal{L}} \le C$ for all $t \in I$. Lemma 5.3.28 yields the pointwise convergence of $\mathcal{R}_\lambda$ to $\mathcal{T}^+$ on $\mathcal{T}(\mathcal{X})$ as $\lambda \to 0+$. Furthermore, $\mathcal{T}(\mathcal{X})$ is dense in $\overline{\mathcal{T}(\mathcal{X})}$ and each $\mathcal{R}_\lambda$ is also a continuous linear operator on $\overline{\mathcal{T}(\mathcal{X})}$, where $\|\mathcal{R}_\lambda\|_{\mathcal{L}(\overline{\mathcal{T}(\mathcal{X})},\mathcal{Y})} \le C$ for all $\lambda \in I$. Hence, we may apply the Banach–Steinhaus theorem (Theorem 2.5.33) such that $\mathcal{R}_\lambda G \to \mathcal{T}^+ G$ for all $G \in \overline{\mathcal{T}(\mathcal{X})}$ and $\mathcal{T}^+$ is continuous, which, however, implies that $\mathcal{T}(\mathcal{X}) = \overline{\mathcal{T}(\mathcal{X})}$, see Theorem 5.3.12. This is a contradiction. □

The following theorem addresses the question of the speed of convergence that we can achieve for a regularization. Unfortunately, the answer is not very pleasant: the convergence can be arbitrarily slow.

Theorem 5.3.30 *Let the inverse problem $\mathcal{T}F = G$ be ill-posed in the sense of Nashed and let $(\{\mathcal{R}_t\}_{t\in I}, \gamma)$ be a regularization of $\mathcal{T}^+$. Then the set of all functions $h\colon \mathbb{R}^+ \to \mathbb{R}^+$ with $\lim_{\varepsilon\to 0+} h(\varepsilon) = 0$ and*

$$\sup_{G^\varepsilon \in \overline{B_\varepsilon(G)},\, G\in\mathcal{T}(\mathcal{X})} \left\|\mathcal{T}^+G - \mathcal{R}_{\gamma(\varepsilon,G^\varepsilon)}G^\varepsilon\right\|_{\mathcal{X}} \le h(\varepsilon) \quad \textit{for all } \varepsilon > 0$$

is the empty set.

Proof The easiest way of proving this theorem is to assume that we do have such a function h. Let now $\mathcal{V} := B_1(0) \subset \mathcal{Y}$, $G \in \mathcal{T}(\mathcal{X}) \cap \mathcal{V}$, and $(G_n)_n \subset \mathcal{T}(\mathcal{X}) \cap \mathcal{V}$ with $\|G_n - G\|_{\mathcal{Y}} \to 0$. Simply by the definition of convergence, we get the following: for every $\varepsilon > 0$ (without loss of generality, $\varepsilon \le 1$), there exists $n_0(\varepsilon)$ such that, for all $n \ge n_0$, we have $\|G_n - G\|_{\mathcal{Y}} \le \varepsilon \le 1$. As a result, $G_n - G \in \mathcal{T}(\mathcal{X}) \cap \mathcal{V}$ for all $n \ge n_0$. Moreover, Corollary 5.3.13 says that $\mathcal{T}\mathcal{T}^+ = \mathcal{P}_{\overline{\mathcal{T}(\mathcal{X})}}$ on $\mathcal{D}(\mathcal{T}^+)$. This projector is continuous with its operator norm bounded by 1 due to Theorem 2.5.71. Hence,

$$\left\|\mathcal{T}\mathcal{T}^+(G_n - G)\right\|_{\mathcal{Y}} \le \left\|\mathcal{T}\mathcal{T}^+\right\|_{\mathcal{L}} \|G_n - G\|_{\mathcal{Y}} \le \varepsilon$$

for all $n \ge n_0$. This implies that, again for all $n \ge n_0$,

$$\left\|\mathcal{T}^+(G_n - G)\right\|_{\mathcal{X}} \le \sup\left\{\|x\|_{\mathcal{X}} : x \in \mathcal{T}^+(\mathcal{T}(\mathcal{X}) \cap \mathcal{V}),\ \|\mathcal{T}x\|_{\mathcal{Y}} \le \varepsilon\right\}.$$

By choosing $H^\varepsilon = 0$ and remembering that $\mathcal{R}_t 0 = 0$ for all t (see Definition 5.3.25), we can conclude next that

$$\begin{aligned}\left\|\mathcal{T}^+(G_n - G)\right\|_{\mathcal{X}} &\le \sup\Big\{\left\|x - \mathcal{R}_{\gamma(\varepsilon,H^\varepsilon)}H^\varepsilon\right\|_{\mathcal{X}} : x \in \mathcal{T}^+(\mathcal{T}(\mathcal{X}) \cap \mathcal{V}), \\ &\qquad\qquad H^\varepsilon \in \mathcal{Y},\ \left\|\mathcal{T}x - H^\varepsilon\right\|_{\mathcal{Y}} \le \varepsilon\Big\} \\ &= \sup\Big\{\left\|\mathcal{T}^+H - \mathcal{R}_{\gamma(\varepsilon,H^\varepsilon)}H^\varepsilon\right\|_{\mathcal{X}} : H \in \mathcal{T}(\mathcal{X}) \cap \mathcal{V}, \\ &\qquad\qquad H^\varepsilon \in \mathcal{Y},\ \left\|H - H^\varepsilon\right\|_{\mathcal{Y}} \le \varepsilon\Big\}.\end{aligned}$$

For the latter identity, we replaced x by $x = \mathcal{T}^+H \in (\ker\mathcal{T})^\perp$ (see Theorem 5.3.12) with $H \in \mathcal{T}(\mathcal{X}) \cap \mathcal{V}$ such that $\mathcal{T}x = \mathcal{T}\mathcal{T}^+H = \mathcal{P}_{\overline{\mathcal{T}(\mathcal{X})}}H = H$.

What we derived so far means that, for every $\varepsilon > 0$, there exists $n_0(\varepsilon)$ such that, for all $n \geq n_0$, $\|\mathcal{T}^+(G_n - G)\|_{\mathcal{X}} \leq h(\varepsilon)$. However, if $\lim_{\varepsilon\to 0+} h(\varepsilon) = 0$, then $\mathcal{T}^+G_n \to \mathcal{T}^+G$ as $n \to \infty$ and $\mathcal{T}^+$ is continuous first of all on $\mathcal{T}(\mathcal{X}) \cap \mathcal{V}$ but, due to its linearity, also on the whole space $\mathcal{Y}$. This contradicts the ill-posedness and Theorem 5.3.12. □

This means that no matter which speed of convergence h we hope to get (and no matter how humble our expectations are), we will always find a pair (G, G^ε), where our hope will be frustrated. This fact is underlined by the following theorem. The proof is due to Schock (1985) and is omitted here.

Theorem 5.3.31 *Let $(\{\mathcal{R}_t\}_{t>0}, \gamma)$ be a regularization of $\mathcal{T}^+$, where the underlying inverse problem is ill-posed in the sense of Nashed, and let the function $h\colon \mathbb{R}^+ \to \mathbb{R}^+$ be monotonically increasing with $\lim_{\varepsilon\to 0+} h(\varepsilon) = 0$. Then the set $\{F \in \mathcal{X} : \|F - \mathcal{R}_{\gamma(\varepsilon,\mathcal{T}F)}\mathcal{T}F\|_{\mathcal{X}} = \mathcal{O}(h(\varepsilon))$ as $\varepsilon \to 0+\}$ is of the first category in $\mathcal{X}$.*

In other words, if we choose a speed of convergence h and a regularization method, then it is not only impossible to achieve this speed universally for all right-hand sides, but it is even worse: there is not just one single counterexample for the right-hand side. On the contrary, it is rather rare that a solution F (also without the presence of noise) allows the chosen speed of convergence. In most cases, it will fail. Note that $\mathcal{X}$ is of the second category (see Definition 2.5.17 and Theorem 2.5.18).

The reason for this arbitrarily slow convergence is hidden in the ill-posedness. If $\mathcal{T}^+$ is discontinuous, then small changes ε to the data can have severe effects on the solution. Therefore, one needs certain smoothness assumptions on the right-hand side for proving convergence rates of the regularization. This is achieved by the so-called **source conditions**.

Definition 5.3.32 Let the compact operator $\mathcal{T}$ have the singular system $\{(\sigma_n, u_n, v_n)\}_n$. For each $\nu \geq 0$, we define the space $\mathcal{X}_\nu \subset \mathcal{X}$ as the range $\mathcal{X}_\nu := |\mathcal{T}|^\nu(\mathcal{X})$ and set

$$\|x\|_\nu := \left(\sum_n \sigma_n^{-2\nu} |\langle x, u_n\rangle_{\mathcal{X}}|^2\right)^{1/2} \quad \text{for all } x \in \mathcal{X}_\nu.$$

Note that $|\mathcal{T}|^\nu$ is understood in terms of functional calculus. Moreover, it is easy to show that $\|\cdot\|_\nu$ is a norm on $\mathcal{X}_\nu$.

Theorem 5.3.33 *Let the spaces $\{\mathcal{X}_\nu\}_{\nu\geq 0}$ be given corresponding to the compact operator $\mathcal{T}$ with the singular system $\{(\sigma_n, u_n, v_n)\}_n$. Then the following holds true:*

(a) $x \in \mathcal{X}_\nu \Leftrightarrow x = \sum_n \langle x, u_n\rangle_{\mathcal{X}} u_n$ and $\|x\|_\nu^2 < +\infty$.
(b) $\mathcal{X}_\nu \subset \mathcal{X}_\mu \subset \mathcal{X}_0 = (\ker \mathcal{T})^\perp$ for $\nu \geq \mu$.
(c) $\||\mathcal{T}|^\nu x\|_\nu = \|x\|_{\mathcal{X}}$ for all $x \in (\ker \mathcal{T})^\perp$.

Before we prove the theorem, let us look at an example.

Example 5.3.34 Though the notation is different, we have already encountered these spaces. In the case of the sphere with spherical harmonics as an orthonormal basis, these are the **spherical Sobolev spaces**; compare Definition 4.7.1 with part a of

Theorem 5.3.33. The coefficients A_n in the definition of the Sobolev spaces correspond to the singular values σ_n^ν of $|\mathcal{T}|^\nu$, and the requirement that $\langle F, Y_{n,j}\rangle_{\mathrm{L}^2(\Omega)} = 0$, if $A_n = 0$, corresponds to the expansion of $x \in \mathcal{X}_\nu$ in the basis of $(\ker \mathcal{T})^\perp$. Moreover, part b of Theorem 5.3.33 can be recognized in Theorem 4.7.2.

Proof (of Theorem 5.3.33) By definition, $y \in \mathcal{X}_\nu$, if and only if there exists $\tilde{x} \in \mathcal{X}$ with $y = |\mathcal{T}|^\nu \tilde{x} = \sum_n \sigma_n^\nu \langle \tilde{x}, u_n\rangle_{\mathcal{X}} u_n + 0^\nu \mathcal{P}_{\ker \mathcal{T}} \tilde{x}$. This is equivalent to the existence of $x \in (\ker \mathcal{T})^\perp$ such that $y = \sum_n \sigma_n^\nu \langle x, u_n\rangle_{\mathcal{X}} u_n$. Hence, $y \in \mathcal{X}_\nu$, if and only if $y \in \overline{\operatorname{span}\{u_n\}}$ and $\sum_n |\langle y, u_n\rangle_{\mathcal{X}}|^2 \sigma_n^{-2\nu} < +\infty$. This proves part a and also shows that

$$\|y\|_\nu^2 = \sum_n \sigma_n^{-2\nu} \left|\langle y, u_n\rangle_{\mathcal{X}}\right|^2 = \sum_n |\langle x, u_n\rangle_{\mathcal{X}}|^2 = \|x\|_{\mathcal{X}}^2,$$

which proves part c.

For part b, we observe that, for $\nu \geq \mu$, we have

$$\begin{aligned}\sum_n \sigma_n^{-2\mu} \left|\langle y, u_n\rangle_{\mathcal{X}}\right|^2 &= \sum_{\substack{n \\ \sigma_n \geq 1}} \sigma_n^{-2\mu} \left|\langle y, u_n\rangle_{\mathcal{X}}\right|^2 + \sum_{\substack{n \\ \sigma_n < 1}} \sigma_n^{-2\mu} \left|\langle y, u_n\rangle_{\mathcal{X}}\right|^2 \\ &\leq \sum_{\substack{n \\ \sigma_n \geq 1}} \sigma_n^{-2\mu} \left|\langle y, u_n\rangle_{\mathcal{X}}\right|^2 + \sum_{\substack{n \\ \sigma_n < 1}} \sigma_n^{-2\nu} \left|\langle y, u_n\rangle_{\mathcal{X}}\right|^2 .\end{aligned}$$

Since $\mathcal{T}$ is compact and consequently $(\sigma_n)_n$ tends to zero, the first summation on the right-hand side has only a finite number of summands. The second one is finite if $y \in \mathcal{X}_\nu$. Hence, we also have $y \in \mathcal{X}_\mu$. □

With smoothness properties that we associate to data $G \in \mathcal{Y}$ with $\mathcal{T}^+ G \in \mathcal{X}_\nu$, we are provided with a tool for analyzing the stability of an inversion.

Definition 5.3.35 Let $\mathcal{T} \in \mathcal{K}(\mathcal{X}, \mathcal{Y})$ be given with the corresponding spaces $\{\mathcal{X}_\nu\}_{\nu \geq 0}$. An operator $\mathcal{S} : \mathcal{Y} \to \mathcal{X}$ is called a **(stable) reconstruction method** if $\mathcal{S}$ is continuous and satisfies $\mathcal{S}0 = 0$. For such an operator and given constants $\nu \geq 0$, $\varepsilon, \varrho > 0$, we call

$$\mathrm{E}_\nu(\varepsilon, \varrho, \mathcal{S}) := \sup\left\{\left\|\mathcal{S} G^\varepsilon - \mathcal{T}^+ G\right\|_{\mathcal{X}} : G \in \mathcal{T}(\mathcal{X}), G^\varepsilon \in \overline{B_\varepsilon(G)}, \left\|\mathcal{T}^+ G\right\|_\nu \leq \varrho\right\} \tag{5.34}$$

the **worst-case error** of $\mathcal{S}$ with respect to $\mathcal{T}$ in the case of the noise level ε and the **source condition** $\|\mathcal{T}^+ G\|_\nu \leq \varrho$. Moreover, the **best-possible worst-case error** is defined as

$$\mathrm{e}_\nu(\varepsilon, \varrho) := \inf\left\{\mathrm{E}_\nu(\varepsilon, \varrho, \mathcal{S}) \,|\, \mathcal{S} : \mathcal{Y} \to \mathcal{X} \text{ is continuous and } \mathcal{S}0 = 0\right\}. \tag{5.35}$$

Note that $\mathcal{S}$ may be non-linear.

The worst-case error measures for a specific reconstruction method $\mathcal{S}$ how large the error is between the minimum-norm solution $\mathcal{T}^+ G$ for exact data G and the reconstructed approximation $\mathcal{S} G^\varepsilon$ for noisy data G^ε. Note that we assume here deterministic noise by requiring $\|G^\varepsilon - G\|_{\mathcal{Y}} \leq \varepsilon$. A corresponding theory for the (usually more realistic) case of stochastic noise is also available; see, for example, Lu and Pereverzev (2013). The best-possible worst-case error measures how good a reconstruction method can become maximally. It certainly does not tell us *how* we could get a correspondingly optimal reconstruction method.

Both quantities depend on the operator $\mathcal{T}$ of the inverse problem and thus give us information about how hard it is to regularize the problem and, in the case of E_ν, also how good $\mathcal{S}$ is in achieving this aim.

Theorem 5.3.36 *The formula* (5.34) *is equivalent to*

$$\mathrm{E}_\nu(\varepsilon,\varrho,\mathcal{S}) = \sup\left\{\left\|\mathcal{S}G^\varepsilon - F\right\|_{\mathcal{X}} : F \in \mathcal{X}_\nu,\ G^\varepsilon \in \overline{B_\varepsilon(\mathcal{T}F)},\ \|F\|_\nu \le \varrho\right\} \quad (5.36)$$

and the formula (5.35) *is equivalent to*

$$\mathrm{e}_\nu(\varepsilon,\varrho) = \sup\left\{\|F\|_{\mathcal{X}} : F \in \mathcal{X}_\nu,\ \|\mathcal{T}F\|_{\mathcal{Y}} \le \varepsilon,\ \|F\|_\nu \le \varrho\right\}. \quad (5.37)$$

The proof of (5.36) is analogous to the easy proof of Lemma 5.3.26, whereas the proof of (5.37) is lengthy and, for this reason, omitted here.

Let us assume that we found a 'good' reconstruction method $\mathcal{S}$ for $F^+ = \mathcal{T}^+G \in \mathcal{X}_\nu$, $G \in \mathcal{T}(\mathcal{X})$, in the sense that $\mathcal{S}G \in \mathcal{X}_\nu$ and $\|\mathcal{T}(\mathcal{S}G)-G^\varepsilon\|_{\mathcal{Y}} \le \varepsilon$ for $\|G-G^\varepsilon\|_{\mathcal{Y}} \le \varepsilon$, that is, we can solve the inverse problem with an accuracy which does not exceed the noise level. Then we get the following (note that $\mathcal{T}\mathcal{T}^+G = G$ for $G \in \mathcal{T}(\mathcal{X})$ due to the Moore–Penrose axioms; see Theorem 5.3.14):

$$F^+ - \mathcal{S}G \in \mathcal{X}_\nu, \quad (5.38)$$

$$\left\|\mathcal{T}\left(F^+ - \mathcal{S}G\right)\right\|_{\mathcal{Y}} \le \left\|\mathcal{T}F^+ - G^\varepsilon\right\|_{\mathcal{Y}} + \left\|G^\varepsilon - \mathcal{T}\mathcal{S}G\right\|_{\mathcal{Y}} \le 2\varepsilon, \quad (5.39)$$

$$\left\|F^+ - \mathcal{S}G\right\|_\nu \le \left\|F^+\right\|_\nu + \|\mathcal{S}G\|_\nu \le 2\max\left\{\left\|F^+\right\|_\nu, \|\mathcal{S}G\|_\nu\right\}. \quad (5.40)$$

If we use (5.38)–(5.40) in (5.37), then we directly get the following theorem.

Theorem 5.3.37 *If $\mathcal{T} \in \mathcal{K}(\mathcal{X},\mathcal{Y})$, $G \in \mathcal{T}(\mathcal{X})$, $F^+ = \mathcal{T}^+G \in \mathcal{X}_\nu$, and $\mathcal{S} : \mathcal{Y} \to \mathcal{X}$ is continuous with $\mathcal{S}0 = 0$ and satisfies $\mathcal{S}G \in \mathcal{X}_\nu$ and the inequality $\|\mathcal{T}(\mathcal{S}G) - G^\varepsilon\|_{\mathcal{Y}} \le \varepsilon$ for at least one $G^\varepsilon \in \overline{B_\varepsilon(G)}$, then*

$$\left\|F^+ - \mathcal{S}G\right\|_{\mathcal{X}} \le \mathrm{e}_\nu\left(2\varepsilon, 2\max\left\{\left\|F^+\right\|_\nu, \|\mathcal{S}G\|_\nu\right\}\right).$$

Of particular importance is the dependence of e_ν on ε and ϱ: how quickly does the accuracy still decay if the error increases? How sensitive is the error to the source condition? The probably most relevant answer is given by the next theorem.

Theorem 5.3.38 *For each $\mathcal{T} \in \mathcal{K}(\mathcal{X},\mathcal{Y})$, we have*

$$\mathrm{e}_\nu(\varepsilon,\varrho) \le \varrho^{1/(\nu+1)}\,\varepsilon^{\nu/(\nu+1)} \quad \textit{for all } \varepsilon,\nu,\varrho > 0. \quad (5.41)$$

Moreover, there is no better estimate in the sense that, for each $\varrho > 0$, there exists a sequence $(\varepsilon_k)_k$ with $\varepsilon_k \to 0$ such that

$$\mathrm{e}_\nu(\varepsilon_k,\varrho) = \varrho^{1/(\nu+1)}\,\varepsilon_k^{\nu/(\nu+1)}. \quad (5.42)$$

Proof We first prove the inequality and then construct the sequence $(\varepsilon_k)_k$.

(1) The inequality:

Let $x \in \mathcal{X}_\nu$. From Definition 5.3.32, we know that there exists $z \in \mathcal{X}$ such that $x = |\mathcal{T}|^\nu z$, where we may assume, without loss of generality, that $z \in (\ker \mathcal{T})^\perp$, because $(\ker \mathcal{T})^\perp = (\ker |\mathcal{T}|^\nu)^\perp$. Then, part c of Theorem 5.3.33 tells us that $\|x\|_\nu = \||\mathcal{T}|^\nu z\|_\nu = \|z\|_\mathcal{X}$, while the interpolation inequality (Theorem 5.3.23) leads us to

$$\|x\|_\mathcal{X} = \left\||\mathcal{T}|^\nu z\right\|_\mathcal{X} \leq \left\||\mathcal{T}|^{\nu+1} z\right\|_\mathcal{X}^{\nu/(\nu+1)} \|z\|_\mathcal{X}^{1/(\nu+1)}.$$

Hence, $\|x\|_\mathcal{X} \leq \||\mathcal{T}|x\|_\mathcal{X}^{\nu/(\nu+1)} \|x\|_\nu^{1/(\nu+1)}$. With the svds of $\mathcal{T}$ and $|\mathcal{T}|$,

$$\mathcal{T}x = \sum_n \sigma_n \langle x, u_n\rangle_\mathcal{X} v_n, \qquad |\mathcal{T}|x = \sum_n \sigma_n \langle x, u_n\rangle_\mathcal{X} u_n, \tag{5.43}$$

and the Parseval identity, we easily get that

$$\|\mathcal{T}x\|_\mathcal{Y}^2 = \sum_n \sigma_n^2 |\langle x, u_n\rangle_\mathcal{X}|^2, \quad \||\mathcal{T}|x\|_\mathcal{X}^2 = \sum_n \sigma_n^2 |\langle x, u_n\rangle_\mathcal{X}|^2 = \|\mathcal{T}x\|_\mathcal{Y}^2.$$

Hence, (5.37) allows us to conclude that

$$\mathrm{e}_\nu(\varepsilon, \varrho) \leq \varepsilon^{\nu/(\nu+1)} \varrho^{1/(\nu+1)}.$$

(2) There is no better estimate:

Let $\varrho > 0$ be arbitrary but fixed. We define $\varepsilon_k := \varrho\sigma_k^{\nu+1}$ for all k (if the sequence $(\sigma_k)_k$ is finite, then we extend the sequence $(\varepsilon_k)_k$ by zeros). Since $\mathcal{T}$ is compact, $(\varepsilon_k)_k$ converges to 0. Moreover, if we set $x_k := \varrho|\mathcal{T}|^\nu u_k$ for all k, then Theorem 5.3.33 yields that $x_k \in (\ker \mathcal{T})^\perp$ and $\|x_k\|_\nu = \varrho\|u_k\|_\mathcal{X} = \varrho$ for all k. With the svds (5.43) of $|\mathcal{T}|$ and $\mathcal{T}$, we also get

$$\begin{aligned}\|x_k\|_\mathcal{X} &= \left\|\varrho\sigma_k^\nu u_k\right\|_\mathcal{X} = \varrho\sigma_k^\nu = \varrho\left(\frac{\varepsilon_k}{\varrho}\right)^{\nu/(\nu+1)} \\ &= \varrho^{1-\nu/(\nu+1)} \varepsilon_k^{\nu/(\nu+1)} = \varrho^{1/(\nu+1)} \varepsilon_k^{\nu/(\nu+1)}\end{aligned}$$

and $\|\mathcal{T}x_k\|_\mathcal{Y} = \varrho\|\sigma_k^{\nu+1} v_k\|_\mathcal{Y} = \varepsilon_k$. We use these derivations in (5.37) as particular elements and see that

$$\mathrm{e}_\nu(\varepsilon_k, \varrho) \geq \|x_k\|_\mathcal{X} = \varrho^{1/(\nu+1)} \varepsilon_k^{\nu/(\nu+1)}.$$

In combination with (5.41), we obtain that '=' holds in the latter inequality. □

Let us have a closer look at the estimate (5.41). If the noise level ε increases, then it is obvious that the accuracy of the reconstruction can become worse. The good news is that e_ν increases less than linearly with ε, since $\nu/(\nu+1) < 1$. Moreover, if ϱ increases (i.e., $\|F\|_\nu$ may increase), we include less smooth solutions into the reconstruction. It is clear that this will also have negative effects on the accuracy of the reconstruction. If we increase ν, then $\mathcal{X}_\nu$ becomes 'smaller' and $\|F\|_\nu < \varrho$ becomes a stronger criterion, that is, in total, we consider smoother functions again and an increase of ϱ has a lower influence on e_ν

(as $\varrho^{1/(\nu+1)}$ decreases for increasing ν). Having said this, we need to consider that the influence of the data noise level ε increases with increasing ν (note that $\nu/(\nu+1) = 1 - 1/(\nu+1)$).

The question that certainly arises now is: can we find a method which reaches the behaviour in (5.42)? Commonly, this demand is weakened in the sense that the orders in which the dependencies on ϱ and ε occur can be achieved.

Definition 5.3.39 Let the inverse problem corresponding to the forward operator $\mathcal{T} \in \mathcal{K}(\mathcal{X}, \mathcal{Y})$ be ill-posed in the sense of Nashed. A family $\{\mathcal{S}_\varepsilon\}_{\varepsilon>0}$ of reconstruction methods is called **order optimal** with respect to $\mathcal{X}_\nu$ and $\mathcal{T}$ if there exist constants $C_\nu > 1$ and $\varepsilon_0 > 0$ such that

$$\mathrm{E}_\nu(\varepsilon, \varrho, \mathcal{S}_\varepsilon) \leq C_\nu\, \varrho^{1/(\nu+1)}\, \varepsilon^{\nu/(\nu+1)} \tag{5.44}$$

for all $\varepsilon \in]0, \varepsilon_0]$ and all $\varrho > 0$. If (5.44) also holds true for $C_\nu = 1$, then the family is also called an **optimal** reconstruction method.

Note that every **regularization method** is also a reconstruction method such that the preceding propositions and definitions are also valid for regularization methods. Vice versa and briefly stated, an order optimal reconstruction method is also a regularization. The precise statement is formulated in the next theorem and has a complicated proof, which is omitted here. We refer to the original paper Plato (1990) for a proof and suggest here a name for the theorem.

Theorem 5.3.40 (Plato's Theorem) *Let the inverse problem corresponding to the forward operator $\mathcal{T} \in \mathcal{K}(\mathcal{X}, \mathcal{Y})$ be ill-posed in the sense of Nashed. Moreover, let $\{\mathcal{R}_t\}_{t>0}$ be a family of reconstruction methods and let $\gamma\colon \mathbb{R}^+ \times \mathcal{Y} \to \mathbb{R}^+$ be a function which satisfies* (5.32). *We define for an arbitrary $b > 1$ the function $\gamma_b\colon \mathbb{R}^+ \times \mathcal{Y} \to \mathbb{R}^+$ by $\gamma_b(\varepsilon, y) := \gamma(b\varepsilon, y)$ for all $\varepsilon > 0$ and all $y \in \mathcal{Y}$. If the family $\{\mathcal{R}_{\gamma(\varepsilon,\cdot)}\}_{\varepsilon>0}$ is order optimal with respect to $\mathcal{X}_\nu$ and $\mathcal{T}$, then $(\{\mathcal{R}_t\}_{t>0}, \gamma_b)$ is a regularization method for $\mathcal{T}^+$, which is order optimal with respect to $\mathcal{X}_\mu$ and $\mathcal{T}$ for all $\mu \in]0, \nu]$.*

One of the big questions is: how can we obtain a regularization method, in particular order optimal or even optimal methods? For approaching the answer, we remember that the minimum-norm solution $F^+ = \mathcal{T}^+G$ is the unique solution of the normal equation $\mathcal{T}^*\mathcal{T}F^+ = \mathcal{T}^*G$, $F^+ \in (\ker\mathcal{T})^\perp$ for $G \in \mathcal{D}(\mathcal{T}^+)$ (see Theorem 5.3.10). So, clearly, $F^+ = \left(\mathcal{T}^*\mathcal{T}\big|_{(\ker\mathcal{T})^\perp}\right)^{-1}\mathcal{T}^*G$. Moreover, with Theorem 5.3.17, we have the identity $\mathcal{T}^+G = \sum_n \sigma_n^{-1}\langle G, v_n\rangle_{\mathcal{Y}}\, u_n$. The difficulty is the discontinuity of $\mathcal{T}^+$ if the inverse problem is ill-posed in the sense of Nashed. How can we find stable approximations to $\mathcal{T}^+G$? The functional calculus can provide us with an answer to this question.

Definition 5.3.41 Let $\{F_t\}_{t>0}$ be a family of functions where each F_t satisfies the conditions on φ in Definition 5.3.19. If furthermore

$$\lim_{t\to 0+} F_t(\sigma) = \frac{1}{\sigma} \quad \text{for all } \sigma \in \left]0, \|\mathcal{T}\|_{\mathcal{L}}^2\right],$$

where $\mathcal{T} \in \mathcal{K}(\mathcal{X},\mathcal{Y})$ is a given operator, then the family $\{F_t\}_{t>0}$ is called a **filter**. In this case, we also define the family of operators $\{\mathcal{R}_t\}_{t>0} \subset \mathcal{L}(\mathcal{Y},\mathcal{X})$ by $\mathcal{R}_t := F_t(\mathcal{T}^*\mathcal{T})\mathcal{T}^*$.

We immediately get the svd of $\mathcal{R}_t$ as follows (remember that $\mathcal{P}_{\ker \mathcal{T}}\mathcal{T}^*G = 0$, see Theorem 2.5.72): $\mathcal{R}_t G = \sum_n F_t(\sigma_n^2)\sigma_n \langle G, v_n\rangle_{\mathcal{Y}} u_n$. If we keep in mind that $F_t(\sigma_n^2) \to \sigma_n^{-2}$ as $t \to 0+$ for all n, we can conjecture that $\mathcal{R}_t G$ tends to $\sum_n \sigma_n^{-2}\sigma_n \langle G, v_n\rangle_{\mathcal{Y}} u_n = \mathcal{T}^+ G$. To verify this conjecture, we have a look at the norm of the difference:

$$\left\|\mathcal{R}_t G - \mathcal{T}^+ G\right\|_{\mathcal{X}}^2 = \sum_n \left[F_t(\sigma_n^2)\sigma_n - \sigma_n^{-1}\right]^2 \left|\langle G, v_n\rangle_{\mathcal{Y}}\right|^2. \tag{5.45}$$

However, we need to interchange the limit $t \to 0+$ with the sum over n, which is usually a series. This interchanging is not trivial. It can be achieved, for example, by enforcing a uniform convergence with respect to t in (5.45) with the requirement that $C_F := \sup_{\sigma \in [0, \|\mathcal{T}\|_{\mathcal{L}}^2],\, t\in\mathbb{R}^+} \sigma\, |F_t(\sigma)| < +\infty$. Then

$$\begin{aligned}\left[F_t(\sigma_n^2)\sigma_n - \sigma_n^{-1}\right]^2 \left|\langle G, v_n\rangle_{\mathcal{Y}}\right|^2 &= \left[F_t(\sigma_n^2)\sigma_n^2 - 1\right]^2 \sigma_n^{-2} \left|\langle G, v_n\rangle_{\mathcal{Y}}\right|^2 \\ &\le (C_F + 1)^2 \sigma_n^{-2} \left|\langle G, v_n\rangle_{\mathcal{Y}}\right|^2,\end{aligned}$$

where

$$\sum_n \sigma_n^{-2} \left|\langle G, v_n\rangle_{\mathcal{Y}}\right|^2 = \sum_n \left|\langle \mathcal{T}^+ G, u_n\rangle_{\mathcal{X}}\right|^2 = \left\|\mathcal{T}^+ G\right\|_{\mathcal{X}}^2 < +\infty$$

for all $G \in \mathcal{D}(\mathcal{T}^+)$. Now we are able to conclude that

$$\lim_{t\to 0+} \left\|\mathcal{R}_t G - \mathcal{T}^+ G\right\|_{\mathcal{X}}^2 = \sum_n \lim_{t\to 0+} \left[F_t(\sigma_n^2)\sigma_n - \sigma_n^{-1}\right]^2 \left|\langle G, v_n\rangle_{\mathcal{Y}}\right|^2 = 0.$$

Theorem 5.3.42 *Let $\{F_t\}_{t>0}$ be a filter corresponding to $\mathcal{T} \in \mathcal{K}(\mathcal{X},\mathcal{Y})$. If there exists a constant $C_F > 0$ such that $\sigma|F_t(\sigma)| \le C_F$ for all $\sigma \in [0, \|\mathcal{T}\|_{\mathcal{L}}^2]$ and all $t > 0$, then*

$$\lim_{t\to 0+} \left\|\mathcal{R}_t G - \mathcal{T}^+ G\right\|_{\mathcal{X}} = 0 \quad \textit{for all } G \in \mathcal{D}\left(\mathcal{T}^+\right).$$

*A filter which satisfies the aforementioned conditions is called a **regularizing filter** for $\mathcal{T}$.*

Filters and particularly regularizing filters have analogues in the **wavelet analysis**, where scaling functions Φ_J, $J \in \mathbb{R}^+$, and wavelets Ψ_J, $J \in \mathbb{R}^+$, are constructed and convolutions $\Phi_J * G$ yield approximations to F^+ in the sense that $\Phi_J * G \to F^+$ as $J \to \infty$. Under appropriate constraints, $\mathcal{R}_t G := \Phi_{t^{-1}} * G$ yields a regularization based on regularizing filters. The filters $\mathcal{R}_t$ are (in the language of signal processing) **low-pass filters** because, usually, $F_t(\sigma_n^2)\sigma_n$ is closer to σ_n^{-1} for low n than for large n; see also the examples of regularizing filters that follow. Note in this context that the degree n of $Y_{n,j}$ and of $\sin(nt)$, $\cos(nt)$ can be interpreted as a frequency and the singular values σ_n are usually arranged to a monotonically decreasing sequence.

The larger the scale J, that is, the closer t is to zero, the weaker the filter. In other words, with increasing J (decreasing t), more frequencies may pass the filter. The differences in the resolutions can reveal interesting detail structures in the solution F^+ which are not visible in F^+ or $\mathcal{R}_t G$ themselves. This effect is utilized by constructing detail operators (**band-pass filters**) $\mathcal{S}_t$, which are representable by a convolution $\mathcal{S}_t G := \Psi_{t^{-1}} * G$ with the wavelet $\Psi_{t^{-1}}$. In a continuous approach,

$$\mathcal{R}_{t_2} G = \int_{t_2}^{t_1} \mathcal{S}_\tau G \, \mathrm{d}\tau + \mathcal{R}_{t_1} G \quad \text{for } 0 < t_2 < t_1.$$

The more practicable discrete wavelet analysis uses a strict monotonically decreasing sequence $(t_k)_{k\in\mathbb{N}_0}$ with $t_k \to 0$, for example, $t_k := 2^{-k}$, and corresponding so-called wavelet packets Ψ_k^{P} with operators $\mathcal{S}_k^{\mathrm{P}} G := \Psi_k^{\mathrm{P}} * G$ such that $\mathcal{R}_{t_{k+1}} G = \mathcal{S}_k^{\mathrm{P}} G + \mathcal{R}_{t_k} G$ and, consequently, $F^+ = \sum_{k=-1}^{\infty} \mathcal{S}_k^{\mathrm{P}} G$ with $\mathcal{S}_{-1}^{\mathrm{P}} := \mathcal{R}_{t_0}$. The wavelet analysis $\left\{\mathcal{S}_k^{\mathrm{P}} G\right\}_{k\in\mathbb{N}_0\cup\{-1\}}$ then decomposes F^+ into details of different resolution (e.g., spatial extent). As for the regularizing filters, a theory of regularizing wavelets could also be developed for arbitrary compact operators on Hilbert spaces. For further details, see Michel (2002a). Examples for numerical results of such a wavelet-based regularization can be found in the case of the downward continuation problem in Freeden and Schneider (1998) and Schneider (1997) and for the inverse gravimetric problem in Michel (1999, 2005).

In the theory of inverse problems, two kinds of errors are distinguished. This can be made plausible by considering the following application of the triangle inequality:

$$\left\|\mathcal{T}^+ G - \mathcal{R}_t G^\varepsilon\right\|_{\mathcal{X}} \leq \left\|\mathcal{T}^+ G - \mathcal{R}_t G\right\|_{\mathcal{X}} + \left\|\mathcal{R}_t G - \mathcal{R}_t G^\varepsilon\right\|_{\mathcal{X}}. \tag{5.46}$$

The latter term $\left\|\mathcal{R}_t G - \mathcal{R}_t G^\varepsilon\right\|_{\mathcal{X}}$ is influenced by the noise $G - G^\varepsilon$ and is also called the **data error**, though this error is measured in the solution space $\mathcal{X}$. Since $\mathcal{R}_t$ needs to be continuous, we can hope that $\left\|\mathcal{R}_t G - \mathcal{R}_t G^\varepsilon\right\|_{\mathcal{X}}$ is not too much higher than $\|G - G^\varepsilon\|_{\mathcal{Y}}$. Certainly, in practice, there can still be major differences, particularly if t is close to 0. The other error term $\left\|\mathcal{T}^+ G - \mathcal{R}_t G\right\|_{\mathcal{X}}$ is also called the **approximation error**. It measures how much we have to sacrifice when we replace the exact generalized inverse $\mathcal{T}^+$ by its regularization $\mathcal{R}_t$. While we can expect that, for t close to 0, the data error gets large, the approximation error should become small. Vice versa, for controlling the data error, we would have to move further away from the discontinuous operator $\mathcal{T}^+$, which comes along with a larger approximation error. It is one of the major challenges for practically handling inverse problems to find an 'ideal' regularization parameter t such that both types of errors remain acceptably small. There are numerous parameter choice methods with their own specific pros and cons. It would be beyond the scope of this book to discuss them in detail. The reader is referred to, for example, Bauer et al. (2015), Engl et al. (1996), Gutting et al. (2017), Hanke (2017), and Rieder (2003) and the references therein as well as the very brief discussions on hand after Theorem 5.3.52.

There is a known estimate for the data error – in the solution space and the data space – which will be recapitulated here.

Theorem 5.3.43 *Let $\{F_t\}_{t>0}$ be a regularizing filter for $\mathcal{T} \in \mathcal{K}(\mathcal{X}, \mathcal{Y})$ and let $C_F > 0$ be the constant from the requirements on the filter, that is, $\sigma\, |F_t(\sigma)| \leq C_F$ for all $\sigma \in \left[0, \|\mathcal{T}\|_{\mathcal{L}}^2\right]$ and all $t > 0$. Moreover, let $M(t) := \sup_{\sigma \in [0, \|\mathcal{T}\|_{\mathcal{L}}^2]} |F_t(\sigma)|$ for each $t > 0$. If $G, G^\varepsilon \in \mathcal{Y}$ with $\|G - G^\varepsilon\|_{\mathcal{Y}} \leq \varepsilon$, then*

$$\left\|\mathcal{R}_t G - \mathcal{R}_t G^\varepsilon\right\|_{\mathcal{X}} \leq \varepsilon \sqrt{C_F M(t)}, \tag{5.47}$$

$$\left\|\mathcal{T}\mathcal{R}_t G - \mathcal{T}\mathcal{R}_t G^\varepsilon\right\|_{\mathcal{Y}} \leq C_F \varepsilon. \tag{5.48}$$

Proof We start with the latter estimate.

(1) Estimate in the data space:

With Theorems 5.3.17 and 2.5.72 and the functional calculus, it is easy to see the following commutativity: for all $y \in \mathcal{Y}$, we have

$$\begin{aligned} F_t\left(\mathcal{T}^*\mathcal{T}\right)\mathcal{T}^* y &= F_t\left(\mathcal{T}^*\mathcal{T}\right) \sum_n \sigma_n \langle y, v_n \rangle_{\mathcal{Y}}\, u_n \\ &= \sum_n F_t\left(\sigma_n^2\right) \sigma_n \langle y, v_n \rangle_{\mathcal{Y}}\, u_n + F_t(0) \underbrace{\mathcal{P}_{\ker \mathcal{T}} \mathcal{T}^* y}_{=0} \\ &= \mathcal{T}^* \sum_n F_t\left(\sigma_n^2\right) \langle y, v_n \rangle_{\mathcal{Y}}\, v_n + \underbrace{\mathcal{T}^*\left(F_t(0) \mathcal{P}_{\ker \mathcal{T}^*} y\right)}_{=0} \\ &= \mathcal{T}^* F_t\left(\mathcal{T}\mathcal{T}^*\right) y. \end{aligned} \tag{5.49}$$

Hence, we get

$$\begin{aligned} \left\|\mathcal{T}\mathcal{R}_t G - \mathcal{T}\mathcal{R}_t G^\varepsilon\right\|_{\mathcal{Y}} &= \left\|\mathcal{T} F_t\left(\mathcal{T}^*\mathcal{T}\right)\mathcal{T}^* G - \mathcal{T} F_t\left(\mathcal{T}^*\mathcal{T}\right)\mathcal{T}^* G^\varepsilon\right\|_{\mathcal{Y}} \\ &\leq \left\|\mathcal{T} F_t\left(\mathcal{T}^*\mathcal{T}\right)\mathcal{T}^*\right\|_{\mathcal{L}} \cdot \left\|G - G^\varepsilon\right\|_{\mathcal{Y}} \\ &= \left\|\mathcal{T}\mathcal{T}^* F_t\left(\mathcal{T}\mathcal{T}^*\right)\right\|_{\mathcal{L}} \cdot \left\|G - G^\varepsilon\right\|_{\mathcal{Y}} \\ &= \left\|\tilde{F}_t\left(\mathcal{T}\mathcal{T}^*\right)\right\|_{\mathcal{L}} \cdot \left\|G - G^\varepsilon\right\|_{\mathcal{Y}}, \end{aligned}$$

where $\tilde{F}_t(\sigma) := \sigma F_t(\sigma)$ and Theorem 5.3.20 (together with Theorem 2.5.59) tells us that $\|\tilde{F}_t(\mathcal{T}\mathcal{T}^*)\|_{\mathcal{L}} \leq C_F$. This yields (5.48).

(2) Estimate in the solution space:

By using the underlying inner product, the linearity of $\mathcal{R}_t$, and again (5.49) as well as Theorems 2.5.59 and 2.5.8, we obtain

$$\begin{aligned} \left\|\mathcal{R}_t G - \mathcal{R}_t G^\varepsilon\right\|_{\mathcal{X}}^2 &= \left\langle \mathcal{R}_t G - \mathcal{R}_t G^\varepsilon, \mathcal{R}_t G - \mathcal{R}_t G^\varepsilon \right\rangle_{\mathcal{X}} \\ &= \left\langle \mathcal{R}_t\left(G - G^\varepsilon\right), F_t\left(\mathcal{T}^*\mathcal{T}\right)\mathcal{T}^*\left(G - G^\varepsilon\right)\right\rangle_{\mathcal{X}} \\ &= \left\langle \mathcal{R}_t\left(G - G^\varepsilon\right), \mathcal{T}^* F_t\left(\mathcal{T}\mathcal{T}^*\right)\left(G - G^\varepsilon\right)\right\rangle_{\mathcal{X}} \\ &= \left\langle \mathcal{T}\mathcal{R}_t\left(G - G^\varepsilon\right), F_t\left(\mathcal{T}\mathcal{T}^*\right)\left(G - G^\varepsilon\right)\right\rangle_{\mathcal{Y}} \\ &\leq \left\|\mathcal{T}\mathcal{R}_t\left(G - G^\varepsilon\right)\right\|_{\mathcal{Y}} \cdot \left\|F_t\left(\mathcal{T}\mathcal{T}^*\right)\right\|_{\mathcal{L}} \cdot \left\|G - G^\varepsilon\right\|_{\mathcal{Y}}. \end{aligned}$$

With (5.48) and Theorem 5.3.20, we finally get (5.47), that is,

$$\left\| \mathcal{R}_t G - \mathcal{R}_t G^\varepsilon \right\|_{\mathcal{X}} \le \sqrt{C_F \varepsilon \cdot M(t) \cdot \varepsilon} = \varepsilon \sqrt{C_F M(t)}. \qquad \square$$

Let us come back to the decomposition of the error in (5.46). From Theorem 5.3.42, we know that the approximation error tends to 0 as $t \to 0+$ if $\mathcal{R}_t$ is built with a regularizing filter. The data error remains critical in the limit $t \to 0+$, but Theorem 5.3.43 shows us that it is sufficient to have $\lim_{\varepsilon \to 0+} \varepsilon \sqrt{M(\gamma(\varepsilon))} = 0$ for an a priori parameter choice γ. This leads us immediately to the following corollary.

Corollary 5.3.44 *Provided that all conditions in Theorem 5.3.43 are fulfilled and* $\gamma\colon \mathbb{R}^+ \to \mathbb{R}^+$ *satisfies* $\lim_{\varepsilon \to 0+} \gamma(\varepsilon) = 0$ *and* $\lim_{\varepsilon \to 0+} \varepsilon \sqrt{M(\gamma(\varepsilon))} = 0$, *then* $(\{\mathcal{R}_t\}_{t>0}, \gamma)$ *with* $\mathcal{R}_t := F_t(\mathcal{T}^*\mathcal{T})\mathcal{T}^*$ *is a* ***regularization method*** *for* $\mathcal{T}^+$.

We briefly quote a known result on order optimality in the context of filters and the associated concept of qualifications.

Theorem 5.3.45 *Let* $\{F_t\}_{t>0}$ *be a regularizing filter for* $\mathcal{T} \in \mathcal{K}(\mathcal{X}, \mathcal{Y})$ *and let* $p_t(\sigma) := 1 - \sigma F_t(\sigma)$. *The filter and* $\mu > 0$ *are chosen such that a constant* $t_0 > 0$ *and a function* $\omega_\mu\colon]0, t_0] \to \mathbb{R}$ *exist with*

$$\sup_{\sigma \in [0, \|\mathcal{T}\|_{\mathcal{L}}^2]} \left(\sigma^{\mu/2} \, |p_t(\sigma)|\right) \le \omega_\mu(t) \quad \textit{for all } t \in]0, t_0] \tag{5.50}$$

and $\omega_\mu(t) = \mathcal{O}\left(t^{\mu/2}\right)$ *as* $t \to 0+$. *Moreover,* $M(t) := \sup_{\sigma \in [0, \|\mathcal{T}\|_{\mathcal{L}}^2]} |F_t(\sigma)|$, $t \in \mathbb{R}^+$, *needs to satisfy* $M(t) = \mathcal{O}(t^{-1})$ *as* $t \to 0+$. *Furthermore, we use an a priori parameter choice* $\gamma\colon \mathbb{R}^+ \to \mathbb{R}^+$ *which allows the existence of constants* $C_1, C_2 > 0$ *such that*

$$C_1 \left(\frac{\varepsilon}{\varrho}\right)^{2/(\mu+1)} \le \gamma(\varepsilon) \le C_2 \left(\frac{\varepsilon}{\varrho}\right)^{2/(\mu+1)}$$

for sufficiently small but arbitrary $\varepsilon, \varrho > 0$. *Then* $(\{\mathcal{R}_t\}_{t>0}, \gamma)$ *with the operators* $\mathcal{R}_t := F_t(\mathcal{T}^*\mathcal{T})\mathcal{T}^*$ *is an* ***order optimal regularization method*** *for* $\mathcal{T}^+$ *with respect to* $\mathcal{X}_\mu$.

The phrase 'the filter and $\mu > 0$ are chosen such that ...' should not be forgotten here. Since μ refers to the source condition in $\mathcal{X}_\mu$ in the implication of the theorem, the important question is: for which μ is (5.50) valid? This certainly depends on the filter. We can further guess that, with increasing μ, it gets harder to fulfil this condition. The 'best' μ which we can get in this respect is called the qualification of the filter.

Definition 5.3.46 Let $\{F_t\}_{t>0}$ be a regularizing filter for $\mathcal{T} \in \mathcal{K}(\mathcal{X}, \mathcal{Y})$ for which $M(t) := \sup_{\sigma \in [0, \|\mathcal{T}\|_{\mathcal{L}}^2]} |F_t(\sigma)|$ satisfies $M(t) = \mathcal{O}(t^{-1})$ as $t \to 0+$. Moreover, let $p_t(\sigma) := 1 - \sigma F_t(\sigma)$. Then the number

$$\sup \left\{ \mu_0 \in \mathbb{R}^+ \,\middle|\, \forall \mu \in]0, \mu_0] \colon \sup_{\sigma \in [0, \|\mathcal{T}\|_{\mathcal{L}}^2]} \left(\sigma^{\mu/2} \, |p_t(\sigma)|\right) = \mathcal{O}\left(t^{\mu/2}\right) \text{as } t \to 0+ \right\} \tag{5.51}$$

is called the **qualification** of the filter.

Remember Definition 5.3.39. Order optimality means (here) that we have $\mathrm{E}_\mu(\varepsilon, \varrho, \mathcal{R}_\gamma) \le C_\mu\, \varrho^{1/(\mu+1)}\, \varepsilon^{\mu/(\mu+1)}$ for all $\mu \in]0, \mu_0[$ and all $\varrho > 0$. In other words, the qualification μ_0 means that the convergence rate cannot exceed $\mathcal{O}\left(\varepsilon^{\mu_0/(\mu_0+1)}\right)$ as $\varepsilon \to 0+$. In the case $\mu_0 = +\infty$, the convergence rate can get arbitrarily close to $\mathcal{O}(\varepsilon)$.

We consider now two examples of filters which are commonly used (not only) in Earth sciences.

Theorem 5.3.47 (Truncated Singular-Value Decomposition (TSVD)) *The family of functions*

$$F_t(\sigma) := \begin{cases} \frac{1}{\sigma}, & \text{if } \sigma \ge t, \\ 0, & \text{if } \sigma < t, \end{cases} \qquad \sigma \in \mathbb{R}_0^+,\ t \in \mathbb{R}^+,$$

yields a ***regularizing filter*** for every $\mathcal{T} \in \mathcal{K}(\mathcal{X}, \mathcal{Y})$ *with the qualification* $\mu_0 = +\infty$. *The corresponding regularization is obtained with*

$$\mathcal{R}_t G = \sum_{\substack{n, \\ \sigma_n^2 \ge t}} \sigma_n^{-1} \langle G, v_n \rangle_{\mathcal{Y}}\, u_n, \quad G \in \mathcal{Y}, \tag{5.52}$$

where the summation in (5.52) *always consists of only a finite number of summands.*

Proof Clearly, each F_t is piecewise continuous with only one discontinuity, while the left- and right-hand limits exist at this point $\sigma_0 = t$. Moreover, for each fixed $\sigma \in \mathbb{R}^+$, we have $\lim_{t \to 0+} F_t(\sigma) = \sigma^{-1}$. Furthermore, $\sigma |F_t(\sigma)| \le 1$ for all $\sigma \in \mathbb{R}_0^+$ and all $t \in \mathbb{R}^+$. Hence, $\{F_t\}_{t>0}$ is a regularizing filter for every $\mathcal{T} \in \mathcal{K}(\mathcal{X}, \mathcal{Y})$. The corresponding regularization is constructed via

$$\mathcal{R}_t G = F_t\left(\mathcal{T}^*\mathcal{T}\right)\mathcal{T}^* G = \sum_n F_t\left(\sigma_n^2\right)\sigma_n \langle G, v_n \rangle_{\mathcal{Y}}\, u_n = \sum_{\substack{n, \\ \sigma_n^2 \ge t}} \frac{\sigma_n}{\sigma_n^2} \langle G, v_n \rangle_{\mathcal{Y}}\, u_n,$$

which yields (5.52). For a compact operator $\mathcal{T}$, the singular values $(\sigma_n)_n$ are either a finite sequence or tend to zero. In both cases, there are only finitely many σ_n with $\sigma_n^2 \ge t$ for a fixed $t \in \mathbb{R}^+$.

Eventually, we determine the qualification, which is also an easy task for the TSVD. We have $M(t) = \sup_{\sigma \in [0, \|\mathcal{T}\|_{\mathcal{L}}^2]} |F_t(\sigma)| = \sup_{\sigma \in [t, \|\mathcal{T}\|_{\mathcal{L}}^2]} \sigma^{-1} = t^{-1}$ for all $t \in \mathbb{R}^+$ such that $M(t) = \mathcal{O}(t^{-1})$ as $t \to 0+$ is obviously true. For

$$\sigma^{\mu/2} |p_t(\sigma)| = \sigma^{\mu/2} |1 - \sigma F_t(\sigma)| = \begin{cases} 0, & \text{if } \sigma \ge t, \\ \sigma^{\mu/2}, & \text{if } \sigma < t, \end{cases}$$

we observe that

$$\sup_{\sigma \in \left[0, \|\mathcal{T}\|_{\mathcal{L}}^2\right]} \left(\sigma^{\mu/2} |p_t(\sigma)|\right) \le \sup_{\sigma \in [0, t]} \sigma^{\mu/2} = t^{\mu/2}$$

for all $\mu > 0$. Hence, the qualification is $\mu_0 = +\infty$. □

Equation (5.52) explains the name of this regularization method. We take the svd of the Moore–Penrose inverse $\mathcal{T}^+ G = \sum_n \sigma_n^{-1} \langle G, v_n \rangle_{\mathcal{Y}}\, u_n$, $G \in \mathcal{D}(\mathcal{T}^+)$, and cut out all summands (i.e., we truncate the expansion) where the coefficients σ_n^{-1} are larger than the threshold $t^{-1/2}$. The closer t gets to 0, the fewer summands are omitted, which can be expected to reduce the approximation error. On the other hand, including terms with very large σ_n^{-1} means that noise in the associated coefficients $\langle G^\varepsilon, v_n\rangle_\mathcal{Y}$ of the data G^ε gains essentially more influence and the data error might grow.

Let us have a look now at the other regularization method which we discuss here. While the TSVD is a 'hard' way of regularizing, since some coefficients $\langle \mathcal{T}^+ G, u_n\rangle_\mathcal{X} = \sigma_n^{-1}\langle G, v_n\rangle_\mathcal{Y}$ are taken just as they are and the others are replaced by 0, the Tikhonov–Phillips regularization provides us with a 'smoother' alternative.

Theorem 5.3.48 (Tikhonov–Phillips (TP) Regularization) *The family of functions* $F_t(\sigma) := (\sigma + t)^{-1}$, $\sigma \in \mathbb{R}_0^+$, $t \in \mathbb{R}^+$, *yields a* ***regularizing filter*** *for every* $\mathcal{T} \in \mathcal{K}(\mathcal{X},\mathcal{Y})$, *where the filter has the qualification* $\mu_0 = 2$. *The corresponding regularization is obtained with*

$$\mathcal{R}_t G = \sum_n \frac{\sigma_n}{\sigma_n^2 + t} \langle G, v_n\rangle_\mathcal{Y}\, u_n, \quad G \in \mathcal{Y}. \tag{5.53}$$

Proof Every F_t is continuous and, for each $\sigma \in \mathbb{R}^+$, we get $\lim_{t\to 0+} F_t(\sigma) = \sigma^{-1}$. Furthermore, $\sigma\,|F_t(\sigma)| = \sigma(\sigma+t)^{-1} \le 1$ for all $\sigma \in \mathbb{R}_0^+$ and all $t \in \mathbb{R}^+$. Hence, we have a regularizing filter, where

$$\mathcal{R}_t G = \sum_n F_t\left(\sigma_n^2\right)\sigma_n \langle G, v_n\rangle_\mathcal{Y}\, u_n = \sum_n \frac{\sigma_n}{\sigma_n^2 + t} \langle G, v_n\rangle_\mathcal{Y}\, u_n$$

for all $t \in \mathbb{R}^+$, which proves (5.53). All these conclusions are possible for every $\mathcal{T} \in \mathcal{K}(\mathcal{X},\mathcal{Y})$.

Concerning the qualification, we first observe that, for all $\mathcal{T} \in \mathcal{K}(\mathcal{X},\mathcal{Y})$,

$$M(t) = \sup_{\sigma\in\left[0,\|\mathcal{T}\|_\mathcal{L}^2\right]} |F_t(\sigma)| = F_t(0) = \frac{1}{t} = \mathcal{O}\left(t^{-1}\right) \quad \text{as } t \to 0+$$

and

$$p_t(\sigma) = 1 - \sigma F_t(\sigma) = \frac{\sigma + t - \sigma}{\sigma + t} = \frac{t}{\sigma + t}.$$

We now consider the quotient $\sigma^{\mu/2}\,|p_t(\sigma)| = \sigma^{\mu/2} t/(\sigma + t)$, which is relevant for (5.51), and distinguish two cases.

If $\mu > 2$, then

$$\frac{\partial}{\partial\sigma}\frac{\sigma^{\mu/2} t}{\sigma + t} = \frac{1}{2}\,\frac{\mu\sigma^{\mu/2-1}t(\sigma+t) - 2\sigma^{\mu/2}t}{(\sigma+t)^2} = \frac{\sigma^{\mu/2}}{2}\,\frac{t(\mu-2) + \sigma^{-1}\mu t^2}{(\sigma+t)^2} > 0.$$

The implied strict monotonicity with respect to σ yields

$$\sup_{\sigma\in\left[0,\|\mathcal{T}\|_\mathcal{L}^2\right]} \left(\sigma^{\mu/2}\,|p_t(\sigma)|\right) = \frac{\|\mathcal{T}\|_\mathcal{L}^\mu}{\|\mathcal{T}\|_\mathcal{L}^2 + t}\, t \begin{cases} = \mathcal{O}\left(t^{2/2}\right), & \\ \neq \mathcal{O}\left(t^\alpha\right), & \text{if } \alpha > 1 \end{cases}$$

as $t \to 0+$ *for all* $\mu > 2$. Hence, the qualification cannot be larger than 2.

If now $\mu \leq 2$, then

$$\sup_{\sigma \in [0, \|\mathcal{T}\|_{\mathcal{L}}^2]} \left(\sigma^{\mu/2}\, |p_t(\sigma)|\right) = \sup_{\sigma \in [0, \|\mathcal{T}\|_{\mathcal{L}}^2]} \left[t^{\mu/2}\, \frac{(\sigma/t)^{\mu/2}}{\sigma/t+1}\right] \leq t^{\mu/2} \underbrace{\sup_{r \in \mathbb{R}^+} \frac{r^{\mu/2}}{r+1}}_{=:C_\mu}.$$

Since $\mu \leq 2$, we have that C_μ is finite.

Combining both cases, we are able to conclude that the filter has the qualification $\mu_0 = 2$. □

This result implies that we can expect a convergence rate of $\mathcal{O}(\varepsilon^{2/3})$ as $\varepsilon \to 0+$ for the TP regularization, while the TSVD can approach $\mathcal{O}(\varepsilon)$. However, we have here an example where theoretical results need not coincide with practical numerical experiences. The convergence rate $\mathcal{O}(\varepsilon^{\mu_0/(\mu_0+1)})$ is only one criterion for the behaviour of a regularization method. The various real-world inverse problems and different requirements on the solution – depending, for example, on the way how the error $F^+ - \mathcal{R}_t G^\varepsilon$ is quantified and which specific properties (such as smoothness, interpretability, or more particular aspects such as known characteristics of the real but in general unknown solution F) are expected – often paint a different picture. There are many cases where the TP regularization outperforms the TSVD. Besides, there are many other regularization methods which are preferred for specific inverse problems and requirements on the solution. Some of them are constructed by imposing side conditions such as the minimization of a certain functional. We (maybe recklessly) neglect such approaches here and apologize this again with reasons of brevity. However, we remark, because this will also be important later on, that the TP regularization is equivalent to imposing a particular minimization side condition.

Theorem 5.3.49 *Let $\mathcal{T} \in \mathcal{K}(\mathcal{X}, \mathcal{Y})$. While $F^+ = \mathcal{T}^+ G$ for $G \in \mathcal{D}(\mathcal{T}^+)$ is known to be (see Theorems 5.3.5 and 5.3.8) the minimum-norm solution of the normal equation*

$$\mathcal{T}^*\mathcal{T} F = \mathcal{T}^* G$$

and one minimizer of the residual $\|\mathcal{T}F - G\|_{\mathcal{Y}}$, the approximation $\mathcal{R}_t G^\varepsilon$ obtained by the TP regularization for arbitrary right-hand sides $G^\varepsilon \in \mathcal{Y}$ is the unique solution of the ***Tikhonov–Phillips (TP)-regularized normal equation***

$$\left(\mathcal{T}^*\mathcal{T} + t\mathcal{I}\right) F = \mathcal{T}^* G^\varepsilon, \tag{5.54}$$

where $\mathcal{I}$ is the identity operator on $\mathcal{X}$. Moreover, $(\mathcal{T}^*\mathcal{T} + t\mathcal{I})^{-1}$ is continuous for every $t > 0$ and $F_t^\varepsilon := \mathcal{R}_t G^\varepsilon$ is the unique minimizer of the ***Tikhonov–Phillips (TP) functional***

$$J_{t, G^\varepsilon}(F) := \left\|\mathcal{T}F - G^\varepsilon\right\|_{\mathcal{Y}}^2 + t\|F\|_{\mathcal{X}}^2.$$

Proof Let $t > 0$ throughout the whole proof and let $\mathcal{X}$ and $\mathcal{Y}$ be Hilbert spaces with respect to the field $\mathbb{K} \in \{\mathbb{R}, \mathbb{C}\}$.

(1) The regularized normal equation:

From (5.53) and Theorem 5.3.17, we see that

$$(\mathcal{T}^*\mathcal{T} + t\mathcal{I})\,\mathcal{R}_t G^\varepsilon = \sum_n (\sigma_n^2 + t)\frac{\sigma_n}{\sigma_n^2 + t}\langle G^\varepsilon, v_n\rangle_{\mathcal{Y}}\, u_n = \mathcal{T}^* G^\varepsilon$$

for all $G^\varepsilon \in \mathcal{Y}$. Moreover, if $(\mathcal{T}^*\mathcal{T} + t\mathcal{I})F = 0$ for an $F \in \mathcal{X}$, then

$$0 = \langle(\mathcal{T}^*\mathcal{T} + t\mathcal{I})\,F, F\rangle_{\mathcal{X}} = \langle\mathcal{T}^*\mathcal{T}F, F\rangle_{\mathcal{X}} + t\langle F, F\rangle_{\mathcal{X}} = \|\mathcal{T}F\|_{\mathcal{Y}}^2 + t\|F\|_{\mathcal{X}}^2$$

due to Theorem 2.5.59. Hence, $F = 0$ and $\mathcal{T}^*\mathcal{T} + t\mathcal{I}$ is injective. Furthermore, if $H \in \mathcal{X}$ is arbitrary, then

$$\left|\frac{1}{\sigma_n^2 + t}\langle H, u_n\rangle_{\mathcal{X}}\right| \le \frac{1}{t}\,|\langle H, u_n\rangle_{\mathcal{X}}| \quad \text{for all } n,$$

while $\sum_n |\langle H, u_n\rangle_{\mathcal{X}}|^2 < +\infty$. Thus,

$$\tilde{F}_t := \sum_n \frac{1}{\sigma_n^2 + t}\langle H, u_n\rangle_{\mathcal{X}}\, u_n + \frac{1}{t}\,\mathcal{P}_{\ker\mathcal{T}} H$$

exists in $\mathcal{X}$ and satisfies the equation $(\mathcal{T}^*\mathcal{T} + t\mathcal{I})\tilde{F}_t = H$. Hence, the operator $\mathcal{T}^*\mathcal{T} + t\mathcal{I} : \mathcal{X} \to \mathcal{X}$ is bijective, and consequently $(\mathcal{T}^*\mathcal{T} + t\mathcal{I})^{-1}$ exists and is continuous due to the inverse mapping theorem (Theorem 2.5.41).

(2) The minimization of the Tikhonov–Phillips functional:

For the functional J_{t,G^ε}, we can derive an analogous formula as follows (where $\alpha \in \mathbb{K}$ and $d \in \mathcal{X}$):

$$\begin{aligned}
&J_{t,G^\varepsilon}\left(F_t^\varepsilon + \alpha d\right)\\
&\quad= \langle\mathcal{T}\left(F_t^\varepsilon + \alpha d\right) - G^\varepsilon, \mathcal{T}\left(F_t^\varepsilon + \alpha d\right) - G^\varepsilon\rangle_{\mathcal{Y}} + t\langle F_t^\varepsilon + \alpha d, F_t^\varepsilon + \alpha d\rangle_{\mathcal{X}}\\
&\quad= \langle\mathcal{T}F_t^\varepsilon - G^\varepsilon, \mathcal{T}F_t^\varepsilon - G^\varepsilon\rangle_{\mathcal{Y}} + t\langle F_t^\varepsilon, F_t^\varepsilon\rangle_{\mathcal{X}}\\
&\qquad+ |\alpha|^2\left(\langle\mathcal{T}d, \mathcal{T}d\rangle_{\mathcal{Y}} + t\langle d, d\rangle_{\mathcal{X}}\right)\\
&\qquad+ \bar{\alpha}\langle\mathcal{T}F_t^\varepsilon - G^\varepsilon, \mathcal{T}d\rangle_{\mathcal{Y}} + \alpha\langle\mathcal{T}d, \mathcal{T}F_t^\varepsilon - G^\varepsilon\rangle_{\mathcal{Y}}\\
&\qquad+ t\bar{\alpha}\langle F_t^\varepsilon, d\rangle_{\mathcal{X}} + t\alpha\langle d, F_t^\varepsilon\rangle_{\mathcal{X}}\\
&\quad= \underbrace{\|\mathcal{T}F_t^\varepsilon - G^\varepsilon\|_{\mathcal{Y}}^2 + t\,\|F_t^\varepsilon\|_{\mathcal{X}}^2}_{=J_{t,G^\varepsilon}(F_t^\varepsilon)} + |\alpha|^2\left(\|\mathcal{T}d\|_{\mathcal{Y}}^2 + t\|d\|_{\mathcal{X}}^2\right)\\
&\qquad+ 2\Re\left[\alpha\langle d, \mathcal{T}^*\mathcal{T}F_t^\varepsilon - \mathcal{T}^*G^\varepsilon + tF_t^\varepsilon\rangle_{\mathcal{X}}\right],
\end{aligned} \tag{5.55}$$

where $\Re$ stands for the real part and $\bar{\alpha}$ is the complex conjugate of α. Let us now only consider coefficients $\alpha \in \mathbb{R}$ for a moment to ensure the existence of the following derivative. If F_t^ε is a minimizer of J_{t,G^ε}, then

$$0 = \left.\left(\frac{\partial}{\partial\alpha} J_{t,G^\varepsilon}\left(F_t^\varepsilon + \alpha d\right)\right)\right|_{\alpha=0} = 2\Re\left[\langle d, \mathcal{T}^*\mathcal{T}F_t^\varepsilon - \mathcal{T}^*G^\varepsilon + tF_t^\varepsilon\rangle_{\mathcal{X}}\right]$$

must hold true for all $d \in \mathcal{X}$. If $\mathbb{K} = \mathbb{C}$, then we add the following: since the latter identity must remain true, if we replace d by $\mathrm{i}d$, we also have for the imaginary part that $2\Im[\langle d, \mathcal{T}^*\mathcal{T}F_t^\varepsilon - \mathcal{T}^*G^\varepsilon + tF_t^\varepsilon\rangle_{\mathcal{X}}] = 0$ for all $d \in \mathcal{X}$. Hence, if $\mathbb{K} = \mathbb{R}$ or $\mathbb{K} = \mathbb{C}$, then $\langle d, \mathcal{T}^*\mathcal{T}F_t^\varepsilon - \mathcal{T}^*G^\varepsilon + tF_t^\varepsilon\rangle_{\mathcal{X}} = 0$ for all $d \in \mathcal{X}$, that is, $\mathcal{T}^*\mathcal{T}F_t^\varepsilon - \mathcal{T}^*G^\varepsilon + tF_t^\varepsilon = 0$, which equals (5.54).

Vice versa, if F_t^ε solves the regularized normal equation (5.54), then (5.55), now again with $\alpha \in \mathbb{K}$, simplifies to

$$J_{t,G^\varepsilon}\left(F_t^\varepsilon + \alpha d\right) = J_{t,G^\varepsilon}\left(F_t^\varepsilon\right) + |\alpha|^2\left(\|\mathcal{T}d\|_{\mathcal{Y}}^2 + t\|d\|_{\mathcal{X}}^2\right).$$

Obviously, $J_{t,G^\varepsilon}(F) \geq J_{t,G^\varepsilon}(F_t^\varepsilon)$ for all $F \in \mathcal{X}$, where '$=$' holds true if and only if $d := F - F_t^\varepsilon \in \mathcal{X}$ (with $\alpha = 1$) satisfies $\|\mathcal{T}d\|_{\mathcal{Y}}^2 + t\|d\|_{\mathcal{X}}^2 = 0$, where the latter is equivalent to $d = 0$, since $t > 0$. □

Note that a **generalized Tikhonov–Phillips functional** of the form

$$J_{t,G^\varepsilon}(F) := \left\|\mathcal{T}F - G^\varepsilon\right\|_{\mathcal{Y}}^2 + t\|\mathcal{S}F\|_{\mathcal{Z}}^2$$

can be found in the literature (see, e.g., Rieder, 2003, section 4.1), where $\mathcal{S} \in \mathcal{L}(\mathcal{X},\mathcal{Z})$ is an a priori chosen operator with a continuous inverse. In this case, the regularized normal equation becomes $(\mathcal{T}^*\mathcal{T} + t\mathcal{S}^*\mathcal{S})F = \mathcal{T}^*G^\varepsilon$. We will restrict our considerations here to the case $\mathcal{S} = \mathcal{I}$, which was discussed in Theorem 5.3.49.

There is a very elaborated theory on the Tikonov–Phillips regularization. We only cite here a small selection of theorems and omit their proofs for reasons of brevity.

Theorem 5.3.50 *Let $\mathcal{T} \in \mathcal{K}(\mathcal{X},\mathcal{Y})$ and let $(\{\mathcal{R}_t\}_{t>0},\gamma)$ be given by the Tikhonov–Phillips filter (see Theorem 5.3.48) together with a parameter choice $\gamma\colon \mathbb{R}^+ \to \mathbb{R}^+$ which satisfies*

$$\lim_{\varepsilon\to 0+}\gamma(\varepsilon) = 0 \qquad \textit{and} \qquad \lim_{\varepsilon\to 0+}\frac{\varepsilon}{\sqrt{\gamma(\varepsilon)}} = 0.$$

Then $(\{\mathcal{R}_t\}_{t>0},\gamma)$ is a regularization method for $\mathcal{T}^+$.

Theorem 5.3.51 *Let $\tau, c > 0$ be constants and let $G \in \mathcal{D}(\mathcal{T}^+)$ be the right-hand side of the inverse problem $\mathcal{T}F = G$, $\mathcal{T} \in \mathcal{K}(\mathcal{X},\mathcal{Y})$, with the restriction that its minimum-norm solution needs to satisfy $F^+ \in \mathcal{X}_2$ with $\|F^+\|_2 \leq \tau$. If we define an a priori parameter choice by $\gamma(\varepsilon) := c(\varepsilon/\tau)^{2/3}$, then the Tikonov–Phillips regularized approximation*

$$F_{\gamma(\varepsilon)}^\varepsilon := \left(\mathcal{T}^*\mathcal{T} + \gamma(\varepsilon)\mathcal{I}\right)^{-1}\mathcal{T}^*G^\varepsilon$$

for $\|G^\varepsilon - G\|_{\mathcal{Y}} \leq \varepsilon$ satisfies

$$\|F_{\gamma(\varepsilon)}^\varepsilon - F^+\|_{\mathcal{X}} \leq \left(\frac{1}{2\sqrt{c}} + c\right)\tau^{1/3}\varepsilon^{2/3},$$

that is, order optimality is achieved.

See Kirsch (1996, Theorem 2.12c) for a proof. Note that $F^+ \in \mathcal{X}_2$ and $\|F^+\|_2 \leq \tau$ represent **source conditions** on G here; see Definition 5.3.32 and Theorem 5.3.33. This is, however, a rather theoretical result confirming the optimal convergence rate $\mathcal{O}(\varepsilon^{2/3})$ for $\varepsilon \to 0+$, since we will, in practice, often not be able to estimate τ before the calculation of F^+. For this reason, we present here a short look at some parameter choice strategies for the TP regularization.

Theorem 5.3.52 *Let $G \in \mathcal{T}(\mathcal{X})$, $t > 0$, and $\mathcal{T} \in \mathcal{K}(\mathcal{X},\mathcal{Y})$. If*

$$F_t := \left(\mathcal{T}^*\mathcal{T} + t\mathcal{I}\right)^{-1} \mathcal{T}^* G \quad \textit{for all } t > 0,$$

then the dependence of F_t on t is continuous. Moreover, as $t \to 0+$, the norm $\|F_t\|_{\mathcal{X}}$ is monotonically increasing and the data misfit $\|\mathcal{T}F_t - G\|_{\mathcal{Y}}$ is monotonically decreasing, where $\lim_{t\to 0+} \mathcal{T}F_t = G$ *and* $\lim_{t\to +\infty} F_t = 0$. *If $\mathcal{T}^*G \neq 0$, then both monotonicities are strict.*

The latter theorem is useful for determining a regularization parameter t in practice. It tells us that, by choosing t very close to 0, we can fulfil the equation $\mathcal{T}F = G$ very accurately, while F_t might have a very large norm, that is, it might be very oscillatory and we might have under-regularized the problem. On the other hand, by choosing t too large, we might over-regularize the problem, that is, the data misfit can be large, F_t might be too smooth, and interesting details in F could not be visible any more in F_t.

At this point, two common parameter choice strategies should be mentioned. One strategy is the **L-curve method** (see Hansen, 1992, 1998; and Hansen and O'Leary, 1993), where a diagram with $\log \|F_t^\varepsilon\|_{\mathcal{X}}$ on one axis and $\log \|\mathcal{T}F_t^\varepsilon - G^\varepsilon\|_{\mathcal{Y}}$ on the other axis are plotted (sometimes, the logarithms are omitted). In an ideal situation such as in Figure 5.2, one observes bottom-left a 'kink'. At this point, $\|F_t^\varepsilon\|_{\mathcal{X}}$ and $\|\mathcal{T}F_t^\varepsilon - G^\varepsilon\|_{\mathcal{Y}}$ are both acceptably small. The corresponding parameter t is then chosen. However, one does not always obtain a clear kink, as shown in Figure 5.3. Moreover, it turned out to be problematic to find theoretical results on a general convergence behaviour of the L-curve method; see, for example, Vogel (1996). Nevertheless, the L-curve method is popular and turned out to produce useful results for a series of practical applications. Long-term experiences, also by the author, suggest that there is no 'perfect and universal' parameter choice method. It can significantly vary from problem to problem which particular parameter choice method is eventually the best performing one.

The **discrepancy principle** (see Morozov, 1966, 1967, 1984; and Phillips, 1962) recommends to search for a parameter t which satisfies

$$\left\|\mathcal{T}F_t^\varepsilon - G^\varepsilon\right\|_{\mathcal{Y}} = \varepsilon \tag{5.56}$$

(or, at least, approximately). The idea is that, in the presence of a noise level $\|G - G^\varepsilon\|_{\mathcal{Y}} \leq \varepsilon$, it does not make sense to fulfil the equation $\mathcal{T}F = G^\varepsilon$ with a higher accuracy than ε. Due to Theorem 5.3.52, we have the monotonic limits

$$0 = \left\|G^\varepsilon - G^\varepsilon\right\|_{\mathcal{Y}} \xleftarrow{t\to 0+} \left\|\mathcal{T}F_t^\varepsilon - G^\varepsilon\right\|_{\mathcal{Y}} \xrightarrow{t\to +\infty} \left\|\mathcal{T}0 - G^\varepsilon\right\|_{\mathcal{Y}} = \left\|G^\varepsilon\right\|_{\mathcal{Y}}.$$

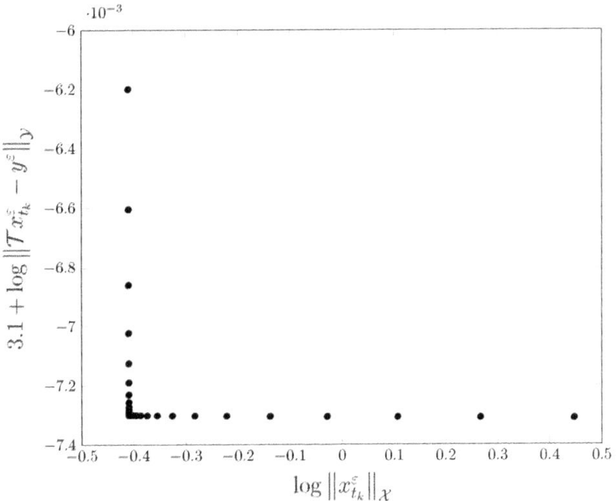

Figure 5.2 As a simple example, the system of linear equations $\mathcal{T}x = y$ is used for $\mathcal{T} = \begin{pmatrix} -1 & 1 \\ -2 & 2-10^{-6} \end{pmatrix}$, $y = (-1, -2)^{\mathrm{T}} \in \mathbb{R}^2 =: \mathcal{Y}$, and the noise vector $\tilde{\varepsilon} := (0.1, 0.1)^{\mathrm{T}}$ such that $y^{\varepsilon} := y + \tilde{\varepsilon}$. Then the TP-regularized approximation $x^{\varepsilon}_{t_k} := \left(\mathcal{T}^*\mathcal{T} + t_k\mathcal{I}\right)^{-1}\mathcal{T}^*y^{\varepsilon} \in \mathbb{R}^2 =: \mathcal{X}$ is computed for $t_k = 10^{-8+k/10}$, $k = 0, \ldots, 60$, and the values $\log \left\|x^{\varepsilon}_{t_k}\right\|_{\mathcal{X}}$ and $\log \left\|\mathcal{T}x^{\varepsilon}_{t_k} - y^{\varepsilon}\right\|_{\mathcal{Y}}$ are plotted. A typical L-shape can be seen in the diagram. The L-curve method suggests the use of the parameter t_k which corresponds to the kink in the 'L'.

If we have $\|G^{\varepsilon}\|_{\mathcal{Y}} > \varepsilon$, that is, the signal-to-noise ratio is larger than 1, then, assuming $\mathcal{T}^*G^{\varepsilon} \neq 0$ (otherwise, $F^{\varepsilon}_t = 0$ for all t), we can use the intermediate value theorem to conclude that there is exactly one $t > 0$ which satisfies (5.56). It can be shown that, under some conditions, the discrepancy principle for the TP regularization can achieve an order of convergence $\mathcal{O}(\varepsilon^{1/2})$ as $\varepsilon \to 0+$ but not more (see, e.g., Kirsch, 1996, theorems 2.17 and 2.18). Often, the discrepancy principle is modified for practical purposes in the sense that a monotonically decreasing sequence $(t_k)_k$ with $t_k \to 0$ is used and an index k^* is chosen such that

$$\left\|\mathcal{T}F^{\varepsilon}_{t_{k^*}} - G^{\varepsilon}\right\|_{\mathcal{Y}} \leq \alpha\varepsilon < \left\|\mathcal{T}F^{\varepsilon}_{t_{k^*-1}} - G^{\varepsilon}\right\|_{\mathcal{Y}}. \tag{5.57}$$

Here, $\alpha > 1$ is a factor which is selected based on a particular heuristic. Under certain constraints, this procedure yields order optimality with respect to X_{μ} for $\mu \leq \mu_0 - 1$ in the case of filters with a qualification $\mu_0 > 1$ (including the TP filter). The best possible convergence order in this respect is here, again, $\mathcal{O}\left(\varepsilon^{1/2}\right)$ if we set $\mu = \mu_0 - 1$ and $\mu_0 = 2$ for the TP regularization (see Theorem 5.3.48). Indeed, one can show that, in the case of the TP filter, a stronger convergence order than $\mathcal{O}\left(\varepsilon^{1/2}\right)$ in the presence of the discrepancy principle (5.57) can only be achieved if $\dim \mathcal{T}(\mathcal{X}) < +\infty$. For further details, see, for example, Rieder (2003, section 3.4).

For this reason, alternatives to the discrepancy principle were considered in the literature. One of these options is the **generalized discrepancy principle** (see Engl and Gfrerer,

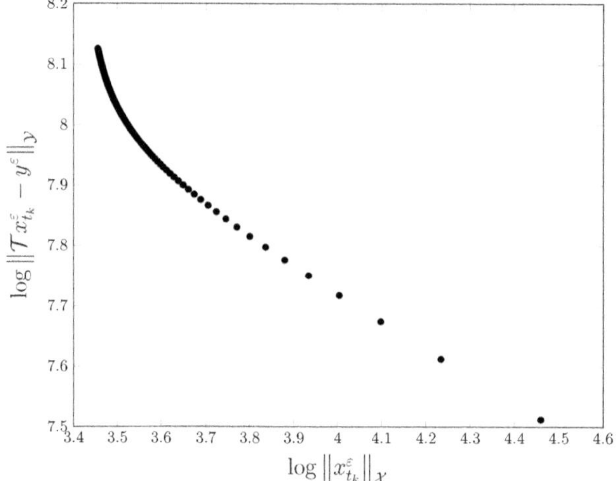

Figure 5.3 Points $\xi_1, \ldots, \xi_n$ with $n := 2{,}026$ are chosen on a sphere with radius 1 together with contrived masses $x := (x_1, \ldots, x_n)^{\mathrm{T}} \in \mathbb{R}^n =: \mathcal{X}$. Then the norm of the gravitational force vector (up to a constant factor) $y_i = \sum_{j=1}^{n} x_j/|\xi_j - \eta_i|^2$, $i = 1, \ldots, m := 2{,}502$ is computed at some points η_j on a sphere with radius 6,971/6,371 (i.e., an orbit of 600 km altitude). The data $y = \mathcal{T}x \in \mathbb{R}^m =: \mathcal{Y}$ are contaminated with normally distributed random noise $\tilde{\varepsilon} \in \mathcal{Y}$ such that $y^{\varepsilon} := y + \tilde{\varepsilon}$. Note that, due to the finite dimensions, $\mathcal{T}$ can be represented by a matrix. Eventually, the TP-regularized approximation $x^{\varepsilon}_{t_k} := (\mathcal{T}^*\mathcal{T} + t_k\mathcal{I})^{-1}\mathcal{T}^* y^{\varepsilon}$ is computed for $t_k = 0.0025 + 0.0025\,k$, $k = 0, \ldots, 199$ and the values $\log \|x^{\varepsilon}_{t_k}\|_{\mathcal{X}}$ and $\log \|\mathcal{T}x^{\varepsilon}_{t_k} - y^{\varepsilon}\|_{\mathcal{Y}}$ are plotted. The identification of the location of the kink is here not obvious.

1988; Hanke and Engl, 1994; Raus, 1985). In this case, a family of positive and piecewise continuous functions $s_t : \left[0, \|\mathcal{T}\|^2_{\mathcal{L}}\right] \to \mathbb{R}$, $t \in \mathbb{R}^+$, is chosen such that

$$C_1 \left(\frac{t}{\sigma^2 + t}\right)^{\mu_0+1} \le s_t(\sigma^2) \le C_2 \left(\frac{t}{\sigma^2 + t}\right)^{\mu_0+1}$$

for all $t \in \mathbb{R}^+$ and all $\sigma \in [0, \|\mathcal{T}\|_{\mathcal{L}}]$ and some positive constants C_1 and C_2. Then, based again on a sequence $(t_k)_k$ and a factor $\tau > \sup_{\sigma,t} s_t(\sigma^2)$, an index k^* is chosen such that

$$t_{k^*} = \sup\left\{t_k \,\middle|\, \left\langle G^{\varepsilon}, s_{t_k}\left(\mathcal{T}\mathcal{T}^*\right) G^{\varepsilon}\right\rangle_{\mathcal{Y}} \le \tau\varepsilon^2\right\}.$$

One can show that, again under some constraints, this parameter choice yields order optimality but now with respect to X_μ for all $\mu \in]0, \mu_0]$ if $\mu_0 > 0$ is the qualification of the filter. Hence, the convergence order $\mathcal{O}(\varepsilon^{2/3})$ can be achieved for the TP regularization in the case $\mu = 2\,(= \mu_0)$ if the generalized discrepancy principle is applied.

As we previously mentioned, there are numerous other parameter choice strategies, which we cannot discuss here for reasons of brevity. Nevertheless, they all have their own particular use and justification; see the references listed in this section for further details.

5.4 Best Bases for Ill-Posed Inverse Problems

In Section 4.9, the best basis algorithms FMP, OFMP, and WFMP were explained. These algorithms start with a dictionary $\mathcal{D} \subset \mathcal{X} \setminus \ker \mathcal{T}$ of possibly suitable functions and choose an (in some sense) optimal selection $d_1, \ldots, d_N \in \mathcal{D}$ and corresponding coefficients $\alpha_1, \ldots, \alpha_N \in \mathbb{R}$ such that $F_N = \sum_{k=1}^{N} \alpha_k d_k$ is a sufficiently close approximation to a solution of an inverse problem $\mathcal{T}F = y$. These algorithms are appropriate if the inverse problem is well-posed. For ill-posed problems, we need to introduce a regularization into the algorithms. Moreover, it can also be reasonable for well-posed problems to include a regularization, for example if large amounts of data are used or if the data are sampled at a very scattered point set.

The Tikhonov–Phillips-regularized counterparts, or better generalizations (the parameter choice $\lambda = 0$ yields the unregularized versions), of the aforementioned algorithms, namely the RFMP, the ROFMP, and the RWFMP, were introduced in Fischer (2011), Fischer and Michel (2012), Kontak (2018), Kontak and Michel (2019), Michel and Telschow (2016), and Telschow (2014). In Kontak (2018) and Kontak and Michel (2019), it was pointed out that an elegant way of including such a regularization is to consider an appropriately modified problem $\widetilde{\mathcal{T}}_\lambda F = y$. For elaborating this in detail, we first need to do a little preliminary work.

We assume again throughout this section that $(\mathcal{X}, \langle\cdot,\cdot\rangle_{\mathcal{X}})$ and $(\mathcal{Y}, \langle\cdot,\cdot\rangle_{\mathcal{Y}})$ are arbitrary Hilbert spaces. The considered inverse problem is $\mathcal{T}F = y$ with $\mathcal{T} \in \mathcal{L}(\mathcal{X}, \mathcal{Y})$.

Lemma 5.4.1 *Corresponding to two arbitrary Hilbert spaces $(\mathcal{X}, \langle\cdot,\cdot\rangle_{\mathcal{X}})$ and $(\mathcal{Y}, \langle\cdot,\cdot\rangle_{\mathcal{Y}})$, we define, for all $x_1, x_2 \in \mathcal{X}$ and $y_1, y_2 \in \mathcal{Y}$ the bilinear mapping $\left\langle (x_1, y_1)^{\mathrm{T}}, (x_2, y_2)^{\mathrm{T}} \right\rangle_{\mathcal{X}\times\mathcal{Y}} := \langle x_1, x_2\rangle_{\mathcal{X}} + \langle y_1, y_2\rangle_{\mathcal{Y}}$. This mapping is an inner product on $\mathcal{X} \times \mathcal{Y}$, which becomes a Hilbert space this way.*

The validity of Lemma 5.4.1 is easy to verify. Note that the induced norm is the same as the norm in Lemma 2.5.37.

Theorem 5.4.2 *Let $\mathcal{T} \in \mathcal{L}(\mathcal{X}, \mathcal{Y})$ and $\lambda \geq 0$. We define the operator $\widetilde{\mathcal{T}}_\lambda \colon \mathcal{X} \to \mathcal{X} \times \mathcal{Y}$ by $\widetilde{\mathcal{T}}_\lambda F := \left(\sqrt{\lambda}\, F, \mathcal{T}F\right)^{\mathrm{T}}$, $F \in \mathcal{X}$. If $\lambda > 0$, then the image $\widetilde{\mathcal{T}}_\lambda(\mathcal{X})$ is closed, that is, the inverse problem $\widetilde{\mathcal{T}}_\lambda F = G$, $G \in \mathcal{X} \times \mathcal{Y}$ given, is well-posed in the sense of Nashed (see Definition 5.3.3).*

Proof Obviously, the image $\widetilde{\mathcal{T}}_\lambda(\mathcal{X})$ of $\widetilde{\mathcal{T}}_\lambda$ is the graph of the operator $\lambda^{-1/2}\mathcal{T}$, because $\mathcal{T}$ is linear. Since $\lambda^{-1/2}\mathcal{T}$ is linear and continuous, its graph is closed (see, *nomen est omen*, the closed graph theorem, i.e., Theorem 2.5.38). □

We now replace the (possibly ill-posed) inverse problem $\mathcal{T}F = y$ by the well-posed inverse problem $\widetilde{\mathcal{T}}_\lambda F = (0, y)^{\mathrm{T}}$. As a consequence, instead of minimizing $\|\mathcal{T}F - y\|_{\mathcal{Y}}^2$, we now seek to minimize

$$\left\|\widetilde{\mathcal{T}}_\lambda F - (0, y)^{\mathrm{T}}\right\|_{\mathcal{X}\times\mathcal{Y}}^2 = \lambda \|F\|_{\mathcal{X}}^2 + \|\mathcal{T}F - y\|_{\mathcal{Y}}^2. \tag{5.58}$$

This also shows again very simply the basic principle of the Tikhonov–Phillips regularization.

Let us now go through the theory of the algorithms FMP, OFMP, and WFMP, where we will study the consequences of replacing the inverse problem accordingly.

5.4.1 Regularized Functional Matching Pursuit (RFMP)

For transferring the FMP (Algorithm 4.9.1) to the problem $\widetilde{\mathcal{T}}_\lambda F = (0, y)^{\mathrm{T}}$, we need to have a look at the terms

$$\left\langle \begin{pmatrix} 0 \\ y \end{pmatrix} - \widetilde{\mathcal{T}}_\lambda F_n, \widetilde{\mathcal{T}}_\lambda d_{n+1} \right\rangle_{\mathcal{X}\times\mathcal{Y}} = \left\langle -\sqrt{\lambda}\, F_n, \sqrt{\lambda}\, d_{n+1} \right\rangle_{\mathcal{X}} + \langle y - \mathcal{T} F_n, \mathcal{T} d_{n+1} \rangle_{\mathcal{Y}}$$

and

$$\|\widetilde{\mathcal{T}}_\lambda d_{n+1}\|_{\mathcal{X}\times\mathcal{Y}} = \sqrt{\lambda\ \|d_{n+1}\|_{\mathcal{X}}^2 + \|\mathcal{T} d_{n+1}\|_{\mathcal{Y}}^2}. \tag{5.59}$$

Moreover, for $\lambda > 0$, $\widetilde{\mathcal{T}}_\lambda$ is injective, that is, $\ker \widetilde{\mathcal{T}}_\lambda = \{0\}$, while $\ker \widetilde{\mathcal{T}}_0 = \ker \mathcal{T}$. This leads us to the following algorithm.

Algorithm 5.4.3 (Regularized Functional Matching Pursuit (RFMP))
Let an operator $\mathcal{T} \in \mathcal{L}(\mathcal{X}, \mathcal{Y}) \setminus \{0\}$ and a right-hand side $y \in \mathcal{Y}$ be given.

(1) Set $n := 0$ and $F_0 := 0$ and choose a stopping criterion. Select a dictionary $\mathcal{D} \subset \mathcal{X} \setminus \{0\}$. Choose a regularization parameter $\lambda \in \mathbb{R}^+$.
(2) Find $d_{n+1} \in \mathcal{D}$ such that

$$\frac{\left(\langle y - \mathcal{T} F_n, \mathcal{T} d_{n+1}\rangle_{\mathcal{Y}} - \lambda \langle F_n, d_{n+1}\rangle_{\mathcal{X}}\right)^2}{\|\mathcal{T} d_{n+1}\|_{\mathcal{Y}}^2 + \lambda\ \|d_{n+1}\|_{\mathcal{X}}^2} \geq \frac{\left(\langle y - \mathcal{T} F_n, \mathcal{T} d\rangle_{\mathcal{Y}} - \lambda \langle F_n, d\rangle_{\mathcal{X}}\right)^2}{\|\mathcal{T} d\|_{\mathcal{Y}}^2 + \lambda\ \|d\|_{\mathcal{X}}^2} \tag{5.60}$$

for all $d \in \mathcal{D}$ and set

$$\alpha_{n+1} := \frac{\langle y - \mathcal{T} F_n, \mathcal{T} d_{n+1}\rangle_{\mathcal{Y}} - \lambda \langle F_n, d_{n+1}\rangle_{\mathcal{X}}}{\|\mathcal{T} d_{n+1}\|_{\mathcal{Y}}^2 + \lambda\ \|d_{n+1}\|_{\mathcal{X}}^2} \tag{5.61}$$

as well as $F_{n+1} := F_n + \alpha_{n+1} d_{n+1}$.
(3) If the stopping criterion is fulfilled, then F_{n+1} is the output as an approximate and regularized solution of $\mathcal{T} F = y$. Otherwise, increase n by 1 and go to step 2.

In the case $\lambda = 0$, the RFMP would become the previously discussed FMP (Algorithm 4.9.1), except that $\mathcal{D} \subset \mathcal{X} \setminus \ker \mathcal{T}$ is needed for the FMP.

We have to reconsider what an image-normalized dictionary is in this case, due to (5.59).

Definition 5.4.4 Let $\lambda \in \mathbb{R}_0^+$ be a chosen regularization parameter. A dictionary $\mathcal{D} \subset \mathcal{X} \setminus \ker \widetilde{\mathcal{T}}_\lambda$ is called an **image-normalized dictionary** (or, briefly, **i.n.d.**), if $\|\mathcal{T} d\|_{\mathcal{Y}}^2 + \lambda \|d\|_{\mathcal{X}}^2 = 1$, that is, $\|\widetilde{\mathcal{T}}_\lambda d\|_{\mathcal{X}\times\mathcal{Y}} = 1$, for all $d \in \mathcal{D}$.

Note that this definition includes our previous definition for the case $\lambda = 0$; see page 239.

If we use an i.n.d., then (5.60) and (5.61) simplify to the requirement that

$$\left(\langle y - \mathcal{T} F_n, \mathcal{T} d_{n+1}\rangle_{\mathcal{Y}} - \lambda\langle F_n, d_{n+1}\rangle_{\mathcal{X}}\right)^2 \geq \left(\langle y - \mathcal{T} F_n, \mathcal{T} d\rangle_{\mathcal{Y}} - \lambda\langle F_n, d\rangle_{\mathcal{X}}\right)^2 \tag{5.62}$$

for all $d \in \mathcal{D}$ and the definition

$$\alpha_{n+1} := \langle y - \mathcal{T} F_n, \mathcal{T} d_{n+1}\rangle_{\mathcal{Y}} - \lambda\langle F_n, d_{+1}\rangle_{\mathcal{X}}. \tag{5.63}$$

Example 5.4.5 We consider the same scenario of an interpolation problem from Example 4.9.2 again, but this time we use the RFMP. The regularization term uses the norm of $\mathcal{X} = \mathcal{H}_2(\Omega)$; see Definition 4.7.11. Several regularization parameters were tested. For some parameters, the truncation after the maximal number of iterations applied because of an insufficient decay of the data error. In the following, we only considered those parameters for which the truncation based on the data error occurred. Eventually that parameter was used which yielded the lowest relative approximation error, that is, we assume an ideal parameter choice rule. This optimal parameter is $\lambda = 10^{-10}\|y\|_{\mathcal{Y}}$. Like the FMP, the RFMP also stops at a residual with $\|R^n\|_{\mathcal{Y}}/\|y\|_{\mathcal{Y}} \approx 0.05$, where 46, 131 iterations were necessary here. The relative approximation error is 0.000205 and is, therefore, slightly lower than in the case of the FMP. The analogues of Figures 4.26 and 4.27 are shown in Figures 5.4 and 5.5. Again we obtain a very good approximation. By construction, the regularization as part of the RFMP has a smoothing effect to the calculated approximations, which, however, is not visible here in this theoretically stable problem. According to the experiences of the Geomathematics Groups, for such typical geomathematical problems, it nevertheless appears to be the best practice to use the RFMP instead of the FMP, at least as soon as the problem becomes relatively high dimensional or ill-posed.

Example 5.4.6 Since the RFMP is specifically designed for ill-posed inverse problems, we also demonstrate its performance for such a scenario. We consider the downward continuation problem, which is exponentially ill-posed and therefore particularly challenging, as discussed in Section 5.1. The example is designed in analogy to the interpolation problem in Examples 4.9.2 and 5.4.5. The differences are as follows: the data y_i are obtained by evaluating EGM2008 at the same Reuter grid as before, but now at an orbit height of 500 km (for the formula of upward continuation, see Definition 5.1.1), that is, we assume grid-based space-borne data. Then noise is added as in the case of the interpolation problem. With the upward continuation operator $\mathcal{T}^{\mathrm{up}}$ (more precisely, the version with a discretized image: $F \mapsto \left(\mathcal{T}^{\mathrm{up}}\left(\sigma\xi^{(i)}\right)\right)_{i=1,\ldots,l}$) as the forward operator $\mathcal{T}$, we run the RFMP in analogy to the previous interpolation tests. We pick again the optimal regularization parameter out of a finite selection and choose $\lambda = 10^{-9}\|y\|_{\mathcal{Y}}$.

Also here, the algorithm stops when the residual reaches the 5% threshold, which occurs after 957 iterations. Due to the ill-posed nature of the inverse problem, the relative approximation error 0.000466 is higher than in the case of terrestrial data, but it is still very low, considering the exponential ill-posedness. The result is shown in Figures 5.6 and 5.7. Also these images show that locally the error increases in

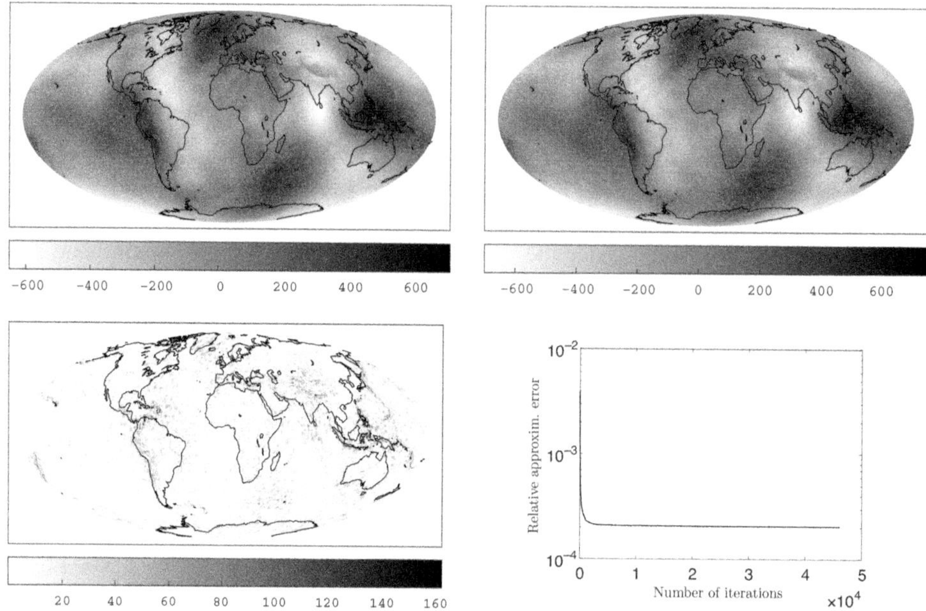

Figure 5.4 The RFMP is used to interpolate grid-based data from EGM2008. The exact solution (given by EGM2008) is shown at the top on the left-hand side. The obtained approximation by the RFMP (top-right) is very close to the exact solution. The absolute difference (bottom-left) shows that only small deviations, particularly in areas with fine local details, occur. These deviations become visible on the finer plot grid but are not captured by the coarser data grid, in particular in the presence of noise, as it was the case for this demonstrating example. All values of these plots are in $\mathrm{m}^2\mathrm{s}^{-2}$. Moreover, the curve on the right-hand side at the bottom shows the rapid decay of the relative approximation error during the iterations.

comparison to the pure interpolation problem, which can be expected. It is remarkable that these errors remain local. They occur mainly in regions with a very fine, 'low-wavelength' signal. Such 'high-frequency' parts are particularly smoothened by $\mathcal{T}^{\mathrm{up}}$ and are, therefore, hardly extractable from the data at the orbit, especially if the data are contaminated with noise. In this respect, the approximation of the RFMP can be considered as very good. Moreover, the solution is smoother in comparison to the solution of the interpolation problem: the ill-posedness also resulted in a larger regularization parameter such that non-smooth updates of the approximations are more strongly penalized for the sake of the stability of the numerical result. The smoothness of the approximation can particularly be seen if the contributions of the radial basis functions are compared, as shown in the bottom images in Figures 5.5 and 5.7. Fewer anomalies are visible now.

Let us now investigate the theoretical properties of the RFMP. Since the residual satisfies (5.58), we can apply Lemmata 4.9.3 and 4.9.4 in order to immediately obtain the following propositions.

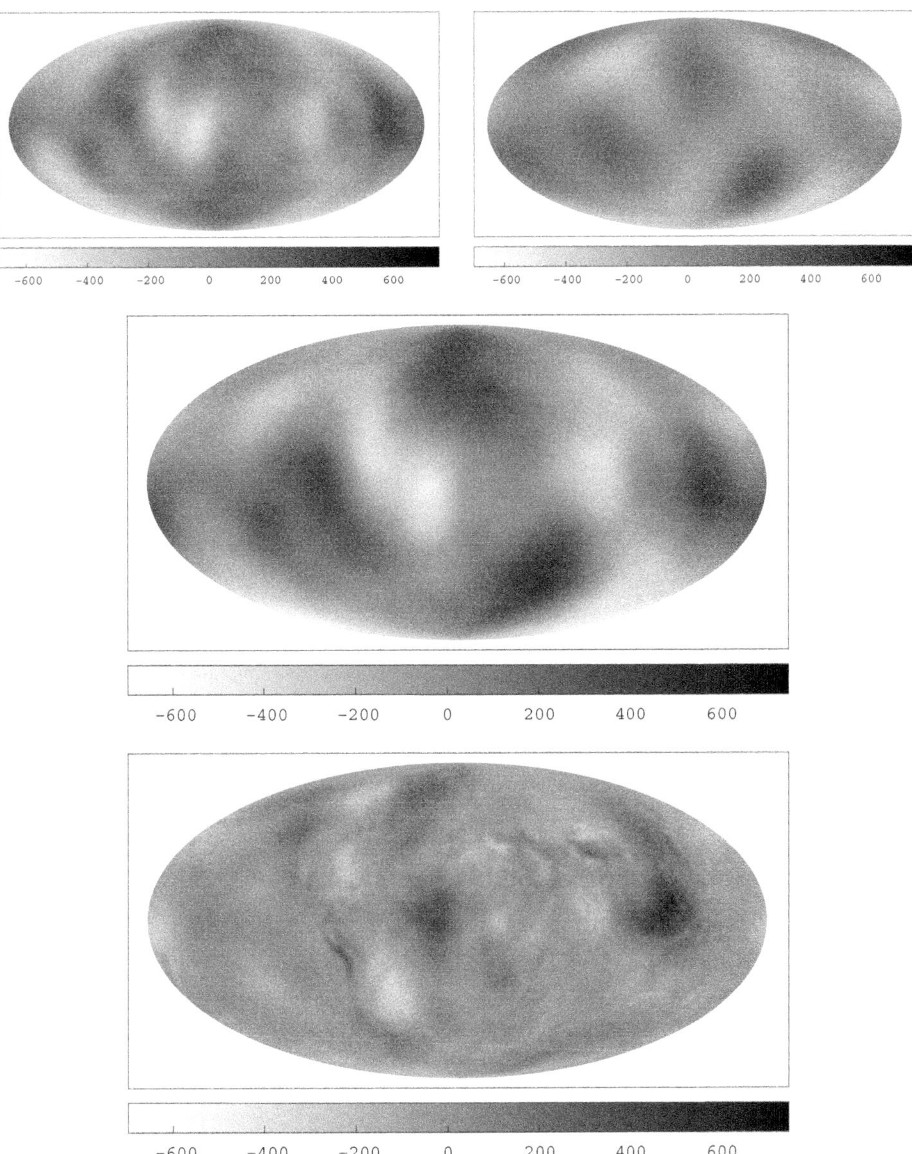

Figure 5.5 The dictionary elements which are chosen by the RFMP in Example 5.4.5 (in analogy to Example 4.9.2) are separated: we have a look at the spherical harmonics alone (top-left), the Slepian functions alone (top-right) and their sum (middle). These basis functions provide a coarse approximation of the large global structures of the solution, while the sum of all chosen radial basis functions (bottom), that is, the Abel–Poisson kernels and wavelets, reveals more localized structures in the gravitational field such as the signals due to the Andes or the Himalayas.

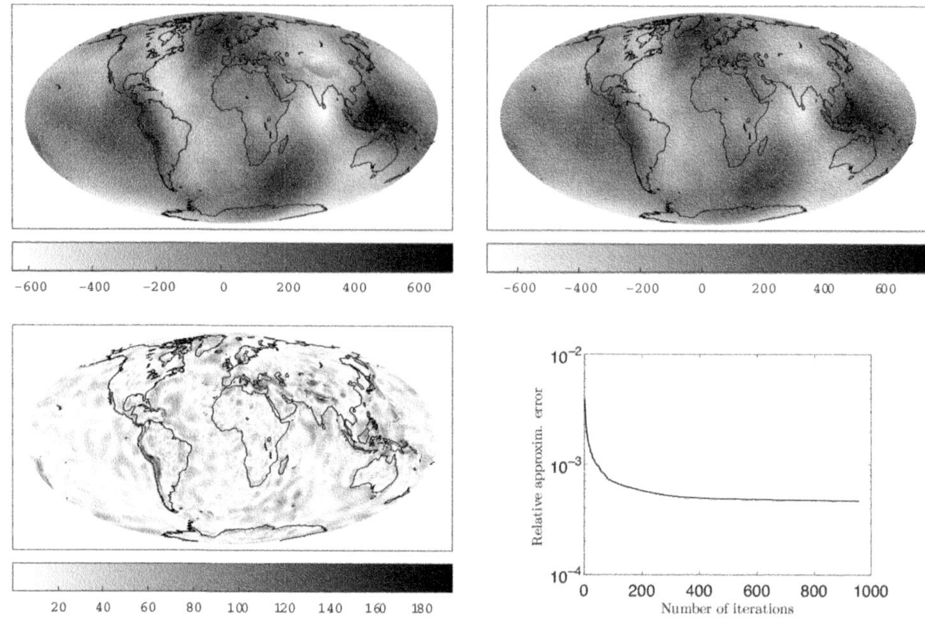

Figure 5.6 The RFMP is used for the downward continuation of grid-based data from EGM2008 at an orbit height of 500 km. The images are arranged as in Figure 5.4. Considering the exponentially ill-posed nature of downward continuation and the artificial noise on the data, the approximation by the RFMP is very close to the exact solution.

Lemma 5.4.7 *Let the sequence* $(F_n)_n$ *be produced by the RFMP (Algorithm 5.4.3). Then the corresponding sequence of norms of the regularized residuals* $\left(\|y - \mathcal{T} F_n\|_{\mathcal{Y}}^2 + \lambda \|F_n\|_{\mathcal{X}}^2\right)_n$ *is monotonically decreasing.*

Lemma 5.4.8 *Let the sequence* $(F_n)_n$ *be produced by the RFMP (Algorithm 5.4.3). Then the sequence of corresponding norms of the regularized residuals* $\left(\|y - \mathcal{T} F_n\|_{\mathcal{Y}}^2 + \lambda \|F_n\|_{\mathcal{X}}^2\right)_n$ *is convergent.*

Also Lemma 4.9.7 is applicable to $\widetilde{\mathcal{T}}_\lambda$, and it provides us with a very valuable result.

Lemma 5.4.9 *Let the sequence* $(F_n)_n$ *be produced by the RFMP (Algorithm 5.4.3). Then the sequence of regularized residuals* $\left(\left(\sqrt{\lambda}\, F_n, \mathcal{T} F_n\right)^{\mathrm{T}} - (0, y)^{\mathrm{T}}\right)_n$ *is convergent in* $\mathcal{X} \times \mathcal{Y}$*. Moreover, the sequence* $(\alpha_n)_n$ *is square-summable.*

Theorem 5.4.10 *Every sequence* $(F_n)_n$ *which is produced by the RFMP (Algorithm 5.4.3) with* $\lambda > 0$ *is convergent in* $\mathcal{X}$*.*

Proof Lemma 5.4.9 says that the sequence $\left(\left(\sqrt{\lambda}\, F_n, \mathcal{T} F_n - y\right)^{\mathrm{T}}\right)_n$ strongly converges in $\mathcal{X} \times \mathcal{Y}$, that is, there is a so far unknown limit $(\tilde{F}, \tilde{y}) \in \mathcal{X} \times \mathcal{Y}$ such that

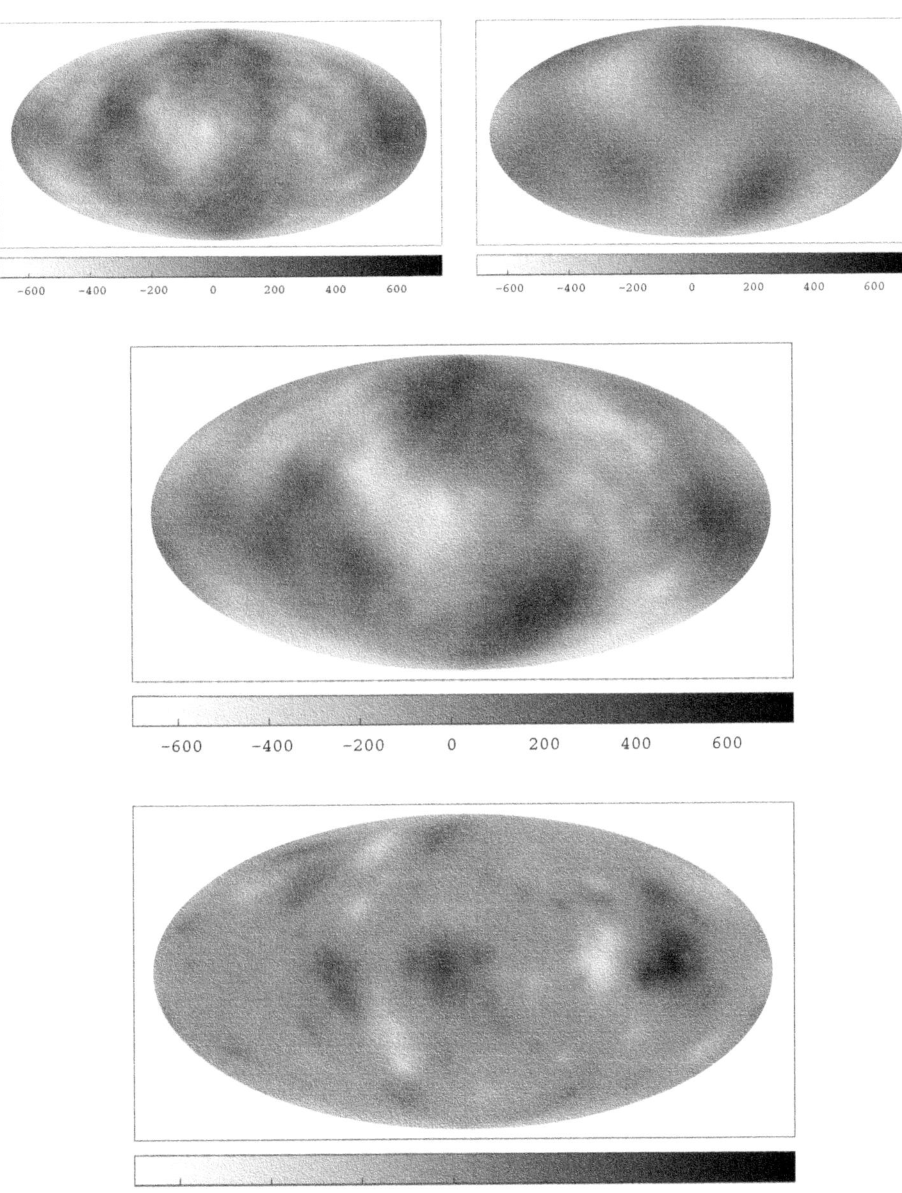

Figure 5.7 In analogy to Figure 5.5, we decompose the approximate solution of the downward continuation problem in Example 5.4.6 and Figure 5.6 into bandlimited functions (top-left: spherical harmonics; top-right: Slepian functions; middle: sum of both) and non-bandlimited radial basis functions (bottom).

$$0 = \lim_{n\to\infty} \left\| \begin{pmatrix} \sqrt{\lambda}\, F_n \\ \mathcal{T} F_n - y \end{pmatrix} - \begin{pmatrix} \tilde{F} \\ \tilde{y} \end{pmatrix} \right\|^2_{\mathcal{X}\times\mathcal{Y}}$$
$$= \lim_{n\to\infty} \left(\left\| \sqrt{\lambda}\, F_n - \tilde{F} \right\|^2_{\mathcal{X}} + \|\mathcal{T} F_n - y - \tilde{y}\|^2_{\mathcal{Y}} \right).$$

A necessary conclusion is the convergence of $\left(\sqrt{\lambda} F_n\right)_n$ to $\tilde{F}$. □

Note that we *do not* need the semi-frame condition in the regularized case!

In Theorem 4.9.8, we required that the span of all $\mathcal{T}d$, $d \in \mathcal{D}$, is dense in the closure of $\mathcal{T}(\mathcal{X})$. In the case of $\widetilde{\mathcal{T}}_\lambda$, the image is already closed (see Theorem 5.4.2). Moreover, we have $\widetilde{\mathcal{T}}_\lambda d = \left(\sqrt{\lambda}\, d, \mathcal{T}d\right)^{\mathrm{T}}$ for all $d \in \mathcal{D}$ by the definition of $\widetilde{\mathcal{T}}_\lambda$. Hence, it suffices if span $\mathcal{D}$ is dense in $\mathcal{X}$: if $\left(\sqrt{\lambda}\, F, \mathcal{T}F\right)^{\mathrm{T}}$ is an arbitrary element of $\widetilde{\mathcal{T}}_\lambda(\mathcal{X})$ with $F \in \mathcal{X}$, then we find, for every $\varepsilon > 0$, a corresponding $G \in \operatorname{span} \mathcal{D}$ such that $\|F - G\|_{\mathcal{X}} < \varepsilon$. Since $\mathcal{T}$ is linear and continuous, we obtain $\|\mathcal{T}F - \mathcal{T}G\|_{\mathcal{Y}} \le \|\mathcal{T}\|_{\mathcal{L}(\mathcal{X},\mathcal{Y})} \|F - G\|_{\mathcal{X}} \le \|\mathcal{T}\|_{\mathcal{L}(\mathcal{X},\mathcal{Y})}\, \varepsilon$. Hence,

$$\left\| \left(\sqrt{\lambda}\, F, \mathcal{T}F\right)^{\mathrm{T}} - \widetilde{\mathcal{T}}_\lambda G \right\|^2_{\mathcal{X}\times\mathcal{Y}} = \lambda \|F - G\|^2_{\mathcal{X}} + \|\mathcal{T}F - \mathcal{T}G\|^2_{\mathcal{Y}}$$
$$\le \left(\lambda + \|\mathcal{T}\|^2_{\mathcal{L}(\mathcal{X},\mathcal{Y})}\right) \varepsilon^2,$$

where $\widetilde{\mathcal{T}}_\lambda G$ is an element of span $\{\widetilde{\mathcal{T}}_\lambda d \mid d \in \mathcal{D}\}$ because of the linearity of $\mathcal{T}$.

Nevertheless, we cannot immediately apply Theorem 4.9.8, because it requires (formulated for the replaced problem with $\widetilde{\mathcal{T}}_\lambda$) that $(0, y)^{\mathrm{T}} \in \widetilde{\mathcal{T}}_\lambda(\mathcal{X})$. Provided that $\lambda > 0$, this is valid if and only if $y = 0$. Therefore, we cannot proceed analogously to the FMP-case any more.

With some building blocks from the convergence proof in Michel and Orzlowski (2017), we are able to prove that the limit $F_\infty \in \mathcal{X}$ of the RFMP, which must exist due to Theorem 5.4.10, is consistent with the choice of the TP functional as the objective function (see also Theorem 5.3.49).

Theorem 5.4.11 *Let a dictionary $\mathcal{D}$ with $\overline{\operatorname{span} \mathcal{D}}^{\|\cdot\|_{\mathcal{X}}} = \mathcal{X}$ be used in the RFMP (Algorithm 5.4.3) with a given right-hand side $y \in \mathcal{Y}$ and a regularization parameter $\lambda > 0$. Then the limit $F_\infty := \lim_{n\to\infty} F_n$ of the RFMP satisfies the TP-regularized normal equation*

$$\left(\mathcal{T}^*\mathcal{T} + \lambda\mathcal{I}\right) F_\infty = \mathcal{T}^* y. \tag{5.64}$$

In particular, the unregularized algorithm, that is the FMP (Algorithm 4.9.1), yields, if convergent, a limit F_∞ which satisfies the normal equation

$$\mathcal{T}^*\mathcal{T} F_\infty = \mathcal{T}^* y. \tag{5.65}$$

Note that no restrictions apply for the right-hand side y in both cases (FMP and RFMP) and the propositions for the RFMP also do not require the semi-frame condition. For the FMP, we still need to know that (F_n) converges and (5.65) requires $y \in \mathcal{D}(\mathcal{T}^+)$, see Theorem 5.3.6 and Definition 5.3.9, that is, this is necessary for the convergence of the FMP. Let us now prove Theorem 5.4.11.

Proof Again, we may assume that $\mathcal{D}$ is an i.n.d., because this does not influence the sequence (F_n). We first have a look at the residual of the inverse problem $\widetilde{\mathcal{T}}_\lambda F = (0, y)^{\mathrm{T}}$. We define $\tilde{R}_\lambda^n := (0, y)^{\mathrm{T}} - \widetilde{\mathcal{T}}_\lambda F_n = \left(-\sqrt{\lambda}\, F_n, y - \mathcal{T} F_n\right)^{\mathrm{T}}$. Note that Lemma 5.4.9 tells us that $(\alpha_n)_n$ is square-summable.

(1) A weak limit:
The square-summability can only be valid if $\alpha_n \to 0$ as $n \to \infty$. Hence, (5.63) yields $\langle y - \mathcal{T} F_n, \mathcal{T} d_{n+1}\rangle_{\mathcal{Y}} - \lambda \langle F_n, d_{n+1}\rangle_{\mathcal{X}} \to 0$ as $n \to \infty$. With (5.62), we are also able to conclude that $\langle y - \mathcal{T} F_n, \mathcal{T} d\rangle_{\mathcal{Y}} - \lambda \langle F_n, d\rangle_{\mathcal{X}} \to 0$ as $n \to \infty$ for all $d \in \mathcal{D}$. With the adjoint operator (see Theorem 2.5.59), we can equivalently write that

$$\left\langle \mathcal{T}^* y - \mathcal{T}^* \mathcal{T} F_n - \lambda F_n, d\right\rangle_{\mathcal{X}} \to 0 \tag{5.66}$$

as $n \to \infty$ for all $d \in \mathcal{D}$ (and, due to the bilinearity of an inner product, also for all $d \in \operatorname{span} \mathcal{D}$). The sequence $(F_n)_n$ converges in $\mathcal{X}$ due to Theorem 5.4.10, and $\mathcal{T}$ as well as $\mathcal{T}^*$ are continuous. Hence, the left-hand argument in the inner product of (5.66) is a bounded sequence. Consequently, we also get for every convergent sequence $(\tilde{d}_m)_m \subset \operatorname{span} \mathcal{D}$ with $d := \lim_{m\to\infty} \tilde{d}_m \in \mathcal{X}$ the identity

$$\begin{aligned}
&\lim_{n\to\infty} \left\langle \mathcal{T}^* y - \mathcal{T}^* \mathcal{T} F_n - \lambda F_n, d\right\rangle_{\mathcal{X}} \\
&\quad = \lim_{m\to\infty} \lim_{n\to\infty} \left\langle \mathcal{T}^* y - \mathcal{T}^* \mathcal{T} F_n - \lambda F_n, \tilde{d}_m\right\rangle_{\mathcal{X}} = 0.
\end{aligned}$$

(2) A strong limit:
Since the span of $\mathcal{D}$ is dense in $\mathcal{X}$, the latter result means that the sequence $\left(\mathcal{T}^* y - \mathcal{T}^* \mathcal{T} F_n - \lambda F_n\right)_n$ weakly converges to zero. As we know that the strong limit $\mathcal{T}^* y - \mathcal{T}^* \mathcal{T} F_\infty - \lambda F_\infty$ of this sequence exists, we have $\mathcal{T}^* y - \mathcal{T}^* \mathcal{T} F_\infty - \lambda F_\infty = 0$, which is equivalent to (5.64).

(3) The unregularized case:
The RFMP with $\lambda = 0$ coincides with the FMP. The only difference is that we do not automatically get the convergence of the FMP, which, therefore, occurs as an additional condition. Eventually, (5.64) with $\lambda = 0$ is the normal equation (5.65). □

Theorem 5.4.11 in combination with the well-known theory of the TP regularization (see Theorems 5.3.42, 5.3.48, and 5.3.49) yields the following essential properties of the RFMP.

Theorem 5.4.12 *Let $\mathcal{D}$ be a dictionary with $\overline{\operatorname{span} \mathcal{D}}^{\|\cdot\|_{\mathcal{X}}} = \mathcal{X}$ and let $F_{\infty,\lambda}$ be the limit obtained by the RFMP (Algorithm 5.4.3) for the right-hand side $y \in \mathcal{D}(\mathcal{T}^+)$ in dependence of the chosen regularization parameter $\lambda > 0$, where $\mathcal{T} \in \mathcal{K}(\mathcal{X}, \mathcal{Y})$. Then $\lim_{\lambda\to 0+} \| F_{\infty,\lambda} - \mathcal{T}^+ y\|_{\mathcal{X}} = 0$.*

Note that $\mathcal{D}(\mathcal{T}^+)$ is the notation for the domain of the Moore–Penrose inverse (see Definition 5.3.9) and has nothing to do with the dictionary $\mathcal{D}$.

Theorem 5.4.13 *Let $\mathcal{D}$ be a dictionary with $\overline{\operatorname{span} \mathcal{D}}^{\|\cdot\|_{\mathcal{X}}} = \mathcal{X}$ and $\mathcal{T} \in \mathcal{K}(\mathcal{X}, \mathcal{Y})$. Then the limit F_∞ of the RFMP (Algorithm 5.4.3) with $\lambda > 0$ depends continuously on the right-hand side $y \in \mathcal{Y}$.*

Hence, noise on the data has only a small influence on the approximation (i.e., the RFMP is stable). Moreover, depending on the noise level, the regularization parameter can be chosen small for small noise levels. Note also the results on TP regularization in Theorems 5.3.50 and 5.3.51 concerning the parameter choice. Furthermore, first experiments where different parameter choice strategies for the RFMP are compared were presented in Gutting et al. (2017).

5.4.2 Regularized Weak Functional Matching Pursuit (RWFMP)

In analogy to the WFMP (see Section 4.9.2), a weak formulation of the RFMP was also developed in Kontak (2018) and Kontak and Michel (2019). We will also summarize the state of the art on this algorithm here. The basic principle is again that the selection criterion for d_{n+1}, here given by (5.60), is weakened by trying to get a value of the objective function which is at least ϱ-times the supremum of it. This also has the advantage that one need not worry about the existence of a maximizer.

Algorithm 5.4.14 (Regularized Weak Functional Matching Pursuit)
Let an operator $\mathcal{T} \in \mathcal{L}(\mathcal{X},\mathcal{Y}) \setminus \{0\}$ and a right-hand side $y \in \mathcal{Y}$ be given.

(1) *Set $n := 0$ and $F_0 := 0$ and choose a stopping criterion. Select a dictionary $\mathcal{D} \subset \mathcal{X} \setminus \{0\}$. Choose a regularization parameter $\lambda \in \mathbb{R}^+$ and a real number (weakness parameter) $\varrho \in]0,1]$.*
(2) *Find $d_{n+1} \in \mathcal{D}$ such that*

$$\frac{\left(\langle y - \mathcal{T}F_n, \mathcal{T}d_{n+1}\rangle_{\mathcal{Y}} - \lambda\langle F_n, d_{n+1}\rangle_{\mathcal{X}}\right)^2}{\|\mathcal{T}d_{n+1}\|_{\mathcal{Y}}^2 + \lambda\,\|d_{n+1}\|_{\mathcal{X}}^2} \geq \varrho \frac{\left(\langle y - \mathcal{T}F_n, \mathcal{T}d\rangle_{\mathcal{Y}} - \lambda\langle F_n, d\rangle_{\mathcal{X}}\right)^2}{\|\mathcal{T}d\|_{\mathcal{Y}}^2 + \lambda\,\|d\|_{\mathcal{X}}^2}$$

for all $d \in \mathcal{D}$ and set

$$\alpha_{n+1} := \frac{\langle y - \mathcal{T}F_n, \mathcal{T}d_{n+1}\rangle_{\mathcal{Y}} - \lambda\langle F_n, d_{n+1}\rangle_{\mathcal{X}}}{\|\mathcal{T}d_{n+1}\|_{\mathcal{Y}}^2 + \lambda\,\|d_{n+1}\|_{\mathcal{X}}^2}$$

as well as $F_{n+1} := F_n + \alpha_{n+1}d_{n+1}$.
(3) *If the stopping criterion is fulfilled, then F_{n+1} is the output as an approximate and regularized solution of $\mathcal{T}F = y$. Otherwise, increase n by 1 and go to step 2.*

Some of the previous results can again be obtained analogously by simply introducing the factor ϱ in the case of the proofs for the RFMP (e.g., when we use (5.62) in the proof of Theorem 5.4.11) or by replacing $\mathcal{T}$ with $\widetilde{\mathcal{T}}_\lambda$ in the proofs for the WFMP. We then obtain the following propositions rather immediately.

Lemma 5.4.15 *Let the sequence $(F_n)_n$ be produced by the RWFMP (Algorithm 5.4.14). Then the sequence of norms of the regularized residuals $\left(\|y - \mathcal{T}F_n\|_{\mathcal{Y}}^2 + \lambda\|F_n\|_{\mathcal{X}}^2\right)_n$ is monotonically decreasing.*

Lemma 5.4.16 *Let the sequence $(F_n)_n$ be produced by the RWFMP (Algorithm 5.4.14). Then the sequence of norms of the regularized residuals $\left(\|y - \mathcal{T} F_n\|_{\mathcal{Y}}^2 + \lambda \|F_n\|_{\mathcal{X}}^2\right)_n$ is convergent.*

Lemma 5.4.17 *Let the sequence $(F_n)_n$ be produced by the RWFMP (Algorithm 5.4.14). Then the sequence of regularized residuals $\left(\left(\sqrt{\lambda}\, F_n, \mathcal{T} F_n\right)^{\mathrm{T}} - (0, y)^{\mathrm{T}}\right)_n$ is convergent in $\mathcal{X} \times \mathcal{Y}$.*

Theorem 5.4.18 *Every sequence $(F_n)_n$ which is produced by the RWFMP (Algorithm 5.4.14) with $\lambda > 0$ is convergent in $\mathcal{X}$.*

Theorem 5.4.19 *Let a dictionary $\mathcal{D}$ with $\overline{\operatorname{span} \mathcal{D}}^{\|\cdot\|_{\mathcal{X}}} = \mathcal{X}$ be used in the RWFMP (Algorithm 5.4.14) with a given right-hand side $y \in \mathcal{Y}$ and a regularization parameter $\lambda > 0$. Then the limit $F_\infty := \lim_{n\to\infty} F_n$ of the RWFMP satisfies the TP-regularized normal equation*

$$\left(\mathcal{T}^*\mathcal{T} + \lambda\mathcal{I}\right) F_\infty = \mathcal{T}^* y.$$

In particular, the unregularized algorithm, that is, the WFMP (Algorithm 4.9.17), yields, if convergent, a limit F_∞ which satisfies the normal equation

$$\mathcal{T}^*\mathcal{T} F_\infty = \mathcal{T}^* y.$$

Theorem 5.4.20 *Let $\mathcal{D}$ be a dictionary with $\overline{\operatorname{span} \mathcal{D}}^{\|\cdot\|_{\mathcal{X}}} = \mathcal{X}$ and let $F_{\infty,\lambda}$ be the limit obtained by the RWFMP (Algorithm 5.4.14) for the right-hand side $y \in \mathcal{D}(\mathcal{T}^+)$ and the regularization parameter $\lambda > 0$, where $\mathcal{T} \in \mathcal{K}(\mathcal{X}, \mathcal{Y})$. Then*

$$\lim_{\lambda\to 0+} \left\| F_{\infty,\lambda} - \mathcal{T}^+ y \right\|_{\mathcal{X}} = 0.$$

Note again that $\mathcal{D}(\mathcal{T}^+)$ is the notation for the domain of the Moore–Penrose inverse (see Definition 5.3.9) and has nothing to do with the dictionary $\mathcal{D}$.

Theorem 5.4.21 *Let $\mathcal{D}$ be a dictionary with $\overline{\operatorname{span} \mathcal{D}}^{\|\cdot\|_{\mathcal{X}}} = \mathcal{X}$ and $\mathcal{T} \in \mathcal{K}(\mathcal{X}, \mathcal{Y})$. If $\lambda > 0$, then the limit F_∞ of the RWFMP (Algorithm 5.4.14) depends continuously on the right-hand side $y \in \mathcal{Y}$.*

5.4.3 Regularized Orthogonal Functional Matching Pursuit (ROFMP)

We present here the regularized version of the OFMP, which was developed in Section 4.9.3. Much like the RFMP, the ROFMP, by updating the coefficients α_i, avoids choosing too many dictionary elements multiple times. The numerical experiments which have been done so far for some inverse problems in Earth sciences and medical imaging revealed that the ROFMP usually yields the most satisfactory numerical results among all matching pursuits.

While the FMP seeks to minimize $\| R^n - \alpha \mathcal{T} d \|_{\mathcal{Y}}$, the OFMP minimizes

$$\left\| R^n - \alpha \mathcal{P}_{\mathcal{W}_n} \mathcal{T} d \right\|_{\mathcal{Y}} = \left\| R^n - \alpha \left(\mathcal{T} d - \mathcal{P}_{\mathcal{V}_n} \mathcal{T} d \right) \right\|_{\mathcal{Y}}$$
$$= \left\| R^n - \alpha \left(\mathcal{T} d - \sum_{k=1}^{n} \beta_k^{(n)}(d) \mathcal{T} d_k \right) \right\|_{\mathcal{Y}} .$$

See Section 4.9.3 for details. Since the RFMP has the objective function $\| R^n - \alpha \mathcal{T} d \|_{\mathcal{Y}}^2 + \lambda \| F_n + \alpha d \|_{\mathcal{X}}^2$, the ROFMP now seeks to minimize

$$\left\| R^n - \alpha \left(\mathcal{T} d - \sum_{k=1}^{n} \beta_k^{(n)}(d) \mathcal{T} d_k \right) \right\|_{\mathcal{Y}}^2 + \lambda \left\| F_n + \alpha \left(d - \sum_{k=1}^{n} \beta_k^{(n)}(d) d_k \right) \right\|_{\mathcal{X}}^2 . \tag{5.67}$$

The derivation of the algorithm for choosing α_{n+1} and d_{n+1} as minimizers of this objective function is analogous to the derivation of the FMP.

Algorithm 5.4.22 (ROFMP) *Let an operator $\mathcal{T} \in \mathcal{L}(\mathcal{X}, \mathcal{Y}) \setminus \{0\}$ and a right-hand side $y \in \mathcal{Y}$ be given.*

(1) Set $n := 0$ and $F_0 := 0$ and choose a stopping criterion. Select a dictionary $\mathcal{D} \subset \mathcal{X} \setminus \{0\}$ and a regularization parameter $\lambda > 0$. Moreover, set $\mathcal{V}_0 := \{0\}$ and $\mathcal{W}_0 := \mathcal{Y}$.

(2) Calculate

$$\mathcal{P}_{\mathcal{W}_n} \mathcal{T} d = \mathcal{T} d - \sum_{j=1}^{n} \beta_j^{(n)}(d)\, \mathcal{T} d_j, \quad W^{(n)}(d) := d - \sum_{j=1}^{n} \beta_j^{(n)}(d)\, d_j \tag{5.68}$$

for each $d \in \mathcal{D}$. Find $d_{n+1} \in \mathcal{D}$ such that

$$\frac{\left(\langle y - \mathcal{T} F_n, \mathcal{P}_{\mathcal{W}_n} \mathcal{T} d_{n+1} \rangle_{\mathcal{Y}} - \lambda \langle F_n, W^{(n)}(d_{n+1}) \rangle_{\mathcal{X}} \right)^2}{\left\| \mathcal{P}_{\mathcal{W}_n} \mathcal{T} d_{n+1} \right\|_{\mathcal{Y}}^2 + \lambda \left\| W^{(n)}(d_{n+1}) \right\|_{\mathcal{X}}^2}$$
$$\geq \frac{\left(\langle y - \mathcal{T} F_n, \mathcal{P}_{\mathcal{W}_n} \mathcal{T} d \rangle_{\mathcal{Y}} - \lambda \langle F_n, W^{(n)}(d) \rangle_{\mathcal{X}} \right)^2}{\left\| \mathcal{P}_{\mathcal{W}_n} \mathcal{T} d \right\|_{\mathcal{Y}}^2 + \lambda \left\| W^{(n)}(d) \right\|_{\mathcal{X}}^2}$$

(where only $d \in \mathcal{D}$ are considered for which the preceding denominator is non-zero) for all $d \in \mathcal{D}$ and set

$$\alpha_{n+1}^{(n+1)} := \frac{\langle y - \mathcal{T} F_n, \mathcal{P}_{\mathcal{W}_n} \mathcal{T} d_{n+1} \rangle_{\mathcal{Y}} - \lambda \langle F_n, W^{(n)}(d_{n+1}) \rangle_{\mathcal{X}}}{\left\| \mathcal{P}_{\mathcal{W}_n} \mathcal{T} d_{n+1} \right\|_{\mathcal{Y}}^2 + \lambda \left\| W^{(n)}(d_{n+1}) \right\|_{\mathcal{X}}^2}.$$

With the coefficients $\beta_j^{(n)}$ from (5.68), *we also set*

$$\alpha_j^{(n+1)} := \alpha_j^{(n)} - \alpha_{n+1}^{(n+1)} \beta_j^{(n)}(d_{n+1}), \quad j = 1, \ldots, n,$$
$$F_{n+1} := \sum_{j=1}^{n+1} \alpha_j^{(n+1)} d_j,$$

$$
\begin{aligned}
R^{n+1} &:= R^n - \alpha_{n+1}^{(n+1)} \mathcal{P}_{\mathcal{W}_n} \mathcal{T} d_{n+1}, \\
\mathcal{V}_{n+1} &:= \mathcal{V}_n + \{ r \mathcal{T} d_{n+1} \,|\, r \in \mathbb{R} \}, \\
\mathcal{W}_{n+1} &:= \mathcal{V}_{n+1}^{\perp} \quad in\ \mathcal{Y}.
\end{aligned}
\tag{5.69}
$$

(3) If the stopping criterion is fulfilled, then F_{n+1} is the output as an approximate and regularized solution of $\mathcal{T}F = y$. Otherwise, increase n by 1 *and go to step 2.*

For an efficient calculation of the β_j, see Theorems 4.9.31 and 4.9.32, which are also valid for the ROFMP (where $\mathcal{T}$ remains $\mathcal{T}$ and is *not* $\widetilde{\mathcal{T}}_\lambda$).

Example 5.4.23 We apply the ROFMP to the interpolation problem in Examples 4.9.2 and 5.4.5. We pick again the optimal regularization parameter out of a finite selection and choose $\lambda = 10^{-10}\|y\|_{\mathcal{Y}}$. A restart was performed after 250 iterations each.

The truncation criteria are only modified in comparison to FMP and RFMP in the sense that the maximum number of iterations is now 6, 000. Nevertheless, also here, the algorithm stops when the residual reaches the 5% threshold. The relative approximation error for the obtained solution is 0.000212. The result is shown in Figures 5.8 and 5.9. The obtained function F_N is approximately of the same quality as

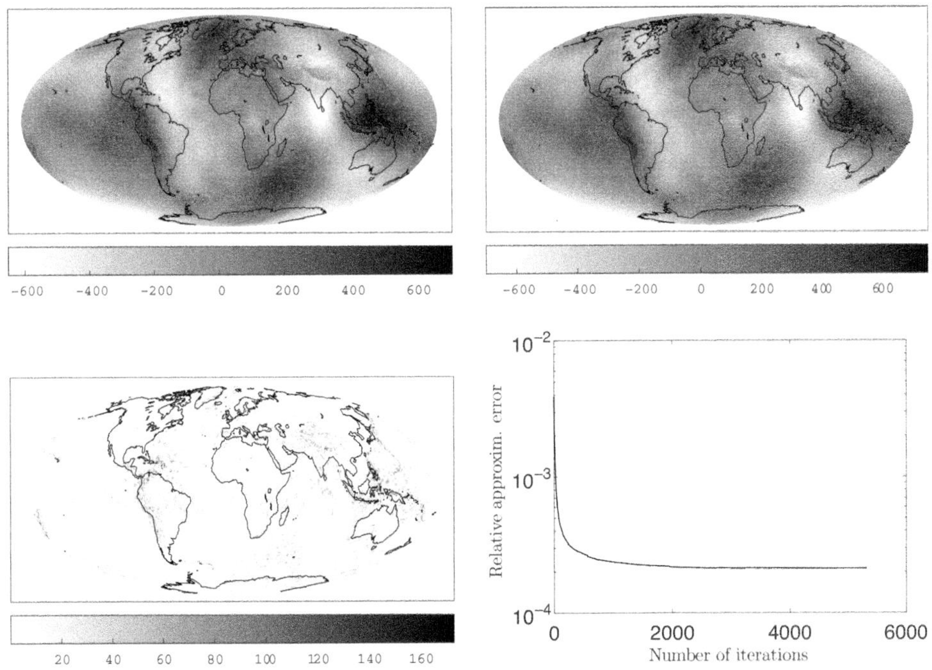

Figure 5.8 The ROFMP is used for the interpolation of grid-based data from EGM2008 (top-left) at the surface. The approximation is shown on the right-hand side of the first row. The absolute approximation error is shown on the left-hand side of the second row. The decay of the relative approximation error during the iterations is demonstrated on the right-hand side of the second row.

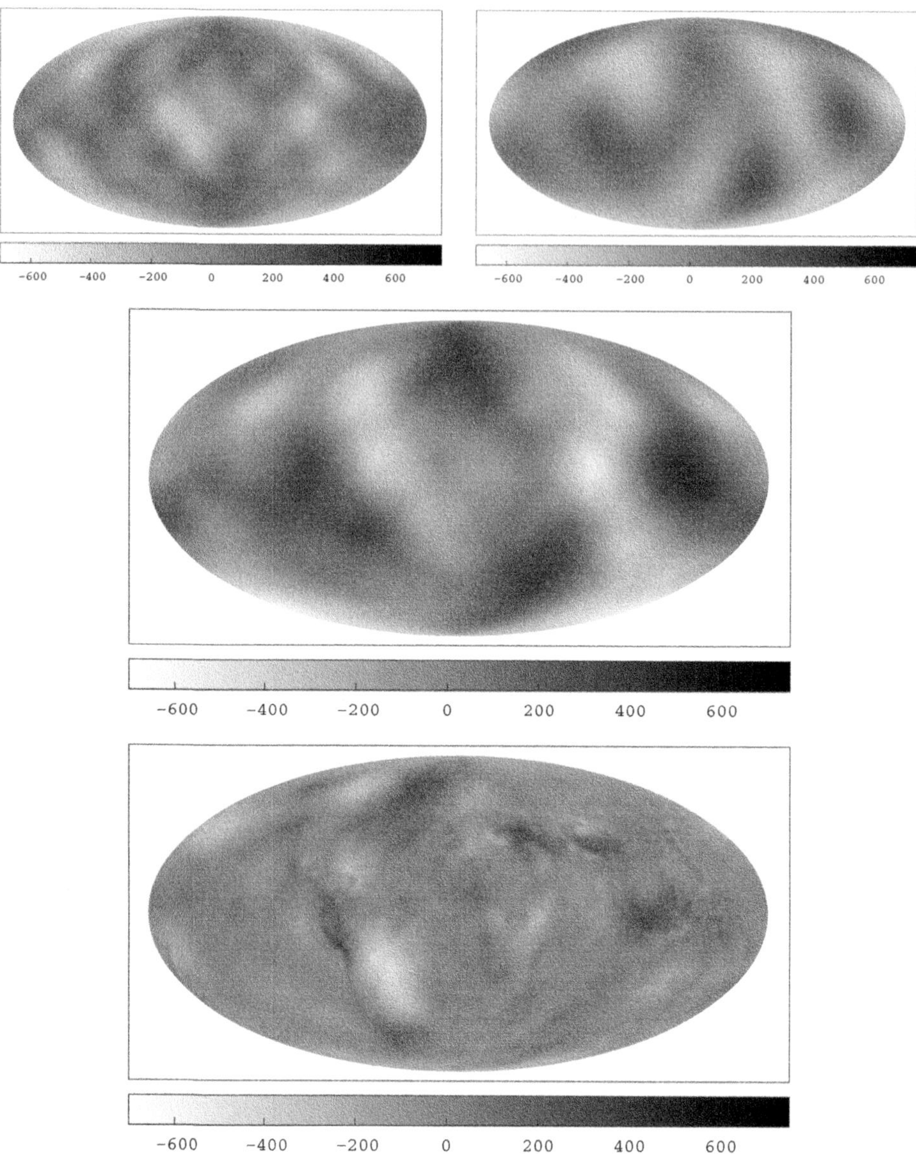

Figure 5.9 In analogy to Figure 5.5, we decompose the approximate solution of the interpolation problem in Example 5.4.23 and Figure 5.8 into bandlimited functions (top-left: spherical harmonics; top-right: Slepian functions; middle: sum of both); and non-bandlimited radial basis functions (bottom).

the result for the RFMP (see Example 5.4.5), with slight local differences. However, the RFMP needed 46, 131 iterations to obtain the approximation which reaches the 5% residual threshold, while the ROFMP only needed 5, 337 iterations, which is a reduction by 88%.

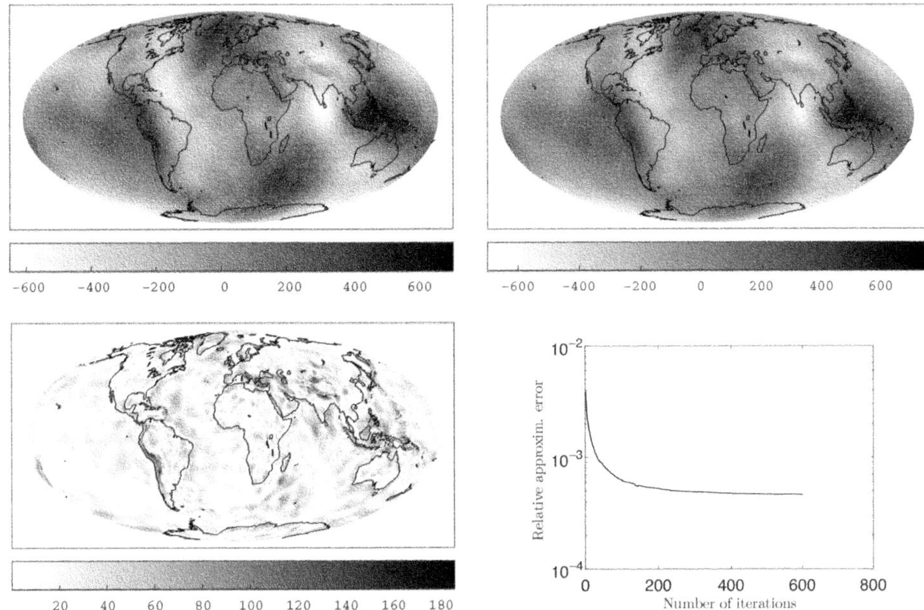

Figure 5.10 This figure shows EGM2008 (top-left), its approximation (top-right) by the ROFMP for the downward continuation of synthetic, noisy data at a 500 km orbit, and the corresponding absolute difference (bottom-left). The obtained approximation is close to the exact, noise-free solution (local structures get lost due to the exponential ill-posedness and the noise) and is obtained with essentially fewer iterations than the similar result of the RFMP; see Example 5.4.6. The decay of the relative approximation error during the iterations is demonstrated at the bottom-right.

Figure 5.9 shows also in this case that a multi-resolution analysis of the signal is possible: local details can be revealed by looking at the contributions of the radial basis functions which were chosen by the ROFMP.

Example 5.4.24 We apply the ROFMP to the same downward continuation problem as in Example 5.4.6. The choice of the regularization parameter led us to $\lambda = 10^{-9}\|y\|_{\mathcal{Y}}$. Restarts were performed after each 250th iteration. The maximum number of iterations was set to 6,000. The algorithm stopped again at the 5% threshold for the residual. The result has the relative approximation error 0.000463, which is slightly lower than for the RFMP. More importantly, the RFMP needed 957 iterations, while the ROFMP needed only 600 iterations, which is a reduction by 37%. The result is illustrated in Figures 5.10 and 5.11.

From the construction of the algorithm as a stepwise minimization of (5.67), we immediately get the following result in analogy the derivations for the previously discussed matching pursuits.

Theorem 5.4.25 *The sequence* $\left(\|R^n\|_{\mathcal{Y}}^2 + \lambda\|F_n\|_{\mathcal{X}}^2\right)_n$ *of norms of the regularized residuals produced by the ROFMP (Algorithm 5.4.22) is monotonically decreasing and convergent.*

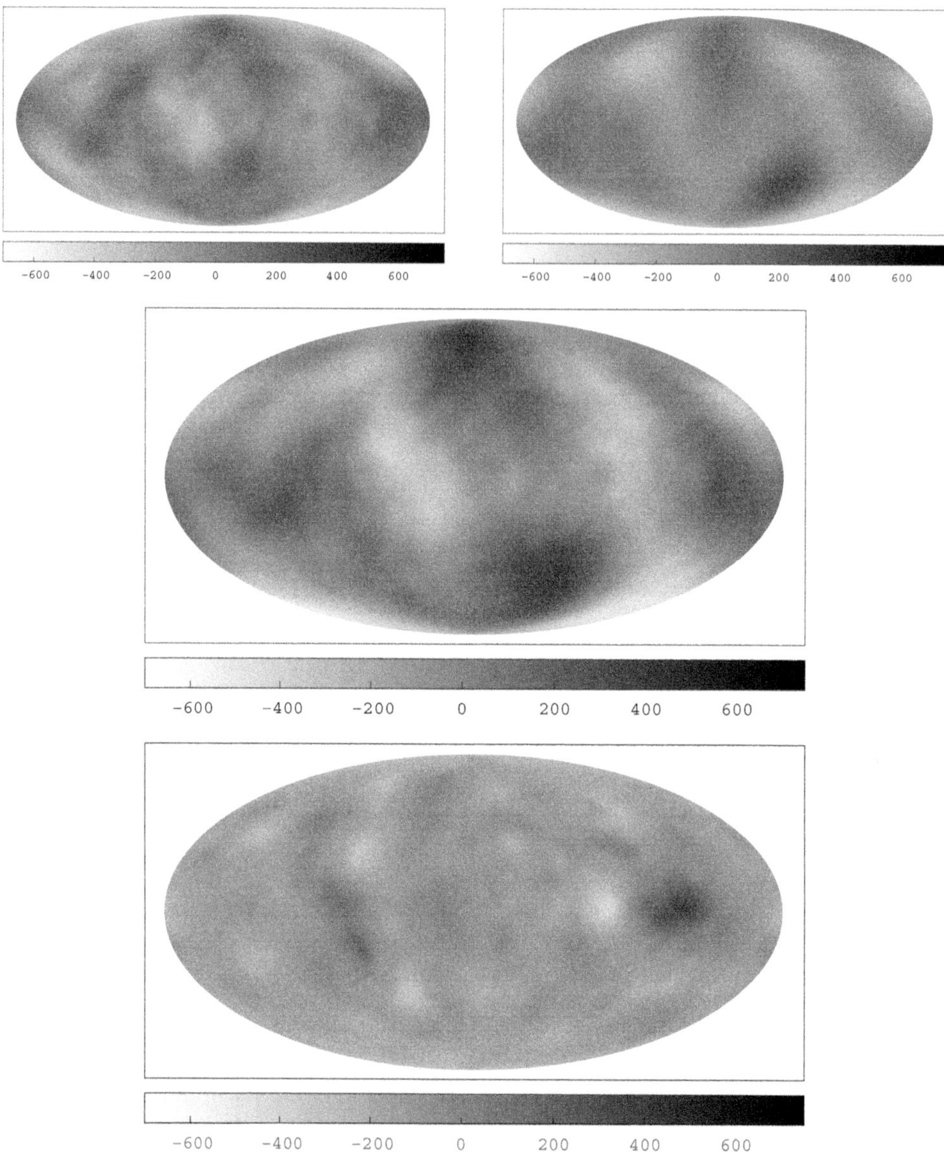

Figure 5.11 We decompose the approximate solution of the downward continuation problem in Example 5.4.24 and Figure 5.10 into bandlimited functions (top-left: spherical harmonics; top-right: Slepian functions; middle: sum of both) and non-bandlimited radial basis functions (bottom).

There is a drawback which needs to be discussed here. Due to the regularization term, we do not necessarily get $R^n \in \mathcal{W}_n$ as it was the case for the OFMP. Hence, we also cannot expect that the ROFMP yields a vanishing residual after a finite number of steps. This is, however, also not intended here, because $R^n = 0$ would mean that we get the/an exact, that is, unstable, solution of the ill-posed problem $\mathcal{T}F = y$. Let us have a closer look at these circumstances.

In the case of the OFMP, we had that $\mathcal{T} F_{n+1} = \mathcal{P}_{\mathcal{V}_{n+1}} y$, which resulted in $R^n \in \mathcal{W}_n$ and $\mathcal{P}_{\mathcal{W}_n} \mathcal{T} d_{n+1} \neq 0$ (for $R^n \neq 0$ and a sufficiently large dictionary). The latter implied $\dim \mathcal{V}_{n+1} = \dim \mathcal{V}_n + 1$ and the convergence in finite steps (if $\dim \mathcal{Y} < +\infty$). In the case of the ROFMP, it is, however, possible that $\mathcal{P}_{\mathcal{W}_n} \mathcal{T} d_{n+1} = 0$. If this situation occurs, then $\mathcal{T} d_{n+1} \in \mathcal{V}_n$, that is, $\mathcal{T} d_{n+1}$ is a linear combination of $\mathcal{T} d_1, \ldots, \mathcal{T} d_n$. In contrast to the OFMP, the ROFMP might choose such dictionary elements d_{n+1} due to the regularization. As the residual is updated by $R^{n+1} = R^n - \alpha_{n+1}^{(n+1)} \mathcal{P}_{\mathcal{W}_n} \mathcal{T} d_{n+1}$, we obtain $R^{n+1} = R^n$ in the discussed case. This implies that in such a step where $\mathcal{P}_{\mathcal{W}_n} \mathcal{T} d_{n+1} = 0$, we do not make any progress regarding the accuracy of fulfilling the equation $\mathcal{T} F = y$. This step is thus completely devoted to the regularization. The objective function in (5.67) is minimized by reducing the norm of the approximation only.

If $\mathcal{P}_{\mathcal{W}_n} \mathcal{T} d_{n+1} \neq 0$, then automatically $\dim \mathcal{V}_{n+1} = \dim \mathcal{V}_n + 1$ holds again. Therefore, we have (almost) derived the following property of the ROFMP.

Theorem 5.4.26 *Let $(R^n)_n$ be the sequence of the residuals produced by the ROFMP (Algorithm 5.4.22). If $\dim \mathcal{Y} < +\infty$, then there exists an index $N \in \mathbb{N}$ such that $R^n = R^N$ for all $n \geq N$.*

Proof There can only be a finite number of steps (namely $\dim \mathcal{Y}$ steps) with $\mathcal{P}_{\mathcal{W}_n} \mathcal{T} d_{n+1} \neq 0$. Either this maximal number of steps occurs after a finite number $N-1$ of iterations, then $\mathcal{W}_N = \{0\}$ and $\mathcal{V}_N = \mathcal{Y}$, or less than $\dim \mathcal{Y}$ steps of this type occur. In both cases, there is an ultimate N with $\mathcal{P}_{\mathcal{W}_{N-1}} \mathcal{T} d_N \neq 0$. After this, $R^n = R^N$ for all $n \geq N$. □

This stagnation of the residual of the ROFMP might cause the obtained result to be not as desired. The reason behind this is the regularization term in the objective function (5.67). We do not entirely concentrate on minimizing the residual such that $R^n \not\perp \mathcal{V}_n$ might occur. Then, as a consequence, there is a non-vanishing projection of R^n onto $\mathcal{V}_n$, which cannot be approximated anymore, because we only reduce R^n with elements of $\mathcal{W}_n$; see (5.69). However, after a large number of iterations, when we have already significantly reduced the residual, it might indeed be reasonable to take care of this part of the residual which has been left aside before. For this reason, it is recommended to use the whole dictionary again after an appropriately chosen number K of iterations. This procedure is called the **restart of the ROFMP**. It has a positive side effect: the orthogonal projections become more and more expensive with increasing n (in particular, the terms $\langle F_n, W^{(n)}(d) \rangle_{\mathcal{X}}$ and $\| W^{(n)}(d) \|_{\mathcal{X}}^2$ are expensive to calculate for large n) while the restart also re-initializes $\mathcal{V}_n$ to $\{0\}$.

We follow here Michel and Telschow (2016) and Telschow (2014), to which the reader is also referred for further details. We now equip the approximation and the residual (and a couple of related objects) with two indices: the first one (n) counts the iterations of the ROFMP between restarts, and the second one (j) counts the restarts. Hence, we start with $F_{0,0}$, which was previously F_0, and perform K iterations, which yield

$$F_{n,0} = \sum_{k=1}^{n} \alpha_{k,0}^{(n)} d_{k,0}, \quad 0 \leq n \leq K,$$

and the corresponding residuals $R^{n,0} := y - \mathcal{T} F_{n,0}$. After the K iterations, we set $F_{0,1} := F_{K,0}$, $R^{0,1} := R^{K,0}$, $\mathcal{V}_{0,1} := \{0\}$, and $\mathcal{W}_{0,1} := \mathcal{Y}$. With this setting, we run another K

iterations of the ROFMP, and so on. Note that we always work with the complete regularization term ($F_{0,1} = F_{K,0}$), that is, we remember the previously chosen approximation in the regularization term. This means that (5.67) is replaced by

$$\left\| R^{n,j} - \alpha \left(\mathcal{T} d - \sum_{k=1}^{n} \beta_{k,j}^{(n)}(d) \mathcal{T} d_{k,j} \right) \right\|_{\mathcal{Y}}^{2} + \lambda \left\| F_{n,j} + \alpha \left(d - \sum_{k=1}^{n} \beta_{k,j}^{(n)}(d) d_{k,j} \right) \right\|_{\mathcal{X}}^{2},$$

where

$$F_{n,j} = \sum_{i=0}^{j-1} \sum_{k=1}^{K} \alpha_{k,i}^{(K)} d_{k,i} + \sum_{k=1}^{n} \alpha_{k,j}^{(n)} d_{k,j}.$$

We now focus on the subsequence of those approximations which are obtained immediately before each restart, that is, we look at the sequence $(F_{K,j})_j$. We obtain the following result.

Theorem 5.4.27 *For the residuals and the approximations produced by the ROFMP with restart, the sequence* $(\|R^{K,j}\|_{\mathcal{Y}}^2 + \lambda \|F_{K,j}\|_{\mathcal{X}}^2)_j$ *is monotonically decreasing and convergent.*

Proof We use the formulae of the restart procedure according to which we have

$$\left\| R^{K,j-1} \right\|_{\mathcal{Y}}^{2} + \lambda \left\| F_{K,j-1} \right\|_{\mathcal{X}}^{2} = \left\| R^{0,j} \right\|_{\mathcal{Y}}^{2} + \lambda \left\| F_{0,j} \right\|_{\mathcal{X}}^{2}. \tag{5.70}$$

For every j, the procedure from $R^{0,j}$ to $R^{K,j}$ and from $F_{0,j}$ to $F_{K,j}$ is the same as in the ROFMP without restart. Hence, we may utilize Theorem 5.4.25 and conclude that

$$\left\| R^{0,j} \right\|_{\mathcal{Y}}^{2} + \lambda \left\| F_{0,j} \right\|_{\mathcal{X}}^{2} \geq \left\| R^{K,j} \right\|_{\mathcal{Y}}^{2} + \lambda \left\| F_{K,j} \right\|_{\mathcal{X}}^{2}. \tag{5.71}$$

Equations (5.70) and (5.71) together prove the monotonicity with respect to j. In combination with the non-negativity of the sequence, we obtain the convergence. □

Under some technical conditions, the convergence of the sequence $(F_{K,j})_j$ in $\mathcal{X}$ could be shown in Michel and Telschow (2016) and Telschow (2014). The details are omitted here. However, properties of the limit of this sequence have not been derived so far. Though there is this lack of a theoretical fundament, the previous numerical experiments revealed that this algorithm can be very useful for ill-posed inverse problems, for example in Earth sciences but also in medical imaging.

Exercises

5.1 Let $V : \mathbb{R}^3 \setminus \overline{B_R(0)} \to \mathbb{R}$ be a gravitational potential such that the inverse gravimetric problem $\mathcal{T}^{\mathrm{gr}} F = V$ is solvable. Prove that with the additional requirement $\Delta_x(F(x)|x|^{-p}) = 0$ in $B_R(0)$ for a fixed $p \in \mathbb{R}_0^+$, there is a unique solution F of $\mathcal{T}^{\mathrm{gr}} F = V$. Derive also an expansion of F in the following form $F(x) = \sum_{n=0}^{\infty} \sum_{j=-n}^{n} \gamma_{n,j} |x|^{\alpha_n} Y_{n,j}(x/|x|)$ and determine the coefficients $\gamma_{n,j}$ and α_n.

5.2 Let a spherical shell be given by $0 \leq \tau \leq |x| \leq \tau + \delta \leq R$ and $\delta > 0$. Moreover, let $F^{\mathrm{L}}\colon B_R(0) \to \mathbb{R}$ be a sufficiently harmless function of the form $F^{\mathrm{L}}(x) := \sum_{n=0}^{\infty} \sum_{j=-n}^{n} F_{n,j}^{\mathrm{L}} Y_{n,j}(x/|x|)$ for x inside the shell and $F^{\mathrm{L}}(x) := 0$ elsewhere. Eventually, let V be the gravitational potential to the mass density function F^{L}. Derive an explicit formula for $F_{n,j}^{\mathrm{L}}$ in dependence of the spherical harmonic coefficients of $V|_{S_{R+\varepsilon}(0)}$ for $\varepsilon > 0$.

5.3 Prove Lemma 5.4.1.

5.4 Let $\mathcal{T} \in \mathcal{L}(\mathcal{H})$ where $(\mathcal{H}, \langle\cdot,\cdot\rangle)$ is a Hilbert space. The Galerkin method for approximately solving an equation $\mathcal{T}x = y$ with given $y \in \mathcal{H}$ and unknown $x \in \mathcal{H}$ works as follows: let $\mathcal{H}_n := \operatorname{span}\{z_1, \ldots, z_n\}$ be an n-dimensional subspace of $\mathcal{H}$. We seek $\xi = \sum_{j=1}^{n} \alpha_j z_j \in \mathcal{H}_n$ such that

$$\sum_{j=1}^{n} \alpha_j \left\langle \mathcal{T} z_j, z_k \right\rangle = \langle y, z_k \rangle \quad \text{for all } k = 1, \ldots, n. \tag{5.72}$$

Let $\mathcal{P} := \mathcal{P}_{\mathcal{H}_n}\colon \mathcal{H} \to \mathcal{H}_n$ be the orthogonal projection onto $\mathcal{H}_n$. Prove that (5.72) is equivalent to $\mathcal{P}\mathcal{T}\mathcal{P}\xi = \mathcal{P}y$.

5.5 Prove Theorem 5.3.14 on the Moore–Penrose axioms.

5.6 Let $(\mathcal{X}, \langle\cdot,\cdot\rangle)$ be a Hilbert space and $\mathcal{A} \in \mathcal{L}(\mathcal{X})$ be an injective and self-adjoint operator.

(a) Why is the image $\mathcal{A}(\mathcal{X})$ dense in $\mathcal{X}$?

(b) Assume the existence of a constant $\gamma > 0$ such that $\langle \mathcal{A}x, x\rangle \geq \gamma \|x\|^2$ for all $x \in \mathcal{X}$. Prove that $\mathcal{A}$ has a continuous inverse $\mathcal{A}^{-1}\colon \mathcal{X} \to \mathcal{X}$ with $\|\mathcal{A}^{-1}\| \leq \gamma^{-1}$.

5.7 Let $\mathcal{T}\colon \mathrm{L}^2(D) \to \mathrm{L}^2(D)$ be a Fredholm integral operator of the first kind, that is, $(\mathcal{T}F)(y) = \int_D k(y,x)F(x)\,\mathrm{d}x$ for (almost all) $y \in D$ with $k \in \mathrm{L}^2(D \times D)$, and let $\lambda \neq 0$ be a constant. Prove the following propositions, which are altogether also known as the Fredholm alternative:

(a) $\mathcal{T}$ has at most a countable number of eigenvalues $\{\lambda_n\}_n$. The non-vanishing eigenvalues of $\mathcal{T}^*$ are their complex conjugates $\{\overline{\lambda_n}\}_n \setminus \{0\}$.

(b) The solution spaces of each of the two equations

$$\int_D k(y,x)F(x)\,\mathrm{d}x - \lambda F(y) = 0, \qquad y \in D, \tag{5.73}$$

$$\int_D \overline{k(y,x)}H(y)\,\mathrm{d}y - \overline{\lambda}H(x) = 0, \qquad x \in D, \tag{5.74}$$

are both finite dimensional and they have the same dimensions.

(c) The equation

$$\int_D k(y,x)F(x)\,\mathrm{d}x - \lambda F(y) = G(y), \qquad y \in D,$$

has a solution F if and only if G is $\mathrm{L}^2(D)$-orthogonal to all solutions H of (5.74).

(d) The equation

$$\int_D \overline{k(y,x)} H(y)\,\mathrm{d}y - \overline{\lambda} H(x) = G(x), \qquad x \in D,$$

has a solution H if and only if G is $\mathrm{L}^2(D)$-orthogonal to all solutions F of (5.73).

5.8 Prove the interpolation inequality in Theorem 5.3.23.

5.9 Let $\mathcal{T} \in \mathcal{K}(\mathcal{X},\mathcal{Y})$ have the singular system $\{(\sigma_n, u_n, v_n)\}_{n\in\mathbb{N}}$ with monotonically decreasing $(\sigma_n)_n$, where $\mathcal{X}$ and $\mathcal{Y}$ are Hilbert spaces. Furthermore, let $\varphi\colon \mathbb{R}_0^+ \to \mathbb{R}$ and $\psi\colon \mathbb{R}_0^+ \to \mathbb{R}$ satisfy the conditions stated in Definition 5.3.19. Prove the following propositions:

(a) $\|\mathcal{T}\| = \sigma_1$.

(b) $\varphi(\mathcal{T}^*\mathcal{T})\mathcal{T}^* = \mathcal{T}^*\varphi(\mathcal{T}\mathcal{T}^*)$.

(c) $\left\|\varphi(\mathcal{T}^*\mathcal{T})\psi(\mathcal{T}^*\mathcal{T})\right\| = \sup\left\{\left|\varphi(\sigma_j^2)\,\psi(\sigma_j^2)\right| \,:\, j \in \mathbb{N}\right\} \le \sup\left\{|\varphi(\lambda)\psi(\lambda)| \,:\, 0 \le \lambda \le \|\mathcal{T}\|^2\right\}$.

(d) $\left\|\varphi(\mathcal{T}^*\mathcal{T})\mathcal{T}^*\right\| = \sup\left\{\sigma_j\left|\varphi(\sigma_j^2)\right| \,:\, j \in \mathbb{N}\right\} \le \sup\left\{\sqrt{\lambda}\,|\varphi(\lambda)| \,:\, 0 \le \lambda \le \|\mathcal{T}\|^2\right\}$.

5.10 Let $\mathcal{T}\colon \mathcal{X} \to \mathcal{Y}$ be a compact linear operator between Hilbert spaces and let $\{\mathcal{X}_\nu\}_{\nu\ge 0}$ be the associated spaces. Prove the inequalities:

$$\|x\|_{\vartheta\nu+(1-\vartheta)\mu} \le \|x\|_\nu^\vartheta \|x\|_\mu^{1-\vartheta} \qquad \text{for all } x \in X_{\max\{\nu,\mu\}},\ \mu,\nu \ge 0,\ \vartheta \in [0,1],$$

$$\|x\|_\mu \le \|\mathcal{T}\|^{\nu-\mu}\|x\|_\nu \qquad \text{for all } x \in X_\nu,\ \nu \ge \mu \ge 0.$$

5.11 Let $\mathcal{T}\colon \mathcal{X} \to \mathcal{Y}$ be a compact linear operator between Hilbert spaces and let $\{\mathcal{X}_\nu\}_{\nu\ge 0}$ be the associated spaces. Moreover, let $f_1, f_2 \in \mathcal{X}_\nu$ for some $\nu \ge 0$ be approximate solutions of $\mathcal{T}f = g$ for a given $g \in \mathcal{Y}$ such that $\|\mathcal{T}f_j - g\|_{\mathcal{Y}} \le \varepsilon$, $j = 1,2$, for some $\varepsilon > 0$. Prove that

$$\|f_1 - f_2\|_{\mathcal{X}} \le \mathrm{e}_\nu\left(2\varepsilon, 2\max\left\{\|f_1\|_\nu, \|f_2\|_\nu\right\}\right).$$

5.12 Prove Theorem 5.3.45.

6
The Magnetic Field

We will discuss here some mathematical tools which are useful for modelling and analyzing the geomagnetic field. Handling this topic completely would exceed the available capacity for such a book. Therefore, we will focus on essentials which are expected to be of interest for a broad audience. For further aspects and details, we refer to the publications Backus (1986), Backus et al. (1996), Bayer (2000), and Blakely (1996) on particular geomagnetic topics as well as, from a general mathematical point of view, the publications Freeden and Gerhards (2013), Freeden and Gutting (2013), Leweke (2018), and Leweke et al. (2020). These works have also been consulted for the production of this chapter.

6.1 The Governing Equations

Maybe, the best starting point of our discussions is represented by the famous **Maxwell's equations**

$$\nabla \times E = -\frac{\partial}{\partial t} B, \tag{6.1}$$

$$\nabla \cdot E = \frac{\varrho}{\varepsilon_0}, \tag{6.2}$$

$$\nabla \times B = \mu_0 \left(J + \varepsilon_0 \frac{\partial}{\partial t} E \right), \tag{6.3}$$

$$\nabla \cdot B = 0, \tag{6.4}$$

where $E, B \in \mathrm{c}^{(1)}(\mathbb{R}^3 \times \mathbb{R})$ are the electric and magnetic field, respectively, which depend on space and time; $\varrho \in \mathrm{C}(\mathbb{R}^3 \times \mathbb{R})$ is the charge density; ε_0, $\mu_0 \in \mathbb{R}$ are the capacity and permeability of vacuum, respectively; and $J \in \mathrm{c}(\mathbb{R}^3 \times \mathbb{R})$ is the current density. The derivatives represented by the nabla operator all apply on the spatial variables only.

The preceding equations are mathematically the same for the vacuum and for polarized media. In the latter case, the functions are to be understood as integral mean values over a ball of small radius (much larger than the distance of atoms). The spatial dependence occurs then with respect to the centre of the ball.

If we consider the static case where $\frac{\partial}{\partial t} B \equiv 0$, then (6.1) reduces to the equation $\nabla \times E = 0$. Also, if we only assume that $\frac{\partial}{\partial t} B \equiv 0$ on a subdomain of $\mathbb{R}^3$, as long as this domain is simply connected, we obtain (in the subdomain) the existence of an **electric**

potential ϕ with $E = -\nabla\phi$. From (6.2), we get $-\nabla \cdot \nabla\phi = -\Delta\phi = \varrho\varepsilon_0^{-1}$, which is the **Poisson equation**. We may now consult Theorem 3.1.11 to get a solution for ϕ. We will come back to this soon.

For considering (6.3) and (6.4), we first need a very useful tool from mathematical physics: the Helmholtz decomposition.

Theorem 6.1.1 (Helmholtz Decomposition) *Let $f: \overline{D} \to \mathbb{R}^3$ be (componentwise) Hölder continuous on $\overline{D}$ with $D = \Sigma_{\text{int}}$, the interior of a regular surface, that is $f \in \mathrm{c}^{(0,\alpha)}(\overline{D})$ for some $\alpha > 0$. Then there exist functions $\phi \in \mathrm{C}^{(1)}(D)$ and $A \in \mathrm{c}^{(1)}(D)$ such that*

$$f = \text{grad } \phi + \text{curl } A = \nabla\phi + \nabla \times A \quad \text{on } D. \tag{6.5}$$

Moreover, if additionally $f \in \mathrm{c}^{(1,\alpha)}(\overline{D})$, then we can choose $\phi \in \mathrm{C}^{(2)}(D)$ and $A \in \mathrm{c}^{(2)}(D)$ with

$$\phi(y) = -\frac{1}{4\pi}\int_D \frac{\text{div } f(x)}{|x-y|}\,\mathrm{d}x \quad \text{and} \tag{6.6}$$

$$A(y) = \frac{1}{4\pi}\int_D \frac{\text{curl } f(x)}{|x-y|}\,\mathrm{d}x \quad \text{for all } y \in D, \tag{6.7}$$

if we additionally require that div $A = 0$ *on D (**Coulomb gauge**).*

Proof We define the vectorial function

$$v(y) := -\frac{1}{4\pi}\int_D \frac{f(x)}{|x-y|}\,\mathrm{d}x, \quad y \in D.$$

Due to Theorem 3.1.2 we know that v exists. Moreover, Theorem 3.1.11 tells us that $v \in \mathrm{c}^{(2)}(D)$ and $\Delta v(y) = f(y)$ for all $y \in D$. Now we use (2.5) from Theorem 2.3.4. It tells us that $f = \text{grad div } v - \text{curl curl } v$. Let now $\phi := \text{div } v$ and $A := -\text{curl } v$. Then $f = \text{grad } \phi + \text{curl } A$, where $\phi \in \mathrm{C}^{(1)}(D)$ and $A \in \mathrm{c}^{(1)}(D)$.

Let now the stronger condition on f hold true. Then $\phi \in \mathrm{C}^{(2)}(D)$ and $A \in \mathrm{c}^{(2)}(D)$ due to Theorem 3.1.12. We apply now the divergence and the curl operators, respectively, to (6.5) and use (2.2)–(2.5) from Theorem 2.3.4. This leads us to div $f = \Delta\phi$ and

$$\text{curl } f = \text{curl curl } A = \text{grad div } A - \Delta A.$$

With the assumption div $A = 0$, the latter identity becomes $-\text{curl } f = \Delta A$. If we now consult Theorem 3.1.11 again, then we obtain (6.6) and (6.7) as possible choices for ϕ and A. □

Let us come back to the discussion of the static case. We had $E = -\nabla\phi$ and $\frac{\partial}{\partial t} B \equiv 0$ on a subdomain D. Let us assume that $D = \Sigma_{\text{int}}$ is the interior of a regular surface (which is then also simply connected). Then the Helmholtz decomposition and (6.4) yield $B = \nabla \times A$ with (6.7) for A. We can insert (6.3), which becomes in the static case: $\nabla \times B = \mu_0 J$. Hence, $\nabla \cdot J = \mu_0^{-1}\nabla \cdot (\nabla \times B) = 0$ due to (2.3).

Theorem 6.1.2 (Static Case) *Let $D = \Sigma_{\mathrm{int}}$ be the interior of a regular surface $\Sigma \subset \mathbb{R}^3$. Moreover, let $E, J \in \mathrm{c}^{(1)}(\overline{D})$ and $B \in \mathrm{c}^{(2)}(\overline{D}) \cap \mathrm{c}^{(1,\alpha)}(\overline{D})$ be independent of the time t. Furthermore, let ϱ be Hölder continuous on $\overline{D}$. If Maxwell's equations hold true, then $E = -\nabla\phi$, $B = \nabla \times A$, and $\nabla \cdot J = 0$ on D, where, for all $y \in D$,*

$$\phi(y) = \frac{1}{4\pi\varepsilon_0} \int_D \frac{\varrho(x)}{|x-y|}\, \mathrm{d}x, \tag{6.8}$$

$$A(y) = \frac{\mu_0}{4\pi} \int_D \frac{J(x)}{|x-y|}\, \mathrm{d}x. \tag{6.9}$$

The quantity ϕ in (6.8) is called the **Coulomb potential**. Concerning its mathematical properties, no further work has to be done by us, because its formula coincides with the formula of Newton's gravitational potential. Hence, the discussions starting in Section 3.1 are also valid here. Eventually, it could be added that $E = -\nabla\phi$ in combination with (6.8) is known as **Coulomb's law**.

Moreover, the vectorial quantity A in (6.9) is, from the mathematical point of view, componentwise the same as Newton's gravitational potential. Hence, we may apply Corollary 3.1.4 and get

$$\frac{\partial}{\partial y_j} A_k(y) = \frac{\mu_0}{4\pi} \int_D J_k(x) \frac{x_j - y_j}{|x-y|^3}\, \mathrm{d}x,$$

where A_k and J_k are the kth components of the vectors A and J. Hence,

$$\operatorname{curl} A(y) = \nabla \times A(y) = \frac{\mu_0}{4\pi} \int_D \frac{x-y}{|x-y|^3} \times J(x)\, \mathrm{d}x.$$

By changing the order in the cross product inside the integrand and taking into account the associated change of sign, we obtain another fundamental physical law.

Corollary 6.1.3 (Biot–Savart Law) *If the conditions of Theorem 6.1.2 are fulfilled, then the magnetic field satisfies*

$$B(y) = \frac{\mu_0}{4\pi} \int_D J(x) \times \frac{y-x}{|y-x|^3}\, \mathrm{d}x \quad \textit{for all } y \in D.$$

6.2 The Gauß Representation and Inner and Outer Sources

As we have seen before, the equations for the magnetic field B in the static case are

$$\nabla \times B = \mu_0 J \quad \text{and} \quad \nabla \cdot B = 0.$$

In the Gauß representation of the geomagnetic field, the presence of electric currents is neglected. This is valid, for example, in the lower parts of the atmosphere (where, at Gauß's era, the only way of measuring the magnetic field existed). However, in the case of satellite measurements from, for instance, the ionosphere, this assumption is not realistic.

If $J = 0$, then we have

$$\nabla \times B = 0 \quad \text{and} \quad \nabla \cdot B = 0.$$

A layer in the atmosphere is, geometrically speaking, a spherical shell. Such a domain is simply connected. Hence, in the Gauß representation, also the magnetic field possesses a potential. We set $B = -\nabla U$ and immediately get $0 = \nabla \cdot B = -\nabla \cdot \nabla U = -\Delta U$. Thus, the **magnetic potential** U is harmonic. Harmonic functions are something with which we are very familiar now at this stage of the book. We have discussed this topic in detail, when we studied the gravitational field. However, there is a difference here: remember that, when we developed a basis system in Section 3.4, we came across the inner harmonics and the outer harmonics (see Definition 3.4.26):

$$H_{n,j}^{\mathrm{int}}(x) := |x|^n Y_{n,j}\left(\frac{x}{|x|}\right), \qquad H_{-n-1,j}^{\mathrm{ext}}(y) := |y|^{-n-1} Y_{n,j}\left(\frac{y}{|y|}\right),$$

$x \in \Omega_{\mathrm{int}}$, $y \in \Omega_{\mathrm{ext}}$. The former were only appropriate for interior problems (the IDP and INP), because we know that the gravitational potential is regular at infinity while the inner harmonics are not. The latter are only admissible for exterior problems (the EDP and ENP), because they cannot be continued to $y = 0$. In contrast to the gravitational field modelling, which takes place in the exterior of a regular surface (usually approximated by a sphere), we now have a spherical shell as the domain D for the modelling of the magnetic field. This provides us neither with arguments against the inner harmonics nor against the outer harmonics, because the domain is bounded but it does not contain the zero point. Therefore, we have to deal with both basis functions:

$$U(x) = \sum_{n=0}^{\infty} \sum_{j=-n}^{n} \left(U_{n,j} H_{n,j}^{\mathrm{int}}(x) + U_{-n-1,j} H_{-n-1,j}^{\mathrm{ext}}(x) \right),$$

$x \in D$. This reflects the physical reality: there are **inner sources** due to the core field and the crustal field of the Earth and there are **outer sources** from space. If we remember what we learned about gravitational field modelling, then it is evident that the harmonicity caused by a source is given outside the source. Therefore, the coefficients $U_{n,j}$, $n \in \mathbb{N}_0$, which accompany the *inner* harmonics are associated to *outer* sources, while the coefficients $U_{-n-1,j}$, $n \in \mathbb{N}_0$, of the *outer* harmonics belong to the *inner* sources – just like outer harmonics correspond to the gravitational field caused by the masses inside the Earth.

We have learned before that the Edmonds vector spherical harmonics are a good tool for representing gradient fields such as the gradient of U; see Definition 4.3.6 and (4.30). We get, with $x = r\xi$, $r = |x|$, $\xi \in \Omega$,

$$B(x) = -\nabla U(x) = \sum_{n=0}^{\infty} \sum_{j=-n}^{n} U_{-n-1,j} \sqrt{(n+1)(2n+1)}\, r^{-n-2} \tilde{y}_{n,j}^{(1)}(\xi)$$
$$- \sum_{n=1}^{\infty} \sum_{j=-n}^{n} U_{n,j} \sqrt{n(2n+1)}\, r^{n-1} \tilde{y}_{n,j}^{(2)}(\xi).$$

Therefore, the orthonormality of the Edmonds vector spherical harmonics allows a complete separation of the fields due to the internal and the external sources.

However, as we noted previously, the Gauß representation requires that the magnetic field is curl-free, that is, there are no electric currents in the region of interest. This assumption is not valid everywhere. Therefore, we also need to consider the more general Mie representation, which we will do in the next section.

6.3 The Spherical Helmholtz Decomposition and the Mie Representation

For finding an appropriate way of modelling magnetic data, we have to exploit the fact that $\operatorname{div} B = 0$. The investigation of divergence-free fields leads to the concept of solenoidal functions and the Mie representation. We first need some definitions. Before we proceed, it should be mentioned that the following discussions are based on Backus (1986) and Backus et al. (1996).

Definition 6.3.1 Let $U \subset \mathbb{R}^3$ be an open subset and $f : U \to \mathbb{R}^3$ be a given vector field.

(a) f is called **solenoidal** if, for every regular surface $\Sigma \subset U$, the surface integral $\int_\Sigma f(x) \cdot \nu(x) \, \mathrm{d}\omega(x)$, where ν is the outer unit normal to Σ, exists and vanishes.
(b) The operator $\mathrm{L} \colon \mathrm{C}^{(1)}(U) \to \mathrm{c}(U)$ is given by $\mathrm{L}_x Q(x) := x \times \nabla_x Q(x)$, $x \in D$, for all $Q \in \mathrm{C}^{(1)}(U)$. The vector field f is called **toroidal** if there exists a scalar field $Q \in \mathrm{C}^{(1)}(U)$ such that $f = \mathrm{L}Q$. In this case, Q is called a **toroidal scalar** of f.
(c) f is called **poloidal**, if there exists a scalar field $P \in \mathrm{C}^{(2)}(U)$ such that $f = \operatorname{curl} \mathrm{L}P$. In this case, P is called a **poloidal scalar** of f.

Obviously, every poloidal field is the curl of a toroidal field. Moreover, the poloidal and toroidal scalars are never unique: let us assume that we have a function $R \colon U \to \mathbb{R}$ which is sufficiently often differentiable and which is only radially dependent, that is, there is another function $F \colon \mathbb{R}_0^+ \to \mathbb{R}$ such that $R(x) = F(|x|)$ for all $x \in U$. Then, using Theorem 3.4.1 and $x = r\xi$ with $r = |x|$ and $\xi \in \Omega$, we obtain

$$\begin{aligned}\mathrm{L}_x(P(x) + R(x)) &= r\xi \times \left(\xi \frac{\partial}{\partial r} + \frac{1}{r} \nabla_\xi^* \right) (P(r\xi) + F(r)) \\ &= \xi \times \nabla_\xi^*(P(r\xi) + F(r)) = \xi \times \nabla_\xi^* P(r\xi).\end{aligned}$$

Hence, $\mathrm{L}P = \mathrm{L}(P + R)$, and consequently $\operatorname{curl} \mathrm{L}P = \operatorname{curl} \mathrm{L}(P + R)$. The preceding considerations also show us that $\mathrm{L} = \mathrm{L}^*$ (for the surface curl gradient; see Definition 4.1.1) in the sense that

$$\mathrm{L}_x G(x) = \mathrm{L}_\xi^* G(r\xi), \quad x = r\xi, \, \xi \in \Omega, \tag{6.10}$$

for all differentiable functions G on open subsets of $\mathbb{R}^3$.

There are several further properties of solenoidal, toroidal, and poloidal vector fields, which we will discuss here. Let us first have a look at the link to divergence-free fields.

Theorem 6.3.2 *Every continuously differentiable vector field which is poloidal or toroidal is also divergence-free (provided that the toroidal scalar is in* $\mathrm{C}^{(2)}(U)$ *and the poloidal scalar is in* $\mathrm{C}^{(3)}(U)$*, respectively).*

Proof From Theorem 2.3.4, we get immediately that $\operatorname{div}\operatorname{curl}\mathrm{L}P = 0$. Let now $f = \mathrm{L}Q$ be toroidal. Then we saw earlier that $\operatorname{div}\mathrm{L}Q = \nabla\cdot\left(\mathrm{L}_\xi^* Q(r\xi)\right) = \left(\xi\,\frac{\partial}{\partial r} + r^{-1}\nabla_\xi^*\right)\cdot\mathrm{L}_\xi^* Q(r\xi)$. Finally, we only have to consult Theorem 4.1.6, which tells us that $\operatorname{div}\mathrm{L}Q = 0$ must hold consequently. □

Theorem 6.3.3 *Let* $U \subset \mathbb{R}^3$ *be an open subset and* $f\colon U \to \mathbb{R}^3$ *be a solenoidal and continuously differentiable vector field. Then* $\operatorname{div} f = 0$ *on the whole set* U.

Proof Let $x \in U$ be an arbitrary point. Since U is open, there exists an $\varepsilon > 0$ such that $B_\varepsilon(x) \subset U$. Since $S_\varrho(x)$ with $0 < \varrho < \varepsilon$ is a regular surface inside U and f is solenoidal, we have $\int_{S_\varrho(x)} f(y)\cdot\nu(y)\,\mathrm{d}\omega(y) = 0$, where ν is the outer unit normal to $S_\varrho(x)$. Hence, Gauß's law (see Theorem 2.4.2) implies that $\int_{B_\varrho(x)} \operatorname{div} f(y)\,\mathrm{d}y = 0$ for all $\varrho \in]0,\varepsilon[$. By taking the limit $\varrho \to 0+$ and using the continuity of $\operatorname{div} f$, we get $\operatorname{div} f(x) = 0$. □

The converse of Theorem 6.3.3 is false, as the following counterexample shows.

Example 6.3.4 If $U := \mathbb{R}^3 \setminus \{0\}$ and $f(x) := |x|^{-3}x$ for all $x \in U$, then

$$\operatorname{div} f(x) = \sum_{j=1}^{3} \frac{\partial}{\partial x_j}\frac{x_j}{\left(x_1^2+x_2^2+x_3^2\right)^{3/2}} = \sum_{j=1}^{3}\frac{1\cdot|x|^3 - x_j|x|\cdot 2x_j\cdot 3/2}{|x|^6}$$

$$= \frac{1}{|x|^5}\left(3|x|^2 - 3\sum_{j=1}^{3} x_j^2\right) = 0.$$

Hence, f is divergence-free. However, it is not solenoidal, because

$$\int_{S_r(0)} f(x)\cdot\nu(x)\,\mathrm{d}\omega(x) = \int_{S_r(0)} \frac{x}{|x|^3}\cdot\frac{x}{|x|}\,\mathrm{d}\omega(x) = \frac{1}{r^2}\cdot 4\pi r^2 = 4\pi \neq 0.$$

The question that occurs here is: what is the difference between the proof of Theorem 6.3.3 and the counterexample in Example 6.3.4? A closer look reveals the answer: for the opposite conclusion in Theorem 6.3.3, we would have to be able to apply Gauß's law to $\int_{\Sigma_{\text{int}}} \operatorname{div} f(y)\,\mathrm{d}y = 0$ for *every* regular surface $\Sigma \subset U$. However, in Example 6.3.4, regular surfaces Σ with $0 \in \Sigma_{\text{int}}$, like all spheres $S_r(0)$, do *not* satisfy $\Sigma_{\text{int}} \subset U$. Therefore, the converse in Theorem 6.3.3 is true if $\Sigma_{\text{int}} \subset U$ for every regular surface $\Sigma \subset U$. This property is connected to the topological concept of a contractible set (see also Ballmann, 2015; Jänich, 2005; Laures and Szymik, 2015; and Munkres, 2000). In this context, the author gratefully acknowledges valuable discussions with his colleague Mohamed Barakat.

Definition 6.3.5 A set $U \subset \mathbb{R}^n$, $n \in \mathbb{N}$, is called **contractible** if there exists a point $x_0 \in U$ and a continuous mapping $F\colon U \times [0,1] \to U$ such that $F(x,0) = x$ and $F(x,1) = x_0$ for all $x \in U$. In other words, the set U can be continuously shrunk to a single point.

We can easily deduce that contractible sets have the desired property. If $\Sigma \subset U$ and $y \in \Sigma_{\text{int}}$ but $y \notin U$, then one can only deform Σ to the point x_0 without leaving the set U by 'tearing' the set Σ apart, that is, in a discontinuous manner.

Examples of contractible sets are all **star-shaped sets**, that is, all sets U for which a point $x_0 \in U$ exists such that every line segment $\overline{x_0 x}$ with $x \in U$ is contained in U. This includes all balls but, for instance, *not* spherical shells.

Theorem 6.3.6 *Let $U \subset \mathbb{R}^3$ be an open and contractible subset and let $f \colon U \to \mathbb{R}^3$ be a continuously differentiable vector field. If $\operatorname{div} f = 0$ on the whole set U, then f is solenoidal.*

Proof Let $\Sigma \subset U$ be an arbitrary regular surface. Since U is contractible, we have $\Sigma_{\text{int}} \subset U$ such that we apply Gauß's law (Theorem 2.4.2) to

$$\int_\Sigma f(x) \cdot \nu(x) \, \mathrm{d}\omega(x) = \int_{\Sigma_{\text{int}}} \operatorname{div} f(x) \, \mathrm{d}x = 0.$$

Hence, f is solenoidal. □

Since the magnetic field is divergence-free everywhere in $\mathbb{R}^3$ (due to the fact that the non-existence of magnetic monopoles is a fundamental component of the classical electromagnetic theory), we can assume that the magnetic field is also solenoidal. This is an important conclusion, because the Mie representation for solenoidal vector fields is a common tool in geomagnetic modelling.

In the case of a spherical shell, which is a non-contractible domain, an alternative criterion is sometimes helpful.

Theorem 6.3.7 *Let $U := \Omega_{]\alpha,\beta[} := \{x \in \mathbb{R}^3 \mid \alpha < |x| < \beta\}$ be a spherical shell, where $0 < \alpha < \beta < +\infty$. If $f \in \mathrm{c}^{(1)}(U)$ is divergence-free, that is, $\operatorname{div} f = 0$ in U, and if additionally*

$$\int_{S_\varrho(0)} f(x) \cdot \frac{x}{|x|} \, \mathrm{d}\omega(x) = 0 \quad \textit{for at least one } \varrho \in]\alpha, \beta[,$$

then f is solenoidal.

Proof Let Σ be an arbitrary regular surface inside U and let ν be its unit normal. If $\Sigma_{\text{int}} \subset U$, then we can follow the proof of Theorem 6.3.6 and obtain $\int_\Sigma f(x) \cdot \nu(x) \, \mathrm{d}\omega(x) = 0$. Otherwise, as shown in Figure 6.1, only $\Sigma_{\text{int}} \setminus \overline{B_\alpha(0)}$ is a subset of U. Since Σ is a compact set, there exists $\sigma_{\min} := \min\{|x| \colon x \in \Sigma\}$, where $\sigma_{\min} > \alpha$, because $\Sigma \subset U$. Let now $\alpha < r < \sigma_{\min}$.

If $r \neq \varrho$, then we need an additional discussion: let S be the spherical shell which is given by $\Omega_{]r,\varrho[}$ or $\Omega_{]\varrho,r[}$, depending on if ϱ or r is larger. Clearly, $S \subset U$ and Gauß's law (Theorem 2.4.2) implies that

$$\int_S \operatorname{div} f(x) \, \mathrm{d}x = \pm \left(\int_{S_\varrho(0)} f(x) \cdot \frac{x}{|x|} \, \mathrm{d}\omega(x) - \int_{S_r(0)} f(x) \cdot \frac{x}{|x|} \, \mathrm{d}\omega(x) \right),$$

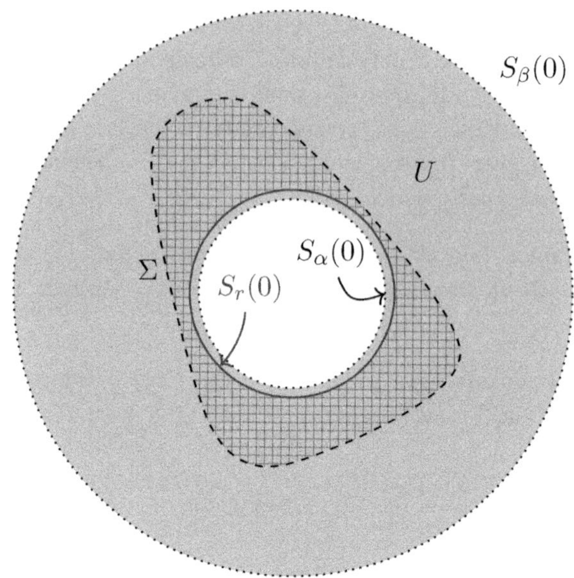

Figure 6.1 Two-dimensional analogue of the situation discussed in the proof of Theorem 6.3.7: the regular surface Σ (dashed) is located inside the spherical shell U (light grey with dotted boundaries $S_\alpha(0)$ and $S_\beta(0)$) in a way such that $\Sigma_{\text{int}} \not\subset U$. In this case, we choose a sphere $S_r(0)$ with $\alpha < r < \min\{|x| : x \in \Sigma\}$ (solid, dark grey line). Then $\overline{\Sigma_{\text{int}}} \setminus B_r(0)$ (crosshatched) is contained in U, and Gauß's law can be applied on this set.

where '+' occurs if and only if $r < \varrho$. Due to the conditions on f, the first two integrals vanish. Hence, $\int_{S_r(0)} f(x) \cdot x\, |x|^{-1} \,\mathrm{d}\omega(x) = 0$, no matter if $r = \varrho$ or not.

We have a look now at $V := \overline{\Sigma_{\text{int}}} \setminus B_r(0)$. Obviously, $V \subset U$ and Gauß's law (Theorem 2.4.2) yields

$$\int_V \operatorname{div} f(x)\,\mathrm{d}x = \int_\Sigma f(x) \cdot \nu(x)\,\mathrm{d}\omega(x) + \int_{S_r(0)} f(x) \cdot \left(-\frac{x}{|x|}\right) \mathrm{d}\omega(x).$$

The volume integral and the latter surface integral vanish. Hence, also the integral over Σ must vanish. □

Before we can prove the Mie representation, we first have to show the spherical version of the Helmholtz decomposition, which is also called the **Hodge decomposition** or the **Hansen decomposition**.

Theorem 6.3.8 (Spherical Helmholtz Decomposition) *Let* $0 \le a < c$ *and let* $U := \Omega_{]a,c[} := \{x \in \mathbb{R}^3 \,|\, a < |x| < c\}$ *be a spherical shell. If* $f \in \mathrm{c}^{(1,\alpha)}(U)$ *for* $0 < \alpha < 1$*, then there exist scalar functions* $F, G, H\colon U \to \mathbb{R}$ *such that* $F \in \mathrm{C}^{(1,\alpha)}(U)$*,* $G \in \mathrm{C}^{(1)}(U)$*, and* $G(r\cdot), H(r\cdot) \in \mathrm{C}^{(2,\alpha)}(\Omega)$ *for all* $r \in]a,c[$ *and*

$$f(r\xi) = \xi F(r\xi) + \nabla^*_\xi G(r\xi) + \mathrm{L}^*_\xi H(r\xi) \tag{6.11}$$

for all $r \in]a,c[$ and all $\xi \in \Omega$. The function F is unique, where G and H are unique up to summands which only depend on the radial coordinate.

Here, ∇^* and L^* are the surface gradient and the surface curl gradient (see Definitions 3.4.2 and 4.1.1).

Proof This will get a bit lengthy.

(1) Normal and tangential part:

Let us first observe that we can decompose f into a radial vector field and a field which is tangential to all concentric spheres around 0:

$$f(r\xi) = \underbrace{[\xi \cdot f(r\xi)]\xi}_{\text{radial/normal}} + \underbrace{f(\xi) - [\xi \cdot f(r\xi)]\xi}_{\text{tangential}}, \quad a < r < c,\, \xi \in \Omega.$$

Remember that the outer unit normal vector to $S_r(0)$ in $x = r\xi$ is given by ξ. Since $\nabla^* G$ and $\mathrm{L}^* H$ are tangential in this respect (see Theorem 4.1.6), it is clear that $F(r\xi) = \xi \cdot f(r\xi)$ for all $r \in]a,c[$ and all $\xi \in \Omega$. Therefore, we 'only' have to deal now with the tangential part. Let now $\widetilde{f}$ represent this tangential part.

(2) Uniqueness:

Due to the differentiation which is included in ∇^* and L^*, it is clear that additive constants do not change the outcome. Let us assume that f is decomposed like in (6.11).

We now refer to Theorem 4.1.6 and conclude that

$$\nabla_\xi^* \cdot \widetilde{f}(r\xi) = \nabla_\xi^* \cdot \nabla_\xi^* G(r\xi) + \nabla_\xi^* \cdot \mathrm{L}_\xi^* H(r\xi) = \Delta_\xi^* G(r\xi).$$

Hence, if there is another scalar $\widehat{G}$ which could be used instead of G in (6.11), then $\Delta_\xi^*(G(r\xi) - \widehat{G}(r\xi)) = 0$ for all $r \in]a,c[$ and all $\xi \in \Omega$. From Theorem 4.2.19, we know that only the constants are eigenfunctions of Δ^* to the eigenvalue 0. Hence, for *each* $r \in]a,c[$, there exists a constant $C(r) \in \mathbb{R}$ such that $G(r\xi) = \widehat{G}(r\xi) + C(r)$ for all $r \in]a,c[$ and all $\xi \in \Omega$.

Analogously, $\mathrm{L}_\xi^* \cdot \widetilde{f}(r\xi) = \Delta_\xi^* H(r\xi)$ yields that H is also unique up to summands which only depend on r.

(3) Existence:

Since $\widetilde{f}$ is a tangential vector field, we can use Corollary 4.1.13 to conclude that

$$\left(\int_\Omega \nabla_\xi^* \cdot \widetilde{f}(r\xi)\, \mathrm{d}\omega(\xi) = \right) 0 = \int_\Omega \mathrm{L}_\xi^* \cdot \widetilde{f}(r\xi)\, \mathrm{d}\omega(\xi)$$

for all $r \in]a,c[$. Hence, Theorems 4.7.27 and 4.7.30 tell us that there is, for each $r \in]a,c[$, a function $H(r\cdot) \in \mathrm{C}^{(2,\alpha)}(\Omega)$ such that $\Delta_\xi^* H(r\xi) = \mathrm{L}_\xi^* \cdot \widetilde{f}(r\xi)$ for all $r \in]a,c[$ and all $\xi \in \Omega$. Our next step is now to use Theorem 4.1.6, because it allows us to conclude that

$$\mathrm{L}_\xi^* \cdot \left(\widetilde{f}(r\xi) - \mathrm{L}_\xi^* H(r\xi)\right) = \mathrm{L}_\xi^* \cdot \widetilde{f}(r\xi) - \Delta_\xi^* H(r\xi) = 0$$

for all $r \in]a,c[$ and all $\xi \in \Omega$. Hence, Theorem 4.1.14 and its derivation yield the existence of a continuously differentiable function $G\colon \Omega_{]a,c[} \to \mathbb{R}$ such that

$$\widetilde{f}(r\xi) - \mathrm{L}_\xi^* H(r\xi) = \nabla_\xi^* G(r\xi)$$

for all $\xi \in \Omega$ and all $r \in]a,c[$. Note that $\nabla_\xi^* \cdot \widetilde{f}(r\xi) = \Delta_\xi^* G(r\xi)$ such that also $G(r\cdot) \in \mathrm{C}^{(2,\alpha)}(\Omega)$ for all $r \in]a,c[$. □

In view of the preceding proof and of the used Theorem 4.7.27 (see also Definition 4.7.28), we can also further specify what F, G, and H are.

Corollary 6.3.9 *The function F in Theorem 6.3.8 is uniquely determined by $F(r\xi) = \xi \cdot f(r\xi)$, $r \in]a,c[$, $\xi \in \Omega$. Moreover, the functions G and H in the same theorem are uniquely determined if*

$$\int_{S_r(0)} G(x)\,\mathrm{d}\omega(x) = \int_{S_r(0)} H(x)\,\mathrm{d}\omega(x) = 0$$

is required for all $r \in]a,c[$. In this case,

$$G(r\xi) = \left(\Delta_\xi^*\right)^{-1}\left(\nabla_\xi^* \cdot \widetilde{f}(r\xi)\right), \qquad H(r\xi) = \left(\Delta_\xi^*\right)^{-1}\left(\mathrm{L}_\xi^* \cdot \widetilde{f}(r\xi)\right)$$

for all $r \in]a,c[$ and all $\xi \in \Omega$, where $\widetilde{f}(r\xi) := f(r\xi) - \xi F(r\xi)$. Moreover, $G(r\cdot), H(r\cdot) \in \mathcal{H}_{\Delta^*}$ for all $r \in]a,c[$.

Note that we have *not* proved that $H \in \mathrm{C}^{(2)}(\Omega_{]a,c[})$. For us it will be important that $\mathrm{L}_\xi^* H(r\xi)$ is well defined, which is the case for the spherical Helmholtz decomposition. In view of (6.10), we will also write $\mathrm{L}_x H(x)$ instead of $\mathrm{L}_\xi^* H(r\xi)$, independent of the existence of a radial derivative of H.

Lemma 6.3.10 *Let $P \in \mathrm{C}^{(2)}(D)$, where $D \subset \mathbb{R}^3$ is an open set. Then the poloidal field* curl LP *can be represented as*

$$\begin{aligned}(\operatorname{curl} \mathrm{L}P)(x) &= x\Delta_x P(x) - \nabla_x P(x) - \nabla_x(x \cdot \nabla_x P(x)) \\ &= \frac{1}{r}\,\xi\Delta_\xi^* P(r\xi) + \nabla_\xi^*\left(-\frac{1}{r} - \frac{\partial}{\partial r}\right) P(r\xi),\end{aligned} \tag{6.12}$$

where $x = r\xi$, $r = |x|$, $\xi \in \Omega$. Instead of $P \in \mathrm{C}^{(2)}(D)$, it also suffices if P is twice partially differentiable of all kinds and $\frac{\partial}{\partial r}\nabla^* P = \nabla^* \frac{\partial}{\partial r} P$ in D.

Proof We use the Levi-Cività symbol (see Theorem 2.1.4) to write

$$(\mathrm{L}P)(x) = (x \times \nabla_x)P(x) = \left(\sum_{l,m=1}^{3} \varepsilon_{klm}\, x_l \frac{\partial P}{\partial x_m}(x)\right)_{k=1,2,3},$$

$$\begin{aligned}(\operatorname{curl} \mathrm{L}P)(x) &= \left(\sum_{j,k=1}^{3} \varepsilon_{ijk} \frac{\partial}{\partial x_j}(\mathrm{L}P)_k(x)\right)_{i=1,2,3} \\ &= \left(\sum_{j,k,l,m=1}^{3} \varepsilon_{ijk}\,\varepsilon_{klm} \frac{\partial}{\partial x_j}\left(x_l \frac{\partial P}{\partial x_m}(x)\right)\right)_{i=1,2,3}\end{aligned}$$

$$= \left(\sum_{j,k,l,m=1}^{3} \varepsilon_{ijk}\,\varepsilon_{klm} \left(\delta_{jl}\, \frac{\partial P}{\partial x_m}(x) + x_l\, \frac{\partial^2 P}{\partial x_j \partial x_m}(x) \right) \right)_{i=1,2,3},$$

where δ_{jl} is the Kronecker delta. Since $\varepsilon_{ijk} = \varepsilon_{kij}$ (the permutation from (i,j,k) to (k,i,j) is even and, therefore, does not change the sign here), Theorem 2.1.4 yields

$$(\operatorname{curl} \mathrm{L} P)(x) = \left(\sum_{j,l,m=1}^{3} \left(\delta_{il}\delta_{jm} - \delta_{im}\delta_{jl} \right) \left(\delta_{jl}\, \frac{\partial P}{\partial x_m}(x) + x_l\, \frac{\partial^2 P}{\partial x_j \partial x_m}(x) \right) \right)_{i=1,2,3}$$

$$= \left(\frac{\partial P}{\partial x_i}(x) + \sum_{j=1}^{3} \left(x_i\, \frac{\partial^2 P}{\partial x_j^2}(x) - \frac{\partial P}{\partial x_i}(x) - x_j\, \frac{\partial^2 P}{\partial x_j \partial x_i}(x) \right) \right)_{i=1,2,3}.$$

Note that Theorems 3.4.1 and 4.1.6 yield

$$\left(\sum_{j=1}^{3} x_j\, \frac{\partial^2 P}{\partial x_j \partial x_i}(x) \right)_{i=1,2,3} = (x \cdot \nabla)\nabla P(x) = r\, \frac{\partial}{\partial r} \left(\xi\, \frac{\partial}{\partial r} + \frac{1}{r}\, \nabla_\xi^* \right) P(r\xi)$$

$$= r\xi\, \frac{\partial^2}{\partial r^2} P(r\xi) - \frac{1}{r}\, \nabla_\xi^* P(r\xi) + \frac{\partial}{\partial r} \nabla_\xi^* P(r\xi)$$

$$= \xi\, \frac{\partial}{\partial r} P(r\xi) + r\xi\, \frac{\partial^2}{\partial r^2} P(r\xi) + \nabla_\xi^* \frac{\partial}{\partial r} P(r\xi) - \nabla_{r\xi} P(r\xi)$$

$$= \nabla_{r\xi} \left(r\, \frac{\partial}{\partial r} P(r\xi) \right) - \nabla_{r\xi} P(r\xi)$$

$$= \nabla_x \left(x \cdot \nabla_x P(x) \right) - \nabla_x P(x).$$

Hence, we can represent the poloidal function by

$$(\operatorname{curl} \mathrm{L} P)(x) = -\nabla_x P(x) + x \Delta_x P(x) - \nabla_x \left(x \cdot \nabla_x P(x) \right),$$

which proves the first part of (6.12). With the radial-angular decomposition of ∇ and Δ (see Theorems 3.4.1 and 3.4.3 as well as Definition 3.4.2) and Theorem 4.1.6, we obtain

$$(\operatorname{curl} \mathrm{L} P)(r\xi) = -\left(\xi\, \frac{\partial}{\partial r} + \frac{1}{r}\, \nabla_\xi^* \right) P(r\xi) + r\xi \left(\frac{\partial^2}{\partial r^2} + \frac{2}{r}\, \frac{\partial}{\partial r} + \frac{1}{r^2}\, \Delta_\xi^* \right) P(r\xi)$$

$$- \left(\xi\, \frac{\partial}{\partial r} + \frac{1}{r}\, \nabla_\xi^* \right) \left(r\, \frac{\partial}{\partial r} P(r\xi) \right)$$

$$= -\frac{1}{r}\, \nabla_\xi^* P(r\xi) - \nabla_\xi^*\, \frac{\partial}{\partial r}\, P(r\xi) + \frac{1}{r}\, \xi \Delta_\xi^* P(r\xi).$$

This completes the proof. □

This lemma enables us now to prove the Mie representation of solenoidal vector fields, which goes back to Mie (1908).

Theorem 6.3.11 (Mie Representation) *Let* $0 < a < c$ *and let* $\Omega_{]a,c[} := \{x \in \mathbb{R}^3 \mid a < |x| < c\}$ *be an open spherical shell. If* $f \in \mathrm{c}^{(j,\alpha)}(\Omega_{]a,c[})$ *for a fixed* $0 < \alpha < 1$ *and for all* $j = 0, \ldots, k$ *(with a fixed* $k \in \mathbb{N} \setminus \{1\}$*) is a solenoidal vector field, then there exist scalar functions* $P, Q\colon \Omega_{]a,c[} \to \mathbb{R}$ *with* $P(r\cdot) \in \mathrm{C}^{(2+k,\alpha)}(\Omega)$ *and* $Q(r\cdot) \in \mathrm{C}^{(2,\alpha)}(\Omega)$ *for all* $r \in]a,c[$*, existing second-order partial derivatives of* P *on* $\Omega_{]a,c[}$*, and*

$$f = \operatorname{curl} \mathrm{L} P + \mathrm{L} Q. \tag{6.13}$$

Moreover, the functions P *and* Q *are uniquely determined by* (6.13) *up to summands which only depend on the radial coordinate. They are called* ***Mie scalars****. In addition, the poloidal part* $\operatorname{curl} \mathrm{L} P$ *and the toroidal part* $\mathrm{L} Q$ *are unique.*

Proof Again, it is better to subdivide a lengthy proof.

(1) Existence:

We have already proved a decomposition of functions such as f, namely the spherical Helmholtz decomposition (Theorem 6.3.8):

$$f(r\xi) = \xi F(r\xi) + \nabla_\xi^* G(r\xi) + \mathrm{L}_\xi^* H(r\xi), \tag{6.14}$$

where $r \in]a,c[$ and $\xi \in \Omega$. Here F, G, and H are scalar fields, where we use Corollary 6.3.9 and require

$$\int_{S_r(0)} G(x)\, \mathrm{d}\omega(x) = \int_{S_r(0)} H(x)\, \mathrm{d}\omega(x) = 0 \tag{6.15}$$

for all $r \in]a,c[$ to obtain unique functions G and H. Considering (6.10), we see that it makes sense to set $Q := H$. Moreover, Theorem 4.1.6 implies that

$$0 = r \int_{S_r(0)} f(x) \cdot \nu(x)\, \mathrm{d}\omega(x) = \int_{S_r(0)} r F(x)\, \mathrm{d}\omega(x)$$

for all $r \in]a,c[$, because f is solenoidal. Hence, Theorems 4.7.27 and 4.7.30 guarantee that, for each arbitrary but fixed $r \in]a,c[$, there is a function $P_r \in \mathrm{C}^{(2+k,\alpha)}(\Omega)$ such that

$$\int_\Omega P_r(\xi)\, \mathrm{d}\omega(\xi) = 0 \tag{6.16}$$

and

$$\Delta_\xi^* P_r(\xi) = r F(r\xi) \quad \text{for all } \xi \in \Omega, \tag{6.17}$$

where the function $r\xi \mapsto \Delta_\xi^* P_r(\xi)$ is in $\mathrm{C}^{(k,\alpha)}(\Omega_{]a,c[})$. If we use this in (6.14), we arrive at

$$f(r\xi) = \frac{1}{r}\, \xi \Delta_\xi^* P_r(\xi) + \nabla_\xi^* G(r\xi) + \mathrm{L}_\xi^* Q(r\xi) \tag{6.18}$$

for all $r \in]a,c[$ and $\xi \in \Omega$. From Theorem 4.7.29, we know that

$$P_r(\xi) = \int_\Omega K(\xi \cdot \eta) r F(r\eta)\, \mathrm{d}\omega(\eta) \tag{6.19}$$

for all $r \in]a,c[$ and $\xi \in \Omega$, where

$$K(\xi \cdot \eta) = \sum_{n=1}^{\infty} \frac{-2n-1}{n(n+1)4\pi} P_n(\xi \cdot \eta); \quad \xi, \eta \in \Omega.$$

It is possible to derive that the preceding series converges to the function $K(t) = \{\log[(1-t)/2]+1\}/(4\pi)$ if $\xi \neq \eta$; see, for example, Backus et al. (1996, pp. 165–166) or Freeden and Gutting (2013, pp. 159–161), where K is also called Green's function of the Beltrami operator. Moreover, for the Lebesgue integral, it is clear that it does not make any difference if $P_r(\xi)$ in (6.19) is, for each fixed $\xi \in \Omega$, determined by integrating over $\Omega \setminus \{\xi\}$. Furthermore, the orthogonality of the Legendre polynomials and the absence of degree 0 in the expansion of K imply that $\int_\Omega K(\xi \cdot \eta)\, \mathrm{d}\omega(\eta) = 0$ for all $\xi \in \Omega$ such that we have $\int_\Omega |K(\xi \cdot \eta)|\, \mathrm{d}\omega(\eta) < +\infty$ for all $\xi \in \Omega$.

Now let $r \in]a,c[$ be arbitrary. Clearly, there exist $\gamma, \delta \in \mathbb{R}$ with $a < \gamma < r < \delta < c$. Then F is k-times continuously differentiable on the compact set $\Omega_{[\gamma,\delta]} := \{x \in \mathbb{R}^3 \mid \gamma \le |x| \le \delta\}$ and Theorem 2.4.9 allows us to conclude that

$$\frac{\partial^j}{\partial r^j} P_r(\xi) = \int_\Omega K(\xi \cdot \eta) \frac{\partial^j}{\partial r^j} (r F(r\eta))\, \mathrm{d}\omega(\eta), \quad \xi \in \Omega,$$

exists and holds true for all $j = 0, \ldots, k$. Moreover, the continuity of these derivatives follows from Theorems 2.4.5 (regarding r) and 4.7.30 (regarding ξ). For general limits $r_k \xi^{(k)} \to \varrho\zeta$ within $\Omega_{[\gamma,\delta]}$, we get with $A(r\eta) := \frac{\partial^j}{\partial r^j}(r F(r\eta))$ that

$$\begin{aligned}
&\left| \int_\Omega K\left(\xi^{(k)} \cdot \eta\right) A\left(r_k \eta\right) \mathrm{d}\omega(\eta) - \int_\Omega K(\zeta \cdot \eta) A(\varrho\eta)\, \mathrm{d}\omega(\eta) \right| \\
&\quad \le \left| \int_\Omega K\left(\xi^{(k)} \cdot \eta\right) \left(A\left(r_k\eta\right) - A(\varrho\eta)\right) \mathrm{d}\omega(\eta) \right| \\
&\qquad + \left| \int_\Omega \left(K\left(\xi^{(k)} \cdot \eta\right) - K(\zeta \cdot \eta) \right) A(\varrho\eta)\, \mathrm{d}\omega(\eta) \right| \\
&\quad \le \left(2\pi \int_{-1}^{1} K(t)^2\, \mathrm{d}t \right)^{1/2} \left(\int_\Omega \left(A\left(r_k\eta\right) - A(\varrho\eta)\right)^2 \mathrm{d}\omega(\eta) \right)^{1/2} \\
&\qquad + \left(\int_\Omega \left(K\left(\xi^{(k)} \cdot \eta\right) - K(\zeta \cdot \eta) \right)^2 \mathrm{d}\omega(\eta) \right)^{1/2} \left(\int_\Omega (A(\varrho\eta))^2\, \mathrm{d}\omega(\eta) \right)^{1/2}.
\end{aligned}$$

Note that $K \in \mathrm{L}^2[-1,1]$, A is continuous, and Ω as well as $\Omega_{[\gamma,\delta]}$ are compact. For given $\varepsilon > 0$, we can now find a k_0 such that, for all $k \ge k_0$, we have $|A(r_k\eta) - A(\varrho\eta)| \le \varepsilon$ for all $\eta \in \Omega$. Moreover,

$$\begin{aligned}
\left\| K\left(\xi^{(k)} \cdot\right) - K(\zeta \cdot) \right\|_{\mathrm{L}^2(\Omega)}^2 &= \sum_{n=1}^{\infty} \sum_{j=-n}^{n} \frac{1}{n^2(n+1)^2} \left(Y_{n,j}\left(\xi^{(k)}\right) - Y_{n,j}(\zeta) \right)^2 \\
&\le \sum_{n=1}^{\infty} \frac{2n+1}{n^2(n+1)^2\pi}
\end{aligned}$$

is uniformly convergent such that we can conclude that there is also a k_1 such that, for all $k \geq k_1$, this norm is less than ε. In total, we obtain that $r\xi \mapsto \frac{\partial^j}{\partial r^j} P_r(\xi)$ is continuous on $\Omega_{]a,c[}$. We also see that

$$\Delta_\xi^* \frac{\partial^j}{\partial r^j} P_r(\xi) = \frac{\partial^j}{\partial r^j} (r F(r\xi)) = \frac{\partial^j}{\partial r^j} \Delta_\xi^* P_r(\xi)$$

for all $j = 0, \ldots, k$, all $r \in]a,c[$, and all $\xi \in \Omega$ due to Theorem 4.7.29 and (6.17). Note that the term in the middle has a vanishing spherical integral, because the radial derivative may be interchanged with the integration due to reasons which are analogous to the preceding case.

Theorem 4.7.30 tells us that $\frac{\partial^j}{\partial r^j} P_r \in \mathrm{C}^{(k-j+2)}(\Omega)$ for $j = 0, \ldots, k$ and each $r \in]a,c[$. In particular, $\nabla^* \frac{\partial^j}{\partial r^j} P_r$ exists and is continuous in the angular variables. From Freeden and Schreiner (2009, Theorem 4.5), we know that ∇^* may be interchanged with the integral over K times a continuous function. Hence,

$$\nabla_\xi^* \frac{\partial^j}{\partial r^j} P_r(\xi) = \int_\Omega \nabla_\xi^* K(\xi \cdot \eta) \frac{\partial^j}{\partial r^j} (r F(r\eta)) \, \mathrm{d}\omega(\eta)$$

for all $j = 0, \ldots, k$ and $r\xi \in \Omega_{]a,c[}$. It is easy to verify that $\nabla_\xi^* K(\xi \cdot \eta) = -(\eta - (\xi \cdot \eta)\xi)/[4\pi(1 - \xi \cdot \eta)]$ and $|\nabla_\xi^* K(\xi \cdot \eta)|^2 = (1 + \xi \cdot \eta)/[16\pi^2(1 - \xi \cdot \eta)]$ for $\xi \neq \eta$. Since Theorem 4.1.9 yields

$$\int_\Omega \left| \nabla_\xi^* K(\xi \cdot \eta) \right| \mathrm{d}\omega(\eta) = \frac{1}{2} \int_{-1}^{1} \sqrt{\frac{1+t}{1-t}} \, \mathrm{d}t = \frac{\pi}{2},$$

we can apply the dominated convergence theorem (Theorem 2.4.5) similarly to how we argued previously and get the continuity of $r \mapsto \nabla_\xi^* \frac{\partial^j}{\partial r^j} P_r(\xi)$. Let now $r_k \xi^{(k)} \to \varrho\zeta$ in $\Omega_{[\gamma,\delta]}$. Then, for all $k \geq k_0(\varepsilon)$ (as introduced earlier in this proof), we have (with a slightly modified derivation) that

$$\begin{aligned}
&\left| \int_\Omega \nabla_\xi^* K\left(\xi^{(k)} \cdot \eta\right) A(r_k \eta) \, \mathrm{d}\omega(\eta) - \int_\Omega K(\zeta \cdot \eta) A(\varrho\eta) \, \mathrm{d}\omega(\eta) \right| \\
&\leq \varepsilon \int_\Omega \left| \nabla_\xi^* K\left(\xi^{(k)} \cdot \eta\right) \right| \mathrm{d}\omega(\eta) \\
&\quad + \left| \int_\Omega \left(\nabla_\xi^* K\left(\xi^{(k)} \cdot \eta\right) - \nabla_\xi^* K(\zeta \cdot \eta) \right) A(\varrho\eta) \, \mathrm{d}\omega(\eta) \right| \\
&= \frac{\varepsilon\pi}{2} + \left| \nabla_\xi^* \frac{\partial^j}{\partial r^j} P_r\left(\xi^{(k)}\right) \Big|_{r=\varrho} - \nabla_\xi^* \frac{\partial^j}{\partial r^j} P_r(\zeta) \Big|_{r=\varrho} \right|,
\end{aligned}$$

where the latter term is less than ε for sufficiently large k. Hence, we get continuous derivatives $\nabla^* \frac{\partial^j}{\partial r^j} P_r$ for $j = 0, \ldots, k$ with $k \geq 2$ on $\Omega_{]a,c[}$. Since we know already that $\frac{\partial^j}{\partial r^j} P_r$ exists and is continuous on the spherical shell $\Omega_{]a,c[}$ for $j = 0, \ldots, k$ and $\nabla^* \otimes \nabla^* P_r$ exists on the shell, the function $r\xi \mapsto P_r(\xi)$ is twice partially differentiable in all combinations of variables and $\nabla^* \frac{\partial}{\partial r} P_r = \frac{\partial}{\partial r} \nabla^* P_r$ on $\Omega_{]a,c[}$.

We continue with some elementary calculations to get a useful formula for the divergence of f (remember Theorems 3.4.1 and 4.1.6):

$$\begin{aligned}\nabla_{r\xi} \cdot f(r\xi) &= \left(\xi \frac{\partial}{\partial r} + \frac{1}{r} \nabla_\xi^*\right) \cdot f(r\xi) \\ &= -\frac{1}{r^2} \Delta_\xi^* P_r(\xi) + \frac{1}{r} \Delta_\xi^* \frac{\partial}{\partial r} P_r(\xi) \\ &\quad + \frac{1}{r^2} \nabla_\xi^* \cdot \left(\xi \Delta_\xi^* P_r(\xi)\right) + \frac{1}{r} \Delta_\xi^* G(r\xi).\end{aligned} \tag{6.20}$$

We have a closer look at the third term of the right-hand side:

$$\begin{aligned}\nabla_\xi^* \cdot \left(\xi \Delta_\xi^* P_r(\xi)\right) &= \sum_{j=1}^{3} \left(\nabla_\xi^*\right)_j \left(\xi_j \Delta_\xi^* P_r(\xi)\right) \\ &= \left(\nabla_\xi^* \cdot \xi\right) \Delta_\xi^* P_r(\xi) + \underbrace{\xi \cdot \nabla_\xi^* \Delta_\xi^* P_r(\xi)}_{=0},\end{aligned}$$

where $\nabla_\xi^* \cdot \xi = 2$ due to Theorem 4.1.3. Hence, (6.20) becomes

$$\begin{aligned}\nabla_{r\xi} \cdot f(r\xi) &= -\frac{1}{r^2} \Delta_\xi^* P_r(\xi) + \frac{1}{r} \Delta_\xi^* \frac{\partial}{\partial r} P_r(\xi) + \frac{2}{r^2} \Delta_\xi^* P_r(\xi) + \frac{1}{r} \Delta_\xi^* G(r\xi) \\ &= \Delta_\xi^* \left(\frac{1}{r^2} P_r(\xi) + \frac{1}{r} \frac{\partial}{\partial r} P_r(\xi) + \frac{1}{r} G(r\xi)\right) \\ &= \Delta_\xi^* \left[\frac{1}{r^2} \frac{\partial}{\partial r} (r P_r(\xi)) + \frac{1}{r} G(r\xi)\right].\end{aligned}$$

Since f is solenoidal, Theorems 4.2.19 and 6.3.3 imply that the term in square brackets is, for each fixed r, constant. In other words, there is a function $A\colon]a,c[\to \mathbb{R}$ such that

$$\frac{1}{r^2} \frac{\partial}{\partial r} (r P_r(\xi)) + \frac{1}{r} G(r\xi) = A(r) \tag{6.21}$$

for all $r \in]a,c[$ and all $\xi \in \Omega$. Remember that we required in (6.15) that

$$\int_{S_r(0)} G(x) \, \mathrm{d}\omega(x) = 0 \tag{6.22}$$

for all $r \in]a,c[$. Remember also that P_r originates from (6.17). Since we have already seen that $r\xi \mapsto \frac{\partial^j}{\partial r^j} P_r(\xi)$, $j = 0, \ldots, k$, is continuous on each compact shell $\Omega_{[\gamma,\delta]}$, we may apply Theorem 2.4.9 and (6.16) in the sense that (with $x = r\xi$)

$$\begin{aligned}\int_{S_r(0)} \frac{1}{r^2} \frac{\partial}{\partial r} (r P_r(\xi)) \, \mathrm{d}\omega(x) &= \int_\Omega \frac{\partial}{\partial r} (r P_r(\xi)) \, \mathrm{d}\omega(\xi) \\ &= \frac{\partial}{\partial r} \left(r \int_\Omega P_r(\xi) \, \mathrm{d}\omega(\xi)\right) = 0\end{aligned} \tag{6.23}$$

for all $r \in]a,c[$. However, (6.21) in combination with (6.22) and (6.23) implies that $A \equiv 0$, which means that

$$G(r\xi) = -\frac{1}{r}\frac{\partial}{\partial r}(r P_r(\xi)) \tag{6.24}$$

for all $r \in]a,c[$ and all $\xi \in \Omega$. Let now $P(r\xi) := P_r(\xi)$ for all $r\xi \in \Omega_{]a,c[}$. It is now the right moment to summarize our results: from (6.10), (6.18), and (6.24) as well as Lemma 6.3.10, we obtain

$$\begin{aligned} f(r\xi) &= \frac{1}{r}\,\xi\,\Delta_\xi^* P(r\xi) - \frac{1}{r}\,\nabla_\xi^* \frac{\partial}{\partial r}(r P(r\xi)) + \mathrm{L}_{r\xi}\, Q(r\xi) \\ &= \frac{1}{r}\,\xi\,\Delta_\xi^* P(r\xi) - \nabla_\xi^* \left(\frac{1}{r} + \frac{\partial}{\partial r}\right) P(r\xi) + \mathrm{L}_{r\xi}\, Q(r\xi) \\ &= (\operatorname{curl} \mathrm{L} P)(r\xi) + (\mathrm{L} Q)(r\xi) \end{aligned} \tag{6.25}$$

for all $r \in]a,c[$ and all $\xi \in \Omega$. This is the Mie representation (6.13) of f.

(2) Uniqueness:

Let us assume that f is decomposed as in (6.13). Then Lemma 6.3.10 implies the representation in (6.25), which means that (remember Theorems 4.1.3 and 4.1.6)

$$\xi \cdot f(r\xi) = \frac{1}{r}\,\Delta_\xi^* P(r\xi), \tag{6.26}$$

$$\begin{aligned} \nabla_\xi^* \cdot f(r\xi) &= \frac{2}{r}\,\Delta_\xi^* P(r\xi) + \frac{1}{r}\,\underbrace{\xi \cdot \nabla_\xi^* \Delta_\xi^* P(r\xi)}_{=0} - \Delta_\xi^* \left(\frac{1}{r} + \frac{\partial}{\partial r}\right) P(r\xi) \\ &= \Delta_\xi^* \left(\frac{1}{r} - \frac{\partial}{\partial r}\right) P(r\xi), \end{aligned} \tag{6.27}$$

$$\mathrm{L}_\xi^* \cdot f(r\xi) = \Delta_\xi^* Q(r\xi) \tag{6.28}$$

for all $r \in]a,c[$ and all $\xi \in \Omega$. Hence, if there are further Mie scalars $\widetilde{P}$ and $\widetilde{Q}$, then $\Delta_\xi^*(P - \widetilde{P})(r\xi) = 0 = \Delta_\xi^*(Q - \widetilde{Q})(r\xi)$ such that Theorem 4.2.19 implies also here that, for fixed r, $(P - \widetilde{P})(r\xi)$ and $(Q - \widetilde{Q})(r\xi)$ are constant with respect to $\xi \in \Omega$. Therefore, let $A, B\colon]a,c[\to \mathbb{R}$ be functions such that $P(r\xi) = \widetilde{P}(r\xi) + A(r)$ and $Q(r\xi) = \widetilde{Q}(r\xi) + B(r)$ for all $r \in]a,c[$ and $\xi \in \Omega$. Then

$$\begin{aligned} \mathrm{L}_x \widetilde{Q}(x) = \mathrm{L}_\xi^* \widetilde{Q}(r\xi) &= \left(\xi \times \nabla_\xi^*\right) \widetilde{Q}(r\xi) \\ &= (\xi \times \nabla_\xi^*)\left(\widetilde{Q}(r\xi) + B(r)\right) = \mathrm{L}_x Q(x) \end{aligned}$$

and analogously $\mathrm{L}_x \widetilde{P}(x) = \mathrm{L}_x P(x)$ for all $x \in \Omega_{]a,c[}$. This shows that, independent of the chosen Mie scalars P and Q, the poloidal part $\operatorname{curl} \mathrm{L} P$ and the toroidal part $\mathrm{L} Q$ are unique. □

In view of (6.26)–(6.28) and Theorem 4.7.27, we are immediately enabled to deduce the following additional result.

Corollary 6.3.12 *The Mie scalars P and Q in Theorem 6.3.11 are uniquely determined if one additionally requires that*

$$\int_{S_r(0)} P(x)\,\mathrm{d}\omega(x) = 0 = \int_{S_r(0)} Q(x)\,\mathrm{d}\omega(x) \quad \textit{for all } r \in]a,c[.$$

An orthogonal polynomial basis system on the foundations of the spherical Helmholtz decomposition and the Mie representation for functions on a ball was constructed in Kazantsev and Kardakov (2019) by the use of (generalized) Zernike polynomials.

The Mie representation also allows the following short characterizations of poloidal and toroidal fields.

Corollary 6.3.13 *The only vector field f with $f \in \mathrm{c}^{(j,\alpha)}(\Omega_{]a,c[})$ for all $j = 0,1,2$ on a spherical shell $\Omega_{]a,c[}$ which is poloidal and toroidal is the zero function.*

Proof Since the zero function is obviously poloidal and toroidal, f has the Mie representation as 'poloidal + toroidal' in two ways: $f = f + 0 = 0 + f$. However, the Mie representation is unique (see Theorem 6.3.11). Hence, both decompositions must be identical: $f = 0$. □

Corollary 6.3.14 *Let $\Omega_{]a,c[}$ be a spherical shell and let $p \in \mathrm{c}^{(j,\alpha)}(\Omega_{]a,c[})$ for all $j = 0,1,2$ and a fixed $\alpha \in]0,1[$. Then the following holds true: p is poloidal with a poloidal scalar in $\mathrm{C}^{(3)}(\Omega_{]a,c[}) \Leftrightarrow p$ is solenoidal with poloidal Mie scalar in $\mathrm{C}^{(3)}(\Omega_{]a,c[})$ and toroidal Mie scalar in $\mathrm{C}^{(2)}(\Omega_{]a,c[})$ and its curl is tangential (i.e., $\xi \cdot \mathrm{curl}_{r\xi}\, p(r\xi) = 0$ for all $r \in]a,c[$ and all $\xi \in \Omega$).*

Proof We prove both implications separately.

(1) '⇒':

Let p be poloidal such that there is a scalar function $P \in \mathrm{C}^{(3)}(\Omega_{]a,c[})$ with $p = \mathrm{curl}\,\mathrm{L}P$. Hence, Theorem 2.3.4 yields that p is divergence-free. Furthermore, the representation of a poloidal field in (6.12) in Lemma 6.3.10 in combination with Theorem 4.1.6 leads us to $\xi \cdot (\mathrm{curl}\,\mathrm{L}P)(r\xi) = r^{-1}\Delta^*_\xi P(r\xi)$ for all $r \in]a,c[$ and all $\xi \in \Omega$. Hence, we have, for all $r \in]a,c[$, the result

$$\int_{S_r(0)} \xi \cdot p(r\xi)\,\mathrm{d}\omega(\xi) = \frac{1}{r}\int_{S_r(0)} \Delta^*_\xi P(r\xi)\,\mathrm{d}\omega(\xi),$$

where Green's first surface identity (see Theorem 4.1.11) implies that we obviously have $\int_{S_r(0)} 1 \cdot \Delta^*_\xi P(r\xi)\,\mathrm{d}\omega(\xi) = 0$. Therefore, we may apply Theorem 6.3.7 and conclude that p is solenoidal. Moreover, due to Theorem 2.3.4, we have

$$\mathrm{curl}\, p = \mathrm{curl}\,\mathrm{curl}\,\mathrm{L}P = \mathrm{grad}\,\mathrm{div}\,\mathrm{L}P - \Delta\mathrm{L}P. \tag{6.29}$$

From Theorem 6.3.2, we learn that the toroidal field $\mathrm{L}P$ is divergence-free. Moreover, it is easy to verify that Δ and L commute on $\mathrm{C}^{(3)}(\Omega_{]a,c[})$, that is, we have $\Delta\mathrm{L}P = \mathrm{L}\Delta P$. Hence, $\mathrm{curl}\, p = -\mathrm{L}\Delta P$ and, hence, $\xi \cdot \mathrm{curl}\, p(r\xi) = -\xi \cdot \mathrm{L}^*_\xi \Delta P(r\xi) = 0$, where we used (6.10) and Theorem 4.1.6.

(2) '$\Leftarrow$':

If p is solenoidal, then it allows the Mie representation $p = \operatorname{curl} \mathrm{L}P + \mathrm{L}Q$ with

$$\int_{S_r(0)} P(r\xi)\,\mathrm{d}\omega(\xi) = 0 = \int_{S_r(0)} Q(r\xi)\,\mathrm{d}\omega(\xi) \tag{6.30}$$

for all $r \in]a,c[$, where we assume $P \in \mathrm{C}^{(3)}(\Omega_{]a,c[})$. In analogy to our considerations in the context of (6.29), we get that $\operatorname{curl}\operatorname{curl} \mathrm{L}P = -\mathrm{L}\Delta P$. Furthermore, we consult Lemma 6.3.10 and conclude that

$$\operatorname{curl} p(r\xi) = -\mathrm{L}\Delta P(r\xi) + \frac{1}{r}\,\xi \Delta_\xi^* Q(r\xi) + \nabla_\xi^* \left(-\frac{1}{r} - \frac{\partial}{\partial r}\right) Q(r\xi)$$

for all $r \in]a,c[$ and all $\xi \in \Omega$. For the converse implication, which we want to prove here, we also have the requirement that the curl of p is tangential. In view of (6.10) and Theorem 4.1.6, we now obtain $0 = r^{-1}\Delta_\xi^* Q(r\xi)$ for all $r \in]a,c[$ and all $\xi \in \Omega$, which is only valid if Q has no angular dependence (remember the eigenfunctions of Δ^*; see Theorem 4.2.19), that is, $\xi \mapsto Q(r\xi)$ is constant for each fixed $r \in]a,c[$. However, (6.30) only allows that this constant is zero. Hence, $Q \equiv 0$ on the spherical shell. Thus, $p = \operatorname{curl} \mathrm{L}P$. □

In the context of (6.29) we have also proved the following identity.

Corollary 6.3.15 *Every function $F \in \mathrm{C}^{(3)}(\Omega_{]a,c[})$ on a spherical shell $\Omega_{]a,c[}$ satisfies* $\operatorname{curl}\operatorname{curl} \mathrm{L}F = -\mathrm{L}\Delta F$.

We prove some further properties of poloidal and toroidal fields.

Corollary 6.3.16 *Let $\Omega_{]a,c[}$ be a spherical shell and let $p \in \mathrm{c}^{(j,\alpha)}(\Omega_{]a,c[})$ for all $j = 0,1,2$ and a fixed $\alpha \in]0,1[$. Then the following holds true: p is poloidal with a poloidal scalar in $\mathrm{C}^{(3)}(\Omega_{]a,c[})$ $\Leftrightarrow$ p is solenoidal with poloidal Mie scalar in $\mathrm{C}^{(3)}(\Omega_{]a,c[})$ and toroidal Mie scalar in $\mathrm{C}^{(2)}(\Omega_{]a,c[})$ and* curl *p is toroidal.*

Proof If p is poloidal with $\mathrm{C}^{(3)}$-scalar, then Corollary 6.3.14 tells us that p is solenoidal and Corollary 6.3.15 shows that curl p is a toroidal vector field.

Vice versa, let p be solenoidal and let there be a scalar field Q with $\operatorname{curl} p = \mathrm{L}Q$. We know that L corresponds to L^*, as shown in (6.10), and L^* produces tangential vector fields, as a result of Theorem 4.1.6. Eventually, it is only left to consult Corollary 6.3.14 to see that p is poloidal. □

Corollary 6.3.17 *Let $\Omega_{]a,c[}$ be a given shell and let $q \in \mathrm{c}^{(j,\alpha)}(\Omega_{]a,c[})$ for all $j = 0, 1,2$ and a fixed $\alpha \in]0,1[$. Then the following holds true: q is toroidal $\Leftrightarrow$ q is solenoidal and tangential, that is, $\xi \cdot q(r\xi) = 0$ for all $r \in]a,c[$ and all $\xi \in \Omega$.*

Proof We have already seen that toroidal fields q are tangential. Moreover, Theorem 6.3.2 tells us that q is also divergence-free. If we combine both properties of toroidal fields q, then Theorem 6.3.7 yields that q is also solenoidal.

Vice versa, let q be solenoidal and tangential. Then we can apply the Mie representation and get scalar fields P and Q with $q = \operatorname{curl} \mathrm{L}P + \mathrm{L}Q$ and

$$\int_{S_r(0)} P(x)\,\mathrm{d}\omega(x) = \int_{S_r(0)} Q(x)\,\mathrm{d}\omega(x) = 0 \tag{6.31}$$

for all $r \in]a,c[$; see Theorem 6.3.11 and Corollary 6.3.12. Due to the first part of this proof, the toroidal field $\mathrm{L}Q$ is tangential such that $\xi \cdot q(r\xi) = \xi \cdot (\operatorname{curl} \mathrm{L}P)(r\xi)$ for all $r \in]a,c[$ and all $\xi \in \Omega$. Moreover, Lemma 6.3.10 tells us that $\xi \cdot (\operatorname{curl} \mathrm{L}P)(r\xi) = r^{-1}\, \Delta_\xi^* P(r\xi)$.

Since we assume that q is tangential, we see now that $\Delta_\xi^* P(r\xi) = 0$ for all $r \in]a,c[$ and all $\xi \in \Omega$, which means that $\Omega \ni \xi \mapsto P(r\xi)$ is constant for each fixed $r \in]a,c[$, see Theorem 4.2.19. Due to (6.31), this constant is zero for each r. Hence, $q = \mathrm{L}Q$ is toroidal. □

Let us now apply the Mie representation to the geomagnetic field modelling. The magnetic field B is solenoidal due to (6.4), which holds true in the whole $\mathbb{R}^3$, and Theorem 6.3.6. Hence, we find scalar fields P^B and Q^B, where the superscript is a reference to the magnetic field, such that

$$B = \operatorname{curl} \mathrm{L}P^B + \mathrm{L}Q^B \tag{6.32}$$

and

$$\int_{S_r(0)} P^B(x)\,\mathrm{d}\omega(x) = 0 = \int_{S_r(0)} Q^B(x)\,\mathrm{d}\omega(x) \tag{6.33}$$

for all r in a subinterval of $\mathbb{R}_0^+$ which may be chosen arbitrarily. We assume sufficient regularity for all occurring functions here. In the static case, which we assume to be valid in a spherical shell $\Omega_{]a,c[}$, (6.3) reduces to $\nabla \times B = \mu_0 J$. The left-hand side, that is, the curl of B, can be expressed, provided that the Mie representation (6.32) is used, by applying Corollary 6.3.15. We get

$$\mu_0 J = \operatorname{curl} B = -\mathrm{L}\Delta P^B + \operatorname{curl} \mathrm{L}Q^B. \tag{6.34}$$

Hence, J is solenoidal and has a unique Mie representation

$$J = \operatorname{curl} \mathrm{L}P^J + \mathrm{L}Q^J \tag{6.35}$$

with

$$\int_{S_r(0)} P^J(x)\,\mathrm{d}\omega(x) = 0 = \int_{S_r(0)} Q^J(x)\,\mathrm{d}\omega(x) \tag{6.36}$$

for all $r \in]a,c[$. Let us compare the Mie scalars in (6.34) and (6.35). It appears that $P^J = \mu_0^{-1} Q^B$ and $Q^J = -\mu_0^{-1} \Delta P^B$, but we need to check, if (6.36) is satisfied by these scalars. For P^J, this is immediately clear because of (6.33). For Q^J, we utilize the decomposition of the Laplace operator (see Theorem 3.4.3). Furthermore, we observe that the following interchanging of the radial differentiation and the angular integration is allowed, because we assumed a sufficient regularity of the occurring functions and because a sphere is compact. We get

$$\begin{aligned}\int_{S_r(0)} \Delta P^B(x)\, d\omega(x) &= \int_{S_r(0)} \left(\frac{\partial^2}{\partial r^2} + \frac{2}{r}\frac{\partial}{\partial r} + \frac{1}{r^2}\Delta_\xi^*\right) P^B(r\xi)\, d\omega(r\xi)\\ &= \left(\frac{\partial^2}{\partial r^2} + \frac{2}{r}\frac{\partial}{\partial r}\right)\int_{S_r(0)} P^B(r\xi)\, d\omega(r\xi)\\ &\quad + \frac{1}{r^2}\int_{S_r(0)} \Delta_\xi^* P^B(r\xi)\, d\omega(r\xi).\end{aligned}$$

The first integral on the right-hand side vanishes due to (6.33). The latter one can be handled with one of Green's surface identities (where the other function is constant), see Theorem 4.1.11, which yields a vanishing integral. Hence, we have, indeed,

$$P^J = \frac{1}{\mu_0} Q^B \quad \text{and} \quad Q^J = -\frac{1}{\mu_0}\Delta P^B. \tag{6.37}$$

6.4 Internal and External Fields under the Mie Representation

We have seen in Section 6.2 how external and internal fields can be separated if currents are neglected, that is, if the Gauß representation is used. The essential argument of the derivation was the harmonicity of the field and its potential. This is no longer the case if currents are present and the Mie representation needs to be used.

Remember Theorem 6.1.2 and Corollary 6.1.3. If $D = \Sigma_{\text{int}}$ is the interior of a regular surface and represents a planet, for example the Earth, then currents $J \in \mathrm{c}^{(1)}(\overline{D})$ produce the magnetic field $B^{\mathrm{I}} = \nabla \times A^{\mathrm{I}}$ with

$$A^{\mathrm{I}}(y) = \frac{\mu_0}{4\pi}\int_D \frac{J(x)}{|x-y|}\, dx \quad \text{and} \quad B^{\mathrm{I}}(y) = \frac{\mu_0}{4\pi}\int_D J(x) \times \frac{y-x}{|y-x|^3}\, dx, \tag{6.38}$$

where the latter is the Biot–Savart law. We use here the superscript 'I' in order to stress that this is the magnetic field caused by inner sources, which is called the internal field. If J is Hölder continuous, then

$$\Delta A^{\mathrm{I}} = -\mu_0 J \text{ in } D \quad \text{and} \quad \Delta A^{\mathrm{I}} = 0 \text{ in } \mathbb{R}^3 \setminus \overline{D}$$

due to Theorems 3.1.5 and 3.1.11. Hence, with Theorem 2.3.4, we can deduce that $\nabla \times B^{\mathrm{I}} = \nabla \times \nabla \times A^{\mathrm{I}} = \nabla(\nabla \cdot A^{\mathrm{I}}) - \Delta A^{\mathrm{I}}$ and we get

$$\nabla \times B^{\mathrm{I}} = \nabla\left(\nabla \cdot A^{\mathrm{I}}\right) + \mu_0 J \text{ in } D \quad \text{and} \quad \nabla \times B^{\mathrm{I}} = \nabla\left(\nabla \cdot A^{\mathrm{I}}\right) \text{ in } \mathbb{R}^3 \setminus \overline{D}. \tag{6.39}$$

The integral for A^{I} in (6.38) has componentwise the form of a gravitational potential, where the components of J play the role of the mass density function. Hence, Corollary 3.1.4 yields

$$\frac{\partial}{\partial y_j} A_j^{\mathrm{I}}(y) = \frac{\mu_0}{4\pi}\int_D J_j(x)\frac{\partial}{\partial y_j}\frac{1}{|x-y|}\, dx.$$

Since $\nabla_x |x-y|^{-1} = -\nabla_y |x-y|^{-1}$, we can conclude, by also using the product rule, that

$$\frac{\partial}{\partial y_j} A_j^{\mathrm{I}}(y) = -\frac{\mu_0}{4\pi} \int_D J_j(x) \frac{\partial}{\partial x_j} \frac{1}{|x-y|} \,\mathrm{d}x$$

$$= -\frac{\mu_0}{4\pi} \int_D \frac{\partial}{\partial x_j} \frac{J_j(x)}{|x-y|} - \frac{1}{|x-y|} \frac{\partial}{\partial x_j} J_j(x) \,\mathrm{d}x.$$

Hence, the divergence of A^{I} takes the form

$$\nabla \cdot A^{\mathrm{I}}(y) = -\frac{\mu_0}{4\pi} \int_D \nabla_x \cdot \frac{J(x)}{|x-y|} - \frac{\nabla \cdot J(x)}{|x-y|} \,\mathrm{d}x.$$

In the static case, we have $\nabla \cdot J = 0$ (see Theorem 6.1.2). Moreover, we apply Gauß's law (Theorem 2.4.2). This leads us to

$$\nabla \cdot A^{\mathrm{I}}(y) = -\frac{\mu_0}{4\pi} \int_\Sigma \frac{1}{|x-y|} \nu(x) \cdot J(x) \,\mathrm{d}\omega(x), \tag{6.40}$$

where ν is the outer unit normal on Σ. Note that this has the form of a single-layer potential. The formula shows us that, if the currents on the surface are only tangential, then $\nabla \cdot A^{\mathrm{I}} \equiv 0$ and, due to (6.39), $\nabla \times B^{\mathrm{I}}(y) = 0$ for all $y \in \mathbb{R}^3 \setminus \overline{D}$. In general, we will not have a purely tangential J on Σ. For this reason, external sources also need to be modelled.

For mathematical reasons, it is easier to assume that the external sources are in a bounded domain. This is not a serious restriction, because we can take a ball with an arbitrarily large radius. So, let $E := \Omega_{]a,c[} \subset \mathbb{R}^3$ be the domain which includes all external sources but not the internal sources, that is, we require $E \subset \mathbb{R}^3 \setminus \overline{D}$. Obviously, E is measurable and bounded. Moreover, we choose $D := B_a(0)$, that is, $\Sigma = S_a(0)$. An established ansatz is to define the vector fields

$$p^{\mathrm{I}}(y) := \frac{\mu_0}{4\pi} \int_D \left(\mathrm{L}Q^J\right)(x) \times \frac{y-x}{|y-x|^3} \,\mathrm{d}x,$$

$$p^{\mathrm{E}}(y) := \frac{\mu_0}{4\pi} \int_E \left(\mathrm{L}Q^J\right)(x) \times \frac{y-x}{|y-x|^3} \,\mathrm{d}x,$$

$y \in \mathbb{R}^3$, where Q^J is the toroidal scalar of the current J; see (6.35). If we compare this ansatz with (6.38), then we see that these vector fields satisfy, in analogy to (6.39) and (6.40), the partial differential equation

$$\nabla \times p^{\mathrm{I}}(y) = -\frac{\mu_0}{4\pi} \nabla \left(\int_\Sigma \frac{1}{|x-y|} \underbrace{\nu(x) \cdot \left(\mathrm{L}Q^J\right)(x)}_{=0} \,\mathrm{d}\omega(x) \right) + \mu_0 \left(\mathrm{L}Q^J\right)(y) \tag{6.41}$$

for all $y \in D$, where we used Corollary 6.3.17 (the fact that toroidal fields are tangential is independent of the domain). As a consequence and by analogous considerations, we get

$$\nabla \times p^{\mathrm{I}} = \mu_0 \mathrm{L} Q^J \qquad \text{on } D, \tag{6.42}$$

$$\nabla \times p^{\mathrm{I}} = 0 \qquad \text{on } \mathbb{R}^3 \setminus \overline{D} \supset E, \tag{6.43}$$

$$\nabla \times p^{\mathrm{E}} = 0 \qquad \text{on } D, \tag{6.44}$$

$$\nabla \times p^{\mathrm{E}} = \mu_0 \mathrm{L} Q^J \qquad \text{on } E. \tag{6.45}$$

For p^{E}, we get two surface integrals in the analogue of (6.41), because ∂E consists of two spheres. Both integrals vanish also in this case.

Hence, p^{I} and p^{E} have toroidal curls on D and E (separately). Moreover, both are curls of a field like A^{I} in (6.38), but with $\mathrm{L}Q^J$ instead of J; see also the derivation of the Biot–Savart law (Corollary 6.1.3). Since p^{I} and p^{E} are defined on the whole $\mathbb{R}^3$, they are consequently solenoidal due to Theorem 6.3.6. Now we have all properties which are required by Corollary 6.3.16 such that we may conclude that p^{I} and p^{E} are poloidal (on D and E). Furthermore, we have $\nabla \times \left(p^{\mathrm{I}} + p^{\mathrm{E}}\right) = \mu_0 \mathrm{L} Q^J$ due to (6.42) and (6.45), and consequently $\nabla \times \left(p^{\mathrm{I}} + p^{\mathrm{E}} + \mathrm{L} Q^B\right) = \mu_0 \mathrm{L} Q^J + \mu_0 \nabla \times \mathrm{L} P^J = \mu_0 J$; see (6.37) and (6.35). In view of (6.32), the static Maxwell equation $\nabla \times B = \mu_0 J$, and the uniqueness of the Mie representation, we see that

$$B = p^{\mathrm{I}} + p^{\mathrm{E}} + \mathrm{L} Q^B \tag{6.46}$$

works, where p^{I} and p^{E} are poloidal fields associated to internal and external sources, respectively, while these sources are toroidal currents. This justifies the ansatz.

For the toroidal part of the magnetic field, the situation is different. From (6.37), we know that $\mathrm{L} Q^B = \mu_0 \mathrm{L} P^J$. We combine this with (6.17) from the proof of the Mie representation and obtain

$$\mathrm{L} Q^B(r\xi) = \mu_0 \mathrm{L}\left(\Delta_\xi^*\right)^{-1}(r\xi \cdot J(r\xi)), \tag{6.47}$$

because P^J is the poloidal scalar of J, and F in (6.17) is the scalar of the normal part of the decomposed function, which is in this case the current J. The preceding identity is true for all $\xi \in \Omega$ and all $r \in]a,c[$, where we use again $\Sigma = S_a(0)$ as the Earth's surface and $c > a$ is a radius which we can choose based on the practical situation such that $\Omega_{]a,c[}$ covers the interesting domain. Referring again to our assumption that all occurring functions are sufficiently regular, we obtain that (6.47) still holds true in the limit $r \to a+$. Hence, based on (6.46), we can decompose B at the Earth's surface as follows:

$$B(a\xi) = p^{\mathrm{I}}(a\xi) + p^{\mathrm{E}}(a\xi) + q(a\xi),$$

where the toroidal part $q(r\xi) := \mathrm{L} Q^B(r\xi)$, $r \in [a,c[$, $\xi \in \Omega$, at $r = a$, is produced by the normal component of the current crossing the surface $S_a(0)$. Therefore, we do not have to (and cannot) subdivide the toroidal magnetic field into contributions of inner and outer sources. Instead, it corresponds to a process taking place at the very same sphere where we also consider the magnetic field.

We have seen in Section 6.2 that, in the absence of currents, the Gauß representation allows an orthogonal decomposition of the field based on outer and inner sources. In the more general situation of the Mie representation, this is still possible. Due to our previous considerations and analogous conclusions, the inner-source part p^{I} of the poloidal magnetic field has a poloidal scalar P^{I}, which satisfies

$$\Delta P^{\mathrm{I}} = -\mu_0 Q^J \quad \text{on } D, \qquad\qquad \Delta P^{\mathrm{I}} = 0 \quad \text{on } \mathbb{R}^3 \setminus \overline{D} \supset E.$$

Analogously, we get for the poloidal scalar P^{E} of p^{E}, due to (6.44) and (6.45),

$$\Delta P^{\mathrm{E}} = 0 \quad \text{on } D, \qquad\qquad \Delta P^{\mathrm{E}} = -\mu_0 Q^J \quad \text{on } E.$$

Since $D = B_a(0)$, we know, due to our previous work in this book, how to expand these poloidal scalars in outer and inner harmonics (see the discussions after Theorem 3.4.25):

$$P^{\mathrm{I}}(r\xi) = \sum_{n=0}^{\infty} \sum_{j=-n}^{n} \left\langle P^{\mathrm{I}}|_{S_a(0)}, Y_{n,j}^a \right\rangle_{\mathrm{L}^2(S_a(0))} \left(\frac{a}{r}\right)^{n+1} Y_{n,j}^a(r\xi)$$

for $r > a$ and $\xi \in \Omega$, while

$$P^{\mathrm{E}}(r\xi) = \sum_{n=0}^{\infty} \sum_{j=-n}^{n} \left\langle P^{\mathrm{E}}|_{S_a(0)}, Y_{n,j}^a \right\rangle_{\mathrm{L}^2(S_a(0))} \left(\frac{r}{a}\right)^{n} Y_{n,j}^a(r\xi)$$

for $r < a$ and $\xi \in \Omega$. Our assumption on a sufficient regularity allows us to extend these equations to $r = a$. Furthermore, (6.43) implies the existence of a scalar potential ψ^{I} such that $p^{\mathrm{I}} = -\nabla \psi^{\mathrm{I}}$ on $\mathbb{R}^3 \setminus \overline{D}$. We compare now the decomposition of the gradient (see Theorem 3.4.1) with the structure of poloidal fields (see Lemma 6.3.10) in the case of p^{I}:

$$\begin{aligned} p^{\mathrm{I}}(r\xi) &= -\xi \frac{\partial}{\partial r} \psi^{\mathrm{I}}(r\xi) - \frac{1}{r} \nabla_\xi^* \psi^{\mathrm{I}}(r\xi) \\ &= \frac{1}{r} \xi \Delta_\xi^* P^{\mathrm{I}}(r\xi) + \nabla_\xi^* \left(-\frac{1}{r} - \frac{\partial}{\partial r}\right) P^{\mathrm{I}}(r\xi) \end{aligned}$$

for all $r > a$ and all $\xi \in \Omega$. Since surface gradients are tangential, we obtain two equations out of this identity:

$$-r \frac{\partial}{\partial r} \psi^{\mathrm{I}}(r\xi) = \Delta_\xi^* P^{\mathrm{I}}(r\xi) \quad \text{and} \quad \nabla_\xi^* \psi^{\mathrm{I}}(r\xi) = \nabla_\xi^* \left(1 + r \frac{\partial}{\partial r}\right) P^{\mathrm{I}}(r\xi).$$

Since ψ^{I} is unique up to a constant, we can choose this constant in a way such that $\psi^{\mathrm{I}}(r\xi) = \left(1 + r \frac{\partial}{\partial r}\right) P^{\mathrm{I}}(r\xi)$ at $r = a$ for all $\xi \in \Omega$. With a slight simplification of the right-hand side, we obtain

$$\begin{aligned} \psi^{\mathrm{I}}(r\xi) &= \frac{\partial}{\partial r} \left(r P^{\mathrm{I}}(r\xi)\right) \\ &= -\sum_{n=0}^{\infty} \sum_{j=-n}^{n} \left\langle P^{\mathrm{I}}|_{S_a(0)}, Y_{n,j}^a \right\rangle_{\mathrm{L}^2(S_a(0))} n \left(\frac{a}{r}\right)^{n+1} Y_{n,j}^a(r\xi) \end{aligned}$$

at $r = a$, but, since p^{I} is divergence-free (see Theorem 6.3.2) and $\nabla \cdot p^{\mathrm{I}} = -\Delta \psi^{\mathrm{I}}$ on $\mathbb{R}^3 \setminus \overline{D}$, the potential ψ^{I} allows the preceding expansion for all $r \geq a$ (due to the definition of p^{I}, it needs to be regular at infinity such that only outer harmonics can be used).

Analogously, we have a potential ψ^{E} of p^{E} on D due to (6.44) such that

$$\begin{aligned}\psi^{\mathrm{E}}(r\xi) &= \frac{\partial}{\partial r}\left(rP^{\mathrm{E}}(r\xi)\right)\\ &= \sum_{n=0}^{\infty}\sum_{j=-n}^{n}\left\langle P^{\mathrm{E}}\left|_{S_a(0)}, Y^a_{n,j}\right.\right\rangle_{\mathrm{L}^2(S_a(0))}(n+1)\left(\frac{r}{a}\right)^n Y^a_{n,j}(r\xi)\end{aligned}$$

for all $r \leq a$ and all $\xi \in \Omega$. By calculating the gradients of these potentials and taking into accout (4.30) and the consequent discussions, we see that

$$p^{\mathrm{I}}(r\xi) = \sum_{n=1}^{\infty}\sum_{j=-n}^{n} p^{\mathrm{I}}_{n,j}\left(\frac{a}{r}\right)^{n+2}\tilde{y}^{(1)}_{n,j}(\xi), \quad r \geq a, \quad \xi \in \Omega, \tag{6.48}$$

$$p^{\mathrm{E}}(r\xi) = \sum_{n=1}^{\infty}\sum_{j=-n}^{n} p^{\mathrm{E}}_{n,j}\left(\frac{r}{a}\right)^{n-1}\tilde{y}^{(2)}_{n,j}(\xi), \quad r \leq a, \quad \xi \in \Omega, \tag{6.49}$$

where

$$\begin{aligned}p^{\mathrm{I}}_{n,j} &= -\left\langle P^{\mathrm{I}}\left|_{S_a(0)}, Y^a_{n,j}\right.\right\rangle_{\mathrm{L}^2(S_a(0))}\frac{n\sqrt{(n+1)(2n+1)}}{a^2},\\ p^{\mathrm{E}}_{n,j} &= -\left\langle P^{\mathrm{E}}\left|_{S_a(0)}, Y^a_{n,j}\right.\right\rangle_{\mathrm{L}^2(S_a(0))}\frac{(n+1)\sqrt{n(2n+1)}}{a^2}\end{aligned}$$

for all $n \geq 1$ and all $j = -n, \ldots, n$. Eventually, the toroidal magnetic field is also expandable in Edmonds vector spherical harmonics (see also Definition 4.3.6 and the preceding discussions):

$$q(a\xi) = \sum_{n=1}^{\infty}\sum_{j=-n}^{n}\underbrace{\left\langle Q^B\left|_{S_a(0)}, Y^a_{n,j}\right.\right\rangle_{\mathrm{L}^2(S_a(0))}\frac{\sqrt{n(n+1)}}{a}}_{=:q_{n,j}}\tilde{y}^{(3)}_{n,j}(\xi), \quad \xi \in \Omega. \tag{6.50}$$

Hence, (6.48)–(6.50) give us an expansion of B at the Earth's surface $S_a(0)$ in the form

$$B(a\xi) = \sum_{n=1}^{\infty}\sum_{j=-n}^{n}\left(p^{\mathrm{I}}_{n,j}\tilde{y}^{(1)}_{n,j}(\xi) + p^{\mathrm{E}}_{n,j}\tilde{y}^{(2)}_{n,j}(\xi) + q_{n,j}\tilde{y}^{(3)}_{n,j}(\xi)\right)$$

for all $\xi \in \Omega$. Since the Edmonds vector spherical harmonics are orthonormal with respect to $\mathrm{l}^2(\Omega)$, the three parts of B, namely,

- the poloidal field p^{I}, which is produced by the toroidal current inside the Earth,
- the poloidal field p^{E}, which originates from external currents, and
- the toroidal field q, which is produced by the normal component of the current crossing the surface,

are orthogonal in the same sense and can, therefore, be completely decoupled if B is known on the surface $\Sigma = S_a(0)$.

Exercises

6.1 Determine a Helmholtz decomposition of the functions $f(x) = x$ and $g(x) = (x_2, x_3, x_1)^{\mathrm{T}}$.

6.2 Let $f(x) := \varepsilon^{\varphi}(\varphi)$ on the spherical shell $\Omega_{]\alpha,\beta[}$, where φ is the longitude of x. Is f solenoidal, toroidal, or poloidal? Determine corresponding Mie, toroidal, and poloidal scalars, as far as they exist.

6.3 What do you get if f in Exercise 6.2 is replaced by ε^{t} or ε^{r}?

6.4 Let again f be one of the vector fields ε^{r}, ε^{φ}, and ε^{t}. Determine a spherical Helmholtz decomposition in each case.

6.5 Use the Levi-Cività alternating symbol to prove that Δ and L commute on $\mathrm{C}^{(3)}(D)$, $D \subset \mathbb{R}^3$ open.

7
Mathematical Models in Seismology

The field of mathematical models in seismology is large and versatile. For this reason, this book would not be able to cover this whole area. We will focus here on some selected topics which are important as theoretical foundations of the science of earthquakes and which are related in parts to mathematical tools we have come across in earlier chapters. For more detailed discussions, for example in cases where we restrict our considerations to a spherical Earth and where the reader is interested in the 'whole picture', but also regarding more general theories on topics such as elasticity or waves, we recommend the standard books and relevant research papers. A small selection of such works is Aki and Richards (2002), Ben-Menahem and Singh (1981), Berkel and Michel (2010), Billingham and King (2000), Dahlen and Tromp (1998), Marsden and Hughes (1994), Michel (2003), and Nolet (2008), which also partially served as the basis for this chapter.

7.1 Continuum Mechanics with Focus on Elasticity

Briefly speaking, the existence of earthquakes is a consequence of the elasticity of the Earth. Tectonic plates move relative to each other, but this motion sometimes gets blocked at the boundaries, where different plates meet, the seismic faults. Since the driving forces are still there but a motion is not possible, the Earth starts to deform at the faults. This continues until the stress exceeds a certain point and the motion continues all of a sudden as an earthquake – like a rubber band ruptures if we pull too strongly. This nutshell explanation of the cause of an earthquake shows us that understanding earthquakes first requires the understanding of the mathematical methods for elastic bodies, which is a part of the mathematical theory of continuum mechanics. This section adopts concepts from Marsden and Hughes (1994), which is also recommended for further reading.

We start with some notations which will be used throughout this chapter.

Definition 7.1.1 A **simple body** is a bounded and open subset of $\mathbb{R}^3$ and is usually denoted by $\mathcal{B}$. A **configuration** of a simple body $\mathcal{B}$ is a mapping $F\colon \mathcal{B} \to \mathbb{R}^3$. Configurations are supposed to model a deformed state of a simple body. The set of all configurations of $\mathcal{B}$ is denoted by $\mathcal{C}(\mathcal{B})$ or simply $\mathcal{C}$, if the reference to the simple body is clear from the context. A mapping which associates configurations to every point of time t, that is, an operator of the kind $\phi\colon \mathbb{R} \ni t \mapsto \phi_t \in \mathcal{C}$, is called a **motion**

of $\mathcal{B}$. Elements (i.e., points) in $\mathcal{B}$ will be denoted by capital letters $X, Y, \ldots$, while elements of $\phi_t(\mathcal{B})$, that is, positions to which a point $X \in \mathcal{B}$ has moved at time t, will be denoted with lower-case letters $x, y, \ldots$, where usually $x = \phi_t(X)$, $y = \phi_t(Y)$, ... Moreover, we will use the notations $\phi_t(X) =: \phi(X,t) =: \phi_X(t)$.

The idea of this modelling is that a particle $X \in \mathcal{B}$ moves at time t to the position $x = \phi_t(X) \in \phi_t(\mathcal{B})$. In this sense, $\phi_t(\mathcal{B})$ represents the deformed state of the body $\mathcal{B}$ at time t. We will assume later in this chapter that $\mathcal{B}$ is the form of the body at time $t = 0$ such that $\phi_0(X) = X$ for all $X \in \mathcal{B}$. Moreover, the function ϕ_X follows the trajectory of particle X over the observed period of time. It is comparable to what is called a word line in relativity theory.

Example 7.1.2 As an example, we consider the motion

$$\phi(X,t) := \begin{pmatrix} X_1 + tX_2 \\ X_2 \\ X_3 + \frac{1}{2}tX_1 \end{pmatrix},$$

which is applied to the cube $]0,1[^3 =: \mathcal{B}$ as a simple body. Figure 7.1 shows the states $\phi_t(\mathcal{B})$ of the cube at the times $t = 0, 0.5, 1$, where $t = 0$ stands for the undeformed state, that is, $\phi_0(\mathcal{B}) = \mathcal{B}$.

Definition 7.1.3 Let $\mathcal{B}$ be a simple body. A motion ϕ of $\mathcal{B}$ is called **regular** or **invertible** if, for each time t, the set $\phi_t(\mathcal{B})$ is open and each mapping $\phi_t : \mathcal{B} \to \phi_t(\mathcal{B})$ has an inverse $\phi_t^{-1} : \phi_t(\mathcal{B}) \to \mathcal{B}$. If the motion additionally satisfies the requirement $\phi \in \mathrm{C}^{(r)}(\overline{\mathcal{B}} \times \mathbb{R}, \mathbb{R}^3)$ for $r \in \mathbb{N}_0$, then we call the motion ϕ a $\mathrm{C}^{(r)}$-regular motion.

Regularity is a reasonable assumption for motions in seismology. To understand this, let us assume that a motion is not regular, then two different particles $X, Y \in \mathcal{B}$ would move at time t to the same position $x = \phi_t(X) = y = \phi_t(Y)$. This would be the case, if, for example, a chemical reaction is taking place, where two molecules form a chemical compound. This is not a relevant situation in the modelling which we are interested in.

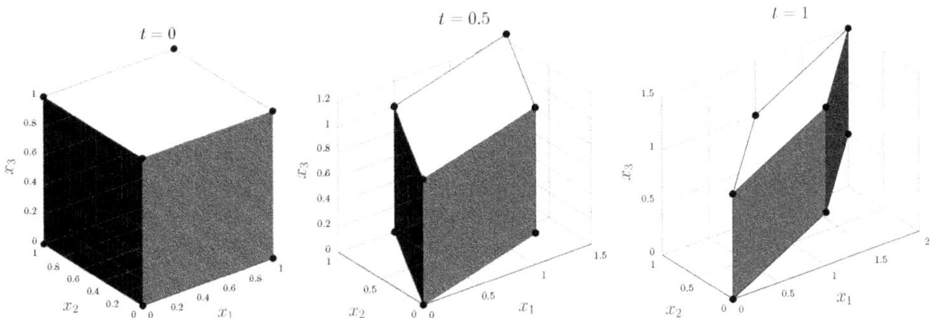

Figure 7.1 Example of a simple body (here: the unit cube) and a motion.

Based on our initial modelling, we can now state a general assumption, which shall be valid throughout the whole chapter.

Assumption 7.1.4 We assume that the following is valid unless anything different is stated.

(a) An initial state of the deformation is given at time $t = 0$, that is, $\phi(X,0) = X$ for all $X \in \mathcal{B}$.
(b) $\mathcal{B}$ represents a simple body and ϕ is a $\mathrm{C}^{(2)}$-regular motion, where the Jacobian satisfies $\det(\nabla_X \phi_t(X)) \neq 0$ for all times t and all $X \in \mathcal{B}$.

In continuum mechanics, one distinguishes two points of view how a motion can be observed or represented. The **Lagrangian description** or **material description** follows a fixed particle while it moves in space. Since particles are, in our modelling, elements of $\mathcal{B}$ and are represented by capital letters, functions corresponding to the material description are represented with capital letters as well according to a common notation, that is, we write, for example, $Q(X,t)$ for such a quantity.

On the other hand, the **Eulerian description** or **spatial description** observes the motion from a fixed point in space. Spatial quantities, which are accordingly denoted by lower-case letters, are transferred from material quantities by the formula $Q(X,t) = q(\phi_t(X),t)$ or, equivalently, $q(x,t) = Q\left(\phi_t^{-1}(x),t\right)$. For a better understanding, we first define the velocity and the acceleration in their Lagrangian and their Eulerian version. In this context, it should be mentioned that we will use 'Lagrangian' and 'material' as equivalent terms, that is, it does not matter if we speak about a Lagrangian velocity or a material velocity. The same holds true for the terms 'Eulerian' and 'spatial'.

Definition 7.1.5 Let ϕ be a motion which satisfies Assumption 7.1.4.

(a) The **material velocity** is defined by

$$V_t(X) := V_X(t) := V(X,t) := \frac{\partial}{\partial t}\phi(X,t), \quad X \in \mathcal{B}, t \in \mathbb{R}.$$

(b) The **material acceleration** is defined by

$$A_t(X) := A_X(t) := A(X,t) := \frac{\partial}{\partial t}V(X,t), \quad X \in \mathcal{B}, t \in \mathbb{R}.$$

(c) The **spatial velocity** is defined by

$$v_t(x) := v(x,t) := V\left(\phi_t^{-1}(x),t\right), \quad x \in \phi_t(\mathcal{B}), t \in \mathbb{R}.$$

(d) The **spatial acceleration** is defined by

$$a_t(x) := a(x,t) := A\left(\phi_t^{-1}(x),t\right), \quad x \in \phi_t(\mathcal{B}), t \in \mathbb{R}.$$

Note that we will often simply state that $t \in \mathbb{R}$, as we did it here, though certainly the considered quantities need not and will not be available on such a time scale.

A simple example can illustrate here the difference between the material description and the spatial description. If you drive around in your car, then your car is something like a particle X. The velocity which your speedometer shows at time t is the material velocity $V(X,t)$. At every time t, the material velocity $V(X,t)$ gives us the velocity of the very same object/particle/atom/…X, which we observe. If the police install a speed trap, then the position where they install it is a fixed position x in space. The speed which the police's instrument indicates at different times is associated to different cars. If you pass by the speed trap with your car at a particular point of time t_0, then the speed $v(x,t_0)$ which the police measure with their instrument at the position x at the time t_0 should be the same as the speed $V(X,t_0)$ which the speedometer of your car X measures at the time t_0. Hence, $V(X,t_0) = v(x,t_0)$, while $x = \phi_{t_0}(X)$, that is the car X has moved to the position $x = \phi_{t_0}(X)$ at time t_0. The motion ϕ itself would describe how all vehicles move around. If X and Y are different vehicles, then V_X and V_Y deliver the time series of the velocity of vehicle X and vehicle Y, respectively. If x and y are different positions in space, then $v(x,\cdot)$ and $v(y,\cdot)$ represent the velocities of vehicles which pass by these points at different times.

There is a helpful formula which allows us to transfer temporal derivatives from material quantities to spatial quantities.

Lemma 7.1.6 *Let $Q\colon \mathcal{B}\times\mathbb{R}\to\mathbb{R}^n$ with $n\in\mathbb{N}$ be a differentiable Lagrangian quantity and let q be the corresponding Eulerian quantity, that is, $q(x,t) = Q(\phi_t^{-1}(x),t)$ for all $x\in\phi_t(\mathcal{B})$ and all $t\in\mathbb{R}$. Then the following identity holds true for all $X\in\mathcal{B}$ and all $t\in\mathbb{R}$:*

$$\frac{\partial}{\partial t}\,Q(X,t) = \frac{\partial}{\partial x}\,q(\phi(X,t),t)\,V(X,t) + \frac{\partial}{\partial t}\,q(\phi(X,t),t). \tag{7.1}$$

Proof The proof consists basically of the application of the chain rule.

$$\begin{aligned}\frac{\partial}{\partial t}\,Q(X,t) &= \frac{\mathrm{d}}{\mathrm{d}t}\,q(\phi(X,t),t) = \frac{\partial}{\partial x}\,q(\phi(X,t),t)\,\frac{\partial}{\partial t}\,\phi(X,t) + \frac{\partial}{\partial t}\,q(\phi(X,t),t)\\ &= \frac{\partial}{\partial x}\,q(\phi(X,t),t)\,V(X,t) + \frac{\partial}{\partial t}\,q(\phi(X,t),t).\end{aligned}$$

Here, $\frac{\mathrm{d}}{\mathrm{d}t}$ stands for the total time derivative with respect to all explicit and implicit time dependencies, where $\frac{\partial}{\partial t}$ is the common partial time derivative, which only applies to the explicit temporal argument of $Q(\cdot,t)$ and $q(\cdot,t)$. Note also that $\frac{\partial}{\partial x}\,q$ is here an $n\times 3$ Jacobian matrix, which is multiplied with the three-component vector V. □

For example, we get for the material acceleration

$$\begin{aligned}A(X,t) &= \frac{\partial}{\partial t}\,V(X,t) = \frac{\partial}{\partial x}\,v(\phi(X,t),t)\,V(X,t) + \frac{\partial}{\partial t}\,v(\phi(X,t),t)\\ &= \frac{\partial}{\partial x}\,v(x,t)\,v(x,t) + \frac{\partial}{\partial t}\,v(x,t),\end{aligned}$$

where $x = \phi_t(X)$. Note that $\frac{\partial}{\partial x}\,v\,v = \left(\frac{\partial}{\partial x}\,v\right)v$ in the usual mathematical syntax.

Definition 7.1.7 Let ϕ be a motion which satisfies Assumption 7.1.4.

(a) The Jacobian matrix which is given by $g := \left(\frac{\partial v}{\partial x}\right)^{\mathrm{T}}$ is called the **spatial deformation-rate tensor**.

(b) The operator D_t which is defined by

$$\frac{\mathrm{D}}{\mathrm{D}t}\, q(x,t) := \mathrm{D}_t q(x,t) := \frac{\partial}{\partial x}\, q(x,t)\, v(x,t) + \frac{\partial}{\partial t}\, q(x,t)$$

is called the **material time derivative**. We will rarely also use the notation $D_t := v \bullet \frac{\partial}{\partial x} + \frac{\partial}{\partial t}$ for this operator.

Hence, a shorter version of (7.1) can be written as

$$\frac{\partial}{\partial t}\, Q(X,t) = \mathrm{D}_t q(x,t), \quad x = \phi_t(X).$$

Definition 7.1.8 The second-rank-tensor-valued function F is a material quantity which is defined by $F(X,t) := \nabla_X \phi(X,t)$ for all $X \in \mathcal{B}$ and all $t \in \mathbb{R}$ and is called the **deformation gradient** of ϕ. As for other quantities, the alternative notations $F_t(X) := F_X(t) := F(X,t)$ may be used. Moreover, the second-rank-tensor-valued material quantity C is defined by $C(X,t) := (F(X,t))^{\mathrm{T}}\, F(X,t)$ for all $X \in \mathcal{B}$ and all $t \in \mathbb{R}$ and is called the **Green deformation tensor** or the **right Cauchy–Green tensor**.

Assumption 7.1.4 guarantees that, for each $X \in \mathcal{B}$ and $t \in \mathbb{R}$, the second-rank tensors (or matrices) $F(X,t)$ and $C(X,t)$ are invertible. We can, therefore, always define the following additional quantity.

Definition 7.1.9 The second-rank-tensor-valued material quantity B is defined by $B(X,t) := (C(X,t))^{-1}$ for all $X \in \mathcal{B}$ and all $t \in \mathbb{R}$ and is called the **Piola deformation tensor**.

The proof of the following lemma is an easy exercise in linear algebra.

Lemma 7.1.10 *For each $X \in \mathcal{B}$ and $t \in \mathbb{R}$, the matrix $C(X,t)$ is invertible, symmetric, and positive definite.*

From the principal axis theorem, which is a basic theorem in linear algebra, we know that symmetric matrices always have only real eigenvalues (which need to be positive if the matrix is positive definite) and an orthonormal basis of eigenvectors. Moreover, there is another well-known theorem from linear algebra in this context: if M is a symmetric and positive definite matrix, then there is one and only one symmetric and positive definite matrix N such that $N^2 = M$. This matrix N is often also called the **square root** of matrix M and denoted by $N =: M^{1/2}$.

Definition 7.1.11 Let C be the right Cauchy–Green tensor. Then the symmetric and positive definite matrix U which satisfies $U^2 = C$ is called the **right stretch tensor**. Its eigenvalues are called the **principal stretches**.

We continue now with the formulation of essential physical conservation laws. The first one, which is also the easiest to formulate, is the conservation of mass. Moreover, we also introduce a notation for the domain of a spatial quantity and adopt a known terminology for a particular class of sets.

Definition 7.1.12 Let ϕ be an arbitrary motion of the simple body $\mathcal{B}$. Then the domain of a spatial quantity is denoted by

$$\mathfrak{D}_\phi := \{(x,t) \mid x \in \phi_t(\mathcal{B}) \text{ and } t \in \mathbb{R}\}.$$

Furthermore, a subset $\mathcal{U} \subset \mathcal{B}$ is called **nice** if $\mathcal{U}$ is open and has a boundary with a piecewise continuously differentiable parameterization.

Definition 7.1.13 A time-dependent mass density function $\varrho \colon \overline{\mathfrak{D}_\phi} \to \mathbb{R}$ is said to obey **conservation of mass** if, for every nice set $\mathcal{U} \subset \mathcal{B}$, the following identity holds true:

$$\frac{\mathrm{d}}{\mathrm{d}t} \int_{\phi_t(\mathcal{U})} \varrho(x,t)\,\mathrm{d}x = 0 \quad \text{for all } t \in \mathbb{R}. \tag{7.2}$$

Definition 7.1.13 says not only that the total mass $\int_{\phi_t(\mathcal{B})} \varrho(x,t)\,\mathrm{d}x$ of the simple body $\mathcal{B}$ does not change over time. In (7.2), we require that every arbitrary part $\mathcal{U} \subset \mathcal{B}$ keeps its mass, that is, the masses of $\phi_t(\mathcal{U})$ are constant with respect to the time t. This is realistic, since all particles move together with the motion. If tectonic plates collide, then parts of the plates might deform and, for instance, build fold mountains, but the total masses of the affected region and all of its local areas do not change during this deformation.

There is a property of motions which should not be mixed up with conservation of mass: while conservation of mass is a fundamental physical law, which is always true, volume preservation is a property which some motions have but others do not. For seismological applications, we cannot, in general, assume that motions are volume preserving, while they must always satisfy conservation of mass.

Definition 7.1.14 A motion ϕ is called **volume preserving**, alternatively also **isochoric** or **incompressible**, if, for every nice region $\mathcal{U} \subset \mathcal{B}$, the following holds true:

$$\frac{\mathrm{d}}{\mathrm{d}t} \int_{\phi_t(\mathcal{U})} 1\,\mathrm{d}x = 0 \quad \text{for all } t \in \mathbb{R}. \tag{7.3}$$

Note that the integral in (7.3) yields the volume of $\phi_t(\mathcal{U})$. Hence, the requirement is here that each (connected) part $\mathcal{U} \subset \mathcal{B}$ keeps its volume during the motion. Thus, compression and expansion are excluded.

At this point, we should realize that conditions such as (7.2) and (7.3) are not practicable. We would not want to verify them for a specific motion. We would also not prefer to use them as starting points for the derivation of theorems. Therefore, we need a different, but equivalent, statement. The preceding definitions have the purpose of an easy interpretation. We see that they claim what we understand by conversation of mass and volume preservation, respectively. Now they need to be turned into something more practicable. We will see that such conservation laws can be modified into equivalent partial differential equations. However, we first need some preliminary work.

Definition 7.1.15 The Jacobian associated to a motion ϕ is denoted by $J(X,t) := \det F(X,t) = \det(\nabla_X \phi(X,t))$, $X \in \mathcal{B}$, $t \in \mathbb{R}$.

Lemma 7.1.16 *We have $J(X,t) > 0$ for all $X \in \mathcal{B}$ and all $t \in \mathbb{R}$.*

Proof From Assumption 7.1.4, we immediately get that J is continuous and $J(X,t) \neq 0$ for all $X \in \mathcal{B}$ and all $t \in \mathbb{R}$. The assumption also requires that $\phi(X,0) = X$ for all $X \in \mathcal{B}$ such that $F(X,0)$ is the identity matrix for each $X \in \mathcal{B}$. Hence, $J(X,0) = 1$ for every $X \in \mathcal{B}$. Thus, the intermediate value theorem implies that J must remain positive. □

Lemma 7.1.17 *Let $A \in \mathrm{C}^{(1)}(D,\mathbb{R}^{n\times n})$ with $D \subset \mathbb{R}$ (open) be an $n\times n$-matrix-valued function whose components are denoted by a_{jk}, $j,k = 1,\dots,n$. Then the derivative of the determinant of A satisfies*

$$\frac{\mathrm{d}}{\mathrm{d}t} \det A = \sum_{j,k=1}^{n} (-1)^{j+k} \frac{\mathrm{d}a_{jk}}{\mathrm{d}t} \det A_{jk},$$

where A_{jk} represents the $(n-1)\times(n-1)$-matrix which is obtained from A by removing the jth row and the kth column.

The proof of Lemma 7.1.17 is left to the reader as an exercise.

Lemma 7.1.18 *The Jacobian J of the motion ϕ satisfies*

$$\frac{\partial}{\partial t} J(X,t) = (\mathrm{div}_x\, v\,(\phi_t(X),t))\; J(X,t) \tag{7.4}$$

for all $X \in \mathcal{B}$ and $t \in \mathbb{R}$, where v is the spatial velocity (see Definition 7.1.5) and div_x stands for the divergence with respect to the spatial variable $x = \phi_t(X)$.

Proof We apply Lemma 7.1.17 to $J(X,t) = \det(\nabla_X \phi(X,t))$ and obtain

$$\frac{\partial}{\partial t} J(X,t) = \sum_{j,k=1}^{3} (-1)^{j+k} \left(\frac{\partial^2}{\partial t\, \partial X_k} \phi_j(X,t)\right) \det\left(\frac{\partial \phi_i(X,t)}{\partial X_l}\right)_{\substack{i=1,2,3;\ i\neq j\\ l=1,2,3;\ l\neq k}}, \tag{7.5}$$

where ϕ_j is the jth component of ϕ and X_k is the kth material coordinate. Moreover, we get from the definition of the spatial velocity and the chain rule the identity

$$\frac{\partial v}{\partial x}(x,t) = \frac{\partial V}{\partial x}(\phi_t^{-1}(x),t) = \left(\frac{\partial V}{\partial X}(\phi_t^{-1}(x),t)\right) \frac{\partial \phi_t^{-1}}{\partial x}(x),$$

where the right-hand side is a multiplication of matrices. Remember also that $x = \phi_t(X)$. We now apply the inverse function theorem from differential calculus and observe that

$$\left(\frac{\partial v}{\partial x}(x,t)\right) \frac{\partial \phi_t(X)}{\partial X} = \left(\frac{\partial V}{\partial X}(X,t)\right) \frac{\partial \phi_t^{-1}(x)}{\partial x} \frac{\partial \phi_t(X)}{\partial X} = \frac{\partial V}{\partial X}(X,t). \tag{7.6}$$

Keeping in mind that $\frac{\partial V}{\partial X}(X,t) = \frac{\partial^2}{\partial X \partial t}\phi(X,t) = \frac{\partial^2}{\partial t \partial X}\phi(X,t)$ holds true due to Schwarz's theorem, we insert (7.6) into (7.5) and get (with '$\cdot$' denoting, as usual, the dot product)

$$\begin{aligned}\frac{\partial}{\partial t} J(X,t) &= \sum_{j,k=1}^{3} (-1)^{j+k} \frac{\partial v_j(x,t)}{\partial x} \cdot \frac{\partial \phi(X,t)}{\partial X_k} \det\left(\frac{\partial \phi_i(X,t)}{\partial X_l}\right)_{\substack{i=1,2,3;\ i\neq j\\ l=1,2,3;\ l\neq k}} \\ &= \sum_{j,k,p=1}^{3} (-1)^{j+k} \frac{\partial v_j(x,t)}{\partial x_p} \frac{\partial \phi_p(X,t)}{\partial X_k} \det\left(\frac{\partial \phi_i(X,t)}{\partial X_l}\right)_{\substack{i=1,2,3;\ i\neq j\\ l=1,2,3;\ l\neq k}}.\end{aligned} \tag{7.7}$$

We have a look now at those summands which we obtain for each fixed pair (j, p), but we factor out $\frac{\partial v_j}{\partial x_p}$, that is, we consider

$$\Psi_{jp} := \sum_{k=1}^{3}(-1)^{j+k}\frac{\partial \phi_p(X,t)}{\partial X_k} \det\left(\frac{\partial \phi_i(X,t)}{\partial X_l}\right)_{\substack{i=1,2,3;\ i\neq j\\ l=1,2,3;\ l\neq k}}. \tag{7.8}$$

We distinguish two cases. If $p = j$, then Laplace's formula for a determinant yields

$$\Psi_{jj} = \sum_{k=1}^{3}(-1)^{j+k}\frac{\partial \phi_j(X,t)}{\partial X_k} \det\left(\frac{\partial \phi_i(X,t)}{\partial X_l}\right)_{\substack{i=1,2,3;\ i\neq j\\ l=1,2,3;\ l\neq k}} = \det(\nabla_X \phi(X,t)).$$

If $p \neq j$, we also obtain Laplace's formula, but for a different matrix. In (7.8), the determinant of the underlying matrix is expanded along the jth row, which is

$$\left(\frac{\partial \phi_p(X,t)}{\partial X_1}, \frac{\partial \phi_p(X,t)}{\partial X_2}, \frac{\partial \phi_p(X,t)}{\partial X_3}\right).$$

The other rows are identical to those of $\nabla_X \phi(X,t)$. This means that the j-th row has the same entries as the pth row, which is, however, not the same row, because $j \neq p$. Hence, the underlying matrix has two identical rows and the corresponding determinant must consequently vanish. Thus, in total, $\Psi_{jp} = \delta_{jp} J(X,t)$. If we use this conclusion in (7.7), then we get

$$\frac{\partial}{\partial t} J(X,t) = \sum_{p=1}^{3} \frac{\partial v_p(x,t)}{\partial x_p} J(X,t) = (\operatorname{div}_x v(\phi_t(X),t))\ J(X,t). \qquad \square$$

Lemma 7.1.19 *Let $\mathcal{B}$ be a simple body and let $F, G\colon \mathcal{B} \to \mathbb{R}$ be two continuous functions. Then $\int_{\mathcal{U}} F(X)\,\mathrm{d}X = \int_{\mathcal{U}} G(X)\,\mathrm{d}X$ holds true for every nice region $\mathcal{U} \subset \mathcal{B}$ if and only if $F \equiv G$.*

The proof mainly uses the continuity of both functions and is left to the reader as an easy exercise.

Lemma 7.1.20 *Let ϕ_t be a configuration of the body $\mathcal{B}$ which originates from a motion ϕ which fulfils Assumption 7.1.4, and let $f, g : \phi_t(\mathcal{B}) \to \mathbb{R}$ be two continuous functions. Then $\int_{\phi_t(\mathcal{U})} f(x)\,\mathrm{d}x = \int_{\phi_t(\mathcal{U})} g(x)\,\mathrm{d}x$ holds true for every nice region $\mathcal{U} \subset \mathcal{B}$ if and only if $f \equiv g$.*

Proof With the substitution $x = \phi_t(X)$, we get

$$\int_{\mathcal{U}} f(\phi_t(X))\ |\det(\nabla_X \phi_t(X))|\ \mathrm{d}X = \int_{\mathcal{U}} g(\phi_t(X))\ |\det(\nabla_X \phi_t(X))|\ \mathrm{d}X$$

for all nice regions $\mathcal{U} \subset \mathcal{B}$. With Lemmata 7.1.16 and 7.1.19, we obtain then $f(\phi_t(X))J(X,t) = g(\phi_t(X))J(X,t)$ for all $X \in \mathcal{B}$. Keeping in mind that $J(X,t) \neq 0$ for all $X \in \mathcal{B}$, we only have to resubstitute $x = \phi_t(X)$ and get the desired result. $\square$

Theorem 7.1.21 *The following properties of a motion ϕ are equivalent:*

(i) *ϕ is volume preserving.*
(ii) *$J(X,t) = 1$ for all $X \in \mathcal{B}$ and all $t \in \mathbb{R}$.*
(iii) *$\operatorname{div}_x v(x,t) = 0$ for all $x \in \phi_t(\mathcal{B})$ and all $t \in \mathbb{R}$.*

Proof We prove the equivalence in three steps.

(1) (i) $\Leftrightarrow$ (ii):
Since $\Phi_0(X) = X$ for all $X \in \mathcal{B}$ due to Assumption 7.1.4, ϕ is volume preserving if and only if $\int_{\phi_t(\mathcal{U})} 1\,\mathrm{d}x = \int_{\mathcal{U}} 1\,\mathrm{d}X$ for every nice region $\mathcal{U} \subset \mathcal{B}$ and every $t \in \mathbb{R}$. We substitute now $x = \phi_t(X)$ in the left-hand integral. Then volume preservation is equivalent to

$$\int_{\mathcal{U}} |\det(\nabla_X \phi_t(X))|\,\mathrm{d}X = \int_{\mathcal{U}} 1\,\mathrm{d}X \quad \text{for all } t \in \mathbb{R} \text{ and nice regions } \mathcal{U}.$$

With Lemmata 7.1.16 and 7.1.19, we get that the latter holds true, if and only if $J(X,t) = 1$ for all $X \in \mathcal{B}$ and all $t \in \mathbb{R}$.

(2) (ii) $\Rightarrow$ (iii):
If (ii) holds true, then J is constant and $\frac{\partial J}{\partial t}(X,t) = 0$ for all $X \in \mathcal{B}$ and all $t \in \mathbb{R}$. If we insert this into (7.4) from Lemma 7.1.18 and if we take into account that $J(X,t) \neq 0$ for all $X \in \mathcal{B}$ and all $t \in \mathbb{R}$ due to Assumption 7.1.4, then we get $\operatorname{div}_x v(x,t) = 0$ for all $x \in \phi_t(\mathcal{B})$ and all $t \in \mathbb{R}$.

(3) (iii) $\Rightarrow$ (ii):
If we use the vanishing divergence of the spatial velocity in (7.4), then $\frac{\partial J}{\partial t}(X,t) = 0$ must be valid for all $X \in \mathcal{B}$ and all $t \in \mathbb{R}$, that is, J is constant in time. Due to Assumption 7.1.4, $J(X,0) = 1$ for all $X \in \mathcal{B}$. Hence, $J(X,t) = 1$ is also valid for all times $t \in \mathbb{R}$. □

Theorem 7.1.22 *Let $\varrho \in \mathrm{C}^{(1)}(\overline{\mathfrak{D}_\phi})$ be a spatial mass density function. Then the following properties are equivalent:*

(i) *ϱ obeys conservation of mass.*
(ii) *$\varrho(\phi(X,t),t)\,J(X,t) = \varrho(X,0)$ for all $X \in \mathcal{B}$ and all $t \in \mathbb{R}$.*
(iii) *The **equation of continuity** is valid: for all $x \in \phi_t(\mathcal{B})$ and all $t \in \mathbb{R}$,*

$$\mathrm{D}_t\varrho(x,t) + \varrho(x,t)\operatorname{div}_x v(x,t) = 0, \tag{7.9}$$

or, equivalently,

$$\frac{\partial}{\partial t}\varrho(x,t) + \operatorname{div}_x(\varrho(x,t)\,v(x,t)) = 0. \tag{7.10}$$

Proof We will need two parts to prove this theorem.

(1) (i) $\Leftrightarrow$ (ii):
Due to the vanishing time derivative in (7.2), conservation of mass is given if and only if

$$\int_{\phi_t(\mathcal{U})} \varrho(x,t)\,\mathrm{d}x = \int_{\mathcal{U}} \varrho(X,0)\,\mathrm{d}X \quad \text{for all } t \in \mathbb{R} \text{ and nice } \mathcal{U} \subset \mathcal{B}.$$

The substitution $x = \phi_t(X)$ yields the equivalent identity

$$\int_{\mathcal{U}} \varrho\,(\phi_t(X),t)\ |\det\,(\nabla_X \phi_t(X))|\ \mathrm{d}X = \int_{\mathcal{U}} \varrho(X,0)\,\mathrm{d}X$$

for all $t \in \mathbb{R}$ and nice subsets $\mathcal{U} \subset \mathcal{B}$. In analogy to the previous proof, we also apply here Lemmata 7.1.16 and 7.1.19. It turns out that ϱ obeys conservation of mass if and only if

$$\varrho\,(\phi_t(X),t)\ J(X,t) = \varrho(X,0) \quad \text{for all } X \in \mathcal{B} \text{ and all } t \in \mathbb{R}.$$

(2) (ii) ⇔ (iii):

Remember that $\Phi_0(X) = X$ for all $X \in \mathcal{B}$ due to Assumption 7.1.4. For this reason, (ii) states that the left-hand side is independent of time and therefore equals its value at $t = 0$, which is the right-hand side of (ii). Hence, (ii) is equivalently represented as

$$\frac{\mathrm{d}}{\mathrm{d}t}\,(\varrho(\phi_t(X),t)\,J(X,t)) = 0 \quad \text{for all } X \in \mathcal{B} \text{ and } t \in \mathbb{R}. \tag{7.11}$$

Let us now apply the total differentiation operator on the left-hand side and use Lemma 7.1.18:

$$\begin{aligned}
&\frac{\mathrm{d}}{\mathrm{d}t}\,(\varrho(\phi_t(X),t)\,J(X,t)) \\
&\quad = \frac{\mathrm{d}}{\mathrm{d}t}\varrho(\phi_t(X),t)\,J(X,t) + \varrho(\phi_t(X),t)\,\frac{\partial}{\partial t}J(X,t) \\
&\quad = \frac{\mathrm{d}}{\mathrm{d}t}\varrho(\phi_t(X),t)\,J(X,t) + \varrho(\phi_t(X),t)\,\mathrm{div}_x\, v\,(\phi_t(X),t)\ J(X,t).
\end{aligned}$$

We insert this into (7.11) and use the fact that J never vanishes (see Assumption 7.1.4). As a consequence, (ii) is equivalent to

$$\frac{\mathrm{d}}{\mathrm{d}t}\varrho(\phi_t(X),t) + \varrho(\phi_t(X),t)\,\mathrm{div}_x\, v\,(\phi_t(X),t) = 0 \text{ for all } X \in \mathcal{B} \text{ and } t \in \mathbb{R}.$$

We need to change now from material coordinates to spatial coordinates, where $x = \phi_t(X)$. The total time derivative in the first summand transforms then into a material time derivative (see Lemma 7.1.6 and Definition 7.1.7). Hence, (ii) holds true if and only if

$$\mathrm{D}_t\varrho(x,t) + \varrho(x,t)\,\mathrm{div}_x\, v(x,t) = 0 \quad \text{for all } x \in \phi_t(\mathcal{B}) \text{ and } t \in \mathbb{R}. \tag{7.12}$$

This is the first version (7.9) of the equation of continuity. If we apply the definition of D_t, we get that (7.12) is the same as

$$\frac{\partial}{\partial t}\varrho(x,t) + \frac{\partial \varrho(x,t)}{\partial x}\cdot v(x,t) + \varrho(x,t)\,\mathrm{div}_x\, v(x,t) = 0, \quad x \in \phi_t(\mathcal{B}),\ t \in \mathbb{R}. \tag{7.13}$$

Finally, we apply Theorem 2.3.4 to see that (7.13) can also be written as

$$\frac{\partial}{\partial t}\varrho(x,t) + \operatorname{div}_x(\varrho v)(x,t) = 0 \quad \text{for all } x \in \phi_t(\mathcal{B}) \text{ and } t \in \mathbb{R},$$

which is (7.10). □

We will see that all conservation laws which we need have a common mathematical structure. This is also the reason why Marsden and Hughes (1994) introduce and discuss a master balance law. We will adhere to this concept in this chapter. Therefore, it is reasonable if we first concentrate on some general propositions.

Theorem 7.1.23 (Transport Theorem) *Let* $f \in \mathrm{C}^{(1)}(\overline{\mathfrak{D}_\phi})$ *be a given spatial quantity and let* $\mathcal{U} \subset \mathcal{B}$ *be an arbitrary nice set. Then the following identity holds true:*

$$\begin{aligned}\frac{\mathrm{d}}{\mathrm{d}t}\int_{\phi_t(\mathcal{U})} f(x,t)\,\mathrm{d}x &= \int_{\phi_t(\mathcal{U})} \mathrm{D}_t f(x,t) + f(x,t)\operatorname{div}_x v(x,t)\,\mathrm{d}x \\ &= \int_{\phi_t(\mathcal{U})} \frac{\partial}{\partial t} f(x,t) + \operatorname{div}_x(fv)(x,t)\,\mathrm{d}x.\end{aligned} \tag{7.14}$$

Proof We first apply the familiar substitution $x = \phi_t(X)$.

$$\frac{\mathrm{d}}{\mathrm{d}t}\int_{\phi_t(\mathcal{U})} f(x,t)\,\mathrm{d}x = \frac{\mathrm{d}}{\mathrm{d}t}\int_{\mathcal{U}} f(\phi(X,t),t)\,J(X,t)\,\mathrm{d}X.$$

Since f and its first-order derivatives are bounded due to the compact domain, we may apply Theorem 2.4.9 and get

$$\frac{\mathrm{d}}{\mathrm{d}t}\int_{\phi_t(\mathcal{U})} f(x,t)\,\mathrm{d}x = \int_{\mathcal{U}} \frac{\mathrm{d}}{\mathrm{d}t} f(\phi(X,t),t)\,J(X,t) + f(\phi(X,t),t)\,\frac{\partial J}{\partial t}(X,t)\,\mathrm{d}X.$$

Then we use Lemma 7.1.18, the resubstitution $X = \phi_t^{-1}(x)$, the material time derivative (see Lemma 7.1.6 and Definition 7.1.7), and Theorem 2.3.4:

$$\begin{aligned}&\frac{\mathrm{d}}{\mathrm{d}t}\int_{\phi_t(\mathcal{U})} f(x,t)\,\mathrm{d}x \\ &= \int_{\mathcal{U}} \frac{\mathrm{d}}{\mathrm{d}t} f(\phi(X,t),t)\,J(X,t) + f(\phi(X,t),t)\operatorname{div}_x v\,(\phi_t(X),t)\;J(X,t)\,\mathrm{d}X \\ &= \int_{\phi_t(\mathcal{U})} \mathrm{D}_t f(x,t) + f(x,t)\operatorname{div}_x v(x,t)\,\mathrm{d}x \\ &= \int_{\phi_t(\mathcal{U})} \frac{\partial f(x,t)}{\partial t} + \frac{\partial f(x,t)}{\partial x}\cdot v(x,t) + f(x,t)\operatorname{div}_x v(x,t)\,\mathrm{d}x \\ &= \int_{\phi_t(\mathcal{U})} \frac{\partial f(x,t)}{\partial t} + \operatorname{div}_x(fv)(x,t)\,\mathrm{d}x.\end{aligned}$$

Hence, both right-hand sides of (7.14) are valid. □

Definition 7.1.24 Spatial quantities $a, b\colon \mathfrak{D}_\phi \to \mathbb{R}$ and $c\colon \mathfrak{D}_\phi \times \Omega \to \mathbb{R}$ are said to satisfy the **master balance law** if, for every nice set $\mathcal{U} \subset \mathcal{B}$ and each $t \in \mathbb{R}$, all following terms (i.e., the integrals and the derivative) exist such that the identity

$$\frac{\mathrm{d}}{\mathrm{d}t}\int_{\phi_t(\mathcal{U})} a(x,t)\,\mathrm{d}x = \int_{\phi_t(\mathcal{U})} b(x,t)\,\mathrm{d}x + \int_{\partial\phi_t(\mathcal{U})} c(x,t,n(x))\,\mathrm{d}\omega(x)$$

holds true, where n is the outer unit normal to $\partial\phi_t(\mathcal{U})$.

Remember that Ω is the unit sphere in $\mathbb{R}^3$. Moreover, note that a is here an arbitrary function and not the spatial acceleration.

In the following, we show now the announced equivalence of conservation laws of the type previously described to a partial differential equation.

Theorem 7.1.25 (Spatial Localization Theorem) *Let $a \in \mathrm{C}^{(1)}(\overline{\mathfrak{D}_\phi})$, $b \in \mathrm{C}(\overline{\mathfrak{D}_\phi})$, and $c \in \mathrm{C}^{(1)}(\overline{\mathfrak{D}_\phi} \times \Omega)$ be given functions, where the latter has a particular form: there exists $d \in \mathrm{c}^{(1)}(\overline{\mathfrak{D}_\phi})$ such that*

$$c(x,t,\xi) = d(x,t)\cdot\xi \quad \textit{for all } x \in \overline{\phi_t(\mathcal{B})},\ t \in \mathbb{R},\ \textit{and } \xi \in \Omega. \tag{7.15}$$

Then the following holds true: a, b, and c satisfy the master balance law, if and only if

$$\frac{\partial}{\partial t}a + \operatorname{div}_x(av) = b + \operatorname{div}_x d \quad \textit{on } \mathfrak{D}_\phi, \tag{7.16}$$

where v is the spatial velocity (see Definition 7.1.5).

Proof We apply (7.14) from the transport theorem to the left-hand side of the master balance law. Then we see that the master balance law is satisfied if and only if

$$\int_{\phi_t(\mathcal{U})}\left(\frac{\partial}{\partial t}a(x,t) + \operatorname{div}_x(av)(x,t)\right)\mathrm{d}x = \int_{\phi_t(\mathcal{U})} b(x,t)\,\mathrm{d}x + \int_{\partial\phi_t(\mathcal{U})} d(x,t)\cdot n(x)\,\mathrm{d}\omega(x) \tag{7.17}$$

for all nice $\mathcal{U} \subset \mathcal{B}$ and all $t \in \mathbb{R}$, where Gauß's law (see Theorem 2.4.2) allows us to write the surface integral as

$$\int_{\partial\phi_t(\mathcal{U})} d(x,t)\cdot n(x)\,\mathrm{d}\omega(x) = \int_{\phi_t(\mathcal{U})} \operatorname{div}_x d(x,t)\,\mathrm{d}x.$$

With this replacement in mind, we can apply Lemma 7.1.20 to (7.17). The immediate conclusion is then: the master balance law is satisfied by a, b, and c, if and only if (7.16) holds true. □

Example 7.1.26 Conservation of mass, see Definition 7.1.13, is an example of the master balance law, where $a \equiv \varrho$, $b \equiv 0$, and $c \equiv 0$ (and, therefore, also $d \equiv 0$). The spatial localization theorem, see above, yields the partial differential equation $\frac{\partial}{\partial t}\varrho + \operatorname{div}_x(\varrho v) = 0$ in $\mathfrak{D}_\phi$. This is exactly the equation of continuity (7.10), see Theorem 7.1.22.

In Theorem 7.1.25, the representation of c via (7.15) was formulated as a condition. However, we will see now that this is not an additional requirement if the master balance law is already satisfied.

Theorem 7.1.27 (Cauchy's Theorem) *Let $a \in \mathrm{C}^{(1)}(\overline{\mathfrak{D}_\phi})$, $b \in \mathrm{C}(\overline{\mathfrak{D}_\phi})$, and $c \in \mathrm{C}(\overline{\mathfrak{D}_\phi} \times \Omega)$ satisfy the master balance law. Then there exists a unique vector field $d\colon \overline{\mathfrak{D}_\phi} \to \mathbb{R}^3$ such that*

$$c(x,t,\xi) = d(x,t) \cdot \xi \quad \textit{for all } x \in \overline{\phi_t(\mathcal{B})},\ t \in \mathbb{R},\ \textit{and } \xi \in \Omega.$$

Proof To make a long story short: *divide et impera.*

(1) Application of the transport theorem:

With the first part in (7.14), we get

$$\int_{\phi_t(\mathcal{U})} \mathrm{D}_t a(x,t) + a(x,t)\,\mathrm{div}_x\, v(x,t)\,\mathrm{d}x \tag{7.18}$$
$$= \int_{\phi_t(\mathcal{U})} b(x,t)\,\mathrm{d}x + \int_{\partial\phi_t(\mathcal{U})} c(x,t,n(x))\,\mathrm{d}\omega(x)$$

as the master balance law (for all nice $\mathcal{U} \subset \mathcal{B}$ and all $t \in \mathbb{R}$). We choose now arbitrary $t \in \mathbb{R}$ and $x^0 \in \phi_t(\mathcal{B})$ and keep them fixed. Moreover, we move our coordinate system, without loss of generality, such that $x^0 = 0$. This is possible since an affine transformation $x' := x - x^0$ does not harm (7.18) notably.

(2) Some geometrical framework:

In the first octant (given by $x_1, x_2, x_2 \geq 0$), we draw a closed tetrahedron T_ℓ (or, maybe more precisely, a 3-simplex) with the following specifications: each vertex lies on a coordinate axis, and three edges lie on the three-coordinate axis and have the length $\ell > 0$, as shown in Figure 7.2.

We assume that ℓ is sufficiently small such that $T_\ell \subset \phi_t(\mathcal{B})$. The interior of T_ℓ is open and, as a consequence of Assumption 7.1.4, so is $\phi_t^{-1}(\mathrm{int}\, T_\ell) =: \mathcal{U}$, which is

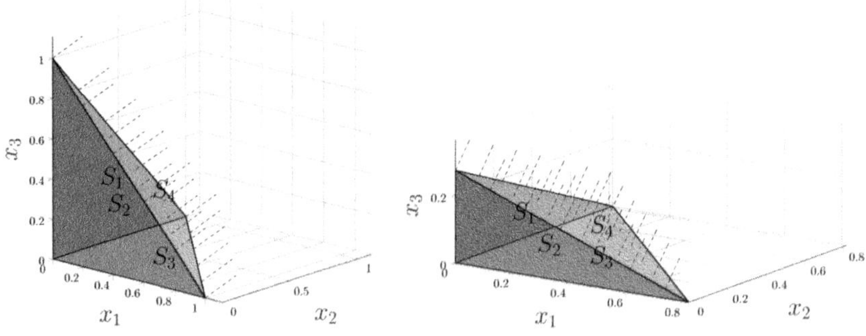

Figure 7.2 A tetrahedron in the first octant with three vertices along the coordinate axis: if these vertices have equal length (left), then the (constant) outer unit normal vector on S_4 is $(1,1,1)^{\mathrm{T}}/\sqrt{3}$. By varying the vertex lengths, different normal vectors can be generated (right).

a nice subset of $\mathcal{B}$. Hence, we get for $\operatorname{int} T_\ell = \phi_t(\mathcal{U})$ and $\partial T_\ell = \partial\phi_t(\mathcal{U})$ that (7.18) yields (note that $\partial T_\ell = T_\ell \setminus \operatorname{int} T_\ell$ is a set of measure zero)

$$\int_{T_\ell} \underbrace{\mathrm{D}_t a(x,t) + a(x,t)\operatorname{div}_x v(x,t) - b(x,t)}_{=: f(x,t)} \mathrm{d}x = \int_{\partial T_\ell} c(x,t,n(x))\,\mathrm{d}\omega(x)$$

for all sufficiently small $\ell > 0$. Remember that f is a continuous function. We denote (again) the Lebesgue measure (i.e., the volume measure) by λ and the corresponding surface measure by ω. Then we can see that

$$\lim_{\ell\to 0+}\left|\frac{1}{\omega(\partial T_\ell)}\int_{\partial T_\ell} c(x,t,n(x))\,\mathrm{d}\omega(x)\right| \le \lim_{\ell\to 0+}\left(\frac{\lambda(T_\ell)}{\omega(\partial T_\ell)}\max_{x\in T_\ell}|f(x,t)|\right). \tag{7.19}$$

Both sides are zero, because $\lambda(T_\ell) = \ell^3/6$ and $\omega(\partial T_l) = C\ell^2$ for a constant $C > 0$ (the precise formula is presented later).

(3) The faces of the tetrahedron/simplex:

T has four faces, which are triangles (or 2-simplices). We denote them as follows: S_j, $j = 1,2,3$, is the face which is orthogonal to the x_j-axis and S_4 is the remaining face (remember that they depend on ℓ, though this is not indicated here). The associated (constant) outer unit normal vectors are called $n^{(j)}$, $j = 1,\dots,4$. Clearly, S_1, S_2, and S_3 have the area $l^2/2$ each and S_4 is an equilateral triangle with side length $\sqrt{2}\,\ell$ and area $\sqrt{3}\,\ell^2/2$. Hence, $\omega(\partial T_\ell) = (3+\sqrt{3})\ell^2/2$. We get now for the surface integral in (7.19) that

$$\min_{x\in\partial T_\ell} c(x,t,n(x)) \le \frac{1}{\omega(\partial T_\ell)}\int_{\partial T_\ell} c(x,t,n(x))\,\mathrm{d}\omega(x) \le \max_{x\in\partial T_\ell} c(x,t,n(x))$$

and

$$\begin{aligned}
&\frac{1}{\omega(\partial T_\ell)}\int_{\partial T_\ell} c(x,t,n(x))\,\mathrm{d}\omega(x)\\
&= \sum_{j=1}^{4}\frac{1}{\omega(\partial T_\ell)}\int_{S_j} c\big(x,t,n^{(j)}\big)\,\mathrm{d}\omega(x)\\
&\le \sum_{j=1}^{4}\frac{\omega(S_j)}{\omega(\partial T_\ell)}\max_{x\in S_j} c\big(x,t,n^{(j)}\big)\\
&= \frac{1}{3+\sqrt{3}}\sum_{j=1}^{3}\max_{x\in S_j} c\big(x,t,n^{(j)}\big) + \frac{\sqrt{3}}{3+\sqrt{3}}\max_{x\in S_4} c\big(x,t,n^{(4)}\big),
\end{aligned}$$

where the analogous statement with '$\geq$' and 'min' holds. Hence, the limit $\ell \to 0+$ yields, due to the continuity of c, the identity

$$0 = \frac{1}{3+\sqrt{3}}\sum_{j=1}^{3} c\big(x^0,t,n^{(j)}\big) + \frac{\sqrt{3}}{3+\sqrt{3}} c\big(x^0,t,n^{(4)}\big). \tag{7.20}$$

(4) A closer look at the normal vectors:

For $j = 1,2,3$, we obviously have $n^{(j)} = -\varepsilon^j$. The fourth outer unit normal vector is $n^{(4)} = (1,1,1)^{\mathrm{T}}/\sqrt{3}$. Hence, (7.20) leads us to

$$0 = \frac{\sqrt{3}}{3+\sqrt{3}}\left(\frac{1}{\sqrt{3}}\sum_{j=1}^{3} c\big(x^0,t,n^{(j)}\big) + c\big(x^0,t,n^{(4)}\big)\right)$$
$$= \frac{\sqrt{3}}{3+\sqrt{3}}\left(\sum_{j=1}^{3} c\big(x^0,t,-\varepsilon^j\big)n_j^{(4)} + c\big(x^0,t,n^{(4)}\big)\right).$$

Consequently,

$$c\big(x^0,t,n^{(4)}\big) = -\sum_{j=1}^{3} c\big(x^0,t,-\varepsilon^j\big)n_j^{(4)}. \tag{7.21}$$

(5) Deforming the tetrahedron:

Let now $\xi \in \Omega$ be an arbitrary direction in the interior of the first octant, which means that all components ξ_j are positive. We build a modification of the tetrahedron (or simplex) T_ℓ with the vertices $v^{(0)} := (0,0,0)^{\mathrm{T}}$, $v^{(1)} := 3\,\xi_2\xi_3\ell\,\varepsilon^1$, $v^{(2)} := 3\,\xi_1\xi_3\ell\,\varepsilon^2$, and $v^{(3)} := 3\xi_1\xi_2\ell\,\varepsilon^3$. Then one can easily verify that ξ is the outer unit normal vector to S_4. Note that we previously had $\xi_j = 1/\sqrt{3}$ for all j.

The volume of T_ℓ is obtained as $1/6$ times the lengths of the axis-parallel vertices, that is, $\lambda(T_\ell) = 9\xi_1^2\xi_2^2\xi_3^2\ell^3/2$. For the vertices which are right-angled triangles, the area is easy to calculate: $\omega(S_1) = 9\xi_1^2\xi_2\xi_3\ell^2/2$, $\omega(S_2) = 9\xi_1\xi_2^2\xi_3\ell^2/2$, and $\omega(S_3) = 9\xi_1\xi_2\xi_3^2\ell^2/2$. For S_4, we get

$$\omega\,(S_4) = \frac{1}{2}\left|\big(v^{(2)} - v^{(1)}\big) \times \big(v^{(3)} - v^{(1)}\big)\right| = \frac{9\ell^2}{2}\left|\begin{pmatrix}-\xi_2\xi_3\\ \xi_1\xi_3\\ 0\end{pmatrix} \times \begin{pmatrix}-\xi_2\xi_3\\ 0\\ \xi_1\xi_2\end{pmatrix}\right|$$
$$= \frac{9\ell^2}{2}\left|\begin{pmatrix}\xi_1^2\xi_2\xi_3\\ \xi_1\xi_2^2\xi_3\\ \xi_1\xi_2\xi_3^2\end{pmatrix}\right| = \frac{9\ell^2}{2}\sqrt{\xi_1^4\xi_2^2\xi_3^2 + \xi_1^2\xi_2^4\xi_3^2 + \xi_1^2\xi_2^2\xi_3^4}$$
$$= \frac{9\ell^2}{2}\sqrt{\xi_1^2\xi_2^2\xi_3^2\left(\xi_1^2 + \xi_2^2 + \xi_3^2\right)} = \frac{9\ell^2}{2}\,\xi_1\xi_2\xi_3.$$

Hence, $\omega(\partial T_\ell) = 9\xi_1\xi_2\xi_3(\xi_1 + \xi_2 + \xi_3 + 1)\ell^2/2$ such that, like before, we obtain $\lim_{\ell\to 0+} \lambda(T_\ell)(\omega(\partial T_\ell))^{-1} = 0$ and

$$\frac{\omega(S_j)}{\omega(\partial T_\ell)} = \frac{\xi_j}{\xi_1 + \xi_2 + \xi_3 + 1} = \xi_j\,\frac{\omega\,(S_4)}{\omega(\partial T_\ell)} \quad \text{for all } j = 1,2,3.$$

Consequently, if we follow the preceding argumentation with $n^{(4)} = \xi$, then we arrive again at (7.21).

Let us harvest from our work. We can now set $d_j(x^0,t) := -c(x^0,t,-\varepsilon^j)$, $j = 1,2,3$. Then

$$c(x^0,t,\xi) = d(x^0,t)\cdot\xi \tag{7.22}$$

holds true for all $\xi \in \Omega$ with $\xi_1,\xi_2,\xi_3 > 0$. For symmetry reasons, this must also work in the other octants. Moreover, all occurred functions, in particular c, are continuous such that the relevant identities can be continuously extended to the cases where components of ξ may vanish. Eventually, (7.22) holds true for all $\xi \in \Omega$. The extension of the identity to $x \in \overline{\phi_t(\mathcal{B})}$ is also possible due to the continuity of c. Furthermore, note that it is clear that d is uniquely defined by (7.22). □

We are now in the position to take care of further conservation laws. We start with the conservation of (linear) momentum, which is basically **Newton's second law**. It states that the temporal changes of the linear momentum (in terms of a time derivative) equal the sum of all external forces. In particular, in elasticity it is important to subdivide these forces into body forces, which act on the whole volume $\phi_t(\mathcal{U})$, and surface forces, which are exerted on surfaces $\partial\phi_t(\mathcal{U})$. These forces are represented by (force) density functions, in the sense of a force per volume or surface area, and are denoted by b and $\mathfrak{t}$ here. The latter symbol helps us avoid confusion with the time t. Note that conservation laws are sometimes also called balance laws, that is, mass balance, balance of momentum, etc.

Definition 7.1.28 The motion ϕ, the mass density $\varrho\colon \mathfrak{D}_\phi \to \mathbb{R}$, and the **force densities** $b\colon \mathfrak{D}_\phi \to \mathbb{R}^3$, $\mathfrak{t}\colon \mathfrak{D}_\phi \times \Omega \to \mathbb{R}^3$ are said to satisfy **conservation of (linear) momentum** if, for each nice set $\mathcal{U} \subset \mathcal{B}$ and all $t \in \mathbb{R}$, the following terms exist such that the identity

$$\frac{\mathrm{d}}{\mathrm{d}t}\int_{\phi_t(\mathcal{U})} (\varrho v)(x,t)\,\mathrm{d}x = \int_{\phi_t(\mathcal{U})} (\varrho b)(x,t)\,\mathrm{d}x + \int_{\partial\phi_t(\mathcal{U})} \mathfrak{t}(x,t,n(x))\,\mathrm{d}\omega(x) \tag{7.23}$$

holds true, where n is again the outer unit normal to $\partial\phi_t(\mathcal{U})$ and v is the spatial velocity.

Theorem 7.1.29 *If the motion ϕ and the densities $\varrho \in \mathrm{C}^{(1)}(\overline{\mathfrak{D}_\phi})$, $b \in \mathrm{c}(\overline{\mathfrak{D}_\phi})$, and $\mathfrak{t} \in \mathrm{c}(\overline{\mathfrak{D}_\phi} \times \Omega)$ satisfy conservation of momentum, then there exists a unique tensor field $\sigma \in \mathbf{c}(\overline{\mathfrak{D}_\phi})$ such that $\mathfrak{t}(x,t,\xi) = \sigma(x,t)\,\xi$ for all $x \in \overline{\phi_t(\mathcal{B})}$, $t \in \mathbb{R}$, and $\xi \in \Omega$. This tensor field σ is called the* ***Cauchy stress tensor****.*

Proof Equation (7.23) is a vectorial equation. Each component yields a scalar equation, which has the form of the master balance law. We may now apply Cauchy's theorem (Theorem 7.1.27), which provides us, for each index $j = 1,2,3$, with a vectorial function $d^{(j)}\colon \overline{\mathfrak{D}_\phi} \to \mathbb{R}^3$ such that

$$\mathfrak{t}_j(x,t,\xi) = d^{(j)}(x,t)\cdot\xi \quad \text{for all } x \in \overline{\phi_t(\mathcal{B})},\ t \in \mathbb{R}, \text{ and } \xi \in \Omega,$$

where $d^{(j)}$ inherits its continuity from $\mathfrak{t}$. If we use the $d^{(j)}$ as the rows of σ, then we get the desired result. The uniqueness of σ is a consequence of the obtained linearity of $\mathfrak{t}$ with respect to ξ. □

Theorem 7.1.30 (Euler's Equation of Motion) *Let a motion ϕ and density functions $\varrho \in C^{(1)}(\overline{\mathfrak{D}_\phi})$, $b \in c(\overline{\mathfrak{D}_\phi})$, and $\mathfrak{t} \in c^{(1)}(\overline{\mathfrak{D}_\phi} \times \Omega)$ be given such that conservation of mass and momentum are fulfilled. The (consequently existing) Cauchy stress tensor is denoted, as usual, by $\sigma \in \mathbf{c}^{(1)}(\overline{\mathfrak{D}_\phi})$. Then* ***Euler's equation of motion*** *holds true on $\mathfrak{D}_\phi$:*

$$\varrho \, \mathrm{D}_t v = \varrho b + \operatorname{div}_x \sigma. \tag{7.24}$$

Proof First, we remark that the regularity of σ (continuously differentiable) is inherited from $\mathfrak{t}$. Then we remember that conservation of mass yields the equation of continuity

$$\mathrm{D}_t \varrho + \varrho \operatorname{div}_x v = 0 \quad \text{on } \mathfrak{D}_\phi; \tag{7.25}$$

see Theorem 7.1.22. After that, we apply the spatial localization theorem (Theorem 7.1.25) to conservation of momentum, which we have already identified as a componentwise particular case of the master balance law:

$$\frac{\partial}{\partial t}\left(\varrho v_j\right) + \operatorname{div}_x \left(\varrho v_j v\right) = \varrho b_j + \operatorname{div}_x \sigma_j \quad \text{on } \mathfrak{D}_\phi \text{ for all } j = 1,2,3,$$

where σ_j is the jth row of σ. We apply now the product rule and Theorem 2.3.4 and get

$$\frac{\partial \varrho}{\partial t} v_j + \varrho \frac{\partial v_j}{\partial t} + v_j \left(\nabla_x \varrho\right) \cdot v + \varrho \left(\nabla_x v_j\right) \cdot v + \varrho v_j \operatorname{div}_x v = \varrho b_j + \operatorname{div}_x \sigma_j.$$

With a well-chosen factorization, we obtain

$$v_j \left(\frac{\partial \varrho}{\partial t} + \left(\nabla_x \varrho\right) \cdot v + \varrho \operatorname{div}_x v \right) + \varrho \left(\frac{\partial v_j}{\partial t} + \left(\nabla_x v_j\right) \cdot v \right) = \varrho b_j + \operatorname{div}_x \sigma_j.$$

We remember Definition 7.1.7 regarding the material time derivative and recognize (7.25) in the latter identity. Hence, we get $\varrho \, \mathrm{D}_t v_j = \varrho b_j + \operatorname{div}_x \sigma_j$ for all $j = 1,2,3$, which is the desired equation of motion. □

Euler's equation of motion is maybe the most important governing equation in seismology (but also in many other applications). We will later discuss what this particularly means if b represents the body force density which is exerted to the Earth and if σ is further modelled regarding elasticity properties.

We continue now with the conservation of angular momentum. It says that the temporal changes of the angular momentum are due to the external torques which are produced by the body and surface forces.

Definition 7.1.31 The motion ϕ, the mass density $\varrho \colon \mathfrak{D}_\phi \to \mathbb{R}$, the body force density $b \colon \mathfrak{D}_\phi \to \mathbb{R}^3$, and the Cauchy stress tensor $\sigma \colon \mathfrak{D}_\phi \to \mathbb{R}^{3\times 3}$ are said to satisfy **conservation of angular momentum** if, for each nice set $\mathcal{U} \subset \mathcal{B}$ and all $t \in \mathbb{R}$, the following terms exist such that

$$\begin{aligned} &\frac{\mathrm{d}}{\mathrm{d}t} \int_{\phi_t(\mathcal{U})} \varrho(x,t)\, x \times v(x,t) \,\mathrm{d}x & (7.26)\\ &\quad = \int_{\phi_t(\mathcal{U})} \varrho(x,t)\, x \times b(x,t) \,\mathrm{d}x + \int_{\partial \phi_t(\mathcal{U})} x \times \left(\sigma(x,t)\, n(x)\right) \mathrm{d}\omega(x). \end{aligned}$$

Before we prove a consequence of this conservation law, we should first notice that something is different here. For the previous two conservation laws (mass and momentum), we had introduced new quantities (ϱ in the first case and $b, \mathfrak{t}$ in the second case). The conservation laws stated rules for these new quantities. However, there is no new quantity in Definition 7.1.31, but we get three new conditions (in terms of the three components of the vectorial equation (7.26)). Therefore, conservation of angular momentum can be expected to reduce the degrees of freedom of our previous quantities by three. Indeed, this conservation law is equivalent to the symmetry of σ, which causes three pairs of equal components.

Theorem 7.1.32 (Symmetry of the Cauchy Stress Tensor) *Let a motion ϕ and density functions $\varrho \in \mathrm{C}^{(1)}(\overline{\mathfrak{D}_\phi})$, $b \in \mathrm{c}(\overline{\mathfrak{D}_\phi})$, and $\mathfrak{t} \in \mathrm{c}^{(1)}(\overline{\mathfrak{D}_\phi} \times \Omega)$ be given such that conservation of mass and momentum are fulfilled. The (consequently existing) Cauchy stress tensor is denoted, as usual, by $\sigma \in \mathbf{c}^{(1)}(\overline{\mathfrak{D}_\phi})$. Then conservation of angular momentum is satisfied if and only if σ is symmetric.*

Proof The proof is a bit lengthy.

(1) The integrand of the surface integral:

If we denote the rows of σ by σ_j, $j = 1,2,3$, and the kth component of σn by $(\sigma n)_k$, then the integrand of the surface integral in (7.26) can be written as

$$x \times (\sigma n) = \begin{pmatrix} x_2(\sigma n)_3 - x_3(\sigma n)_2 \\ x_3(\sigma n)_1 - x_1(\sigma n)_3 \\ x_1(\sigma n)_2 - x_2(\sigma n)_1 \end{pmatrix} = \begin{pmatrix} x_2\sigma_3 \cdot n - x_3\sigma_2 \cdot n \\ x_3\sigma_1 \cdot n - x_1\sigma_3 \cdot n \\ x_1\sigma_2 \cdot n - x_2\sigma_1 \cdot n \end{pmatrix} = (x \times \sigma)n.$$

For vector products like $x \times \sigma$, see Definition 2.2.11.

(2) Spatial localization:

Also conservation of angular momentum has componentwise the form of a master balance law. Therefore, Theorem 7.1.25 tells us that angular momentum is conversed, if and only if

$$\begin{aligned} &\frac{\partial}{\partial t}\left[\varrho(x,t)(x \times v(x,t))_j\right] + \operatorname{div}_x \left[\varrho(x,t)(x \times v(x,t))_j v(x,t)\right] \\ &\quad = \varrho(x,t)(x \times b(x,t))_j + \operatorname{div}_x (x \times \sigma(x,t))_j \end{aligned} \tag{7.27}$$

for all $(x,t) \in \mathfrak{D}_\phi$ and all $j = 1,2,3$. We apply now the product rule and Theorem 2.3.4, but we skip the argument '(x,t)' for the sake of readability:

$$\begin{aligned} &\frac{\partial \varrho}{\partial t}(x \times v)_j + \varrho \frac{\partial}{\partial t}(x \times v)_j + (x \times v)_j (\nabla_x \varrho) \cdot v + \varrho \left[\nabla_x (x \times v)_j\right] \cdot v \\ &\quad + \varrho(x \times v)_j \operatorname{div}_x v = \varrho(x \times b)_j + \operatorname{div}_x (x \times \sigma)_j. \end{aligned}$$

We factor out again in a well-chosen way to get

$$\begin{aligned} &(x \times v)_j \left(\frac{\partial \varrho}{\partial t} + (\nabla_x \varrho) \cdot v + \varrho \operatorname{div}_x v\right) + \varrho \left(\frac{\partial}{\partial t}(x \times v)_j + \left[\nabla_x (x \times v)_j\right] \cdot v\right) \\ &\quad = \varrho(x \times b)_j + \operatorname{div}_x (x \times \sigma)_j. \end{aligned}$$

We can still follow the path of the previous proof and recognize the equation of continuity (which holds due to conservation of mass; see Theorem 7.1.22) and the material time derivative (see Definition 7.1.7):

$$\varrho\,\mathrm{D}_t(x \times v)_j = \varrho(x \times b)_j + \mathrm{div}_x\,(x \times \sigma)_j.$$

Remember that this equation holds on $\mathfrak{D}_\phi$ and for all $j = 1,2,3$. We will now write the equation in vectorial form and apply the product rule to the material time derivative, where $\mathrm{D}_t x = \frac{\partial x}{\partial t} + (\nabla_x x)v = 0 + v = v$. Therefore, we obtain

$$\varrho\,\underbrace{v \times v}_{=0} + \varrho\,x \times \mathrm{D}_t v = \varrho\,x \times b + \mathrm{div}_x\,(x \times \sigma). \tag{7.28}$$

Since ϱ is a scalar function, we can write (7.28) as

$$x \times (\varrho\,\mathrm{D}_t v - \varrho b) = \mathrm{div}_x\,(x \times \sigma). \tag{7.29}$$

We have not used the conservation of momentum yet, but this is the right moment, as we discover with a closer look on the left-hand side. Hence, (7.29) is the same as

$$x \times \mathrm{div}_x\,\sigma = \mathrm{div}_x\,(x \times \sigma). \tag{7.30}$$

Note that all steps from (7.27) to (7.30) are equivalent. We can state, therefore, that conservation of angular momentum is given if and only if (7.30) is valid on $\mathfrak{D}_\phi$.

(3) A few simple calculations at the end:

With the components σ_{jk} of σ, we obtain the following (note Definition 2.3.5 and Theorem 2.3.4):

$$\begin{aligned}
\mathrm{div}_x\,(x \times \sigma) &= \mathrm{div}_x \begin{pmatrix} x_2\sigma_3 - x_3\sigma_2 \\ x_3\sigma_1 - x_1\sigma_3 \\ x_1\sigma_2 - x_2\sigma_1 \end{pmatrix} \\
&= \begin{pmatrix} x_2\,\mathrm{div}_x\,\sigma_3 + \sigma_{32} - x_3\,\mathrm{div}_x\,\sigma_2 - \sigma_{23} \\ x_3\,\mathrm{div}_x\,\sigma_1 + \sigma_{13} - x_1\,\mathrm{div}_x\,\sigma_3 - \sigma_{31} \\ x_1\,\mathrm{div}_x\,\sigma_2 + \sigma_{21} - x_2\,\mathrm{div}_x\,\sigma_1 - \sigma_{12} \end{pmatrix} \\
&= \begin{pmatrix} x_2\,\mathrm{div}_x\,\sigma_3 - x_3\,\mathrm{div}_x\,\sigma_2 \\ x_3\,\mathrm{div}_x\,\sigma_1 - x_1\,\mathrm{div}_x\,\sigma_3 \\ x_1\,\mathrm{div}_x\,\sigma_2 - x_2\,\mathrm{div}_x\,\sigma_1 \end{pmatrix} + \begin{pmatrix} \sigma_{32} - \sigma_{23} \\ \sigma_{13} - \sigma_{31} \\ \sigma_{21} - \sigma_{12} \end{pmatrix} \\
&= x \times \mathrm{div}_x\,\sigma + \begin{pmatrix} \sigma_{32} - \sigma_{23} \\ \sigma_{13} - \sigma_{31} \\ \sigma_{21} - \sigma_{12} \end{pmatrix}.
\end{aligned}$$

Let us summarize what we have: conservation of angular momentum $\Leftrightarrow$ (7.30) $\Leftrightarrow \sigma_{32} = \sigma_{23}$ and $\sigma_{13} = \sigma_{31}$ and $\sigma_{21} - \sigma_{12} \Leftrightarrow \sigma = \sigma^{\mathrm{T}}$. □

One conservation law is still missing. It involves the introduction of three new quantities.

Definition 7.1.33 The motion ϕ, the mass density $\varrho\colon \mathfrak{D}_\phi \to \mathbb{R}$, the body force density $b\colon \mathfrak{D}_\phi \to \mathbb{R}^3$, the surface force density $\mathfrak{t}\colon \mathfrak{D}_\phi \times \Omega \to \mathbb{R}^3$, the **internal energy density** $e\colon \mathfrak{D}_\phi \to \mathbb{R}$, the **heat supply density** $r\colon \mathfrak{D}_\phi \to \mathbb{R}$, and the **heat flux density** $h\colon \mathfrak{D}_\phi \times \Omega \to \mathbb{R}$ are said to satisfy **conservation of energy** if, for each nice set $\mathcal{U} \subset \mathcal{B}$ and all $t \in \mathbb{R}$, the following terms exist such that

$$\begin{aligned}
&\frac{\mathrm{d}}{\mathrm{d}t}\int_{\phi_t(\mathcal{U})} \varrho(x,t)\left(e(x,t)+\frac{1}{2}(v\cdot v)(x,t)\right)\mathrm{d}x \\
&\quad= \int_{\phi_t(\mathcal{U})} \varrho(x,t)\,(b(x,t)\cdot v(x,t)+r(x,t))\,\mathrm{d}x \\
&\qquad+\int_{\partial\phi_t(\mathcal{U})} \mathfrak{t}(x,t,n(x))\cdot v(x,t)+h(x,t,n(x))\,\mathrm{d}\omega(x).
\end{aligned} \tag{7.31}$$

On the left-hand side of (7.31), the temporal changes of two kinds of energies are summed up: the kinetic energy $2^{-1}\int_{\phi_t(\mathcal{U})}(\varrho v\cdot v)(x,t)\,\mathrm{d}x$ due to the motion ϕ and the energy which is stored in any other imaginable way inside the medium (e.g., as chemical binding energy, intermolecular energy, or molecular vibrations). If this temporal derivative is non-zero, then energy must be inserted from outside or gets lost. This can have mechanical reasons due to the power associated to the body forces b and surface forces $\mathfrak{t}$ and thermodynamic reasons: the heat supply r per unit mass represents the heat energy which is transmitted from outside into the volume (this can be caused, e.g., by radiation) and the heat flux h per unit surface area represents energy which is transmitted via heat conduction on a surface or interface.

A simplified illustration can be seen in a kitchen: if you put a pot with water on your cooker, then the hot plate transfers energy (heat) via heat conduction to the bottom of your pot. And heat conduction is also responsible for the fact that the heat finally arrives in the water. Hence, this kind of bringing water to boil acts with the surface integral part on the right-hand side of (7.31). On the left-hand side, the energy density e increases and, at least when the water finally boils, you can also see that there is some kinetic energy in the medium (actually, the convection between the hot parts on the bottom and the colder parts on the top also has some kinetic energy before the boiling). In contrast, if you put a bowl with water into your microwave oven, then electromagnetic radiation enters the water, gets absorbed, and thus increases the total energy of the medium, but via the volume integral on the right-hand side.

We can now apply familiar procedures to this conservation law.

Theorem 7.1.34 (Localized Energy Balance, Spatial Form) *Let the motion ϕ and the densities $\varrho \in \mathrm{C}^{(1)}(\overline{\mathfrak{D}_\phi})$, $b \in \mathrm{c}(\overline{\mathfrak{D}_\phi})$, $\mathfrak{t} \in \mathrm{c}^{(1)}(\overline{\mathfrak{D}_\phi} \times \Omega)$, $e \in \mathrm{C}^{(1)}(\overline{\mathfrak{D}_\phi})$, $r \in \mathrm{C}(\overline{\mathfrak{D}_\phi})$, and $h \in \mathrm{C}^{(1)}(\overline{\mathfrak{D}_\phi} \times \Omega)$ be given such that all four conservation laws (mass, momentum, angular momentum, and energy) are satisfied. Moreover, let $\sigma \in \mathbf{c}^{(1)}(\overline{\mathfrak{D}_\phi})$ be the Cauchy stress tensor. Then there exists a unique vector field $q \in \mathrm{c}^{(1)}(\overline{\mathfrak{D}_\phi})$, which is called the **heat flux vector**, such that*

$$h(x,t,\xi) = -q(x,t)\cdot\xi \quad \textit{for all } (x,t) \in \overline{\mathfrak{D}_\phi} \textit{ and all } \xi \in \Omega \tag{7.32}$$

and

$$\varrho\, \mathrm{D}_t e + \operatorname{div}_x q = \sigma : d + \varrho r, \tag{7.33}$$

where $d = \left(\nabla_x v + (\nabla_x v)^{\mathrm{T}}\right)/2$.

For the definition of the double-dot product, see Definition 2.2.8 or Example 2.5.2, part c.

Proof Obviously, conservation of energy also has the form of the master balance law.

(1) Existence of the heat flux vector:

We first apply Cauchy's theorem (Theorem 7.1.27), which gives us a unique vector field $f \in \mathrm{c}^{(1)}(\overline{\mathfrak{D}_\phi})$ with the property that

$$\mathfrak{t}(x,t,\xi) \cdot v(x,t) + h(x,t,\xi) = (v(x,t))^{\mathrm{T}} \sigma(x,t)\xi + h(x,t,\xi) = f(x,t) \cdot \xi$$

for all $(x,t) \in \overline{\mathfrak{D}_\phi}$ and all $\xi \in \Omega$. Note that we have already used here the conservation of momentum, which gave us the Cauchy stress tensor σ. With the symmetry of σ (equivalent to conservation of angular momentum), we obtain

$$h(x,t,\xi) = f(x,t) \cdot \xi - \left[(\sigma(x,t))^{\mathrm{T}} v(x,t)\right]^{\mathrm{T}} \xi = [f(x,t) - \sigma(x,t)\, v(x,t)] \cdot \xi.$$

Hence, $q(x,t) := -f(x,t) + \sigma(x,t)\, v(x,t)$ fulfils the desired property (7.32), which cannot be satisfied by more than one function.

(2) The differential equation:

In the next step, we apply the spatial localization theorem (Theorem 7.1.25):

$$\frac{\partial}{\partial t}\left[\varrho\left(e + \frac{1}{2} v \cdot v\right)\right] + \operatorname{div}_x \left[\varrho\left(e + \frac{1}{2} v \cdot v\right) v\right] = \varrho(b \cdot v + r) + \operatorname{div}_x f.$$

In analogy to what we have already done before, we use the material time derivative (see Lemma 7.1.6 and Definition 7.1.7) and the formula for the divergence of a scalar times a vectorial function (see Theorem 2.3.4) and apply these to the left-hand side of the latter equation:

$$\mathrm{D}_t\left[\varrho\left(e + \frac{1}{2} v \cdot v\right)\right] + \varrho\left(e + \frac{1}{2} v \cdot v\right) \operatorname{div}_x v = \varrho(b \cdot v + r) + \operatorname{div}_x f. \tag{7.34}$$

Let us now have a closer look at the divergence of f and use the symmetry of σ. This leads us to

$$\begin{aligned}
\operatorname{div}_x f &= \operatorname{div}_x (\sigma v - q) \\
&= \sum_{i=1}^{3} \frac{\partial}{\partial x_i}\left(\sum_{j=1}^{3} \sigma_{ij} v_j - q_i\right) \\
&= \sum_{i=1}^{3}\left[\sum_{j=1}^{3}\left(\frac{\partial \sigma_{ij}}{\partial x_i} v_j + \sigma_{ij} \frac{\partial v_j}{\partial x_i}\right) - \frac{\partial q_i}{\partial x_i}\right]
\end{aligned}$$

$$
\begin{aligned}
&= \sum_{i,j=1}^{3} \frac{\partial \sigma_{ji}}{\partial x_i} v_j + \sum_{i=1}^{3} \frac{1}{2} \left[\sum_{j=1}^{3} \left(\sigma_{ij} \frac{\partial v_j}{\partial x_i} + \sigma_{ji} \frac{\partial v_j}{\partial x_i} \right) \right] - \operatorname{div}_x q \\
&= \sum_{j=1}^{3} \sum_{i=1}^{3} \frac{\partial \sigma_{ji}}{\partial x_i} v_j + \frac{1}{2} \sum_{i,j=1}^{3} \left(\sigma_{ij} \frac{\partial v_j}{\partial x_i} + \sigma_{ij} \frac{\partial v_i}{\partial x_j} \right) - \operatorname{div}_x q \\
&= \sum_{j=1}^{3} \left(\operatorname{div}_x \sigma_j \right) v_j + \frac{1}{2} \sum_{i,j=1}^{3} \left(\sigma_{ij} \left(\nabla_x v \right)_{ij}^{\mathrm{T}} + \sigma_{ij} \left(\nabla_x v \right)_{ij} \right) - \operatorname{div}_x q \\
&= \left(\operatorname{div}_x \sigma \right) \cdot v + \sigma : d - \operatorname{div}_x q,
\end{aligned} \tag{7.35}
$$

where σ_j is the jth row of σ and $(\nabla_x v)_{ij}, (\nabla_x v)_{ij}^{\mathrm{T}}$ denote the (i, j)th component of the corresponding Jacobian matrix and transposed Jacobian matrix, respectively.

Our next auxiliary calculation considers the material time derivative of the inner product $v \cdot v$:

$$
\begin{aligned}
\mathrm{D}_t(v \cdot v) &= \frac{\partial}{\partial t}(v \cdot v) + \left[\nabla_x (v \cdot v) \right] \cdot v \\
&= \left(\frac{\partial}{\partial t} v \right) \cdot v + v \cdot \left(\frac{\partial}{\partial t} v \right) + \left(\frac{\partial}{\partial x_i} \sum_{j=1}^{3} v_j^2 \right)_{i=1,2,3} \cdot v \\
&= 2 \left(\frac{\partial}{\partial t} v \right) \cdot v + \left(\sum_{j=1}^{3} 2 v_j \frac{\partial v_j}{\partial x_i} \right)_{i=1,2,3} \cdot v \\
&= 2 \left(\frac{\partial}{\partial t} v \right) \cdot v + 2 \left(v^{\mathrm{T}} \nabla_x v \right) v = 2 \left[\frac{\partial}{\partial t} v + \left(\nabla_x v \right) v \right] \cdot v \\
&= 2 \left(\mathrm{D}_t v \right) \cdot v.
\end{aligned} \tag{7.36}
$$

We can insert (7.35) and (7.36) into (7.34) and get

$$
\begin{aligned}
&\mathrm{D}_t(\varrho e) + \frac{1}{2} \left(\mathrm{D}_t \varrho \right) (v \cdot v) + \varrho \left(\mathrm{D}_t v \right) \cdot v + \varrho e \operatorname{div}_x v + \frac{1}{2} \varrho (v \cdot v) \operatorname{div}_x v \\
&\quad = \varrho (b \cdot v + r) + \left(\operatorname{div}_x \sigma \right) \cdot v + \sigma : d - \operatorname{div}_x q.
\end{aligned}
$$

We apply the product rule to the first summand and rearrange the terms:

$$
\begin{aligned}
&\left(\mathrm{D}_t \varrho + \varrho \operatorname{div}_x v \right) e + \varrho \mathrm{D}_t e + \left(\mathrm{D}_t \varrho + \varrho \operatorname{div}_x v \right) \frac{1}{2} (v \cdot v) + \varrho \left(\mathrm{D}_t v \right) \cdot v \\
&\quad = \varrho (b \cdot v + r) + \left(\operatorname{div}_x \sigma \right) \cdot v + \sigma : d - \operatorname{div}_x q.
\end{aligned}
$$

The identical factors in front of e and $(v \cdot v)/2$ on the left-hand side are well known to us from the equation of continuity (7.9). Since conservation of mass holds, these two factors vanish. We are left with

$$
\varrho \mathrm{D}_t e + \varrho \left(\mathrm{D}_t v \right) \cdot v = \varrho b \cdot v + \varrho r + \left(\operatorname{div}_x \sigma \right) \cdot v + \sigma : d - \operatorname{div}_x q. \tag{7.37}
$$

Since also conservation of momentum is valid, Euler's equation of motion (7.24) holds true, namely

$$\varrho \mathrm{D}_t v = \varrho b + \mathrm{div}_x\, \sigma\,. \tag{7.38}$$

If we take the dot product of (7.38) with v, then we recognize the terms in (7.37). Hence, our equation is now reduced to

$$\varrho \mathrm{D}_t e = \varrho r + \sigma : d - \mathrm{div}_x\, q\,. \qquad \square$$

Definition 7.1.35 For a motion ϕ, we define the **first Piola–Kirchhoff stress tensor** by $P(X,t) := J(X,t)\,\sigma(\phi_t(X),t)[(F(X,t))^{-1}]^{\mathrm{T}}$ and the **second Piola–Kirchhoff stress tensor** by $S(X,t) := (F(X,t))^{-1}P(X,t)$ for all $(X,t) \in \mathcal{B} \times \mathbb{R}$, where F is the deformation gradient (see Definition 7.1.8) and J is the determinant of F (see Definition 7.1.15).

Theorem 7.1.36 (Localized Energy Balance, Material Form) *If the conditions of Theorem 7.1.34 are satisfied, in particular all four conservation laws are given, then the localized material form of* (7.33) *is*

$$\varrho_{\mathrm{Ref}} \frac{\partial E}{\partial t} + \mathrm{div}_X\, Q = \varrho_{\mathrm{Ref}}\, R + S : D, \tag{7.39}$$

where the occurring functions of (X,t) *with* $X \in \mathcal{B}$, $t \in \mathbb{R}$ *are based on the coordinate transformation* $x = \phi_t(X)$ *in the sense that* $E(X,t) := e(x,t)$, $\varrho_{\mathrm{Ref}}(X,t) := J(X,t)\varrho(x,t)$, $Q(X,t) := J(X,t)(F(X,t))^{-1}q(x,t)$,

$$D(X,t) := \frac{1}{2}\left[(\nabla_X V(X,t))^{\mathrm{T}} F(X,t) + (F(X,t))^{\mathrm{T}} \nabla_X V(X,t)\right],$$

$R(X,t) := r(x,t)$, *and* S *is the second Piola–Kirchhoff stress tensor. Note that* $\varrho_{\mathrm{Ref}}(X,t) = \varrho(X,0) = \varrho_0(X)$ *for all* $X \in \mathcal{B}$ *and all* $t \in \mathbb{R}$.

Proof The fact that ϱ_{Ref} is the mass density at the time $t = 0$ is an immediate consequence of the equivalent statements for conservation of mass; see Theorem 7.1.22. For proving (7.39), we multiply (7.33) with $J(X,t)$ and apply the transformation of x into the material coordinate X such that

$$\begin{aligned} &\varrho_{\mathrm{Ref}}(X,t)\frac{\partial E}{\partial t}(X,t) + J(X,t)\,\mathrm{div}_x\, q\,(\phi_t(X),t) \\ &\quad = \varrho_{\mathrm{Ref}}(X,t)R(X,t) + J(X,t)\sum_{i,j=1}^{3} \sigma_{ij}(\phi_t(X),t)\,d_{ij}(\phi_t(X),t) \end{aligned} \tag{7.40}$$

holds true for all $X \in \mathcal{B}$ and all $t \in \mathbb{R}$. For reasons of brevity, we omit here the derivation of the following **Piola identity**:

$$\mathrm{div}_X\, Q(X,t) = J(X,t)\,\mathrm{div}_x\, q(\phi_t(X),t) \quad \text{for all } X \in \mathcal{B} \text{ and all } t \in \mathbb{R}. \tag{7.41}$$

It can be shown by using Gauß's law and arguments from differential geometry, see, for example, Marsden and Hughes (1994, p. 117).

From the definition of the first Piola–Kirchhoff stress tensor P, we know that $J\sigma = PF^{\mathrm{T}}$ with the usual relation between material and spatial coordinates. Since the left-hand side is symmetric (see Theorem 7.1.32), this must also be the case for the right-hand side. Note that we will omit the arguments of the functions for the remaining proof, but keep in mind that we will handle spatial and material quantities in the same context here. We get

$$
\begin{aligned}
J\sum_{i,j=1}^{3}\sigma_{ij}d_{ij} &= \sum_{i,j=1}^{3}\left(PF^{\mathrm{T}}\right)_{ij}\frac{1}{2}\left(\frac{\partial v_i}{\partial x_j}+\frac{\partial v_j}{\partial x_i}\right)\\
&= \frac{1}{2}\sum_{i,j=1}^{3}\left(\sum_{k=1}^{3}P_{ik}F_{jk}\right)\frac{\partial v_i}{\partial x_j}+\frac{1}{2}\sum_{i,j=1}^{3}\underbrace{\left(\sum_{k=1}^{3}P_{ik}F_{jk}\right)}_{=\sum_{k=1}^{3}P_{jk}F_{ik}}\frac{\partial v_j}{\partial x_i}\\
&= \sum_{i,j=1}^{3}\left(\sum_{k=1}^{3}P_{ik}F_{jk}\right)\frac{\partial v_i}{\partial x_j} = \sum_{i,k=1}^{3}P_{ik}\sum_{j=1}^{3}F_{jk}\frac{\partial v_i}{\partial x_j}.
\end{aligned}
$$

Briefly, we have just seen that $J\sigma : d = P : [(\nabla_x v)F]$. On the other hand, $\nabla_X V(X,t) = [\nabla_x v(\phi_t(X),t)]\nabla_X \phi_t(X) = [\nabla_x v(\phi_t(X),t)]F(X,t)$ holds due to the chain rule such that $J\sigma : d = P : \nabla_X V = (FS) : \nabla_X V$, where $S = F^{-1}P = JF^{-1}\sigma(F^{-1})^{\mathrm{T}}$ is the second Piola–Kirchhoff stress tensor. Obviously, S is symmetric, which allows us to derive that

$$
\begin{aligned}
J\sigma : d = (FS) : \nabla_X V &= \sum_{i,j=1}^{3}(FS)_{ij}\frac{\partial V_i}{\partial X_j} \qquad (7.42)\\
&= \sum_{i,j=1}^{3}\left(\sum_{k=1}^{3}F_{ik}S_{kj}\right)\frac{\partial V_i}{\partial X_j} = \sum_{k,j=1}^{3}S_{kj}\left(\sum_{i=1}^{3}F_{ik}\frac{\partial V_i}{\partial X_j}\right)\\
&= \frac{1}{2}\sum_{k,j=1}^{3}S_{kj}\underbrace{\sum_{i=1}^{3}F_{ik}\frac{\partial V_i}{\partial X_j}}_{=(F^{\mathrm{T}}\nabla_X V)_{kj}}+\frac{1}{2}\sum_{k,j=1}^{3}S_{jk}\underbrace{\sum_{i=1}^{3}F_{ik}\frac{\partial V_i}{\partial X_j}}_{=((\nabla_X V)^{\mathrm{T}}F)_{jk}} = S : D.
\end{aligned}
$$

If we now insert (7.41) and (7.42) into (7.40), then we obtain the desired identity (7.39). □

In a similar manner, we can also transfer Euler's equation of motion (see Theorem 7.1.30) into a Lagrangian form, that is, material form. The proof is omitted here – it is actually a merely moderately difficult exercise – and we only state the result, the, so to speak, **Lagrangian equation of motion**.

Theorem 7.1.37 (Lagrangian Equation of Motion) *Provided that the conditions of Theorem 7.1.30 are satisfied, in particular conservation of mass and momentum are given, then*

$$\varrho_{\mathrm{Ref}} A = \varrho_{\mathrm{Ref}} B + \operatorname{div}_X P \tag{7.43}$$

holds true in all $(X,t) \in \mathcal{B} \times \mathbb{R}$*, where* $B(X,t) = b(\phi_t(X),t)$ *is the body force density and* P *is the first Piola–Kirchhoff stress tensor. Remember that* A *is the material acceleration and* $\varrho_{\mathrm{Ref}} = \varrho_0$.

7.2 Specifics and Simplifications for the Elastic Body Earth

The derivations of the general equations of continuum mechanics cover a wide range of motions of a body, which may even be fluid and not solid. We need to keep in mind that such general equations never result in general solution formulae. Therefore, we need to have a look at the specifics which occur in the case of the Earth as the body $\mathcal{B}$, where we are particularly interested in modelling seismological phenomena.

Having a look at the specifics has here a double meaning in this section: on the one hand, we will look at the formulae of the body forces which occur if $\mathcal{B}$ is the Earth. This will at least make the equations more lengthy. On the other hand, we have to simplify terms by omitting them if they are too small to have any influence or by replacing them with approximations which are easier to handle. One omission can already be discussed now: thermal effects and radiation are considered to be negligible in seismology such that we may ignore h (and consequently also q) as well as r.

Let us first start with a general assumption, which saves us the repetition of recurring conditions (Assumption 7.1.4 remains valid).

Assumption 7.2.1 If nothing different is required, we always assume that $\mathcal{B}$ is a simple body (representing the shape of the Earth) and the functions introduced in Section 7.1 belong to the function classes $\phi \in \mathrm{c}^{(2)}(\overline{\mathcal{B}} \times \mathbb{R})$, $\varrho \in \mathrm{C}^{(1)}(\overline{\mathfrak{D}_\phi})$, $b \in \mathrm{c}(\overline{\mathfrak{D}_\phi})$, $\mathrm{t} \in \mathrm{c}^{(1)}(\overline{\mathfrak{D}_\phi} \times \Omega)$, and $e \in \mathrm{C}^{(2)}(\overline{\mathfrak{D}_\phi})$. Moreover, these functions are assumed to obey conservation of mass, momentum, angular momentum, and energy. The corresponding (symmetrical) Cauchy stress tensor is denoted by σ. Furthermore, we assume that the state at time $t = 0$, where $\phi_0(X) = X$ for all $X \in \mathcal{B}$, is a state of equilibrium with minimal internal energy. This minimal energy is gauged to be zero.

The latter gauging is possible, since we will only use derivatives of the energy density E.

Regarding the body force density b, there are three forces which play a role for the body Earth:

Gravitation: More precisely, if we include gravitation into the equations, then we model the fact that every small subset (every grain of sand, so to speak) of the Earth is attracted by the rest. This is also called **self-gravitation**. At this stage of the book, this is an easy exercise for us. Every point mass m at the position $x \in \phi_t(\mathcal{B})$ is attracted by the Earth with the force

$$f(x) = Gm\nabla_x \int_{\phi_t(\mathcal{B})} \frac{\varrho(y,t)}{|x-y|}\,\mathrm{d}y = -Gm \int_{\phi_t(\mathcal{B})} \frac{\varrho(y,t)}{|x-y|^3}\,(x-y)\,\mathrm{d}y,$$

where G is the gravitational constant. Thus, if $\mathcal{U}$ is a nice subset of $\mathcal{B}$, then the gravitational attraction of this part of the Earth by the planet itself is given by

$$-G \int_{\phi_t(\mathcal{U})} \varrho(x,t) \int_{\phi_t(\mathcal{B})} \frac{\varrho(y,t)}{|x-y|^3}\,(x-y)\,\mathrm{d}y\,\mathrm{d}x.$$

Since we represented the body forces exerted on $\mathcal{U}$ at time t by $\int_{\phi_t(\mathcal{U})} \varrho(x,t)\,b(x,t)\,\mathrm{d}x$, the contribution of self-gravitation to the total body force (per volume) term b, is given by

$$g(x,t) := -G \int_{\phi_t(\mathcal{B})} \frac{\varrho(y,t)}{|x-y|^3}\,(x-y)\,\mathrm{d}y, \quad x \in \phi_t(\mathcal{B}),\ t \in \mathbb{R}.$$

Centrifugal force: If $\tilde{\omega} \in \mathbb{R}^3$ is the angular velocity in the sense that $\tilde{\omega} \approx \frac{2\pi}{24\,\mathrm{h}}\,\varepsilon^3$, then the contribution of the centrifugal force to b is given by the well-known formula $b_{\mathrm{cen}}(x) := -\tilde{\omega} \times (\tilde{\omega} \times x)$. We neglect here the minor variations of $\tilde{\omega}$ in time, since they most likely do not have an essential influence on earthquakes and their waves.

Coriolis force: Also for the Coriolis force, we have a well-known formula at our disposal. By removing also here the factor corresponding to the mass, we obtain the contribution to b in the form $b_{\mathrm{Cor}}(x,t) := -2\tilde{\omega} \times v(x,t)$. Note that this force term implicitly depends on the time via the spatial velocity v.

Theorem 7.2.2 *The gravitational force and the centrifugal force are gradient fields. Their primitive functions (or potentials) in the sense that $g = \nabla_x \varphi$ and $b_{\mathrm{cen}} = \nabla_x \psi$ are given, for all $x \in \phi_t(\mathcal{B})$ and all $t \in \mathbb{R}$, by*

$$\varphi(x,t) = G \int_{\phi_t(\mathcal{B})} \frac{\varrho(y,t)}{|x-y|}\,\mathrm{d}y, \qquad \psi(x) = \frac{1}{2}\big[\,|\tilde{\omega}|^2\,|x|^2 - (\tilde{\omega}\cdot x)^2\,\big].$$

However, the Coriolis force is not a gradient field.

Proof The proof is very easy and therefore omitted here. For the gravitational force, this is old hat for us; see Section 3.1. The gradient of the centrifugal potential ψ can be calculated without effort, which yields a verification of the proposition. Eventually, it is also an easy exercise to determine curl b_{Cor} to see that it is, in general, not identical to zero. □

For a **self-gravitating and rotating Earth model**, the ansatz would be $b := g + b_{\mathrm{cen}} + b_{\mathrm{Cor}}$. **Euler's equation of motion** (7.24) then becomes

$$\begin{aligned}\varrho(x,t)\,\mathrm{D}_t v(x,t) &= \varrho(x,t)\Big[-G \int_{\phi_t(\mathcal{B})} \frac{\varrho(y,t)}{|x-y|^3}\,(x-y)\,\mathrm{d}y - \tilde{\omega}\times(\tilde{\omega}\times x) - 2\tilde{\omega}\times v(x,t)\Big]\\ &\quad + \mathrm{div}_x\,\sigma(x,t)\\ &= \varrho(x,t)\big[\nabla_x\,(\varphi(x,t)+\psi(x)) - 2\tilde{\omega}\times v(x,t)\big] + \mathrm{div}_x\,\sigma(x,t)\end{aligned}$$

for all $x \in \phi_t(\mathcal{B})$ and all $t \in \mathbb{R}$. This is indeed an equation which we cannot expect to be solvable in such a general constellation. What we would be looking for would be the motion ϕ or its (in some sense) derivative v. However, the equation above is non-linear in the motion and we have no imagination yet how σ should be handled here. Therefore, we need to continue with our modelling by applying also some simplifications.

Typically, in such situations, equations are simplified by linearizing them. For reasons of brevity, the mathematical background of this procedure is not explained in detail here. What is done is essentially the following: each quantity is considered as an operator depending on the motion. For example, each configuration ϕ_t yields a corresponding gravitational potential φ and a Coriolis force density b_{Cor} (as functions). Moreover, for operators between Banach spaces, there also exists a kind of a Taylor's theorem which allows an expansion into terms with multi-linear operators (with increasing order of multi-linearity) and higher-order derivatives (so-called Fréchet derivatives). This expansion can be truncated after, for example, the linear term to get a simplified relation between the motion on the one hand and the considered quantity on the other hand.

First, we do the following: we do not look for the whole motion ϕ but for the deviation from the initial configuration ϕ_0. This deviation is also called the **displacement** and is denoted by U, that is, we have $\phi(X,t) = \phi_0(X) + U(X,t)$ for all $X \in \mathcal{B}$ and all $t \in \mathbb{R}$. Clearly, the material velocity and acceleration satisfy $V(X,t) = \frac{\partial}{\partial t}\phi(X,t) = \frac{\partial}{\partial t}U(X,t)$ and $A(X,t) = \frac{\partial^2}{\partial t^2}\phi(X,t) = \frac{\partial^2}{\partial t^2}U(X,t)$. Moreover, the material time derivative of the spatial velocity satisfies, due to Lemma 7.1.6, the identity

$$\mathrm{D}_t v(x,t) = \frac{\mathrm{d}}{\mathrm{d}t} V\left(\phi_t^{-1}(x),t\right) = \frac{\partial}{\partial t} V(X,t) = \frac{\partial^2}{\partial t^2} U(X,t),$$

where $x = \phi_t(X)$, $X \in \mathcal{B}$, and $t \in \mathbb{R}$. We assume now that the displacement is 'small', which is definitely the case if we compare the amplitude of U with the size of the Earth. This assumption has a consequence: the borderline between the Lagrangian formalism and the Eulerian formalism is not entirely sharp any more. If the displacement is small, then the difference between $\frac{\partial}{\partial t}$ and D_t is also small. Since we know that this difference is represented by $v \bullet \frac{\partial}{\partial x}$, this means, for example in the case of v, that the term $(\nabla_x v)v$ in Euler's equation of motion is neglected. This is a non-linear part of the equation, which would then be removed.

Anyway, we stick to the Lagrangian formalism in the following discussions and remember the Lagrangian equation of motion (7.43), which we can now write as

$$\varrho_0 \frac{\partial^2}{\partial t^2} U = \varrho_0 B + \mathrm{div}_X\, P, \tag{7.44}$$

and the Lagrangian energy balance equation (see Theorem 7.1.36), which has the form

$$\varrho_0 \frac{\partial}{\partial t} E = S : D, \tag{7.45}$$

where Q and R were omitted as we discussed previously. Remember also that we saw in the derivation of the latter equation that

$$S : D = P : \nabla_X V = P : \left(\nabla_X \frac{\partial}{\partial t}\phi(X,t)\right) = P : \left(\frac{\partial}{\partial t} F\right), \tag{7.46}$$

where we used here Schwarz's theorem.

Our next task is to discuss the interdependence of some of the quantities which play a role in our modelling. Alternatively worded: can there be particular different motions (or simply configurations) which nevertheless produce certain identical quantities? Let us understand what is meant by this by having a look at the Lagrangian internal energy density E, which occurs in (7.45). Let $X_0 \in \mathcal{B}$ be a fixed point in material coordinates and let ϕ^0 and ϕ^1 be two configurations (the superscripts are merely enumerations) which may be arbitrary with the exception of one requirement: they need to produce the same deformation gradient F in X_0, that is, $\nabla_X \phi^0(X_0) = \nabla_X \phi^1(X_0)$. Then we define a contrived motion ϕ as follows:

$$\phi(X,t) := \phi^0(X) + t\big(\phi^1(X) - \phi^0(X)\big), \quad X \in \mathcal{B}.$$

We assume that ϕ^0 and ϕ^1 are chosen such that ϕ satisfies our assumptions on motions. Obviously, $\phi_0(X) = \phi(X,0) = \phi^0(X)$ and $\phi_1(X) = \phi(X,1) = \phi^1(X)$ for all $X \in \mathcal{B}$. We easily see that the deformation gradient corresponding to this motion satisfies

$$\begin{aligned} F(X,t) &= \nabla_X \phi^0(X) + t\big(\nabla_X \phi^1(X) - \nabla_X \phi^0(X)\big), \\ \frac{\partial}{\partial t} F(X,t) &= \nabla_X \phi^1(X) - \nabla_X \phi^0(X), \qquad \frac{\partial}{\partial t} F(X_0,t) = 0. \end{aligned}$$

Hence, (7.45) and (7.46) imply that $\frac{\partial}{\partial t} E(X_0,t) = 0$ for all t, because $\varrho_0 > 0$ everywhere. In particular, we see that $E(X_0,0) = E(X_0,1)$, that is, at time $t = 0$, when the motion equals the configuration ϕ^0, and at time $t = 1$, when the motion equals ϕ^1, the internal energy at X_0 is the same. In other words, only changing the deformation gradient F in X_0 can change the internal energy there.

Theorem 7.2.3 *The internal energy density E depends on the material coordinate X and the deformation gradient F only. Moreover, this dependence obeys the identity*

$$P_{ij} = \varrho_0 \frac{\partial E}{\partial F_{ij}} \quad \textit{for all } i,j = 1,2,3. \tag{7.47}$$

Proof We have already proved the reduced dependence. Regarding (7.47), we have a look at the Lagrangian energy balance equation, which states that

$$\varrho_0 \frac{\partial}{\partial t} E = P : \left(\frac{\partial}{\partial t} F\right) = \sum_{k,l=1}^{3} P_{kl} \left(\frac{\partial}{\partial t} F_{kl}\right); \tag{7.48}$$

see (7.45) and (7.46). Since E only depends on X and F, and X is independent of t, we are allowed to apply the chain rule to the left-hand side in the following manner:

$$\varrho_0 \frac{\partial}{\partial t} E = \varrho_0 \sum_{k,l=1}^{3} \frac{\partial E}{\partial F_{kl}} \frac{\partial F_{kl}}{\partial t}. \tag{7.49}$$

Note that (7.48) and (7.49) must be valid for all motions and, therefore, also if $\phi(X,t) := X + t^2 X_j \varepsilon^i$ for arbitrary $i,j \in \{1,2,3\}$. For each of these nine motions, we get $F(X,t) = I + t^2 \varepsilon^i \otimes \varepsilon^j$, where I is the identity matrix. Note that $F(X,t)$ is an invertible matrix for each choice of (i,j) and Assumption 7.1.4 is satisfied. Moreover, we have $\frac{\partial F}{\partial t}(X,t) = 2t\varepsilon^i \otimes \varepsilon^j$, that is, $\frac{\partial F_{kl}}{\partial t}(X,t) = 2t\delta_{ik}\delta_{jl}$, which we may insert in the preceding equations. This reveals that (7.48) can only be true if (7.47) is valid. □

These discussions of interdependencies can be put on an abstract, more mathematically rigorous basis by using so-called constitutive equations. However, for our purposes this would be a bit of overshooting the mark. Readers who are interested in such a background theory, can, for example, consult Marsden and Hughes (1994, chapter 3).

Now we come to an assumption which is often stated as an axiom. Though its name may be somehow misleading, it is not a material property. This axiom basically states that the properties of an elastic material (in this particular case, its internal energy) are independent of how we set our coordinate system. More precisely, if we rotate or mirror the coordinate system, then this should not make any difference. It appears to be intuitively clear that this is a feasible assumption.

Axiom 7.2.4 (Axiom of Material Frame Indifference) *If $X \in \mathcal{B}$ and F is a deformation gradient of a motion, then the internal energy density E which is given by X and F remains the same if F is replaced by RF for any orthogonal matrix R (i.e., $R^{\mathrm{T}} = R^{-1}$, briefly $R \in \mathrm{O}(3)$).*

This axiom is assumed to be valid throughout the remaining modelling.

Theorem 7.2.5 *The internal energy density E depends on the material coordinate X and the right Cauchy–Green tensor C only. Moreover, this dependence obeys the identity*

$$S_{ij} = 2\varrho_0 \frac{\partial E}{\partial C_{ij}} \quad \text{for all } i,j = 1,2,3. \tag{7.50}$$

Remember Definition 7.1.8, where we defined the right Cauchy–Green tensor by $C := F^{\mathrm{T}}F$. Theorem 7.2.5 is indeed a restriction of the degrees of freedom of the variability of E. Formerly, Theorem 7.2.3 stated a dependence on X (three independent components) and F (nine independent components). However, C is symmetric and has, thus, only six independent components. So, the variability is reduced by three degrees of freedom. Let us now prove Theorem 7.2.5.

Proof Let F^1 and F^2 be two deformation gradients (the superscripts are again only enumerations and no powers) with the same right Cauchy–Green tensor $C = (F^1)^{\mathrm{T}}F^1 = (F^2)^{\mathrm{T}}F^2$. We define now the second-rank tensor (or, let us simply say, the matrix) $R := F^2(F^1)^{-1}$. Note that F^1 must be invertible due to Assumption 7.1.4. Then we have $RF^1 = F^2$. Consequently, the right Cauchy–Green tensor has the representations

$$C = (F^1)^{\mathrm{T}}F^1 = (F^2)^{\mathrm{T}}F^2 = (RF^1)^{\mathrm{T}}\left(RF^1\right) = (F^1)^{\mathrm{T}}R^{\mathrm{T}}RF^1.$$

We focus now on the second term and on the last term of the latter identity. We multiply them with $\left[(F^1)^{\mathrm{T}}\right]^{-1}$ from the left-hand side and with $(F^1)^{-1}$ from the right-hand side. Then we obtain $I = R^{\mathrm{T}}R$, where I is the identity matrix. Hence, R is an orthogonal matrix. Since $F^2 = RF^1$, the axiom of material frame indifference requires that F^2 and F^1 cause the same internal energy density.

For proving (7.50), we remember (7.45) and (7.46), which tell us that $\varrho_0 \frac{\partial E}{\partial t} = P : \frac{\partial F}{\partial t}$. Moreover, the Piola–Kirchhoff stress tensors are related by $P = FS$; see Definition 7.1.35. We apply now the chain rule, the symmetry of S, an index exchange, and the product rule and get

$$\begin{aligned}
\varrho_0 \sum_{k,l=1}^{3} \frac{\partial E}{\partial C_{kl}} \frac{\partial C_{kl}}{\partial t} &= \sum_{k,l=1}^{3} P_{kl} \frac{\partial F_{kl}}{\partial t} = \sum_{k,l,m=1}^{3} F_{km} S_{ml} \frac{\partial F_{kl}}{\partial t} \\
&= \sum_{l,m=1}^{3} S_{ml} \sum_{k=1}^{3} F_{km} \frac{\partial F_{kl}}{\partial t} \\
&= \frac{1}{2} \left(\sum_{l,m=1}^{3} S_{ml} \sum_{k=1}^{3} F_{km} \frac{\partial F_{kl}}{\partial t} + \sum_{l,m=1}^{3} S_{lm} \sum_{k=1}^{3} F_{km} \frac{\partial F_{kl}}{\partial t} \right) \\
&= \frac{1}{2} \sum_{l,m=1}^{3} S_{ml} \sum_{k=1}^{3} \left(F_{km} \frac{\partial F_{kl}}{\partial t} + F_{kl} \frac{\partial F_{km}}{\partial t} \right) \\
&= \frac{1}{2} \sum_{l,m=1}^{3} S_{ml} \frac{\partial}{\partial t} \sum_{k=1}^{3} F_{km} F_{kl} = \frac{1}{2} \sum_{l,m=1}^{3} S_{ml} \frac{\partial}{\partial t} C_{ml}.
\end{aligned}$$

Hence, respecting the symmetries of C and S, we conclude that

$$2\varrho_0 \sum_{k=1}^{3} \sum_{l=1}^{k} \frac{\partial E}{\partial C_{kl}} \frac{\partial C_{kl}}{\partial t} = \sum_{k=1}^{3} \sum_{l=1}^{k} S_{kl} \frac{\partial C_{kl}}{\partial t}.$$

With analogous arguments to those which we used at the end of the proof of Theorem 7.2.3, we obtain (7.50). □

Corollary 7.2.6 *Energy balance can also be represented by*

$$2\varrho_0 \frac{\partial E}{\partial t} = S : \frac{\partial C}{\partial t}.$$

Due to the observations which we made earlier, we can consider the first Piola–Kirchhoff stress tensor P as a function of X and F and its relative S as a function of X and C. This justifies the following definition.

Definition 7.2.7 The fourth-order tensor fields Λ and Ξ are defined by their components as follows:

$$\Lambda_{ijkl} := \frac{\partial P_{ij}}{\partial F_{kl}}, \qquad \Xi_{ijkl} := \frac{\partial S_{ij}}{\partial C_{kl}} \qquad \text{for all } i,j,k,l = 1,2,3.$$

They are called the first and the second **elasticity tensor**, respectively.

Theorem 7.2.8 *The elasticity tensors satisfy the following identities:*

$$\Lambda_{ijkl} = \varrho_0 \frac{\partial^2 E}{\partial F_{kl} \partial F_{ij}}, \tag{7.51}$$

$$\Lambda_{ijkl} = \Lambda_{klij}, \tag{7.52}$$

$$\Xi_{ijkl} = 2\varrho_0 \frac{\partial^2 E}{\partial C_{kl} \partial C_{ij}}, \tag{7.53}$$

$$\Xi_{ijkl} = \Xi_{jikl} = \Xi_{ijlk} = \Xi_{klij}. \tag{7.54}$$

Proof Identities (7.51) and (7.53) are immediate consequences of the definition of the tensors and of (7.47) and (7.50), respectively. The first two symmetry properties of Ξ in (7.54) are implied by the symmetry of C. The third symmetry of Ξ and the only symmetry of Λ in (7.52) are given by Schwarz's theorem. □

Theorem 7.2.9 *Λ has 45 independent components while Ξ has 21 independent components.*

Proof There are nine pairs $(i,j) \in \{1,2,3\}^2$. The same holds true for the pairs $(k,l) \in \{1,2,3\}^2$. Let us arrange these pairs in some way, for example by $I := 3(i-1)+j \in \{1,\ldots,9\}$ and $K := 3(k-1)+l$. Due to the symmetry, the component of Λ corresponding to (I,K) is the same as the component corresponding to (K,I). Thus, for each I, we have I independent components, namely those belonging to $(I,1),\ldots,(I,I)$. Hence, the total number of independent components of Λ is $\sum_{I=1}^{9} I = 45$.

In the case of Ξ, the argumentation is similar. The only difference is that the additional symmetries reduce the number of relevant pairs (i,j) as well as (k,l) to six each. The reason is actually the same as before: for each i, we only need to know i components with the initial index pairs: $(i,1),\ldots,(i,i)$, these are $1+2+3=6$. Hence, in analogy to the earlier case, the tensor Ξ has and $\sum_{I=1}^{6} I = 21$ independent components. □

At this point, we should allow ourselves a short break before we continue. The study of the elastic behaviour of materials with this amount of independent components is certainly interesting, important, and challenging. There are indeed publications which investigate the theory and numerics for elastic waves under such general conditions such as Chattopadhyay (2004), just to mention one example. However, regarding the scope of this book and the broadness of its topics, this is beyond its aims. Therefore, we consider a simpler case, which is nevertheless a very useful and realistic assumption for most of the geophysical applications, namely the investigation of isotropic materials.

Definition 7.2.10 Whenever we assume in the modelling of a simply body $\mathcal{B}$ that the underlying material is **isotropic** in $X_0 \in \mathcal{B}$, then this means: regarding E as depending on X and C only, then the value of E for $X = X_0$ and an arbitrary C does not change if C is replaced by $R^{\mathrm{T}}CR$ for any choice of an orthogonal matrix R with positive determinant, briefly $R \in \mathrm{SO}(3)$. We say that the material is isotropic if it is isotropic in every $X \in \mathcal{B}$.

Isotropy literally means 'same direction'. An isotropic material behaves independently of the direction of the motion. At first sight, the condition in Definition 7.2.10 sounds like the axiom of material frame indifference (Axiom 7.2.4). However, there is an essential difference. Material frame indifference means that the behaviour of the material cannot depend on our choice of setting the coordinate system, which certainly needs to be always the case. Isotropy means something different: let $x = \phi_t(X)$ and $\tilde{x} = \phi_t(RX)$ for an arbitrary $R \in \mathrm{SO}(3)$. Elements of O(3) are rotations and mirrorings, the absolute value of their determinant needs to be 1. The subset SO(3) contains the rotations only. If we compare the x-motion with the rotated $\tilde{x}$-motion (we will denote now all quantities belonging to $\tilde{x}$ with a tilde), then $\tilde{F} = \nabla_X \phi_t(RX) = [(\nabla_X \phi_t)(RX)]R = FR$. Hence, the right Cauchy–Green tensor satisfies $\tilde{C} = (FR)^{\mathrm{T}} FR = R^{\mathrm{T}} F^{\mathrm{T}} FR = R^{\mathrm{T}} CR$. This is exactly the replacement which is made in the condition for isotropy.

The difference to material frame indifference becomes clear, if we compare the motion with $\hat{x} := R\phi_t(X)$, where $\hat{F} = \nabla_X(R\phi_t(X)) = R\nabla_X \phi_t(X) = RF$. The latter is the replacement which was made in the axiom of material frame indifference. It leads us to $\hat{C} = \hat{F}^{\mathrm{T}} \hat{F} = F^{\mathrm{T}} R^{\mathrm{T}} RF = C$. The 'hat'-motion means that we have a body and let something happen to it, that is, it is deformed or a wave propagates through it, etc. *After that*, we rotate the deformed object and observe its state, particularly its internal energy, which results, as we know now, in a distribution of stresses. Certainly, the rotation will only cause the same stresses to now to occur at different coordinates, but the body itself does not change its deformation only because we rotate it after everything is over. However, in the case of the 'tilde'-motion, we rotate the body *before* something happens. Imagine a part of the Earth which is isotropic. Then it will behave the same way if a wave passes through it, no matter how we rotated it beforehand. Or equivalently worded, no matter from which direction the wave arrives, it will always propagate in the same way through an isotropic material. This is certainly not a natural law but a specific property which some materials have and others do not have.

Now, if we have an isotropic material, then the degrees of freedom for the variability of E (and consequently also the stress) are further decreased. This can be seen as follows: since C is symmetric, the principal axis theorem allows us to choose an orthonormal basis of eigenvectors as the columns of the matrix R, which is then automatically orthogonal (and also with a positive determinant if we arrange the column vectors in an appropriate order). Then $R^{\mathrm{T}} CR$ is a diagonal matrix with the eigenvalues of C as the diagonal entries. These eigenvalues are the squares of the principal stretches; see Definition 7.1.11. In other words, E depends, in general, on X (three independent components) and C (six independent components). In the case of isotropic materials, E still depends on the three components of X and additionally now only on the three independent principal stretches. The same must hold true for the second Piola–Kirchhoff stress tensor.

This section was also dedicated to simplifications of the model. So, we should now take care of this again. For this purpose, we have a closer look at the dependence of E on C. We assumed small deformations, that is, the deviations from the initial state $\phi_0(X) = X$ must be small in comparison to ϕ_0. This initial state yields the identity matrix as deformation gradient: $F = I$. Consequently, $C = I$. Thus, if we want to make a Taylor expansion of E with respect to C, then $C = I$ would be the appropriate choice for the centre of the expansion, from which C should not be too far away. Actually, $(C - I)/2$ is also called the **material (or Lagrangian) strain tensor**.

And here is the simplification: we truncate this Taylor expansion after the quadratic term. We will denote the resulting approximated internal energy by W. The change of the notation of the energy function is recommended here, because we change the variables and because it is only an approximation. The link between E and W is as follows: if we have a motion ϕ, then the energy at point X and time t is given by $E(X,t) = e(\phi_t(X),t)$ and is approximated by $W(X,(\nabla_X\phi_t(X))^{\mathrm{T}}\nabla_X\phi_t(X))$. We also avoid writing '$\approx$' for the cases where we use such kinds of approximation but simply write '$=$'. We get

$$\begin{aligned}
W(X,C) &= W(X,I) + \sum_{i,j=1}^{3} \frac{\partial W}{\partial C_{ij}}(X,I)\left(C_{ij}-\delta_{ij}\right) \\
&\quad + \frac{1}{2}\sum_{i,j,k,l=1}^{3} \frac{\partial^2 W}{\partial C_{ij}\partial C_{kl}}(X,I)\left(C_{ij}-\delta_{ij}\right)\left(C_{kl}-\delta_{kl}\right) \\
&= W(X,I) + \left(\frac{\partial W}{\partial C}(X,I)\right) : (C-I) \\
&\quad + \frac{1}{2}\left(\frac{\partial^2 W}{\partial C^2}(X,I)\right) : \left[(C-I)\otimes(C-I)\right].
\end{aligned} \tag{7.55}$$

The next step is to remember Assumption 7.2.1, where we required that the initial state ϕ_0 corresponds to an equilibrium of minimal energy, where we set the latter to zero. This implies that $W(X,I) = 0$ and $\frac{\partial W}{\partial C}(X,I) = 0$. Moreover, if we have a look at (7.53), then we eventually get

$$W = \frac{1}{4\varrho_0}\,\Xi : \left[(C-I)\otimes(C-I)\right].$$

We use now (7.50) and one of the symmetries of Ξ in (7.54) and represent the previous double-dot product via components:

$$\begin{aligned}
S_{mn} &= \frac{1}{2}\frac{\partial}{\partial C_{mn}} \sum_{i,j,k,l=1}^{3} \Xi_{ijkl}\left(C_{ij}-\delta_{ij}\right)\left(C_{kl}-\delta_{kl}\right) \\
&= \frac{1}{2}\left[\sum_{k,l=1}^{3} \Xi_{mnkl}\left(C_{kl}-\delta_{kl}\right) + \sum_{i,j=1}^{3} \Xi_{ijmn}\left(C_{ij}-\delta_{ij}\right)\right] \\
&= \frac{1}{2}\left[\sum_{k,l=1}^{3} \Xi_{mnkl}\left(C_{kl}-\delta_{kl}\right) + \sum_{i,j=1}^{3} \Xi_{mnij}\left(C_{ij}-\delta_{ij}\right)\right] \\
&= \sum_{i,j=1}^{3} \Xi_{mnij}\left(C_{ij}-\delta_{ij}\right).
\end{aligned} \tag{7.56}$$

Hence,

$$S = \Xi : (C-I). \tag{7.57}$$

Now, we need to add a linearization. If a small perturbation h is added to the initial state ϕ_0, then the difference for the right Cauchy–Green tensor is as follows:

$$\begin{aligned}
&[\nabla_X(\phi_0+h)]^{\mathrm{T}}[\nabla_X(\phi_0+h)]-(\nabla_X\phi_0)^{\mathrm{T}}(\nabla_X\phi_0)\\
&=(\nabla_X\phi_0)^{\mathrm{T}}(\nabla_X h)+(\nabla_X h)^{\mathrm{T}}(\nabla_X\phi_0)+(\nabla_x h)^{\mathrm{T}}(\nabla_X h)\\
&=(\nabla_X h)+(\nabla_X h)^{\mathrm{T}}+(\nabla_x h)^{\mathrm{T}}(\nabla_X h)\,.
\end{aligned}$$

The latter term is quadratic in h and therefore neglected in the further discussions (with an appropriate norm, a small norm of h implies that $(\nabla_X h)^{\mathrm{T}}(\nabla_X h)$ is also small). This means that the right Cauchy–Green tensor which corresponds to a motion ϕ_0+U is approximated by $I+\nabla_X U+(\nabla_X U)^{\mathrm{T}}$. For (7.57), this means

$$S=\Xi:\left[\nabla_X U+(\nabla_X U)^{\mathrm{T}}\right]. \tag{7.58}$$

Writing this in components again and using the second symmetry of Ξ in (7.54), we arrive at

$$S_{ij}=\sum_{k,l=1}^{3}\Xi_{ijkl}\left(\frac{\partial U_k}{\partial X_l}+\frac{\partial U_l}{\partial X_k}\right)=\sum_{k,l=1}^{3}\Xi_{ijkl}\frac{\partial U_k}{\partial X_l}+\sum_{k,l=1}^{3}\Xi_{ijlk}\frac{\partial U_l}{\partial X_k}$$

such that

$$S=2\,\Xi:(\nabla_X U)\,. \tag{7.59}$$

Equation (7.58) and its equivalent form (7.59) are known as **Hooke's law**. It states, that, for small displacements U (with small derivatives as well), the relation between the stress (S) and the strain ($\nabla_X U+(\nabla_X U)^{\mathrm{T}}$) can be linearized and therefore be regarded as proportional. The associated proportionality factor (in the sense of a double-dot product) is the (second) elasticity tensor.

In analogy to (7.55), we can also consider the internal energy density E as a function of X and F and expand it around $F=I$ in a Taylor series, which is truncated after the quadratic term. We obtain then an approximation which we denote by $\tilde{W}$. Since the deformation gradient $F=I$ corresponds to the equilibrium with minimal and also vanishing energy, we arrive at

$$\tilde{W}(X,F)=\frac{1}{2}\left(\frac{\partial^2\tilde{W}}{\partial F^2}(X,I)\right):\left[(F-I)\otimes(F-I)\right]=\frac{1}{2\varrho_0}\Lambda:\left[(F-I)\otimes(F-I)\right].$$

In the case of $W(X,C)$, we continued by calculating S in (7.56). We can do this completely analogously if we seek P out of $\tilde{W}(X,F)$, because (7.47) gives us a corresponding formula for P as a derivative of the internal energy with respect to F and the particular symmetry of Ξ, which was used in (7.56), is also available for Λ; see (7.52). Eventually, we get

$$P=\Lambda:(F-I)=\Lambda:(\nabla_X U)\,. \tag{7.60}$$

This is **Hooke's law** for the first Piola–Kirchhoff stress tensor and the first elasticity tensor as the proportionality factor.

Let us summarize what we have done here. After introducing the small displacement U, we turned the equation of motion into (7.44). If we also use Hooke's law (7.60), then we get

$$\varrho_0 \frac{\partial^2 U}{\partial t^2} = \varrho_0 B + \operatorname{div}_X \left(\Lambda : \nabla_X U\right).$$

This equation can also be transformed into a spatial form, for which the derivation is omitted here (see, e.g., Marsden and Hughes, 1994, chapter 4). See also Gurtin (1984), where a consistent linear theory of elasticity is constructed from the beginning. Independent of the way the modelling is started and in which mathematical sense simplifications such as linearizations are applied, one usually arrives at this point at the following equation

$$\varrho \frac{\partial^2 u}{\partial t^2} = \varrho b + \operatorname{div}_x \left(\Gamma : \nabla_x u\right) \tag{7.61}$$

for small displacements u and an **elasticity tensor** Γ with $\Gamma_{ijkl} = \Gamma_{klij} = \Gamma_{jikl}$. Equation (7.61) is the **equation of linear elasticity**. It should be noted that the postulated symmetries of Γ also imply that $\Gamma_{ijkl} = \Gamma_{ijlk}$, because $\Gamma_{ijkl} = \Gamma_{klij} = \Gamma_{lkij} = \Gamma_{ijlk}$. Besides, the use of a capital letter for the elasticity tensor, although we stick here to the spatial formalism, could be a bit confusing, but Γ indeed also occurs in some of the literature. Moreover, note that the additional symmetries of Γ in comparison to Λ should be rather regarded as a postulation in our context of modelling here.

Remember how we modelled the internal energy density as a quadratic function with respect to one of the elasticity tensors and also that we linearized the Lagrangian strain tensor to $[\nabla_X U + (\nabla_X U)^{\mathrm{T}}]/2$. In this context, the following definition should be seen.

Definition 7.2.11 The **linearized strain tensor** $\overline{e}$ as well as the **elastic stored energy function** ϵ is defined by

$$\overline{e}(x,t) := \frac{1}{2}\left[\nabla_x u(x,t) + (\nabla_x u(x,t))^{\mathrm{T}}\right], \qquad \epsilon(x,t) := \frac{1}{2}\,\Gamma(x,t) : (\overline{e} \otimes \overline{e})(x,t).$$

Moreover, the **linearized Cauchy stress tensor** is defined by $\sigma := \Gamma : \overline{e}$. Since the pre-linearization version has the same notation, we declare that, from now on, σ denotes the linearized Cauchy stress tensor only, unless anything different is stated.

Indeed, the state of equilibrium where $\overline{e} \equiv 0$ is associated again to the minimal energy. Moreover, we obviously get the following identities.

Theorem 7.2.12 *In every* $(x,t) \in \mathfrak{D}_\phi$ *and for all* $i,j,k,l \in \{1,2,3\}$, *we have* $\sigma_{ij} = \frac{\partial \epsilon}{\partial \overline{e}_{ij}}$ *and* $\Gamma_{ijkl} = \frac{\partial \epsilon}{\partial \overline{e}_{ij} \partial \overline{e}_{kl}}$.

In analogy to Definition 7.2.10, where isotropy was defined, we also specify what isotropy means in the linearized model. Certainly, this needs to be done in a consistent way.

Definition 7.2.13 Whenever we assume in the context of the linearized model for motions ϕ of the simply body $\mathcal{B}$ that the underlying material is **isotropic** in $x_0 \in \phi_t(\mathcal{B})$, then this means that regarding ϵ as depending on x and $\overline{e}$ only, then the value of ϵ for $x = x_0$ and an arbitrary $\overline{e}$ does not change, if $\overline{e}$ is replaced by $R^{\mathrm{T}}\overline{e}R$ for any choice of an orthogonal matrix $R \in \mathrm{SO}(3)$. We say that the material is isotropic if it is isotropic in every $x_0 \in \phi_t(\mathcal{B})$ at all times t.

Definition 7.2.14 A material is called **homogeneous** in a non-void open subset $\mathcal{U} \subset \mathcal{B}$ if the elasticity tensor Γ and the mass density ϱ are constant in $\phi_t(\mathcal{U})$ at all times t. The material is called homogeneous if it is homogeneous in $\mathcal{B}$.

By applying analytical methods to some advanced identities from linear algebra including the invariants of symmetric matrices, one can derive that in the homogeneous isotropic case, the elasticity tensor Γ only depends on two constants. The proof for this is omitted here. Readers who are interested in further details are referred to Marsden and Hughes (1994, sections 3.5 and 4.2). We state here the result, which will be very important for our further considerations.

Theorem 7.2.15 *If Γ is isotropic and homogeneous in a non-void open subset $\mathcal{U} \subset \mathcal{B}$, then there exist constants $\lambda, \mu \in \mathbb{R}$ such that the (constant) components of Γ on $\mathcal{U}$ are representable by*

$$\Gamma_{ijkl} = \lambda\, \delta_{ij}\delta_{kl} + \mu\left(\delta_{ik}\delta_{jl} + \delta_{il}\delta_{jk}\right). \tag{7.62}$$

Definition 7.2.16 The constants λ and μ in Theorem 7.2.15 are called the **Lamé parameters** or the **Lamé moduli**. The parameter μ is also called the **rigidity** or the **shear modulus**. Furthermore, instead of using the parameter pair (λ, μ), the pair (κ, μ) is also often used in the literature, where $\kappa := \lambda + 2\mu/3$ is the **incompressibility** and is also called the **bulk modulus** or the **modulus of compression**.

For physical reasons, only $\lambda, \mu \in \mathbb{R}_0^+$ make sense.

Let us see what isotropy implies for the Cauchy stress tensor. First of all, we observe that one of the symmetries of Γ allows us to show that $\Gamma : \overline{e} = \Gamma : (\nabla_x u)$, which can be done in complete analogy to the derivation of (7.59). With (7.62), we obtain then

$$\begin{aligned}\sigma_{ij} &= \sum_{k,l=1}^{3} \left[\lambda\, \delta_{ij}\delta_{kl} + \mu\left(\delta_{ik}\delta_{jl} + \delta_{il}\delta_{jk}\right)\right] \frac{\partial u_k}{\partial x_l} \\ &= \lambda\, \delta_{ij} \sum_{k=1}^{3} \frac{\partial u_k}{\partial x_k} + \mu\left(\frac{\partial u_i}{\partial x_j} + \frac{\partial u_j}{\partial x_i}\right).\end{aligned} \tag{7.63}$$

This leads us to the following identity.

Corollary 7.2.17 *On homogeneous and isotropic subsets, the (linearized) Cauchy stress tensor is given by*

$$\sigma = \lambda(\operatorname{div}_x u)I + \mu\left[\nabla_x u + (\nabla_x u)^{\mathrm{T}}\right] = \lambda(\operatorname{div}_x u)I + 2\mu\overline{e}, \tag{7.64}$$

where I is the second-rank identity tensor.

Remark 7.2.18 A **perfect fluid** is usually modelled by the requirement that the Cauchy stress tensor satisfies that $\sigma\xi$ is a vector which is parallel to ξ for all $\xi \in \mathbb{R}^3 \setminus \{0\}$ (due to the linearity of the condition, it suffices to consider only unit vectors ξ), which means that stresses at surfaces are normal and have no tangential component. It is easy to verify that this constraint can only be fulfilled if σ is diagonal with equal diagonal entries. We can, therefore, write $\sigma = -pI$, where I is the identity matrix and p is a scalar function, which is called the **pressure**. In comparison with (7.64), we see that a perfect fluid is associated to a vanishing shear modulus μ. Then we have $p = -\lambda \operatorname{div}_x u$.

7.3 Propagation of Body Waves and Normal Modes

Within this section, we consider the following scenario: the Earth (more precisely, the part of the Earth which will serve as the medium for wave propagation) consists of a finite number of homogeneous layers, where 'layer' does not refer to any specific geometry. There could be some open subsets $\mathcal{E}_i \subset \mathcal{B}$, $i = 1, \ldots, N$, such that $\bigcup_{i=1}^{N} \overline{\mathcal{E}_i}$ is the considered part of the Earth. We will only investigate here waves which propagate through the Earth's body (unlike waves which propagate across its surface). These waves are called **body waves**.

For solving the equation of linear elasticity (7.61), we will use the ansatz of a plane progressive wave here (remember that Ω is the unit sphere).

Definition 7.3.1 A **plane progressive wave** is a vector field $u\colon \overline{\mathfrak{D}_\phi} \to \mathbb{R}^3$ which has the form

$$u(x,t) = \alpha\varphi(x \cdot k - ct), \tag{7.65}$$

where $\alpha \in \mathbb{R}^3$, $k \in \Omega$, and $c \in \mathbb{R}$ are constant (in each layer $\mathcal{E}_i$) and $\varphi \in \mathrm{C}^{(2)}(\mathbb{R})$ is a function whose second derivative is not identical to zero. We call α the **amplitude of the wave**, k the **direction of wave propagation**, and c the **speed of wave propagation**.

Example 7.3.2 Let, for instance, $k = 5^{-1/2}(2,-1,0)^{\mathrm{T}}$. The vanishing third component makes a graphical illustration in the x_1-x_2-plane possible, which is done in Figure 7.3. Obviously, the shape of the wave depends on the function φ. More precisely, if we choose an arbitrary $\tau \in \mathbb{R}$, then $u(x,t)$ is constant on the set of all (x,t) for which $x \cdot k - ct = \tau$. From a slightly different point of view, we could say: at a fixed time t, the vectorial value $u(x,t)$ is the same on the set of all x for which $x \cdot k = \tau + ct$. The latter identity is the normal form of a plane, which explains why we talk about *plane* waves. So, whatever the shape of the wave is which $\varphi(\tau)$ indicates – let it, for example, be the crest of the wave or its trough – it occurs on a plane. A closer look at the normal form of the plane shows us that k is the normal vector, which does not depend on the time. This is why it appears that the wave propagates in the direction of k. Moreover, the right-hand side $\tau + ct$ determines the distance of the plane to the origin, because $x = (\tau + ct)k$ is the point of the plane which has the minimal distance to 0 (this is the well-known fact that the plumbline gives the shortest distance from a plane to a given point).

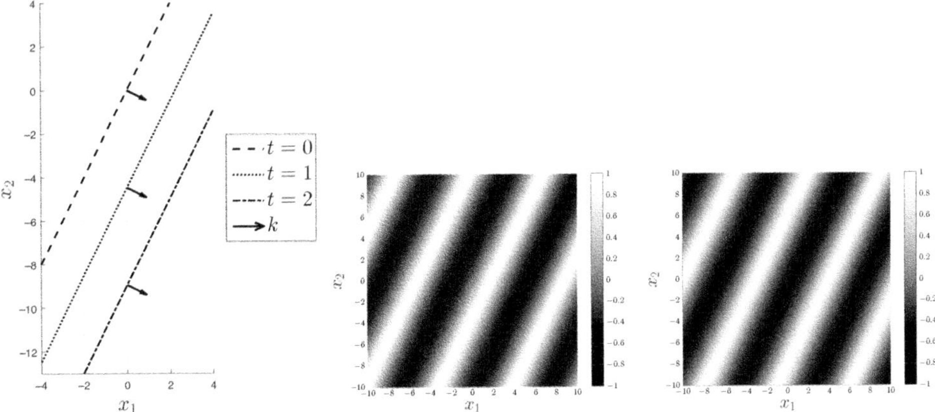

Figure 7.3 Illustration of the propagation of a plane progressive wave, restricted to the x_1-x_2-plane: on the left-hand side, planes are shown which are given by $x \cdot k = \tau + ct$ for a fixed value of τ and varying times t. At each time t, the wave has on the corresponding plane the state which is given by $\alpha\varphi(\tau)$. The other two pictures show $\sin(x \cdot k - ct)$ for the same k and c and the times $t = 0$ (middle) and $t = 1$ (right).

Before we can insert the ansatz (7.65) into the equation of linear elasticity (7.61), we need to do some simple calculations: if u is given by (7.65), then

$$\frac{\partial u_i}{\partial t}(x,t) = -c\alpha_i\varphi'(x \cdot k - ct), \qquad \frac{\partial u_i}{\partial x_j}(x,t) = k_j\alpha_i\varphi'(x \cdot k - ct),$$

$$\frac{\partial^2 u_i}{\partial t^2}(x,t) = c^2\alpha_i\varphi''(x \cdot k - ct), \qquad \frac{\partial^2 u_i}{\partial x_j \partial x_l}(x,t) = k_j k_l \alpha_i\varphi''(x \cdot k - ct).$$

Hence, respecting the symmetries of the elasticity tensor Γ, which is assumed to be constant inside a layer (see the previous discussions), we obtain

$$\begin{aligned}\operatorname{div}_x\left(\Gamma : \nabla_x u\right) &= \operatorname{div}_x \left(\sum_{l,m=1}^{3} \Gamma_{ijlm} \frac{\partial u_l}{\partial x_m}\right)_{i,j=1,2,3} \\ &= \left(\sum_{j,l,m=1}^{3} \Gamma_{ijlm} \frac{\partial^2 u_l}{\partial x_j \partial x_m}\right)_{i=1,2,3} = \left(\sum_{j,l,m=1}^{3} \Gamma_{ijlm} k_j k_m \alpha_l\right)_{i=1,2,3} \varphi'' \\ &= \left[\Gamma : (k \otimes \alpha \otimes k)\right]\varphi'',\end{aligned}$$

where the argument $x \cdot k - ct$ of φ'' was omitted here.

It is common to consider a modified version of (7.61) in this context. It means basically that the body forces are neglected in the equation of linear elasticity. There are alternative modellings (see, e.g., Dahlen and Tromp, 1998, section 3.6.3) which arrive at essentially the same result, and there is also a (let us say) typically mathematical way of just defining that there is this new equation whose solutions are called elastic waves (see, e.g., Marsden and Hughes, 1994, section 4.3). For our purposes, let us try to get accustomed to the following imagination: if a seismic wave propagates, then neither its reason nor its way of propagating is predominantly a consequence of gravitation or the Earth's rotation. The latter two certainly have influence on the wave, but it is primarily caused and determined by the elasticity of the Earth. Thus, what is called an elastic wave is the wave as it would occur as the pure result of elasticity. The body forces will cause some (likely minor) perturbations.

Definition 7.3.3 A function u of the form (7.65) which solves the equation

$$\varrho \frac{\partial^2 u}{\partial t^2} = \operatorname{div}_x (\Gamma : \nabla_x u) \tag{7.66}$$

is called an **elastic plane progressive wave**. In general, solutions of (7.66) are called **elastic waves**.

Taking into account that $\varphi'' \not\equiv 0$, our preceding calculations immediately lead us to the following result.

Theorem 7.3.4 (Fresnel–Hadamard Theorem) *A plane progressive wave is elastic if and only if we have inside each layer the identity*

$$\Gamma : (k \otimes \alpha \otimes k) = \varrho c^2 \alpha. \tag{7.67}$$

With the definition of the second-rank tensor Υ *as* $\Upsilon_{pq} := \sum_{i,j=1}^{3} \Gamma_{pijq} k_i k_j$, $p, q = 1, 2, 3$, (7.67) *is equivalent to the* ***Christoffel equation***:

$$\Upsilon \alpha = \varrho c^2 \alpha. \tag{7.68}$$

Proof Equation (7.67) follows from the preceding derivations. For the Christoffel equation, we have a closer look at the left-hand side of (7.67):

$$\begin{aligned}
\Gamma : (k \otimes \alpha \otimes k) &= \left(\sum_{j,l,m=1}^{3} \Gamma_{ijlm} k_j \alpha_l k_m \right)_{i=1,2,3} \\
&= \left(\sum_{j,l,m=1}^{3} \Gamma_{ijml} k_j \alpha_l k_m \right)_{i=1,2,3} = \left(\sum_{j,m=1}^{3} \Gamma_{ijml} k_j k_m \right)_{i,l=1,2,3} \alpha \\
&= \Upsilon \alpha,
\end{aligned}$$

where we used one of the symmetries of Γ and the latter two terms stand for matrix-vector multiplications. □

It is not difficult to see that the Christoffel equation (7.68) represents an eigenvalue problem. For solving this, the next proposition will be helpful.

Corollary 7.3.5 *The second-rank tensor* Υ *in the Christoffel equation* (7.68) *is symmetric.*

Proof We use the symmetries of Γ to obtain, for all $p, q = 1, 2, 3$,

$$\Upsilon_{pq} = \sum_{i,j=1}^{3} \Gamma_{pijq} k_i k_j = \sum_{i,j=1}^{3} \Gamma_{jqpi} k_i k_j = \sum_{i,j=1}^{3} \Gamma_{qjip} k_i k_j = \Upsilon_{qp}. \qquad \square$$

Hence, the principal axis theorem from linear algebra tells us that Υ has only real eigenvalues and we can find a corresponding system of three mutually orthonormal eigenvectors. We will calculate these eigenvalues and eigenvectors for the particular case of an isotropic material.

Corollary 7.3.6 *In the case of an* isotropic material, *the previous second-rank tensor* Υ *has the form* $\Upsilon = (\lambda + \mu) k \otimes k + \mu I$, *where* λ *and* μ *are the Lamé parameters and* I *is the second-rank identity tensor (i.e., the identity matrix).*

Proof Theorem 7.2.15 provides us with a formula for the elasticity tensor in the case of an isotropic material. By inserting this into the definition of the components of Υ and remembering that $k \in \Omega$, that is, $|k| = 1$, we obtain

$$\begin{aligned}\Upsilon_{p,q} &= \sum_{i,j=1}^{3} \left[\lambda\, \delta_{pi} \delta_{jq} + \mu \left(\delta_{pj} \delta_{iq} + \delta_{pq} \delta_{ij}\right)\right] k_i k_j \\ &= \lambda k_p k_q + \mu k_q k_p + \mu \delta_{pq} \sum_{i=1}^{3} k_i k_i = (\lambda + \mu) k_p k_q + \mu \delta_{pq}\end{aligned}$$

for all $p, q = 1, 2, 3$. $\square$

The eigenvalues and eigenvectors of Υ in the isotropic case are easy to find. First, we observe that

$$\Upsilon k = (\lambda + \mu) k k^{\mathrm{T}} k + \mu k = (\lambda + \mu) k + \mu k = (\lambda + 2\mu) k$$

such that the vector k of the direction of propagation is an eigenvalue (including its non-trivial multiples). The associated eigenvalue can be compared to eigenvalue problem (7.68), where we see that $\varrho c^2 = \lambda + 2\mu$. Moreover, all vectors α which are orthogonal to k need to satisfy

$$\Upsilon \alpha = (\lambda + \mu) k k^{\mathrm{T}} \alpha + \mu \alpha = \mu \alpha.$$

Thus, the plane of all vectors which are orthogonal to k is a two-dimensional eigenspace associated to the eigenvalue $\mu = \varrho c^2$. This leads us to the following fundamental property of body waves.

Theorem 7.3.7 *In the* isotropic case, *the Christoffel equation only has two kinds of solutions: first, all multiples of* k, *the direction of wave propagation, are solutions. In this case, the speed of propagation is* $c = \sqrt{(\lambda + 2\mu)\varrho^{-1}}$. *Every other solution is given by all vectors* $\alpha \in \mathbb{R}^3 \setminus \{0\}$ *with* $k \cdot \alpha = 0$. *For these waves, the speed of propagation is* $c = \sqrt{\mu \varrho^{-1}}$.

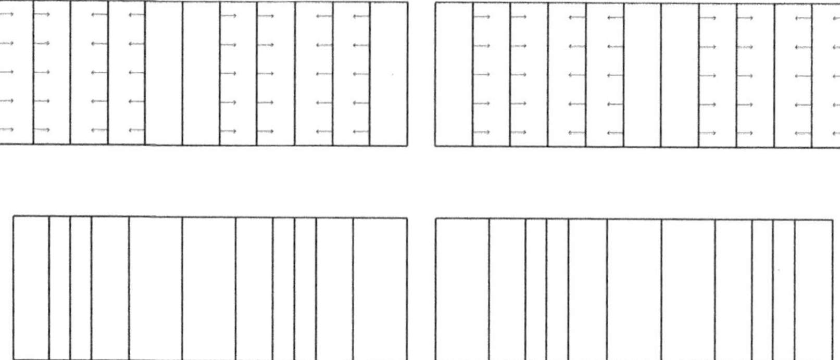

Figure 7.4 Illustration of a P-wave: a beam is provided with equidistant vertical lines (top row). An example of a displacement vector u corresponding to a P-wave is shown by the arrows in the first row, where the right column belongs to a point of time slightly later than in the case of the left column. The effect of the displacement is indicated by the moved vertical lines in the second row. The wave propagates to the right, while the displacement due to the oscillating wave is horizontal, that is, parallel to the direction of propagation.

Thus, there are two kinds of waves. The first type is a wave where the direction of the oscillations (i.e., the amplitude α) is parallel to the direction of propagation k, as shown in Figure 7.4. For this reason, these waves are called **longitudinal waves**. Oscillations parallel to the direction of propagation cause compressions and extensions of the material. So, from the physical point of view, they are nothing else than **compressional waves** or, simply put, they are **sound waves**. Since their speed is obviously larger than the speed of the other type, they are always the first to arrive at a seismic station. For this reason, the longitudinal waves are also called the primary waves, which is commonly abbreviated as **P-waves**.

The second type of waves, which consequently obtains the name **S-waves** for secondary waves, has in common that the direction of oscillation is orthogonal to the direction of propagation. These waves, which are therefore called the **transversal waves**, are not associated to compressions and extensions. Instead, they cause deformations in the sense of a shear strain, that is, S-waves are **shear waves**; see Figure 7.5. Remember Remark 7.2.18: shear waves can obviously not occur in perfect fluids, because the formula for the speed would yield zero.

However, looking at a seismic wave from a very local perspective can allow us to consider the wave as a plane progressive wave, while this is not realistic from a global point of view. In the latter perspective, we would expect something like spherical wavefronts in a homogeneous medium and wavefronts with more complicated geometries in more general media. For this reason, we use now a different ansatz to solve (7.66), but we will see that the speeds of wave propagation will obey the same formulae that we obtained for the plane waves. For reasons of simplicity, we reduce our discussions also here to a homogeneous and isotropic medium (such as a layer of the Earth with this property). From (7.64), we obtain, by using Schwarz's theorem,

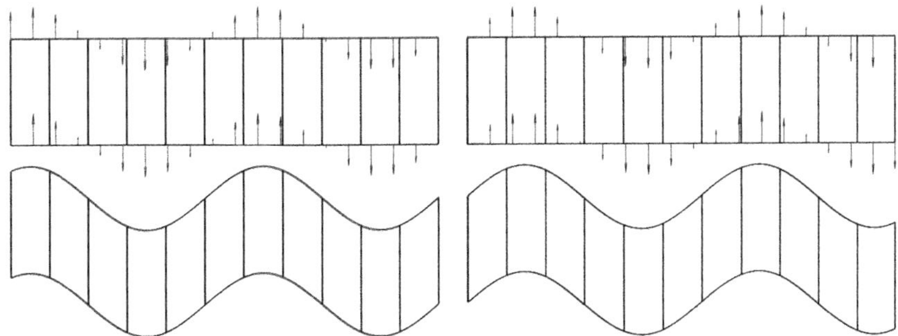

Figure 7.5 Illustration of an S-wave: a delineation in analogy to Figure 7.4 is used, but now the displacement u corresponds to an S-wave and is therefore orthogonal to the direction of propagation. That is, here a vertical displacement occurs which is indicated by the arrows and the moved horizontal boundaries of the beam. Unlike for P-waves, the material is not compressed or extended, but shear strains occur.

$$\begin{aligned}
\operatorname{div}_x \sigma &= \left(\sum_{j=1}^{3} \frac{\partial}{\partial x_j} (\lambda \operatorname{div}_x u) \delta_{ij} \right)_{i=1,2,3} + \mu \sum_{j=1}^{3} \frac{\partial}{\partial x_j} \left(\frac{\partial u_i}{\partial x_j} + \frac{\partial u_j}{\partial x_i} \right)_{i=1,2,3} \\
&= \lambda \operatorname{grad}_x \operatorname{div}_x u + \mu \Delta_x u + \mu \left(\frac{\partial}{\partial x_i} \sum_{j=1}^{3} \frac{\partial u_j}{\partial x_j} \right)_{i=1,2,3} \\
&= (\lambda + \mu) \operatorname{grad}_x \operatorname{div}_x u + \mu \Delta_x u. \qquad (7.69)
\end{aligned}$$

Theorem 7.3.8 *In isotropic and homogeneous media, a twice continuously differentiable function u is an elastic wave if and only if the* **Cauchy–Navier equation** *holds true, which has the following equivalent forms:*

$$(\lambda + \mu) \operatorname{grad}_x \operatorname{div}_x u + \mu \Delta_x u = \varrho \frac{\partial^2 u}{\partial t^2} \qquad (7.70)$$

$$\Leftrightarrow \quad (\lambda + 2\mu) \operatorname{grad}_x \operatorname{div}_x u - \mu \operatorname{curl}_x \operatorname{curl}_x u = \varrho \frac{\partial^2 u}{\partial t^2} \qquad (7.71)$$

$$\Leftrightarrow \quad \alpha^2 \operatorname{grad}_x \operatorname{div}_x u - \beta^2 \operatorname{curl}_x \operatorname{curl}_x u = \frac{\partial^2 u}{\partial t^2}, \qquad (7.72)$$

where $\alpha := \sqrt{(\lambda + 2\mu)\varrho^{-1}}$ *and* $\beta := \sqrt{\mu \varrho^{-1}}$ *are the velocities of P- and S-waves.*

Proof Equation (7.70) is obtained by combining (7.66) with (7.69). The modification into (7.71) is a consequence of the application of (2.5) from Theorem 2.3.4. Moreover, (7.72) is derived by dividing (7.71) by ϱ. □

To get rid of the time derivative, we first apply the **Fourier transform**

$$\mathcal{F}(f)(\omega) := \int_{-\infty}^{+\infty} f(t) \mathrm{e}^{-\mathrm{i}\omega t} \, \mathrm{d}t, \quad f \in \mathrm{L}^1(\mathbb{R}), \, \omega \in \mathbb{R}.$$

It is well known that $\mathcal{F}\left(\frac{\partial^k}{\partial t^k} u(x,\cdot)\right)(\omega) = (\mathrm{i}\omega)^k \mathcal{F}(u(x,\cdot))$ if the Fourier transform applies to the temporal dependence, provided that the 0th to kth derivatives are in $\mathrm{L}^1(\mathbb{R})$. Hence, the term $\frac{\partial^2 u}{\partial t^2}$, which occurs in the Cauchy–Navier equation, can be replaced by its Fourier transform $-\omega^2 \mathcal{F}(u(x,\cdot))(\omega)$. For reasons of better readability of the formulae, we will abbreviate the Fourier transform $\mathcal{F}(u(x,\cdot))(\omega)$ by $u(x,\omega)$, and similar assertions hold for related terms, such as the spatial derivatives of u. We arrive now at the following equation, where u stands for the function $(x,\omega) \mapsto u(x,\omega)$, which will always be the case from now on, unless anything different is stated:

$$\alpha^2 \operatorname{grad}_x \operatorname{div}_x u - \beta^2 \operatorname{curl}_x \operatorname{curl}_x u + \omega^2 u = 0. \tag{7.73}$$

An alternative to the use of the Fourier transform is a separation ansatz, where we assume that $u(x,t) = v(x)w(t)$. The insertion into (7.72) yields

$$\left(\alpha^2 \operatorname{grad}_x \operatorname{div}_x v(x) - \beta^2 \operatorname{curl}_x \operatorname{curl}_x v(x)\right) w(t) = v(x) w''(t). \tag{7.74}$$

Since we are not looking for solutions which are identical to zero, we divide (7.74) by vw outside their zeros and arrive at the equation

$$\left(\alpha^2 \operatorname{grad}_x \operatorname{div}_x v(x) - \beta^2 \operatorname{curl}_x \operatorname{curl}_x v(x)\right)(v(x))^{-1} = (w(t))^{-1} w''(t), \tag{7.75}$$

whose left-hand side only depends on x while the right-hand side only depends on t. Hence, the identity in (7.75) can only be guaranteed if both sides are constant. Consequently, there is a constant C such that

$$\alpha^2 \operatorname{grad}_x \operatorname{div}_x v(x) - \beta^2 \operatorname{curl}_x \operatorname{curl}_x v(x) = C\, v(x), \qquad w''(t) = C\, w(t) \tag{7.76}$$

must hold true. The right-hand equation is easy to solve: because we are expecting oscillations and not an exponential behaviour with respect to t, the constant C must be negative. We write it as $C = -\omega^2$ and get

$$w(t) = \gamma_1 \cos(\omega t) + \gamma_2 \sin(\omega t) \tag{7.77}$$

for arbitrary constants $\gamma_1, \gamma_2 \in \mathbb{R}$. It is clear that every constant C (or every eigenfrequency ω) has its own solutions w and v. To take this into account, we denote the solution $x \mapsto v(x)$ corresponding to ω by $(x,\omega) \mapsto u(x,\omega)$. The left-hand equation in (7.76) becomes then the same equation as (7.73).

The next essential step for solving the Cauchy–Navier equation is the application of the Helmholtz decomposition (see Theorem 6.1.1). Provided that u satisfies the corresponding conditions, it can be decomposed into $u = u_\alpha + u_\beta$, where u_α is a gradient field (and, thus, curl-free) and u_β is a curl field (and, thus, divergence-free), see also Theorem 2.3.4. We obtain

$$\alpha^2 \operatorname{grad}_x \operatorname{div}_x u_\alpha - \beta^2 \operatorname{curl}_x \operatorname{curl}_x u_\beta + \omega^2 (u_\alpha + u_\beta) = 0. \tag{7.78}$$

It is clear that, if

$$\alpha^2 \operatorname{grad}_x \operatorname{div}_x u_\alpha + \omega^2 u_\alpha = 0 \qquad \text{and} \qquad -\beta^2 \operatorname{curl}_x \operatorname{curl}_x u_\beta + \omega^2 u_\beta = 0,$$

then (7.78) is also satisfied. By applying (2.5) from Theorem 2.3.4, we can equivalently state the following: it is also clear that, if

$$\Delta_x u_\alpha + \frac{\omega^2}{\alpha^2} u_\alpha = 0 \qquad \text{and} \qquad \Delta_x u_\beta + \frac{\omega^2}{\beta^2} u_\beta = 0,$$

then (7.78) is also satisfied. Certainly, the converse is not automatically true (if a sum vanishes, then the summands need not vanish). Nevertheless, the following can be proved (see Michel, 2003).

Theorem 7.3.9 *For each three-times continuously differentiable vectorial function u, which is, regarding its spatial dependence, Hölder continuous up to the third differentiation order and given on $D := \overline{\Sigma_{\text{int}}}$, where Σ is a regular surface, the following holds true: u satisfies the (Fourier transformed) Cauchy–Navier equation (7.78) if and only if u can be decomposed into a sum $u = u_\alpha + u_\beta$ where the summands satisfy*

$$\Delta_x u_\alpha + \frac{\omega^2}{\alpha^2} u_\alpha = 0, \qquad \operatorname{curl}_x u_\alpha = 0, \tag{7.79}$$

$$\Delta_x u_\beta + \frac{\omega^2}{\beta^2} u_\beta = 0, \qquad \operatorname{div}_x u_\beta = 0. \tag{7.80}$$

Proof We have already derived that every sum $u_\alpha + u_\beta$ whose summands obey (7.79) and (7.80) must satisfy (7.78).

Now let u be an arbitrary solution of (7.78) with the regularity requirements stated previously. The Helmholtz decomposition still allows us to decompose it into a curl-free part u_α and a divergence-free part u_β. However, the left-hand sides of (7.79) and (7.80) need not hold true. Therefore, we set $f := \Delta_x u_\alpha + \omega^2\alpha^{-2} u_\alpha$. The Cauchy–Navier equation now implies that $\Delta_x u_\beta + \omega^2\beta^{-2} u_\beta = -\alpha^2\beta^{-2} f$.

From Theorems 3.1.12 and 6.1.1, we know that u_α and u_β are three-times continuously differentiable. Therefore, we may apply Schwarz's theorem in

$$\operatorname{curl}_x f = \Delta_x (\operatorname{curl}_x u_\alpha) + \frac{\omega^2}{\alpha^2} \operatorname{curl}_x u_\alpha = 0, \tag{7.81}$$

$$\operatorname{div}_x f = -\frac{\beta^2}{\alpha^2} \Delta_x \operatorname{div}_x u_\beta - \frac{\omega^2}{\alpha^2} \operatorname{div}_x u_\beta = 0. \tag{7.82}$$

Equation (7.81) implies that f is a gradient field. By inserting $f = \operatorname{grad}_x F$ into (7.82), we see that the potential F is harmonic due to (2.4). Using Schwarz's theorem again, $\Delta_x f = \Delta_x \operatorname{grad}_x F = \operatorname{grad}_x \Delta_x F = 0$, we get immediately that f is also harmonic. Thus,

$$\Delta_x \left(u_\alpha - \frac{\alpha^2}{\omega^2} f \right) + \frac{\omega^2}{\alpha^2} \left(u_\alpha - \frac{\alpha^2}{\omega^2} f \right) = 0,$$

$$\Delta_x \left(u_\beta + \frac{\alpha^2}{\omega^2} f \right) + \frac{\omega^2}{\beta^2} \left(u_\beta + \frac{\alpha^2}{\omega^2} f \right) = 0.$$

By setting $\widetilde{u}_\alpha := u_\alpha - \alpha^2\omega^{-2} f$ and $\widetilde{u}_\beta := u_\beta + \alpha^2\omega^{-2} f$, we obtain $u = \widetilde{u}_\alpha + \widetilde{u}_\beta$, where the summands satisfy (7.79) and (7.80). □

Equations of the type $\Delta_x\varphi(x,\omega) + \omega^2 c^{-2}\varphi(x,\omega) = 0$ are summarized under the **Helmholtz equation**. The Helmholtz equation is the Fourier transform of the **wave equation** $\Delta_x\varphi(x,t) = c^{-2}\frac{\partial^2}{\partial t^2}\varphi(x,t)$. It describes waves which propagate with the speed c. This means that the use of the Helmholtz decomposition provides us with a more flexible way of concluding that there must be waves with the speeds of P- and S-waves.

It is known that solutions of the Helmholtz equation, under conditions which are satisfied in our modelling, are always analytical; see Müller (1969, §4) and Smirnow (1977, pp. 592–593). We can, therefore and with good conscience, restrict our quest to infinitely differentiable solutions from now on.

Our next task is therefore to find $\mathrm{c}^{(\infty)}$-solutions of (7.79) and (7.80). This is now feasible due to the preliminary work which we have done so far. Regarding u_α, we know that we have a gradient field, and for u_β, we will apply the Mie representation. We will write here $\mathrm{c}_{\mathrm{H}}^{(\infty)}(D)$ for all functions in $\mathrm{c}^{(\infty)}(D)$ where all derivatives are Hölder continuous. Theorem 6.3.6 indeed suggests that u_β can be assumed to be solenoidal. However, the regularity which we get for the Mie scalars in Theorem 6.3.11 does not suffice for the argumentation which we will use within the following proof, and we do not have a spherical shell here. For this reason, we will use a slightly cautious statement in the following theorem for the sake of the mathematical rigour. In fact, the application of the Mie representation in this context is a common and well-working tool and should not pose a notable restriction to the modelling. Moreover, we could imagine that the considered domain is a union of overlapping spherical shells. The overlap should be sufficiently large such that the Mie scalars in each shell are sufficiently smooth continuations of each other. Besides, if we focus, for example, on waves in the mantle, that is, on a subset of the domain, we would have a structure which is approximately a spherical shell (where, however, divergence-free fields need not be solenoidal).

Note also that we consider, for a moment, the angular frequency ω as a constant and only look for functions of the spatial variables. Certainly, every ω causes a different Helmholtz equation and, consequently, different solutions. The resulting ω-parameterized family of functions represents the general dependence on space and frequency.

Theorem 7.3.10 *Let $D := \overline{\Sigma_{\mathrm{int}}}$ be the closed interior of the regular surface Σ. Every $\mathrm{c}_{\mathrm{H}}^{(\infty)}(D)$-solution u_α of (7.79) is representable by*

$$u_\alpha = \nabla\varphi_\alpha, \qquad \Delta\varphi_\alpha + \frac{\omega^2}{\alpha^2}\varphi_\alpha = 0. \tag{7.83}$$

Furthermore, every $\mathrm{c}_{\mathrm{H}}^{(\infty)}(D)$-solution u_β of (7.80) possesses a Mie representation on each spherical shell $E := \Omega_{]a,b[} \subset D$. If the corresponding Mie scalars are elements of $\mathrm{C}^{(\infty)}(E)$, then u_β is representable on E by

$$u_\beta = \mathrm{L}^*\varphi_{\beta,1} + \nabla\times\mathrm{L}^*\varphi_{\beta,2}, \qquad \Delta\varphi_{\beta,j} + \frac{\omega^2}{\beta^2}\varphi_{\beta,j} = 0,\ j = 1,2. \tag{7.84}$$

In both cases (7.83) and (7.84), all possible, infinitely differentiable solutions $\varphi_\alpha, \varphi_{\beta,1}, \varphi_{\beta,2}$ of the right-hand equations can be inserted into the left-hand terms. We set $u_{\beta,1} := \mathrm{L}^*\varphi_{\beta,1}$ and $u_{\beta,2} := \nabla\times\mathrm{L}^*\varphi_{\beta,2}$.

Proof The proof of (7.83) is easy. The unknown solution u_α is curl-free and D is simply connected. Thus, there exists a scalar potential $\widetilde{\varphi}_\alpha$ such that $u_\alpha = \nabla\widetilde{\varphi}_\alpha$. Hence, Schwarz's theorem and (7.79) yield

$$0 = \Delta\left(\nabla\widetilde{\varphi}_\alpha\right) + \frac{\omega^2}{\alpha^2}\nabla\widetilde{\varphi}_\alpha = \nabla\left(\Delta\widetilde{\varphi}_\alpha + \frac{\omega^2}{\alpha^2}\widetilde{\varphi}_\alpha\right). \tag{7.85}$$

As a consequence, the term in the large brackets on the right-hand side must be constant on D. We multiply this constant by $\alpha^2\omega^{-2}$ and denote the resulting constant by γ. Eventually, we set $\varphi_\alpha := \widetilde{\varphi}_\alpha - \gamma$. Then $\nabla\varphi_\alpha = \nabla\widetilde{\varphi}_\alpha = u_\alpha$ and

$$\Delta\varphi_\alpha + \frac{\omega^2}{\alpha^2}\varphi_\alpha = 0. \tag{7.86}$$

Hence, every solution u_α of (7.79) has the form of (7.83). Vice versa, we get, in analogy to (7.85), that the gradient of each solution φ_α of (7.86) also solves (7.79).

Let now $P, Q \in \mathrm{C}^{(\infty)}(E)$ such that $u_\beta = \mathrm{L}^* Q + \operatorname{curl}\mathrm{L}^* P$ solves (7.80) on E. We know that Q and P are uniquely determined up to summands which only depend on the radial coordinate. Due to Schwarz's theorem, Δ and curl commute. Moreover, it is also easy to verify that $\Delta\mathrm{L}^* = \mathrm{L}^*\Delta$. Hence, the Helmholtz equation for u_β leads us to

$$\operatorname{curl}\mathrm{L}^*\left(\Delta P + \frac{\omega^2}{\beta^2}P\right) + \mathrm{L}^*\left(\Delta Q + \frac{\omega^2}{\beta^2}Q\right) = 0 \tag{7.87}$$

on the spherical shell E. This is a Mie representation of the zero function. From Theorem 6.3.11, we know that the Mie scalars are unique up to summands which only depend on the radial coordinate. Since we could replace both bracketed terms in (7.87) by the zero function, the bracketed terms themselves may only depend on the radial coordinate r.

We focus first on Q and observe that consequently there must exist a scalar function $G \in \mathrm{C}^{(\infty)}(\,]a,b[\,)$ such that $\Delta Q(x) + \omega^2\beta^{-2}Q(x) = G(|x|)$ for all $x \in E$. This is an inhomogeneous Helmholtz equation. According to Smirnow (1977, p. 588), one solution of this equation is given by

$$\widetilde{Q}(x) = -\frac{1}{4\pi}\int_E \frac{\mathrm{e}^{-\mathrm{i}\omega^2\beta^{-2}|x-y|}}{|x-y|}G(|y|)\,\mathrm{d}y, \quad x \in E.$$

If we write $x = |x|\xi$, $y = r\eta$ with $\xi, \eta \in \Omega$ and use Theorem 4.1.9, then the formula above becomes

$$\begin{aligned}\widetilde{Q}(x) &= -\frac{1}{4\pi}\int_a^b r^2 G(r)\int_\Omega \frac{\mathrm{e}^{-\mathrm{i}\omega^2\beta^{-2}\sqrt{|x|^2+r^2-2|x|r\xi\cdot\eta}}}{\sqrt{|x|^2+r^2-2|x|r\xi\cdot\eta}}\,\mathrm{d}\omega(\eta)\,\mathrm{d}r\\ &= -\frac{1}{2}\int_a^b r^2 G(r)\int_{-1}^1 \frac{\mathrm{e}^{-\mathrm{i}\omega^2\beta^{-2}\sqrt{|x|^2+r^2-2|x|rt}}}{\sqrt{|x|^2+r^2-2|x|rt}}\,\mathrm{d}t\,\mathrm{d}r.\end{aligned}$$

Hence, $\widetilde{Q}$ also only depends on the radial coordinate. We now define $\varphi_{\beta,1} := Q - \widetilde{Q}$. This new function satisfies the homogeneous Helmholtz equation and the property of $\widetilde{Q}$ yields $\mathrm{L}^*\varphi_{\beta,1} = \mathrm{L}^*(Q - \widetilde{Q}) = \mathrm{L}^*Q - \mathrm{L}^*\widetilde{Q} = \mathrm{L}^*Q$, that is, $\mathrm{L}^*\varphi_{\beta,1}$ is indeed the (unique) toroidal part $u_{\beta,1}$ of u_β on E, and the toroidal scalar $\varphi_{\beta,1}$ solves the right-hand equation in (7.84).

Regarding the poloidal part, the conclusions are completely analogous. Equation (7.87) also yields that P satisfies a Helmholtz equation with a merely radially dependent inhomogeneity. For this equation, we find an also purely radially dependent solution $\widetilde{P}$. Eventually, $\varphi_{\beta,2} := P - \widetilde{P}$ yields the desired result and (7.84) holds true.

Vice versa, every u_β which is constructed via (7.84) satisfies the corresponding Helmholtz equation – this follows in analogy to (7.87) – and is divergence-free – this is a consequence of Theorem 6.3.2. Hence, u_β obeys (7.80). □

Definition 7.3.11 The vectorial solutions u_α, $u_{\beta,1}$, and $u_{\beta,2}$ are called the **Hansen vectors**. In the literature, they are also denoted (in this order) as L, M, and N. Moreover, solutions of the type $u_{\beta,1}$ are called **toroidal** or **torsional oscillations**, while linear combinations of u_α and $u_{\beta,2}$ are called **poloidal** or **spheroidal oscillations**. All kinds of solutions of these types are called **normal modes**.

We will now derive a basis system for the Hansen vectors. We still ignore the angular frequency, though we are aware that this ignorance cannot last forever. For finding a basis, we approach the scalar potentials $\varphi_{\ldots}$ by using a separation ansatz of the kind $\varphi(r\xi) = F(r)Y(\xi)$, $r \in]a,b[$, $\xi \in \Omega$. Note that all these potentials are given as solutions of the Helmholtz equation with only differing speeds. The easiest starting point is the investigation of the toroidal oscillations, where we get $u_{\beta,1}(r\xi) = \mathrm{L}^*_\xi(F(r)Y(\xi)) = F(r)\mathrm{L}^*_\xi Y(\xi)$. Therefore, it appears to be reasonable to start with vector spherical harmonics of type 3 (see Section 4.3). We set $M_{n,j}(r\xi) := F_{n,j}(r)y^{(3)}_{n,j}(\xi)$, $r \in]a,b[$, $\xi \in \Omega$, $n \in \mathbb{N}$, $j \in \{-n,\ldots,n\}$. Actually, if we set $\varphi_{\beta,1}(r\xi) = F_{n,j}(r)Y_{n,j}(\xi)$ here, then a normalization factor would occur in $M_{n,j}$. However, since we are dealing with a linear and homogeneous equation, we can skip this factor (more precisely, we hide it by using the following potential $\varphi_{\beta,1}(r\xi) = [n(n+1)]^{-1/2}F_{n,j}(r)Y_{n,j}(\xi)$).

Let us continue with the poloidal oscillations of type u_α. For this purpose, we have (4.29) available, which yields for $\varphi_\alpha(r\xi) = \widetilde{F}_{n,j}(r)Y_{n,j}(\xi)$,

$$u_\alpha(r\xi) = \widetilde{F}'_{n,j}(r)y^{(1)}_{n,j}(\xi) + \frac{\widetilde{F}_{n,j}(r)}{r}\sqrt{n(n+1)}\,y^{(2)}_{n,j}(\xi) =: L_{n,j}(r\xi), \tag{7.88}$$

$r \in]a,b[$, $\xi \in \Omega$, $n \in \mathbb{N}_0$, $j \in \{-n,\ldots,n\}$. This also means that we will stick here to the use of Morse–Feshbach vector spherical harmonics. For $n = 0$, the second summand in (7.88) is considered to be vanishing due to the factor $\sqrt{n(n+1)}$ such that the non-existence of the vector spherical harmonic of type 2 and degree 0 does not matter here. Note that we cannot avoid the occurrence of this normalizing factor here. Furthermore, the tilde on the radial part takes into account that φ_α solves a Helmholtz equation with a different speed.

Finally, for $\varphi_{\beta,2}$, we need a formula for $\operatorname{curl}_{r\xi}\mathrm{L}^*_\xi(F_{n,j}(r)Y_{n,j}(\xi))$. We have already solved this problem in Lemma 6.3.10 such that we have

$$u_{\beta,2}(r\xi) = \frac{F_{n,j}(r)}{r}\,\xi\Delta^*_\xi Y_{n,j}(\xi) - \left(\frac{F_{n,j}(r)}{r} + F'_{n,j}(r)\right)\nabla^*_\xi Y_{n,j}(\xi).$$

Due to the linearity of the Helmholtz equation, we permit ourselves to multiply the previous function with -1. Moreover, we use the Helmholtz equation in combination with Theorem 3.4.3, which yields

$$F''_{n,j}(r)Y_{n,j}(\xi) + \frac{2}{r}\, F'_{n,j}(r)Y_{n,j}(\xi) + \frac{F_{n,j}(r)}{r^2}\Delta^*_\xi Y_{n,j}(\xi) + \frac{\omega^2}{\beta^2}\, F_{n,j}(r)Y_{n,j}(\xi) = 0. \tag{7.89}$$

This leads us eventually to

$$\begin{aligned}-u_{\beta,2}(r\xi) &= \left(r F''_{n,j}(r) + 2F'_{n,j}(r) + \frac{\omega^2}{\beta^2}\, r F_{n,j}(r)\right) y^{(1)}_{n,j}(\xi)\\ &\quad + \left(\frac{F_{n,j}(r)}{r} + F'_{n,j}(r)\right)\sqrt{n(n+1)}\, y^{(2)}_{n,j}(\xi) =: N_{n,j}(r\xi),\end{aligned}$$

$r \in]a,b[$, $\xi \in \Omega$, $n \in \mathbb{N}$, $j \in \{-n,\ldots,n\}$. Note that the coefficient of $y^{(1)}_{n,j}$ vanishes for $n = 0$ because of (7.89) and $\Delta^* Y_{n,j} = -n(n+1)Y_{n,j}$.

In view of the properties which we obtained earlier for the vector spherical harmonics (see also Theorem 4.1.6), we can already see here that Hansen vectors of types L and N indeed belong together, while the type M oscillations have a different mathematical nature. The latter are surface-divergence-free and they are tangential to concentric spheres around 0. This explains why these oscillations are also characterized as torsional.

We need to take care of the radially dependent part of the Hansen vectors now. Since this radial part originates from the scalar potentials φ, it suffices to use the Helmholtz equation here. Due to the different speeds, we write now c for the speed and $\widehat{F}_{n,j}$ for the radial part, which may then be replaced appropriately. We start with the use of Theorems 3.4.3 and 4.2.19:

$$\begin{aligned}0 &= \Delta_{r\xi}\left(\widehat{F}_{n,j}(r)Y_{n,j}(\xi)\right) + \frac{\omega^2}{c^2}\, \widehat{F}_{n,j}(r)Y_{n,j}(\xi)\\ &= \widehat{F}''_{n,j}(r)Y_{n,j}(\xi) + \frac{2}{r}\, \widehat{F}'_{n,j}(r)Y_{n,j}(\xi) + \frac{\widehat{F}_{n,j}(r)}{r^2}\, \Delta^*_\xi Y_{n,j}(\xi) + \frac{\omega^2}{c^2}\, \widehat{F}_{n,j}(r)Y_{n,j}(\xi)\\ &= \left[\widehat{F}''_{n,j}(r) + \frac{2}{r}\, \widehat{F}'_{n,j}(r) + \left(\frac{\omega^2}{c^2} - \frac{n(n+1)}{r^2}\right)\widehat{F}_{n,j}(r)\right] Y_{n,j}(\xi)\end{aligned}$$

is valid for all $r \in]a,b[$ and all $\xi \in \Omega$. Obviously, we can cancel out the spherical harmonic here such that we get an ordinary differential equation for the radial part alone:

$$\widehat{F}''_{n,j}(r) + \frac{2}{r}\, \widehat{F}'_{n,j}(r) + \left(\frac{\omega^2}{c^2} - \frac{n(n+1)}{r^2}\right)\widehat{F}_{n,j}(r) = 0, \quad r \in]a,b[. \tag{7.90}$$

In addition, we see that we can skip the dependence of the unknown function on the order j of the spherical harmonics. We substitute now $\widehat{F}_n(r) =: r^{-1/2}G_n(\omega c^{-1}r)$ and insert this into (7.90). We obtain

$$-\frac{1}{2}\left(-\frac{3}{2}\right)r^{-5/2}G_n\left(\frac{\omega}{c}r\right)+2\left(-\frac{1}{2}\right)r^{-3/2}G_n'\left(\frac{\omega}{c}r\right)\frac{\omega}{c}$$

$$+r^{-1/2}G_n''\left(\frac{\omega}{c}r\right)\frac{\omega^2}{c^2}+\frac{2}{r}\left[-\frac{1}{2}r^{-3/2}G_n\left(\frac{\omega}{c}r\right)+r^{-1/2}G_n'\left(\frac{\omega}{c}r\right)\frac{\omega}{c}\right]$$

$$+\frac{\omega^2}{c^2}r^{-1/2}G_n\left(\frac{\omega}{c}r\right)-n(n+1)r^{-5/2}G_n\left(\frac{\omega}{c}r\right)=0$$

$$\Leftrightarrow \qquad \left(\frac{\omega}{c}r\right)^2 G_n''\left(\frac{\omega}{c}r\right)+\frac{\omega}{c}r\,G_n'\left(\frac{\omega}{c}r\right)$$

$$+\left[\frac{\omega^2}{c^2}r^2+\left(\frac{3}{4}-1-n(n+1)\right)\right]G_n\left(\frac{\omega}{c}r\right)=0,$$

$r \in]a,b[$, where $3/4-1-n(n+1)=-n^2-n-1/4=-(n+1/2)^2$. With $s:=\omega c^{-1}r$, we obtain for G_n the **Bessel differential equation**:

$$s^2G_n''(s)+s\,G_n'(s)+\left[s^2-\left(n+\frac{1}{2}\right)^2\right]G_n(s)=0, \quad s\in \left]\frac{\omega}{c}a,\frac{\omega}{c}b\right[.$$

This equation is well known, and its solutions can be found in the standard literature on ordinary differential equations; see, for example, Heuser (1991, sections 27 and 28). We briefly summarize what we need from the available theory.

Definition 7.3.12 The function $J_\nu\colon \mathbb{R}^+ \to \mathbb{R}$ with $\nu \in \mathbb{R}\setminus(-\mathbb{N})$ and

$$J_\nu(x):=\sum_{k=0}^{\infty}\frac{(-1)^k}{k!\;\Gamma(k+\nu+1)}\left(\frac{x}{2}\right)^{2k+\nu}, \quad x\in\mathbb{R}^+, \tag{7.91}$$

is called the **Bessel function of the first kind** and the order ν.

Definition 7.3.13 The function $Y_\nu\colon \mathbb{R}^+ \to \mathbb{R}$ with $\nu \in \mathbb{N}_0$ is defined by

$$Y_\nu(x):=\frac{2}{\pi}\left[(C+\log x-\log 2)J_\nu(x)-\frac{1}{2}\sum_{k=0}^{\nu-1}\frac{(\nu-k-1)!}{k!}\left(\frac{x}{2}\right)^{2k-\nu}\right.$$
$$\left.-\frac{1}{2}\sum_{k=0}^{\infty}\frac{(-1)^k\,(h_k+h_{k+\nu})}{k!\;(k+\nu)!}\left(\frac{x}{2}\right)^{2k+\nu}\right], \quad x\in\mathbb{R}^+,$$

where $h_k := \sum_{j=1}^{k}\frac{1}{j}$ and $C := \lim_{k\to\infty}(h_k - \log k)$ is the **Euler–Mascheroni constant**. The function Y_ν is called the **Bessel function of the second kind** and the order ν as well as the **Neumann function** of the order ν.

There is a large amount of literature on Bessel functions including the common books on special functions, such as Abramowitz and Stegun (1972), Beals and Wong (2016), Freeden and Gutting (2013), and Magnus et al. (1966). Moreover, the monograph Watson (1944) is entirely dedicated to the Bessel functions. For us, the following property is primarily important.

Theorem 7.3.14 *The general solution of the Bessel differential equation*

$$x^2\, y'' + x\, y' + (x^2 - \nu^2)\, y = 0$$

of the order $\nu \in \mathbb{R}_0^+$ *is representable on* $\mathbb{R}^+$ *by*

(a) $y(x) = C_1\, J_\nu(x) + C_2\, J_{-\nu}(x)$, $x \in \mathbb{R}^+$, *with arbitrary coefficients* $C_1, C_2 \in \mathbb{R}$, *if* $\nu \notin \mathbb{N}_0$,

(b) $y(x) = C_1\, J_\nu(x) + C_2\, Y_\nu(x)$, $x \in \mathbb{R}^+$, *with arbitrary coefficients* $C_1, C_2 \in \mathbb{R}$, *if* $\nu \in \mathbb{N}_0$.

In our case, $\nu = n + 1/2$ is definitely not an integer, that is, we need the linear combination of two Bessel functions of the first kind.

We only discuss here the basic essentials for the modelling of normal modes. This includes that we restrict our derivations to the simplified case of a homogeneous ball. This implies that we need to be able to calculate a limit $a \to 0+$ for the domain of the derived functions. Remember that a and b are the radii of the bounding spheres of the spherical shell. For b, we can simply take the radius of the homogeneous ball. The limit $a \to 0+$ restricts the possibilities of how we can choose the radial part $\widehat{F}_n$, which we are still seeking. From (7.91), we deduce that the Bessel functions with negative order do not have a finite limit if their argument tends to zero. Therefore, we can only choose $\widehat{F}_n(r) = r^{-1/2} G_n(\omega c^{-1} r) = \gamma r^{-1/2} J_{n+1/2}(\omega c^{-1} r)$, where γ is an arbitrary constant. We can now summarize what we have derived on the previous pages and redefine our preliminary basis. Before we do this, we quote a common abbreviation (we add a tilde to the notation to avoid confusion with the order of a spherical harmonic).

Definition 7.3.15 The nth **spherical Bessel function** $\widetilde{j}_n : \mathbb{R}^+ \to \mathbb{R}$ with $n \in \mathbb{N}_0$ is given by $\widetilde{j}_n(s) := [\pi/(2s)]^{1/2} J_{n+1/2}(s)$, $s \in \mathbb{R}^+$.

Theorem 7.3.16 *The normal modes which solve the (Fourier-transformed) Cauchy–Navier equation* $\alpha^2 \operatorname{grad}_x \operatorname{div}_x u - \beta^2 \operatorname{curl}_x \operatorname{curl}_x u + \omega^2 u = 0$ *on a homogeneous ball of radius* b *and with centre* 0 *can be expressed as linear combinations of the Hansen basis vectors:*

$$L_{n,j}(r\xi,\omega) := \frac{\mathrm{d}}{\mathrm{d}r} \widetilde{j}_n\left(\frac{\omega}{\alpha} r\right) y^{(1)}_{n,j}(\xi) + \frac{1}{r}\, \widetilde{j}_n\left(\frac{\omega}{\alpha} r\right) \sqrt{n(n+1)}\, y^{(2)}_{n,j}(\xi),$$

$$M_{n,j}(r\xi,\omega) := \widetilde{j}_n\left(\frac{\omega}{\beta} r\right) y^{(3)}_{n,j}(\xi),$$

$$N_{n,j}(r\xi,\omega) := \left(r \frac{\mathrm{d}^2}{\mathrm{d}r^2} + 2\frac{\mathrm{d}}{\mathrm{d}r} + \frac{\omega^2}{\beta^2} r\right) \widetilde{j}_n\left(\frac{\omega}{\beta} r\right) y^{(1)}_{n,j}(\xi)$$
$$+ \left(\frac{1}{r} + \frac{\mathrm{d}}{\mathrm{d}r}\right) \widetilde{j}_n\left(\frac{\omega}{\beta} r\right) \sqrt{n(n+1)}\, y^{(2)}_{n,j}(\xi),$$

$r \in \,]0,b]$, $\xi \in \Omega$, *where* $j \in \{-n,\ldots,n\}$ *and* $n \in \mathbb{N}_0$ *for* $L_{n,j}$ *and* $n \in \mathbb{N}$ *for the others.*

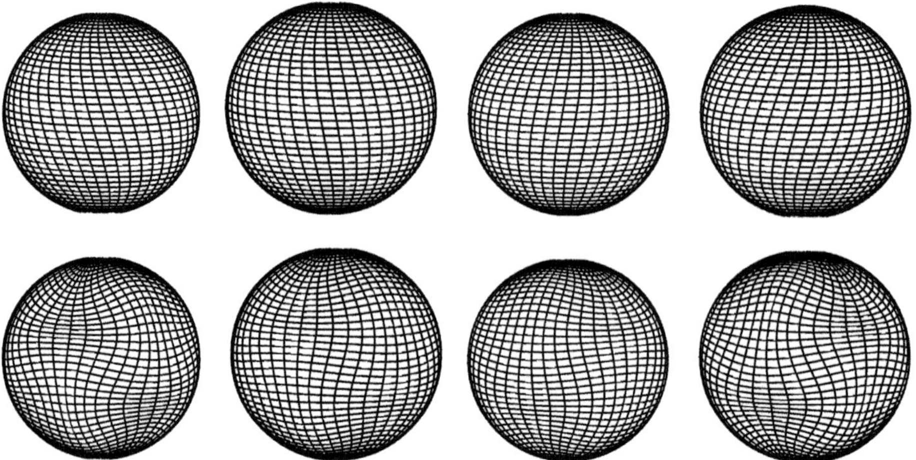

Figure 7.6 Qualitative illustration of the toroidal normal modes $M_{n,j}w$ for different points of time (left to right) and different degrees and orders (top: $(n, j) = (2, 0)$; bottom: $(n, j) = (5, 3)$): type M Hansen vectors are purely tangential and surface-divergence-free.

There are two ways of getting time-dependent solutions in the end. One possibility is the application of the inverse Fourier transform, which was done in Michel (2003). A drawback of this choice is that we cannot automatically conclude that the inverse Fourier transform also solves the original time-dependent equation (7.72). Remember, therefore, that we saw that a separation ansatz is an alternative. It easily gives us solutions on the space-time domain by multiplying the previous solutions (for arbitrary but fixed ω) with the function w from (7.77).

In the preceding formulae, we can see that the toroidal oscillations are purely tangential, where these tangential motions are described by the vector spherical harmonics of type 3, that is, by surface-divergence-free basis functions. The spheroidal oscillations have a normal component and a tangential part. The former is (naturally) described by the Morse–Feshbach vector spherical harmonics of type 1 and the latter by those of type 2, that is, by surface-curl-free basis functions. If we plotted qualitatively those Hansen vectors and their normal and tangential projections, we would not get essentially different images than those for the Morse–Feshbach vector spherical harmonics (see Figures 4.8–4.13 on pages 172–177). Therefore, we show an alternative illustration in Figures 7.6–7.8. There, longitudes and latitudes are deformed based on the tangential part of the normal modes times the function w for the time dependence.

There are two issues which should keep us from relaxing and leaning back at this point: first, people who are familiar with partial differential equations would pose the question if there are any boundary conditions. Second, seismologists would object that ω cannot be arbitrary, since there are discrete eigenfrequencies which can be observed. Indeed, the boundary conditions are the reasons why the frequencies are discrete. So, this becomes our next topic.

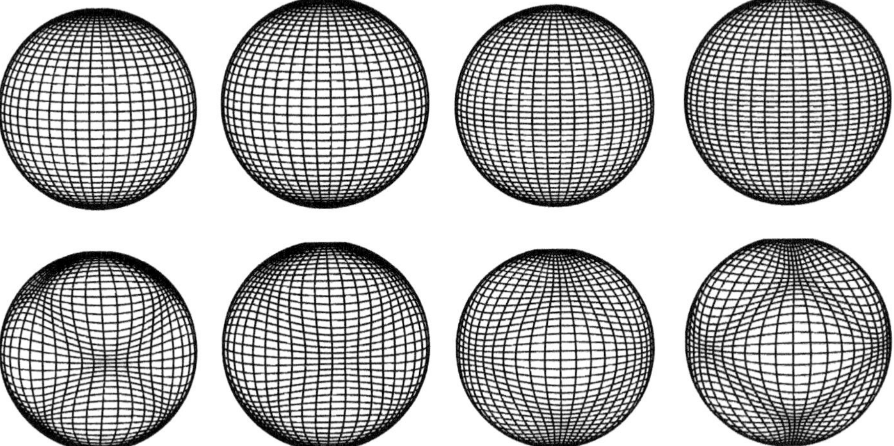

Figure 7.7 Qualitative illustration of the tangential part of the spheroidal normal modes $L_{n,j}w$ and $N_{n,j}w$ for different points of time (left to right) and different degrees and orders (top: $(n, j) = (2, 0)$; bottom: $(n, j) = (5, 3)$): the tangential part of type L and N Hansen vectors is surface-curl-free. All lines in the figure and in Figure 7.6 are located on the sphere, since only horizontal motions are shown. Due to an optical illusion, it might look as if there were vertical motions.

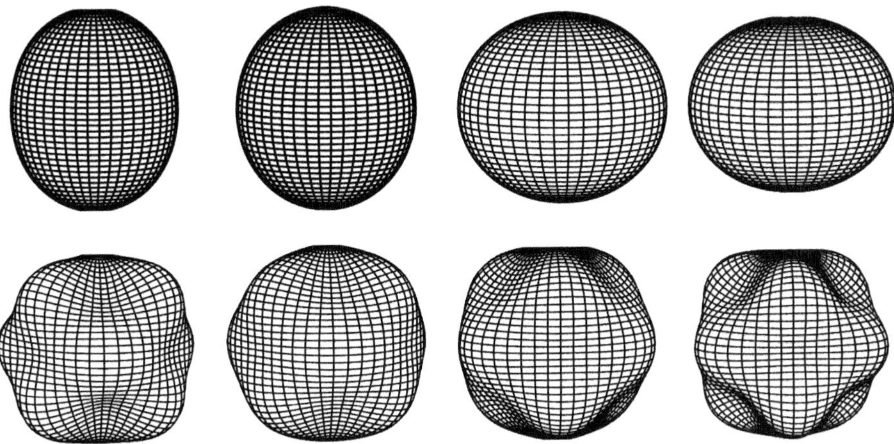

Figure 7.8 Qualitative illustration of the normal part of the spheroidal normal modes $L_{n,j}w$ and $N_{n,j}w$ for different points of time (left to right) and different degrees and orders (top: $(n, j) = (2, 0)$; bottom: $(n, j) = (5, 3)$): on a sphere, the normal part is proportional to the spherical harmonic of the same degree and order. As it is the case for all normal modes, the number of crests and troughs increases with the polynomial degree n, where modes corresponding to $j = 0$ have a rotational symmetry. The spheroidal normal mode for $n = 2$ and $j = 0$ is also called the football mode due to the resemblance of the top-left image to an American football.

In our simple model of a homogeneous and isotropic ball, we will use the (hydrostatic) boundary condition $\sigma(b\xi,\omega)\xi = 0$ for all $\xi \in \Omega$, where we respected the frequency dependence in this formula. Remember that b is the radius of the ball and σ is the (linearized) Cauchy stress tensor, which obeys Corollary 7.2.17. For handling this boundary condition, we need some preparatory work.

Lemma 7.3.17 *The transposed* ***surface Jacobian matrix*** *of a function* $y \in \mathrm{c}^{(1)}(\Omega)$ *satisfies, for all* $\xi \in \Omega$,

$$\left(\nabla_\xi^* y(\xi)\right)^{\mathrm{T}}\xi = \left(\nabla_\xi^* \otimes y(\xi)\right)\xi = \nabla_\xi^*(y(\xi)\cdot\xi) + (\xi\otimes\xi - I)y(\xi), \tag{7.92}$$

$$\xi^{\mathrm{T}}\left(\nabla_\xi^* y(\xi)\right)^{\mathrm{T}} = \xi^{\mathrm{T}}\left(\nabla_\xi^* \otimes y(\xi)\right) = 0, \tag{7.93}$$

where I *is the identity matrix. In particular, the Morse–Feshbach vector spherical harmonics fulfil*

$$\left(\nabla_\xi^* \otimes y_{n,j}^{(1)}(\xi)\right)\xi = \sqrt{n(n+1)}\, y_{n,j}^{(2)}(\xi),$$

$$\left(\nabla_\xi^* \otimes y_{n,j}^{(2)}(\xi)\right)\xi = -y_{n,j}^{(2)}(\xi), \qquad \left(\nabla_\xi^* \otimes y_{n,j}^{(3)}(\xi)\right)\xi = -y_{n,j}^{(3)}(\xi).$$

Proof Remember that y_k represents the kth component of the vector y and, for example, $y = (y_k)_{k=1,2,3}$ represents the vector by its components. For tensors of higher orders, the notations are analogous. We start the proof by calculating the terms, as always here for all $\xi \in \Omega$ and with $x = r\xi$:

$$\begin{aligned}
\nabla_\xi^*(y(\xi)\cdot\xi) &= \left(\left(\nabla_\xi^*\right)_i\left(\sum_{k=1}^{3} y_k(\xi)\xi_k\right)\right)_{i=1,2,3}\\
&= \left(\sum_{k=1}^{3}\left\{\left[\left(\nabla_\xi^*\right)_i y_k(\xi)\right]\xi_k + y_k(\xi)\left(\nabla_\xi^*\right)_i \xi_k\right\}\right)_{i=1,2,3}\\
&= \left(\nabla_\xi^* \otimes y(\xi)\right)\xi + \left(\nabla_\xi^* \otimes \xi\right)y(\xi).
\end{aligned}$$

From Theorem 4.1.3, we know that $\nabla_\xi^* \otimes \xi = I - \xi\otimes\xi$ such that we have proved (7.92). The insertion of the Morse–Feshbach vector spherical harmonics (see Section 4.3) leads us to

$$\begin{aligned}
\left(\nabla_\xi^* \otimes y_{n,j}^{(1)}(\xi)\right)\xi &= \nabla_\xi^*\left(Y_{n,j}(\xi)\xi\cdot\xi\right) + \xi\xi^{\mathrm{T}}\xi Y_{n,j}(\xi) - \xi Y_{n,j}(\xi)\\
&= \sqrt{n(n+1)}\, y_{n,j}^{(2)}(\xi)
\end{aligned}$$

and for $i \in \{2,3\}$ with the use of Theorem 4.1.6 to

$$\left(\nabla_\xi^* \otimes y_{n,j}^{(i)}(\xi)\right)\xi = \nabla_\xi^*\left(y_{n,j}^{(i)}(\xi)\cdot\xi\right) + \xi\left(\xi\cdot y_{n,j}^{(i)}(\xi)\right) - y_{n,j}^{(i)}(\xi) = -y_{n,j}^{(i)}(\xi).$$

For proving (7.93), we only need to consider that, due to Theorem 4.1.6,

$$\xi^{\mathrm{T}}\left(\nabla_\xi^* y(\xi)\right)^{\mathrm{T}} = \left(\sum_{k=1}^{3} \xi_k \left(\nabla_\xi^* y(\xi)\right)_{l,k}\right)_{l=1,2,3} = \left(\xi \cdot \left(\nabla_\xi^* y_l(\xi)\right)\right)_{l=1,2,3}$$

is the constant zero vector. □

With these formulae at hand, we are now able to calculate the stress $\sigma(b\xi,\omega)\xi$ in the case of a toroidal oscillation. For this purpose, we insert $M_{n,j}(r\xi,\omega) = \widetilde{j}_n(\omega\beta^{-1}r)y_{n,j}^{(3)}(\xi)$ into (7.64) from Corollary 7.2.17, that is $\sigma = \lambda(\nabla_x \cdot u)I + \mu[(\nabla_x \otimes u)^{\mathrm{T}} + \nabla_x \otimes u]$. We already know that $M_{n,j}$ is divergence-free. For the Jacobian matrix and its transpose, we use Theorems 3.4.1 and 4.1.6 as well as Lemma 7.3.17 and obtain, with $x = r\xi$,

$$\begin{aligned}
\left(\nabla_x \otimes M_{n,j}(x,\omega)\right)\xi &= \left[\left(\xi\frac{\partial}{\partial r} + \frac{1}{r}\nabla_\xi^*\right) \otimes \left(\widetilde{j}_n\left(\frac{\omega}{\beta}r\right) y_{n,j}^{(3)}(\xi)\right)\right]\xi \\
&= \left[\frac{\mathrm{d}}{\mathrm{d}r}\widetilde{j}_n\left(\frac{\omega}{\beta}r\right)\xi \otimes y_{n,j}^{(3)}(\xi) + \frac{1}{r}\widetilde{j}_n\left(\frac{\omega}{\beta}r\right)\nabla_\xi^* \otimes y_{n,j}^{(3)}(\xi)\right]\xi \\
&= \frac{\mathrm{d}}{\mathrm{d}r}\widetilde{j}_n\left(\frac{\omega}{\beta}r\right)\xi\left(y_{n,j}^{(3)}(\xi)\right)^{\mathrm{T}}\xi + \frac{1}{r}\widetilde{j}_n\left(\frac{\omega}{\beta}r\right)\left(-y_{n,j}^{(3)}(\xi)\right) \\
&= -\frac{1}{r}\widetilde{j}_n\left(\frac{\omega}{\beta}r\right)y_{n,j}^{(3)}(\xi)
\end{aligned}$$

and, with an analogous beginning, additionally

$$\begin{aligned}
\left(\nabla_x \otimes M_{n,j}(x,\omega)\right)^{\mathrm{T}}\xi &= \frac{\mathrm{d}}{\mathrm{d}r}\widetilde{j}_n\left(\frac{\omega}{\beta}r\right)\left(y_{n,j}^{(3)}(\xi)\right)\xi^{\mathrm{T}}\xi + \frac{1}{r}\widetilde{j}_n\left(\frac{\omega}{\beta}r\right)\left(\nabla_\xi^* \otimes y_{n,j}^{(3)}(\xi)\right)^{\mathrm{T}}\xi \\
&= \frac{\mathrm{d}}{\mathrm{d}r}\widetilde{j}_n\left(\frac{\omega}{\beta}r\right)y_{n,j}^{(3)}(\xi).
\end{aligned}$$

Theorem 7.3.18 *Let a toroidal oscillation be represented by the series*

$$M(r\xi,\omega) = \sum_{n=1}^{\infty}\sum_{j=-n}^{n}\gamma_{n,j}\widetilde{j}_n\left(\frac{\omega}{\beta}r\right)y_{n,j}^{(3)}(\xi), \quad r \in\,]0,b],\ \xi \in \Omega,$$

where this series and its first-order term-by-term derivative are assumed to be uniformly convergent. Then the (linearized) Cauchy stress which is caused by this oscillation obeys the formula

$$\sigma(r\xi,\omega)\xi = \mu\sum_{n=1}^{\infty}\left(\frac{\mathrm{d}}{\mathrm{d}r}\widetilde{j}_n\left(\frac{\omega}{\beta}r\right) - \frac{1}{r}\widetilde{j}_n\left(\frac{\omega}{\beta}r\right)\right)\sum_{j=-n}^{n}\gamma_{n,j}y_{n,j}^{(3)}(\xi) \tag{7.94}$$

*for all $r \in\,]0,b]$ and $\xi \in \Omega$. The boundary condition $\sigma(r\xi,\omega)\xi = 0$ for $r = b$ and all $\xi \in \Omega$ then implies the following two equivalent **toroidal frequency equations***

$$\frac{\omega}{\beta}b\,\widetilde{j}_n'\left(\frac{\omega}{\beta}b\right) = \widetilde{j}_n\left(\frac{\omega}{\beta}b\right) \quad\Leftrightarrow\quad \frac{\omega}{\beta}b\,\widetilde{j}_{n+1}\left(\frac{\omega}{\beta}b\right) = (n-1)\widetilde{j}_n\left(\frac{\omega}{\beta}b\right) \tag{7.95}$$

which hold for all $n \in \mathbb{N}$ (provided that $\mu > 0$).

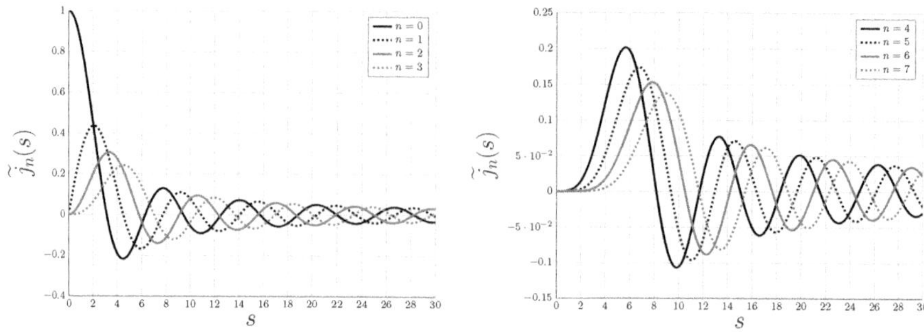

Figure 7.9 Spherical Bessel functions $\widetilde{j}_n$ for different orders n: the functions show a structure reminiscent of typical oscillations.

Proof Equation (7.94) is an immediate consequence of the preceding derivations. Regarding the boundary condition, we mention that certainly some $\gamma_{n,j}$ might vanish, but our point of view is different here: if one of the modes $M_{n,j}$ occurs, then its corresponding $\gamma_{n,j}$ does not vanish and this imposes a condition on the frequency. So, if for some $n \in \mathbb{N}$ at least one $\gamma_{n,j}$ is non-zero, then the boundary condition and the linear independence of the $y_{n,j}^{(i)}$ imply that

$$\frac{\mathrm{d}}{\mathrm{d}r}\widetilde{j}_n\left(\frac{\omega}{\beta}r\right)-\frac{1}{r}\widetilde{j}_n\left(\frac{\omega}{\beta}r\right)=0 \quad \text{at } r=b.$$

A simple application of the chain rule yields the left-hand frequency equation. The right-hand version in (7.95) is a result of a particular property of spherical Bessel functions, which can be found in the literature (see, e.g., Ben-Menahem and Singh, 1981), namely $\widetilde{j}_n'(s)=s^{-1}n\widetilde{j}_n(s)-\widetilde{j}_{n+1}(s)$ for all $n \in \mathbb{N}_0$ and all $s>0$. Then the first equation becomes

$$n\widetilde{j}_n\left(\frac{\omega}{\beta}b\right)-\frac{\omega}{\beta}b\,\widetilde{j}_{n+1}\left(\frac{\omega}{\beta}b\right)=\widetilde{j}_n\left(\frac{\omega}{\beta}b\right),$$

which immediately results in the right-hand equation of (7.95). □

The following property of spherical Bessel functions, which is an immediate consequence of a theorem for Bessel functions, which is proved in Beals and Wong (2016, Theorem 9.2.1), tells us something about the solutions of (7.95).

Theorem 7.3.19 *Each spherical Bessel function $\widetilde{j}_n$, $n \in \mathbb{N}_0$, has a countable number of positive zeros $(s_k)_{k\in\mathbb{N}_0}$. The difference $s_{k+1}-s_k$ of consecutive zeros is at least π. Moreover, $s_{k+1}-s_k=\pi+\mathcal{O}(k^{-2})$ as $k\to\infty$.*

Hence, spherical Bessel functions show some kind of oscillatory behaviour, which is also confirmed by Figures 7.9–7.10. The frequency equation restricts ω to a discrete, that is, infinite but countable, set of **toroidal eigenfrequencies**. They can be calculated numerically, and they obviously depend on the degree n of the vector spherical harmonics but not on their order j (i.e., there are $2n+1$ modes with the same eigenfrequency). So, if ${}_k\omega_n$ denotes the kth solution (sorted as an increasing sequence) of (7.95) corresponding to a particular

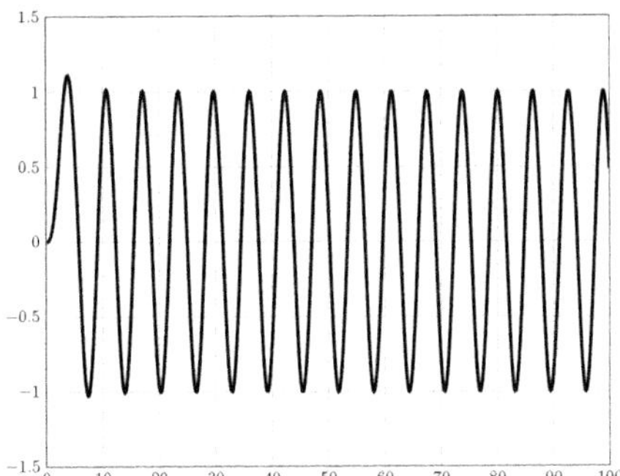

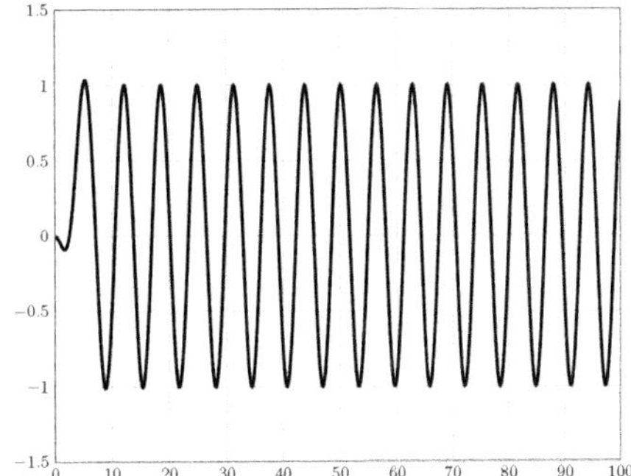

Figure 7.10 The plots show the function $g_n(s) := s\widetilde{j}_{n+1}(s) - (n-1)\widetilde{j}_n(s)$ for $n = 1$ (left) and $n = 2$ (right). The zeros of g_n (multiplied by βb^{-1}) are the solutions of the toroidal frequency equation (7.95) to the degree n.

$n \in \mathbb{N}$, then the **general representation of a toroidal normal mode for a homogeneous and isotropic ball** is, in the space-time domain,

$$M(r\xi,t) = \sum_{n=1}^{\infty}\sum_{k=0}^{\infty}\widetilde{j}_n\left(\frac{{}_k\omega_n}{\beta}r\right)\sum_{j=-n}^{n}\left(\gamma_{n,j,k}^{\mathrm{c}}\cos\left({}_k\omega_n t\right) + \gamma_{n,j,k}^{\mathrm{s}}\sin({}_k\omega_n t)\right)y_{n,j}^{(3)}(\xi), \tag{7.96}$$

$r \in]0,b]$, $\xi \in \Omega$, $t \in \mathbb{R}$ for arbitrary coefficients $\gamma_{n,j,k}^{\mathrm{c}}, \gamma_{n,j,k}^{\mathrm{s}}$ (as long as a proper convergence, see previous discussions, is guaranteed – but in practice we will only have a finite number of modes).

The discussion of the **spheroidal frequency equation** is lengthy and omitted here for reasons of brevity (see, e.g., Ben-Menahem and Singh, 1981). It also yields a discrete set of **spheroidal eigenfrequencies** ${}_k\omega_n$ such that the **general representation of a spheroidal normal mode for a homogeneous and isotropic ball** is, in the space-time domain, analogous to (7.96), where $M_{n,j}$ needs to be replaced by an arbitrary linear combination of $L_{n,j}$ and $N_{n,j}$ – note the two different speeds which are associated to these two kinds of modes.

Certainly, the Earth is not a homogeneous ball. The preceding derivations can essentially also be used for unions of homogeneous spherical shells, which are also not the truth but a better approximation of it. A more realistic Earth model has to take into account the inhomogeneity (in the radial but also the angular coordinates) and the (at least in parts) anisotropy of it as well as the occurring body forces, including those caused by the rotation. This leads to much more complicated problems, which are far beyond the scope of such a general book, whose author struggles to fulfil the given page limitations and tries to give a broad overview of some highly exciting mathematical problems in Earth sciences. In fact, more sophisticated models lead to perturbations of the modes and eigenfrequencies which theoretically occur for a homogeneous ball. Such more accurate models also yield the practically proved fact of the splitting of the normal modes, that is, the eigenfrequencies depend on the order j of the vector spherical harmonics. The degenerate case of $2n+1$ identical eigenfrequencies given earlier is actually split into $2n+1$ (slightly) different

Table 7.1. Approximate solutions of the toroidal frequency equation given as $\chi := b\beta^{-1}\omega$. The numerical solutions were obtained with Newton's method.

n	${}_0\chi_n$	${}_1\chi_n$	${}_2\chi_n$	${}_3\chi_n$	${}_4\chi_n$	${}_5\chi_n$
1	5.76346	9.09501	12.32294	15.51460	18.68904	21.85387
2	2.50113	7.13601	10.51460	13.77168	16.98306	20.17171
n	${}_6\chi_n$	${}_7\chi_n$	${}_8\chi_n$	${}_9\chi_n$	${}_{10}\chi_n$	${}_{11}\chi_n$
1	25.01280	28.16783	31.32014	34.47049	37.61937	40.76712
2	23.34729	26.51462	29.67640	32.83426	35.98925	39.14206

eigenfrequencies per degree n. For further details, the readers are referred to, for example, Aki and Richards (2002), Ben-Menahem and Singh (1981), and Dahlen and Tromp (1998).

Remark 7.3.20 An easy way of finding the solutions of the frequency equations is available with Newton's method. For example, the solutions of (7.95) in terms of $s := \omega\beta^{-1}b$ are the zeros of $g_n(s) := s\,\widetilde{j}_{n+1}(s) - (n-1)\widetilde{j}_n(s)$; see also Figure 7.10. The derivative of this function is given by

$$\begin{aligned} g_n'(s) &= \widetilde{j}_{n+1}(s) + s\,\widetilde{j}_{n+1}'(s) - (n-1)\widetilde{j}_n'(s) \\ &= \widetilde{j}_{n+1}(s) + (n+1)\widetilde{j}_{n+1}(s) - s\,\widetilde{j}_{n+2}(s) \\ &\quad - (n-1)ns^{-1}\widetilde{j}_n(s) + (n-1)\widetilde{j}_{n+1}(s) \\ &= -s\,\widetilde{j}_{n+2}(s) + (2n+1)\widetilde{j}_{n+1}(s) - (n-1)ns^{-1}\widetilde{j}_n(s). \end{aligned}$$

After an initial approximation s_0 was chosen (e.g., by looking at the graph of g such as in Figure 7.10), we can determine an approximation to a zero by the well-known iteration $s_{p+1} := s_p - g_n(s_p)(g_n'(s_p))^{-1}$. Table 7.1 shows some examples of approximate solutions. Note that these solutions are also denoted by ${}_k\chi_n := b\beta^{-1}{}_k\omega_n$ in the literature.

7.4 An Inverse Problem in Seismology

In this section, we will have a look at the mathematical and theoretical basics of an inverse problem which can be worded as follows: we have some seismic data (more precisely, some travel times of seismic waves) and we want to find out which velocity such waves have at different points inside the Earth. Such a velocity model then allows conclusions on the material which occurs in the various layers and substructures in the interior of our planet. For reasons of brevity, we reduce the discussions here (in some parts but not in general) to the case that we seek a model which only depends on the radial coordinates, such as the well-known Preliminary Reference Earth Model (PREM; see Dziewonski and Anderson, 1981). Moreover, we reduce our discussions to the infinite-frequency limit (also called the high-frequency limit).

This is only a selection out of several inverse problems about which we could talk in this section. However, a detailed and all-embracing discussion of inverse problems in seismology could fill entire books or major parts of suchlike, which it has already done indeed; see Aki and Richards (2002), Ben-Menahem and Singh (1981), Dahlen and Tromp (1998), and Nolet (2008) as examples. For instance, the analysis of the frequency anomalies which can be observed for normal modes allows a partial inference of structures inside the Earth; see also Berkel (2009), Berkel and Michel (2010), Berkel et al. (2011), and the references therein. Moreover, in Section 7.3, we have already restricted our attention to the propagation of body waves. Certainly, surface waves also have a very interesting mathematical modelling, and the analysis of their travel times is another challenging inverse problem, which has been investigated and solved in numerous papers, including Amirbekyan (2007), and Amirbekyan et al. (2008) about the theory and numerics, where Amirbekyan (2007) and Amirbekyan and Michel (2008) also consider the inversion of body wave travel times.

Moreover, the finite-frequency modelling is more accurate than the infinite-frequency limit but (as usual in such cases) more complicated and page consuming. It plays an important role in regions with a very large number of data and a dense grid of measurements. The inverse problem is then represented by an integral equation which relates the deviations of the travel times δT_q to the P-wave and S-wave velocities as well as the mass density ϱ. In detail, we have

$$\delta T_q = \int_{B_R(0)} K_{\alpha,q}(x)\,\frac{\delta\alpha}{\alpha_{\mathrm{M}}}(x) + K_{\beta,q}(x)\,\frac{\delta\beta}{\beta_{\mathrm{M}}}(x) + K_{\varrho,q}(x)\,\frac{\delta\varrho}{\varrho_{\mathrm{M}}}(x)\,\mathrm{d}x,$$

$q = 1, \ldots, N$, where $B_R(0)$ is a ball as an approximation of the Earth's shape. Moreover, $K_{\alpha,q}$, $K_{\beta,q}$, and $K_{\varrho,q}$ are given kernels, which depend, inter alia, on the source and the receiver. Furthermore, $\delta\alpha$, $\delta\beta$, and $\delta\varrho$ are deviations from a reference model $(\alpha_{\mathrm{M}}, \beta_{\mathrm{M}}, \varrho_{\mathrm{M}})$. It should be noted that the travel times are more strongly correlated with the velocities than with the mass density. More details on this modelling can be found, for instance, in Dahlen and Tromp (1998), Marquering et al. (1998, 1999), and Yomogida (1992).

Let us now come back to the case which we consider in this book. Since we are interested in the detection of velocity variations inside the Earth, we have to go back in our modelling to the point before we started to assume that we have a homogeneous ball. We now have an inhomogeneous, but still isotropic, medium. More precisely, we assume that we have the equation of linear elasticity (7.61) with an elasticity tensor which obeys (7.62), but with variable λ and μ. Remember that the speeds of body waves depend on these two parameters.

The derivation of the succeeding equation (7.103) is based on a particular ansatz which occurs in varying versions, among others, in Aki and Richards (2002, section 4.4), Ben-Menahem and Singh (1981, section 7.1.3), Billingham and King (2000, section 5.4.3), Dahlen and Tromp (1998, sections 15.1 and 15.2), and Nolet (2008, section 2.5). The underlying principle is a so-called JWKB approximation (which also occurs in the literature with permutations or omissions of some letters but also as Liouville–Green method): we assume that u (as a function of space and angular frequency) is representable as $u(x,\omega) = A(x)\exp(-\mathrm{i}\omega T(x))$, where $A(x)$ represents the (vectorial) amplitude of the considered wave at x and $T(x)$ stands for the **travel time**, that is, the time at which the wave (more precisely, a certain state of the wave) arrives at x. Moreover, it is assumed that A is expandable as $A(x) = \sum_{n=0}^{\infty}(\mathrm{i}\omega)^{-n} A_n(x)$ with vectorial A_n. This expansion also shows that somehow 'large' frequencies are required for a reasonable modelling here. The first two derivative orders of this ansatz for u are given by

$$\frac{\partial u}{\partial x_j} = \frac{\partial A}{\partial x_j} \exp(-\mathrm{i}\omega T) - A \exp(-\mathrm{i}\omega T)\, \mathrm{i}\omega \frac{\partial T}{\partial x_j}, \tag{7.97}$$

$$\begin{aligned} \frac{\partial^2 u}{\partial x_k \partial x_j} = & \frac{\partial^2 A}{\partial x_k \partial x_j} \exp(-\mathrm{i}\omega T) - \frac{\partial A}{\partial x_j} \exp(-\mathrm{i}\omega T)\, \mathrm{i}\omega \frac{\partial T}{\partial x_k} \\ & - \frac{\partial A}{\partial x_k} \exp(-\mathrm{i}\omega T)\, \mathrm{i}\omega \frac{\partial T}{\partial x_j} - A \exp(-\mathrm{i}\omega T)\, \omega^2 \frac{\partial T}{\partial x_k} \frac{\partial T}{\partial x_j} \\ & - A \exp(-\mathrm{i}\omega T)\, \mathrm{i}\omega \frac{\partial^2 T}{\partial x_k \partial x_j}. \end{aligned} \tag{7.98}$$

Furthermore, we need to consider the equation of linear elasticity (7.61) in the space-frequency domain (i.e., we either use the Fourier transform or a separation ansatz; see the discussions before and after (7.73)). We also insert (7.63), omit the body force term, and use (2.6) such that we obtain

$$\begin{aligned} -\varrho\omega^2 u &= \operatorname{div}_x \left[\lambda\, (\operatorname{div}_x u)\, I + \mu \nabla_x u + \mu\, (\nabla_x u)^{\mathrm{T}}\right] \\ &= (\nabla_x \lambda) \operatorname{div}_x u + \lambda \operatorname{grad}_x \operatorname{div}_x u + \left[\nabla_x u + (\nabla_x u)^{\mathrm{T}}\right] \nabla_x \mu \\ &\quad + \mu\left(\Delta_x u + \operatorname{grad}_x \operatorname{div}_x u\right) \\ &= (\nabla_x \lambda) \operatorname{div}_x u + (\lambda + \mu) \operatorname{grad}_x \operatorname{div}_x u + \left[\nabla_x u + (\nabla_x u)^{\mathrm{T}}\right] \nabla_x \mu + \mu\, \Delta_x u. \end{aligned} \tag{7.99}$$

In the next step, we insert (7.97) and (7.98) as well as the expansion of A into (7.99), while we keep an eye on the powers of ω which occur in the resulting equation. It is clear that we get exponents which consist of all integers up to 2 (the exponential term is not included in this point of view, because it is located on the unit circle line in $\mathbb{C}$). We can reason now that the coefficients of each power of ω must fulfil the equation on their own and then only look at the coefficients of ω^2. Or we can say that all terms which do not include ω^2 are $\mathcal{O}(\omega)$ as $\omega \to \infty$ (here is again the high/infinite frequency modelling). Certainly, this ansatz has its drawbacks and inaccuracies: we cannot be sure that the applied formula for u covers all possible solutions, and we certainly produce a modelling error by ignoring all terms that are not coefficients of ω^2 (in particular, those of ω, if we assume high or infinite frequencies). Nevertheless, this modelling has served as a foundation for many acceptably good approximations to the reality for quite a long time. To make a long story short, the second-order terms in ω which we get are obviously as follows:

$$\begin{aligned} -\varrho\omega^2 A_0 \exp(-\mathrm{i}\omega T) = & -(\lambda + \mu)\omega^2 \exp(-\mathrm{i}\omega T)\, (A_0 \cdot \nabla_x T)\, \nabla_x T \\ & - \mu A_0 \exp(-\mathrm{i}\omega T)\, \omega^2\, |\nabla_x T|^2 \end{aligned}$$

such that

$$\varrho A_0 = (\lambda + \mu)\,(A_0 \cdot \nabla_x T)\,\nabla_x T + \mu\,|\nabla_x T|^2\,A_0. \tag{7.100}$$

We need to get a geometrical imagination of what we are doing here: if the function T describes the travel time, then a surface across which T is constant contains all points in space where the wave (or, again, a particular state of the wave) arrives at the same time. It is a basic result of real analysis that the gradient of a level set, which here is $\nabla_x T$, is a normal vector to this surface. Comparing this with the propagation of planar waves and realizing that $\nabla_x T$ is the normal vector to the tangential plane, we get the interpretation that $\nabla_x T(x)$ points into the direction of wave propagation at point x. In view of the experience which we have made before regarding P-waves and S-waves, we distinguish now two cases: either A_0 is orthogonal to $\nabla_x T$ (S-wave) or A_0 is parallel to $\nabla_x T$ (P-wave). In the former case, (7.100) becomes

$$\varrho = \mu\,|\nabla_x T|^2\,. \tag{7.101}$$

In the latter case, where $A_0 = \pm c|A_0|\nabla_x T$, $c \in \mathbb{R}$, we obtain

$$\varrho = (\lambda + \mu)\,|\nabla_x T|^2 + \mu\,|\nabla_x T|^2\,. \tag{7.102}$$

We abbreviate the speeds of P-waves and S-waves with c and summarize (7.101) and (7.102) with the famous **eikonal equation**, which we assume to be valid from now on:

$$|\nabla_x T(x)| = \frac{1}{c(x)}. \tag{7.103}$$

Definition 7.4.1 The vectorial function $x \mapsto \mathfrak{p}(x) := c(x)\nabla_x T(x)$ is called the **unit slowness vector** and the reciprocal of the speed function, that is, $x \mapsto (c(x))^{-1} =: S(x)$, is called the **slowness**. Moreover, each surface on which T is constant is called a **wavefront**.

We use here the Gothic print letter $\mathfrak{p}$ for the unit slowness vector in order to distinguish it from the ray parameter p, which will be introduced later. Alternatively, one also finds the boldface letter $\mathbf{p}$ for the unit slowness vector in the literature. As we mentioned previously, $\nabla_x T(x)$ is a normal vector to the wavefront through x. Due to the eikonal equation, $\mathfrak{p}(x) = (S(x))^{-1}\nabla_x T(x)$ is a unit normal vector.

Solving the eikonal equation is not an easy task. A common approach is the construction of seismic rays. From the mathematical point of view, this is related to the method of characteristics (see, e.g., Agarwal and O'Regan, 2009, lecture 28). From the physical point of view, the eikonal equation also occurs in optics, where the propagation of light is visualized by light rays, which is basically the same concept. The fundamental idea behind this is to replace the solution of the partial differential equation (7.103) with a family of solutions of a system of ordinary differential equations. The parameterization of this family is achieved by leaving the initial condition open.

Definition 7.4.2 Every arc $I \ni s \mapsto X(s) \in \mathbb{R}^3$, where $I \subset \mathbb{R}$ is an interval and the parameter s is the arc length, which satisfies the system of ordinary differential equations

$$\frac{\mathrm{d}}{\mathrm{d}s}X(s) = \frac{\nabla_x T(X(s))}{|\nabla_x T(X(s))|}, \tag{7.104}$$

is called a **(seismic) ray**.

We should talk about geometry once again. As we know, the right-hand side of (7.104) is the unit slowness vector at $X(s)$, which is a unit normal vector to the wavefront. Regarding the rays, it is a unit tangential vector along the rays – since every derivative of an arc with respect to the arc length has this property. Therefore, rays are everywhere normal to the wavefronts. Note that solutions of systems of differential equations are also called trajectories in mathematics and note also that we use here capital letters X for the rays in order to better distinguish them from points in space. This has nothing to do with material coordinates.

Example 7.4.3 We consider a simple two-dimensional example where all rays start in the origin of coordinates and the travel times fulfil $T(x) = ax_1^2 + bx_2^2$ for $a = 5$ and $b = 6$. Then the wavefronts in the x_1-x_2-plane are obviously ellipses. The rays are defined (in 2D) as solutions of

$$\frac{\mathrm{d}}{\mathrm{d}s}X(s) = \frac{1}{\sqrt{4a^2X_1(s)^2 + 4b^2X_2(s)^2}}\begin{pmatrix}2aX_1(s)\\ 2bX_2(s)\end{pmatrix}. \tag{7.105}$$

We calculate the rays numerically as follows: first, we choose a step size $h := 2.5 \times 10^{-4}$. Then we start in $(0,0)^{\mathrm{T}}$ and use, for each ray, an initial angle out of $0, \pi/25, 2\pi/25, \ldots, 2\pi$. We make a first step into the direction of the initial angle and with the length h. From this new point, we iteratively proceed as follows: we calculate the right-hand side of (7.105) at the current point, make a step in this direction, again with the length h, and reach our new point. The iteration is truncated, if the ray leaves the frame of the image. The result is shown in Figure 7.11. This is certainly a simple way of calculating a ray. For more sophisticated methods, see, for example, Tian et al. (2007).

We will further investigate now the properties of rays, because this will lead us to interesting conclusions on the behaviour of seismic waves and their travel times.

Lemma 7.4.4 *Along an arbitrary ray, the slowness satisfies*

$$\frac{\mathrm{d}}{\mathrm{d}s}(S\mathfrak{p})(X(s)) = \nabla_x S(X(s)), \qquad \frac{\mathrm{d}}{\mathrm{d}s}T(X(s)) = S(X(s)). \tag{7.106}$$

Proof Identities such as (7.106) are always to be understood as valid for all s in the interval which is the domain of the trajectory X. We first prove the second identity by using the chain rule, the defining equation (7.104) of the rays, and the eikonal equation (7.103):

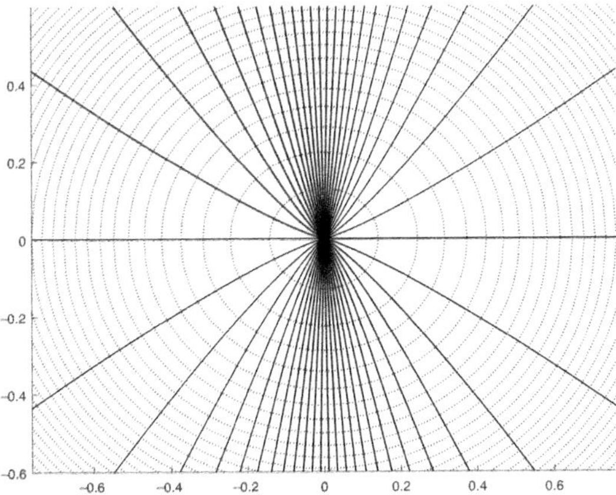

Figure 7.11 Illustration of a simple example of two-dimensional elliptic wavefronts (dotted): the rays (solid lines) are everywhere normal to the wavefronts.

$$\frac{\mathrm{d}}{\mathrm{d}s}T(X(s)) = \nabla_x T(X(s)) \cdot \frac{\mathrm{d}}{\mathrm{d}s}X(s) = \nabla_x T(X(s)) \cdot \frac{\nabla_x T(X(s))}{|\nabla_x T(X(s))|}$$
$$= \frac{1}{c(X(s))} = S(X(s)).$$

For the remaining identity, we first need an auxiliary calculation, which is obtained by using the multivariate differential quotient:

$$\frac{\partial}{\partial x_j}\left[\nabla_x T(X(s)) \cdot \frac{\mathrm{d}X(s)}{\mathrm{d}s}\right]$$
$$= \lim_{h\to 0}\frac{1}{h}\left[\nabla_x T\left(X(s)+h\varepsilon^j\right) \cdot \frac{\mathrm{d}\left(X(s)+h\varepsilon^j\right)}{\mathrm{d}s} - \nabla_x T(X(s)) \cdot \frac{\mathrm{d}X(s)}{\mathrm{d}s}\right]$$
$$= \lim_{h\to 0}\frac{1}{h}\left[\nabla_x T\left(X(s)+h\varepsilon^j\right) - \nabla_x T(X(s))\right] \cdot \frac{\mathrm{d}X(s)}{\mathrm{d}s}$$
$$= \left(\frac{\partial}{\partial x_j}\nabla_x T(X(s))\right) \cdot \frac{\mathrm{d}X(s)}{\mathrm{d}s}.$$

With the previous two identities and the chain rule, we eventually obtain

$$\frac{\mathrm{d}}{\mathrm{d}s}(S\mathfrak{p})(X(s)) = \frac{\mathrm{d}}{\mathrm{d}s}\left[\nabla_x T(X(s))\right] = \left[\nabla_x \otimes \nabla_x T(X(s))\right]\frac{\mathrm{d}X(s)}{\mathrm{d}s}$$
$$= \nabla_x\left[\nabla_x T(X(s)) \cdot \frac{\mathrm{d}X(s)}{\mathrm{d}s}\right] = \nabla_x \frac{\mathrm{d}T(X(s))}{\mathrm{d}s} = \nabla_x S(X(s)).$$ □

Definition 7.4.5 The vectorial function $S\mathfrak{p}$ is called the **ray vector**.

Theorem 7.4.6 *If $X : I \to X(I) =: \gamma$ describes a ray between two points A and B, then the travel time from A to B is given by the line integral*

$$\int_\gamma S(x)\,\mathrm{dl}(x) =: T_{A\curvearrowright B}.$$

Proof The theorem is a rather immediate conclusion of Lemma 7.4.4 and the fundamental theorem of calculus. By evaluating the integral, we obtain

$$\int_\gamma S(x)\,\mathrm{dl}(x) = \int_I S(X(s)) \left|\frac{\mathrm{d}X(s)}{\mathrm{d}s}\right| \mathrm{d}s = \int_I \frac{\mathrm{d}}{\mathrm{d}s} T(X(s))\,\mathrm{d}s = T(B) - T(A)$$

with the help of $\left|\frac{\mathrm{d}}{\mathrm{d}s} X(s)\right| = 1$. □

So far, we have no notable restriction on the Earth model, which could be characterized, for example, by the mass density ϱ and the Lamé parameters λ and μ. We will now introduce a simplification, namely an **SNREI Earth model** (see, e.g., Dahlen and Tromp, 1998, section 8.2), where only the 'S' is new here. SNREI stands for *spherically symmetric*, *non-rotating*, perfectly *elastic*, and *isotropic*. The spherical symmetry means that the material parameters only depend on the radial coordinate r, which implies that the speeds c of P-waves and S-waves also only depend on r. We will, therefore, represent c and S as univariate functions of r now. The spherical symmetry leads us, indeed, to some very interesting results.

Theorem 7.4.7 *In the case of an SNREI Earth model, the ray vector satisfies, along each ray X, the identity*

$$\frac{\mathrm{d}}{\mathrm{d}s}\left[X(s) \times (S\mathfrak{p})(X(s))\right] = 0. \tag{7.107}$$

Proof We use the product rule, the defining equation (7.104) of the rays, and Lemma 7.4.4 and get

$$\begin{aligned}\frac{\mathrm{d}}{\mathrm{d}s}\left[X(s) \times (S\mathfrak{p})(X(s))\right] &= \frac{\mathrm{d}X(s)}{\mathrm{d}s} \times (S\mathfrak{p})(X(s)) + X(s) \times \frac{\mathrm{d}(S\mathfrak{p})(X(s))}{\mathrm{d}s}\\ &= [\mathfrak{p} \times (S\mathfrak{p})](X(s)) + X(s) \times \nabla_x S(|x|)|_{x=X(s)}\\ &= X(s) \times \left(S'(|X(s)|)\frac{X(s)}{|X(s)|}\right).\end{aligned}$$

Obviously, also the remaining cross product vanishes. □

Theorem 7.4.7 can be exploited in two ways: since the derivative of the cross product in (7.107) vanishes along each ray, its norm and its orientation both have to be constant on each ray. The former conclusion means that $|X(s)|\, S(X(s))\, |\mathfrak{p}(X(s))| \sin\measuredangle(X(s), \mathfrak{p}(X(s)))$ is constant with respect to s, where $|\mathfrak{p}(X(s))| = 1$. This yields the celebrated Snell's law.

Definition 7.4.8 Along an arbitrary ray, the angle between the position vector $X(s)$ and the unit tangential vector $\mathfrak{p}(X(s))$ is denoted by $\mathrm{i}(X(s))$ and is called the **angle of incidence**; see also Figure 7.12.

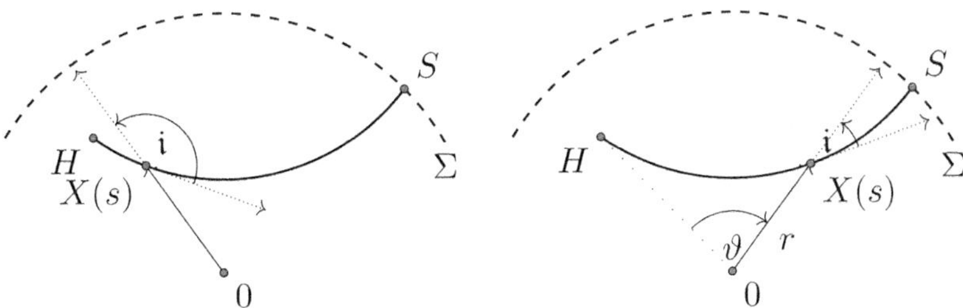

Figure 7.12 Illustration of the angle of incidence $\mathfrak{i}$ as well as the coordinates $r = |X(s)|$ and ϑ, which suffice for a planar ray in an SNREI Earth model. The points H and S represent a hypocentre and a seismic station, respectively, and Σ is the Earth's surface.

We choose also here a Gothic print letter, because i is a common enumeration index and this might cause a confusion otherwise. Note that $\mathfrak{i} \in [0, \pi/2]$ on an 'upgoing leg' and $\mathfrak{i} \in [\pi/2, \pi]$ on a 'downgoing leg'.

Corollary 7.4.9 (Snell's Law) *In the case of an SNREI model, the term* $Sr \sin \mathfrak{i}$ *is constant along every arbitrary ray.*

If we consider the rays caused by one earthquake with a point-shaped source (as in Figure 7.11), then the initial angle at the source defines a unique ray (at least if we ignore splitting of rays due to reflections or other phenomena at sharp changes of the material). This means that if $X(0)$ is the starting point of every ray, then it is exactly $\mathfrak{i}(X(0))$ which determines a specific ray. In this respect, $S(X(s))\,|X(s)|\,\sin \mathfrak{i}(X(s))$ is not only constant along the ray X, its constant value also uniquely identifies the ray itself (up to rotational symmetries). This is the reason for the following common definition.

Definition 7.4.10 If X is an arbitrary ray in an SNREI Earth model, then its associated constant function $s \mapsto S(X(s))\,|X(s)|\,\sin \mathfrak{i}(X(s)) =: p$ is called its **ray parameter**.

The exploitation of (7.107) regarding the constant orientation of the cross product can be done as follows: $X(s)$ is a point on the ray, while $(S\mathfrak{p})(X(s))$ is a tangential vector to the ray at the same point. Both span a plane (if we exclude the degenerate case where they are collinear) and their cross product is a normal vector to this plane. Hence, this normal vector does not change its orientation, which means that this spanned plane is always the same for each point $X(s)$ of the same ray. In other words, we get the following result.

Corollary 7.4.11 *Every ray in an SNREI Earth model is a planar arc.*

Each single ray is, therefore, located in a disc, where the disc certainly can differ from ray to ray. Thus, if we keep an arbitrary ray fixed, then two coordinates suffice to describe each point on the ray: a radial coordinate (naturally the distance to the centre of the ball-shaped Earth) and one angle instead of two. We denote this angle by ϑ and gauge it such that $\vartheta = 0$ corresponds to the starting point of the ray (usually the hypocentre of the earthquake) and ϑ

increases along the ray; see Figure 7.12. Also here, we ignore more complicated geometries of rays where they might be reflected and cross their own curve again or similar scenarios.

Theorem 7.4.12 *If X is a ray in an SNREI Earth model and if it is parameterized by r and ϑ as described in the preceding paragraph, then the eikonal equation* (7.103) *restricted to the corresponding disc is equivalent to the equation*

$$\left(\frac{\partial T}{\partial r}\right)^2 + \frac{1}{r^2}\left(\frac{\partial T}{\partial \vartheta}\right)^2 = \frac{1}{c^2(r)}. \tag{7.108}$$

Proof We use r and ϑ as two of the polar coordinates in the ball-shaped Earth and supplement them appropriately with the angle φ (basically, this means that we put the epicentre in the North pole). From Theorem 3.4.1, we know that the gradient can be decomposed into

$$\nabla T = \frac{\partial T}{\partial r}\varepsilon^r + \frac{1}{r\sqrt{1-t^2}}\frac{\partial T}{\partial \varphi}\varepsilon^\varphi + \frac{\sqrt{1-t^2}}{r}\frac{\partial T}{\partial t}\varepsilon^t,$$

where $t = \cos\vartheta$, $\vartheta \in [0,\pi]$. Since we restrict the eikonal equation to a disc where φ is constant, the derivative with respect to φ does not appear. Due to the chain rule, $\frac{\partial}{\partial t} = -(\sin\vartheta)^{-1}\frac{\partial}{\partial\vartheta} = -(1-t^2)^{-1/2}\frac{\partial}{\partial\vartheta}$ for all $\vartheta \in]0,\pi[$. Due to the orthonormality of ε^r, ε^φ, and ε^t, we immediately obtain (7.108). □

Definition 7.4.13 The radial coordinate of a hypocentre will be denoted by r_{h}.

Based on our modelling, it is reasonable that we initialize the travel time as vanishing at the hypocentre coordinates, which are $r = r_{\mathrm{h}}$ and $\vartheta = 0$.

Theorem 7.4.14 *Some solutions of* (7.108) *together with the boundary condition* $T(x(r_{\mathrm{h}},0)) = 0$ *are given by*

$$T(x(r,\vartheta)) = p\vartheta \pm \int_{r_{\mathrm{h}}}^{r}\sqrt{S^2(\varrho)\varrho^2 - p^2}\,\frac{1}{\varrho}\,\mathrm{d}\varrho, \tag{7.109}$$

where p is the ray parameter and $S = c^{-1}$ is the slowness. Moreover, $x(r,\vartheta)$ represents the Cartesian point x in terms of the disc coordinates r and ϑ.

Proof For proving this theorem, we only have to calculate the derivatives $\left(\frac{\partial T}{\partial r}\right)^2 = \left(\sqrt{S^2(r)r^2 - p^2}\,r^{-1}\right)^2 = S^2(r) - p^2r^{-2}$ and $\left(\frac{\partial T}{\partial\vartheta}\right)^2 = p^2$. Inserting both identities into the left-hand side of (7.108) yields $S^2(r)$. □

We immediately get the following equation.

Corollary 7.4.15 (Benndorf's Relation) *Let X be a ray in an SNREI Earth model and with the travel time function* (7.109), *then* $\frac{\partial T}{\partial\vartheta} = p$.

Hence, the ratio between the increasing travel time along a ray and the passed angle is constant along each ray. The fact that this constant is the ray parameter gives us, hence, another interpretation of p.

We will now have a look at the inverse problem of travel time tomography. We apply the limitations announced earlier: we only investigate the problem for the infinite frequency model. Moreover, we first discuss the case of an SNREI Earth model while we will also have a look at more general models afterwards.

Definition 7.4.16 The **inverse problem of travel time tomography** (in the infinite frequency model) is represented here as follows: for a given finite set of hypocentre locations A_j^{h} and positions of seismic stations B_j^{s}, $j = 1, \ldots, N$, we know the travel times $T_{A_j^{\mathrm{h}} \curvearrowright B_j^{\mathrm{s}}}$ of seismic waves of a particular type (P or S). The task is to find a slowness function S such that

$$\int_{A_j^{\mathrm{h}}}^{B_j^{\mathrm{s}}} S(x)\,\mathrm{dl}(x) = T_{A_j^{\mathrm{h}} \curvearrowright B_j^{\mathrm{s}}} \quad \text{for all } j = 1, \ldots, N; \tag{7.110}$$

see also Theorem 7.4.6.

For the SNREI model, we continue with the formula (7.109), where we need to have a closer look at the '$\pm$' which is there. Clearly, if r increases, then T also increases in the '$+$'-case, while in the '$-$'-case, a decreasing r causes an increasing T. Naturally, we want to follow the propagation of the wave (or the ray), which means that we need a monotonically increasing T along the ray. Hence, the '$+$'-case belongs to the upgoing leg and '$-$' occurs on the downgoing leg. We also change our point of view. We consider (7.109) on a wavefront, that is, T is now a constant, but the ray parameter p is a variable, which means that we compare different rays which reach the wavefront at time T. The derivative of (7.109) with respect to p becomes then

$$0 = \vartheta \pm \int_{r_{\mathrm{h}}}^{r} \frac{1}{2}\left(S^2(\varrho)\varrho^2 - p^2\right)^{-1/2}(-2p)\frac{1}{\varrho}\,\mathrm{d}\varrho$$

such that

$$\vartheta = \pm p \int_{r_{\mathrm{h}}}^{r} \left(S^2(\varrho)\varrho^2 - p^2\right)^{-1/2}\frac{1}{\varrho}\,\mathrm{d}\varrho.$$

For reasons of simplicity, we only look at rays which start with a downgoing leg and then have an upgoing leg which finally reaches the surface (as in Figure 7.12) – more precisely, we pick out such a ray among those rays which reach our wavefront. Moreover, we assume a shallow earthquake, that is, $r_{\mathrm{h}} = a$, where a is now the Earth's radius. Therefore, we introduce δ as the total angle between the hypocentre and the seismic station, as shown in Figure 7.13. The constancy of the ray parameter $S(r)r \sin \mathrm{i}$ implies a symmetry of the rays: if two different points of the same ray have equal radial coordinates, then the angles of incidence are either equal or are symmetric to $\pi/2$ – in our case of one downgoing leg and one upgoing leg, it needs to be the latter case. The symmetry axis is the line through the centre of the Earth and the deepest point of the ray, whose radial coordinate is, therefore, minimal along the ray and represented by r_{m}.

Certainly, we have reached very strong limitations here. Indeed, this is less exciting from the point of view of today's research, where one seeks lateral heterogeneities in the

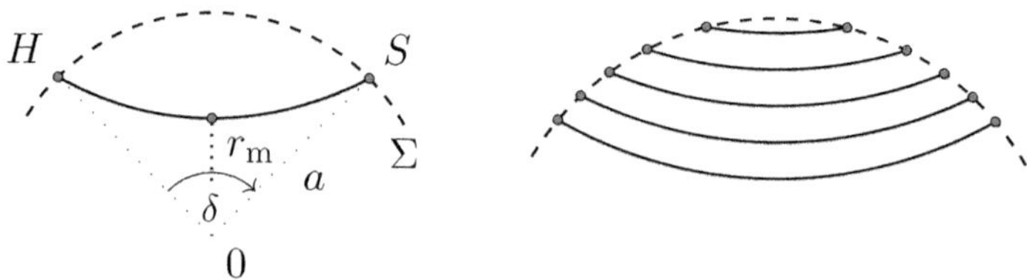

Figure 7.13 In the simplified case of a hypocentre on the surface of an SNREI Earth model, the ray is symmetric with respect to its deepest point. The total angle between hypocentre H and seismic station S is denoted by δ. The radial coordinate at the deepest point is represented by r_{m}. The radius of the Earth is a. The right-hand graphic illustrates that the limit $\mathrm{i} \to \pi/2$ for the angle of incidence at the surface points corresponds to the limit $\delta \to 0$.

Earth's structure. However, as a textbook-like starting point into the theory, it is nevertheless interesting to have a closer look at this specific scenario, because we gain a simple way of learning something about the coarse structures of the Earth.

Taking into account the aforementioned derivations and assumptions, we obtain the following formula for the total angle (note that δ depends on the particular ray):

$$\delta(p) = 2p \int_{r_{\mathrm{m}}}^{a} \left(S^2(r)r^2 - p^2\right)^{-1/2} \frac{1}{r}\,\mathrm{d}r.$$

The factor 2 occurs from the combination of the downgoing leg and the upgoing leg. We substitute now $y := (S(r)r/a)^2$ and use $\frac{\mathrm{d}}{\mathrm{d}y} \log r(y) = (r(y))^{-1} \frac{\mathrm{d}r}{\mathrm{d}y}$. We should not ignore that this substitution requires an injective relation between r and y. In other words, $S(r)r$ needs to depend in a monotonical way on r. We will discuss this condition later on. At the moment, we assume its validity (at least, in an upper layer where the ray is located) and obtain

$$\delta(p) = 2p \int_{(S(r_{\mathrm{m}})r_{\mathrm{m}}/a)^2}^{S^2(a)} \left(a^2 y - p^2\right)^{-1/2} \frac{\mathrm{d}}{\mathrm{d}y} \log r(y)\,\mathrm{d}y.$$

Since $p \neq 0$ (a ray with a vanishing ray parameter would have an angle of incidence which is constantly 0 or π), we can also write

$$\frac{a\delta(p)}{2p} = \int_{(S(r_{\mathrm{m}})r_{\mathrm{m}}/a)^2}^{S^2(a)} \left(y - p^2 a^{-2}\right)^{-1/2} \frac{\mathrm{d}}{\mathrm{d}y} \log r(y)\,\mathrm{d}y.$$

At $r = r_{\mathrm{m}}$, we have $\mathrm{i} = \pi/2$ (see Figure 7.13). Hence, $p = S(r_{\mathrm{m}})r_{\mathrm{m}}$ holds true for each ray. The lower integral limit is thus the term $p^2 a^{-2} =: y_{\mathrm{m}}$, which also occurs elsewhere. With $G(y) := \frac{\mathrm{d}}{\mathrm{d}y} \log r(y)$, we can write

$$\frac{\delta\left(a\sqrt{y_{\mathrm{m}}}\right)}{2\sqrt{y_{\mathrm{m}}}} = \int_{y_{\mathrm{m}}}^{S^2(a)} (y - y_{\mathrm{m}})^{-1/2} G(y)\,\mathrm{d}y.$$

We substitute now $z := y_\mathrm{m} + w - y$, where $w := S^2(a) - y_\mathrm{m}$ (such that $z = S^2(a) - y$), and get

$$\frac{\delta\left(a\sqrt{S^2(a) - w}\right)}{2\sqrt{S^2(a) - w}} = \int_0^w (w - z)^{-1/2}\, G\big(S^2(a) - z\big)\, \mathrm{d}z. \tag{7.111}$$

Equation (7.111) has the form of an **Abel integral equation**. For further details on this type of integral equation, we refer, as an example of related literature, to Gorenflo and Vessella (1991), whose theorem 1.2.1 tells us that, provided that the left-hand side of (7.111) is absolutely continuous with respect to w (i.e., actually, with respect to p), there is one and only one solution, and this is given by

$$G\big(S^2(a) - z\big) = \frac{1}{\pi}\frac{\mathrm{d}}{\mathrm{d}z}\int_0^z (z - w)^{-1/2}\, \frac{\delta\left(a\sqrt{S^2(a) - w}\right)}{2\sqrt{S^2(a) - w}}\, \mathrm{d}w.$$

Resubstituting back to y and y_m, we arrive at

$$\frac{\mathrm{d}}{\mathrm{d}y}\log r(y) = G(y) = \frac{1}{2\pi}(-1)\frac{\mathrm{d}}{\mathrm{d}y}\int_{S^2(a)}^y (y_\mathrm{m} - y)^{-1/2}\, \frac{\delta\left(a\sqrt{y_\mathrm{m}}\right)}{\sqrt{y_\mathrm{m}}}\, \mathrm{d}y_\mathrm{m}\cdot(-1).$$

We integrate now with respect to y and insert $S^2(a)$ as the lower limit. This leads us to

$$\log r(y) - \log r\big(S^2(a)\big) = -\frac{1}{2\pi}\int_y^{S^2(a)} (y_\mathrm{m} - y)^{-1/2}\, \frac{\delta\left(a\sqrt{y_\mathrm{m}}\right)}{\sqrt{y_\mathrm{m}}}\, \mathrm{d}y_\mathrm{m}. \tag{7.112}$$

The integration variable $y_\mathrm{m} = p^2a^{-2}$ is still artificial. So, actually we have a hidden integration with respect to the ray parameter p. The limits of the integration satisfy $(S(r)r/a)^2 = y \le p^2a^{-2} \le S^2(a) = (S(a)a/a)^2$. Hence, we can rewrite (7.112) as

$$\begin{aligned}\log\frac{r}{a} &= -\frac{1}{2\pi}\int_{S(r)r}^{S(a)a}\left(\frac{p^2}{a^2} - \left(S(r)\frac{r}{a}\right)^2\right)^{-1/2}\frac{a\delta(p)}{p}\frac{2p}{a^2}\,\mathrm{d}p\\ &= -\frac{1}{\pi}\int_{S(r)r}^{S(a)a}\left(p^2 - (S(r)r)^2\right)^{-1/2}\delta(p)\,\mathrm{d}p.\end{aligned}$$

From here on, we consider r as an arbitrary but fixed value. We integrate now by parts, which requires that $p \mapsto \delta(p)$ is continuously differentiable. For the solution of the Abel integral equation, we have already required that δ is absolutely continuous with respect to p, which is (on compact intervals) necessary but not sufficient for continuous differentiability (see Theorem 2.4.11), which is now our stronger condition. We get

$$\log\frac{r}{a} = -\frac{1}{\pi}\,\delta(p)\operatorname{arcosh}\frac{p}{S(r)r}\bigg|_{p=S(r)r}^{p=S(a)a} + \frac{1}{\pi}\int_{S(r)r}^{S(a)a}\delta'(p)\operatorname{arcosh}\frac{p}{S(r)r}\,\mathrm{d}p. \tag{7.113}$$

The insertion $p = S(a)a$ should be regarded as a limit $p \to S(a)a$, which is obtained by letting $\mathrm{i}|_{r=a}$ tend to $\pi/2$. In fact, the closer the hypocentre and station are, the closer is the initial angle of incidence to $\pi/2$. The limit yields $\delta(p) \to 0$ (see Figure 7.13). Thus, (7.113) becomes

$$\log \frac{r}{a} = \frac{1}{\pi} \int_{S(r)r}^{S(a)a} \delta'(p) \operatorname{arcosh} \frac{p}{S(r)r} \, \mathrm{d}p = \frac{1}{\pi} \int_{\widetilde{\delta}(r)}^{0} \operatorname{arcosh} \left(\frac{p(\delta)}{S(r)r} \right) \mathrm{d}\delta,$$

where we used integration by substitution, which causes the condition that $\delta'(p) \neq 0$ for all p (which implies the invertibility of $p \mapsto \delta(p)$). Here, $\widetilde{\delta}(r)$ represents the total angle δ, which corresponds to the ray with ray parameter $p(\widetilde{\delta}(r)) = S(r)r = r/c(r)$, that is, the ray for which our fixed value r is the radial coordinate r_{m} of the deepest point.

After the application of the exponential function and the multiplication with $a[p(\widetilde{\delta}(r))]^{-1}$, we obtain the following famous result.

Theorem 7.4.17 (Herglotz–Wiechert Formula) *Let a be the Earth's radius in an SNREI Earth model, whose upper layer satisfies the following requirements: $r \mapsto S(r)r$ is monotonically increasing and $p \mapsto \delta(p)$ is continuously differentiable with a non-vanishing derivative. If r is a radial coordinate which belongs to this layer and if $\widetilde{\delta}(r)$ is the total angle of the ray with r as lowest radial coordinate, such that all rays with total angles $\leq \widetilde{\delta}(r)$ are also inside the layer, then the wave speed at this coordinate satisfies*

$$c(r) = \frac{a}{p\left(\widetilde{\delta}(r)\right)} \exp\left(-\frac{1}{\pi} \int_0^{\widetilde{\delta}(r)} \operatorname{arcosh} \frac{p(\delta)}{p\left(\widetilde{\delta}(r)\right)} \, \mathrm{d}\delta \right). \tag{7.114}$$

The vigilant reader might get confused at this point: if one chooses a radius r and is interested in $c(r)$, then (7.114) requires the knowledge $\widetilde{\delta}(r)$. However, this knowledge is only available if the geometry of the ray with lowest radial coordinate r is known, which is only accessible via a velocity model. So, this is a catch-22. The way out is, fortunately, very simple: we start in a different way. What we actually have is a table of travel times between hypocentres and stations. In an SNREI model with our further assumptions, the travel time actually only depends on the total angle, no matter where the hypocentre and station are located. So, what we have is a finite sequence $\{(\delta_j, T_j)\}_{j=1,\ldots,N}$, where the total angles are out of a range $[0, \delta_{\max}]$. Benndorf's relation (Corollary 7.4.15) suggests that we can approximately calculate the ray parameters $p(\delta) = \frac{\mathrm{d}T}{\mathrm{d}\delta}$ by numerical differentiation. If we replace $\widetilde{\delta}(r)$ in (7.114) by $\delta_{\max}$, then we know everything on the right-hand side and can numerically calculate the formula, which gives a velocity value at an unknown radius r. However, since $p(\delta_{\max})c(r) = r$, we also know at which radius this velocity occurs. Then we proceed with different values for $\delta_{\max}$ (within the data range) and obtain in this way a collection of velocity values for different radial coordinates.

Even more interesting than the derivation of Theorem 7.4.17 is the discussion of the cases where its preconditions are violated. We will have a look at two particular cases:

Low-Velocity Zones (LVZs): A layer in which $r \mapsto S(r)r$ is monotonically decreasing is called an LVZ. If we follow a downgoing leg, then r decreases and, in an LVZ, $r/c(r)$

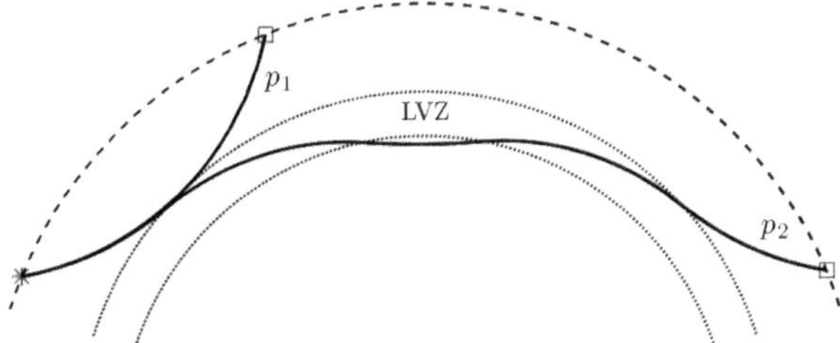

Figure 7.14 (Exaggerated) illustration of an LVZ: rays which enter the LVZ are refracted towards the centre of the Earth. Their angle of incidence cannot decrease on the downgoing leg as long as the ray is inside the LVZ. The consequence is a shadow zone on the surface, where no waves occur.

increases, that is, the velocity decreases more than linearly, which explains the nomenclature. For understanding what is so particular about an LVZ, we assume that the top layer of the Earth is not an LVZ, but somewhere deeper there is an LVZ. Let p_1 be the ray parameter of that ray which just touches the upper boundary of the LVZ but does not enter it, as shown in Figure 7.14. Rays with lower r_{m} have a larger initial angle of incidence (i.e., at $r = a$) and (since this angle is in $[\pi/2, \pi]$ and $p = S(a)a \sin \mathrm{i}|_{r=a}$) a smaller ray parameter p. So let $p_2 := p_1 - \varepsilon$ for a sufficiently small $\varepsilon > 0$. As long as this ray propagates through the upper (non-LVZ) layer downwards, $r/c(r)$ decreases and the constancy of the ray parameter implies that $\sin \mathrm{i}$ must increase, which means that i decreases. In the case of p_1, the angle of incidence eventually reaches $\pi/2$ and the upgoing leg starts, which is already determined due to the symmetry of the rays under our assumptions.

For the ray with parameter p_2, the situation becomes, however, different as soon as it enters the LVZ. It did not reach $\mathrm{i} = \pi/2$ so far, but now $r/c(r)$ starts increasing, and therefore $\sin \mathrm{i}$ must decrease. Hence, the angle of incidence increases, that is, it moves away from $\pi/2$, as long as the ray propagates through the LVZ. After that, i can decrease again and the upgoing leg comes in sight. The consequence is a **shadow zone** where no waves arrive. This shadow zone has the form of a belt which is defined by the total angles δ (with respect to the hypocentre) between $\delta(p_1)$ and $\lim_{p \to p_1-} \delta(p)$. This also shows us that an LVZ causes a discontinuity in the function $p \mapsto \delta(p)$. A continuous differentiability is thus by far not given. Interestingly, the requirement of a monotonical function $r \mapsto S(r)r$ is also violated in the very same context, as we have seen.

Another cause of a shadow zone is a discontinuity of c. If the velocity jumps from high to low, regarded from above a layer boundary to below it, then the angle of incidence needs to jump as well to keep the ray parameter constant. On a downgoing leg, this means that i jumps to a larger value and the ray is refracted towards the centre of the Earth, as shown in Figure 7.15. As a consequence, the path of such a ray is much longer with an essentially larger total angle. Also here, we get a shadow zone. LVZs and jump discontinuities occur, indeed, in the Earth; see Dziewonski and Anderson (1981). A famous jump discontinuity, where waves are decelerated if they move below it and a large shadow zone occurs, is the core–mantle boundary between the solid mantle and the liquid outer core.

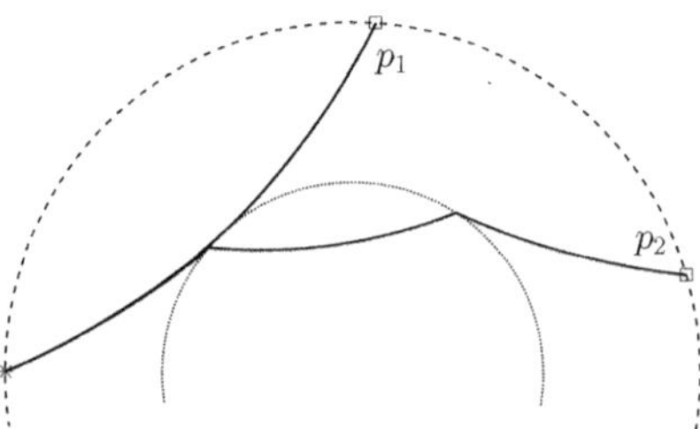

Figure 7.15 A refraction of the rays towards the centre and a resulting shadow zone also occur at velocity discontinuities, where the velocity jumps on the downgoing leg to a lower value. The discontinuity surface is dotted in this figure.

Zones of Rapid Velocity Increase (ZRVI): These zones are defined by the property that $r \mapsto c(r)|c'(r)|^{-1}$ is monotonically increasing. From the point of view of a downgoing leg, where r decreases, the latter term decreases as well along the leg in the ZRVI. In the majority of the Earth's layers, the velocity increases with depth. Hence, the derivative of the velocity (i.e. the acceleration) increases faster than the velocity does in a ZRVI.

For understanding wave propagation in a ZRVI, we need a little bit of basic differential geometry. The vector $\frac{\mathrm{d}}{\mathrm{d}s}X(s) = \mathfrak{p}(X(s))$ is the unit tangential vector to the ray in $X(s)$, its derivative with respect to the arc length, that is, $\frac{\mathrm{d}^2}{\mathrm{d}s^2}X(s) = \frac{\mathrm{d}}{\mathrm{d}s}\mathfrak{p}(X(s)) =: \mathfrak{m}(X(s))$, is a normal vector to the ray; and the Euclidean norm of this derivative is the curvature of the ray at the very same point. By definition of $\mathfrak{p}$, we have $S\mathfrak{p} = \nabla T$ and thus $\operatorname{curl}(S\mathfrak{p}) = 0$ such that

$$S \operatorname{curl} \mathfrak{p} + (\nabla_x S) \times \mathfrak{p} = 0. \tag{7.115}$$

Due to Lemma 7.4.4, we obtain for the second term

$$(\nabla_x S) \times \mathfrak{p} = \frac{\mathrm{d}}{\mathrm{d}s}(S\mathfrak{p}) \times \mathfrak{p} = \left(\frac{\mathrm{d}S}{\mathrm{d}s}\mathfrak{p} + S\frac{\mathrm{d}\mathfrak{p}}{\mathrm{d}s} \right) \times \mathfrak{p} = S\,\mathfrak{m} \times \mathfrak{p} \tag{7.116}$$

along each ray. Combining (7.115) and (7.116), we see that $\operatorname{curl} \mathfrak{p} = -\mathfrak{m} \times \mathfrak{p}$. With the antisymmetry of the cross product and the expansion theorem (see Theorem 2.1.5), we come to the conclusions

$$(\operatorname{curl} \mathfrak{p}) \times \mathfrak{p} = \mathfrak{p} \times (\mathfrak{m} \times \mathfrak{p}) \quad \Rightarrow \quad (\operatorname{curl} \mathfrak{p}) \times \mathfrak{p} = \mathfrak{m}.$$

The curvature of a ray must consequently obey

$$\begin{aligned} |\mathfrak{m}| &= |\mathfrak{p} \times \mathfrak{m}| = |\mathfrak{p} \times [(\operatorname{curl} \mathfrak{p}) \times \mathfrak{p}]| = |\operatorname{curl} \mathfrak{p} - (\mathfrak{p} \cdot \operatorname{curl} \mathfrak{p})\mathfrak{p}| \\ &= |\operatorname{curl} \mathfrak{p} - [\mathfrak{p} \cdot (-\mathfrak{m} \times \mathfrak{p})]\mathfrak{p}| = |\operatorname{curl} \mathfrak{p}|. \end{aligned}$$

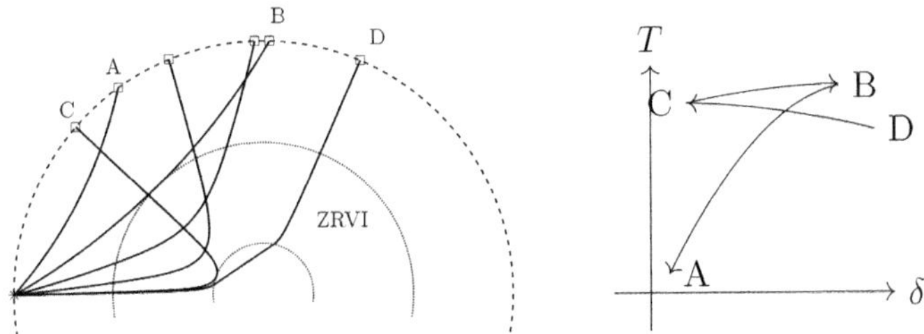

Figure 7.16 In a ZRVI, rays which enter deeper into the Earth obtain a higher curvature and have a smaller total angle. The result is a triplication of the travel time curve (right, parameterized with the ray parameter): there is an interval of total angles δ where three rays (with mostly different but similar travel times) occur. The letters A to D refer to special rays and their (δ, T)-pairs. The edges of the reverse segment, which correspond to B and C, are called cusps.

We insert (7.115) and use the spherical symmetry of our Earth model to obtain the following (remember that $S = c^{-1}$), again along a ray $x = X(s)$:

$$|\mathfrak{m}| = S^{-1}|(\nabla_x S) \times \mathfrak{p}| = S^{-1} \left| \frac{\mathrm{d}S}{\mathrm{d}r} \frac{x}{|x|} \times \mathfrak{p} \right| = S^{-1} \left| S' \right| \sin \mathfrak{i} = c^{-1} \left| c' \right| \sin \mathfrak{i}.$$

Let us now compare different rays which arrive at the same radial coordinate. Their only difference is their angle of incidence. The deeper this radius r is, provided it is in a ZRVI, the smaller is $c(r)|c'(r)|^{-1}$, and thus the larger is the curvature of the ray. The effect becomes particularly clear if we look at the case $r = r_{\mathrm{m}}$ for each ray. The deeper this 'turning point' is, the larger is the curvature there, which is simply $(c(r_{\mathrm{m}}))^{-1}|c'(r_{\mathrm{m}})|$. Therefore, rays going down at a steeper angle (and hence reaching deeper into the Earth) arrive at a surface point which is closer to the hypocentre, as shown in Figure 7.16. Hence, the condition $\delta'(p) \neq 0$ is violated at those ray parameters which correspond to rays which touch tangentially one of the bounding spheres of the ZRVI. The mapping $p \mapsto \delta(p)$ loses its injectivity, which leads to a **triplication of the travel time curve**.

A way out is available with the so-called **intercept time**, which is defined by $\tau(p) := T(p) - p\delta(p)$. With Benndorf's relation (Corollary 7.4.15), we easily see that

$$\tau'(p) = T'(p) - \delta(p) - p\delta'(p) = \frac{\mathrm{d}T(\delta(p))}{\mathrm{d}\delta} \frac{\mathrm{d}\delta(p)}{\mathrm{d}p} - \delta(p) - p\delta'(p) = -\delta(p) < 0$$

outside the cusps (there is no $T(\delta)$, but there is a $T(\delta(p))$ in the sense that p gives the information in which segment of the travel time curve we are, e.g., between A and B, such that in each segment there is a well-defined $T(\delta)$, while at the cusps B and C there is no neighbourhood which allows this). Hence, $p \mapsto \tau(p)$ is injective, that is, different rays have different intercept times. For further details on the use of the intercept time for the inversion of seismic data, also in the case of an LVZ, see Aki and Richards (2002, chapter 9).

If we do not have an SNREI Earth model, then the Herglotz–Wiechert formula and many of the other derivations in this section are not applicable. In this case, we have to use Equation (7.110) as it was stated in Definition 7.4.16. In other words, we know the values of line integrals $\int_{A_j^{\mathrm{h}}}^{B_j^{\mathrm{s}}} S(x)\,\mathrm{d}l(x)$ along selected curves and we seek the integrand S. This is related to the Radon transform as it occurs in computerized tomography (CT), but it is now more complicated, since the CT case is associated to integrations along straight lines in a planar disc. The scope and available size for this book do not allow a comprehensive discussion of the methodologies and the theory for this problem, but we will give a short summary of some important mathematical issues and references.

First of all, we need to be aware that the inverse problem is actually non-linear, because the ray (i.e., the domain of integration) depends on the slowness itself. In practice, this is avoided by taking a reference model which is already a good approximation, a priori calculating the rays based on this model, and keeping them fixed such that only the integrand remains unknown. This adds a slight inaccuracy into the inversion. However, the solution of non-linear inverse problems is often extremely challenging, while linear inverse problems can be regularized very well with a number of methods such that it is better to add an error in the modelling than to obtain severe limitations in the resolution for the original problem.

A survey on the theory of the linearized problem is given in the thesis Amirbekyan (2007). In particular, Bernstein and Gerver (1978) showed a sufficient condition for a unique solution: if the set of rays along which travel times are given is large enough such that through each point of the domain (e.g., the Earth) rays exist with all possible tangent vectors in this point and some other criteria are satisfied, then there is not more than one solution. Certainly, this is not possible in practice, and it suggests that we will have to deal with a severe non-uniqueness of the solution when we deal with real data. Moreover, Amirbekyan (2007) proved that, if a unique solution is available, it must be unstable, that is, the slowness function does not continuously depend on the travel times. Hence, no matter if it is due to a non-uniqueness or due to the instability, the seismic travel time tomography problem is for sure an ill-posed inverse problem.

For the numerical solution of travel time inversion at the global scale, several methodologies exist. Often, the problem is discretized by subdividing the Earth or the region into small volumes (blocks) such as *tetrahedra*. This is combined with the assumption that the solution has a very simple structure within each single block. For example, the values of the solution at the vertices are considered as unknowns and are linearly interpolated in between. In this context, there also exist methods where the structure of the blocks is more or less manually adapted, for example, by a subdivision into smaller subblocks or by melting several blocks into a larger block. Examples of such works can be found in Humphreys and Clayton (1988), Nolet (2008), Sigloch (2008), Tanabe and Ogata (1990), and Tian et al. (2009), as well as Bianco and Gerstoft (2018), where the latter uses a subdivision of a two-dimensional domain into pixels and considers dictionaries of Haar wavelets and discrete cosine transforms together with a dictionary learning technique. For the similar inverse problem of ocean acoustic tomography, a related approach to the latter is introduced in Wang and Xu (2016), where the well-known K-SVD algorithm is used for dictionary learning. Also for ultrasound tomography in medical imaging, such kinds of techniques exist; see, for example, Tosic et al. (2010). These dictionary-based approaches have sparsity as a main objective and the methods rely on an a priori discretization of the solution into a finite set of function values. Several alternative approaches have been discussed and applied so far for seismic travel time tomography. This includes the use of localized basis functions such as splines in Amirbekyan (2007) and Amirbekyan and Michel (2008).

Exercises

7.1 Let $\phi_t(X) := \mathrm{e}^{-(X_1^2+X_2^2)t}(X_1, X_2, (t+1)X_3)^{\mathrm{T}}$ be defined on the set $B_1(0)$. Verify if Assumption 7.1.4 is fulfilled and determine the material velocity and acceleration.

7.2 Prove Lemma 7.1.17.

7.3 Prove Lemma 7.1.19.

7.4 Let $\phi_t(X) := \left(X_1(1+t), X_2(1+t^2), X_3(1+2t)\right)^{\mathrm{T}}$ on the set $B_1(0)$ for $t \in \mathbb{R}_0^+$. Check if Assumption 7.1.4 is fulfilled and verify the validity of (7.4) by calculating both sides of the equation.

7.5 Prove that the deformation gradient obeys $\frac{\partial F}{\partial t}(X,t) = \frac{\partial v}{\partial x}(\phi_t(X),t)\, F(X,t)$ for all $X \in \mathcal{B}$ and all $t \in \mathbb{R}$, where $\frac{\partial v}{\partial x}$ stands for a Jacobian matrix.

7.6 Show that the definition of a perfect fluid in Remark 7.2.18 leads, indeed, to the particular form of σ and verify that $\mathrm{div}_x\, \sigma = -\nabla_x p$.

7.7 Let U be a displacement. We define $\mathcal{E}(U) := \left(\nabla_X U + (\nabla_X U)^{\mathrm{T}}\right)/2$ and $\mathcal{R}(U) := \left(\nabla_X U - (\nabla_X U)^{\mathrm{T}}\right)/2$. Prove that $\mathcal{R}(U) = (1/2)(\mathrm{curl}_X U) \times I$ and give a justification of the approximation $F \simeq I + \mathcal{E} + \mathcal{R}$ for the deformation gradient.

7.8 Consider the linearized model and the case of a homogeneous and isotropic medium, where Ξ has a form as in (7.62). Show that $\mathrm{tr} S \simeq (6\lambda + 4\mu)\mathrm{div}_X U$ and $\mathcal{E} \simeq (4\mu)^{-1}(S - \lambda(3\lambda + 2\mu)^{-1}(\mathrm{tr} S)I)$.

7.9 A long, thin, and cylinder-shaped wire is located along the X_1-axis and is being stretched along this axis. Its second Piola–Kirchhoff stress tensor has then vanishing components S_{ij} for $(i,j) \neq (1,1)$. The assumptions of Exercises 7.7 and 7.8 also apply here. Determine $\mathcal{E}$, Young's modulus $S_{11}\mathcal{E}_{11}^{-1}$, and Poisson's ratio $-\mathcal{E}_{jj}\mathcal{E}_{11}^{-1}$, $j = 2, 3$.

7.10 Let u be a plane progressive wave in a homogeneous and isotropic medium in the case of the linearized model. Show that the linearized Cauchy stress tensor $\Gamma : (\nabla_x u)$ is given in the case of a P-wave and an S-wave, respectively, by

$$\sigma_{\mathrm{P}}(x,t) = \varphi'(x \cdot k - ct)(\lambda\, \alpha \cdot k\, I + 2\mu\, \alpha \otimes k),$$
$$\sigma_{\mathrm{S}}(x,t) = \varphi'(x \cdot k - ct)\mu(\alpha \otimes k + k \otimes \alpha).$$

7.11 Use the separation ansatz $U(x) = F(r)G(\varphi)$ in polar coordinates on the disc $B_R(0) \subset \mathbb{R}^2$ to solve the Helmholtz equation $\Delta U + \omega^2 U = 0$ for a fixed $\omega \in \mathbb{R}^+$. Determine the general solution for G and show that all F with $F(r) = J_k(\omega r)$, $k \in \mathbb{N}_0$, can be used.

7.12 Use the substitution $\widehat{F}_{n,j}(r) := \widetilde{j}_n(\omega r/c)$ in (7.90) to derive a differential equation for the nth spherical Bessel function. Verify that $\widetilde{j}_0(x) = x^{-1}\sin x$ and $\widetilde{j}_1(x) = x^{-1}(x^{-1}\sin x - \cos x)$ solve this equation for $x > 0$ in the cases $n = 0$ and $n = 1$, respectively.

7.13 Show that the toroidal frequency equation (7.95) can be written in the case $n = 1$ as $\chi^{-1}\tan\chi = 3(3 - \chi^2)^{-1}$, where $\chi := \omega b \beta^{-1}$ is the unknown. Determine approximately the smallest positive solution. (You may use that the 2nd spherical Bessel function is given by $\widetilde{j}_2(x) = (3x^{-2} - 1)x^{-1}\sin x - 3x^{-2}\cos x$ for $x > 0$.)

7.14 Show that, in the case of a constant speed c, the rays are straight lines.

Appendix A

Hints for the Exercises

In this appendix, some short hints are given for the exercises which have been stated throughout the book.

3.1 Basically, one needs to apply the chain rule to the formula for the occurring Euclidean norm.

3.2 The proof is analogous to the case $n = 3$.

3.3 For example, take $|U(y)| = \mathcal{O}(|y|^{2-n})$, $|\nabla U(y)| = \mathcal{O}(|y|^{1-n})$ as $|y| \to \infty$.

3.4 The proof is analogous to the case $n = 3$.

3.5 Harmonicity in 1D means $U'' = 0$, which essentially restricts the structure of U. Also regions in $\mathbb{R}^1$ have a simple structure. Think about what their boundary would be.

3.6 Start with the fundamental theorem on the ball and apply Green's second identity.

3.7 Calculate $\nabla^*|\nabla V|$, which leads you to the calculation of $\nabla|\nabla V|$ via the comment in the exercise. The rest is basically the application of differentiation rules.

3.8 Part (a): use Exercise 3.6. Part (b): there is a property in real analysis regarding the curvature of the graph.

3.9 For the first part, use the triangle inequality for integrals and estimate the integrand. For the second part, apply the Ascoli–Arzelà theorem to $\mathcal{T}\mathcal{U}$, where $\mathcal{U}$ is the unit sphere in $\mathrm{C}(D)$.

3.10 Use the Cauchy–Schwarz inequality and the dominated convergence theorem.

3.11 Use polar coordinates. The occurrence of an absolute value in $P_{\mathrm{s}}(0,0,x_3)$ and $P_{\mathrm{d}}(0,0,x_3)$ is responsible for the jumps.

3.12 Use the Poisson integral formula for the IDP, find an appropriate estimate for the Abel–Poisson kernel, and use Gauß's mean value theorem.

3.13 Use Harnack's inequality from Exercise 3.12.

3.14 Have a look at the solution of the IDP: $\Delta V = 0$ in $B_r(x_0)$, $V = U$ on $S_r(x_0)$.

3.15 Use Exercise 3.14. Note that the uniform convergence does not automatically allow the interchanging of the limit and the Laplace operator, but it suffices for the interchanging of the limit and an integration.

3.16 Your result should basically look like Figure 3.9.

3.17 The derivation basically uses the chain rule and is a simpler analogue of the 3D case.

3.18 We get $Y'' = -CY$ and $r^2F''(r) + rF'(r) = CF(r)$. Use the periodicity of Y and (e.g.) a power series ansatz for F.

4.1 The proof is similar to the proof of $\mathrm{L}^* \cdot \nabla^* = 0$ in Theorem 4.1.6.

4.2 For verifying your results, note that all functions are algebraic polynomials in the Cartesian coordinates ξ_1, ξ_2, and ξ_3.

4.3 Use the addition theorem for spherical harmonics.

4.4 Use the mean value theorem of differentiation and the Cauchy–Schwarz inequality.

4.5 Insert $(n+1/2)^s(n+1/2)^{-s}$ into the spherical harmonics expansion of $F(\xi) - F(\eta)$ and use the Cauchy–Schwarz inequality and Exercise 4.4.

4.6 Use the formulae for ∇^* and L^* to represent $\nabla^* \otimes \nabla^* H - \mathrm{L}^* \otimes \mathrm{L}^* H$ in terms of the spherical tensor basis $\varepsilon^\varphi \otimes \varepsilon^r$, $\varepsilon^\varphi \otimes \varepsilon^t$, $\varepsilon^\varphi \otimes \varepsilon^\varphi$, $\varepsilon^t \otimes \varepsilon^r$, ... This allows you to determine the transposition, which can then be easily subtracted.

5.1 Since $B_R(0) \ni x \mapsto F(x)|x|^{-p}$ is harmonic, it can be expanded in inner harmonics.

5.2 Use Theorem 5.2.3.

5.3 Separate the properties into their analogues in $\mathcal{X}$ and $\mathcal{Y}$ and use that these spaces are known to be Hilbert spaces.

5.4 Use that $\mathcal{P} z_k = z_k$ and $\mathcal{P}$ is self-adjoint.

5.5 Use the orthogonal decomposition of $\mathcal{X}$ into the null space of $\mathcal{T}$ and its orthogonal complement.

5.6 For part (a), use Theorem 2.5.72. For part (b), use the Cauchy–Schwarz inequality and Theorems 2.5.39 and 5.3.2.

5.7 Fredholm's theorem and its implications, that is, Theorem 2.5.78 and Corollary 2.5.79, are helpful here. Moreover, for part (a), prove by contradiction that $\mathcal{T} - \lambda\mathcal{I}$ is not surjective if λ is an eigenvalue. Then have a look at the orthogonal complement of the image of $\mathcal{T} - \lambda\mathcal{I}$. For part (b), use orthonormal bases $\{x_j\}_{j=1,\ldots,n}$ of $\ker(\mathcal{T} - \lambda\mathcal{I})$ and $\{y_k\}_{k=1,\ldots,m}$ of $\ker(\mathcal{T}^* - \overline{\lambda}\mathcal{I})$ and investigate the operator $\mathcal{S}$ with $\mathcal{S}F := \mathcal{T}F - \sum_{j=1}^n \langle F, x_j\rangle y_j$.

5.8 Use the Hölder inequality $\sum_{n=1}^\infty x_n y_n \leq (\sum_{n=1}^\infty x_n^p)^{1/p}(\sum_{n=1}^\infty y_n^q)^{1/q}$ for $p^{-1} + q^{-1} = 1$.

5.9 Use the singular-value decompositions for the occurring operators and the corresponding analogues in the functional calculus. It is also helpful to solve the parts (a) to (d) consecutively.

5.10 Use the Hölder inequality (see the hint for Exercise 5.8) for the first inequality and use Exercise 5.9a for the second one.

5.11 Use (5.37) in Theorem 5.3.36.

5.12 Start with an arbitrary $f^+ \in \mathcal{X}_\mu$ and let $f_t := \mathcal{R}_t\mathcal{T}f^+$. Then estimate $\|f^+ - f_t\|_\mathcal{X}$ via Exercise 5.9c. Furthermore, use Theorem 5.3.43.

6.1 For example, $f(x) = \operatorname{grad}(|x|^2/2) + \operatorname{curl} 0$, because $\operatorname{curl} f = 0$, and $g(x) = \operatorname{grad} 0 + \operatorname{curl}[(x_3^2, x_1^2, x_2^2)^\mathrm{T}/2]$, because $\operatorname{div} g = 0$.

6.2 Use Theorem 6.3.7 and Corollaries 6.3.17 and 6.3.13 to show that f is solenoidal and toroidal but not poloidal. Use Theorem 2.1.5 to find a toroidal scalar such as $Q(x) = -\arcsin\xi_3 = -\arcsin(x_3/|x|)$. Possible Mie scalars are, hence, the same Q and $P \equiv 0$.

6.3 Both fields are not divergence-free such that they have none of the properties (see Theorems 6.3.2 and 6.3.3).

6.4 $\varepsilon^r = \xi 1 + \nabla^* 0 + \mathrm{L}^* 0$, $\varepsilon^\varphi = \xi 0 + \nabla^* 0 + \mathrm{L}^*(-\arcsin \xi_3)$, and $\varepsilon^t = \xi 0 + \nabla^* \arcsin \xi_3 + \mathrm{L}^* 0$.

6.5 Write LU with the Levi-Cività alternating symbol, elaborate all derivatives, and use Schwarz's theorem.

7.1 The Jacobian J has zeros. For the material velocity and acceleration, use Definition 7.1.5.

7.2 Use induction and the chain rule.

7.3 Take an arbitrary point $X \in \mathcal{B}$, use balls $B_{r_n}(X)$ with $r_n \to 0+$ as nice regions, and use the continuity of the integrands.

7.4 The assumption is fulfilled. Note that v needs to be differentiated with respect to x.

7.5 Use Schwarz's theorem, replace X by $\phi_t^{-1}(\phi_t(X))$, and apply the chain rule.

7.6 For the form of σ, use that also the standard basis vectors are eigenvectors. The formula of $\operatorname{div}_x \sigma$ is a simple calculation.

7.7 Use the Levi-Cività alternating symbol and its properties to calculate $(\operatorname{curl}_X U) \times I$. Moreover, note that F can be replaced by $I + \nabla_X U$.

7.8 Use Hooke's law to calculate S.

7.9 Use the formula for $\mathcal{E}$ in Exercise 7.8.

7.10 Use (7.64) and the formula for the spatial derivative of a plane progressive wave.

7.11 Use the representation of Δ in $\mathbb{R}^2$ from Exercise 3.17 and note the periodicity of G.

7.12 The equation is $\widetilde{j}_n''(x) + 2x^{-1}\widetilde{j}_n'(x) + (1 - n(n+1)x^{-2})\widetilde{j}_n(x) = 0$.

7.13 The equation results from $\widetilde{j}_2(\chi) = 0$ and the observation that the zeros of cos do not solve this. The smallest solution is $\chi \approx 5.763459197$.

7.14 Use Lemma 7.4.4 and the definition of rays.

Appendix B

Questions for Understanding

The following selection of questions refers to some essential topics of the book. The answers can be found by a thorough study of this book (with a few indicated exceptions where outlooks to different topics are given). Readers, in particular students, who want to learn the contents which were derived and explained in the previous chapters are recommended to use them as a self test.

- What is the formula of Newton's gravitational potential?
- Which properties does it have? What are conditions on the mass density function for each of these properties of the potential?
- Which properties do harmonic functions have in general?
- Where does the Poisson equation play a direct or indirect role within this book?
- What does Hölder continuity mean and how is it related to other properties of functions?
- What is the Kelvin transform and what is it good for?
- Outlook question: there is a terminus in the theory of functions of a complex variable which has strikingly similar properties in comparison to harmonic functions. What is this? Can you imagine (or find in the literature) how this can be mathematically explained?
- We defined four common boundary-value problems for the Laplace equation. What was their precise definition?
- Outlook question: if we measure the gravitational force at the Earth's surface and want to determine the potential outside the Earth, then this also leads us to a boundary-value problem. Why is it not precisely one of the four preceding problems? What is different?
- Which answers connected to the three questions behind well-posedness in the sense of Hadamard have been given for each of the four boundary-value problems?
- What is a Green's function for Δ? What does it have to do with the preceding boundary-value problems?
- Which properties of Green's functions were mentioned in the book?
- What is the Green's function in the case of a spherical boundary? To which famous formula did this lead us?
- What is the definition of the single and double layer potentials?

- Which properties do the layer potentials have? What are the associated conditions on the numerator of the integrand (here usually denoted by F)?
- In particular: what are the differences between these two potentials regarding discontinuity properties?
- What do the layer potentials have in common with Newton's (volume) potential? What is different?
- What is a Fredholm integral operator? Which kinds are distinguished?
- Where did Fredholm integral operators occur within the book?
- How can the gradient and the Laplace operator be decomposed into polar coordinates? What is the surface curl gradient? What did we learn about the Beltrami operator?
- We used a separation ansatz to determine a basis for harmonic functions in polar coordinates. What was the ansatz and what were the essential steps which eventually led us to the $Y_{n,j}$-functions?
- What are the Legendre polynomials and which properties do they have?
- What is a spherical harmonic?
- What are inner and outer harmonics?
- Which properties of spherical harmonics did we encounter throughout this book?
- What are, in particular, the properties of the $Y_{n,j}$?
- Outlook question: how could spherical harmonics and a $Y_{n,j}$-basis be constructed in $\mathbb{R}^d$? What would this basis be particularly in $\mathbb{R}^2$, that is, for a circle line?
- What is the Levi-Cività alternating symbol and what can it be used for?
- What is a zonal function? Why is the integration of a zonal function rather simple?
- What do Gauß's surface law and the surface theorem of Stokes look like precisely?
- From basic courses on real analysis, it is well known that gradient fields are curl-free, while the opposite holds true under certain sufficient conditions on the domain. What is, in some sense, the surface analogue on the sphere?
- What is the Poisson integral formula? Note that it occurs twice. How are these two versions connected? Which other question in this list refers to this formula?
- What is the Christoffel–Darboux formula?
- What is an efficient way of calculating the $Y_{n,j}$?
- What is a vector spherical harmonic? Which two kinds of systems are distinguished? What do these two kinds have in common and what is different? Which functions of these two kinds are identical?
- For which kind is it easier to represent the gradient of inner and outer harmonics?
- What is the definition of the Freeden–Gervens–Schreiner tensor spherical harmonics? Which properties do they have?
- In which (broadly geometrical) sense do some of the types of tensor spherical harmonics differ from others?
- What is the difference between a bandlimited and a spacelimited function? Which alternative name to 'spacelimited' is also common?

- What is an optimally space-localizing Slepian function on the sphere (here now just called Slepian function)? How can it be calculated? For which domain and how can this calculation be done in an efficient way?
- What do the vectors of the spherical harmonic coefficients of Slepian functions have in common with the Slepian functions themselves? Why is this the case?
- Which equivalent statements to the definition of Slepian functions exist?
- What is a radial basis function?
- What is a Sobolev space on the sphere (here now briefly Sobolev space)?
- When does it occur that a Sobolev space is embedded into another one? Which particular nested sequence of Sobolev spaces did we use?
- Which further properties do Sobolev spaces and their elements have? What are the requirements on the corresponding sequence $(A_n)_{n\in\mathbb{N}_0}$?
- What is a multi-resolution analysis?
- What is a reproducing kernel?
- When do radial basis functions deserve their name, that is, when do they really constitute a basis?
- Which examples of radial basis functions occurred throughout the book?
- The equation $\Delta^* U = F$ can be considered the spherical version of the Poisson equation. When is it solvable and what is known about its solutions?
- Which basis functions on the three-dimensional ball were considered in the book?
- What is the definition of a Sobolev space on the ball?
- Which best basis algorithms were discussed here? What are their differences regarding their objective function and regarding their properties? Are there algorithmic differences?
- What is the downward continuation problem? What do we know about it? Is it well-posed?
- What is the inverse gravimetric problem? What do we know about it? Is it well-posed?
- For an inverse problem with a known svd, which criterion is available for its solvability? If you want to prove this criterion, then make it so.
- Which two concepts of well-posedness are used in the book? What are their definitions? Is there a connection?
- Moreover, there is a way of classifying how severely ill-posed an inverse problem is. How is this done? What is the case for downward continuation and inverse gravimetry?
- What is the normal equation and what does it have to do with inverse problems?
- What is a minimum-norm solution? What do you know about it?
- What is the Moore–Penrose inverse? Which properties does it have?
- What is a regularization? Which examples occurred in the book?
- What is the benefit of a regularization?
- What is a source condition?
- What is meant by order optimality?

- Tikhonov–Phillips regularization is represented by what? Which properties does it have?
- What are the four Maxwell equations?
- In the chapter on the magnetic field, we encountered three different decomposition principles for particular kinds of functions. What are these and what are their precise propositions? Which of them was used to prove which other one among them?
- For each decomposition principle, where did it, throughout the book, become useful?
- Under which conditions is a function solenoidal, toroidal, or poloidal by definition? Which sufficient, necessary, or equivalent criteria exist?
- We had a counterexample which showed that 'solenoidal' is only a sufficient but not necessary criterion for a differential equation. What is the corresponding theorem and what is the counterexample?
- How can the magnetic field B and the current J be decomposed?
- The distinction of internal and external fields is easy if the absence of currents is assumed. Why is this the case and how can this be done?
- Which alternative exists if currents cannot be neglected? Which kind of vector spherical harmonics is useful in this case?
- What is a simple body and a motion in continuum mechanics?
- What is the difference between an Eulerian and a Lagrangian description? Which other names exist? Explain the difference for the velocity and the acceleration.
- What is the right Cauchy–Green tensor?
- Which conservation laws occurred in the seismology chapter? What is their precise definition? Which implications or equivalences exist? What are the basic steps of proving them?
- What does 'volume preserving' mean? Are there also implications or equivalences? Why is this not a conservation law?
- Which body forces need to be taken into account in the case of the Earth (in a precise modelling)? What is the difference between these forces?
- Which kinds of stress tensors did we encounter? What are their properties?
- What are elasticity tensors?
- Which equations were used in the linearized model?
- What does isotropy and homogeneity mean? Which simplifications occur in the isotropic (and homogeneous) case?
- What does the ansatz of a plane progressive wave stand for? What does it yield and why (for the case of an elastic wave)?
- What is the Cauchy–Navier equation and how is it derived? How can it be solved?
- What is the Helmholtz equation?
- Which kinds of basis vectors did we get for the solutions of the Cauchy–Navier equation? How are they composed? Which kind of vector spherical harmonics is useful here? Which type is needed for which basis vectors?

- We also had boundary conditions in this context. What are they and what did they imply?
- What is a seismic ray?
- What is the inverse problem of travel time tomography?
- Which special properties do rays have in the case of an SNREI model?
- In the derivation of the Herglotz–Wiechert formula, several requirements on the velocity field had to be made. We saw that these are not satisfied, for example, in two kinds of zones. What are these zones and how are they characterized? What happens to rays propagating through such a zone? Which way out was briefly mentioned?

References

Abramowitz, M., and Stegun, I. A. 1972. *Handbook of Mathematical Functions with Formulas, Graphs, and Mathematical Tables*. New York: Dover Publications Inc.

Abrikosov, O., and Schwintzer, P. 2004. Recovery of the Earth's gravity field from GOCE satellite gravity gradiometry: a case study. In: Lacoste, H. (ed.), *GOCE. The Geoid and Oceanography, Proc. of the 2nd International Workshop, Held 8–10 March in Frascati, Italy*. https://ui.adsabs.harvard.edu/abs/2004ESASP.569E..18A/abstract.

Agarwal, R. P., and O'Regan, D. 2009. *Ordinary and Partial Differential Equations. With Special Functions, Fourier Series, and Boundary Value Problems*. New York: Springer.

Aki, K., and Richards, P. G. 2002. *Quantitative Seismology*. 2nd ed. Sausalito: University Science Books.

Akram, M. 2008. *Constructive Approximation on the 3-Dimensional Ball with Focus on Locally Supported Kernels and the Helmholtz Decomposition*. Ph.D. thesis, University of Kaiserslautern, Department of Mathematics, Geomathematics Group. Shaker, Aachen.

Akram, M., and Michel, V. 2010. Locally supported approximate identities on the unit ball. *Revista Matemática Complutense*, **23**, 233–249.

Akram, M., Amina, I., and Michel, V. 2011. A study of differential operators for particular complete orthonormal systems on a 3D ball. *International Journal of Pure and Applied Mathematics*, **73**, 489–506.

Albertella, A., Sansò, F., and Sneeuw, N. 1999. Band-limited functions on a bounded spherical domain: the Slepian problem on the sphere. *Journal of Geodesy*, **73**, 436–447.

Amann, H., and Escher, J. 2008. *Analysis III*. 2nd ed. Basel: Birkhäuser.

Amirbekyan, A. 2007. *The Application of Reproducing Kernel Based Spline Approximation to Seismic Surface and Body Wave Tomography: Theoretical Aspects and Numerical Results*. Ph.D. thesis, University of Kaiserslautern, Department of Mathematics, Geomathematics Group. https://nbn-resolving.org/urn:nbn:de:hbz:386-kluedo-21039.

Amirbekyan, A., and Michel, V. 2008. Splines on the three-dimensional ball and their application to seismic body wave tomography. *Inverse Problems*, **24**, 1–25.

Amirbekyan, A., Michel, V., and Simons, F. J. 2008. Parameterizing surface-wave tomographic models with harmonic spherical splines. *Geophysical Journal International*, **174**, 617–628.

Atiyah, M., et al. 2001a. *Lexikon der Mathematik*, vol. 2. Heidelberg: Spektrum Akademischer Verlag.

Atiyah, M., et al. 2001b. *Lexikon der Mathematik*, vol. 3. Heidelberg: Spektrum Akademischer Verlag.

Atkinson, K., and Han, W. 2012. *Spherical Harmonics and Approximations on the Unit Sphere: An Introduction*. Berlin, Heidelberg: Springer.

Aubin, T. 1982. *Nonlinear Analysis on Manifolds, Monge–Ampère Equations*. Grundlagen der mathematischen Wissenschaften, vol. 252. New York: Springer.

Backus, G. 1986. Poloidal and toroidal fields in geomagnetic field modeling. *Reviews of Geophysics*, **24**, 75–109.

Backus, G., Parker, R., and Constable, C. 1996. *Foundations of Geomagnetism*. Cambridge: Cambridge University Press.

Bagherbandi, M. 2012. A comparison of three gravity inversion methods for crustal thickness modelling in Tibet plateau. *Journal of Asian Earth Sciences*, **43**, 89–97.

Balandin, A. L., Ono, Y., and You, S. 2012. 3D vector tomography using vector spherical harmonics decompositon. *Computers and Mathematics with Applications*, **63**, 1433–1441.

Ballani, L., Engels, J., and Grafarend, E. W. 1993. Global base functions for the mass density in the interior of a massive body (Earth). *Manuscripta Geodaetica*, **18**, 99–114.

Ballmann, W. 2015. *Einführung in die Geometrie und Topologie*. Mathematik Kompakt. Basel: Springer.

Barrera, R. G., Estévez, G. A., and Giraldo, J. 1985. Vector spherical harmonics and their application to magnetostatics. *European Journal of Physics*, **6**, 287–294.

Bauer, F., Gutting, M., and Lukas, M. A. 2015. Evaluation of parameter choice methods for regularization of ill-posed problems in geomathematics. Pages 1713–1774 of: Freeden, W., Nashed, M. Z., and Sonar, T. (eds.), *Handbook of Geomathematics*, 2nd ed. Berlin, Heidelberg: Springer.

Bauer, H. 1990. *Maß- und Integrationstheorie*. Berlin: Walter de Gruyter.

Baur, O., and Sneeuw, N. 2011. Assessing Greenland ice mass loss by means of point-mass modeling: a viable methodology. *Journal of Geodesy*, **85**, 607–615.

Bayer, M. 2000. *Geomagnetic Field Modelling from Satellite Data by First and Second Generation Vector Wavelets*. Ph.D. thesis, University of Kaiserslautern, Department of Mathematics, Geomathematics Group. Shaker, Aachen.

Beals, R., and Wong, R. 2016. *Special Functions and Orthogonal Polynomials*. Cambridge Studies in Advanced Mathematics, vol. 153. Cambridge: Cambridge University Press.

Ben-Menahem, A., and Singh, S. J. 1981. *Seismic Waves and Sources*. New York: Springer.

Berkel, P. 2009. *Multiscale Methods for the Combined Inversion of Normal Mode and Gravity Variations*. Ph.D. thesis, University of Kaiserslautern, Department of Mathematics, Geomathematics Group. Shaker, Aachen.

Berkel, P., and Michel, V. 2010. On mathematical aspects of a combined inversion of gravity and normal mode variations by a spline method. *Mathematical Geosciences*, **42**, 795–816.

Berkel, P., Fischer, D., and Michel, V. 2011. Spline multiresolution and numerical results for joint gravitation and normal mode inversion with an outlook on sparse regularisation. *GEM: International Journal on Geomathematics*, **1**, 167–204.

Bernstein, I. N., and Gerver, M. L. 1978. On the problem of integral geometry for a set of geodesic lines and on the inverse kinematic problem of seismology. *Doklady Akademii Nauk SSSR*, **243**, 302–305.

Bianco, M. J., and Gerstoft, P. 2018. Travel time tomography with adaptive dictionaries. *IEEE Transactions on Computational Imaging*, **4**, 499–511.

Billingham, J., and King, A. C. 2000. *Wave Motion*. Cambridge Texts in Applied Mathematics. Cambridge: Cambridge University Press.

Blakely, R. J. 1996. *Potential Theory in Gravity and Magnetic Applications*. Cambridge: Cambridge University Press.

Blick, C., Freeden, W., and Nutz, H. 2017. Feature extraction of geological signatures by multiscale gravimetry. *GEM: International Journal on Geomathematics*, **8**, 57–83.

Bolton, S., Levin, S., and Bagenal, F. 2017. Juno's first glimpse of Jupiter's complexity. *Geophysical Research Letters*, **44**, 7663–7667. Special Section Early Results: Juno at Jupiter.

Boulanger, O., and Chouteau, M. 2001. Constraints in 3D gravity inversion. *Geophysical Prospecting*, **49**, 265–280.

Buchheim, W. 1975. Zur geophysikalischen Inversionsproblematik. Pages 305–310 of: Maaz, R. (ed.), *Seismology and Solid-Earth-Physics: Proc. Int. Symp. on the Occasion of 50 Years of Seismological Research and 75 Years of Seismic Registration at Jena 1974.*

Candès, E. J., Romberg, J., and Tao, T. 2006. Robust uncertainty principles: exact signal reconstruction from highly incomplete frequency information. *IEEE Transactions on Information Theory*, **52**, 489–509.

Canuto, C., and Tabacco, A. 2010. *Mathematical Analysis II*. Milan: Springer.

Carrascal, B, Estévez, G. A., Lee, P., and Lorenzo, V. 1991. Vector spherical harmonics and their application to classical electrodynamics. *European Journal of Physics*, **12**, 184–191.

Chattopadhyay, A. 2004. Wave reflection and refraction in triclinic crystalline media. *Archive of Applied Mechanics*, **73**, 568–579.

Chen, J. L., Wilson, C. R., and Tapley, B. D. 2006. Satellite gravity measurements confirm accelerated melting of Greenland ice sheet. *Science*, **313**, 1958–1960.

Chihara, T. S. 1978. *An Introduction to Orthogonal Polynomials*. New York: Gordon and Breach.

Chui, C. K. 1992. *An Introduction to Wavelets*. San Diego: Academic Press.

Clapp, R. E., and Li, H. T. 1970. Six integral theorems for vector spherical harmonics. *Journal of Mathematical Physics*, **11**, 4–9.

Craven, B. D. 1982. *Lebesgue Measure and Integral*. Boston: Pitman.

Dahlen, F. A., and Tromp, J. 1998. *Theoretical Global Seismology*. Princeton: Princeton University Press.

Dahlen, F. A., and Simons, F. J. 2008. Spectral estimation on a sphere in geophysics and cosmology. *Geophysical Journal International*, **174**, 774–807.

Dai, F., and Xu, Y. 2014. The Hardy–Rellich inequality and uncertainty principle on the sphere. *Constructive Approximation*, **40**, 141–171.

Dang, P., Qian, T., and Chen, Q. 2017. Uncertainty principle and phase-amplitude analysis of signals on the unit sphere. *Advances in Applied Clifford Algebras*, **27**, 2985–3013.

Davis, P. J. 1975. *Interpolation and Approximation*. New York: Dover Publications.

Dieudonné, J. 1960. *Foundations of Modern Analysis*. New York and London: Academic Press.

Donoho, D. L. 2006. Compressed sensing. *IEEE Transactions on Information Theory*, **52**, 1289–1306.

Dufour, H. M. 1977. Fonctions orthogonales dans la sphère. Résolution théorique du problème du potentiel terrestre. *Bulletin Géodésique*, **51**, 227–237.

Dunkl, C. F., and Xu, Y. 2001. *Orthogonal Polynomials of Several Variables*. Encyclopedia of Mathematics and Its Applications. Cambridge: Cambridge University Press.

Dziewonski, A. M., and Anderson, D. L. 1981. Preliminary reference Earth model. *Physics of the Earth and Planetary Interiors*, **25**, 297–356.

Edmonds, A. R. 1957. *Angular Momentum in Quantum Mechanics*. Princeton: Princeton University Press.

Efthimiou, C., and Frye, C. 2014. *Spherical Harmonics in p Dimensions*. Singapore: World Scientific.

Eicker, A., Mayer-Gürr, T., and Ilk, K. H. 2005. Global gravity field solutions based on a simulation scenario of GRACE SST data and regional refinements by GOCE SGG observations. Pages 66–71 of: Jekeli, C., Bastos, L., and Fernandes, J. (eds.), *Gravity, Geoid and Space Missions. International Association of Geodesy Symposia*, vol. 129. Berlin, Heidelberg: Springer.

Engl, H. W., and Gfrerer, H. 1988. A posteriori parameter choice for general regularization methods for solving linear ill-posed problems. *Applied Numerical Mathematics*, **4**, 395–417.

Engl, H. W., Hanke, M., and Neubauer, A. 1996. *Regularization of Inverse Problems*. Dordrecht, Boston, London: Kluwer Academic Publishers.

Eshagh, M. 2009. Spatially restricted integrals in gradiometric boundary value problems. *Artificial Satellites*, **44**, 131–148.

Fengler, M. J. 2005. *Vector Spherical Harmonic and Vector Wavelet Based Non-Linear Galerkin Schemes for Solving the Incompressible Navier–Stokes Equation on the Sphere*. Ph.D. thesis, University of Kaiserslautern, Department of Mathematics, Geomathematics Group. Shaker, Aachen.

Fengler, M. J., Michel, D., and Michel, V. 2006. Harmonic spline-wavelets on the 3-dimensional ball and their application to the reconstruction of the Earth's density distribution from gravitational data at arbitrarily shaped satellite orbits. *Zeitschrift für Angewandte Mathematik und Mechanik*, **86**, 856–873.

Fengler, M. J., Freeden, W., Kohlhaas, A., Michel, V., and Peters, T. 2007. Wavelet modelling of regional and temporal variations of the Earth's gravitational potential observed by GRACE. *Journal of Geodesy*, **81**, 5–15.

Ferrers, N. M. 1877. *Spherical Harmonics*. London: Macmillan and Co.

Fersch, B., Kunstmann, H., Bárdossy, A., Devaraju, B., and Sneeuw, N. 2012. Continental-scale basin water storage variation from global and dynamically downscaled atmospheric water budgets in comparison with GRACE-derived observations. *Journal of Hydrometeorology*, **13**, 1589–1603.

Fischer, D. 2011. *Sparse Regularization of a Joint Inversion of Gravitational Data and Normal Mode Anomalies*. Ph.D. thesis, University of Siegen, Department of Mathematics, Geomathematics Group. Dr. Hut, Munich, https://nbn-resolving.org/urn:nbn:de:hbz:467-5448.

Fischer, D., and Michel, V. 2012. Sparse regularization of inverse gravimetry – case study: spatial and temporal mass variations in South America. *Inverse Problems*, **28**, 065012 (34pp).

Fischer, D., and Michel, V. 2013a. Automatic best-basis selection for geophysical tomographic inverse problems. *Geophysical Journal International*, **193**, 1291–1299.

Fischer, D., and Michel, V. 2013b. Inverting GRACE gravity data for local climate effects. *Journal of Geodetic Science*, **3**, 151–162.

Flechtner, F., Morton, P., Watkins, M., and Webb, F. 2014. Status of the GRACE Follow-On mission. Pages 117–121 of: Marti, U. (ed.), *Gravity, Geoid and Height Systems. International Association of Geodesy Symposia*, vol. 141. Cham: Springer.

Folland, G. B. 1976. *Introduction to Partial Differential Equations*. Princeton: Princeton University Press.

Fox, A. J., and Johnson, F. A. 1966. On finding the eigenvalues of real symmetric tridiagonal matrices. *The Computer Journal*, **9**, 98–105.

Freeden, W. 1981a. On approximation by harmonic splines. *Manuscripta Geodaetica*, **6**, 193–244.

Freeden, W. 1981b. On spherical spline interpolation and approximation. *Mathematical Methods in the Applied Sciences*, **3**, 551–575.

Freeden, W., Gervens, T., and Schreiner, M. 1994. Tensor spherical harmonics and tensor spherical splines. *Manuscripta Geodaetica*, **19**, 70–100.

Freeden, W., and Schneider, F. 1998. Regularization wavelets and multiresolution. *Inverse Problems*, **14**, 225–243.

Freeden, W., Gervens, T., and Schreiner, M. 1998. *Constructive Approximation on the Sphere with Applications to Geomathematics*. Oxford: Oxford University Press.

Freeden, W. 1999. *Multiscale Modelling of Spaceborne Geodata*. Stuttgart, Leipzig: B. G. Teubner.

Freeden, W., and Hesse, K. 2002. On the multiscale solution of satellite problems by use of locally supported kernel functions corresponding to equidistributed data on spherical orbits. *Studia Scientiarum Mathematicarum Hungarica*, **39**, 37–74.

Freeden, W., Michel, V., and Nutz, H. 2002. Satellite-to-satellite tracking and satellite gravity gradiometry (advanced techniques for high-resolution geopotential field determination). *Journal of Engineering Mathematics*, **43**, 19–56.

Freeden, W., and Michel, V. 2004. *Multiscale Potential Theory (with Applications to Geoscience)*. Boston: Birkhäuser.

Freeden, W., and Gutting, M. 2008. On the completeness and closure of vector and tensor spherical harmonics. *Integral Transforms and Special Functions*, **19**, 713–734.

Freeden, W., and Schreiner, M. 2009. *Spherical Functions of Mathematical Geosciences, a Scalar, Vectorial, and Tensorial Setup*. Berlin: Springer.

Freeden, W., and Gerhards, C. 2013. *Geomathematically Oriented Potential Theory*. Boca Raton: CRC Press.

Freeden, W., and Gutting, M. 2013. *Special Functions of Mathematical (Geo-) Physics*. Basel: Birkhäuser.

Freeden, W., Michel, V., and Simons, F. J. 2018. Spherical harmonics based special function systems and constructive approximation methods. Pages 753–819 of: Freeden, W., and Nashed, M. Z. (eds.), *Handbook of Mathematical Geodesy*. Geosystems Mathematics. Basel: Birkhäuser.

Fukushima, T. 2012. Numerical computation of spherical harmonics of arbitrary degree and order by extending exponent of floating point numbers. *Journal of Geodesy*, **86**, 271–285.

Gilbarg, D., and Trudinger, N. S. 1977. *Elliptic Partial Differential Equations of Second Order*. Grundlagen der mathematischen Wissenschaften, vol. 224. Berlin: Springer.

Gorenflo, R., and Vessella, S. 1991. *Abel Integral Equations – Analysis and Applications*. Lecture Notes in Mathematics, no. 1461. Berlin: Springer.

Grasmair, M., and Naumova, V. 2016. Conditions on optimal support recovery in unmixing problems by means of multi-penalty regularization. *Inverse Problems*, **32**, 104007 (16pp).

Grünbaum, F. A., Longhi, L., and Perlstadt, M. 1982. Differential operators commuting with finite convolution integral operators: some non-Abelian examples. *SIAM Journal on Applied Mathematics*, **42**, 941–955.

Gubbins, D., Ivers, D., Masterton, S. M., and Winch, D. E. 2011. Analysis of lithospheric magnetization in vector spherical harmonics. *Geophysical Journal International*, **187**, 99–117.

Gurtin, M. E. 1984. The linear theory of elasticity. Pages 1–295 of: Truesdell, C. (ed.), *Linear Theories of Elasticity and Thermoelasticity. Linear and Nonlinear Theories of Rods, Plates, and Shells*. Mechanics of Solids, vol. II. Berlin: Springer.

Gutting, M., Kretz, B., Michel, V., and Telschow, R. 2017. Study on parameter choice methods for the RFMP with respect to downward continuation. *Frontiers in Applied Mathematics and Statistics*, **3**, article 10.

Hadamard, J. 1902. Sur les problèmes aux dérivées partielles et leur signification physique. *Princeton University Bulletin*, **13**, 49–52.

Hanke, M. 2017. *A Taste of Inverse Problems. Basic Theory and Examples*. Other Titles in Applied Mathematics, no. 153. Philadelphia: Society for Industrial and Applied Mathematics (SIAM).

Hanke, M., and Engl, H. W. 1994. An optimal stopping rule for the ν-method for solving ill-posed problems using Christoffel functions. *Journal of Approximation Theory*, **79**, 89–108.

Hansen, P. C. 1992. Analysis of discrete ill-posed problems by means of the L-curve. *SIAM Review*, **34**, 561–580.

Hansen, P. C., and O'Leary, D. P. 1993. The use of the L-curve in the regularization of discrete ill-posed problems. *SIAM Journal on Scientific Computing*, **14**, 1487–1503.

Hansen, P. C. 1998. *Rank-Deficient and Discrete Ill-Posed Problems. Numerical Aspects of Linear Inversion*. Philadelphia: SIAM.

Hardy, G. H. 1916. Weierstrass's non-differentiable function. *Transactions of the American Mathematical Society*, **17**, 301–325.

Harig, C., and Simons, F. J. 2016. Icemass loss in Greenland, the Gulf of Alaska, and the Canadian Archipelago: seasonal cycles and decadal trends. *Geophysical Research Letters*, **43**, 3150–3159.

Heiskanen, W. A., and Moritz, H. 1981. *Physical Geodesy, Reprint*. Technical University Graz/Austria: Institute of Physical Geodesy.

Hettlich, F., and Rundell, W. 1996. Iterative methods for the reconstruction of an inverse potential problem. *Inverse Problems*, **12**, 251–266.

Heuser, H. 1991. *Gewöhnliche Differentialgleichungen*. 2nd ed. Stuttgart: B. G. Teubner.

Heuser, H. 1992. *Funktionalanalysis*. 3rd ed. Stuttgart: B. G. Teubner.

Heuser, H. 2009. *Lehrbuch der Analysis, Teil 1*. 17th ed. Mathematische Leitfäden. Wiesbaden: Vieweg + Teubner.

Higuchi, A. 1987. Symmetric tensor spherical harmonics on the N-sphere and their application to the Sitter group $\mathrm{SO}(N, 1)$. *Journal of Mathematical Physics*, **28**, 1553–1566.

Hill, E. L. 1954. The theory of vector spherical harmonics. *American Journal of Physics*, **22**, 211–214.

Hobson, E. W. 1965. *The Theory of Spherical and Ellipsoidal Harmonics*. New York: Chelsea Publ. Co.

Hofmann, B. 1986. *Regularization for Applied Inverse and Ill-Posed Problems*. Teubner-Texte zur Mathematik. Leipzig: BSB Teubner.

Hofmann, B., Kaltenbacher, B., Pöschl, C., and Scherzer, O. 2007. A convergence rates result for Tikhonov regularization in Banach spaces with non-smooth operators. *Inverse Problems*, **23**, 987–1010.

Hofmann-Wellenhof, B., and Moritz, H. 2005. *Physical Geodesy*. Vienna, New York: Springer.

Holmes, S. A., and Featherstone, W. E. 2002. A unified approach to the Clenshaw summation and the recursive computation of very high degree and order normalised associated Legendre functions. *Journal of Geodesy*, **76**, 279–299.

Humphreys, E., and Clayton, R. W. 1988. Adaptation of back projection tomography to seismic travel time problems. *Journal of Geophysical Research: Solid Earth*, **93**, 1073–1085.

Iglewska-Nowak, I. 2016. Multiresolution on n-dimensional spheres. *Kyushu Journal of Mathematics*, **70**, 353–374.

Ilk, K. H., Feuchtinger, M., and Mayer-Gürr, T. 2005. Gravity field recovery and validation by analysis of short arcs of a satellite-to-satellite tracking experiment as CHAMP and GRACE. Pages 189–194 of: Sansò, F. (ed.), *A Window on the Future of Geodesy. International Association of Geodesy Symposia*, vol. 128. Berlin, Heidelberg: Springer.

Isakov, V. 1990. *Inverse Source Problems*. Providence: American Mathematical Society.

Isakov, V. 2006. *Inverse Problems for Partial Differential Equations*. 2nd ed. New York: Springer.

Ishtiaq, A. 2018. *Grid Points and Generalized Discrepancies on the d-Dimensional Ball*. Ph.D. thesis, University of Siegen, Department of Mathematics, Geomathematics Group. https://nbn-resolving.org/urn:nbn:de:hbz:467-13733.

Ishtiaq, A., and Michel, V. 2017. Pseudo-differential operators, cubature and equidistribution on the 3D ball: an approach based on orthonormal basis systems. *Numerical Functional Analysis and Optimization*, **38**, 891–910.

Ishtiaq, A., Michel, V., and Scheffler, H. P. 2019. Theory of generalized discrepancies on a ball of arbitrary dimensions and algorithms for finding low-discrepancy point sets. *GEM: International Journal on Geomathematics*, **10**, article 21.

Iske, A. 2018. *Approximation*. Berlin: Springer Spektrum.

Jahn, K., and Bokor, N. 2012. Vector Slepian basis functions with optimal energy concentration in high numerical aperture focusing. *Optics Communications*, **285**, 2028–2038.

Jahn, K., and Bokor, N. 2013. Solving the inverse problem of high numerical aperture focusing using vector Slepian harmonics and vector Slepian multipole fields. *Optics Communications*, **288**, 13–16.

James, R. W. 1976. New tensor spherical harmonics, for application to the partial differential equations of mathematical physics. *Philosophical Transactions of the Royal Society A*, **281**, 195–221.

Jänich, K. 2005. *Topologie*. 8th ed. Berlin: Springer.

Jansen, M. J. F., Gunter, B. C., and Kusche, J. 2009. The impact of GRACE, GPS and OBP data on estimates of global mass redistribution. *Geophysical Journal International*, **177**, 1–13.

Jekeli, C. 1999. The determination of gravitational potential differences from satellite-to-satellite tracking. *Celestial Mechanics and Dynamical Astronomy*, **75**, 85–101.

Jensen, J. L. W. V. 1906. Sur les fonctions convexes et les inégalités entre les valeurs moyennes. *Acta Mathematica*, **30**, 175–193.

Jones, F. 1993. *Lebesgue Integration on Euclidean Spaces*. Boston: Jones and Bartlett Publishers.

Jones, L. K. 1987. On a conjecture of Huber concerning the convergence of projection pursuit regression. *The Annals of Statistics*, **15**, 880–882.

Kaban, M. K., Flóvenz, Ó. G., and Pálmason, G. 2002. Nature of the crust-mantle transition zone and the thermal state of the upper mantle beneath Iceland from gravity modelling. *Geophysical Journal International*, **149**, 281–299.

Kant, I. 1786. *Metaphysische Anfangsgründe der Naturwissenschaft*. Riga: Johann Friedrich Hartknoch.

Kant, I. 1883. *Kant's Prolegomena and Metaphysical Foundations of Natural Science*. London: George Bell and Sons. Translated from the original, with a biography and introduction by E. B. Bax.

Kazantsev, S. G., and Kardakov, V. B. 2019. Poloidal-toroidal decomposition of solenoidal vector fields in the ball. *Journal of Applied and Industrial Mathematics*, **13**, 480–499.

Kellogg, O. D. 1967. *Foundations of Potential Theory*. Berlin: Springer.

Khalid, Z., Kennedy, R. A., and McEwen, J. D. 2016. Slepian spatial-spectral concentration on the ball. *Applied and Computational Harmonic Analysis*, **40**, 470–504.

Kirsch, A. 1996. *An Introduction to the Mathematical Theory of Inverse Problems*. Applied Mathematical Sciences, no. 120. New York: Springer.

Kontak, M. 2018. *Novel Algorithms of Greedy-Type for Probability Density Estimation as well as Linear and Nonlinear Inverse Problems*. Ph.D. thesis, University of Siegen, Department of Mathematics, Geomathematics Group. Dr. Hut, Munich, https://nbn-resolving.org/urn:nbn:de:hbz:467-13160.

Kontak, M., and Michel, V. 2018. A greedy algorithm for non-linear inverse problems with an application to non-linear inverse gravimetry. *GEM: International Journal on Geomathematics*, **9**, 167–198.

Kontak, M., and Michel, V. 2019. The regularized weak functional matching pursuit for linear inverse problems. *Journal of Inverse and Ill-Posed Problems*, **27**, 317–340.

Koyré, A., and Cohen, I. B. 1726. *Isaac Newton's Philosophiae Naturalis Principia Mathematica, the 3rd Edition with Variant Readings*. Cambridge: Harvard University Press.

Landau, H. J., and Pollak, H. O. 1961. Prolate spheroidal wave functions, Fourier analysis and uncertainty – II. *Bell System Technical Journal*, **40**, 65–84.

Lang, S. 2001. *Undergraduate Analysis*. 2nd ed. New York: Springer.

Last, B. J., and Kubik, K. 1983. Compact gravity inversion. *Geophysics*, **48**, 713–721.

Laures, G., and Szymik, M. 2015. *Grundkurs Topologie*. 2nd ed. Berlin: Springer Spektrum.

Lauricella, G. 1912. Sulla distribuzione della massa nell'interno dei pianeti. *Rendiconti Accademia Nazionale dei Lincei*, **XXI**, 18–26.

Leweke, S. 2018. *The Inverse Magneto-Encephalography Problem for the Spherical Multiple-Shell Model: Theoretical Investigations and Numerical Aspects*. Ph.D. thesis, University of Siegen, Department of Mathematics, Geomathematics Group. https://nbn-resolving.org/urn:nbn:de:hbz:467-13967.

Leweke, S., Michel, V., and Telschow, R. 2018a. On the non-uniqueness of gravitational and magnetic field data inversion (survey article). Pages 883–919 of: Freeden, W., and Nashed, M. Z. (eds.), *Handbook of Mathematical Geodesy*. Basel: Birkhäuser.

Leweke, S., Michel, V., and Schneider, N. 2018b. Vectorial Slepian functions on the ball. *Numerical Functional Analysis and Optimization*, **39**, 1120–1152.

Leweke, S., Michel, V., and Fokas, A. S. 2020. Electro-magnetoencephalography for a spherical multiple-shell model: novel integral operators with singular-value decompositions. *Inverse Problems*, 035003 (31pp).

Li, Y., and Oldenburg, D. W. 1998. 3-D inversion of gravity data. *Geophysics*, **63**, 109–119.

Louis, A. K. 1989. *Inverse und schlecht gestellte Probleme*. Stuttgart: Teubner.

Lu, S., and Pereverzev, S. V. 2013. *Regularization Theory for Ill-Posed Problems: Selected Topics*. Berlin, Boston: de Gruyter.

MacRobert, T. M. 1927. *Spherical Harmonics. An Elementary Treatise on Harmonic Functions with Applications*. London: E. P. Dutton.

Magnus, W., Oberhettinger, F., and Soni, R. P. 1966. *Formulas and Theorems for the Special Functions of Mathematical Physics*. Berlin: Springer.

Mallat, S. G., and Zhang, Z. 1993. Matching pursuits with time-frequency dictionaries. *IEEE Transactions on Signal Processing*, **41**, 3397–3415.

Maniar, H., and Mitra, P. P. 2005. Local basis expansions for MEG source localization. *International Journal of Bioelectromagnetism*, **7**, 30–33.

Marquering, H., Nolet, G., and Dahlen, F. A. 1998. Three-dimensional waveform sensitivity kernels. *Geophysical Journal International*, **132**, 521–534.

Marquering, H., Dahlen, F. A., and Nolet, G. 1999. Three-dimensional sensitivity kernels for finite-frequency traveltimes: the banana-doughnut paradox. *Geophysical Journal International*, **137**, 805–815.

Marsden, J. E., and Hughes, T. J. R. 1994. *Mathematical Foundations of Elasticity*. New York: Dover Publications.

Martinec, Z. 2003. Green's function solution to spherical gradiometric boundary-value problems. *Journal of Geodesy*, **77**, 41–49.

Marussi, A. 1980. On the density distribution in bodies of assigned outer Newtonian attraction. *Bollettino di Geofisica Teorica ed Applicata*, **XXII**, 83–94.

Mathews, J. 1962. Gravitational multipole radiation. *Journal of the Society for Industrial and Applied Mathematics*, **10**, 768–780.

McShane, E. J. 1974. *Integration*. Princeton: Princeton University Press.

Michel, V. 1999. *A Multiscale Method for the Gravimetry Problem: Theoretical and Numerical Aspects of Harmonic and Anharmonic Modelling*. Ph.D. thesis, University of Kaiserslautern, Department of Mathematics, Geomathematics Group. Shaker, Aachen.

Michel, V. 2002a. *A Multiscale Approximation for Operator Equations in Separable Hilbert Spaces – Case Study: Reconstruction and Description of the Earth's Interior*. Aachen: Shaker. Habilitation thesis.

Michel, V. 2002b. Scale continuous, scale discretized and scale discrete harmonic wavelets for the outer and the inner space of a sphere and their application to an inverse problem in geomathematics. *Applied and Computational Harmonic Analysis*, **12**, 77–99.

Michel, V. 2003. Theoretical aspects of a multiscale analysis of the eigenoscillations of the Earth. *Revista Matemática Complutense*, **16**, 519–554.

Michel, V. 2005. Regularized wavelet-based multiresolution recovery of the harmonic mass density distribution from data of the Earth's gravitational field at satellite height. *Inverse Problems*, **21**, 997–1025.

Michel, V., and Fokas, A. S. 2008. A unified approach to various techniques for the non-uniqueness of the inverse gravimetric problem and wavelet-based methods. *Inverse Problems*, **24**, 045019.

Michel, V., and Wolf, K. 2008. Numerical aspects of a spline-based multiresolution recovery of the harmonic mass density out of gravity functionals. *Geophysical Journal International*, **173**, 1–16.

Michel, V. 2013. *Lectures on Constructive Approximation – Fourier, Spline and Wavelet Methods on the Real Line, the Sphere, and the Ball*. New York: Birkhäuser.

Michel, V., and Telschow, R. 2014. A non-linear approximation method on the sphere. *GEM: International Journal on Geomathematics*, **5**, 195–224.

Michel, V. 2015. RFMP – An iterative best basis algorithm for inverse problems in the geosciences. Pages 2121–2147 of: Freeden, W., Nashed, M. Z., and Sonar, T. (eds.), *Handbook of Geomathematics*, 2nd ed. Berlin, Heidelberg: Springer.

Michel, V., and Orzlowski, S. 2016. On the null space of a class of Fredholm integral equations of the first kind. *Journal of Inverse and Ill-Posed Problems*, **24**, 687–710.

Michel, V., and Orzlowski, S. 2017. On the convergence theorem for the Regularized Functional Matching Pursuit (RFMP) algorithm. *GEM: International Journal on Geomathematics*, **8**, 183–190.

Michel, V., and Simons, F. J. 2017. A general approach to regularizing inverse problems with regional data using Slepian wavelets. *Inverse Problems*, **33**, 125016.

Michel, V., and Telschow, R. 2016. The regularized orthogonal functional matching pursuit for ill-posed inverse problems. *SIAM Journal on Numerical Analysis*, **54**, 262–287.

Michel, V., and Schneider, N. 2020. A first approach to learning a best basis for gravitational field modelling. *GEM: International Journal on Geomathematics*, **11**, article 9.

Michel, V., Plattner, A., and Seibert, K. 2021. A unified approach to scalar, vector, and tensor Slepian functions on the sphere and their construction by a commuting operator. *Preprint*, arXiv:2103.14650, accepted for publication in: Analysis and Applications.

Mie, G. 1908. Beiträge zur Optik trüber Medien, speziell kolloidaler Metallösungen. *Annalen der Physik*, **25**, 377–445.

Mikhlin, S. G. 1970. *Mathematical Physics, an Advanced Course*. Amsterdam: North-Holland Publishing Company.

Moritz, H. 1990. *The Figure of the Earth. Theoretical Geodesy of the Earth's Interior*. Karlruhe: Wichmann.

Morozov, V. A. 1966. On the solution of functional equations by the method of regularization. *Soviet Mathematics. Doklady*, **7**, 414–417.

Morozov, V. A. 1967. Choice of parameter for the solution of functional equations by the regularization method. *Soviet Mathematics. Doklady*, **8**, 1000–1003.

Morozov, V. A. 1984. *Methods for Solving Incorrectly Posed Problems*. New York: Springer.

Morse, P. M. C., and Feshbach, H. 1953a. *Methods of Theoretical Physics*. Vol. 1. New York: McGraw–Hill.

Morse, P. M. C., and Feshbach, H. 1953b. *Methods of Theoretical Physics*. Vol. 2. New York: McGraw–Hill.

Müller, C. 1966. *Spherical Harmonics*. Berlin: Springer.

Müller, C. 1969. *Foundations of the Mathematical Theory of Electromagnetic Waves*. Berlin: Springer.

Munkres, J. R. 2000. *Topology*. 2nd ed. Upper Saddle River: Prentice Hall.

Narcowich, F. J., and Ward, J. D. 1996. Nonstationary wavelets on the m-sphere for scattered data. *Applied and Computational Harmonic Analysis*, **3**, 324–336.

Nashed, M. Z. 1987. A new approach to classification and regularization of ill-posed operator equations. Pages 53–75 of: Engl, H. W., and Groetsch, C. W. (eds.), *Inverse and Ill-Posed Problems*. Notes and Reports in Mathematics in Science and Engineering, vol. 4. Boston: Academic Press.

Newton, I. 1687. *Philosophiae Naturalis Principia Mathematica*. London: Royal Society.

Nolet, G. 2008. *A Breviary of Seismic Tomography: Imaging the Interior of the Earth and Sun*. Cambridge: Cambridge University Press.

Novikoff, P. 1938. Sur le problème inverse du potentiel. *Comptes Rendus de l'Académie des Sciences de l'URSS*, **XVIII**, 165–168.

Pail, R., and Wermuth, M. 2003. GOCE SGG and SST quick-look gravity field analysis. *Advances in Geosciences*, **1**, 5–9.

Pail, R., Bruinsma, S., Migliaccio, F., Förste, C, et al. 2011. First GOCE gravity field models derived by three different approaches. *Journal of Geodesy*, **85**, 819–843.

Panet, I., Bonvalot, S., Narteau, C., Remy, D., and Lemoine, J.-M. 2018. Migrating pattern of deformation prior to the Tohoku-Oki earthquake revealed by GRACE data. *Nature Geoscience*, **11**, 367–373.

Pati, Y. C., Rezaiifar, R., and Krishnaprasad, P. S. 1993. Orthogonal matching pursuit: recursive function approximation with applications to wavelet decomposition. Pages 40–44 of: *Asilomar Conference on Signals, Systems and Computers*. Los Alamitos: IEEE Computer Society.

Pavlis, N. K., Holmes, S. A., Kenyon, S. C., and Factor, J. K. 2012. The development and evaluation of the Earth Gravitational Model 2008 (EGM2008). *Journal of Geophysical Research: Solid Earth*, **117**, B04406. Erratum in *Journal of Geophysical Research: Solid Earth*, **118**, 2633–2633.

Peter, S., Artina, M., and Fornasier, M. 2015. Damping noise-folding and enhanced support recovery in compressed sensing. *IEEE Transactions on Signal Processing*, **63**, 5990–6002.

Phillips, D. L. 1962. A technique for the numerical solution of certain integral equations of the first kind. *Journal of the Association for Computing Machinery*, **9**, 84–97.

Pizzetti, P. 1909. Corpi equivalenti rispetto alla attrazione newtoniana esterna. *Accademia dei Lincei, Rendiconti*, **XVIII**, 211–215.

Pizzetti, P. 1910. Intorno alle possibili distribuzioni della massa nell'interno della terra. *Annali di Matematica, Serie III*, **XVII**, 225–258.

Plato, R. 1990. Optimal algorithms for linear ill-posed problems yield regularization methods. *Numerical Functional Analysis and Optimization*, **11**, 111–118.

Plattner, A., Simons, F. J., and Wei, L. 2012. Analysis of real vector fields on the sphere using Slepian functions. Pages 1–4 of: *IEEE Statistical Signal Processing Workshop (SSP), Ann Arbor, MI, USA*.

Plattner, A., and Simons, F. J. 2014. Spatiospectral concentration of vector fields on a sphere. *Applied and Computational Harmonic Analysis*, **36**, 1–22.

Plattner, A., and Simons, F. J. 2017. Internal and external potential field estimation from regional vector data at varying satellite altitude. *Geophysical Journal International*, **211**, 207–238.

Protter, M. H., and Morrey, C. B. 1977. *A First Course in Real Analysis*. 2nd ed. New York: Springer.

Ramillien, G., Lombard, A., Cazenave, A., et al. 2006. Interannual variations of the mass balance of the Antarctica and Greenland ice sheets from GRACE. *Global and Planetary Change*, **53**, 198–208.

Raus, T. 1985. On the discrepancy principle for the solution of ill-posed problems with non-selfadjoint operators. *Acta et Commentationes Universitatis Tartuensis de Mathematica*, **715**, 12–20. In Russian.

Regge, T., and Wheeler, J. A. 1957. Stability of a Schwarzschild singularity. *Physical Review*, **108**, 1063–1069.

Reigber, C., Schmidt, R., Flechtner, F., König, R., and Meyer, U. 2005. An Earth gravity field model complete to degree and order 150 from GRACE: EIGEN-GRACE02S. *Journal of Geodynamics*, **39**, 1–10.

Rennhack, S. 2018. *Der Regularized Weak Functional Matching Pursuit am Beispiel der Erdgravitationsfeldmodellierung*. B.Sc. Thesis, University of Siegen, Department of Mathematics, Geomathematics Group.

Resmerita, E. 2005. Regularization of ill-posed problems in Banach spaces: convergence rates. *Inverse Problems*, **21**, 1303–1314.

Reuter, R. 1982. *Über Integralformeln der Einheitssphäre und harmonische Splinefunktionen*. Ph.D. thesis, RWTH Aachen, Veröffentlichungen des Geodätischen Instituts der Rheinisch-Westfälischen Technischen Hochschule in Aachen, no. 33.

Rieder, A. 2003. *Keine Probleme mit Inversen Problemen*. Wiesbaden: Vieweg.

Riley, K. F., Hobson, M. P., and Bence, S. J. 2008. *Mathematical Methods for Physics and Engineering*. 4th ed. Cambridge: Cambridge University Press.

Roberts, C. E. 2010. *Ordinary Differential Equations. Applications, Models, and Computing*. Boca Raton: CRC Press.

Robin, L. 1957. *Fonctions Sphérique de Legendre et Fonctions Sphéroïdale*. Vol. 1. Paris: Gauthier-Villars.

Robin, L. 1958. *Fonctions Sphérique de Legendre et Fonctions Sphéroïdale*. Vol. 2. Paris: Gauthier-Villars.

Robin, L. 1959. *Fonctions Sphérique de Legendre et Fonctions Sphéroïdale*. Vol. 3. Paris: Gauthier-Villars.

Rubin, M. A., and Ordóñez, C. R. 1984. Eigenvalues and degeneracies for n-dimensional tensor spherical harmonics. *Journal of Mathematical Physics*, **25**, 2888–2894.

Rubincam, D. P. 1979. Gravitational potential energy of the Earth: a spherical harmonics approach. *Journal of Geophysical Research*, **84**, 6219–6225.

Rudin, W. 1987. *Real and Complex Analysis*. New York: McGraw–Hill.

Rummel, R., Yi, W., and Stummer, C. 2001. GOCE gravitational gradiometry. *Journal of Geodesy*, **85**, 777–790.

Rummel, R. 2003. How to climb the gravity wall. In: Beutler, G., Drinkwater, M. R., Rummel, R., and von Steiger, R. (eds.), *Earth Gravity Field from Space – From Sensors to Earth Sciences*. Space Sciences Series of ISSI, vol. 17. Dordrecht: Springer.

Sandberg, V. D. 1978. Tensor spherical harmonics on S^2 and S^3 as eigenvalue problems. *Journal of Mathematical Physics*, **19**, 2441–2446.

Sanna, N. 2000. Vector spherical harmonics: concepts and applications to the single centre expansion method. *Computer Physics Communications*, **132**, 66–83.

Sasgen, I., van den Broeke, M., Bamber, J. L., et al. 2012. Timing and origin of recent regional ice-mass loss in Greenland. *Earth and Planetary Science Letters*, **333–334**, 293–303.

Schmidt, R., Flechtner, F., Meyer, U., et al. 2008. Hydrological signals observed by the GRACE satellites. *Surveys in Geophysics*, **29**, 319–334.

Schneider, F. 1997. *Inverse Problems in Satellite Geodesy and Their Approximate Solution by Splines and Wavelets*. Ph.D. thesis, University of Kaiserslautern, Department of Mathematics, Geomathematics Group. Shaker, Aachen.

Schneider, N. 2020. *Learning Dictionaries for Inverse Problems on the Sphere*. Ph.D. thesis, University of Siegen, Department of Mathematics, Geomathematics Group. http://dx.doi.org/10.25819/ubsi/5431.

Schock, E. 1985. Approximate solution of ill-posed equations: arbitrarily slow convergence vs. superconvergence. Pages 234–243 of: Hämmerlin, G., and Hoffmann, K.-H. (eds.), *Constructive Methods for the Practical Treatment of Integral Equations. Proceedings of the Conference at the Mathematisches Forschungsinstitut Oberwolfach, June 24–30, 1984*. International Series of Numerical Mathematics, vol. 73. Basel, Boston, Stuttgart: Birkhäuser.

Schöpfer, F., Louis, A. K., and Schuster, T. 2006. Nonlinear iterative methods for linear ill-posed problems in Banach spaces. *Inverse Problems*, **22**, 311–329.

Schreiner, M. 1994. *Tensor Spherical Harmonics and Their Application in Satellite Gradiometry*. Ph.D. thesis, University of Kaiserslautern, Department of Mathematics, Geomathematics Group. Shaker, Aachen.

Schuster, T., Kaltenbacher, B., Hofmann, B., and Kazimierski, K. S. 2012. *Regularization Methods in Banach Spaces*. Radon Series on Computational and Applied Mathematics, no. 10. Berlin, Boston: de Gruyter.

Seibert, K. 2018. *Spin-Weighted Spherical Harmonics and Their Application for the Construction of Tensor Slepian Functions on the Spherical Cap*. Ph.D. thesis, University of Siegen, Department of Mathematics, Geomathematics Group. Universi, Siegen, https://nbn-resolving.org/urn:nbn:de:hbz:467-14210.

Sigloch, K. 2008. Two-stage subduction history under North America inferred from multiple-frequency tomography. *Nature Geoscience*, **1**, 458–462.

Simons, F. J., and Dahlen, F. A. 2006. Spherical Slepian functions and the polar gap in geodesy. *Geophysical Journal International*, **166**, 1039–1061.

Simons, F. J., Dahlen, F. A., and Wieczorek, M. A. 2006. Spatiospectral concentration on a sphere. *SIAM Review*, **48**, 504–536.

Simons, F. J. 2010. Slepian functions and their use in signal estimation and spectral analysis. Pages 891–923 of: Freeden, W., Nashed, M. Z., and Sonar, T. (eds.), *Handbook of Geomathematics*. Heidelberg: Springer.

Slepian, D., and Pollak, H. O. 1961. Prolate spheroidal wave functions, Fourier analysis and uncertainty—I. *Bell System Technical Journal*, **40**, 43–63.

Slepian, D. 1964. Prolate spheroidal wave functions, Fourier analysis and uncertainty—IV: extensions to many dimensions; generalized prolate spheroidal functions. *Bell System Technical Journal*, **43**, 3009–3057.

Smirnow, W. I. 1977. *Lehrgang der Höheren Mathematik*. Hochschulbücher für Mathematik, vol. 5. Berlin: VEB Deutscher Verlag der Wissenschaften. Translated from Russian to German by C. Berg and L. Berg.

Stokes, G. G. 1867. On the internal distribution of matter which shall produce a given potential at the surface of a gravitating mass. *Proceedings of the Royal Society*, **16 May 1867**, 482–486. And *Collected Works* IV, pp 277–288.

Szegö, G. 1975. *Orthogonal Polynomials*. 14th ed. Vol. XXIII. Providence: AMS Colloquium Publications.

Tanabe, H. I. Y. F. K., and Ogata, Y. 1990. Whole mantle P-wave travel time tomography. *Physics of the Earth and Planetary Interiors*, **59**, 294–328.

Tapley, B. D., Bettadpur, S., Ries, J. C., Thompson, P. F., and Watkins, M. M. 2004a. GRACE measurements of mass variability in the Earth system. *Science*, **305**, 503–505.

Tapley, B. D., Bettadpur, S., Watkins, M., and Reigber, C. 2004b. The gravity recovery and climate experiment: mission overview and early results. *Geophysical Research Letters*, **31**, L09607.

Telschow, R. 2014. *An Orthogonal Matching Pursuit for the Regularization of Spherical Inverse Problems*. Ph.D. thesis, University of Siegen, Department of Mathematics, Geomathematics Group. Dr. Hut, Munich.

Temlyakov, V. N. 2000. Weak greedy algorithms. *Advances in Computational Mathematics*, **12**, 213–227.

Tesauro, M., Kaban, M. K., and Cloetingh, S. A. P. L. 2008. EuCRUST-07: A new reference model for the European crust. *Geophysical Research Letters*, **35**, L05313.

Tian, Y., Hung, S.-H., Nolet, G., Montelli, R., and Dahlen, F. A. 2007. Dynamic ray tracing and traveltime corrections for global seismic tomography. *Journal of Computational Physics*, **226**, 672–687.

Tian, Y., Sigloch, K., and Nolet, G. 2009. Multiple-frequency SH-wave tomography of the western US upper mantle. *Geophysical Journal International*, **178**, 1384–1402.

Tipler, P. A. 1982. *Physics*. 2nd ed. Vol. 1. New York: Worth Publishers.

Tomita, K. 1982. Tensor spherical and pseudo-spherical harmonics in four-dimensional spaces. *Progress of Theoretical Physics*, **68**, 310–313.

Tosic, I., Jovanovic, I., Frossard, P., Vetterli, M., and Duric, N. 2010. Ultrasound tomography with learned dictionaries. Pages 5502–5505 of: *Proceedings of the IEEE International Conference on Acoustics, Speech, and Signal Processing*.

Trim, D. 1993. *Calculus*. Scarborough: Prentice Hall.

Tscherning, C. C. 1996. Isotropic reproducing kernels for the inner of a sphere or spherical shell and their use as density covariance functions. *Mathematical Geology*, **28**, 161–168.

Velicogna, I., and Wahr, J. 2006. Acceleration of Greenland ice mass loss in spring 2004. *Nature*, **443**, 329–331.

Velicogna, I., and Wahr, J. 2013. Time-variable gravity observations of ice sheet mass balance: precision and limitations of the GRACE satellite data. *Geophysical Research Letters*, **40**, 3055–3063.

Vincent, P., and Bengio, Y. 2002. Kernel matching pursuit. *Machine Learning*, **48**, 169–191.

Vogel, C. R. 1996. Non-convergence of the L-curve regularization parameter selection method. *Inverse Problems*, **12**, 535–547.

Voigt, A., and Wloka, J. 1975. *Hilberträume und elliptische Differentialoperatoren*. Mannheim: Bibliographisches Institut.

von Brecht, J. H. 2016. Localization and vector spherical harmonics. *Journal of Differential Equations*, **260**, 1622–1655.

Wahba, G. 1981. Spline interpolation and smoothing on the sphere. *SIAM Journal on Scientific and Statistical Computing*, **2**, 5–16.

Walter, E., et al. (eds.). 2005. *Cambridge Advanced Learner's Dictionary*. 2nd ed. Cambridge: Cambridge University Press.

Walter, W. 1971. *Einführung in die Potentialtheorie*. Mannheim: Bibliographisches Institut.

Walter, W. 1990. *Analysis II*. Berlin, Heidelberg: Springer.

Wang, T., and Xu, W. 2016. Sparsity-based approach for ocean acoustic tomography using learned dictionaries. Pages 1–6 of: *OCEANS 2016 – Shanghai*. Hoboken: IEEE. https://ieeexplore.ieee.org/abstract/document/7485626.

Watson, G. N. 1944. *A Treatise on the Theory of Bessel functions*. 2nd edn. Cambridge: Cambridge University Press.

Weck, N. 1972. Inverse Probleme der Potentialtheorie. *Applicable Analysis*, **2**, 195–204.

Weinberg, E. J. 1994. Monopole vector spherical harmonics. *Physical Review D*, **49**, 1086–1092.

Wieczorek, M. A., and Simons, F. J. 2005. Localized spectral analysis on the sphere. *Geophysical Journal International*, **162**, 655–675.

Wieczorek, M. A., and Simons, F. J. 2007. Minimum-variance spectral analysis on the sphere. *Journal of Fourier Analysis and Applications*, **13**, 665–692.

Winter, J. 1982. Tensor spherical harmonics. *Letters in Mathematical Physics*, **6**, 91–96.

Wolfers, J. P. 1872. *Sir Isaac Newton's Mathematische Principien der Naturlehre. Mit Bemerkungen und Erläuterungen*. Berlin: R. Oppenheim.

Yegorova, T. P., Kozlenko, V. G., Pavlenkova, N. I, and Starostenko, V. I. 1995. 3D density model for the lithosphere of Europe: construction method and preliminary results. *Geophysical Journal International*, **121**, 873–892.

Yegorova, T. P., Starostenko, V. I., Kozlenko, V. G., and Pavlenkova, N. I. 1997. Three-dimensional gravity modelling of the European Mediterranean lithosphere. *Geophysical Journal International*, **129**, 355–367.

Yegorova, T. P., and Starostenko, V. I. 1999. Large-scale three-dimensional gravity analysis of the lithosphere below the transition zone from Western Europe to the East European Platform. *Tectonophysics*, **314**, 83–100.

Yomogida, K. 1992. Fresnel zone inversion for lateral heterogeneities in the Earth. *Pure and Applied Geophysics*, **138**, 391–406.

Yosida, K. 1995. *Functional Analysis*. 6th ed. Classics in Mathematics. Berlin: Springer.

Zeidler, E. (ed.). 1996. *Teubner-Taschenbuch der Mathematik, originally from I.N. Bronstein and K.A. Semendjajew*. Leipzig: Teubner.

Zerilli, F. J. 1970a. Gravitational field of a particle falling in a Schwarzschild geometry analyzed in tensor harmonics. *Physical Review D*, **2**, 2141–2160.

Zerilli, F. J. 1970b. Tensor harmonics in canonical form for gravitational radiation and other applications. *Journal of Mathematical Physics*, **11**, 2203–2208.

Zhang, S., and Xin, J. 2018. Minimization of transformed L_1 penalty: theory, difference of convex function algorithm, and robust application in compressed sensing. *Mathematical Programming*, **169**, 307–336.

Zuber, M., Smith, D., Lehman, D., Hoffmann, T., Asmar, S., and Watkins, M. 2013. Gravity recovery and interior laboratory (GRAIL): mapping the lunar interior from crust to core. *Space Science Reviews*, **178**, 3–24.

Index